# III International Conference on Laser and Plasma Researches and Technologies 2017



Journal of Physics: Conference Series Volume 941

Moscow, Russia
24 – 27 January 2017

ISBN: 978-1-5108-5846-6
ISSN: 1742-6588

**Printed from e-media with permission by:**

Curran Associates, Inc.
57 Morehouse Lane
Red Hook, NY 12571

**Some format issues inherent in the e-media version may also appear in this print version.**



Printed by Curran Associates, Inc. (2018)

For permission requests, please contact the Institute of Physics
at the address below.

Institute of Physics
Dirac House, Temple Back
Bristol BS1 6BE UK

Phone: 44 1 17 929 7481
Fax: 44 1 17 920 0979

techtracking@iop.org

**Additional copies of this publication are available from:**

Curran Associates, Inc.
57 Morehouse Lane
Red Hook, NY 12571 USA
Phone: 845-758-0400
Fax: 845-758-2633
Email: curran@proceedings.com
Web: www.proceedings.com

# Table of contents

## Volume 941

**III International Conference on Laser and Plasma Researches and Technologies**

**24–27 January 2017, Moscow, Russian Federation**

**Accepted papers received: 14 December 2017**
**Published online: 12 January 2018**

### Preface

III International Conference on Laser and Plasma Researches and Technologies

Peer review statement

### Papers

**Laser physics and its application**

Investigation of hybrid modulator based on nonlinear optical loop mirror and single-walled carbone nanotubes for mode-locking in Yb-doped ultrafast fiber laser

A A Borodkin, D V Khudyakov and S K Vartapetov.....1

Application of direct optical heterodyning methods for studying the processes of chondrite targets destruction by laser radiation

M Gorbashova, I Burdonskiy, K Gubskiy, A Kuznetsov, A Ramazanov, K Lukyanov, A Leonov, K Makarov, I Timofeyev and V Yufa.....6

The technique of investigation of space-time parameters of the master oscillator with fiber optical smoothing based on the slip photo recorder

O I Gorchakov, L A Dushina, D S Korniyenko, D V Sizmin and K V Starodubtsev.....12

Multiline $CO_2$ laser with Q-switching for generation of terahertz radiation

A A Ionin, I O Kinyaevskiy, Yu M Klimachev, V V Stepanishchev and I Zh Khafizov.....16

Numerical simulation of sum frequency generation spectrum in nonlinear crystals with considering dynamics of generation

A A Ionin, I O Kinyaevskiy, Yu M Klimachev, A A Kotkov, D S Kryuchkov and A M Sagitova.....23

Spectral characteristics of Q-switched CO laser

A A Ionin, I O Kinyaevskiy, Yu M Klimachev, D S Kryuchkov and A M Sagitova.....28

**Plasma physics and its application**

Spatiotemporal and spectral characteristics of X-ray radiation emitted by the Z-pinch during the current implosion of quasispherical multiwire arrays

A N Gritsuk.....34

A model of recovering the parameters of fast nonlocal heat transport in magnetic fusion plasmas

A B Kukushkin, A A Kulichenko, P A Sdvizhenskii, A V Sokolov and V V Voloshinov.....39

Greenwald density limit and power balance in tokamaks

D Kh Morozov.....45

Ignition in tokamaks with modulated source of auxiliary heating

D Kh Morozov.....50

Compression of magnetized target in the magneto-inertial fusion

V V Kuzenov.....55

The comparative analysis of particles and films, and the conditions of their formation from arc-discharge plasma and at the CTF devices

N A Smolanov.....60

Spatial separation of particles in modified linear Paul trap

D S Lapitsky.....66

Plasma for environment

G Van Oost.....69

On plasma sheath motion in coaxial electrode system

V V Vikhrev and E I Dodulad.....76

Collective acceleration of laser plasma in a nonstationary and nonuniform magnetic field

A Isaev, K Kozlovskiy, A Shikanov and E Vovchenko.....82

Properties of plasma in case of capillary discharge interaction with metals

D L Kirko and A S Savjolov.....87

Investigation of plasma properties of discharge in electrolyte surface layer

D L Kirko and A S Savjolov.....91

Extracting the differential inverse inelastic mean free path and differential surface excitation probability of Tungsten from X-ray photoelectron spectra and electron energy loss spectra

V P Afanas'ev, A S Gryazev, D S Efremenko, P S Kaplya and A V Kuznetcova.....96

Deuterium diffusion and retention in tungsten coated with barrier layer during ion irradiation

L B Begrambekov, A S Kaplevsky, S S Dovganyuk, A E Evsin and I E Baryshnikova.....102

Stochastic clustering of material surface under powerful load of high-temperature plasma

V P Budaev.....107

On the possibility of thin layers thickness determination with low energy proton scattering

D Bulgadaryan, V Kurnaev, D Sinelnikov and N Efimov.....111

Simulation of adsorbed hydrogen on tungsten surface

N N Degtyarenko and A A Pisarev.....116

The model of thermal field emission from tungsten fuzz

D Sinelnikov, D Bulgadaryan, V Kurnaev and M Lobov.....121

Modelling of lithium transport and its influence on the edge plasma parameters in T-15MD tokamak

A A Pshenov and A S Kukushkin.....125

The image of a nanosecond laser plasma in its own optical radiation

A A. Fronya, N G Borisenko, V N Puzyrev, A T Sahakyan, A N Starodub and O F Yakushev.....130

**Laser, plasma and radiation technologies in industry**

Effect of the laser heat treatment on the formation of the gradient structures in alloys based on Fe – Cr – Ni system

A O Andreev, D P Bykovskiy, A V Osintsev, V N Petrovskiy, I I Ryashko, E N Blinova, M A Libman and A M Glezer.....135

Modification of polyacrylonitrile carbon fibers by highfluence ion irradiation

N N Andrianova, A M Borisov, V A Kazakov, E S Mashkova, M A Ovchinnikov, S V Savushkina and N M Chernenko.....140

Physical and mechanical properties of high-modulus carbon fiber crimped by ion irradiation

V A Anikin, N N Andrianova, A M Borisov, E S Mashkova, M A Ovchinnikov, S V Savushkina, D N Chernenko and N M Chernenko.....146

Calculation and experimental determination of the geometric parameters of the coatings by laser cladding

V P Birukov and A A Fichkov.....150

Direct metal laser deposition of titanium powder Ti-6Al-4V

D P Bykovskiy, V N Petrovskiy, K L Sergeev, A V Osintsev, P S Dzhumaev and V I Polskiy.....154

Laser annealing of porous defects in metals

M S Grigoryeva, I N Zavestovskaya and A P Kanavin.....159

Attenuation of laser radiation by the flame of burning hydrocarbons and efficiency of remote cutting of metals

S V Gvozdev, A F Glova, V Yu Dubrovskii, S T Durmanov, A G Krasyukov, A Yu Lysikov, G V Smirnov and V M Pleshkov.....164

Synthesis of oxide and nitride ceramics in high-power gyrotron discharge

N S Akhmadullina, N N Skvortsova, E A Obraztsova, V D Stepakhin, E M Konchekov, Yu F Kargin and O N Shishilov.....170

Influence of the modes of laser cladding on bond strength and wear resistance of coatings

V P Birukov, D Yu Tatarkin, E V Chriptovish and A A Fichkov.....175

Pulsed abnormal glow discharge with hollow cathode for nitriding of internal cylindrical surfaces

Yu V Borisyuk, V V Kozlova, D V Mozgrin, P S Nenashev, N M Oreshnikova, M M Berdnikova and A A Pisarev.....179

Influence of technological parameters on the geometry of single-track laser clad nickel based alloy on grey cast iron substrate

D S Gusev and A B Lyukhter.....183

Laser-ablative fabrication of nanoparticle inks for 3D inkjetprinting of multifunctional coatings

A A Ionin, A K Ivanova, R A Khmel'nitskii, Yu V Klevkov, S I Kudryashov, N N Mel'nik, A A Nastulyavichus, A A Rudenko, I N Saraeva, N A Smirnov and D A Zayarny.....192

Dimensional processing of composite materials by picosecond pulsed ytterbium fiber laser

S A Kotov.....197

The structure and mechanical properties of parts elaborated by direct laser deposition 316L stainless steel powder obtained in various ways

I S Loginova, A N Solonin, A S Prosviryakov, S B Adisa, A M Khalil, D P Bykovskiy and V N Petrovskiy.....202

The influence of laser hardening on the performance of steel 30KHGSA by the finite elements method

Y A Lyutikov, A B Lyuchter and M A Korobov.....208

Application of low-temperature plasma for the synthesis of hydrogenated graphene (graphane)

M B Shavelkina, R H Amirov, V A Katarzhis and V I Kiselev.....214

**Modern problems of theoretical physics**

Propagation of Circularly Polarized Light Through a Two-Dimensional Random Medium

E E Gorodnichev.....220

Depolarization of an Ultrashort Pulse in a Disordered Ensemble of Mie Particles

E E Gorodnichev, S V Ivliev, A I Kuzovlev and D B Rogozkin.....227

Long-Range Correlations Between Transmitted and Reected Fluxes of Electromagnetic Waves

E E Gorodnichev, A I Kuzovlev and D B Rogozkin.....233

High energy electrons accelerated in the field of tightly focused relativistic laser pulse for peak intensity evaluation

K A Ivanov, I N Tsymbalov, A V Rusakov, R V Volkov and A B Savel'ev.....239

The Kubo-Greenwood formula as a result of the random phase approximation for the electrons of the metal

S V Ivliev.....244

The Scattering of Massive Holes by Supercritical Impurity

V M Kuleshov, V D Mur, A M Fedotov and Yu E Lozovik.....249

Crossover in mesoscopic conductance fluctuations in a quasi-one-dimensional system

V V Marinyuk, K A Kondratiev and D B Rogozkin.....256

Accuracy analysis of automodel solutions for Lévy flight-based transport: from resonance radiative transfer to a simple general model

A B Kukushkin and P A Sdvizhenskii.....262

Hydrodynamic instability at the interface between colliding inhomogeneous mediums

Kirill Gorodnichev, Sergey Kuratov, Pavel Zakharov and Evgenii Gorodnichev.....268

Response of explosive HMX to low-velocity impact: modeling by the crystal plasticity finite element method

Denis Ilnitsky, Nail Inogamov and Vasily Zhakhovsky.....274

Hyperons, Δ resonances and condensate of charged $\rho$ mesons within relativistic mean-field models with scaled hadron masses and couplings

K A Maslov, E E Kolomeitsev and D N Voskresensky.....280

On particle number fluctuations in an interacting pion gas with dynamically fixed number of particles

D N Voskresensky and E E Kolomeitsev.....286

Controlling the light shift of the CPT resonance by modulation technique

E A Tsygankov, S V Petropavlovsky, M I Vaskovskaya, S A Zibrov, V L Velichansky and V P Yakovlev.....292

**Challenges in physics of solid state, functional materials and nanosystems**

Dry cryomagnetic system with $MgB_2$ coil

D A Abin, N A Mineev, M A Osipov, S V Pokrovskii and I A Rudnev.....298

Simulation of magnetization and levitation characteristics of HTS tape stacks

I V Anischenko, S V Pokrovskii and N A Mineev.....303

Low-temperature anomalies of EXAFS at the *K*-edge of As in superconducting $LaFe_{0.89}Co_{0.11}AsO$

A Bianconi, A P Menushenkov, V G Ivanov, A A Ivanov and B Joseph.....309

Temperature dependences of the time of electron-electron interactions in two-dimensional heterojunction

K V Bukhenskyy, A B Dubois, S I Kucheryavyy, S N Mashnina, A S Safoshkin, A A Baukov and E Yu Shchigorev.....313

Metallic atomic hydrogen at a pressure of 300-500 GPa

N N Degtyarenko and E A Mazur.....318

Investigation of the sensitivity of MIS-sensor to thermal decomposition products of cables insulation

D V Filipchuk, A V Litvinov, M O Etrekova and D A Nozdrya.....322

Structure, morphology and electrocatalytic properties of $WO_x$ thin films prepared by reactive pulsed laser deposition

V Y Fominski, S N Grigoriev, R I Romanov, M A Volosova, D V Fominski and A V Irzhak.....327

Magneto-quantum oscillations in $Bi_2Se_3$ nanowires

A V Frolov, A P Orlov, A A Sinchenko, V A Volkov, Ya A Gerasimenko, A Yu Kuntsevich and V M Pudalov.....333

Models of $WO_x$ films growth during pulsed laser deposition at elevated pressures of reactive gas

A G Gnedovets, V Y Fominski, V N Nevolin, R I Romanov, D V Fominski and A A Soloviev.....337

Activation of electrocatalytic properties of a-C films by doping with $MoSe_x$ clusters

S N Grigoriev, V Y Fominski, R I Romanov, M A Volosova and D V Fominski.....343

Momentum distribution in two-orbital Hubbard model: Quantum Monte Carlo approach

Vladimir Kashurnikov, Andrey Krasavin and Yaroslav Zhumagulov.....349

Critical current of layered superconductor with defects in tilted magnetic field

V A Kashurnikov, A N Maksimova, I A Rudnev and D S Odintsov.....354

Pseudospectral method for the quantum state determination of the superconductor in the nonuniform magnetic field

D S Korchagin and P F Kartsev.....358

Evaluation of solution stability for two-component polydisperse systems by small-angle scattering

A E Kryukova, P V Konarev and V V Volkov.....364

Simulation of equilibrium particle distribution of the Bose gas of polaritons using quantum Monte Carlo

I O Kuznetsov and P F Kartsev.....370

Temperature dependence of critical current in $YBa_2Cu_3O_{7-\delta}$ films

A V Kuznetsov, I I Sannikov and A A Ivanov.....374

XMCD and TEM studies of as-cast and rapidly quenched $Fe_{50}Nd_{50}$ alloys

V P Menushenkov, A P Menushenkov, I V Shchetinin, F. Wilhelm, A A Ivanov, I A Rudnev, V G Ivanov, A Rogalev, A G Savchenko, D G Zhukov, A V Rafalskiy and S V Ketov.....379

The influence of the shock treatment under heating on the structure and properties of HTS tapes

B P Mikhailov, A B Mikhailova, I V Borovitskaya, V Ya Nikulin, P V Silin, E N Peregudova, S N Polukhin, S V Shavkin, N A Mineev, V F Shamray, V N Kolokoltsev, N A Krutskih and S Y Alibekov.....385

Influence of strain on transport characteristics of HTSC

A N Moroz, A N Maksimova, V A Kashurnikov and I A Rudnev.....391

Structure and electroconductivity of nanostructured ceramics $BiFeO_3$

Zh Kh Murlieva, D K Palchaev, N M-R Alikhanov, M Kh Rabadanov and S A Sadykov.....397

Structure and conductivity of nanostructured YBCO ceramics

D K Palchayev, S Kh Gadzhimagomedov, Zh Kh Murlieva, M Kh Rabadanov and R M Emirov.....403

Investigation of semiconductor heterostructures HgTe/HgCdTe under magnetic fields up to 50 T

V V Platonov, Yu B Kudasov, I V Makarov, D A Maslov and O M Surdin.....408

Effect of picosecond laser exposure on the transport characteristics of second-generation HTS tapes

S V Pokrovskii, O B Mavritskii, A N Egorov, N A Mineev, A A Timofeev and I A Rudnev.....413

Features of formation and evolution of crystal and local structures in nanocrystalline $Ln_2Zr_2O_7$ (Ln = La - Tb)

V V Popov, A P Menushenkov, B R Gaynanov, Ya V Zubavichus, R D Svetogorov, A A Yastrebtsev, A A Pisarev, L A Arzhatkina and K V Ponkratov.....417

Influence of synthesis conditions on the crystal structure of the powder formed in the "$HfO_2$ - $CeO_2$/$Ce_2O_3$" system

V V Popov, A P Menushenkov, R M Khubbutdinov, A A Yastrebtsev, R D Svetogorov, Ya V Zubavichus, A L Trigub, A S Sharapov, A A Pisarev, V V Kurilkin, N A Tsarenko and L A Arzhatkina.....422

Influence of the concentration and lanthanoide type on the phase transitions in $Ln_2O_3$-$MO_2$ system (Ln = La, Gd, Y; M = Zr, Hf)

V V Popov, A P Menushenkov, A A Yastrebtsev, A A Pisarev, I V Shchetinin, M V Zheleznyi, Ya V Zubavichus, L A Arzhatkina and N A Tsarenko.....428

High-pressure cells for study of condensed matter by diffraction and inelastic neutron scattering at low temperatures and in strong magnetic fields

R A Sadykov, Th Strassle, A Podlesnyak, L Keller, B Fak and J Mesot.....434

**Particle accelerators and radiation technologies**

Statistical precision in charged particle EDM search in storage rings

A E Aksentev and Y V Senichev.....439

Study of multilayer polymer materials after ionization treatment

V T Tarasyuk, A A Semkina, V I Solovyeva, D D Fedotova, N E Strokova, D M Malenko, O V Baranov, A V Bakumenko, S N Puchkov and A V Prokopenko.....444

Design of robust microlinacs for wide replacement of radioisotope sources

A V Smirnov, R A Agustsson, S Boucher, M Harrison, K Junge, E Savin and A Yu Smirnov.....449

Thermometry system development for thermoradiotherapy of deep-seated tumours

A M Fadeev, S M Ivanov, E A Perelstein and S M Polozov.....455

Lagrange Equations of Envelopes

V K Baev.....461

RF design of 324 MHz superconducting (SC) CH cavity for 0.21 beta

K Taletskiy, D Surkov and M Gusarova.....465

Electrodynamic characterisitcs measurements of higher order modes in S-band cavity

R Donetsky, M Lalayan, N P Sobenin, A Orlov and A Bulygin.....470

Numerical simulation of electrons dynamics in a microtron on 6 – 10 MeV

Y A Bashmakov, V S Dyubkov and Y Y Lozeev.....477

Investigation of accelerating ion triode with magnetic insulation for neutron generation

A E Shikanov, K I Kozlovskij, E D Vovchenko, V I Rashchikov, V L Shatokhin and A A Isaev.....481

Intense quasiperiodic beam dynamics in accelerating system: mathematical model and optimization method

I D Rubtsova and D A Ovsyannikov.....486

On finding the analytic dependencies of the external field potential on the control function when optimizing the beam dynamics

A D Ovsyannikov, S A Kozynchenko and V A Kozynchenko.....492

The method of accelerating structure tuning and manufacturing quality control

I V Rybakov and V V Paramonov.....500

Modular compact solid-state modulators for particle accelerators

A A Zavadtsev, D A Zavadtsev and D V Churanov.....505

Compact Storage Ring for an X-Ray Source

L Ovchinnikova, V Shvedunov and K Ivanov.....511

Development of PET projection data correction algorithm

P V Bazhanov and E D Kotina.....520

Electron beam agrobionanotechnologies for agriculture and food industry enabled by electron accelerators

Y S Pavlov, A A Revina, O V Souvorova, N L Voropaeva, D V Chekmar, E V Abkhalimov, M A Zavyalov and V P Filippovich.....524

Quarter-Wave buncher for NICA project

M Trushin, R Fatkullin, A Sitnikov, D Seleznev, V A Koshelev, A S Plastun, S V Barabin, A V Kozlov, V G Kuzmichev, G N Kropachev and T Kulevoy.....534

Progress towards 3-cell superconducting traveling wave cavity cryogenic test

R Kostin, P Avrakhov, A Kanareykin, V Yakovlev and N Solyak.....538

Influence of microwave irradiation of cement mixtures on the strength of cement stone and concrete

M S Dmitriev, M V Dyakonov, R A Krasnokutskiy, A D Kolyaskin and S A Dmitriev.....544

Beam dynamics studies at DAΦNE: from ideas to experimental results

M Zobov on behalf of the DAΦNE Team.....548

Influence of tests results on serial production of EXFEL superconducting cavities

D V Kostin and A A Sulimov.....554

The instrument for investigating magnetic fields of isochronous cyclotrons

N V Avreline.....559

Experience with operation of heavy ion superconducting accelerator ISAC-II and SRF activities at TRIUMF

V Zvyagintsev, R E Laxdal and Z Yao.....566

Cloud-based design of high average power traveling wave linacs

S V Kutsaev, Y Eidelman, D L Bruhwiler, P Moeller, R Nagler and J Barbe Welzel.....571

**Modern trends of quantum metrology**

Method of the production and trapping of thorium ions for nuclear transition investigation

P V Borisyuk, S P Derevyashkin, K Yu Khabarova, N N Kolachevsky, Yu Yu Lebedinsky, S S Poteshin, A A Sysoev, E V Tkalya, D O Tregubov, V I Troyan, O S Vasiliev and V P Yakovlev.....577

Methodical errors caused by superposition of electrical and optical signals in a heterodyne laser interferometer

T V Kazieva, A P Kuznetsov, M V Ponarina, K L Gubskiy, S V Norkin, K S Lukyanov, A A Rusakov, K O Baranova and I I Maslenikov.....581

Formation and study of properties of Ta and Mo nanocluster films

T I Kozlova, O S Vasilyev, P V Borisyuk and Yu Yu Lebedinskii.....587

Investigation of the electronic properties of the surface and bulk forms of gold and palladium

U N Kurelchuk, P V Borisyuk, O S Vasilyev and Yu Yu Lebedinsky.....592

Doppler cooling of thorium ions in a multisectional linear Paul trap

P V Borisyuk, S P Derevyashkin, K YuKhabarova, N N Kolachevsky, Yu Yu Lebedinsky, S S Poteshin, A A Sysoev, E V Tkalya, D O Tregubov, V I Troyan, O S Vasiliev, V P Yakovlev and V I Yudin.....596

Estimation of the operating parameters of miniature radioisotope thermoelectric power unit based on the Th-228 isotope

V V Fetisov, O S Vasilyev, P V Borisyuk and Yu YuLebedinskii.....600

Microwave frequency standard based on $^{25}Mg^{+}$ ions

I Zalivako, I Semerikov, A Borisenko, K Khabarova, V Sorokin and N Kolachevsky.....605

Measurement of the upper clock level lifetime in $^{169}Tm$

A A Golovizin, E S Kalganova, D O Tregubov, G A Vishnyakova, D D Sukachev and V N Sorokin.....611

IOP Conf. Series: Journal of Physics: Conf. Series **941** (2017) 011001 doi:10.1088/1742-6596/941/1/011001

Abstract

A.P. Kuznetsov and S.V. Genisaretskaya

III Conference on Plasma and Laser Research and Technologies took place on January 24th until January 27th, 2017 at the National Research Nuclear University "MEPhI" (NRNU MEPhI). The Conference was organized by the Institute for Laser and Plasma Technologies and was supported by the Competitiveness Program of NRNU MEPhI.

The conference program consisted of nine sections:

- Laser physics and its application
- Plasma physics and its application
- Laser, plasma and radiation technologies in industry
- Physics of extreme light fields
- Controlled thermonuclear fusion
- Modern problems of theoretical physics
- Challenges in physics of solid state, functional materials and nanosystems
- Particle accelerators and radiation technologies
- Modern trends of quantum metrology.

The conference is based on scientific fields as follows:

- Laser, plasma and radiation technologies in industry, energetic, medicine;
- Photonics, quantum metrology, optical information processing;
- New functional materials, metamaterials, “smart” alloys and quantum systems;
- Ultrahigh optical fields, high-power lasers, Mega Science facilities;
- High-temperature plasma physics, environmentally-friendly energetic based on controlled thermonuclear fusion;
- Spectroscopic synchrotron, neutron, laser research methods, quantum mechanical calculation and computer modelling of condensed media and nanostructures.

More than 250 specialists took part in the Conference. They represented leading Russian scientific research centers and universities (National Research Centre "Kurchatov Institute", A.M. Prokhorov General Physics Institute, P.N. Lebedev Physical Institute, Troitsk Institute for Innovation and Fusion Research, Joint Institute for Nuclear Research, Moscow Institute of Physics and Tecnology and others) and leading scientific centers and universities from Germany, France, USA, Canada, Japan.

We would like to thank heartily all of the speakers, participants, organizing and program committee members for their contribution to the conference.

IOP Conf. Series: Journal of Physics: Conf. Series **941** (2017) 011002 doi:10.1088/1742-6596/941/1/011002

# Peer review statement

All papers published in this volume of *Journal of Physics: Conference Series* have been peer reviewed through processes administered by the proceedings Editors. Reviews were conducted by expert referees to the professional and scientific standards expected of a proceedings journal published by IOP Publishing.

IOP Conf. Series: Journal of Physics: Conf. Series **941** (2017) 012001 doi:10.1088/1742-6596/941/1/012001

# Investigation of hybrid modulator based on nonlinear optical loop mirror and single-walled carbone nanotubes for mode-locking in Yb-doped ultrafast fiber laser

**A A Borodkin, D V Khudyakov and S K Vartapetov**

Physics Instrumentation Center of Prokhorov General Physics Institute,Russian Academy of Sciences, Troitsk, Moscow,108840, Russia

E-mail: borodkin_andrey@mail.ru

**Abstract**. Algorithm for the mathematical modelling of a fiber oscillator with a hybrid modulator based on a nonlinear fiber mirror and single-walled carbon nanotubes is suggested. It is based on a numerical solution of the nonlinear Schrödinger equation. Using numerical simulation, optimum NOLM coupler ratio, laser spot diameter on the sample preventing sample degradation and high modulator effect are determined.It is shown that hybrid modulator accelerates the self-starting process and decreases the output pulse duration up to 10%.

## 1. Introduction

Fiber lasers of ultrashort pulses are widely used in many fields of science and technology.The all-normal dispersion (ANDI) oscillator configuration is applied in most modern fiber systems. Such lasers are characterized by a relatively simple design with pulses energies up to several tens of nanojoules directly from the output of the oscillator[1].Therefore, it is important to the investigate and develop the new intracavity modulation methods for the stable self-starting of the pulsed generation without modulator parameters degradation. Presently, the various intracavity nonlinear modulators of radiation intensity, such us effect of nonlinear polarization rotation (NPR) [2,3], semiconductor saturable absorber mirrors (SESAM) [4], nonlinear optical loop mirrors (NOLM) [5] are used for stable pulse generation in ANDI lasers. However, the mode-locking based on the single-walled carbon nanotubes (SWNT) is characterized by self-starting and is more stable than NPR and NOLM modulators. Also polymer films with SWNTs are simple for production, cheap and can be used in ring oscillator configuration, in contrast to SESAMs, which are mainly used in linear oscillator schemes. Therefore, a study of lasers with a SWNT modulator and an investigation of various modulation methods in fiber oscillators providing self-starting of pulse generation without optical degradation of the SWNT modulator are very actual problems.

In this paperthe fiber oscillator with a hybrid modulator based on nonlinear optical loop mirror and SWNT is numerical simulated. The pulse duration and self-starting time dependences on the width of the spectral filter, the length of the fiber mirror, and NOLM coupler ratio are obtained. It is shown that the average power density, which optical degradation limit is 18 kW/cm$^2$,is the critical parameter for single-walled carbon nanotubes as the fiber laser modulator.

## 2. Numerical simulation

The effect of the hybrid modulation method based on a NOLM and a saturable absorption of SWMT was studied. An algorithm based on numerical simulation of the laser radiation propagation in single-mode optical fiber is developedfor numerical simulation of fiber lasers with a hybrid modulator and determination of the optimal configuration of a modulator and a fiber oscillator. The dynamics of temporal and spectral pulses parameters of the fiber laser with an NOLM was modelled using the nonlinear Schrödinger equation.This equation is written in the form of projections onto the selected axes of the optical fiber (1) for polarization maintaining fibers. Analytical solution of this equation exists only for rare cases. The equation was solved by the standard split-step Fourier method [6].

$$\begin{aligned} \frac{\partial A_x}{\partial z} &= -\frac{\Delta\beta_1}{2}\frac{\partial A_x}{\partial T} - i\frac{\beta_2}{2}\frac{\partial^2 A_x}{\partial T^2} + \frac{\beta_3}{6}\frac{\partial^3 A_x}{\partial T^3} + \frac{g_0-\alpha_0}{2}A_x + i\gamma\left(|A_x|^2 + \frac{2}{3}|A_y|^2\right)A_x \\ \frac{\partial A_y}{\partial z} &= +\frac{\Delta\beta_1}{2}\frac{\partial A_y}{\partial T} - i\frac{\beta_2}{2}\frac{\partial^2 A_y}{\partial T^2} + \frac{\beta_3}{6}\frac{\partial^3 A_y}{\partial T^3} + \frac{g_0-\alpha_0}{2}A_y + i\gamma\left(|A_y|^2 + \frac{2}{3}|A_x|^2\right)A_y \end{aligned} \tag{1}$$

where $A_{x,y}(z,\tau)$ is the slowly varying pulse envelope amplitude, which depends on the coordinate z of wave propagation in the fiber and on time t measured in a reference frame moving with the pulse group velocity; g is the gain in the fiber; α corresponds to losses in the fiber; $\Delta\beta_1$ - the first order group velocity dispersion, connected with birefringence of an optical fiber; $\beta_2$, $\beta_3$ are the group velocity dispersion coefficients of the second and third orders, respectively; and $\gamma = n_2\omega_0 / cA_{eff}$ is the nonlinear self-phase modulation coefficient ($n_2$ is the nonlinear refractive index, $\omega_0$ is the central angular frequency, c is the light velocity in vacuum and $A_{eff}$ is the effective beam cross section).The calculations were performed for the following fiber parameters: $\beta_2=2.5\times10^{-26}$ s$^2$/m, $\beta_3=2.4\times10^{-41}$ s$^3$/m, $\gamma=4.7\times10^{-3}$ W$^{-1}\times$m$^{-1}$. The ytterbium gain spectrum was approximated by a Gaussian function with a FWHM of 40 nm and the gain maximum at a wavelength of 1030 nm [7]. Nonlinear transmittance of SWNT-carboxymethyl cellulose (CMC) composite film depending on the peak intensity is described by expression:

$$T_{NT}(I) = T_{NS} - \frac{\Delta T}{1 + I/I_{S,A}}, \tag{2}$$

where $T_{NT}(I)$ - sample transmittance, which depends on the intensity, $\Delta T$ – transmittance change due to saturable absorption, $T_{NS}$ - unsaturated part of transmittance. The saturable absorber parameters for CMC/SWNT are ΔT=2.7 % ,$T_{NS}$=70.5% , $I_{S,A}$=0.043 GW/cm$^2$ [8]. The values of the thresholds of optical destruction of the composite sample: the critical peak intensity (0.75 GW/cm$^2$) and the average intensity (18 ± 1 kW/cm$^2$) are also taken from [8].

The working model of a fiber laser with NOLM (Fig. 1 without element 8) [9] is taken as an initial scheme to optimize the parameters of the hybrid modulator of a fiber laser. Numerical simulation of this laser shows a good agreement with the experimental output characteristics. It is shown that the maximum effect on the pulse duration in the laser is provided by the NOLM and the spectral filter. For this reason, the influence of these components parameters is investigated to reduce the self-starting time.

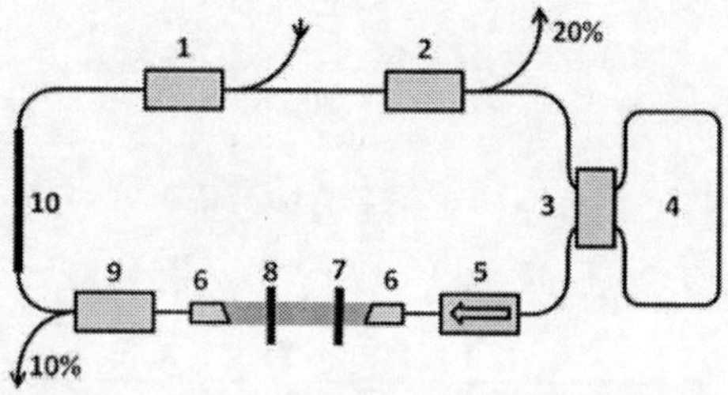


**Figure 1.** Scheme of the fiber laser with NOLM: (1) PM WDM 980/1030 multiplexor; (2) 20/80 output coupler; (3) NOLM coupler; (4) NOLM fiber loop; (5) isolator–polarizer; (6) collimators;(7) spectral filter;(8) SWNT saturable absorber;(9) 10/90 coupler;(10) active Yb-doped fiber.

The numerical simulation starts from a longer pulse 600 ps with a zero spectral chirp. The calculations were continued until the laser radiation parameters (average power and pulse duration) did not change during next ten round trips along the resonator less than 0.1% (stable mode). The value inversed to resulting number of round trips was considered as the rate of the laser self-starting.

The numerical simulation shows the rate of the laser self-starting is derivedfrom the coupling ratio in NOLM (Fig. 2a). The values of the signal part that passed through the coupler in the forward direction are plotted on the ordinate. The coupler with ratio10/90 is marked as 90%. The laser self-starting rate dependences on the NOLM coupling ratio were obtained for the four optical fiber lengths in the loop - 10, 13.3, 16, and 20 m. The optimal ratio of the fiber coupler is in the range from 10/90 to 15/85. However, the pulse duration decreaseswith the coupling ratio (Fig. 2b). This effect is explained by the increasing of the NOLM modulation depth.

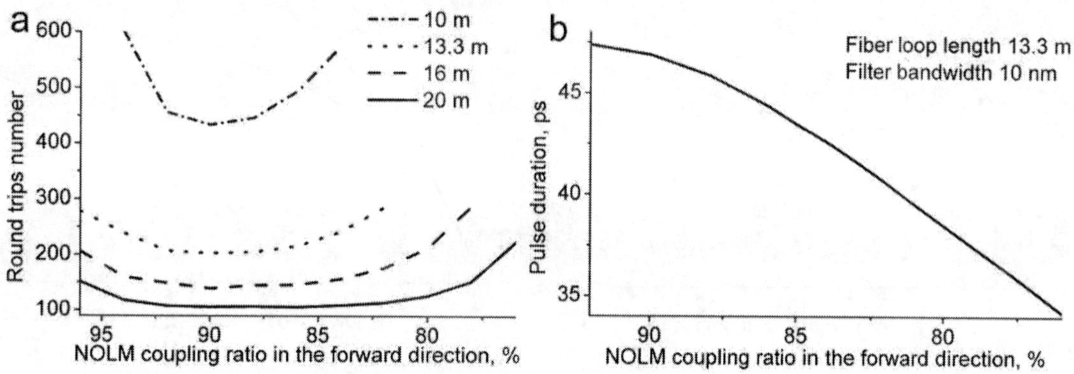


**Figure 2.** a – rate of self-starting and b – pulse durationversus NOLM coupling ratio.

The dependence of the laser self-starting rate on the fiber loop length is shownat Fig. 3a. These results were obtained for the case of a 10/90 NOLM coupling ratio. One can see that the round trips number for self-starting decreases with the fiber loop length increase. However, increasing the length of ANDI fiber oscillator leads to an increase in the pulse duration (Fig. 3b). The optimization of the fiber loop length can not be carried out in oneway and determined by the pulse duration which is chosen for this laser.

IOP Conf. Series: Journal of Physics: Conf. Series **941** (2017) 012001 doi:10.1088/1742-6596/941/1/012001

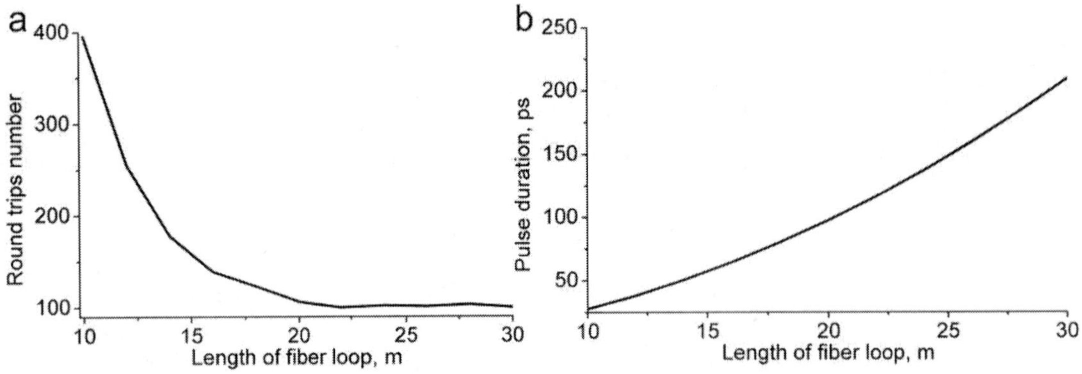


**Figure 3.** a –Rate of self-starting and b – pulse duration versus NOLM fiber length.

The second element, which has a significant effect on the pulse duration in the considered fiber laser scheme, is an intracavity spectral filter. For numerically simulation of the spectral filter bandwidth effect on the laser self-starting rate, the 10/90 NOLM coupling ratio is taken, and the length of the fiber loop is 13.3 m. The dependence of the self-starting rate on the spectral filter bandwidth is shown in Fig. 4a. One can see that filter bandwidth increasing leads to the round trips number for self-starting decreases. The dependence of pulse duration in stable mode on the spectral filter bandwidth is shown at Fig. 4b. In Fig.4b it is shown that the output pulse duration in a stable mode increases with a spectral filter bandwidth.

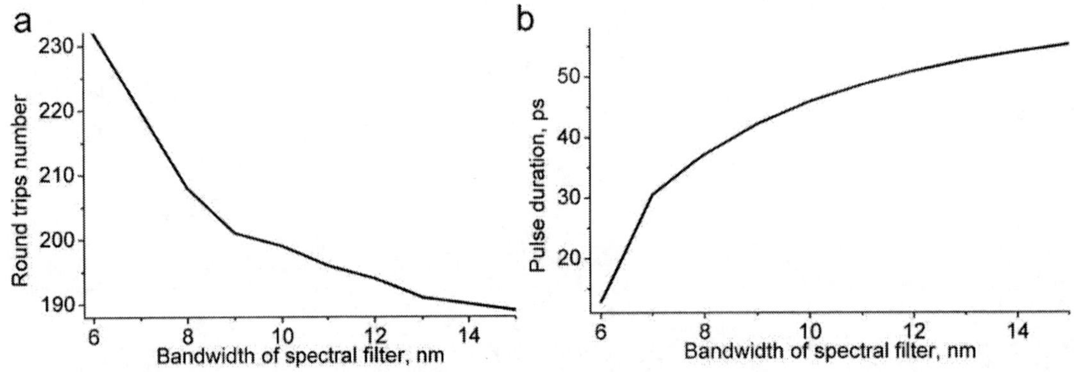


**Figure 4.** a –Rate of self-starting and b – pulse duration versus a bandwidth of the spectral filter.

For a fiber laser with a hybrid modulator (Fig. 1), the dependence of the laser self-starting rate on the peak intensity on the saturable absorber is obtained (Fig.5). In modeling radiation intensity on the sample varied due to the change of the laser beam diameter. The optimum value of the optical intensity on the sample is in the range 1-1.3 GW/cm$^2$, which is 1.5-2 times larger than the optical damage threshold (0.75 GW/cm$^2$). The corresponding value of the average intensity for the sample with SWNTs were 300-500 kW/cm$^2$, which is 20 times higher than the threshold values (18 kW/cm$^2$). Consequently, for fiber lasers with pulse duration more than 30 ps, the damage threshold of a modulator based on SWNT is determined by the thermal damage threshold, i.e. the average power density. The obtained results show that the optimal operation regime of a modulator based on thesaturable absorber is near the thermal destruction threshold. For this fiber oscillator scheme, the threshold average intensityis achieved with a laser beam diameter of 11 μm. In this case, the pulse duration in the stable generation mode is 10% less than for the case of only NOLM modulator.

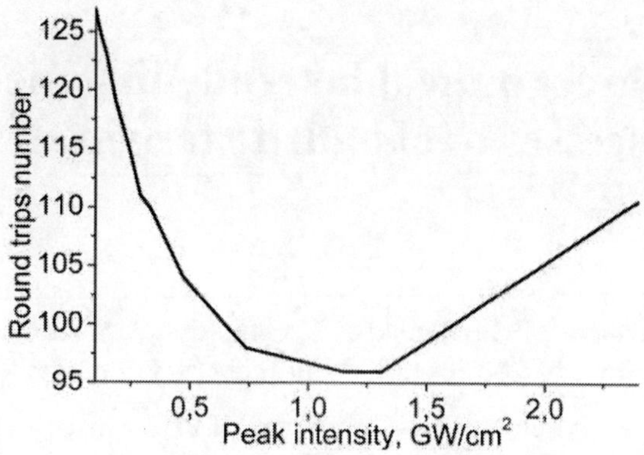


**Figure 5.** Rate of self-starting versus peak radiation intensity on the sample.

## 3. Conclusion

The influence of various hybrid modulator parameters on the output laser characteristics was investigated using the numerical modeling based on the solution of the nonlinear Schrödinger equation. The effect of the length of the NOLM loop, the NOLM coupling ratio, the width of the intracavity spectral filter on alaser self-starting rate and output pulse duration as well as the influence of laser radiation intensity on a modulator based on SWNTs are presented. The optimum NOLM coupling ratio is from 10/90 to 15/85. The critical parameter is found to be the average intensity at the sample, the threshold value of which for the investigated sample is 18 kW/cm$^2$.

The reported study was funded by RFBR, according to the research project No. 16-32-00792 mol_a.

## References

[1] Fermann M, Galvanauskas A, Sucha G and Harter D 1997 Appl. Phys. B **65** 259
[2] Stolen R, Botineau J and Ashkin A 1982 Opt. Lett. **7** 512
[3] Kobtsev S and Smirnov S 2011 Laser Physics **21** 272
[4] Okhotnikov O, Grudinin A and Pessa M 2004 New J. Phys. **6** 177
[5] Aguergaray C, Broderick G, Erkintalo M, Chen S and Kruglov V 2012 Opt. Express**20** 10545
[6] Fisher R and Bischel W 1973 Appl. Phys. Lett. **23** 661
[7] Pask H, Arman R, Hanna D, Tropper A, Mackechnie C, Barber P and Dawes J 1995IEEE J. Selected Top. in Quantum Electron.**1** 2.13
[8] Khudyakov D, Borodkin A, Anatoly S,Lobach A, Ryzhkov A and Vartapetov S 2013 Appl.Optics **52** 150
[9] Borodkin A, Khudyakov D and Vartapetov S 2015 Quantum Electron. **45** 98

IOP Conf. Series: Journal of Physics: Conf. Series **941** (2017) 012002 doi:10.1088/1742-6596/941/1/012002

# Application of direct optical heterodyning methods for studying the processes of chondrite targets destruction by laser radiation

**M Gorbashova[1], I Burdonskiy[2], K Gubskiy[1], A Kuznetsov[1], A Ramazanov[1], K Lukyanov[1], A Leonov[2], K Makarov[2], I Timofeyev[2] and V Yufa[2]**

[1] National Research Nuclear University MEPHI, Moscow, Russia
[2] Moscow Institute of Physics and Technologies (State University), Moscow, Russia

E-mail: gorbashovama@gmail.com

**Abstract**. We have represented a system description used to study of the kinematic parameters of processes of destruction of chondrites under the action of intense shock waves. Interferometric method of measuring speed PDV was the basis for a compact, easy-to-use device intended for the analysis products of target destruction. Using the Fourier transform method of analysis, the dynamic rates of expansion for chondrite targets in interaction with high-power laser radiation and the dependence of the shock wave velocity on the thickness of the target were obtained.

## 1. Introduction

The collision of asteroid-cometary bodies with the Earth poses a significant threat to our planet [1].

After scaling experimental modeling of the destruction of chondritic meteorites by powerful shock waves allows to develop effective mechanisms of antimeteorite protection. Using pulsed laser radiation, it is possible to create shock waves of the necessary intensity for small objects in laboratory conditions.

In present interferometric diagnostic methods are widely used for studying the processes of shock-wave loading of the matter. These methods allow to conduct a non-contact measurement of the object surface velocity under the action of a shock waves. The surface velocity contains information about phase transformations, shock wave parameters, strength properties. It included in the state equation (as mass velocity and pressure).

For problems related to processes of fragmentation and dusting of objects methodic PDV (Photonic Doppler Velocimetry) is usually used. It is based on the fiber interferometers using for the formation of a heterodyne beat signal. This signal is a result of mixing reflected from the object and reference laser radiation [2, 3]. These methods are immune to external influences, accurate, easy to implement. Moreover, the applied algorithms of signal processing allow to analyze fragments and drops distributions (by speed and size). All of the above make the PDV method an optimal tool for studying the kinematic parameters of processes chondritic targets destruction by laser radiation. Experimental installation

Shock wave breaking of chondrite targets experiment was conducted using “Saturn” device in RSRI RAS. “Saturn” device is multistage laser system with automatic stimulation of active medium based on $Nd^{+3}$ - phosphate glass. $Nd^{+3}$ laser operating in Q-switching mode is used as a main generator. Its emission is amplified by preamplifier initially and then by 5 consecutive phosphate glass stages with diameter of 15, 20, 30, 45, 45 mm.

The device is capable of separating beam into two parts which allows to study the glass destruction process by several shock waves. Cuneiform plates made of glass are used for this purpose. Avalanche photodiode, calorimeter and CCD device are used to measure exiting emission characteristics.

Device emission lasts 30 ns with typical energy about 10 Joules which is equivalent to 300 μm spot with power density of $10^{11}$ – $10^{12}$ W/cm$^2$. Targets, which are made of volcanic rock of various composition (chondrite and andesite), are shown in Fig. 1. Some of samples was polished or covered with aluminum (1.5 μm coating) to improve the emission collection.

**Figure 1**. Target photography before and after interaction with emission. Aiming in the center of the target.

New diagnostic scheme was developed in order to explore processes of breaking targets by shock waves. Principal structure of this scheme is shown in Fig. 2.

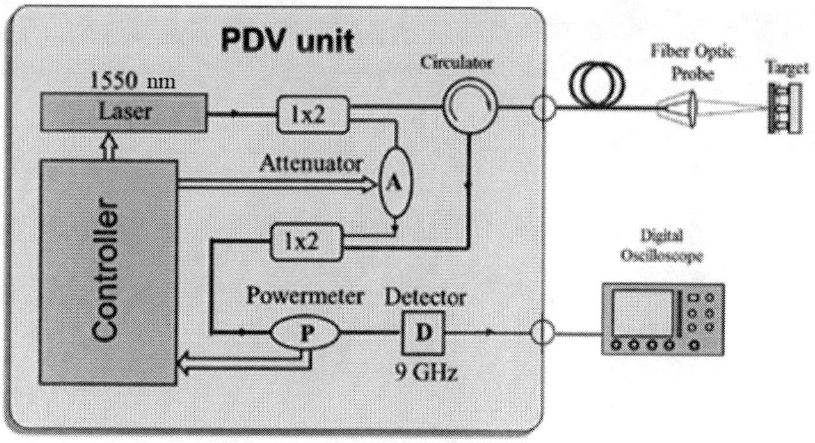


**Figure 2**. Principal scheme fiber interferometer with direct optical signal conversion

Main ideas behind this scheme are listed below. Emission of single frequency continuous laser is conducted to the object of interest through the optical fiber cable. Beam is separated into two parts after passing through the circulator. The first part of the initial beam interacts with the object of interest and is reflected by its surface, then this emission enters the collimator. Both parts combine after the first one passes through the circulator and the second one passes through attenuator. Difference in frequencies of two parts of initial beam caused by Doppler effect produces beats after parts combine. These beats are registered by ET-3500AF photodetector (9 GHz bandwidth) and then the signal with the frequency equal to the Doppler shift is outputted to the LeCroy WaveRunner 640 Zi oscilloscope (4GHz bandwidth). It worth saying that this method requires the usage of single mode fibers and laser with small wideband in order to obtain the necessary contrast [3].

The beat frequency is related to the object's velocity be the following formula:

$$f_b(t) = \frac{2V(t)}{\lambda} \quad (1)$$

Where $\lambda$ is wave length of probing emission.

Relatively low price of this device comes from using of telecommunication equipment. This is also the reason to choose a semiconductor laser with 1.55 μm wavelength as a source of probing emission. Moreover, such choice of wavelength results much lover beat frequencies compared to usage of visible spectrum wave (it uses in devises like VISAR). Thus, it is possible to register the signal on equipment with smaller bandwidth. In this case, maximal velocity which this devise is capable to measure is limited by oscilloscope bandwidth $f_{осц}$ and is equal to $V = \frac{\lambda}{2} f_{осц} = 3100$ m/s.

## 2. Results

A series of measurements of the rear surface velocity for chondrite and andesite targets with different thicknesses values was made. The results are oscillograms describing the dynamics of expansion rates for chondrite targets in interaction with high-power laser radiation. The processing of the received signal was made by software, created in a high-level programming language MATLAB.

To track changes in frequency over time Windowed Fast Fourier Transform was used [4]. The result of its application are spectrograms containing information about the time dynamics of the signal spectrum. The maxima in the spectrum, which stand out against on the noise background, correspond to the target surface velocity and / or the fragments reflecting the laser radiation. To obtain the dynamics of the object speed from the spectrum, it is necessary to perform a recalculation using formula (1). Time and speed resolution are determined by processing parameters. Especially important is the choice of the window type and length.

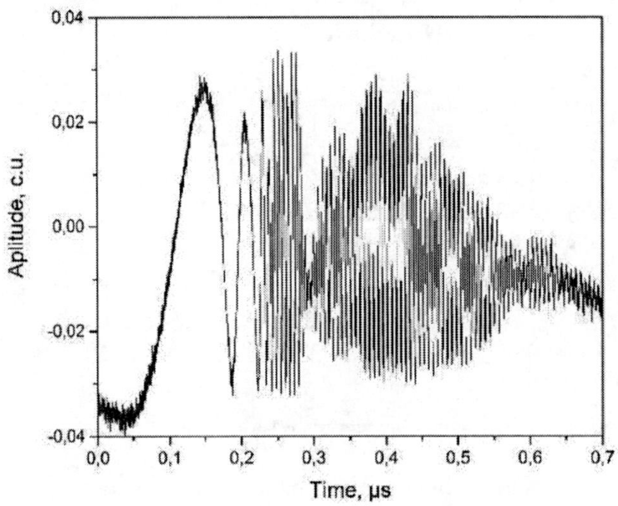


**Figure 3.** Signal recorded with an oscilloscope.

At the Fig. 3 – Fig. 6 the example of result of experimental data processing is given. At the Fig. 3 signal recorded with an oscilloscope is given.

On the spectrogram shown in Fig. 4, there are 2 processes. Velocity which is about 20–50 m/s corresponds to the first process. Oscillation with frequencies, which corresponds to this velocity (dozens MHz) is related to instability in reference laser radiation. Average velocity of target explosion after interaction with laser radiation is equal to 100-130 m/s. Spectral power density cutoff was made to obtain more evident time dependence of shock wave velocity.

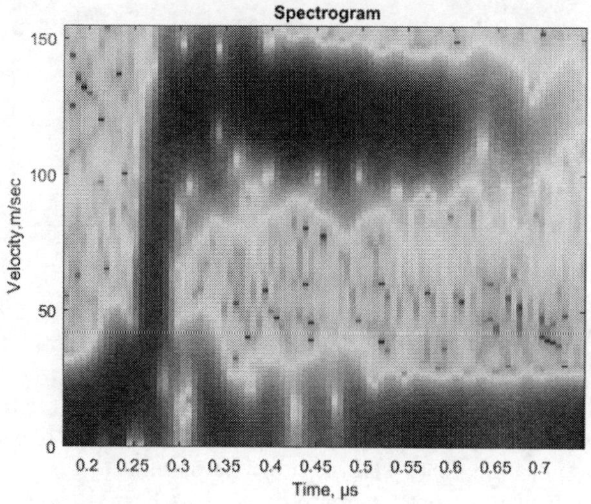


**Figure 4.** Spectrogram of the studied signal

So signals with spectral power density not less than 99% of the maximum was considered. The results of processing are shown in fig. 5.

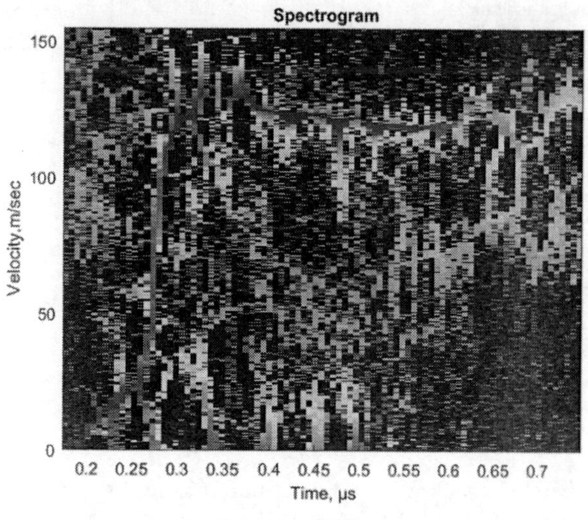


**Figure 5.** Spectrogram of the studied signal after cut-off

So the time dependence of the target surface velocity was obtained (Fig. 6). Obtained dependence is typical for studying processes, its form remains even after radiation energy and target thickness are changed.

Experiment results show that the average shock wave velocity depends weakly on the laser pulse energy. Thus, the beam energy increase by a factor of two leads velocity increase by 5-10%. Perhaps, it is due to the aspect ratio value, since the focusing spot diameter is comparable with the thickness of the target.

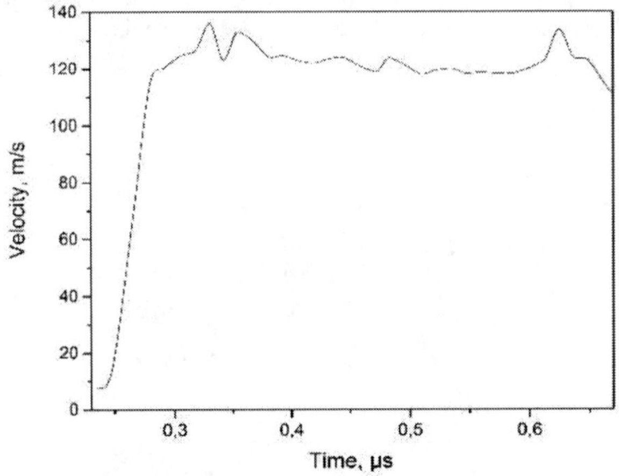


**Figure 6**. Time dependence polished andesite target surface velocity after interaction with pulsed laser radiation (target thickness is 430 µm pulse energy 9. 6 Joules, spot 300 µm, pulse duration 30 ns).

Another oddity is the relation between the shock wave velocity and the thickness of the thick target (Fig. 7). Heating of the target crystalline substance with the energy of plasma flame may explain this oddity. As is well knows, heating the crystalline substances, unlike gases, decreases its speed of sound. The wave moves through matter from the more to the less heated layers that is why it accelerates. Later, it decelerates due to decay of the shock wave at large thicknesses. The error in thickness measurements is determined by the error of used micrometer. The inaccuracy in velocity measurements is related to the averaging of results obtained for same thickness targets.

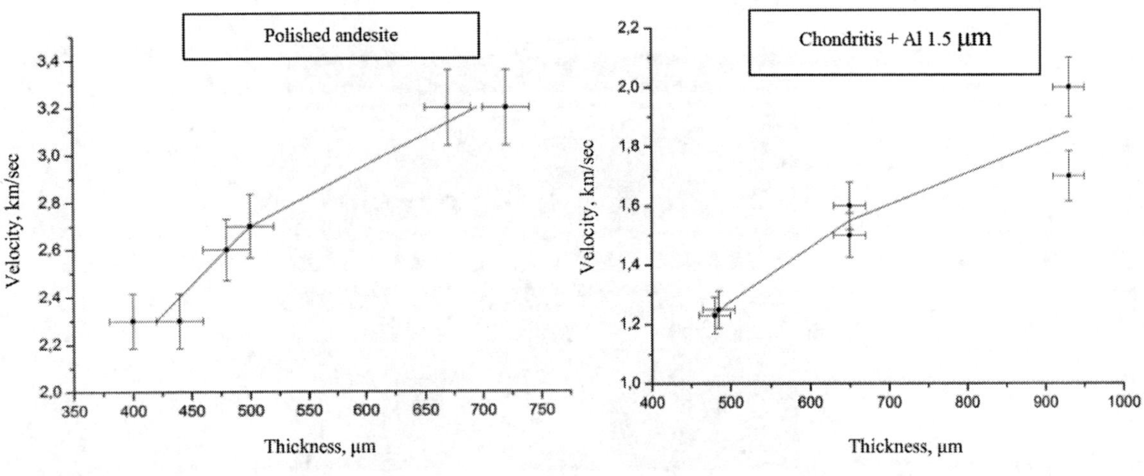


**Figure 7.** Relation between the shock wave velocity and different targets thicknesses. Focusing spot diameter is 300 µm, power density of the laser emission if $4\times10^{11}$ W/cm$^2$.

**Acknowledgments**
Funding: this work was supported by the Russian Science Foundation (RSF) (grant number 17-72-20164).

**References**

[1] Zeleniy L M Esin V I Kokoshin A.A 2013 *Science in Russia* vol 4 p 31-36
[2] O Strand L Berzins D Goosman W Kuhlow P Sargis and T Whitworth 2004 *26 the International Congress on High-Speed Photography and Photonics, edited by D. Paisley SPIE Alexandria* vol. 5580 p. 593.
[3] Strand O T Goosman D R Martinez C Whitworth T L 2006 Compact system for high-speed velocimetry using heterodyne techniques *Rev. Sci. Instum.* vol 77
[4] D. Bergland G 1969 A guided tour of the fast Fourier transform *Bell Telephone Laboratories, Inc*

# The technique of investigation of space-time parameters of the master oscillator with fiber optical smoothing based on the slip photo recorder

**O I Gorchakov, L A Dushina, D S Korniyenko, D V Sizmin, K V Starodubtsev**

Russian Federal Nuclear Center, All-Russia Scientific Research Institute of Experimental Physics, Sarov, Nizhny Novgorod Region, Russian Federation

E-mail: sergeev.vaska@mail.ru

**Abstract**. The basic goal of this work is the development of investigation technique of space-time parameters of broadband oscillator of the powerful Luch Facility of the Russian Federal Nuclear Center designed to smooth radiation profile during passage of fiber optical line. The smoothing is achieved by means of modes mixing because of wave length dispersion in fiber optics, as a result the speckled radiation structure is removed. The optical scheme for recording radiation and the photochronograph with slot scan have been prepared to carry out measurements. The measurements of duration, impulse shape and space-time heterogeneity of laser radiation at output of formation system of reference radiation with fiber optical smoothing were carried out using that developed technique.

## 1. Scheme of diagnostics

The mentioned of broadband oscillator (B.O.) is designed to increase the radiation homogeneity of targets in the powerful laser Luch facility [1]. The width of radiation spectrum of B.O. is 2nm at energy of laser impulse to 50vJ at wave length $\lambda$ =1054nm.

Optical scheme of diagnostics (figure 1) presents reduced two-lens telescope (lenses L1 and L2) corresponding to output aperture of B.O. with the size of photo recorder cathode.

Output plane of fiber optical line of B.O. with scale of order 1:80 for recording of small scale space structure has been changed according to exit slop.

Short focal lens L2 together with L1 focus the radiation to the size comparable to photodiode receiver to record integrated after aperture of impulse shape of laser radiation.

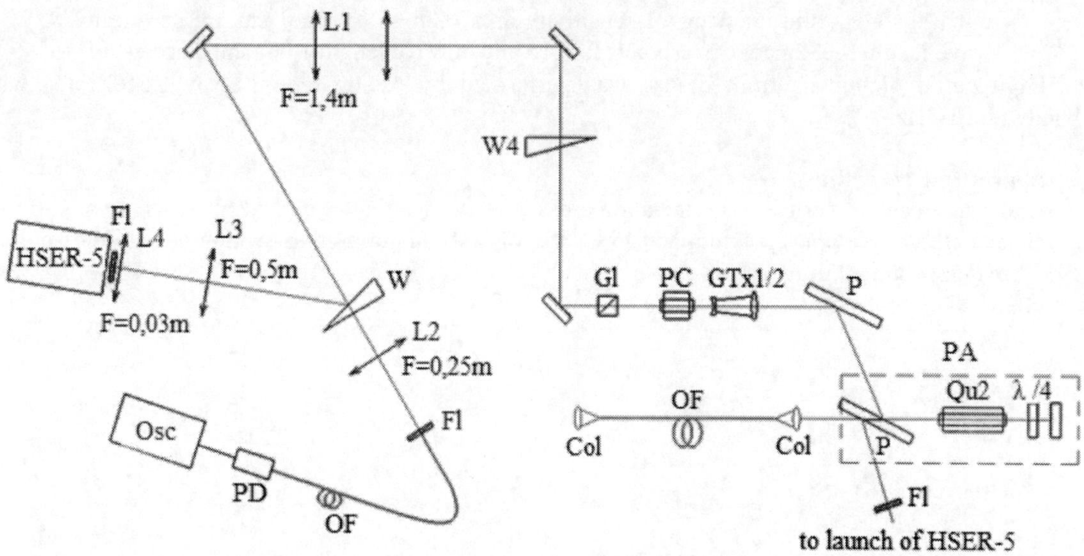


**Figure 1.** Scheme of diagnostics. OF-optical fiber, PA–preamplifier, W–wedge, Qu–quantron, PC–Pockels cell, Gl–Glun prism, λ/4–quarter wave plate, PD–photodiode, P–polarizator, GT–Gallileo's telescope, Col–optical fiber collimator, Fl–filters, HSER-5–photochronograph, Osc-oscillograph

Collecting lens L4 with focal length 3sm was used to record interated shape of laser impulse by photochronograph that allowed fully to focus the whole spot of laser impulse into photochronograph slot.

Synchronization of photochronograph was performed by laser impulse reflected from polarizator surface which is included in two-pass preamplifier and later the radiation is reduced by set of filters.

The photochronograph of type HSER-5 [2] was used. The electronic optical converter of type SPO-23 was used as time analyzing cascade in the device.

This photochronograph has the following metrological characteristics:

- sweep time of recorded – 11ns on screen;
- space resolution – 10 pairs/mm at contrast 10%;
- time resolution of recording – 70ps at with of the installed enter slot 100mcm;
- dynamic range of recording-not less 200 after criterium of duration change 20% in the linear regime of recording.

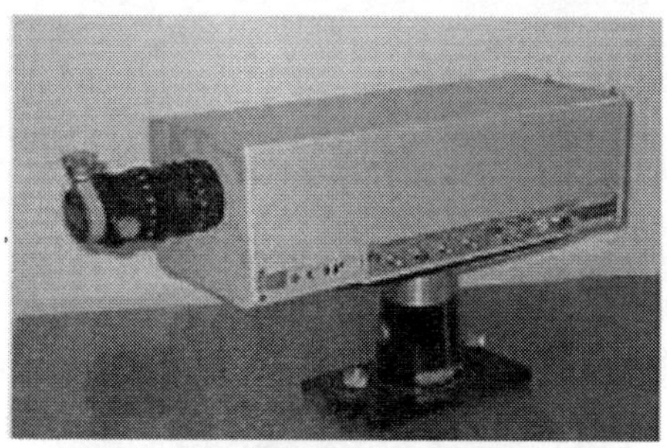

**Figure 2.** Appearance of HSER-5.

Two types of photodetectors were used for integrated recording of impulse shape:

- Photodiode DET08CFC [3]-InGaAs with bandwidth 5GHz and working spectrum range from 800 to 1700nm;

- Photodiode APD2 [4]-avalanche photodiode based on Ge from upper boundary of frequency not less 1GHz and diameter of sensitivity area 0,2mm. Spectral current sensitivity A/W at wave length 0,632mcm corresponds to 0,15 and at wave length 1,06mcm corresponds to 0,5.

Digitization of signals from diodes was performed by oscillograph Lecroy R610Zi [5] with bandwidth 1GHz.

## 2. Space-time recording

Special scan image of vertical laser radiation cross-section at output of optical fiber is shown on figure 3a. The spellings of upper, medium and low parts of laser impulse cross-section at output of optical fiber are demonstrated in figure 3b.

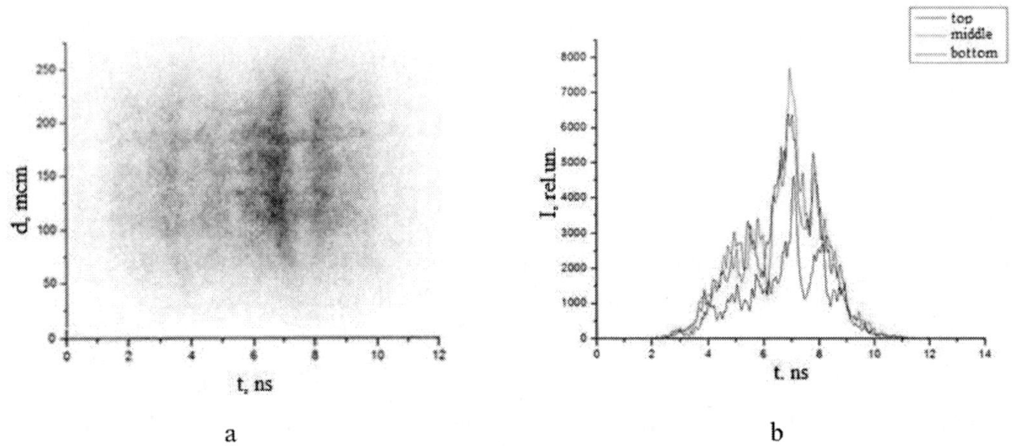


**Figure 3.** Recording of vertical radiation cross-section at output of optical fiber (a) and spellings of upper, medium and low parts of radiation cross-section at output of optical fiber (b).

After receiving the spellings one can judge about local duration of every zone of cross-section (2,1-2,65ns) and space-time heterogeneity of signal (50ps).

Taking into account the scale of optical scheme, space resolution in the plane of optical fiber was of order 1mcm. Heterogeneity with average size of order 5-7mcm with fluctuation intensity of order 7-10% from maximum was recorded in the space structure in the plane of optical fiber.

Recorded enveloping impulses have complex time structure with high-frequency modulation with period from 0,3 to 1ns. Duration of some impulses range from 0,2 to 0,5 ns/ such shape of laser impulse can be explained by stray modes in the initial resonator of B.O. which are not smoothed in fiber line.

The efficiency of obtained data by means of application in the photo recorder the CCD camera will allow to perform in future work setup of B.O. in the regime of real time. Thus, the efficiency of preparation and control of parameters before giving smoothed impulse in amplifying cascades of the Luch facility during powerful experiment can be provided.

## 3. Integrated time characteristics

We carried out the measurements of duration and impulse shape of integrated laser radiation using photochronograph of type HSER -5 and high speed photodiodes of type APD2 and DET08CFC.

The results of measurement demonstrate the advantage of photochronograph both at recording space and time characteristic of laser radiation impulse. In application of high speed slit photo recorder with big dynamic range, the better time resolution was achieved than in application of photodiode recording.

The normalized integrated values of laser radiation impulses recorded using photochronograph of type HSER -5 and photodiodes of type APD2 and DET08CFC are shown in the figure 4.

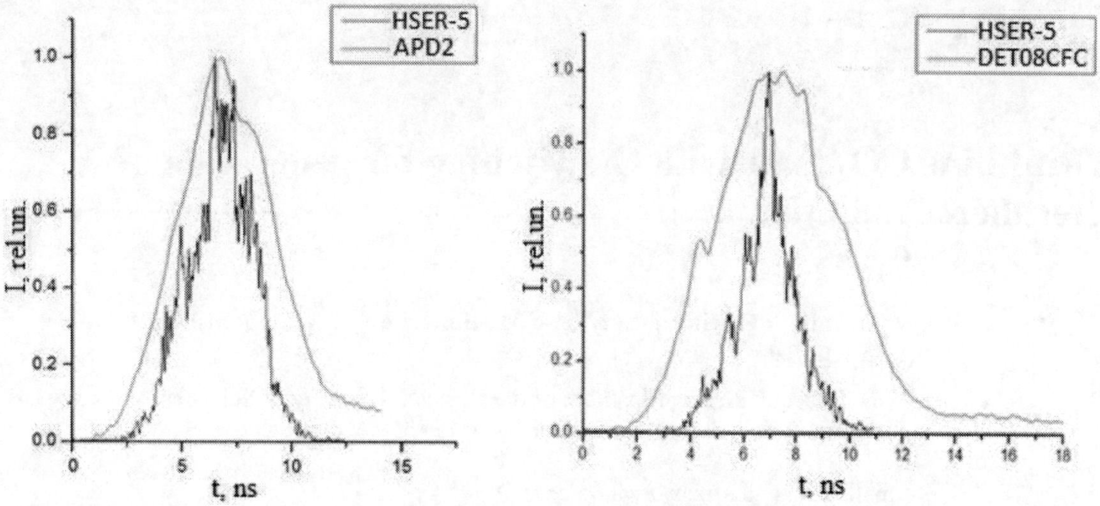


**Figure 4.** Integrated on the whole aperture time shapes of laser impulse, recorded by photo recorded and high speed photodiodes.

It should be noted that time values of laser impulse at half-height, which were recorded by different devices, are different.

Passport characteristics of response time of the used photodiodes are limited by passing band of the applied oscillograph in 1GHz and this must not lead to distortion of the recorded time parameters.

However, duration recorded by photodiodes of type APD2 was 5ns, DET08CFC was 5,7ns. In recording using photochronograph of type HSER-5 durations were 3,1ns and 2,9ns, respectively.

Most likely, prolongation of the recorded duration of laser impulse is connected with its complex structure. As it was reported in Chapter 2, in recording by photochronograph enveloping of output impulse of B.O. presents the train of subnanoseconds peaks moving with high frequency.

The result of recording with photodiodes can be superposition of response to single pulse train that leads to the widening of the recording enveloping by means of dragging of some impulses by semiconductive detectors.

## Conclusion

The device was created for optimization of parameters of broadband oscillator in real time by means of control of space-time characteristics of laser radiation. The advantage of split photochronograph based on electronic optical converter compared to semiconductive detectors by means of high space and time resolution was demonstrated.

## References

[1] Belov I A, Bel'kov S A, Voronich I F 2016 The system for uniform irradiation of targets applying partially coherent light generated in multimode waveguide at Luch facility *Book of abstracts of the 17th International Conference «Laser Optics 2016»* 55

[2] Korniyenko D S, Kravcgenko A G, Litvin D N 2014 Streak cameras for laser fusion experiments *Instruments and Experimental Techniques* **2** 165-175

[3] www.thorlabs.com

[4] Aksenenko M D, Baranochnikov M L 1987 Receivers of optical irradiation *Reference book: Radio and Communications* 66-67

[5] www.lecroy-rus.ru

IOP Conf. Series: Journal of Physics: Conf. Series **941** (2017) 012004 doi:10.1088/1742-6596/941/1/012004

# Multiline $CO_2$ laser with Q-switching for generation of terahertz radiation

**A A Ionin[1], I O Kinyaevskiy[1], Yu M Klimachev[1], V V Stepanishchev[1,2] and I Zh Khafizov[1,2]**

[1]P.N. Lebedev Physical Institute of the Russian Academy of Sciences, Moscow
[2]National Research Nuclear University «MEPhI», Moscow

Email: Ilshat_khafizov@yahoo.com

**Abstract.** In this paper we consider the method of obtaining THz radiation by difference frequency generation (DFG) of multiline $CO_2$ laser. For this purpose a multiline $CO_2$ laser with Q-switching was created. The three strongest lines, 9 R(18), 9 P(20) and 9 P(22) with wavelengths ~9.28, ~9.55 and ~9.57 μm respectively, held 85% of $CO_2$ laser power, and can be used to obtain difference frequency at a wavelength of ~310 μm. DFG of other spectral lines fall within the range of 263 ÷ 8100 μm. Different nonlinear crystals for DFG and filters to separate THz radiation were considered.

## 1. Introduction

The development of the terahertz frequency range is one of the most rapidly developing areas of modern applied physics. Interest in terahertz radiation is due to its following properties: non-ionizing radiation (photon energy 0.04-0.004 eV), does not harm the body, that is, it can be used to scan people and baggage; passes well through turbid environment and finely dispersed materials due to weak Rayleigh scattering.

There are many methods and installations to generate the terahertz radiation: THz quantum cascade lasers (QCL), conversion or titanium-sapphire laser radiation, free-electron lasers (FELs), semiconductor THz signal sources, and the production of THz by plasma filaments. However, almost all these methods give a sufficiently low power of radiation. For example, QCL is the most modern source of THz radiation. But at the moment their potential has not yet been fully revealed. The peak power at 5K is about 248 mW [1].

One of the ways to generate terahertz radiation is to generate the difference frequency (DFG) of $CO_2$ laser radiation in $ZnGeP_2$ [2] or GaSe crystals [3]. Two $CO_2$ lasers with the possibility of wavelength tuning were used in these studies, but with a common active medium and with separate resonators. The radiation of lasers was brought together in a crystal, where the conversion was made. In these studies, the wavelengths of these two lasers were tunable, which made it possible to obtain different wavelengths of THz radiation.

However, publications are known (see, for example, [4]), where a $CO_2$ laser operated simultaneously on several (up to 10) vibrational-rotational transitions. The aim of our work is to create a multiline $CO_2$ laser with a single resonator for DFG in nonlinear crystals. This can be achieved by Q-switching. This method is easier to implement than in the case of two lasers. There is no need to synchronize the radiation of lasers in time and space.

## 2. Optical scheme of measurements and experimental results

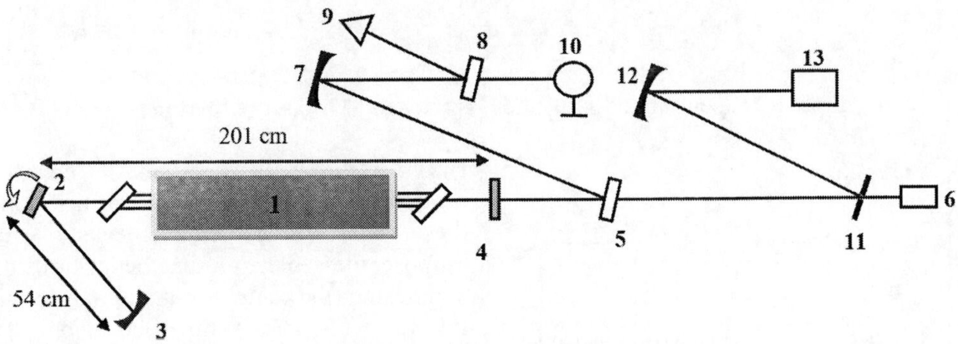


**Figure 1.** The optical scheme of the installation of a multiline Q-switched $CO_2$ laser.

The optical scheme of the installation for the realization of a multiline $CO_2$ laser with Q-switch is shown in figure 1. The active medium 1 had a gas mixture of $CO_2$:$N_2$:He = 1:2:8, excited by a DC discharge at 17kV. The gas discharge tube was covered with $BaF_2$ plates mounted at the Brewster angle. A $CO_2$ resonator with a length of 255 cm is formed by a spherical mirror 3 with a radius of curvature R = 3 m and an output mirror 4. The output mirror 4 was a standard output mirror from an industrial $CO_2$ laser ILGN-704. A part of the radiation coming out of the resonator using plane-parallel plates 5 of $BaF_2$ and a spherical mirror 7 was directed to the thermoelectric sensor 10 (Ophir-3A) for measuring the energy and to the photodetector 9 (PEM-L-3, temporal resolution 0.5 ns) to measure the time behavior of the pulses, using plane-parallel plates 8 of $BaF_2$. The main part of the radiation by a mirror 11 and a spherical mirror 12 was directed into an spectrometer 13 (IKS-31) with the second PEM-L-3 photodetector. The system was adjusted with the help of a semiconductor laser 6, whose radiation was directed to resonator through a millimeter hole in the mirror 11.

$CO_2$ laser can emit in two bands: ~9 μm and ~10 μm. Since the laser tube was covered with plates of $BaF_2$, the laser emitted at a wavelength of 9.2÷9.6 μm. The widest experimental spectrum consisted of 9 lines (Fig. 2). The three strongest lines 9 R(18), 9 P(20) and 9 P(22) with wavelengths ~9.28, ~9.55 and ~9.57 μm respectively, held 85% of $CO_2$ laser power. The line 9 R(18) can give DFG in mixture with 9 P(20) or 9 P(22) line at a wavelength of ~310 μm. DFG wavelengths from other lines fall within the range of 263 - 8100 μm. However, the possibilities of realizing DFG can be substantially limited by the transparent of nonlinear crystals in the terahertz range. Nonlinear crystals of $ZnGeP_2$ [5] and GaSe [6] and PbIn6Te10 [7] have fairly well transparent in the THz wavelength range.

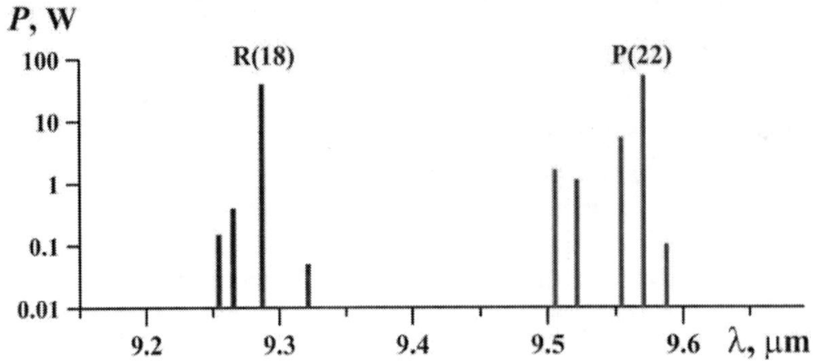


**Figure 2.** Spectrum of multiline Q-switched $CO_2$ laser (experiment).

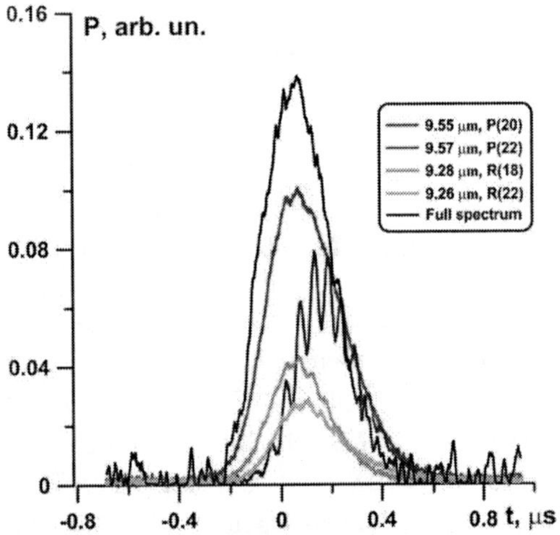


**Figure 3.** Time behaviors of full output spectrum and several spectral lines.

But, for the implementation of the DFG, we needed to make sure that the generation of all these lines occurs simultaneously. For this purpose, the generation time behavior of the lines composing the output radiation spectrum was investigated. These time behaviors for the strongest lines are shown in Figure 3. The analysis showed that the generation of almost all lines occurs at the same time, and if there are deviations, the overlap in time is at least 50 percent (see Figure 3). However, the peak power of these lines can be very different.

Equally important is the stability of the generation of individual lines from pulse to pulse. Figure 4 shows pulsed-periodic generation of the full in the spectrum pulse and on individual lines, spanning ~ 40 consecutive pulses.

An analysis of the results showed that the generation of certain lines does not occur in all pulses (figure 4). For example, generation on the R (22) line is quite rare, on R (18) - more often, and on the lines P (22) and P (20) - almost all the time.

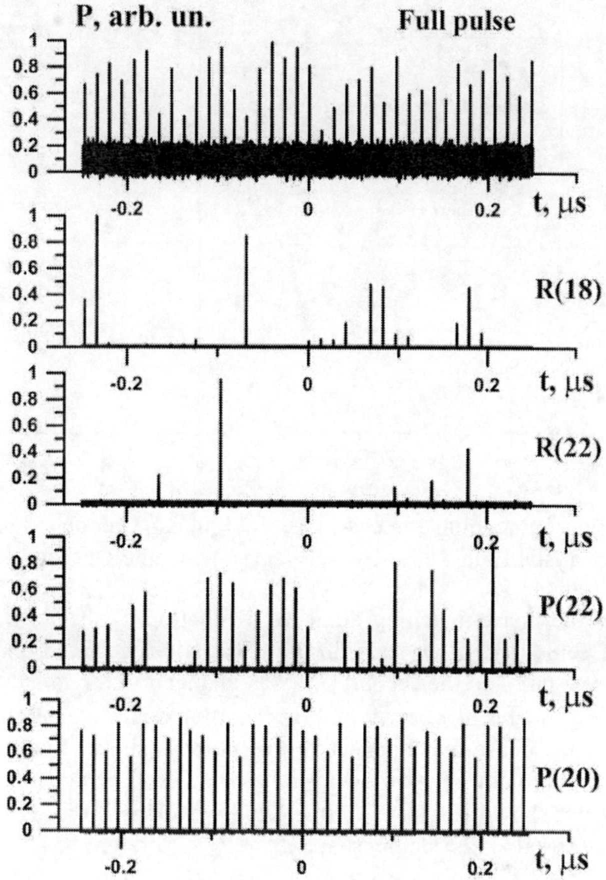


**Figure 4.** Pulse-periodic generation of the full pulse (above) and on individual lines.

Therefore, it was advisable to focus on obtaining DFG between the lines P (20) and P (22). However, the wavelength of DFG in this case is ~ 5380 μm. Registration of such long-wave radiation is not easy. In addition, the quantum efficiency does not exceed ~ 0.1%. If we use DFG between the line R (18) and both lines P (22) and P (20) having near wavelengths, we can immediately obtain two close terahertz lines near ~ 310 μm. In this case, the quantum efficiency can exceed ~ 0.3%.

For further work, it was necessary to solve two problems: which crystal to use for DFG and which filter to use to cut off the pump $CO_2$ laser radiation.

## 3. Nonlinear crystals phase-matching angles

The phase-matching angles of the DFG were calculated by the methods described in [8].

We calculated the phase-matching angles for the $ZnGeP_2$ GaSe and $PbIn_6Te_{10}$ crystals.

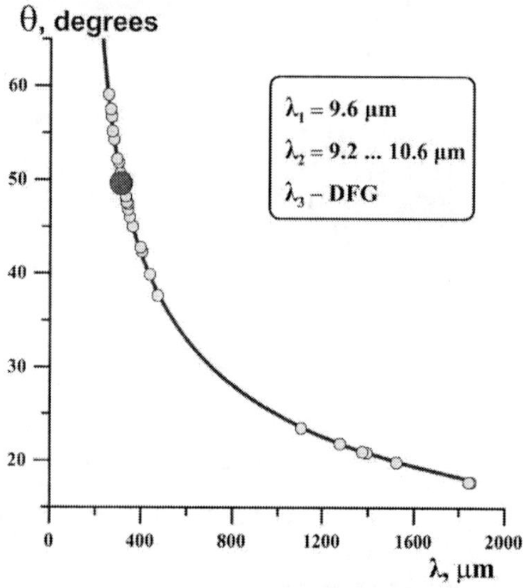


**Figure 5.** Calculated phase-matching angles of the DFG for the crystal $PbIn_6Te_{10}$.

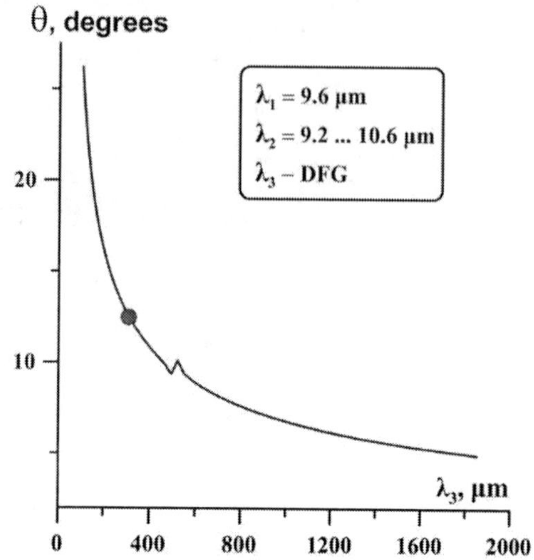


**Figure 6.** Calculated phase-matching angles of the DFG for the crystal GaSe.

Figures 5 and 6 show the phase-matching angle as a function of the DFG wavelength ($\lambda_3$) for the $PbIn_6Te_{10}$ (fig. 5) and GaSe (fig. 6) crystals. In the calculation we fixed one wavelength of the $CO_2$ laser radiation ($\lambda_1$) by 9.6 μm, and the second ($\lambda_2$) was changed in the range from 9.2 μm to 10.6 μm. Since, according to our estimates, the power of THz radiation at a wavelength of ~ 310 μm is expected to be maximal, then on the phase-matching curve this point is shown by a red circle. The remaining circles (yellow) correspond to other possible wavelengths of DFG, which are obtained by using a laser of the $CO_2$ laser spectrum shown in Fig. 2. At the DFG wavelength of ~310 μm, the phase-matching angle was: for a GaSe crystal - 12.5 degrees, for $ZnGeP_2$ - 25 degrees, for $PbIn_6Te_{10}$ - 49.6 degrees. For $ZnGeP_2$ and $PbIn_6Te_{10}$, large phase-matching angles are not an obstacle to the realization of DFG, since these crystals can be cut at the required angle and, as a result, pump radiation can be direct at the normal to the surface of the crystal. But in the case of GaSe this can not be done, since it stratifies. Therefore, a small phase-matching angle for a given crystal shows the possibility of implementing DFG.

## 4. The proposed optical scheme of DFG

In Fig. 5 the proposed installation scheme is presented. We created a multiline Q-switched $CO_2$ laser with one resonator for DFG in nonlinear crystals and selected a crystal. It remains to solve two questions: what detector to choose for recording terahertz radiation and what filter to set before it. The bottom line is that the theoretical efficiency of the conversion is of the order of $10^{-2}$, and in the experiment it is expected to be several orders lower. Therefore, THz radiation should be separated from the radiation of our laser by an optical filter.

IOP Conf. Series: Journal of Physics: Conf. Series **941** (2017) 012004 doi:10.1088/1742-6596/941/1/012004

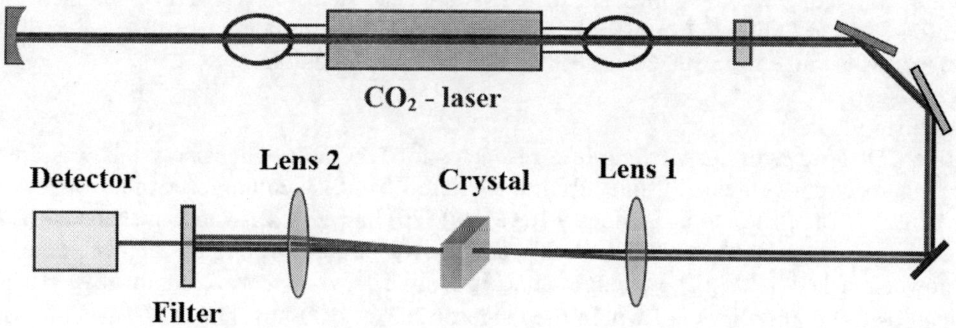


**Figure 7.** The proposed scheme for DFG in nonlinear crystals.

The pump radiation has a wavelength of 9÷10 μm. In figure 5 shows the transmittance and reflection of a teflon (fluoroplastic) sample of a thickness of 0.1 mm in the mid-IR and THz ranges. This material was used in work [2], but it passes a sufficiently large part of the $CO_2$ laser radiation and is not very suitable as a filter, but can be used as a material of lens 2. Another material for filter was considered crystalline quartz. Based on the analysis of figure 6, it was concluded that this material is suitable for further work as a pump radiation filter.

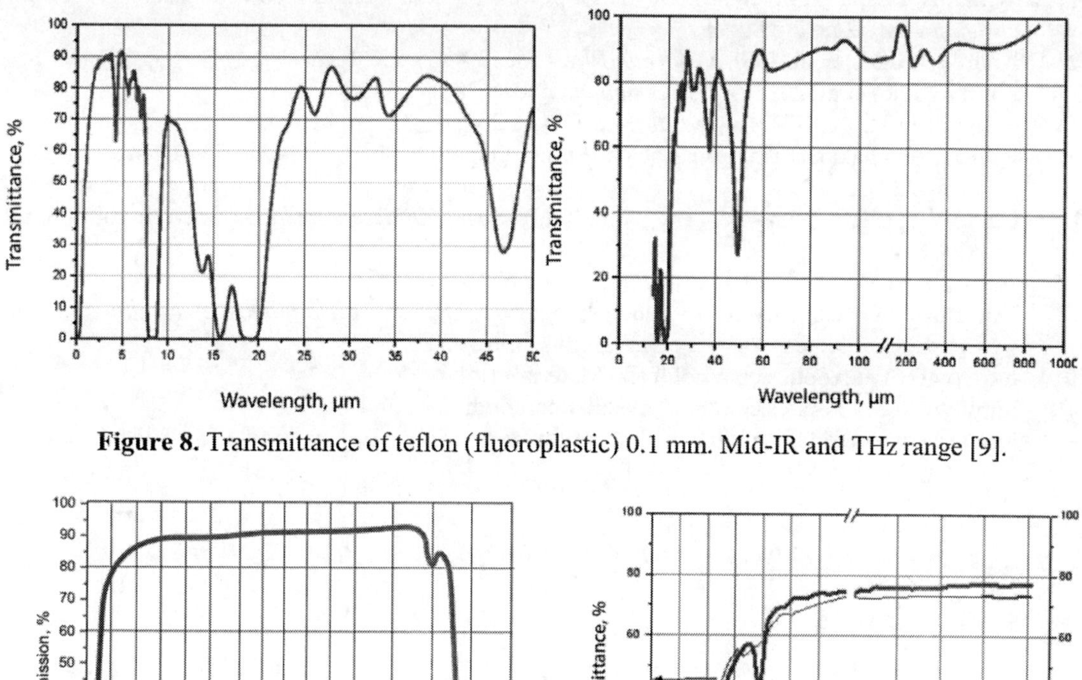


**Figure 8.** Transmittance of teflon (fluoroplastic) 0.1 mm. Mid-IR and THz range [9].

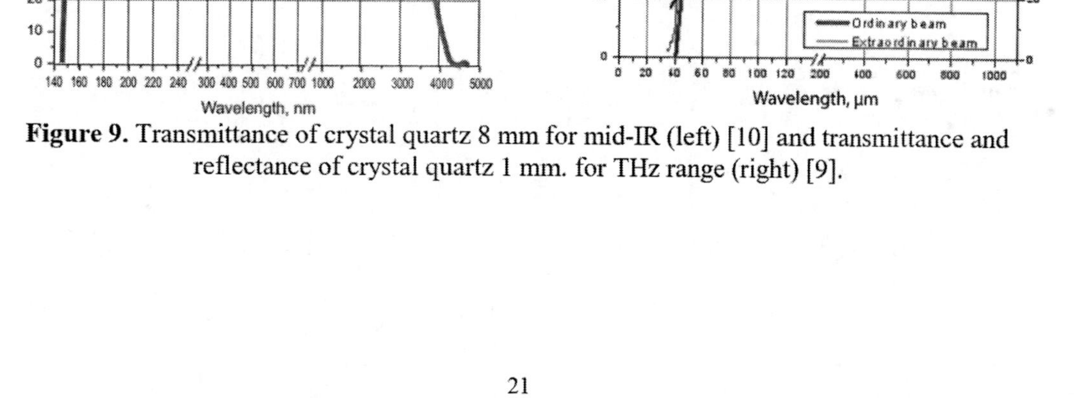


**Figure 9.** Transmittance of crystal quartz 8 mm for mid-IR (left) [10] and transmittance and reflectance of crystal quartz 1 mm. for THz range (right) [9].

Having analyzed the proposals of various manufacturers of medium-IR and THz radiation detectors, we settled on the QMC Instruments Ltd.'s Pyro-electric detector (DLA-TGS element with THz absorption coating) [11]. It has a high sensitivity (1600 V/W at a modulation frequency of the signal 80 Hz) in a wide range from 100 GHz to 100 THz at room temperature.

## 5. Conclusions

Multiline $CO_2$ laser with Q-switching of a resonator for DFG in nonlinear crystals was created. The generation spectrum contained 9 lines from a 9-micron band, since the laser tube was covered with plates of $BaF_2$. The three strongest lines 9 R(18), 9 P(20) and 9 P(22) with wavelengths ~9.28, ~9.55 and ~9.57 μm respectively, held 85% of $CO_2$ laser power. Totally, along all the generation lines, the peak power reached 1 kW. The strongest lines can give DFG at a wavelength of ~ 310 μm. DFG wavelengths from other lines fall within the range of 263 ÷ 8100 μm. The phase matching angles for DFG in the terahertz range in GaSe, $ZnGeP_2$, and $PbIn_6Te_{10}$ crystals were calculated. From these results, it was concluded that all the crystals are suitable for further work. A THz range detector and a filter were selected. The crystalline quartz is best suited for filter.

## Acknowledgments

Work was supported by Russian Science Foundation (16-19-10619).

## References

[1] Williams, Benjamin S., et al. 2006 "High-power terahertz quantum-cascade lasers." *Electronics letters* vol 42 no 2 p 89-91.
[2] Andreev Yu.M., et al. 1998 *J. Korean Phys. Soc.* 33. (3) p 320-325
[3] Zhiming Rao, et al. 2011 *Optics Communications* 284 p 5472-5474
[4] Richardson M. C. 1974 *Applied Physics Letters* 25 p 31-33
[5] Pathik Kumbhakar, Takayoshi Kobayashi and Gopal C. Bhar. 2004 *Applied Optics* vol 43 No. 16 p 3324-3328
[6] Ching-Wei Chen, Tsung-Ta Tang, et al 2009 *Journal Optical Society of America* vol 26 no 9 p A58-A65
[7] Andreev Yu.M., et al. 2016 *Laser Phys. Lett.* vol 13 p 125405
[8] Dmitriev V G, Gurzadyan G G and Nikogosyan D N 1997 *Handbook of Nonlinear Optical Crystals* vol 32, ed. A E Siegman (Berlin: Springer).
[9] http://www.tydexoptics.com/pdf/THz_Materials.pdf
[10] http://www.tydexoptics.com/pdf/Crystal_quartz.pdf
[11] http://www.terahertz.co.uk/index.php?option=com_content&view=article&id=236&Itemid=530

# Numerical simulation of sum frequency generation spectrum in nonlinear crystals with considering dynamics of generation

**A A Ionin[1], I O Kinyaevskiy[1], Yu M Klimachev[1], A A Kotkov[1], D S Kryuchkov[1,3], A M Sagitova[1,2]**

[1]P.N. Lebedev Physical Institute of the Russian Academy of Sciences, 53 Leninskiy prospekt, Moscow 119991, Russia
[2]National research nuclear university MEPhI, 31 Kashirskoe shosse, Moscow 115409, Russia
[3]Moscow Institute of Physics and Technology, 9 Institutskiy per., Dolgoprudny, Moscow Region, 141701, Russia

**Abstract.** Numerical simulation of sum frequency generation spectrum method was elaborated by considering dynamics of generation of every Q-switched CO laser rotational-vibrational lines. It resulted in decreasing of total power by 25% and disappearing approximately 15% lines in comparison with calculation proceeded without taking dynamics into account.

## 1. Introduction

Carbon monoxide laser (CO laser) have an extremely broad spectrum containing about a thousand spectral lines in fundamental (wavelength λ~4.7–8.7μm) [1, 2] and first overtone (λ~2.5-4.2 μm) [1] spectral bands. This laser can run both on a selected single spectral line and in multi-line mode. Efficiency of an electric discharge CO laser being operated in multi-line mode came up to 30-50% for the fundamental band and exceeded 16% for the overtone band [1]. Due to a rich spectrum, multiline CO laser can be successfully applied for multicomponent gas analysis [3]. Multi-line Q-switched CO laser is able to emit dozens of spectral lines (up to 230 lines [4]) in a microsecond pulse. Rich spectrum and high power of multi-line Q-switched CO laser make it very attractive for covering mid-IR range by frequency conversion of radiation in nonlinear crystals. For example broadband two-stage frequency conversion (including both sum and difference frequency generation) was obtained in a single nonlinear $ZnGeP_2$ [5, 6] and $AgGaSe_2$ [7] crystal. As a result the broadband (more than one and half octaves) laser system operating at about 670 lines in the wavelength range from 2.5 to 8.3 μm and based on one the $ZnGeP_2$ crystal and one CO-laser was developed [6]. It should be pointed out that second harmonic generation efficiency of mode-locked CO laser in $ZnGeP_2$ reached 37% [8].

In [5-7] it was considered that the multi-line Q-switched CO laser generates on all rotational-vibrational spectral lines simultaneously. However, as it was noted in [4] the total microsecond CO laser pulse consisted of laser pulses on 230 laser lines with pulse duration about 200 ns each. That can be treated as non-simultaneous generation of lines in microsecond multi-line pulse. Besides, multi-line Q-switched RF discharge CO laser emitted a lot of spectral lines with different pulse duration, which depends on small-signal-gain of different CO laser transition in the active medium [9]. Time behavior of laser output emitted by multi-line Q-switched CO laser with DC discharge was investigated in [10], and a spread of generation beginnings for different spectral lines was 250 ns. However, in [10] only vibrational band 8->7 was considered in details. Time behavior of multi-line Q-switched CO laser on different spectral lines is important for some applications, especially for sum frequency generation

(SFG). Time behavior of the low-pressure multi-line Q-switched CO laser on each spectral line was studied [11]. In present work the influence of spectral-temporal structure of multi-line Q-switched CO laser radiation on broadband SFG in $ZnGeP_2$ crystal was studied with applying data from [11].

## 2. Method of SFG numerical simulation

Calculations were proceeded over experimental data described in detail in [11] when the CO laser generation on each spectral line was measured. For repetition rate 70 Hz the CO laser spectrum consisted of 115 spectral lines in the wavelength range from 4.9 to 6.6 μm (Figure 1a). The laser pulses corresponding to rotational-vibrational transitions of the vibrational band 5->4 are presented on Figure 1(b). It was found that some spectral lines were not simultaneously emitted (see Figure 1(b)). Especially, weak lines with rotational quantum numbers J=6, 7 and J=13 were separated in time. It means that they do not interact with each other under SFG in nonlinear crystals.

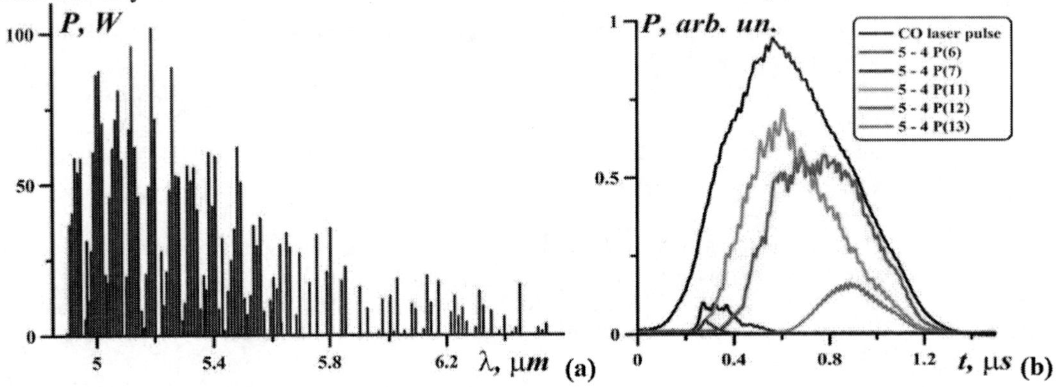


**Figure 1.** Spectrum of multi-line Q-switched CO laser (a) and time behavior of spectral lines in the vibration band 5->4 (b).

In the previous works [5-7], where SFG simulation was carried out, the time behavior of multi-line Q-switched CO laser on different spectral lines was not considered. By other words it was considered that SFG power was proportional to the composition of pulse's amplitudes on two wavelengths *F1* and *F2*:

$$P_{SF}^{(1)} \sim \max(P_{F1}) \cdot \max(P_{F2}) \quad (1)$$

It caused significant mistake for separated in the time pulses (as in Figure 1b). For correct SFG simulation, the time behavior of multi-line Q-switched CO laser must be considered. To do it SFG power must be considered as amplitude of composition of pulse's on two wavelengths *F1* and *F2*:

$$P_{SF}^{(2)} \sim \max(P_{F1} \cdot P_{F2}) \quad (2)$$

Influence of time behavior of multi-line Q-switched CO laser generation on SFG spectrum can be characterized by *K* coefficient, evaluated with:

$$K = \frac{P_{SF}^{(2)}}{P_{SF}^{(1)}} = \frac{\max(P_{F1} \cdot P_{F2})}{\max(P_{F1}) \cdot \max(P_{F2})} \quad (3)$$

*K* has a physical meaning of SFG lines power decreasing due to taking non-simultaneous lasing of different lines into account. Simulation of *K* coefficient for every possible SFG line (for CO laser spectrum Figure 1a) is presented on Figure 2.

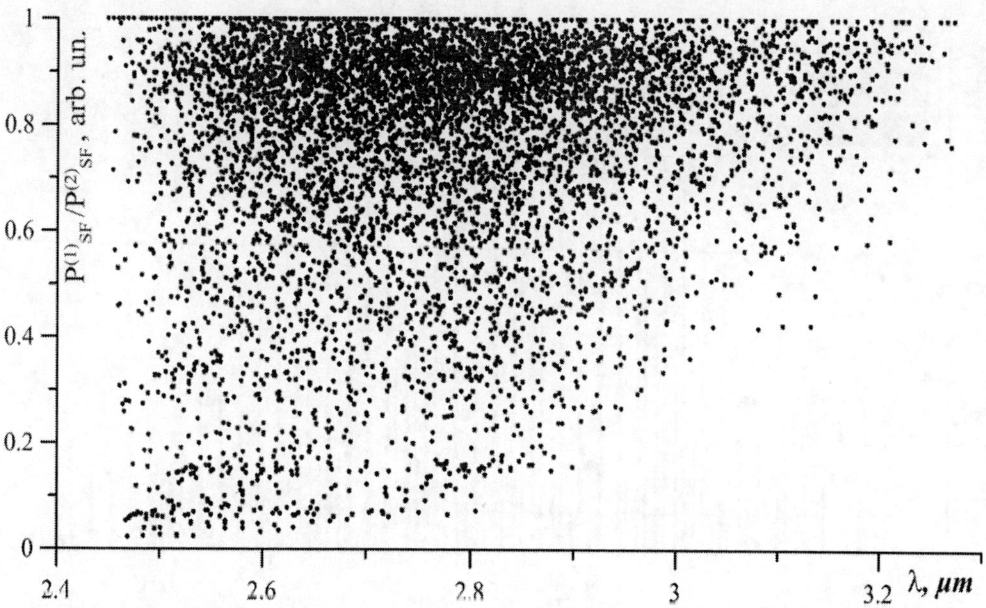


**Figure 2.** Coefficient K, simulated for all possible SFG lines of CO laser spectrum (Fig. 1a).

Figure 2 clearly shows that power of SFG lines significantly decreases, sometimes by more then 10 times, due to non-simultaneous emitting. For second harmonic generation coefficient *K* equales 1. A spread of coefficient *K* for SFG lines with higher wavelength is going less. Average K was about 0.75 that is to say integral power of SFG multiline pulse will be reduced by approximately 25%.

To carry out the numerical simulation of SFG in $ZnGeP_2$ for usual experiment conditions, we applied simulation model presented in [12]. This model is based on a plane wave and low conversion efficiency approximation, where the SFG radiation power $P_{SF}$ can be calculated by expression [13]:

$$P_{SF} = \frac{8\pi^2 d_{eff}^2 L^2 P_{F1} P_{F2}}{\varepsilon_0 c n_{F1} n_{F2} n_{SF} \lambda_{SF}^2 A} sinc^2(\frac{|\Delta k| L}{2}) \quad (4)$$

where $d_{eff}$ is the effective nonlinear coefficient; *n* is refractive index; $\lambda_{SF}$ is SFG wavelength; $\varepsilon_0$ is the dielectric constant; $P_{F1}$ and $P_{F2}$ are the pump laser radiation power for two CO laser lines; $\Delta k$ is the wave mismatch, *L* is the crystal length; *A* is a cross-section of the laser beam. Dispersion equations and effective nonlinear coefficient were taken from [13].

The simulation was proceeded for phase-matching angle 47.4 degree and crystal length of 7 mm. The simulation results with (red bars) and without (bleu bars) considering the time behavior of multi-line Q-switched CO laser generation is presented in Figure 3.

Figure 3 shows that the influence of spectral-temporal structure of multi-line Q-switched CO laser radiation on general shape of broadband SFG spectrum was insignificant. Calculated SFG spectrum with considering the time behavior of multi-line Q-switched CO laser generation consisted of 1291 spectral lines at 0.01 maximal power level from 6670 possible SFG spectral lines. Calculated SFG spectrum without considering consisted of 1487 spectral lines at 0.01 maximal power level. That means that time separation of interacting pulses decreased a number of SFG lines on 13%.

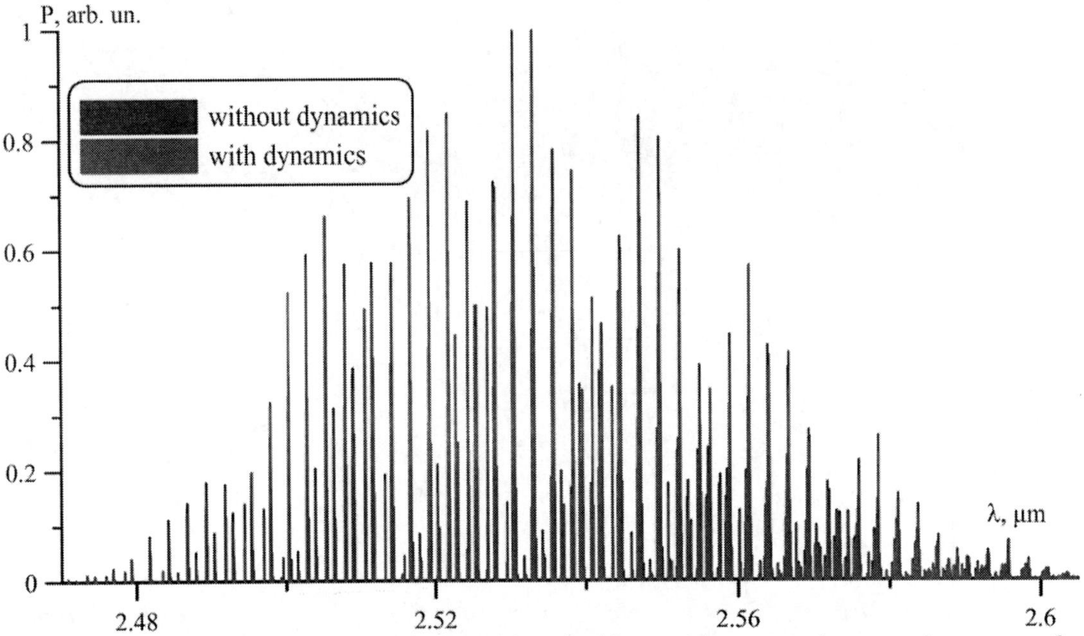


**Figure 3.** The SFG lines power simulation with and without taking spectral-temporal structure of multi-line Q-switched CO laser radiation into consideration.

The integrated over spectrum power of SFG radiation decreased by 18% due to nonsimultaneus generation of CO laser pulses for different spectral lines. Figure 4 presents the enlarged part of SFG spectrum in the region of 2.5μm wavelength.

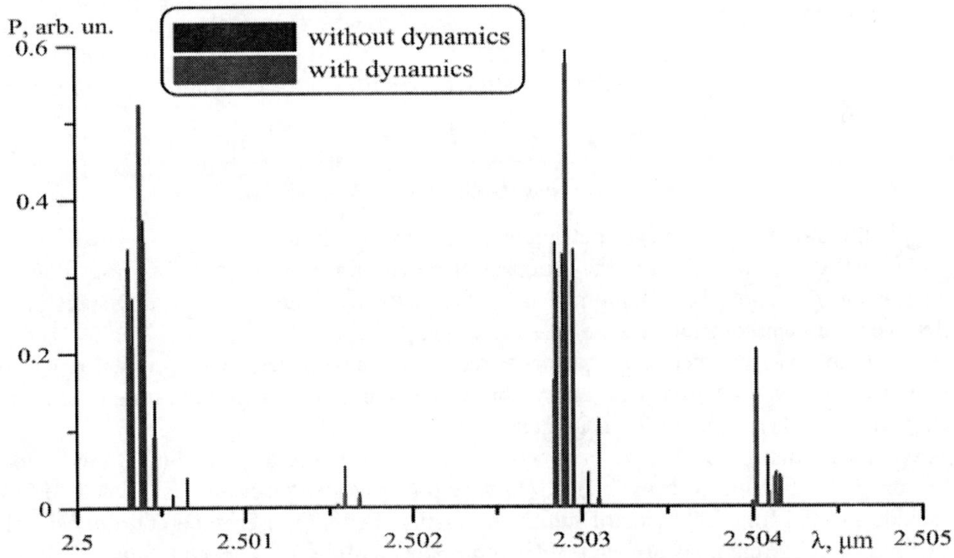


**Figure 4.** Enlarged part of Figure 3 in the region of 2.5025μ wavelength.

Figure 4 shows that the decrease of most powerful spectral lines was small. Significant drop of SFG lines power was observed for weak SFG lines (Figure 4). This fact can be explained as weak CO laser lines at generation were separated significantly on time while powerful CO laser lines generated simultaneously.

## 3. Conclusion

In the first time the influence of time form and duration of Q-switched CO laser rotational-vibrational lines generation was carefully explored. It resulted in elaboration of numerical simulation method used for calculating CO laser SFG spectrum which leaded to decreasing of calculated integral pulse power by 25%. The SFG experiment simulation showed that time separation of interacting pulses of CO laser lines decreased a number of SFG lines by 13% and total spectrum power of SFG radiation by 18%. Future aim is to clarify process of SFG spectrum simulation even more by taking into account atmospheric absorption and compare with upcoming experiments.

**Acknowledgements**
Work was supported by Russian Science Foundation (Project No 16-19-10619).

## References

[1] Ionin A A 2007 Electric discharge CO laser *Gas Lasers* ed M Endo and R Walter (Boca Raton: CRC Press) pp 201-238
[2] Ionin A A, Kinyaevskiy I O, Klimachev Yu M et al 2017 Frequency tunable CO laser operating on the highest vibrational transition with wavelength of 8.7 micron *Optics letters* **42** 498-501
[3] Budilova O V, Ionin A A, Kinyaevskiy I O et al 2016 Absorption of the CO laser sum frequency radiation obtained in a nonlinear crystal $AgGaSe_2$ by molecular gases *Journal of Physics: Conf. Series* **769** 012042
[4] Puerta J, Herrmann W, Bourauel G and Urban W 1979 Extended spectral distribution of lasing transitions in a liquid-nitrogen cooled CO-laser *Applied Physics* **19** 439
[5] Ionin A A, Kinyaevskiy I O, Klimachev Yu M et al 2012 Collinear cascaded mid-infrared frequency conversion in one nonlinear optical crystal *Optics Letters* **37** 2838
[6] Andreev Yu M, Ionin A A, Kinyaevskiy I O et al 2013 Broadband carbon monoxide laser system operating in the wavelength range of 2.5 – 8.3 μm *Quantum Electronics* **43** 139
[7] Budilova O V, Ionin A A, Kinyaevskiy I O et al 2016 Broadband two-stage frequency conversion of CO laser in $AgGaSe_2$ crystal *Optics Letters* **41** 777-780
[8] Andreev Yu M, Budilova O V, Ionin A A et al 2015 Frequency conversion of mode-locked and Q-switched CO laser radiation with efficiency up to 37% *Optics Letters* **40** 2997-3000
[9] Ionin A A, Yu.V. Kochetkov, Kozlov A Yu et al 2017 Q-switched slab RF discharge CO laser *Laser Physics Letters* **14** 055001
[10] Basiev A G, Golubev A A, Gurashvili V A and Izyumov S V 1980 Spectral broadening of Q-switched CO laser generation *Soviet Physics: Technical Physics* **25** 1016
[11] A A Ionin, I O Kinyaevskiy, Yu M Klimachev et al 2017 Spectral characteristics of Q-switched CO laser *Journal of Physics: Conf. Series* (in print).
[12] Ionin A A, Kinyaevskiy I O, Klimachev Yu M et al 2017 Influence of multi-line CO laser focusing on broadband sum frequency generation *Laser physics letters* (in print)
[13] Nikogosyan D N 2005 *Nonlinear optical crystals: a complete survey* (New York: Springer) p 86

IOP Conf. Series: Journal of Physics: Conf. Series **941** (2017) 012006 doi:10.1088/1742-6596/941/1/012006

# Spectral characteristics of Q-switched CO laser

**A A Ionin[1], I O Kinyaevskiy[1], Yu M Klimachev[1], D S Kryuchkov[1,2] and A M Sagitova[1,3]**

[1]P.N. Lebedev Physical Institute of the Russian Academy of Sciences, Moscow 119991, Russia
[2]Moscow Institute of Physics and Technology, Dolgoprudny, Moscow Region, 141701, Russia
[3]National Research Nuclear University MEPhI, Moscow 115409, Russia

Email: adilya.sagitova@gmail.com

**Abstract:** The spectrum of low-pressure cryogenic CO laser operated in Q-switched mode was measured at frequencies of 70 Hz and 120 Hz. CO laser spectrum consisted of 115 spectral lines in the wavelength range from 4.89 μm to 6.54 μm at 70 Hz, and 102 lines in the wavelength range from 4.90 μm to 6.44 μm at 120 Hz. For the first time the dynamics of generation on each spectral line for full CO laser spectrum were measured. We identified rotational-vibrational transitions and analyzed the processes, which affect the formation of the CO laser spectrum. This data were taken into account to determine the peak powers of all the spectral lines correctly.

## 1. Introduction

Research and development of the mid-IR laser source has of great interest for solving many problems of gas analysis, laser chemistry, medicine, isotope separation, and a number of others. For example, many molecules, including pollutants, explosives and toxic substances have absorption lines in this spectral range, and the absorption of human body tissues strongly depends on the wavelength [1].

One of the most effective sources of laser radiation in mid-IR is CO laser. This laser can generate radiation on hundreds of vibrational-rotational lines. The lower vibrational levels of the CO molecule are excited by the electric discharge, and the upper vibrational levels are populated by the vibrational-vibrational exchange due to collisions between anharmonic CO molecules. Thus, a partial inversion population is created on a large number of vibrational-rotational transitions, and due to the cascade mechanism a large number of spectral lines are generated in a wide range of wavelengths [2]. The number of these lines can reach many hundreds [2] for the fundamental vibrational bands in the wavelength range from 4.6 μm [3] to 8.7 μm [4] and the first vibrational overtone in the wavelength range from 2.5 μm to 4, 2 μm [5].

The existence of a cascade mechanism of generation leads to the fact that the CO laser has the highest efficiency among gas-discharge lasers. Output efficiency of fundamental band CO lasers is ~50% [2], first-overtone CO laser efficiency is 16% [6].

However, spectral gap, in which CO laser does not emit, is required for a certain application. One method of expanding and enriching the spectrum of a CO laser is the frequency conversion of the radiation in nonlinear crystals. At the present time sum-frequency conversion of radiation in $ZnGeP_2$ [7], $AgGaSe_2$ [8], GaSe [9], and $PbIn_6Te_{10}$ [10] with an internal conversion coefficient of

6.5%, 1%, 0.3% and 0.01%, respectively, was realized. In order to do this, it is necessary to have a high peak power of radiation, for example, we can use Q-switch laser to achieve this. Actually, the conversion efficiency of broadband CO-laser radiation depends both on the characteristics of used nonlinear crystals and on the characteristics of the CO laser itself.

The purpose of this work is a detailed investigation of the spectral-temporal characteristics of a Q-switched cryogenic CO laser pumped by a direct current discharge. Previously, the spectral-temporal characteristics of such laser were investigated both experimentally and theoretically in [11-12]. Although limited experimental results were given in these publications, but they showed that the generation time on separate vibrational-rotational lines is very different.

## 2. Experiment Set-up

We used Q-switched CO laser for investigating the dynamics of generation on individual vibrational-rotational lines. The optical scheme of our experiment is presented in figure 1. The ends of the CO laser tube 1 were coated with Brewster windows, which are made of $CaF_2$. The gas mixture of active medium 1 consisted of $He:N_2:CO:Air=70:5:1:1$. Tube voltage was 9 kV and a current was 6 mA. The pressure of the mixture was 7.7 Torr.

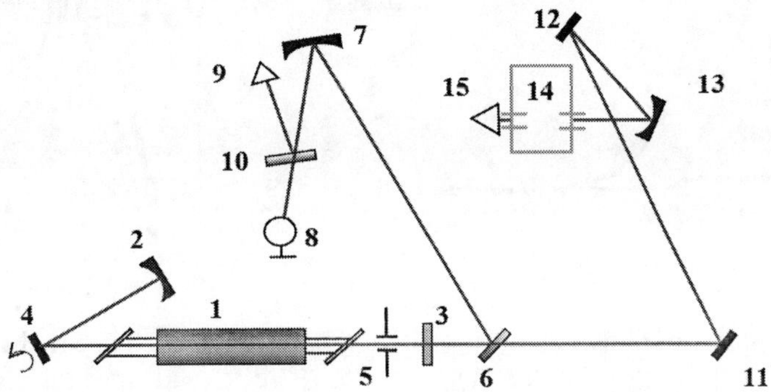


**Figure 1.** Optical scheme. 1 – CO laser tube; 2 – totally reflective spherical mirror; 3 – partially reflective mirror; 4 – rotating mirror; 5 – diaphragm; 6, 10 – $CaF_2$ flat plates; 7, 13, – spherical mirrors; 8 – power meters; 9, 15 – photodetectors; 11, 12 – turning mirrors, 14 – spectrometer IKS-31.

The laser resonator was formed by a reflective spherical mirror 2 (radius of curvature R = 9 m) and a flat output mirror (reflection coefficient ≥90% in the wavelength range ~ 5.0-6.5 μm). A flat rotating mirror 4 was placed inside the resonator, thus the laser operated in Q-switch mode. Diaphragm 5 provided operation of the laser on the fundamental transverse mode. To control the average power of radiation, its part (~ 5%) was directed to a spherical mirror 7 (R = 0.6 m) by a plane-parallel plate 6 of $CaF_2$. Further, the radiation was focused on a power meter 8 (Ophir-10A), and a photodetector 9 (PEM-L-3, temporal resolution 0.5 ns) by another plate 10 of CaF2 to control the temporal shape of the total spectrum pulse.

The main part of the radiation (~ 95%) was directed by turning mirrors 11 and 12, and spherical mirror 13 (R = 0.5 m) into the spectrometer 14 (IKS-31) for measuring the spectral characteristics. The selected radiation was directed to the photodetector 15 (PEM-L-3). Signals from photodetectors 9 and 15 were recorded on an oscilloscope Tektronix TDS 1012 (not shown in figure 1).

## 3. Dynamics of generation of selected spectral lines

A typical oscillogram, which was recorded on an oscilloscope, is presented in figure 2. Signal 1 corresponds to the dynamics of the generation of the total spectrum CO laser pulse, and signal 2 corresponds to the dynamics of generation on a particular vibrational-rotational line. The results were averaged over 4 measurements. Generation durations of 0.1 and 0.5 power level on a particular vibrational-rotational line are shown in blue and red lines, respectively.

The dynamics of generation of the total pulse and individual rotational components of the vibrational band 7-6 is presented in figure 3. One can see from the figure that the strongest lines, for example, P(9) and P(8) completely overlap in time. Consequently, they will interact within a nonlinear crystal and sum-frequency generation occurs the most effectively. There are lines, for example, P(8) and P(11), which partially overlap in time. Consequently, sum-frequency generation occurs less effectively in this case. But there are also such lines, for example, P(7) and P(12), which practically do not overlap in time. It means that sum-frequency generation of these lines doesn't occur at all.

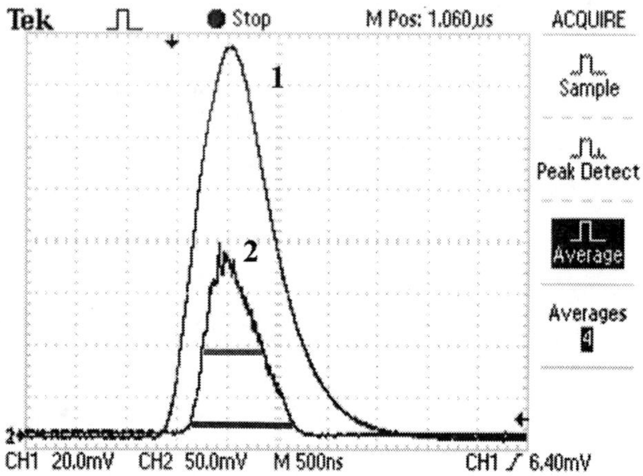


**Figure 2.** A typical oscillogram of generation dynamics of the total spectrum pulse (1), and the dynamics of generation of a specific vibrational-rotational line (2).

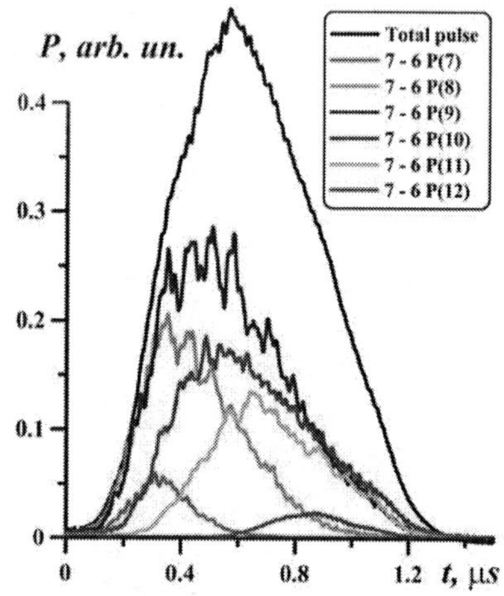


**Figure 3.** Time behavior of of the total pulse and spectral lines in the vibration band 7-6.

Durations of spectral lines generations for Q-switching frequencies of 70 Hz and 120 Hz depending on the wavelength is presented in figure 4. The blue segment shows the duration of the level of 0.1, and the red one shows the duration of the level of 0.5 from the maximum of the pulse. Also, the points indicate the moments when the pulse generation on the each line reached its maximum. In the upper part of figure 4, the horizontal segments show the wavelength ranges corresponding to the vibrational band, the transitions of which are indicated: (V+1) – V P(J), where V is the vibrational quantum number, and J is the rotational quantum number. The generation time at the level of 0.1 at Q-switching frequency of 70 Hz is approximately twice as long the generation time at Q-switching frequency of 120 Hz. It is seen, when vibrational bands are low (V = 3÷13), laser generation begin at low and medium rotational components (J = 6÷10). Whereas the generation at higher rotational components (J = 11÷13) of each vibrational band began in the second half of the entire laser pulse. In general, we can observe that with an increase of the number of the rotational component within a single band the moment of reaching the maximum power is observed at a later time. In middle vibrational bands (V = 14÷23) this analysis was difficult due to a decrease of rotational components. This is apparently due both to a gain decrease of these lines and to the presence of air absorption in this spectral range (5.6÷6.5 µm).

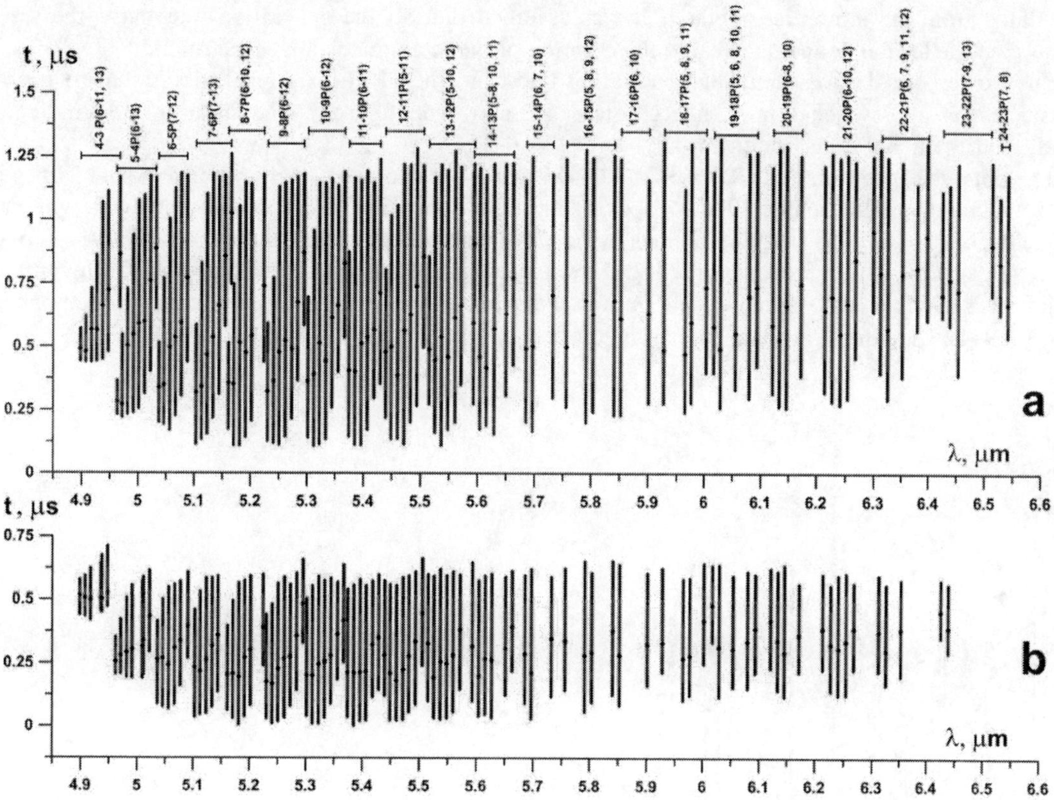


**Figure 4.** Duration of spectral lines generation for Q-switching frequencies of 70 Hz (**a**) and 120 Hz (**b**).

We can observe that the generation at the lowest vibrational band (V = 3) began with a significant delay when compared with generation on higher vibrational bands. It can be concluded that the population inversion in this vibrational transition appears only due to the "falling" of CO molecules from higher levels by generation on higher vibrational bands. In this case, clear advantage of the multilevel generation of CO laser is the cascade mechanism of the appearance of a population inversion.

## 4. Method of calculating the generation peak power of spectral lines and the generation spectrum

The radiation power of individual vibrational-rotational lines was measured in relative units (Volts). The maximum value in these units for each line was multiplied by the generation time for a given line in the level of 0.5. Thus, the generation energy of a given line was obtained in relative units. The sum of the energies over all the lines gave the energy of the generation pulse, which is total in the spectrum, in relative units. Conversion of peak power to Watts in this work was carried out by normalizing the total generation pulse to the real energy. The single pulse energy is equal to the average power of the total radiation (actually measured in the experiment) divided by the Q-switching frequency.

Then the peak power of the spectral lines was refined by the spectral response of the photodetector. According to the dependence, which is known from the photodetector's passport, the actual power of CO laser was calculated.

IOP Conf. Series: Journal of Physics: Conf. Series **941** (2017) 012006 doi:10.1088/1742-6596/941/1/012006

Peak power of the majority of the lines obtained using this method is 20% higher than the power calculated from the simplified method. It was assumed that all the generation lines have the same duration at the level 0.5, and its equal to the duration of the total pulse in the spectrum.

The power distribution along the generation lines and the identification of vibrational-rotational transitions for a Q-switched multi-line CO laser are shown in figure 5. The spectrum also indicates vibrational-rotational transition titles.

The obtained spectrum of a CO laser at a Q-switching frequency of 70 Hz consisted of 115 lines in the wavelength range from 4.89 μm to 6.54 μm. The spectrum of a CO laser at a Q-switching frequency of 120 Hz consisted of 102 lines in the wavelength range from 4.90 to 6.44 microns. In both cases, the strongest lines were observed in the wavelength range from 4.8 to 5.3 μm. The maximum peak power is on the transition 8 - 7 P (9). In each of the vibrational bands, the rotational components with J = 9÷12 were the strongest.

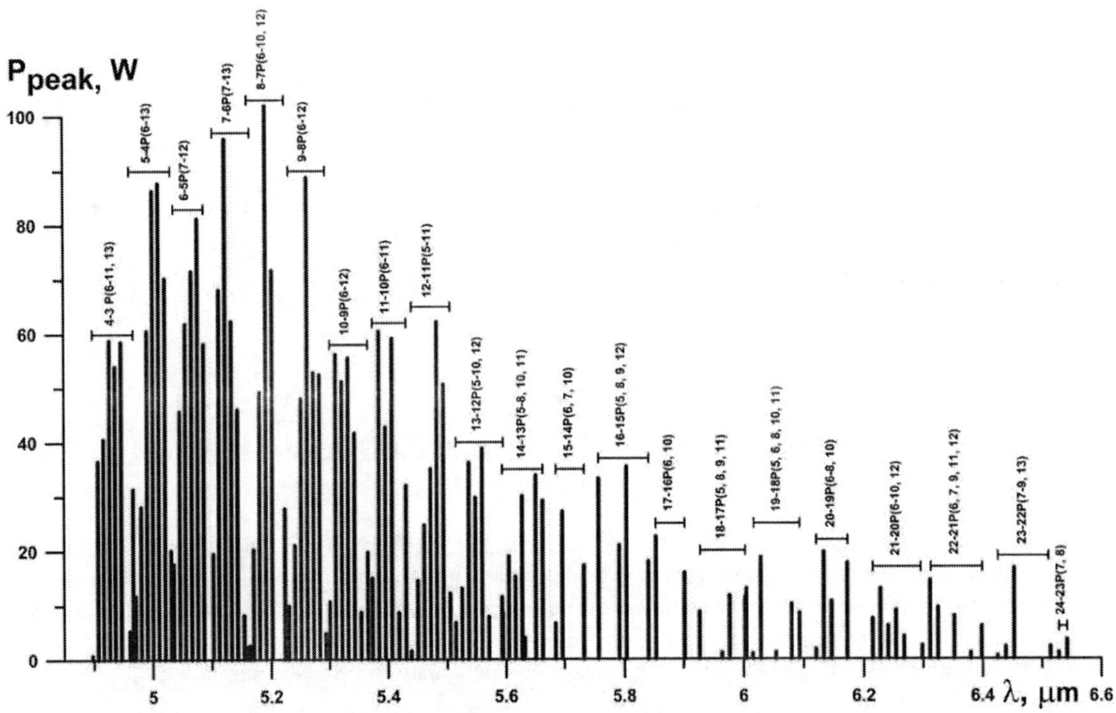


**Figure 5.** CO laser spectrum at a Q-switching frequency of 70 Hz.

Figure 6 presents that the spectral lines are only in those ranges of wavelengths where the air absorption is not very large. Transmission value is indicated by a border between yellow and white.

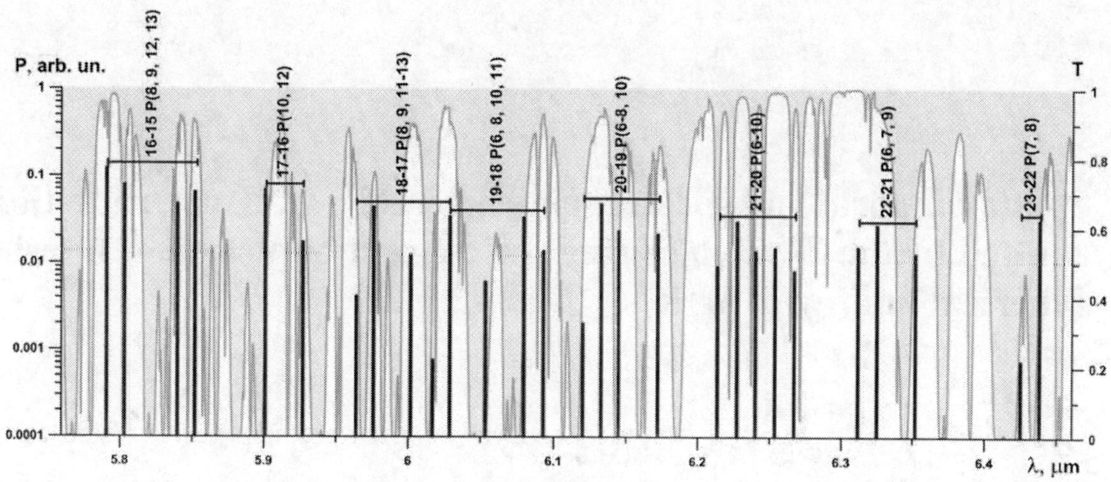


**Figure 6.** Part of the CO laser spectrum at Q-switching frequency of 120 Hz and the air transmittance.

## 5. Conclusions

As a result of experiments, the generation of the Q-switched multi line CO laser spectrum at two different frequencies was obtained. For the first time, the dynamics of the generation of such CO laser on all lines of the spectrum was measured. When Q-switching frequency was 70 Hz the spectrum consisted of 115 spectral lines in the range from 4.89 μm to 6.54 μm. When it was 120 Hz 102 lines were obtained in the wavelength range from 4.90 μm to 6.44 μm. The peak powers of the spectral lines were calculated with normalization for the duration of their generation at a level of 0.5. We identified transitions, which produce every line, and analyzed the processes, which affect the formation of the spectrum.

**Acknowledgments**
Work was supported by Russian Science Foundation (16-19-10619).

## 6. References

[1] Rodin A V, Borisov V M, Kuz'min V N and Mezhevov V S 2004 Pulsed lasers operating from UV to IR at TRINITI for possible medical applications *Proc. of SPIE* **5486** 374
[2] Ionin A A 2007 Electric discharge CO laser *Gas Lasers* ed M Endo and R Walter (Boca Raton: CRC Press) pp 201-238
[3] McCord J E, Ionin A A, Phipps S P, et al. 2000 *IEEE Journal of Quantum Electronics* **36** 1041
[4] Ionin A A, Kinyaevskiy I O, Klimachev Yu M, et al. 2017 *Optics letters* **42** 498
[5] Basov N G, Hager G D, Ionin A A, et al. 2000 Efficient pulsed first-overtone CO laser operating within the spectral range of 2.5-4.2 μm *IEEE J. Quantum Electronics* **36** 810
[6] Ionin A A, Klimachev Yu M, Kozlov A Yu, et al. 2006 *Quantum Electronics* **36** 1153
[7] Andreev Yu M, Ionin A A, Kinyaevskiy I O, et al. 2013 *Quantum Electronics* **43** 139
[8] Budilova O V, Ionin A A, Kinyaevskiy I O, et al. 2016 *Optics Letters* **41** 777
[9] Lai-ming Zh, Ji-Jiang X, Jin G et al 2012 *Optics and Precision Engineering* **20** 277
[10] Ionin A A, Kinyaevskiy I O, Klimachev Yu M, et al. 2016 *Optics Letters* **41** 2390-93.
[11] Basiev A G, Golubev A A, Gurashvili V A and Izyumov S V 1980 *Soviet Physics Technical Physics* **25** 1016
[12] Basiev A G, Galtsev V E, Gurashvili V A, et al. 1981 Features of the spectrum formation of Q-switched CO laser *Preprint of Institute of Atomic Energy* No 3448/12 (in Russian)

IOP Conf. Series: Journal of Physics: Conf. Series **941** (2017) 012007 doi:10.1088/1742-6596/941/1/012007

# Spatiotemporal and spectral characteristics of X-ray radiation emitted by the Z-pinch during the current implosion of quasi-spherical multiwire arrays

**A N Gritsuk**

SRC RF TRINITI, Troitsk, Moscow, Russia

E-mail: griar@triniti.ru

**Abstract**. For the first time, a quasi-spherical current implosion has been experimentally realized on a multimegaampere facility with the peak current of up to 4 MA and a soft X-ray source has been created with high radiation power density on its surface of up to 3 TW/cm$^2$. An increase in the energy density at the centre of the source of soft X-ray radiation (SXR) was experimentally observed upon compression of quasi-spherical arrays with the linear-mass profiling. In this case, the average power density on the surface of the SXR source is three times higher than for implosions of cylindrical arrays of the same mass and close values of the discharge current. Obtained experimental data are compared with the results of modelling the current implosion of multi-wire arrays performed with the help of a three-dimensional radiation-magneto-hydrodynamic code.

## 1. Introduction

To create high density of energy in a substance, the spatial concentration of energy flows generated by powerful energy sources – energy drivers – is used. One such driver is a radiating Z-pinch discharge as the source of a powerful X-ray source. The development of the technology of electric generators of high pulse power made it possible to construct units with a current of up to 26 MA [1]. Further increase in peak currents is associated with significant technical difficulties; therefore, it is extremely important to search for options for more efficient energy concentration, one of which is the three-dimensional compression of spherical arrays.

The transition to spherical arrays will allow us to concentrate the flux of the kinetic energy not only along the radius, but also along the axis of the facility. This will increase the power flow from the driver and make it more symmetrical on the target surface. An additional gain in the concentration of the kinetic energy of the array and the retention of radiation will make it possible to lower the requirements for electrical parameters of the next-generation facilities in the future.

The magnetic field of the current flowing along the load is determined by the distance to the axis of the load and has cylindrical symmetry. The pressure of this field creates a radial acceleration of the load. To maintain the constancy of the acceleration, at all points of the surface of the load its mass must be redistributed so that the sections closer to the axis are heavier, since the pressure of the magnetic field is greater on them. It follows from the equation of the 0-dimensional model of current implosion that for the realization of spherical compression, the angular dependence of the mass per unit surface is chosen in the form $m_s \propto \sin^{-2}\theta$ [2], where $\theta$ is the poloidal angle (Figure 1).

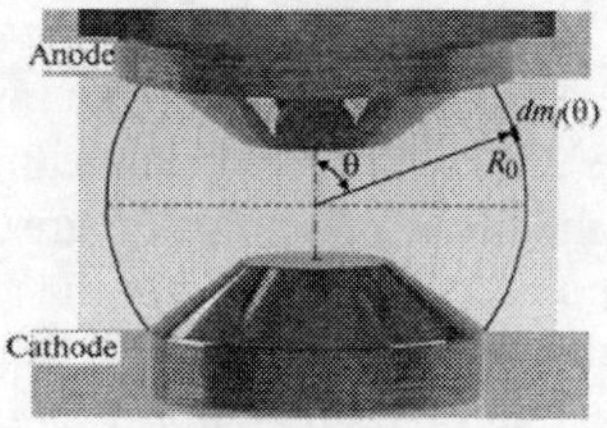


**Figure 1.** Quasi-spherical multiwire array (QSWA) with conical electrodes. A section of a wire with linear mass $dm_l(\theta)$ is shown, located at the initial radius of the assembly $R_0$ at a poloidal angle $\theta$.

In this case, the array will compress while remaining spherical, despite different magnetic pressure at different distances from the array's axis. This will allow to concentrate the array's material in the center and improve the parameters of the X-ray pulse due to the simultaneous arrival of plasma into the center of the quasi-spherical array.

Comparison of the kinetic energy density for a quasi-spherical array with the similar value for a cylindrical array shows that with all other conditions being equal (current amplitude and compression ratio), kinetic energy density for a quasi-spherical implosion, due to axial cumulation, exceeds the corresponding value for the cylindrical one by as many times as $(R_0-R)/R$ is greater than $\ln(R_0/R)$ [2].

## 2. Results

### *2.1. Experimental Layout*

A new method for creating a quasi-spherical multiwire array was developed at the Angara-5-1 facility in 2009 [3]. Parameters of the shape of a spherical surface are determined by the influence of a radial electrostatic field on the wire array. A special ring-shaped electrode is used to create an electric field that affects the array wires. It is located outside the multiwire array coaxially with it in the middle of the anode-cathode gap of the target array. The electric field between the wires and the ring electrode is created by supplying voltage from an additional high-voltage (-10 kV) source. Electrostatic attraction between the ring-shaped electrode and the grounded wires of the array leads to the formation of a stretched quasi-spherical array. Its diameter is determined by the diameters on the electrodes and the length of the wires in the initial state. Such a design makes it possible to form quasi-spherical loads of any diameter.

To create the necessary distribution of mass along the wires of the array, an additional mass of material was deposited on separate parts of a tungsten wire or kapron fiber array. The width of the strip of additional deposited mass near the electrodes is 4.5 mm with constant linear density. This was carried out in a universal vacuum unit VUP-4 by the method of thermal evaporation [4]. To deposit additional mass, materials in a wide range of atomic numbers from 13 to 83 were used: aluminum, tin, indium and bismuth.

The homogeneity of the deposition of additional mass to the surface of the array wires was controlled with the help of optical and electron microscopes, as well as a REMMA-202 spectrograph analyzer with a LiF crystal on control samples [4].

### *2.2. Experimental results and discussion*

The experiments carried out at the Angara-5-1 facility showed that with the implosion of quasi-spherical wire arrays three-dimensional compression of the array material by a magnetic field was indeed experimentally obtained, both radially and axially. To realize a quasi-spherical implosion, it is necessary to profile the mass of the array in accordance with the formula $m_s \propto \sin^{-2}\theta$, compensation of the zippering effect and cone electrode inserts to prevent plasma compression and pinch formation far from the center of the array. A series of experiments was performed with loads confirming the necessity of these conditions [5].

The results of measuring the space-time parameters of the radiation source during the implosion of quasi-spherical metallized fiber arrays (QMFA) in one of the shots are shown in Figure 2. Characteristic time-integral images of the pinch during the implosion of QMFA are shown in the three ranges of photon energy in Figures 2a, 2b, 2c, in which the initial position of the array and the position of the anode and cathode before the onset of implosion are indicated by dashed lines. These images demonstrate that radiating plasma objects are formed in the center of the inter-electrode gap when the QMFA is compressed with cone electrodes. The spatial configuration of these objects is close in shape to the spherical surface. The formation of such compact radiating objects, symmetrical relative to the liner axis, between the anode and the cathode of the array indicates a three-dimensional compression of the liner material during implosion of QMFA.

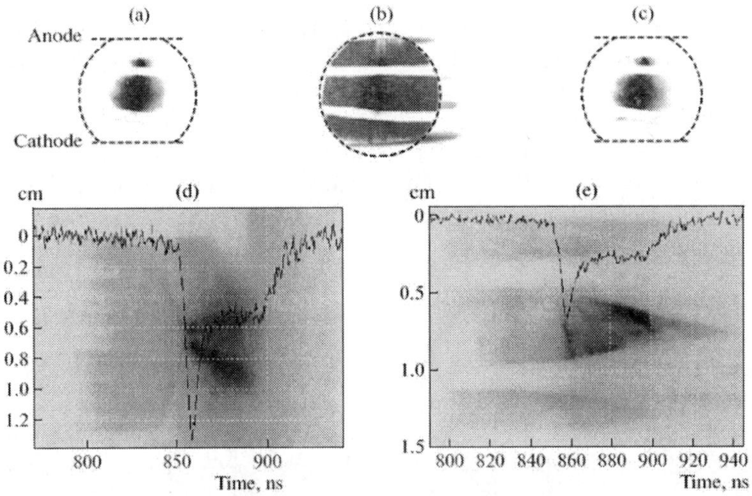


**Figure 2.** Time-integrated and optical streak images in shot No. 5094: QMFA, 40 kapron fibers (Ø25 μm) profiled In–Bi deposition, conical electrodes, total weight of 270 μg, anode-cathode distance 15 mm; a) filter: lavsan - 316 μg / cm$^2$ (> 120 eV); b) without a filter; c) filter Al 3 μm (> 600 eV); d) radial and (e) axial optical streak images of the imploding array synchronized with the SXR pulse with photon energies above 120 eV.

Weak optical radiation from the plasma of the equatorial part of the array converging towards the center is seen from the radial streak (Figure 2d). Then, at the time of the peak of the X-ray pulse (>120 eV), the burst of the glow is observed when the plasma arrives at the center from all areas of the array. After compression, its expansion is observed from the radial streak. On the axial optic streak (Figure 2e), the glow appears at the center of the array only at the peak of the X-ray pulse. In the image of Figure 2d, the contours of the shadows of the stretching electrostatic electrode are visible.

The compactness of compression is manifested in an increase in the power flux from the surface of the pinch. Figure 3a and Figure 3b show the time-integrated X-ray images of a pinch in X-ray radiation with a photon energy above 120 eV obtained by implosion with the current of 2.5 MA amplitude, tungsten wire array (W-CWA) in shot No. 5091 and QMFA in shot No. 5094, respectively.

The diameter of the emitting region during the implosion of the W-CWA from the image of Figure 3a is equal to $d_{CWA}$ = 5 mm. With the array height $H$ = 15 mm, the surface area of the pinch is 2.36 cm$^2$. The X-ray power in shot No. 5091 is $W_{CWA}$=1.22 TW at the current of 2.5 MA, so the X-ray power density per unit surface of the source is 0.52 TW/cm$^2$. In the case of QMFA, the diameter of the radiating region in the image in Figure 3b is equal to $D_{QMFA}$=3 mm. The surface of the compressed plasma is 0.28 cm$^2$. The X-ray power in the shot is $W_{QMFA}$ = 0.4 TW at 2.5 MA current, therefore, the X-ray power, referred to the surface area of the X-ray source, in this case is 1.43 TW/cm$^2$.

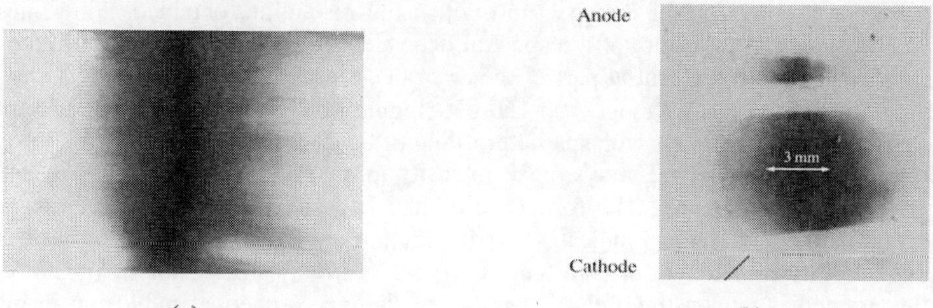


(a) (b)

**Figure 3.** Time-integrated X-ray images of a pinch with photon energy of more than 120 eV (filter lavsan 316 μg/cm$^2$) obtained by implosion of W-CWA and QMFA with close values of the total mass of the arrays: a) shot No. 5091, W-CWA , 40 W wires ∅6 μm, initial diameter of array was 20 mm, total weight 330 μg; b) shot No. 5094, QMFA, 40 kapon fibers (∅25 μm) with In-Bi layers for profiling linear mass and conical electrodes, total mass 270 μg, 3 mm long segment (white arrow) for scale.

Thus, it was found that the average power flux on the surface of the source of X-ray radiation based on QMFA is almost 3 times greater than the corresponding value for W-CWA during the implosion of arrays of the QMFA and CWA type with approximately the same total mass and discharge current. The measured value of the ratio of the average radiation power fluxes corresponds to a simulation of a similar value in [2].

Three-dimensional RMHD modeling of quasi-spherical compression of wire arrays was carried out using the MARPLE-3D code (Keldysh Institute of Applied Mathematics, Russian Academy of Sciences). The numerical calculations of quasi-spherical wire arrays with spatial profiling of the linear mass are in good agreement with the presented experimental data.

A method for measuring the spectrum with spatial resolution perpendicular to and along the axis of multiwire arrays was used to study the spatial characteristics of the plasma radiation of compressed substance. This made it possible to obtain spectra of both the central part of the Z-pinch and the trailing mass. For this purpose, we used a time-integrated grazing incidence spectrometer GIS with a spatial resolution along the radius and height of the array in the wavelength range (20-400) Å. The spatial resolution of the object was 1 mm for the photon energy of 30 eV and, correspondingly, 250 μm for photon energy of the order of 400 eV.

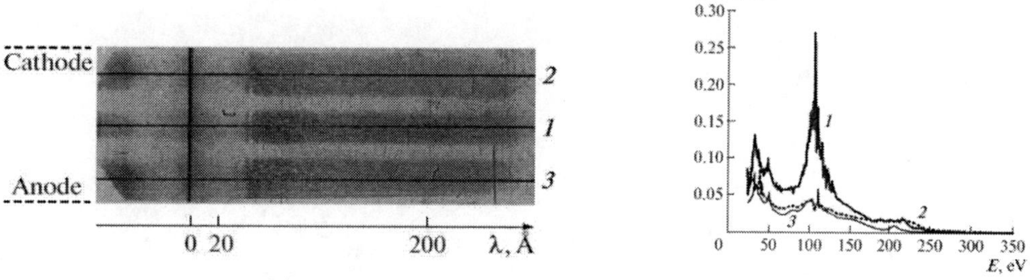


(a) (b)

**Figure 4.** Shot no. 4961, a quasi-spherical array with W wires (total mass of 435 μg). a) SXR spectra recorded with axial resolution and b) emission spectra measured in the directions shown by the horizontal lines in a).

Figure 4b shows the Z-pinch emission intensity distributions at different heights between the anode and the cathode in an image of the spectrum with axial resolution, obtained along lines 1, 2, 3 in Figure 4a. A characteristic feature of the spectral dependence on the wavelength of photons is that in the emission spectrum of the central part of the array (1), a peak in the intensity of X-ray radiation is observed in the photon energy range 100-150 eV (Figure 4b). Note that this intensity maximum is especially pronounced for arrays with spatial profiling of linear mass of the array.

In contrast to quasi-spherical arrays, peak intensity in the range 100-150 eV is absent during the implosion of cylindrical arrays. The formation of this intensity peak in the pinch spectrum in the middle of the inter-electrode gap indicates a redistribution of energy in the spectrum of the radiation source due to the axial motion of the substance during the current compression of W-CWA.

Conducted studies demonstrated the feasibility of the quasi-spherical implosion in terawatt-class facilities. In the experiments on the implosion of QSWA with profiled linear masses and conical electrodes intense and compact quasi-spherical plasma sources of SXR emission has been achieved. An increase in the average power flux density of such a source has been observed experimentally for the first time. The average power flux density in this source is more than 3 times higher than the corresponding value for the radiation source formed at the implosion of a cylindrical wire array with the same total weight. The next stage of the work will be optimization of the parameters of the QSWA to improve the quality of three-dimensional energy cumulation and to achieve significantly higher pulse powers of the Z-pinch X-ray radiation during quasi-spherical implosion.

This work was supported by the Rosatom Science Foundation under Project № 16-12-10487 and in part by the Russian Foundation for Basic Research under Project 16-02-00084, 16-02-00112, 16-02-00491 and 17-02-00167.

## References

[1] Jones M C *et al* 2014 *Rev. Scien. Instr.* **85** 083501

[2] Smirnov V, Zakharov S and Grabovskii E 2005 *JETP Lett.* **81** 442

[3] Grabovskii E, Gritsuk A, Smirnov V, Aleksandrov V, Oleinik G, Frolov I, Laukhin Ya, Gribov A , Samokhin, Sasorov P, Mitrofanov K, and Medovshchikov 2009 *JETP Lett.* **89** 315

[4] Aleksandrov V, Volkov G, Grabovski E, Gribov A, Gritsuk A, Mitrofanov K, Oleinik G, Frolov I, Barsuk V, Medovshchikov S, and Sasorov P 2012 *Plasma Phys. Rep.* **38** 315

[5] Aleksandrov V, Gasilov Grabovski E, Gritsuk, Laukhin Ya, Mitrofanov R, Oleinik G, Ol'khovskaya O, Sasorov P, Smirnov V, Frolov I, and Shevel'ko A 2014 *Plasma Phys. Rep.* **40** 939

IOP Conf. Series: Journal of Physics: Conf. Series **941** (2017) 012008 doi:10.1088/1742-6596/941/1/012008

# A model of recovering the parameters of fast nonlocal heat transport in magnetic fusion plasmas

**A B Kukushkin[1,2], A A Kulichenko[1,2], P A Sdvizhenskii[1], A V Sokolov[3] and V V Voloshinov[3]**

[1] NRC «Kurchatov Institute», Moscow, Russia
[2] National Research Nuclear University MEPhI (Moscow Engineering Physics Institute), Moscow, Russia
[3] Institute for Information Transmission Problems (Kharkevich Institute) of Russian Academy of Science, Moscow, Russia

E-mail: Kukushkin_AB@nrcki.ru

**Abstract.** A model is elaborated for interpreting the initial stage of the fast nonlocal transport events, which exhibit immediate response, in the diffusion time scale, of the spatial profile of electron temperature to its local perturbation, while the net heat flux is directed opposite to ordinary diffusion (i.e. along the temperature gradient). We solve the inverse problem of recovering the kernel of the integral equation, which describes nonlocal (superdiffusive) transport of energy due to emission and absorption of electromagnetic (EM) waves with long free path and strong reflection from the vacuum vessel's wall. To allow for the errors of experimental data, we use the method based on the regularized (in the framework of an ill-posed problem, using the parametric models) approximation of available experimental data. The model is applied to interpreting the data from stellarator LHD and tokamak TFTR. The EM wave transport is considered here in the single-group approximation, however the limitations of the physics model enable us to identify the spectral range of the EM waves which might be responsible for the observed phenomenon.

## 1. Introduction

To describe the anomalous heat transport in the magnetized thermonuclear plasma the superdiffusion formalism was suggested (namely, the equation, integral in space coordinates, with a long-tailed kernel and, respectively, dominance of carriers with a long free path). It includes steady-state transport of heat by electron/ion Bernstein [1] and electron cyclotron waves [2], transport of temperature perturbation by electron Bernstein waves [3] which was possible to describe by Biberman-Holstein equation known from the theory of atomic excitation's transport in spectral lines of atoms/ions; model of fractional derivatives for heat perturbation transport (see references in reviews [4-6]). One of the most interesting examples of fast nonlocal transport is the unusual direction of a heat flux, namely, the instantaneous (in a scale of diffusion time) temperature increase in the plasma center in case of fast cooling the plasma periphery (so called "cold pulse" experiments in tokamaks and stellarators, [4-6]) or an inverse process — the instantaneous temperature fall in the plasma center in case of fast heating the plasma periphery [7].

In this work an inverse problem of recovering the integral equation kernel is formulated, extending/modifying the works [8-10], and solved. The equation describes the nonlocal (superdiffusive) energy transfer caused by emission and absorption of electromagnetic (EM) waves with a long free path and strong reflection of waves from the walls of vacuum chamber. The developed model is used for interpretation of the "cold pulse" experiments in the stellarator LHD [11, 12] and tokamak TFTR [4, 13].

## 2. The fast nonlocal transport model

The model is aimed at recovering the main parameters of the unknown fast heat carriers by solving an inverse problem. The model is based on the following assumptions.

— At the initial stage of the event, unusual behavior of electron temperature $T_e$ (e.g., the instantaneous, within the diffusion time scale, electron temperature increase in the plasma center in case of the instantaneous cooling of the plasma periphery) is completely defined by EM waves. Their group velocity is so high that it is possible to neglect retardation.

— Energy density and total energy of carriers are small in comparison with those of the main plasma, and the total power of energy losses (owing to incomplete reflection of waves from walls) does not exceed a small fraction *f* of the total power of energy losses in plasma.

— Free path length of waves is large, it is not less than the minor radius of toroidal plasma (this condition is satisfied for EM-waves with a frequency significantly higher fundamental cyclotron and plasma frequencies). Waves are strongly reflected from the walls of the vacuum chamber. Then intensity of EM-waves is spatially homogeneous and isotropic in group velocity's angles.

— EM-waves transport is considered in the one-group (monochromatic) approximation.

— Source and sink functions, i.e. the power density of waves' emission and the absorption coefficient of waves, are proportional to electron density and the unknown function depending only on temperature. There is a link between these functions corresponding to the Kirchhoff's law.

— To exclude possible contribution of diffusion processes into the observed evolution of $T_e$ at initial stage, the inverse problem is solved for an energy balance equation only in the central part of plasma.

The first condition allows to obtain the energy balance equation which after simplification takes a form:

$$\left[K(T(\rho,t))I_{\mathrm{EM}}(T(\cdot,t))-K(T_0(\rho))I_{\mathrm{EM}}(T_0(\cdot))\right]=\Pi(\rho,t), \qquad 0<\rho<\rho_{\max}\sim 0,4 \quad (1)$$

where $\rho$ is a radial coordinate normalized on the small radius of a toroidal plasma; $K(T)$ is a temperature-dependent factor of electrons in the absorption coefficient of EM-waves (measured here in the inverse seconds); $T_0(\rho)$ and $T(\rho, t)$ are the spatial electron temperature distributions, respectively, at the initial moment (i.e. at quasi-stationary stage before disturbance of temperature via pellet injection into peripheral plasma) and after that; $I_{\mathrm{EM}}$ is the waves' intensity (energy flux density, measured here in temperature measure units, keV; the point in the argument means that it is a mute variable under the integral),

$$I_{\mathrm{EM}}(T(\cdot,t))=\frac{\int_0^1 n_e(\rho_1,t)Q(T(\rho_1,t))\rho_1 d\rho_1}{F_W+\int_0^1 n_e(\rho_2,t)K(T(\rho_2,t))\rho_2 d\rho_2}, \quad (2)$$

where the power density of EM-waves source (expression under integral in numerator in (2)) and the absorption coefficient (expression under integral in denominator in (2)) are linked by Kirchhoff's law (for the long-wavelength radiation, $\hbar\omega \ll T$, it is the Rayleigh–Jeans's law), and right part of equation (1) is a difference (before and after injection) in balance of plasma thermal energy per one particle (electron density is omitted in the left and right parts in (1) due to its approximate invariance in the central part of plasma),

$$K(T)=Q(T)/T, \qquad \Pi(\rho,t)=A\frac{\partial}{\partial t}\left[T(\rho,t)-T_0(\rho)\right], \quad (3)$$

IOP Conf. Series: Journal of Physics: Conf. Series **941** (2017) 012008 doi:10.1088/1742-6596/941/1/012008

where the coefficient $A$ allows also for the contribution of ions to plasma thermal energy; the constant $F_w$ in (2) is bound to $R_w$, the reflectivity of waves from walls, averaged on the vacuum chamber surface; the constant $F_0$ depends on vacuum chamber's geometry, on the wavelength and the width of spectral range in which the energy transfer takes place (actually, width of the emission and absorption spectrum in central plasma)

$$F_W = (1 - R_W) F_0, \qquad \frac{1}{F_0} = \frac{n_e^* a}{\pi} \left[ \left( \frac{c_w}{\omega} \right)^2 \frac{\pi^2}{\Delta\omega} \right] \frac{2k_{elong}}{1 + k_{elong}} \tag{4}$$

here $a$ — effective minor radius of toroidal plasma, $n_e^*$ — characteristic electron density ($10^{13}$ cm$^{-3}$), $c_w$ — speed of waves equal to velocity of the light, $\omega$ — the wave frequency in monochromatic transport model , $\Delta\omega$ — the above-mentioned spectral width, $k_{elong}$ — vertical elongation of poloidal cross-section of toroidal plasma.

The inverse problem is formulated as a recovery of the function $K(T)$ (which is assumed to be monotonic) and the constant $F_w$ for the known functions $T_0(\rho)$, $T(\rho, t)$ and $n_e(\rho, t)$. It corresponds to recovery of the integral operator kernel in the transport equation (1), (2) for the known dependence of perturbation of electron temperature and density on the time and magnetic surface.

## 3. Method of solving the inverse problem

Numerical calculations are carried out by the method of sequential correction and identification of mathematical model parameters in Eqs. (1) — (4). A method based on a sequential, regularized approximation of the input data by parametric models is used. This approach is ideologically close to the methods of Tikhonov's regularization [14] (in this paper there is also regularization by smoothness). The regularization criterion is the smoothness of the approximating functions. Smoothness is treated as one of the important characteristics of the solution quality. The general scheme for estimating the modeling error is the cross-validation. Similar approaches to data processing can be found in [15]. The general scheme of the method proposed by A.V. Sokolov has already been successfully applied to various problems, including the processing of geophysical data [16].

To estimate the quality of the parametric models and to choose the direction of their improvement, a large number of auxiliary subtasks has to be solved. Each step of the proposed method is a two-level finite-dimensional non-linear problem of mathematical programming. At the "lower" level, it is necessary to solve a large set of independent subtasks, and this phase of calculations can be significantly accelerated when computing is transferred to a distributed computing environment. To do this, the system of services for solving the non-linear problems of mathematical programming on the Everest platform was used [17], where the package pool (Ipopt, https://projects.coin-or.org/Ipopt) was deployed to solve such problems. For the software implementation of the method, including the description of tasks and the data exchange with packages, the Python language and the Pyomo optimization modeling package are used [18, 19]. The general scheme of the Pyomo application in the Everest system of services coincides with that already used for integration of the optimization modeling language AMPL [20] into the Everest. More detailed information on the proposed numerical method can be found in [21].

## 4. Calculation results

In this section we show two typical solutions of the inverse task — recovery the $K(T)$ function in the framework of the model (1) — (4).

Space-time dynamics of the electron temperature and density in shot #48708 in stellarator LHD with auxiliary plasma heating (2 MW power from injection of a neutral beam and 1 MW from injection of electron cyclotron waves) is presented in figure 1 in [11] and figure 1 in [12]. Possible solution of the inverse problem for function $K(T)$ by consideration of plasma dynamics after injection in a time interval from 2.8 s to 2.815 s is presented in figure 1.

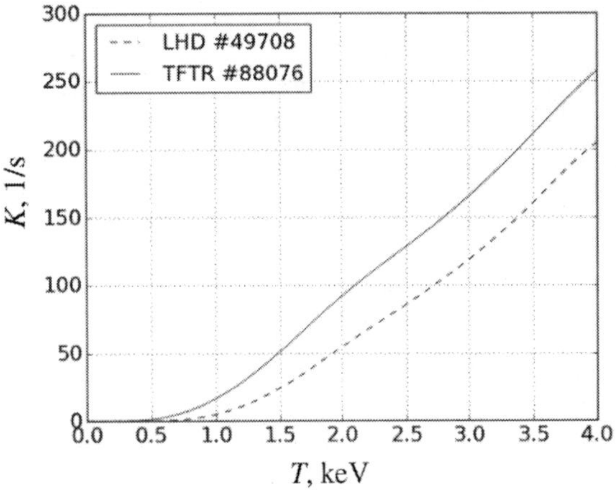


**Figure 1.** *K*(*T*) is a function of temperature recovered by solving the inverse problem in the model of energy balance (1) — (4), for experimental conditions on stellarator LHD in the discharge #49708 [11, 12] (solid curve) and tokamak TFTR in the discharge #88076.

Pellet injection at 2.797 s led to an instantaneous increase of the density in the plasma periphery and to a sharp temperature drop to a minimum at 2.8 s, then that was replaced by a slowly growth. The chosen time interval is characterized by the almost linear increase of temperature and constant density in the central part of plasma (the normalized radius $\rho \leq 0.45$), whereas in the periphery ($0.7 \leq \rho \leq 1$) the temperature grows more slowly after sharp decrease, than in the centre, and dynamics of the density is such that the increase in density, caused by pellet, slowly moves towards plasma center.

Space-time dynamics of electron temperature in tokamak TFTR in the Ohmic discharge #88076 is presented in figures 2 and 3 in [4]. Possible solution of the inverse problem for the *K*(*T*) function if considering the plasma dynamics after injection in a time interval from 3.72 s to 3.74 s, is also presented in figure 1. Pellet injection at 3.71 s produced the same effect, as in case of experiment in stellarator LHD.

## 5. Limitations of physics model

The physics model imposes restrictions for free parameters of the problem: wavelength $\lambda$ of EM waves, relative spectral width $\Delta\omega/\omega$, reflectivity $R_w$ of EM waves from the wall of the vacuum chamber, and the fraction $f$ of power losses, caused by the outgoing EM waves, in the total power losses $P_{\mathrm{loss}}$.

The limitation on the power losses, caused by the outgoing EM waves, takes a form

$$\frac{I_{\mathrm{EM}}(t)}{\mathrm{keV}}\left(\frac{\Delta\omega}{\omega}\right)\left(\frac{10^{-2}\,\mathrm{cm}}{\lambda}\right)^3(1-R_w)\leq f\frac{P_{\mathrm{loss}}}{100\,\mathrm{MW}}. \qquad (5)$$

The condition that free path length is not less than plasma minor radius is as follows:

$$\frac{I_{\mathrm{EM}}(t)}{\mathrm{keV}}\left(\frac{\Delta\omega}{\omega}\right)\left(\frac{10^{-2}\,\mathrm{cm}}{\lambda}\right)^3(1-R_w)\leq f\frac{P_{\mathrm{loss}}}{100\,\mathrm{MW}}. \qquad (6)$$

The total energy of waves has to be much less than the total thermal energy of plasma:

$$\frac{I_{\mathrm{EM}}(t)}{\mathrm{keV}}\left(\frac{\Delta\omega}{\omega}\right)\left(\frac{10^{-2}\,\mathrm{cm}}{\lambda}\right)^3 << 10^6\int_0^1\frac{n_e(\rho,t)}{10^{13}\,\mathrm{cm}^{-3}}\frac{T_e(\rho,t)}{\mathrm{keV}}\rho d\rho. \qquad (7)$$

And at last, for existence of quasi-stationary intensity of EM-waves and notable power exchange between the central and peripheral plasma (i.e. a condition of existence of the studied effect in the considered energy transport model) the contribution to intensity of waves, determined by reflection from the wall and expressed by the first term in denominator of (2), should not exceed the contribution of absorption:

$$\left(\frac{\Delta\omega}{\omega}\right)\left(\frac{10^{-2}\,\mathrm{cm}}{\lambda}\right)^3 (1-R_w) < 2{,}5\cdot 10^{-4}\int_0^1 \frac{n_e(\rho,t)}{10^{13}\,\mathrm{cm}^{-3}}\frac{K(T_e(\rho,t))}{\mathrm{s}^{-1}}\rho d\rho. \tag{8}$$

Restrictions (5) — (8) for applicability of solutions of the inverse problem, shown in figure 1, can be presented in the form of the following system of inequalities:

$$\begin{gathered} a<\left(\frac{\Delta\omega}{\omega}\right)\left(\frac{10^{-2}\,\mathrm{cm}}{\lambda}\right)^3 \le b\frac{P_{\mathrm{loss}}}{100\,\mathrm{MW}}\frac{f}{(1-R_w)}, \\ a<\left(\frac{\Delta\omega}{\omega}\right)\left(\frac{10^{-2}\,\mathrm{cm}}{\lambda}\right)^3 < \frac{d}{1-R_w}, \qquad \left(\frac{\Delta\omega}{\omega}\right)\left(\frac{10^{-2}\,\mathrm{cm}}{\lambda}\right)^3 << g. \end{gathered} \tag{9}$$

The following values of parameters correspond to stellarator LHD: $a = 0.05$, $b = 0.36$, $d = 0.19$, $g = 10^7$, $P_{\mathrm{loss}} \approx 3$ MW, $f < 0.1$. For tokamak TFTR we have parameters $a = 0.0296$, $b = 0.419$, $d = 0.116$, $g = 10^6$, $P_{\mathrm{loss}} \approx 1$ MW, $f < 0.1$. Apparently, the weak and obviously feasible is the condition (7), and the hardest is the first condition in (9).

An important consequence of physical restrictions (5) — (8) for the model (1) — (4) is that the unified parameter in the center of inequalities (9), which depends on the relative spectral width and the wavelength, gives the values of wavelength in the submillimeter range which is studied substantially less than the ambient spectral ranges.

The most important direction of developing the model (1) — (4) seems to be a transition from one-group transport of EM-waves to multi-group ones.

## 6. Conclusions

The model is developed to interpret the initial stage of the phenomenon of fast nonlocal transport in which the response of electron temperature space profile to its local perturbation is almost instantaneous in scales of diffusion time, and the total flux of heat is directed opposite ordinary diffusion (i.e. along temperature gradient). The model is based on dominance of nonlocal (superdiffusive) heat transport by EM-waves with a long free path length and the strong reflection of waves from walls of the vacuum chamber.

The method of regularized optimizing identification is used for solving the inverse problem of recovering the kernel of integral operator in the integral-differential equation. The developed model is used for interpretation of the experiments on a fast cooling of peripheral plasma (so-called "cold pulse" experiments) in stellarator LHD and tokamak TFTR. It is shown that for these experiments, close results of recovering the EM-waves source and coefficient of their absorption are obtained. Despite the transport of EM-waves is considered in the one-group approximation, the restrictions of physics model allow to allocate the spectral range of EM-waves which might be responsible for the observed phenomenon. The found solution of the inverse problem will allow us to look for the concrete physical mechanisms which would be compatible with such dependence of the source and sink of EM-waves on the electron temperature and density.

In general, modeling results indicate possible reality of the hypothesis [8] that the studied phenomenon develops according to the following scenario: fast cooling of the plasma periphery leads to a sharp decrease of absorption of EM-waves in this area; it does possible more free circulation of EM-waves and, under condition of their strong reflection from the wall of vacuum chamber, an increase of the intensity of waves; for monotonic profile of temperature dependence of absorption coefficient, this leads, in turn, to temperature growth in the plasma center at initial stage of perturbation evolution. Note that such scenario is compatible also with an inverse process when on the plasma periphery there is a sharp heating, and in the center the temperature falls. Ability of the model to give possible explanation not only to a unexpected temperature response in the center to perturbations of temperature on the periphery, but also to give a qualitative explanation of the

IOP Conf. Series: Journal of Physics: Conf. Series **941** (2017) 012008 doi:10.1088/1742-6596/941/1/012008

reversibility with respect to the sign of temperature perturbation is, in our opinion, the important advantage of the developed approach.

The realization of the developed physical model was possible only due to carrying out massive numerical calculations in the distributed computing environment.

**Acknowledgments**

This work is partially supported by the Russian Foundation for Basic Research (projects RFBR-15-07-07850-a, 15-07-07428-a and 16-07-01150) and by National Research Nuclear University MEPhI in the framework of the Russian Academic Excellence Project (contract No. 02.a03.21.0005, 27.08.2013). This work has been carried out using computing resources of the federal collective usage center Complex for Simulation and Data Processing for Mega-science Facilities at NRC "Kurchatov Institute", http://ckp.nrcki.ru/.

**References**

[1] Rosenbluth M N, Liu C S 1976 *Phys. Fluids* **19** 815
[2] Tamor S 1988 *Nucl. Instr. and Meth. Phys. Res.* A **271** 37
[3] Kukushkin A B, Lisitsa V S, Saveliev Yu A 1987 *JETP Lett.* **46** 448
[4] Callen J D, Kissick M W 1997 *Plasma Phys. Contr. Fusion* **39** 12B 173
[5] Kukushkin A B, Cherepanov K V 2009 *AIP Conf. Proc.* **1154** 83
[6] Pustovitov V D 2012 *Plasma Phys. Contr. Fusion* **54** 124036
[7] Gentle K W et al 1997 *Phys. Plasmas* **4** 3599
[8] Kukushkin A B 1997 *Proc. 24th EPS Conf. on Plasma Phys. and Contr. Fusion* (Berchtesgaden, Germany) vol **21A** part II pp 849-852
[9] Kukushkin A B, Sdvizhenskii P A, Voloshinov V V, Kulichenko A A 2015 *Proc. 42nd EPS Conf. on Plasma Phys. (Lisbon, Portugal)* vol **39E** P5.182
[10] Kukushkin A B, Kulichenko A A, Sdvizhenskii P A, Sokolov A V, Voloshinov V V 2016 *Proc. 43rd EPS Conference on Plasma Phys. (Leuven, Belgium)* vol **40** P2.028
[11] Tamura N et al 2007 *Nucl. Fusion* **47** 449
[12] Tamura N et al 2010 *Proc. 23rd IAEA Fusion Energy Conf.* (Daejeon, Republic of Korea) EXC/P8-16
[13] Kissick M W, Callen J D, Fredrickson E D, Janos A C, Taylor G 1996 *Nucl. Fusion* **36** 1691
[14] Tikhonov A N 1980 *Problems of Comp. Math.* (Moscow: Moscow State University Press) (in Russian) pp 3-17
[15] Hastie T J, Tibshirani R J, Friedman J 2009 *Unsupervised learning. In: The elements of statistical learning* (Springer) pp 485-585
[16] Linnik V G, Sokolov A V, Mironenko I V, 2016 *Current trends in the development of biogeochemistry* (Moscow: GEOKHI RAS) (in Russian) pp 423-434
[17] Sukhoroslov O, Volkov S, Afanasiev A 2015 A Web-Based Platform for Publication and Distributed Execution of Computing Applications *Parallel and Distributed Computing (ISPDC)* (14th International Symposium on IEEE) pp 175-184
[18] Hart W E 2009 Python Optimization Modeling Objects (Pyomo) *Operations Research and Cyber-Infrastructure* (Springer) eds J W Chinneck, B Kristjansson, M J Saltzman pp 3-19
[19] Hart W E, Laird C, Watson J -P, Woodruff D L 2012 *Pyomo-optimization modeling in python* (Springer) vol 67 p 238
[20] Smirnov S, Voloshinov V, Sukhosroslov O 2016 Distributed Optimization on the Base of AMPL Modeling Language and Everest Platform *Procedia Comp. Science* vol 101 pp 313-322 http://dx.doi.org/10.1016/j.procs.2016.11.037
[21] Voloshinov V, Sokolov A 2016 *Proc. National Supercomputer Forum* (in Russian) http://2016.nscf.ru/TesisAll/04_Reshenie_zadach_optimizatsii/766_VoloshinovVV.pdf

IOP Conf. Series: Journal of Physics: Conf. Series **941** (2017) 012009 doi:10.1088/1742-6596/941/1/012009

# Greenwald density limit and power balance in tokamaks

**D Kh Morozov**[1,2,†]

[1] NRC Kurchatov Institute, Akademika Kurchatova sq. 1, 123182 Moscow, Russia
[2] National Research Nuclear University MEPhI (Moscow Engineering Physics Institute), Kashirskoe shosse 31, 115409 Moscow, Russia

E-mail: Dmitry.morozov.41@mail.ru

**Abstract.** The critical density limit in tokamaks is investigated. It is shown that the equality of the input power and power radiated by impurities corresponds to the Greenwald limit. In Ohmic tokamak plasmas the auxiliary heating may increase the density limit, as it has been shown in experiments. The radiated power threshold for plasmas with heavy impurities, observed in experiments, is derived. Radiation produced by heavy impurities is spread practically uniformly along the plasma radius in contrast to the radiation of light impurities. The effective heating power is decreased by radiation losses and, as a consequence, becomes lower than the threshold for the H-L transition. If the input power is close to the radiated one, the disruption occurs.

## 1. Introduction

Greenwald density limit is known for a long time. (See, for example, the review paper [1]). In most experiments, plasma density in tokamaks cannot exceed the Greenwald upper density limit $\bar{n}_{20} \le n_G = \dfrac{I_p}{\pi a^2}$ using as units $10^{20}$ m$^{-3}$, MA, m. Many theoretical papers are devoted to the problem. Some of them are based on the description of plasma inside the separatrix [2]. Only Ohmic tokamak plasmas are described in [2]. Note that the Greenwald limit describes plasmas in tokamaks with divertors as well as tokamaks with limiters. Other papers are based on the processes in the SOL and divertor [3] and cannot be applied to the limiter tokamak plasmas. The SOL and divertor physics will not be discussed here.

The theory described in [2] and papers cited therein supposes that in the edge plasma the tearing mode develops. It was shown that the island's size rises with the plasma density and saturates in time. However it is not clear why the disruption takes place when the plasma density achieves the Greenwald limit.

It is shown in Section 2 of the present paper that the Greenwald density limit is a consequence of the energy balance. For heavy impurities, like tungsten, the radiation losses are distributed throughout the plasma volume. The radiation losses in tungsten-seeded plasmas are practically uniform inside the separatrix. The difference between the input and radiated power determines the transport processes in tokamaks. The L-H transition also is determined by this difference. It may drop below the L-H threshold. As a consequence, the H-L transition can take place. In the L-regime the input power may

† Deceased 12 September 2017.

not be sufficient to support the discharge. As a result, the current quench may accompany the H-L transition.

The paper is organized as follows. The Ohmic heated plasma in tokamaks with carbon wall is described in Section 2. The auxiliary heated plasmas are investigated in Section 3. Sections 4 is devoted to tungsten-seeded plasmas. The main results are summarized in Section 5.

## 2. Ohmic heating

The thermal balance in tokamak edge plasma may be qualitatively described using the slab geometry (see [4]). Here, the convective term is ignored. Thus, the power balance takes a form

$$\frac{\mathrm{d}}{\mathrm{d}r}\kappa\frac{\mathrm{d}T}{\mathrm{d}r}=\frac{J^2}{\sigma}-nn_I L(T)\ . \tag{1}$$

The first term in the right hand side in (1) determines the Ohmic heating. The second one corresponds to the radiative cooling. The function $L$ depends on the temperature only [5], $\kappa$ is the thermal conductivity, $n$ and $n_I$ are the plasma and impurity densities, respectively, $J$ is the toroidal current density, $\sigma$ is the electrical conductivity. As it is well known, the electrical conductivity has the form

$$\sigma=1.96\frac{0.9\cdot 10^{13}}{(\lambda/10)Z_{ef}}T^{3/2}=\sigma_0\left(\frac{T}{T_0}\right)^{3/2}\ . \tag{2}$$

Here the temperature is expressed in eV, $T_0 = 1$ eV.

The impurity density as well as the plasma density are supposed to be the constants; $n_I=\alpha n$; $\alpha \ll 1$. The boundary conditions yield

$$\left.\frac{\mathrm{d}T}{\mathrm{d}r}\right|_{r=0}=0,\qquad T(a)=0. \tag{3}$$

If the thermal flux at the periphery is equal to zero, $\mathrm{d}T/\mathrm{d}r|_{r=a}=0$, the total input power is radiated completely. This condition defines the density limit.

Multiplying (1) by $\mathrm{d}T/\mathrm{d}r$ and integrating over $r$ from 0 to $a$, where $a$ is the tokamak plasma's minor radius, one can get

$$\int\limits_{T_{min}}^{T(0)}\frac{J^2}{\sigma}\mathrm{d}T=\alpha n_c^2\int\limits_{T_{min}}^{T(0)}L(T)\mathrm{d}T\ . \tag{4}$$

Here $n_c$ is the critical density. The temperature $T_{min}$ is related to the temperature of the ionization threshold and is equal to few eV. With good accuracy, one can put $T_{min}\approx 1\mathrm{eV}$.

The toroidal plasma current density may be estimated as $I_p/(\pi a^2)$, where $I_p$ is the total plasma current. Hence, the left hand side of (3) may be rewritten as

$$\int\limits_{T_{min}}^{T(0)}\frac{J^2}{\sigma}\mathrm{d}T\approx\left(\frac{I_p}{\pi a^2}\right)^2\int\limits_{0}^{T(0)}\frac{1}{\sigma}\mathrm{d}T\approx\left(\frac{I_p}{\pi a^2}\right)^2\frac{T_{min}^{1/2}}{\sigma_0}\ . \tag{5}$$

Here, Eq. (2) is used. The temperature $T$ is expressed in eV.

The value $\int\limits_0^{T(0)}L(T)\mathrm{d}T$ may be calculated for carbon using [5] if one can neglect the bremsstrahlung. Calculations give the following result:

$$g \equiv \int_0^{T(0)} L(T)\mathrm{d}T \approx 7 \cdot 10^{-18}\,\mathrm{erg^2 cm^3 s^{-1}}. \tag{6}$$

The typical impurity concentration may be estimated as $n_I \approx 0.02n$, i.e. $\alpha \approx 0.02$. One can see that the dependence of $n_c$ on $\alpha$ is weak. In order to compare the result with the Greenwald critical density, one has to express the plasma current in MA, and the minor radius in m.

Finally the equation (4) may be transformed to the usual Greenwald's form

$$n_c\left(10^{20}\mathrm{m}^{-3}\right) = 1.3\frac{I_p(\mathrm{MA})}{\pi a^2(\mathrm{m})}. \tag{7}$$

## 3. Auxiliary heating

Auxiliary heating power usually exceeds the Ohmic power significantly in modern tokamaks. If plasma column is heated by the neutral beam as well as by the fusion power, the input power is localized in the plasma center. Hence, the power balance may be presented in the form

$$\frac{\mathrm{d}}{\mathrm{d}r}\kappa\frac{\mathrm{d}T}{\mathrm{d}r} - \frac{J^2}{\sigma} + nn_I L(T) = 0. \tag{8}$$

with the boundary conditions

$$\kappa\left.\frac{\mathrm{d}T}{\mathrm{d}r}\right|_{r=0} = P_{aux},\ T(a) = 0. \tag{9}$$

Here $P_{aux}$ is the energy flux density of auxiliary heating. Again, multiplying (8) by $\mathrm{d}T/\mathrm{d}r$ and integrating over $r$, one can get

$$\kappa\left(\left.\frac{\mathrm{d}T}{\mathrm{d}r}\right|_{r=0}\right)^2 = \int_{T_{min}}^{T(0)} \frac{J^2}{\sigma}\mathrm{d}T - \alpha n_c^2 g. \tag{10}$$

Again replacing the current density by the averaged current density, one can get

$$n_c^2 = \frac{1}{\alpha g}\left(\frac{2\bar{J}^2}{\sigma_0}T_{min}^{1/2} + \frac{P_{aux}^2}{2\kappa}\right). \tag{11}$$

This finally yields:

$$n_c = n_{Gr}\sqrt{1 + \frac{P_{aux}^2}{2\alpha g \kappa n_G^2}}. \tag{12}$$

Here $n_G$ is Greenwald critical density.

Hence, if the auxiliary heating power is comparable with the Ohmic one, the condition (7) should be replaced by (12). The Greenwald density limit may be exceeded, as it has been observed, e.g., in TEXTOR experiments [6]. Nevertheless, the carbon radiative losses have a minimum near the temperature close to 50 eV, which means that the radiation-condensation instability may develop. As a consequence, MARFE appears. However, as it has been shown theoretically [7] as well as in many experiments, the instability may be prevented by the neon injection. On the other hand, the increase of $Z_{eff}$ may produce the L-H transition for tungsten and, as it will be shown below, the disruption.

## 4. Heavy impurities

Tokamak JET operates with the ITER-like divertor plates covered with tungsten. Experiments show that the total radiation losses from tungsten-seeded plasmas do not exceed 70% of the input power,

$P_{rad} < 0.7 P_{aux}$ [8] in contrast to the carbon seeded plasmas. Hence, one can expect some instability to produce the disruption. In particular the tearing instability may develop, as it is shown for Ohmic plasmas by Teng [2] and in the papers cited therein.

However, the modern tokamaks with tungsten divertor use the strong auxiliary heating. The auxiliary power usually significantly exceeds the Ohmic one. Hence, the Ohmic power is ignored in this Section. The auxiliary power is supposed to be localized at the center. Again the plasma temperature is described by the equation (8) with the boundary conditions (9). The variation of tungsten radiation function $L$ in the temperature interval 100 eV $< T <$ 10 keV may be approximated by the constant value of $3\times10^{-17}$ erg cm$^3$/s. The interval $T <$ 100 eV is ignored in the present paper. The boundary condition $T(r = a) = 0$ is used. One can suppose the tungsten radiation losses to be uniform. If the radiation losses increase, the H-regime is replaced by the L-regime [8]. It takes place, if the difference $(P_{aux} - P_{rad})V$ becomes less than the power required for the H-L transition:

$$(P_{aux} - P_{rad})V \le W_{LH} = 2.84 M_i^{-1} B_t^{0.82} \overline{n}_{e20}^{0.58} Ra^{0.81} . \tag{13}$$

Here $V$ is the plasma volume, $M_i$ = 2, $B_t$ is the toroidal magnetic field expressed in $T$, $\overline{n}_{e20}$ is the averaged electron density expressed in $10^{20}$ m$^{-3}$, $R$ and $a$ are the major and minor radii, respectively. For *JET* ($B_t$ = 2.8 T, $\overline{n}_{e20} = 0.28$, $R$ = 3 m, $a$ = 1 m) $W_{aux} - W_{rad} = 2.90\,\mathrm{MW}$. Here $W_{aux} = P_{aux}V$, and $W_{rad} = P_{rad}V$ are the total input and radiated power, respectively.

The power of the neutral beam injection is equal to 14.9 MW. Hence, the relation of the radiated power to the input power in the H–regime cannot exceed the value 80 %.

In this case the inequality (13) may be rewritten as

$$W_{aux} < 5 W_{LH} . \tag{14}$$

If the Ohmic heating may be neglected in comparison with the auxiliary one, the heating equation (8) yields

$$\left[\kappa\left(\frac{dT}{dr}\right)^2\right]_{r=a} - \left[\kappa\left(\frac{dT}{dr}\right)^2\right]_{r=0} = 2nn_I \int_0^{T(0)} L\frac{dr}{dT} dT . \tag{15}$$

Hence, the power balance before the L-H transition takes the form

$$2nn_I \int_0^{T(0)} L\kappa dT = P_{input}^2 . \tag{16}$$

After the *H-L* transition the heat conductivity rises significantly. The equation (16) is not satisfied. The radiated power exceeds the power input. Hence the disruption must take place.

## 5. Summary

Nature of the density limit is analyzed in the present paper. It is shown that different events may lead to the disruption. It is shown that the critical density depends on the heating type and may be different for light and heavy impurities.

As it is shown in Section 2, the Greenwald limit may be explained by the balance of the input power and the impurity radiated power in Ohmically heated tokamak plasmas. As it is shown in [2], one can obtain the same result by analyzing the tearing mode. Some additional investigations must be performed to make a choice between different mechanisms.

As experiments show, the Greenwald limit may be exceeded in tokamaks with auxiliary heating localized at the plasma axis. The experimental results may be explained with the model described in Section 2 at least for carbon, as it is shown in Section 3. The critical density is obtained for the experiments with the auxiliary heating power, exceeding significantly the Ohmic one.

The tungsten seeded plasmas are described in Section 4. In contrast to carbon radiation, variation of the tungsten radiation losses with the temperature is weak, at least, for the temperature less than $T = 10$ keV. Hence, the radiation losses may initiate a significant decrease of the input power assimilation inside the separatrix. The radiated power may achieve approximately 80 % of the input power, and the discharge may transit from H- to L-mode. On the other hand, the heat conductivity in the L-regime increases significantly. Hence, thermal losses may exceed the power input. It means that the thermal quench may take place.

**Acknowledgments**
This work has been partly supported by National Research Nuclear University MEPhI in the framework of the Russian Academic Excellence Project (contract No. 02.a03.21.0005, 27.08.2013).

**References**

[1] Greenwald W 2002 *Plasma Phys. Contr. Fusion* **44** R27
[2] Teng Q, Brennan D P, Delgado-Aparicio L, Gates D A, Sverdlov J and White R B 2016 *Nucl. Fusion* **56** 106001
[3] Borras K, Schneider R and Forengo R 1997 *Nucl. Fusion* **37** 523
[4] Ashby D and Hughes M H 1981 *Nucl. Fusion* **21** 911
[5] Post D E, Jensen R V, Tatler C B, Grasberger W H and Lokke W A 1977 *At. Data Nucl. Data Tables* **20** 397
[6] Rapp J, De Vries R C, Sethuler F C, Tokar M Z et al. 1999 *Nucl. Fusion* **39** 765
[7] Pshenov A A and Morozov D Kh 2010 *Contrib. Plasma Phys.* **50** 380
[8] Huber A, Brezinsek S, Groth M, de Vries P C, Riccardo V, van Rooij G, Sergienko G, Arnoux G, Boboc A, Bilkova P, Calabro G, Clever M, Coenen J W, Beurskens M N A, Eich T, Jachmich S, Lehnen M, Lerche E, Marsen S, Matthews G F, McCormick K, Meigs A G, Mertens Ph, Philipps V, Rapp J, Samm U, Stamp M, Wischmeier M, Wiesen S 2013 *J. Nucl. Mater.* **438** S139

IOP Conf. Series: Journal of Physics: Conf. Series **941** (2017) 012010 doi:10.1088/1742-6596/941/1/012010

# Ignition in tokamaks with modulated source of auxiliary heating

**D Kh Morozov**[1,2,†]

[1] NRC Kurchatov Institute, Akademika Kurchatova sq. 1, 123182 Moscow, Russia
[2] National Research Nuclear University MEPhI (Moscow Engineering Physics Institute), Kashirskoe shosse 31, 115409 Moscow, Russia

E-mail: Dmitry.morozov.41@mail.ru

**Abstract.** It is shown that the ignition may be achieved in tokamaks with the modulated power source. The time-averaged source power may be smaller than the steady-state source power, which is sufficient for the ignition. Nevertheless, the maximal power must be large enough, because the ignition must be achieved within a finite time interval.

## 1. Introduction

In order to achieve the ignition in tokamaks, usage of a stationary source of auxiliary heating has been suggested. On the other hand, the pulsed sources, more simple and cheap, existed earlier. The maximal power of such sources is higher than it is necessary for the ignition but the averaged power is not sufficient to sustain the steady-state fusion. However, the nonlinear dependence of the fusion output power on the temperature can lead to ignition. The harmonic oscillations of the power source produce the nonlinear increase of the temperature and, as the consequence, the additional fusion output. Similar effect has been predicted in [1], where it was shown that the usage of the modulated power source can lead to the decrease of the impurity radiation during the tokamak discharge start-up.

The energy confinement time degradation $\tau_E \sim P_{input}^{-0.69}$ decreases the effect. However, recent investigations show the degradation to be essentially small [2]. The time dependence of the temperature is found in the present paper for the oscillating heating power. Three stages of the temperature growth were found. At the first stage the average temperature rises rapidly. At the second stage the rise is slow. The third stage is related to the thermal instability and extremely rapid rise of the temperature. A simple qualitative model of the temperature evolution is presented below.

## 2. Qualitative model

The zero-dimensional model is used in the present paper:

$$3n\frac{\mathrm{d}T}{\mathrm{d}t} = -\frac{3nT}{\tau_E} + P_{fus} + P_{aux}\ . \tag{1}$$

† Deceased 12 September 2017.

Here $n$ is the plasma density, $\tau_E$ is the energy confinement time, $P_{fus}$ and $P_{aux}$ are the specific fusion heating by $\alpha$-particles and the auxiliary heating power, respectively. The fusion specific power has the form

$$P_{fus} = \frac{n^2}{4} < \sigma v > \varepsilon . \tag{2}$$

Deuterium and tritium densities are supposed to be equal, $n_D = n_T = n/2$, $\varepsilon = 3.52$ MeV. The factor $<\sigma v>$ is presented in [3] with a good accuracy. However, inside the wide temperature interval one can use approximate expression

$$P_{fus} \approx AT^2 . \tag{3}$$

The accuracy of the expression (3) is worse than that of (2) but it is acceptable for qualitative analysis.

The energy confinement time for *H*-regime has the form [3]

$$\tau_E = 0.0562 I_P^{0.98} B_t^{0.15} W^{-0.69} M_i^{0.19} \bar{n}_{e19}^{0.41} \left(\frac{a}{R}\right)^{0.58} k^{0.78} \equiv \Omega \bar{n}^{0.41} W^{-0.69} . \tag{4}$$

Here $I_p$ is the plasma current expressed in MA, $B_t$ is the toroidal magnetic field (T), $W = (P_{fus} + P_{aux})V$ is the input power (MW), $V$ is the plasma volume (m$^3$), $M_i = 2.5$ is the effective atomic mass in a.m.u., $\bar{n}_{e19}$ is the volume-average plasma density ($10^{19}$ m$^{-3}$), $k$ is the ellipticity.

The auxiliary heating power consists of the basic and oscillating parts,

$$W_{aux} = W_{aux}^0 + \tilde{W} . \tag{5}$$

The equation (1) may be rewritten as follows:

$$\frac{dT}{dt} = -\frac{T}{\Omega n^{0.41}} \left(W_{fus} + W_{aux}^0 + \tilde{W}\right)^{\lambda} + \frac{n}{12} AT^2 + \frac{P_{aux}}{3n} . \tag{6}$$

Here $\lambda = 0.69$ for the standard ITER scaling, and may be smaller ($0.25 < \lambda < 0.69$) for the experiments described in [2]. For qualitative evaluation, one can replace the value of $W_{fus}^{max}$ by the value related to the maximum of $<\sigma v>$ (see Figure 1). Usually $W_{aux} << W_{fus}^{max}$. Hence, one can obtain:

$$\Omega \bar{n}^{0.41} (W_{fus} + W_{aux})^{-\lambda} \approx \Omega W_{fus}^{-\lambda} \bar{n}_0^{0.41} \left(\frac{\bar{n}}{\bar{n}_0}\right)^{0.41} \equiv \tau_E^{ef} \left(\frac{\bar{n}}{\bar{n}_0}\right)^{0.41} . \tag{7}$$

Here $\bar{n}_0$ is the reference plasma density. In particular, one can put $\bar{n} = 10^{19}$ m$^{-3}$.

The equation (6) may be rewritten as follows

$$\frac{dT}{dt} = -\frac{T}{\tau_e^{ef}} \left(\frac{n}{n_0}\right)^{2\lambda - 0.41} + \frac{n < \sigma v > E_0}{12} + \frac{P_{aux}}{3n} . \tag{8}$$

Here

$$\tau_E^{ef} = \Omega n_0^{0.41} \left[ W_{fus}^{max} (n = n_0) \left(\frac{n_0}{n}\right)^2 \right]^{\lambda} . \tag{9}$$

The equation (8) takes the form

$$\frac{\mathrm{d}T}{\mathrm{d}\xi} = -\frac{T}{\tau_E^{ef}} + \frac{n_0}{12}\left(\frac{n}{n_0}\right)^{1.41-2\lambda} < \sigma v > E_0 + \frac{P_{aux}}{3n_0}\left(\frac{n_0}{n}\right)^{2\lambda+0.59}. \tag{10}$$

Here $\xi = t\left(n/n_0\right)^{2\lambda-0.41}$.

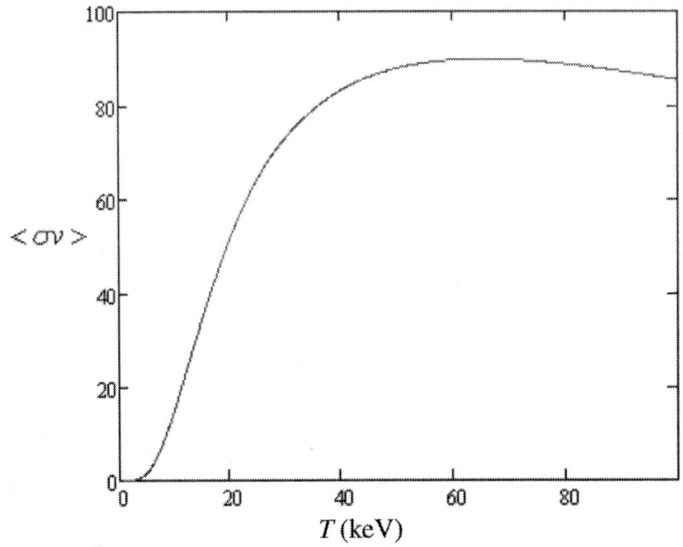


**Figure 1.** $<\sigma v>$ versus the plasma temperature.

Qualitative properties of the solution may be described with a model equation (11)

$$\frac{\mathrm{d}T}{\mathrm{d}t} = -1.5T + T^2 + 0.56 + \eta\sin(10t). \tag{11}$$

Equation (11) for $\eta = 0$ has two solutions, the first, $T = 0.7$ is stable, the second, $T = 0.8$ is unstable. The solution for $\eta = 0.4$ is shown in Figure 2. The average temperature $<T> \approx 0.71$ is stable. It is higher than the equilibrium temperature for the steady-state regime $T = 0.7$. Hence the fusion power output is higher than that in the steady-state regime.

The increase of $\eta$ must increase the nuclear energy output and provide a transition to ignition. The case $\eta = 0.7$ is shown in Figure 3. One can see that the average temperature rises in time. The time scale of the transition into ignition is small. The transition is shown separately in Figure 4.

One can see that for ignition the input power must exceed energy losses not permanently but in a short time interval in contrast to the steady-state regime. Hence, the average auxiliary heating power may be significantly smaller than that in the steady-state regime. The process of the ignition has three stages. The first stage is related to the transition to the quasi-steady state regime with slow rise of the average temperature. The third stage is characterized by a fast increase of the temperature and the transition to the ignition.

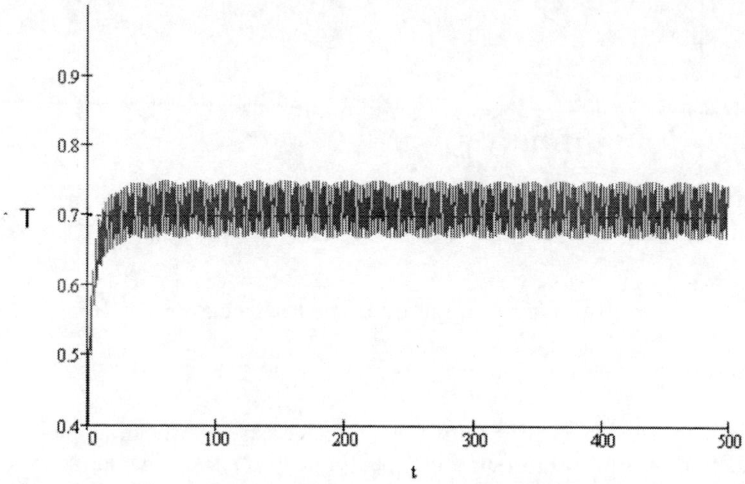


**Figure 2.** Steady-state discharge.

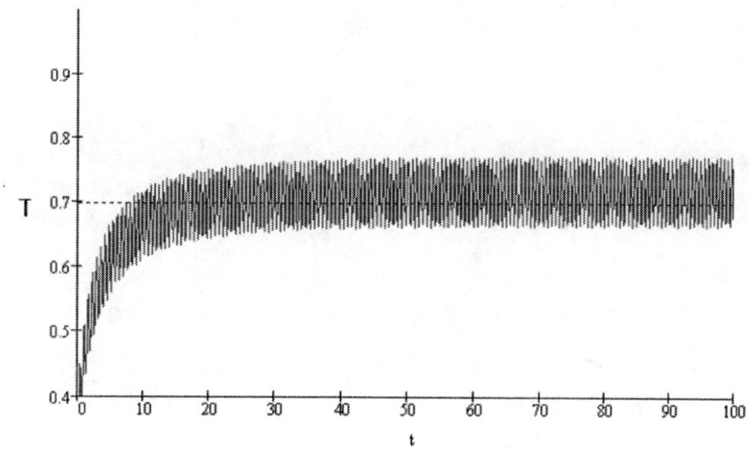


**Figure 3.** Ignition. The first two stages.

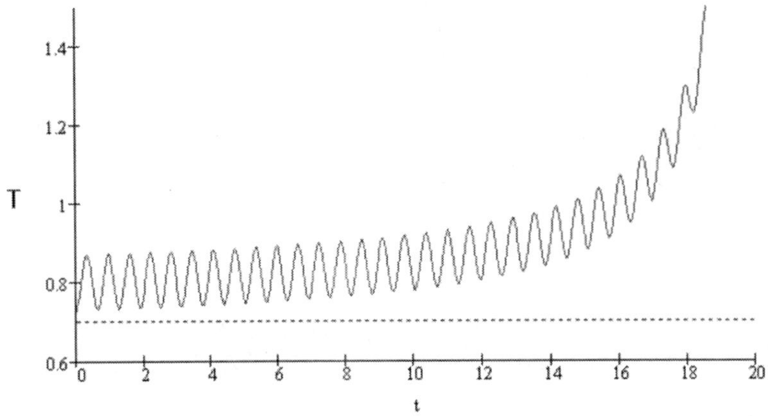


**Figure 4.** Ignition. The final stage.

## 3. Conclusion

In order to increase the fusion energy output and to decrease the heating power required for the ignition the modulation of the power of auxiliary heating is proposed. The harmonic oscillations of the auxiliary heating power produce the fusion power increase. The effect is a consequence of the nonlinear dependence of the fusion power output on the temperature. Two effects are predicted. The first one is the appearance of the additional power output in steady-state regime. Secondly, the ignition is possible even if the averaged power input is not sufficient for ignition in the steady-state regime. Hence, simpler and cheaper power sources may be used for a fusion reactor. The transition to ignition has three stages. The first stage is relatively fast. It is related to the transition from the initial low temperature to the quasi-stationary state with slow temperature rise. The third stage is the extremely fast one. The thermal instability is observed with the ignition as a result. The effect is predicted with the help of a qualitative model based on zero-dimensional approximation. Surely, the problem needs a more detailed analysis.

## Acknowledgments

This work has been partly supported by National Research Nuclear University MEPhI in the framework of the Russian Academic Excellence Project (contract No. 02.a03.21.0005, 27.08.2013).

## References

[1] Morozov D Kh, Pshenov A A and Mineev A B 2008 *Proc. of 35th EPS Conference on Plasma Physics*, ECA v. 32, P-5.067
[2] Challis C D, Gardia J, Beurskens B, Buratti P et al. 2015 *Nucl. Fusion* **55** 053031
[3] Takizuka T and Yamagiwa M 1987 Reactivity for DT Plasma with Beam Injected Tail in Magnetic Confinement System. JAERI-M 87-006.

# Compression of magnetized target in the magneto-inertial fusion

**V V Kuzenov**[1,2]

[1] Department of Thermal Physics, Bauman Moscow State Technical University, 2 Baumanskaya Street 5, 1, Moscow 105005, Russia
[2] N.L. Dukhov All-Russian Research Institute of Automatics (State Corporation «ROSATOM» company), Suschevskaya Street 22, Moscow 127055, Russia

E-mail: vik.kuzenov@gmail.com, svryzhkov@bmstu.ru

**Abstract.** This paper presents a mathematical model, numerical method and results of the computer analysis of the compression process and the energy transfer in the target plasma, used in magneto-inertial fusion. The computer simulation of the compression process of magnetized cylindrical target by high-power laser pulse is presented.

## 1. Introduction

Ongoing experimental and theoretical studies on magneto-inertial confinement fusion (Angara, C-2, CJS-100, General Fusion, MagLIF, MAGPIE, MC-1, YG-1, Omega) and new constructing facilities (Baikal, C-2W, Z300 and Z800) require adequate modeling and description of the physical processes occurring in high-temperature dense plasma in a strong magnetic field [1-7].

Magneto-inertial fusion is an alternative path to the magnetic and inertial fusion, and magnetic field suppresses the thermal conductivity in a magnetized plasma (the target). Thereby, greatly reducing heat losses and increasing the lifetime of the configuration, and hence the thermophysical properties and energy parameters of the system.

This paper presents a detailed description of magneto-plasmadynamic code that allows to perform modeling of X-rays/first wall interaction. This problem is important in assessing the performance of device as the power plant. This mathematical model cannot determine the dynamic response (the evolution of the stress waves and pressures with the strength and compressibility in structural materials) of the first wall on neutron pulse, but the calculations make possible to estimate the interactions between radiation and solid wall, determine the mass of the evaporated material, its temperature, and mechanical impulse.

This work is devoted to the simulation of laser beams interaction with an axisymmetric dense magnetized plasma, e.g., such burning plasma parameters can be produced in magneto-inertial fusion. For this reason the numerical model of the process is based on the multi-component radiation gas dynamics with turbulence model. The radiation transfer is taking into account in multi-group diffusive approach. The main distinguish of this proposed model is the presence of the externally applied magnetic field.

## 2. Mathematical Model

The equation of plasma magnetic field generation is:

$$\frac{\partial \vec{B}}{\partial t} = rot\left[\vec{V}\times\vec{B}\right] - \frac{c^2}{4\pi}\frac{t_*}{L_*^2} rot\left(\frac{rot\vec{B}}{\sigma}\right), \qquad \vec{j} = \sigma\left(\vec{E} + \frac{V_*}{c}\left[\vec{V}\times\vec{B}\right]\right),$$

where $t_*$ is the characteristic exposure time, $L_*$ is the characteristic size, $V_*$ is the characteristic velocity, $\sigma$ is the electrical conductivity coefficient, $\vec{E}$ is the intensity of electric field, $\vec{V}$ is the plasma velocity.

Operator *rot* in the cylindrical coordinate system ($r, z, \varphi$):

$$\left(rot\vec{B}\right)_r = \frac{1}{r}\frac{\partial B_z}{\partial \varphi} - \frac{\partial B_\varphi}{\partial z}, \quad \left(rot\vec{B}\right)_\varphi = \frac{\partial B_r}{\partial z} - \frac{\partial B_z}{\partial r}, \quad \left(rot\vec{B}\right)_z = \frac{1}{r}\frac{\partial rB_\varphi}{\partial r} - \frac{\partial B_r}{\partial \varphi}.$$

It is known, that plasma dynamical processes are strongly affected by irradiation, if laser plasma temperature achieves 1 eV. In that case, the gas dynamic fields of thermophysical variables may be calculated only with taking into account the radiation fields. In this work the radiation transfer equation is used in the form of multi-group diffusion approach:

$$\frac{1}{J}\frac{\partial\left(Jq_{i\xi}\right)}{\partial \xi} + \frac{1}{J}\frac{\partial\left(Jq_{i\eta}\right)}{\partial \eta} + \chi_i c U_i = 4\chi_i\sigma_i T^4,$$

$$\frac{c}{3}\frac{\partial U_i}{\partial \xi} + \chi_i q_{i\xi} = 0, \quad \frac{c}{3}\frac{\partial U_i}{\partial \eta} + \chi_i q_{i\eta} = 0,$$

where $U_i$ ($y$, $z$, $t$) is the radiation power density in $i$-th spectral group, $\chi_i$ is the spectral absorption coefficient. Besides of the method mentioned above, the discrete ordinates method has been used in the work, which gets an opportunity to solve the radiation transfer equation on tetrahedral non–structured mesh. Modified alternatively triangular three-layered iterative scheme is applied for the solution of radiation transport equations, where the time step is selected via the conjugate directions method.

One can use the mechanism of continuum absorption, the opposite to electron bremsstrahlung process under conditions of local thermodynamic equilibrium, to define the plasma absorption coefficient for laser radiation:

$$\chi_\omega = \frac{4{,}97\, gZ_i n_i n_e}{n_c^2\lambda^2\left(kT_e\right)^{3/2}}\frac{1}{\sqrt{1-n_e/n_c}},$$

where $\lambda$ is the laser radiation wavelength (μm), $n_e, n_i$ are the electron and ion densities correspondingly (cm$^{-3}$), $kT_e$ is the electron temperature (keV), $g$ is the Gaunt factor, $n_c = 10^{21}\times\lambda^{-2}$ is the critical electron density (cm$^{-3}$).

The turbulent viscous coefficient $\mu_\Sigma^{\perp,\parallel}$ is calculated by using of the Boussinesq hypothesis, when the effective viscous coefficient is $\mu_\Sigma^{\perp,\parallel} = \mu_m^{\perp,\parallel} + \mu_t$, where $\mu_m^{\perp,\parallel}$ is a molecular viscosity coefficient and $\mu_t$ is a turbulent one, determined from $q-\omega$ turbulence model. The equations of $q-\omega$ turbulence model in curvilinear coordinates $\xi, \eta$ are:

$$\frac{\partial \rho q}{\partial t}+\frac{d\xi}{dt}\frac{\partial \rho q}{\partial \xi}+\frac{d\eta}{dt}\frac{\partial \rho q}{\partial \eta}+\frac{1}{J}\frac{\partial\left(J\rho V_{\xi}q\right)}{\partial \xi}+\frac{1}{J}\frac{\partial\left(J\rho V_{\eta}q\right)}{\partial \eta}+\alpha\frac{\rho q u}{r}=S_q+\frac{\rho q}{2\omega}\left(C_{\mu}fD-\frac{2}{3}\omega div\vec{V}-\omega^2\right),$$

$$\frac{\partial \rho \omega}{\partial t}+\frac{d\xi}{dt}\frac{\partial \rho \omega}{\partial \xi}+\frac{d\eta}{dt}\frac{\partial \rho \omega}{\partial \eta}+\frac{1}{J}\frac{\partial\left(J\rho V_{\xi}\omega\right)}{\partial \xi}+\frac{1}{J}\frac{\partial\left(J\rho V_{\eta}\omega\right)}{\partial \eta}+$$
$$\alpha\frac{\rho\omega u}{r}=S_{\omega}+\rho\left(C_1\left(C_{\mu}D-\frac{2}{3}\omega div\vec{V}\right)-C_2\omega^2\right),$$

$$S_q=\frac{1}{J}\frac{\partial\left\{J\left(\mu^{\perp}_{\Sigma q}\xi_r^2+\mu^{\parallel}_{\Sigma q}\xi_z^2\right)q_{\xi}\right.}{\partial \xi}+\frac{\left.J\left(\mu^{\perp}_{\Sigma q}\xi_r\eta_r+\mu^{\parallel}_{\Sigma q}\xi_z\eta_z\right)q_{\eta}\right\}}{\partial \xi}+$$
$$\frac{1}{J}\frac{\partial\left\{J\left(\mu^{\perp}_{\Sigma q}\eta_r\xi_r+\mu^{\parallel}_{\Sigma q}\eta_z\xi_z\right)q_{\xi}+J\left(\mu^{\perp}_{\Sigma q}\eta_r^2+\mu^{\parallel}_{\Sigma q}\eta_z^2\right)q_{\eta}\right\}}{\partial \eta}+\alpha\frac{\mu^{\perp}_{\Sigma q}}{r}\left\{\xi_r\frac{\partial q}{\partial \xi}+\eta_r\frac{\partial q}{\partial \eta}\right\},$$

$$S_{\omega}=\frac{1}{J}\frac{\partial\left\{J\left(\mu^{\perp}_{\Sigma \omega}\xi_r^2+\mu^{\parallel}_{\Sigma \omega}\xi_z^2\right)\omega_{\xi}\right.}{\partial \xi}+\frac{\left.J\left(\mu^{\perp}_{\Sigma \omega}\xi_r\eta_r+\mu^{\parallel}_{\Sigma \omega}\xi_z\eta_z\right)\omega_{\eta}\right\}}{\partial \xi}+$$
$$\frac{1}{J}\frac{\partial\left\{J\left(\mu^{\perp}_{\Sigma \omega}\eta_r\xi_r+\mu^{\parallel}_{\Sigma \omega}\eta_z\xi_z\right)\omega_{\xi}+J\left(\mu^{\perp}_{\Sigma \omega}\eta_r^2+\mu^{\parallel}_{\Sigma \omega}\eta_z^2\right)\omega_{\eta}\right\}}{\partial \eta}+\alpha\frac{\mu^{\perp}_{\Sigma \omega}}{r}\left\{\xi_r\frac{\partial \omega}{\partial \xi}+\eta_r\frac{\partial \omega}{\partial \eta}\right\},$$

$$\mu^{\perp,\parallel}_{\Sigma q}=\mu^{\perp,\parallel}_{m}+\mu_t,\ \mu^{\perp,\parallel}_{\Sigma \omega}=\mu^{\perp,\parallel}_{m}+1{,}3\mu_t,\ \mu_t=C_{\mu}f(n)\rho\frac{q^2}{\omega},$$

$$C_1=0{,}045+0{,}405f(n),\ C_2=0{,}92,\ C_{\mu}=0{,}09,\ f(n)=1-\exp\left(-0{,}0065\frac{\rho q n}{\mu^{\perp,\parallel}_{m}}\right),$$

where $q$ is a pseudo velocity and $\omega$ is a pseudo vortex, $f(n)$ is a wall function for correct flux parameters evaluation in laminar sub-layer near a solid streamline surface, $n$ is the distance by normal from the given point to the nearest surface.

## 3. Results

Initial values of the intensity of the seed magnetic field in the rarefied medium reach the fraction of $T$. Results of computational modeling [8-10] are presented below. Calculations carried out for the Nd laser radiation for forming rectangular shape (impulse duration = 10 ns). The laser radiation flux is $q_{las}$ = $2\cdot10^{14}$ W/cm$^2$. A thin metallic cylindrical shell material is Al. The target radius is 0.1 cm.

Theoretical results for the case of double-layer target for laser-driven magneto-inertial fusion are presented below. Spatial distribution of the pressure $P$ and temperature $T$, velocity $V$ and density ρ of the plasma at time 5.66 ns (the pulse duration is 10 ns) is shown in figures 1-2. The distributions of the magnetic pressure $P_{mag}$ and laser radiation $W_{las}$ along the radial coordinate are shown in figure 3. These distributions correspond to the "expansion" stage of plasma formation ($t$ = 5.66 ns).

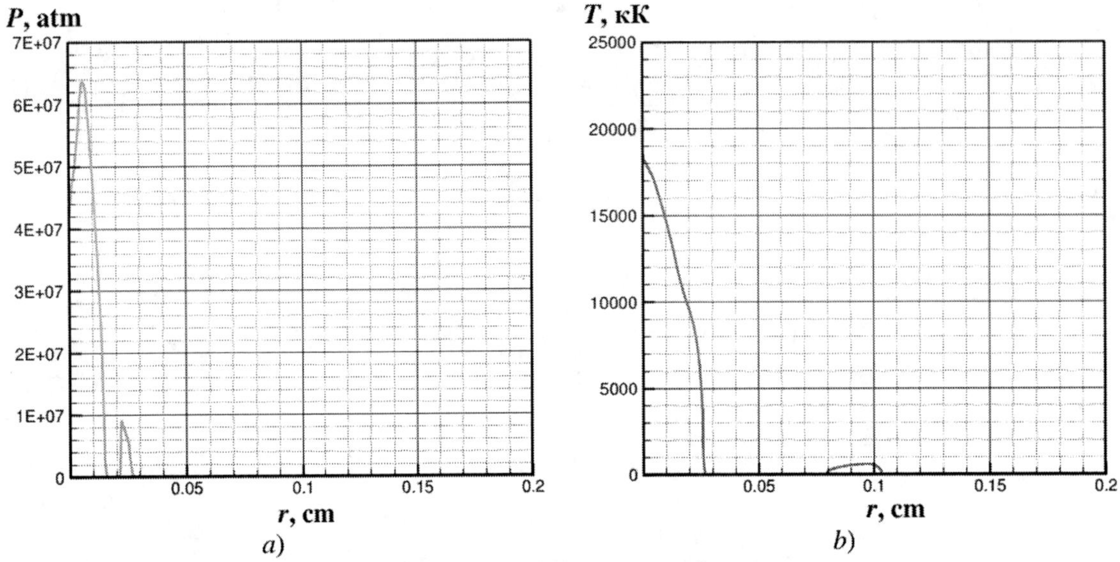


**Figure 1.** Spatial distribution of the plasma pressure (*a*) and temperature (*b*) at time $t = 5.66$ ns. The laser pulse duration $t_{Las} = 10$ ns and the laser radiation flux $q_{las} = 2 \cdot 10^{14}$ W/cm$^2$.

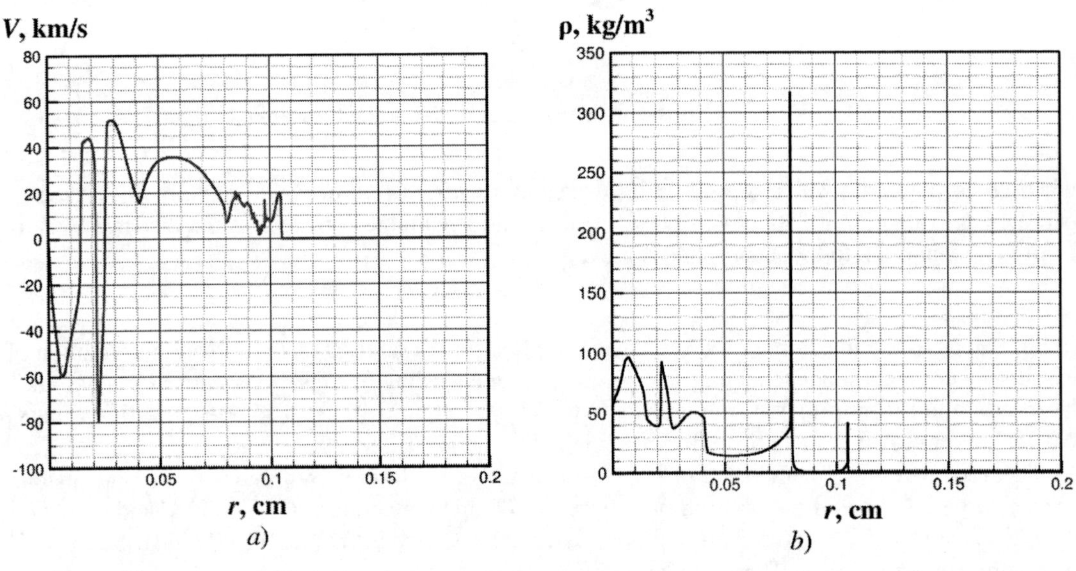


**Figure 2.** Plasma velocity (*a*) and density (*b*) at $t = 5.66$ ns as a function of radial coordinate.

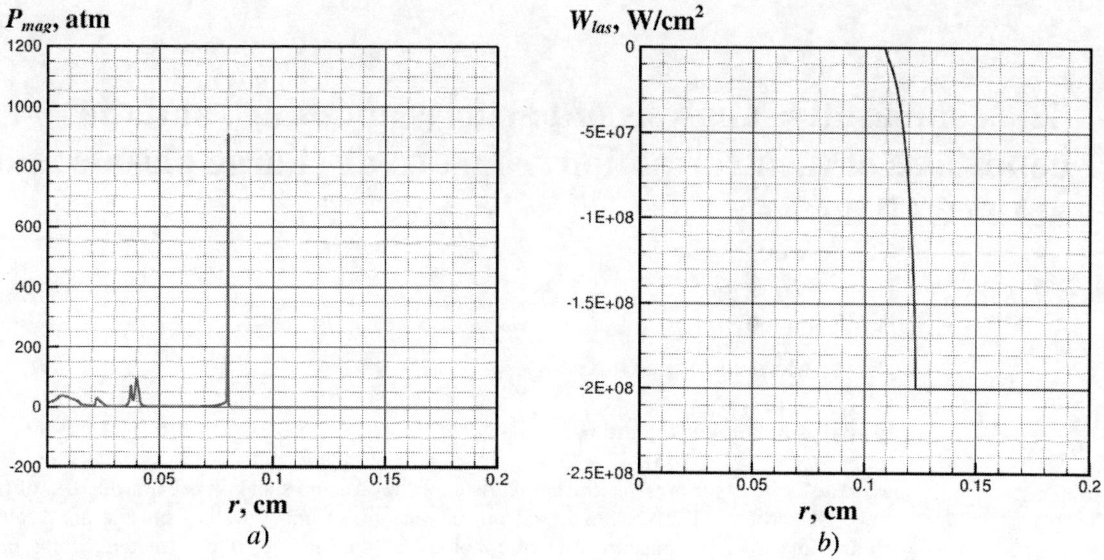


**Figure 3.** Distribution of the magnetic pressure (*a*) and laser radiation (*b*) at $t = 5.66$ ns ($t_{Las} = 10$ ns).

## 4. Conclusion

The non-stationary two dimensional radiation magneto-gas dynamic model is developed. The mathematical model takes into consideration the magnetized plasma compressed by laser beams of high energy pulses and high speed plasma jets. It is based on radiation plasma dynamics equations in arbitrary curvilinear coordinates. An improved two-dimensional radiation-hydrodynamics code which simulates plasmas in cylindrical or spherical geometries is designed. It solves single-fluid, two-temperature equations of motion with contributions from diffusion, convection, heat conduction. Electromagnetic processes are described by the Maxwell-Ohm equations in plasma with final conductivity. The transport coefficients in the given system of the equations taking into account magnetized laser plasma. The model is based on splitting method in terms of physical processes and spatial directions, that in spatially smooth solution allows to get seventh order of accuracy.

## Acknowledgments

This research has been supported by the Ministry of Education and Science of the Russian Federation (Project No. 13.5240.2017/БЧ).

## References

[1] Mima K, Johzaki T, Honrubia J, Nagatomo H, Taguchi T, Sunahara A, Sakagami H, Fujioka S and Logan G 2016 *J. Phys.: Conf. Ser.* **688** 012066

[2] Sinars D B, Campbell E M, Cuneo M E, Jennings C A, Peterson K J, Sefkow A B 2016 *J. Fusion Energy* **35** 78

[3] Ryzhkov S V 2008 *Proc. 35th EPS Conf. on Plasma Physics and Contr. Fusion* ECA **32D** P1.114

[4] Ryzhkov S V 2010 *Problems of Atomic Science and Technology* **4** (7) 105

[5] Ryzhkov S V 2015 *Sustainable Cities and Society* **14** 313

[6] Kostyukov I Yu and Ryzhkov S V 2011 *Plasma Physics Reports* **37** 1092

[7] Ryzhkov S V, Chirkov A Yu and Ivanov A A 2013 *Fusion Science and Technology* **63** (1T) 135

[8] Kuzenov V V and Ryzhkov S V 2013 *Problems of Atomic Science and Technology* **4** (86) 103

[9] Kuzenov V V and Ryzhkov S V 2013 *Problems of Atomic Science and Technology* **1** (83) 12

[10] Kuzenov V V and Ryzhkov S V 2016 *Bulletin of the Russian Academy of Sciences. Physics* **80** 598

# The comparative analysis of particles and films, and the conditions of their formation from arc-discharge plasma and at the CTF devices

**N A Smolanov**

Ogarev Mordovia State University

E-mail: smolanovna@yandex.ru

**Abstract.** The paper presents the comparison of the structures and physical properties of films and dust particles. They are prepared on the vacuum chamber walls from the arc-discharge plasma and tokamak plasma. The morphology of the surface, microstructure, color range, grain-size distribution, and elemental and phase composition of films and micro particles have been compared. It was found that the films and powders have properties of nanocrystal materials. The structure, surface morphology and some properties of the arc-discharge plasma particles are similar to the structures and properties of the particles deposited on the TOKAMAK walls.

## 1. Introduction

The issue of forming films and dust in the plasma is of particular relevance for the safe operation of CTF devices. Technological purity is very important for the devices of plasma etching and spraying. Dust particles ranging in size from tenths of microns to 10-15 microns deposit in the tokamak near-wall areas [1-7]. It is established that dust is the result of drop-and-steam erosion of the walls, the intensity of which increases when current failure occurs. The size of the droplet fraction can increase up to 100 μm. The mechanisms of its formation, transportation and agglomeration remain unexplored. Dropping erosion in tokamaks has not been experimentally studied [5]. When conducting plasma tests of tungsten, we have found dust particles of irregular shape similar to sputtered graphite and other materials. All this underscores the importance of studying all mechanisms of dust formation in the tokamak [3-6]. The tokamak vacuum chamber walls are eroded mainly by near-surface plasma (ions and atoms of the end plasma). The results of the elemental analysis of the spraying products (dust) indicate that dust can also be formed from the limiter (diverter) and from more distant parts. In the SOL area there is intense turbulence, diffusion, and a significant temperature gradient [1-7]. But the conditions in the space behind the (SOL) area and in the diverter, where there is the accumulation of films and dust, are similar to those in the devices for plasma etching and filming [1-11]. The main reason for stimulating the present work was similarity of plasma parameters. The sizes and shapes of the particles near the tokamak walls and in low-temperature arc discharge plasma are also similar (Table 1). The basis of arc discharge plasma is cathode spots [12]. The attention to processes in arc discharges has increased sharply since new carbon structures were discovered [13].

**The objective of this study** is to analyze and compare the structures of films and dust (soot) particles that were deposited on the walls of the vacuum chambers from laboratory arc discharge plasma [14] and tokamak plasma [4-11,16,21-25].

**Table 1.** Comparative analysis of parameters of high-and low-temperature plasma and dust particles.

| No. | Parameter | CTS device | Device for plasma etching and spraying |
|---|---|---|---|
| 1 | Plasma density | $10^{14}$-$10^{15}$ cm$^{-3}$ | $10^{9}$-$10^{11}$ cm$^{-3}$ |
| 2 | Electronic temporary | Up to 1 eV | 2-5 eV |
| 3 | Dust sources | Erosion of the walls | Injection from the walls, condensation of the clusters |
| 4 | Charge of dust participles | $10^{4}$-$10^{5}$ $e$ | $10^{4}$-$10^{5}$ $e$ |
| 5 | Sizes of the dust participles | from 0.1 μm | from 1 μm |
| 6 | Morphology of the participles | Flakes, cauliflowers, fractals | Flakes, cauliflowers, fractals |
| 7 | Ion flows | $2{\cdot}10^{20}$ cm$^{-2}$c$^{-1}$ | Depend on the heat flow on the walls |
| 8 | Mechanisms | Formation of molecular complexes | Formation of molecular complexes $C_xH_y$ (up to y=30) |
| 9 | Peculiarities | Turbulence in the near-wall plasma, short discharge time | Large discharge time |

## 2. Subjects and methods of the study

The film and dust in the arc discharge technological cycle were placed on the underlying of steel 12X18H0T. This steel [3, 6, 10] is used for the tokamak type devices. Plasma supporting gases-nitrogen and acetylene were filed into the area of a spray titanium cathode. The powder deposited on the walls (a fraction of less than 0.075 mm) was extracted with a vacuum cleaner and separated in a magnetic field (1000 Oe) to get magnetic fractions [14].

To research the material deposited from the arc discharge plasma, we used the same set of methods and equipment as in the works, the results of which were taken for comparison [5-8, 12-16]. This ensured objectivity in comparing the structures and properties of the films and dust-like particles from the arc discharge vacuum chamber and tokamak.

## 3. Comparison of color characteristics

Features of hydrocarbon structures of the tokamak T-10 are color range, layer microstructure, and metal impurity (predominantly iron) [3, 7, 10]. The source of the metals in the films are a glow discharge and failures. Yellow-brown smooth films, reaching 0.5 mm in thickness, predominate. Color can be determined not only by iron compounds [3]. For comparison, Figure 3 shows the cladding obtained by us from arc discharge, when a certain proportion of reaction gases (C/N/O) inflows in a vacuum chamber during cathode sputtering of titanium.

The color of the films from the tokamak usually reflects the D/C atomic ratio. In case of arc spraying titanium nitride (carbonitride) of stoichiometric or similar composition gives similar in color film. In the material of yellow (black) films from the tokamak there are traces of spraying tokamak structural components of stainless steel. Metal atoms can be a precursor of forming carbon spheres. This idea [3, 7] deserves further discussion (carbide cycle mechanism) [15].

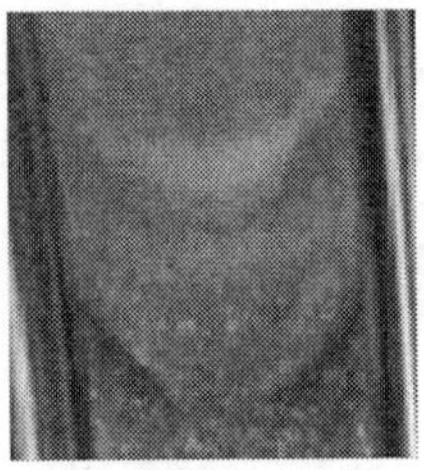

**Figure 1.** Thin smooth films on bellows connection inserts [3].

**Figure 2.** Stainless steel mirror of size 1 × 1 cm with a film up to 300 nm thick [7].

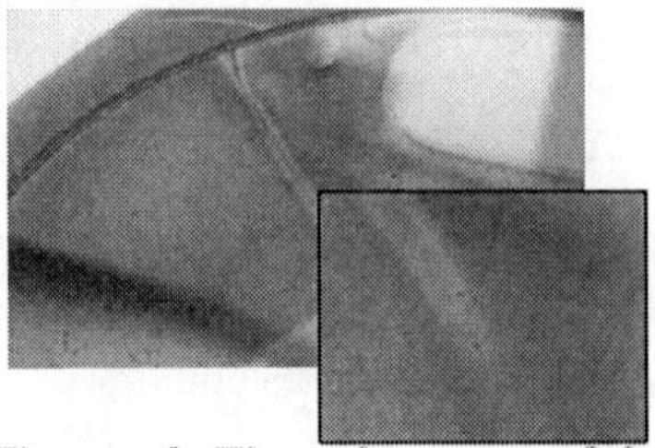

**Figure 3.** The color range of the film from the arc discharge chamber (the substrate is glass).

## 4. Comparison of flakes (scales) from the CTS and LEP cameras

The films of a certain thickness or composed of several layers films in tokamaks begin to flake off stainless steel substrates as a result of internal tensions and the different thermal expansion coefficients with the formation of scales (flakes) [compare Figures 5 and 6]. We have found that in scales (parts of the film) there are trace metals, whose source can be glow discharge being used to clean the camera [3]. For comparison, Figure 5 shows a picture of the scale from the arc discharge chamber, formed after 5-7 cycles of plasma flow deposition with dropping fraction near a spraying titanium cathode. There are obvious matches in shapes and sizes.

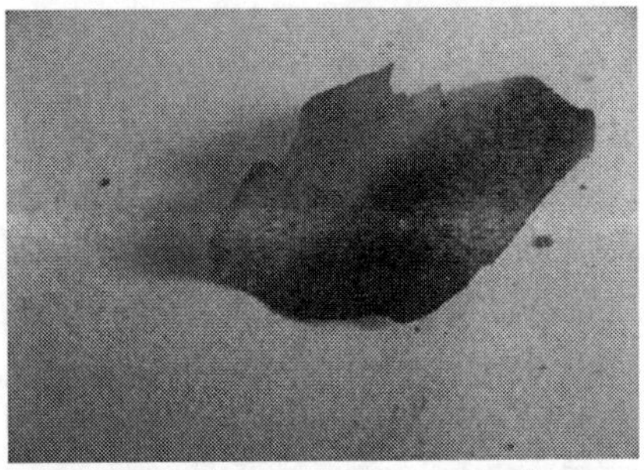

**Figure 4.** A film flake on the lower diagnostic flange [3].

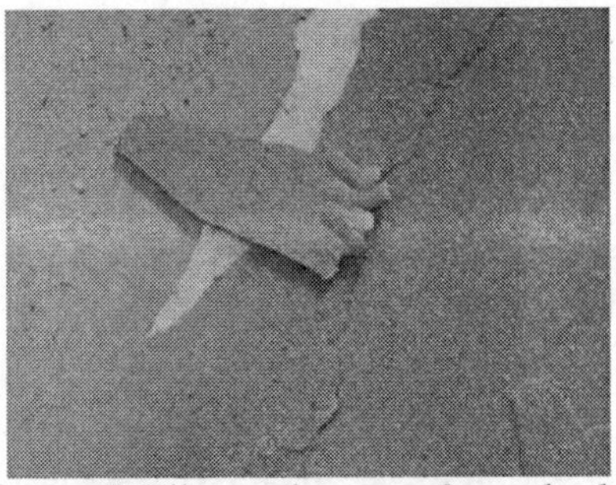

**Figure 5.** Film scales near the cathode sprayed by an arc discharge.

## 5. Globular form of particles

Globules on the films surface in the tokamak are observed in a number of works [10, 11, 16]. Their shapes and sizes match the ones of the globular particles deposited from the dropping fraction of low-temperature plasma (Figures 6, 7). Probably, the mechanisms forming these areas are the same: deposition of carbon on metallic particles and fast catalytic growth in the plasma areas with a temperature of a few electron-volts (near the chamber walls). Another explanation for the formation of spherical particles in the tokamak is based on the fact that a dust particle was inside plasma for a while [7]. In the case of low-temperature plasma, a spherical shape of particles is resulted from their liquid state prior to rapid crystallization (Figures 6, 7).

The globular and columnar structures of the films from tokamaks and low-temperature arc discharge plasma can be presented in a model of structural zones of vacuum condensates [17-19].

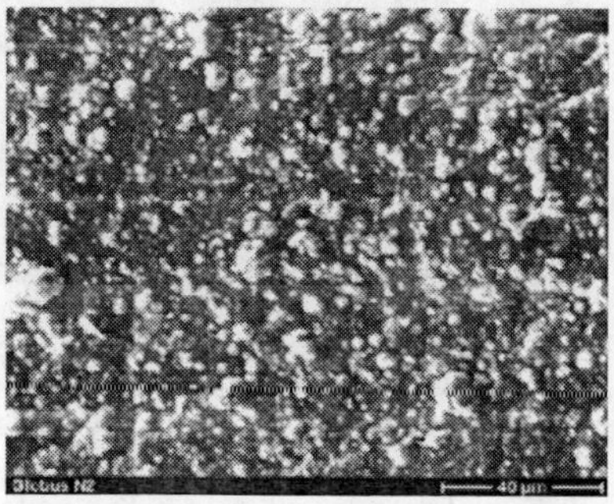

**Figure 6.** Carbon globular deposit formed on the tile [3].

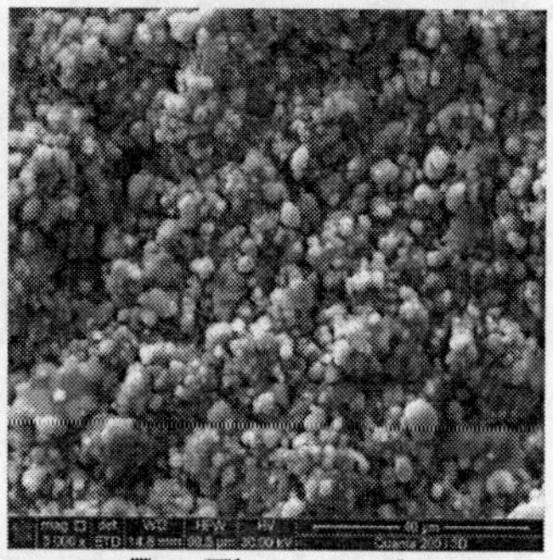

**Figure 7.** The condensate of titanium carbonitride near the cathode pack

## 6. Fractal character of the films

The fractality of deposited films is an important factor for accumulating tritium in a fusion reactor. The authors [3, 7] compare the fractal structure of the films on the tokamak walls with tree rings, which allow retracing the long history of the tokamak T-10 (Figure 8). The similar structures (Figure 9) are formed on the vacuum chamber walls and around cathode sprayed from arc discharge plasma [20]. Fractal growth of films in the thermonuclear facility is caused by a strong turbulence of near-wall plasma, which excites thermal fluctuations in depositing flow. Inhomogeneity of the flow of deposited particles results in interactions similar to those in low-temperature plasma [3-5, 21-23].

**Figure 8.** The multilayer film deposited near the limiter (tokamak T-10), an increase of '2000 [7].

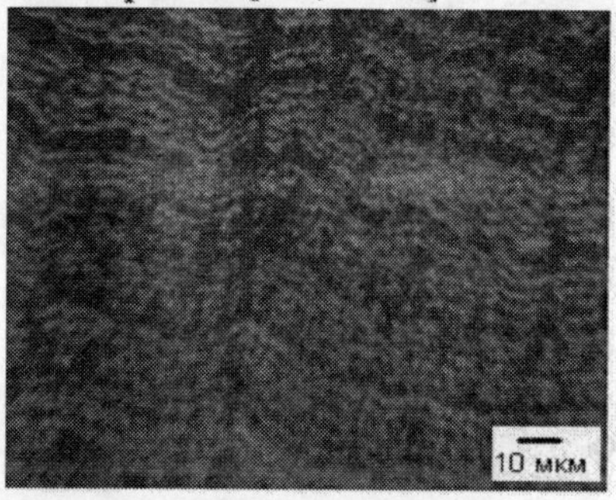


**Figure 9.** The microstructure of a multilayer film near the cathode [20].

## 7. Comparison of dust structures

The structures resulted from the erosion of graphite walls and metal constructions inside the tokamak have been compared with similar materials obtained from low temperature plasma by means of arc sputtering of titanium with spraying nitrogen and acetylene in a combustion zone (Figures 10, 11). In decomposing acetylene in arc discharge for forming titanium carbonitride films, there is growing a-C:H layers [14] similar to the layers depositing while decomposing ethylene in hydrogen plasma [24, 25]. This results not only in increasing the deposition rate, but also in breaking the carbon structure of films caused by forming sp3-carbon bonds [25]. The possibility of reactions of this type is described in a recent review of academician A. K. Rebrov [26].

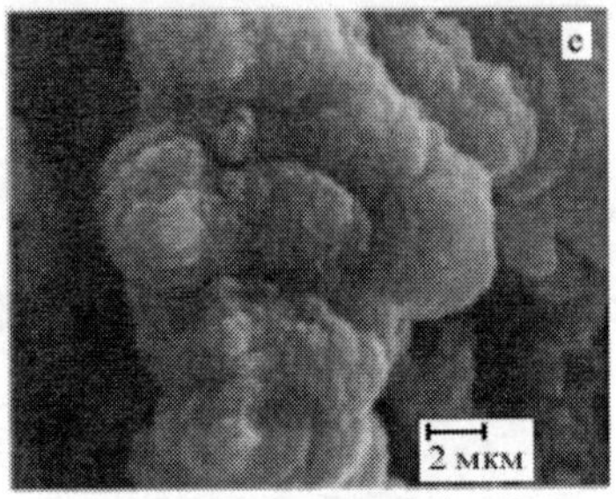


**Figure 10.** Dust particles from the tokamak [4].

**Figure 11.** The particles from arc discharge plasma.

## 8. New (unknown) structures

Of particular interest are new (unknown) structures of CTS plasma and arc discharge plasma. They are resulted from the conditions of their formation: a high speed of steam-liquid-solid or vapor-solid synthesis. The study of the carbon dust structure from the tokomak T-10 [9] has showed that it has a phase that is not in initial graphite. The found phase is characterized by the existence of much larger crystals. The formation of this phase is determined by the pressure of vapor and time of crystallization in the device. Unknown frame structure of carbon nanotubes in sizes from a few to hundreds of nanometers have been also obtained inside the tokamak T-10 [27].

We have also found particles of ASTM structures, which are unknown in databases, in the structures of the carbon black material deposited on the vacuum arc discharge chamber walls (installation of HHB-6). One of the obtained phases corresponds to the point space group Fd3m (diamond) [14].

## 9. Results

1. Compared films and powders obtained through depositing from arc discharge plasma and CTS plasma are nanocrystaline materials.
2. Their structures, surface morphology and some properties are similar.
3. Fractal character of the studied materials can prove the identity of processes starting in the cathode spots of vacuum discharges.
4. The obtained films and microparticles can be considered as a result of the interaction of plasma-dust structures in the electric and magnetic fields.

## 10. Conclusion

Some works note that such studies are important for not only analyzing the hydrocarbon films and particles, but also for studying thermonuclear fuel cycle, modeling and designing new installations. Coupled with the increasing interest in dust plasma, there is a need for more precise and reasonable use of the classification of formed structures, including nanostructures (hierarchy or morphology). In this regard, a Manual of Classification of Nanostructures proposed by material scientists may be very useful [28].

### References

[1] Nedospasov A V 1987 *UFN* **152** 479
[2] Tsitovich V N and Vinter Dzh 1998 *UFN* **168** 899
[3] Kislov A Y, Skovoroda A A and Spitsyn A V 2010 *Problems of Atomic Science and Engineering, ser.: Thermonuclear fusion* **2** 19
[4] Krauz V I, Martynenko Yu V, Svechnikov N Yu, Smirnov V P, Stankevich V G and Khimchenko L N 2010 *UFN* **180** 1055
[5] Budayev V P 2015 *Probl. of Atomic Science andEngineering, ser.: Thermonuclear fusion* **38** 4
[6] Zalavutdinov R Kh, Gorodetsky A E and Zakharov A P 2011 *Problems of Atomic Science and Engineering, ser.: Thermonuclear fusion* **1** 39-47

IOP Conf. Series: Journal of Physics: Conf. Series **941** (2017) 012012 doi:10.1088/1742-6596/941/1/012012


[7] Stankevich V G, Svechnikov N Yu, Zubavichus Ya V and Veligzhanin A A 2011 *Problems of Atomic Science and Engineering, ser.: Thermonuclear* **1** 3

[8] Svechnikov N Yu, Stankevich V G, Lebedev A M, Menshikov K A, Kolbasov B N, Guseva M I and Khimchenko L N 2004 *Problems of Atomic Science and Engineering, ser.: Thermonuclear* **3** 3

[9] Aleksandrov P A, Belova N E, Kolbasov B N and Kuznetsov Yu Yu *Problems of Atomic Science and Engineering, ser.: Thermonuclear* **4** 29

[10] Skovoroda A A, Spitsyn A V, Yanchenkov S V and Zubavichus Ya V 2008 *Problems of Atomic Science and Engineering, ser.: Thermonuclear fusion* **3** 15

[11] Alimov V Kh, Goncharov V L, Komarov D A, Lakhotkin Yu V and Kuzmin V P 2008 *Problems of Atomic Science andEngineering, ser.: Thermonuclear fusion* **4** 31

[12] Anders A 2008 *Cathodic Arcs: From Fractal Spots to Energetic Condensation* (Berckley: Springer Science) 540 p

[13] Kratschmer W, Lamb L D and Fostiropoulos K 1990 *Nature* **347** 354

[14] Smolanov N A 2017 *Journal of Surface Investigation. X-ray, Synchrotron and Neutron Techniques* **4** 72

[15] Buyanov R A and Chesnokov V V 2005 *Khimiya v interesakh ustoychivogo razvitiya* **13** 37

[16] Begrambekov L B 2000 *Entsiklopediya nizkotemperaturnoy plazmy* **3** 126

[17] Movchan B A and Demchishin A V 1969 *The Physics of Metals and Metallography* **28** 23

[18] Cheng Y H, Tay B K and Lau S P 2002 *J. Vac. Sci. Technol. A.* **20** 1270

[19] Thornton J A 1974 *J. Vac. Sci. Technol.* **11** 666

[20] Smolanov N A and Pankin N A 2009 *Plas. Phys. and plas techn* vol. 1 (Minsk: «Polifakt») p. 91-94

[21] Budaev V P and Khimchenko L N 2008 *Problems of Atomic Science and Engineering, ser.: Thermonuclear fusion* **3** 34

[22] Budaev V P, Grashin S A and Karpov A V 2012 *JETP Letters* **95** 84

[23] Budaev V P and Khimchenko L N 2007 *JETP* **131** 711

[24] Markin A V 2004 *Probl. of Atomic Science and Engin., ser.: Thermonuclear fusion* **1** 21

[25] Vukolov K Yu, Mukhammedzyanov T R, Zvonkov S N, Svechnikov N Yu and Stankevich V G 2010 *Problems of Atomic Science and Engineering, ser.: Thermonuclear fusion* **4** 13

[26] Rebrov A K 2017 *UFN* **187** 193

[27] Neverov V S, Kukushkin A B, Marusov N L, Semyonov I B 2010 *Problems of Atomic Science and Engineering, ser.: Thermonuclear fusion* **1** 7

[28] Tretyakov Yu D 2006 *Micro- and nanoworld contemporary materials* (Moscow: Moscow state univ. publ.) 69 p

# Spatial separation of particles in modified linear Paul trap

**D S Lapitsky**

Joint Institute for High Temperatures of the Russian Academy of Sciences, Izhorskayast. 13 Bd. 2, 125412 Moscow, Russia

E-mail: dmitrucho@yandex.ru

**Abstract**. Paper presents theoretical investigations of spatial particle separation in modified linear Paul trap. In the simulations two upper electrodes of four of the trap were angled upwards to study particle separation at different geometries (interelectrode distances) at the same time. To simulate particle motion the Brownian dynamics has been used. Simulations were carried out in the assumption of spherical particles and that particles chargesare proportional to particle surface area. In the simulations particles occupied specific lengths according their charge-to-mass ratio.

## 1. Introduction

Monodisperse particles with known physical parameters are widely used in physics of low-temperature plasma [1–2], in the PIV research methods [3], in nanotechnology [4]. To produce powders of monodisperse particles several methods are used. For an example in [5] monodispese particles are separated from polydisperse powders by organizing of required particles in standing surface acoustic waves. In [6] particle separation is provided by dielectrophoresis. Particle separation in cyclone devices is widespread and often used to separate both solid and liquid particles from gases [7] and particles with different masses. In experimental facilities RF and glow discharges were used for particle separation [8]. In mass spectroscopy ion separation in vacuum is provided by Paul traps [9]. Ion dynamics in mass-spectrometers is described by Mathieu equation. In air at atmospheric pressure due to the energy dissipation in friction of the air the range of parameters for particle and ion capturing becomes wider [10]. In previous works [11–13] the confinement of microparticles by the alternating electric fields has been studied in static gas media and in gas flows.

The aim of this work is theoretical investigations on the spatial separation of charged particles from polydisperse powder in the linear Paul trap. For that task the trap was modified: two upper electrodes were angled upward to simulate particle separation at different interelectrode distances at the same time.

## 2. Mathematical model of charged particle dynamics in Paul trap

To simulate charged particle dynamics in the trap the Brownian dynamics has been used. The simulations took into account stochastic forces of random collisions with neutral particles, viscosity of air, particle interaction with electric fields of the trap, interparticle interaction and the gravitational force. Thus, the microparticle dynamics was described by the following Langevin equation [14]:

$$m_p \frac{d^2 r}{dt^2} = F_t(r) - 6\pi\eta r_p \frac{dr}{dt} + F_b + F_g + F_{\text{int}}, \tag{1}$$

where $m_p$ is particle mass, $r_p$ is particle radius, $r$ is the radius-vector of the particle, $\eta$ is the dynamic viscosity of gas medium (18.2 μPa•s [15]), $F_t(r)$ is the force of trap electrodes, $F_b$ are stochastic delta-correlated forces accounting for stochastic collisions with neutral particles, $F_g$ is the gravitational force, $F_{int}$ is the interparticle force of the Coulomb interaction. To solve the stochastic differential equation (1) the numerical method presented in [16] was used.

The Paul trap [17] consisted of 4 cylindrical electrodes with radius $R_1$ = 1.5 mm and length $L_m$ = 10 cm, arranged in the corners of rectangle. The alternating voltage $U_\omega \sin(\omega t)$ with phase shift $\pi$ was applied to neighboring electrodes. To simulate particle separation at different trap geometries at the same time in the trap two upper electrodes of four were angled, so the vertical interparticle distance varied from $L_b$ = 1.3 cm to 3.5 cm, while horizontal distances were constant 1.3 cm. The magnitude of alternating voltage was $U_\omega$ = 10 kV and the frequency was $f$ = 40 and 50 Hz.

## 3. Results of simulation of charged particle dynamics in Paul trap

In the trap 2000 particles of 9 types were injected. Particles density was 3990 kg/m$^3$. Simulations were carried out in assumption of spherical particles and particles charges are proportional to the surface area. Next types of particles were used: 400 particles with $r_p$ = 1μm and $q_p$ = 2500 e, other species of 200 particles with $r_p$ = 1.25, 1.5, 1.75, 2, 2.25, 2.5, 2.75 and 3 μm and charges correspondingly $q_p$=3906.25, 5625, 7656.25, 10000, 12656.25, 15625, 18906.25 and 22500 e. Results of simulations on spatial particle separation since 30 s after injection are presented in figures 1-3. In figures 1-2 vertical sections of trapped structure 0.2 cm thickness is presented.

In figures 1-2 one can see that particles are spatial separated into layers that correspond to the specific type of particles that can be described by the ratio $\tau = q_p/m_p$. In all figures the bigger $\tau$ the higher layer is situated inside the trap. Increasing the interelectrode distance the phenomenon of compartmentation becomes clearer. In figures 1-2 one can see that even particles with the same $\tau$ arrange in different layers (red dots, $\tau = 10^{17}$ C/kg). The vertical 2 cm cross section of layers in the center of the trap is presented in figure 3.

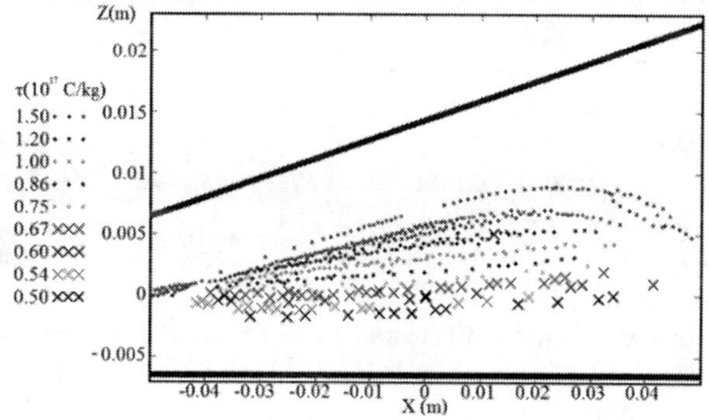


**Figure 1.** Side view of the trap. Spatial particle separation in modified Paul trap, $f$ = 40Hz, $L_b$ = 3.5cm.

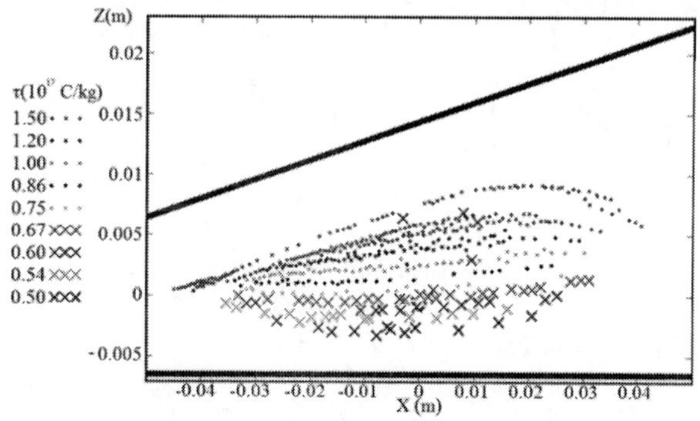


**Figure 2.** Side view of the trap. Spatial particle separation in modified Paul trap, $f$ = 50Hz, $L_b$ = 3.5cm.

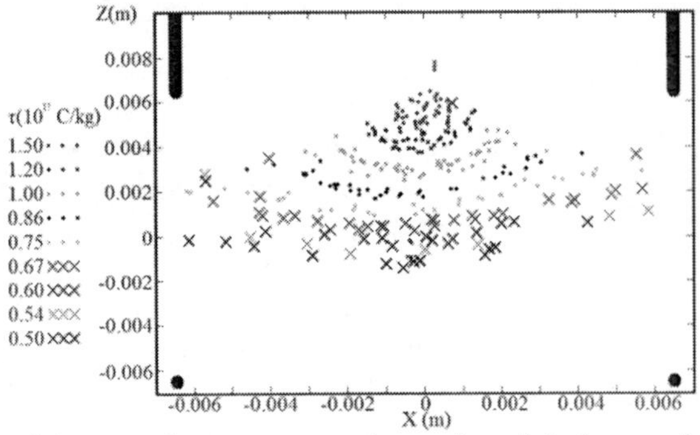


**Figure 3.** End view of the trap. 2 cm cross sections of particle layers, $f$ = 40Hz, $L_b$ = 3.5cm.

## 4. Conclusion

Paper presents theoretical investigations of particle separation in modified linear Paul trap. In assumption of spherical particles and that their charges are proportional to the particle surface area it was shown that particles organize in layers subject to their charge-to-mass ratio: the bigger $\tau$, the higher particles are. The linear dependence of particle charge to its surface area can be achieved by particle charging in the corona discharge [18].

**Acknowledgments**

The work was done by the financial support of the Russian Foundation for Basic Research grant No. 16-32-00031 mol-a.

**References**

[1] Vishnyakov V I and Dragan G S 2003 *Condens. Matter Phys.***6** 687
[2] Fortov V E, Khrapak A G, Khrapak S A, Molotkov V I and Petrov O F 2004 *UFN* **174(5)** 495
[3] Glushniova A V, Savelieva A S, Son E E and Tereshonok D V 2014 *High Temperature* **52** N2 221
[4] PrivmanV 2002 *Mat. Res. Soc. Symp. Proc.* **T3.3**577-585
[5] Rasim Guldiken, Myeong Chan Jo, Nathan D Gallant, Utkan Demirci and Jiang Zhe 2012 *Sensors* **12(1)** 905
[6] Peter R C Gascoyne and Jody Vykoukal 2002 *Electrophoresis* **23(13)**1973
[7] Brouwers J J H1996 *Chem. Eng. Technol.* **19(1)** 1
[8] Pestrikov V V, Grigoriev D A, Vasiliev M M, Petrov O F and Fortov V E 2013 *Granada, Spain: 31st ICPIG* PS2-081
[9] Mihalcea B M, Visan G T, Giurgiu L C and Radan S 2008 *J. Optoel. Adv. Mater.* **10(8)** 1994
[10] Lapitsky D S, Filinov V S, Deputatova L V, Vasilyak L M, Vladimirov V I and Pecherkin V Ya 2015 *High Temperature* **53(1)** 1
[11] Lapitsky D S, Filinov V S, Deputatova L V, Vasilyak L M, Vladimirov V I and Pecherkin V Ya 2015 *EPL* **110** 15001
[12] Deputatova L V, Filinov V S, Lapitsky D S, Pecherkin V Ya, Syrovatka R A, Vasilyak L M and Vladimirov V I 2015 *J. Phys.: Conf. Ser.* **653** 012131
[13] Vasilyak L M, Vladimirov V I, Deputatova L V, Lapitsky D S, Molotkov V I, Pecherkin V Ya, Filinov V S and Fortov V E 2013 *New J. Phys.* **15** 043047
[14] Filinov V S, Lapitsky D S, Deputatova L V, Vasilyak L M, Vladimirov V I and Sinkevich O A 2012 *Contrib. Plasma Phys.* **52(1)** 66
[15] Tsilingiris P T 2008 *Energy Conversion and Management* **49** 1098
[16] Skeel R D and Izaguirre J A 2002 Mol. Phys. **100** 3885
[17] Paul W 1990 *Reviews of Modern Physics***62** 531
[18] Syrovatka R A, Deputatova L V, Filinov V S, Lapitsky D S, Pecherkin V Ya, Vasilyak L M, and Vladimirov V I 2016 *Contrib. Plasma Phys.* **56** 419

IOP Conf. Series: Journal of Physics: Conf. Series **941** (2017) 012014 doi:10.1088/1742-6596/941/1/012014

# Plasma for environment

**G Van Oost**[1,2]

[1] Department of Applied Physics, Ghent University, Belgium
[2] National Research Nuclear University MEPhI (Moscow Engineering Physics Institute), Kashirskoe highway 31, 115409 Moscow, Russia

E-mail: guido.vanoost@ugent.be

**Abstract.** Human activity is associated with the permanent emergence of a very wide range of waste streams. The most widely used treatment of waste is thermal processing such as incineration. An alternative environmentally friendly process is based on thermal plasma technology which is a very flexible tool because it allows to operate in a wide temperature range with almost any chemical composition of waste and chemicals needed for processing this waste. It allows the conversion of organic waste into energy or chemical substances as well as the destruction of toxic organic compounds in a scenario that for each specific type of waste can be considered optimal, both in terms of energy efficiency and environmental safety.

## 1. Introduction

Human activity is associated with the permanent emergence of a very wide range of waste streams. The most widely used treatment is thermal processing such as incineration. However, the combustion conditions do not always match those parameters that are necessary for complete combustion of organic waste, which leads to large emissions of harmful substances into the atmosphere. As a result, exhaust gases may contain dangerous products of incomplete chemical burning.

The increasingly stringent legislation on treatment of waste streams and the limitations of conventional technologies such as thermal incineration, catalytic oxidation and adsorption render plasma technologies more and more attractive. The driving force is to give priority to environmental quality at affordable cost, and to contribute to sustainable development. Thus, the investigation of ways to increase the efficiency of the plasma based treatment processes is very important. In most industrialized countries, integrated solid waste management is mainly governed by the so-called "Ladder of Lansink". This ladder specifies a generally accepted hierarchy of preferred methods for dealing with waste. The highest rung is occupied by waste prevention, followed (in that order) by re-use (of the object as it is), recycling (material re-use without loss of quality) and down-cycling (material re-use with loss of quality). A still lower rung is the energy recuperation from waste (by incineration, gasification or pyrolysis), followed by incineration and/or destruction of waste without energy recuperation while the least preferred option is landfill.

An alternative waste treatment process is based on thermal plasma technology [1]. Plasmas have a great potential to contribute to sustainable development. The present paper deals with the use of thermal plasmas for the treatment of waste. The use of electric-arc plasma (plasma torches or plasmatrons) with mean temperatures of order 5000 K allows to effectively carry out the efficient destruction of organic compounds into atoms and ions with very high speed and a high degree of conversion. In addition, the destruction of complex compounds in the plasma is very efficient. The amount of gaseous products is much smaller than that of products resulting from traditional technologies based on combustion. This leads to a drastic reduction in the amount of flue gas to be cleaned. Plasma technology is a very flexible tool because it allows to operate in a wide temperature range with almost any chemical composition of waste and chemicals needed for processing this waste. It allows the conversion of organic waste into energy or chemical substances in a scenario that for each specific type of waste can be considered optimal, both in terms of energy efficiency and environmental safety. Efficient automation of the processes is possible since for varying waste stream parameters the plasma generators can be easily feedback controlled, and there is a possibility to change the temperature and enthalpy of the plasma flow in real time and, if necessary, the chemical composition of the plasma can be adjusted. Furthermore, the low inertia of the process minimizes the risk of harmful emissions in case of emergency, with the ability to quickly stop the process.

*Thermal plasma pyrolysis and vitrification is a dual simultaneous disintegration process*. The feedstock is treated by thermal plasma in a reactor chamber, whereby organic components are converted into a synthetic gas of high calorific value, and inorganic components are converted into a non-leachable vitrified lava. The treatment of feedstocks is being optimized with respect to the quality of the synthetic gas according to criteria determined by the end use. These criteria can be: the maximum energy content of the syngas for electricity or heat production (*thermal conversion*), or the production or recovery of a valuable by-product from the syngas such as methanol, respectively hydrogen for fuel cells (*chemical conversion*).

The fact that thermal plasma technology is characterized by a high energy consumption has hindered its further development for industrial applications. Since the environmental legislation becomes more and more stringent, for the overall evaluation of the technical and economic feasibility of thermal plasma technology, its substantial environmental advantages have to be taken into account in the comparison with non-plasma methods.

Two examples of thermal plasma treatment of waste will be given from the author's personal experience.

## 2. Gasification and vitrification by thermal plasma

Since the 1980s applications of thermal plasmas experienced an important increase. In the 1990s fundamental research led to great progress in the understanding of the basic phenomena involved, and to a renewed interest in applying thermal plasmas to material processing and waste treatment. The application of plasma torches for environmental purposes is a relatively new process. Generators of thermal plasma (plasma torches) operate simultaneously as a *plasma-chemical and a thermal apparatus*. Plasma is the medium with the highest energy content per particle. Therefore, thermal plasma offers the unique capability of carrying an extremely high energy by a small amount of plasma and ensures high heat transfer to the material to be treated. All materials can be decomposed if they are brought in contact with thermal plasma. The electrical energy of the torches goes into the plasma which transfers its energy to the substances to be treated, thereby triggering a dual simultaneous reaction process in the plasma-chemical reactor: the organic compounds are thermally decomposed into their constituent elements (principally a mixture of carbon monoxide and hydrogen, called *syngas* with more complete conversion of C into gas than in incinerators), and the inorganic materials are melted and converted into a dense, inert, non-leachable *vitrified slag* that does not require controlled disposal. Therefore, it can be viewed as a totally closed treatment system. The main goal of plasma treatment of organic waste is to produce syngas, while the main goal of incineration is material decomposition.

*The use of thermal plasmas has the following advantages:*

- Far higher temperatures than can be reached by conventional heat generators
- Highly reactive and reducing environment
- High energy density and high heat transfer efficiency, allowing shorter residence times and large throughputs
- Low thermal inertia and easy feedback control

- Lower plasma gas input per unit heating power than the gas flow of a classical burner and thus lower energy loss corresponding to the energy necessary for heating of gas to reaction temperature; also lower amount of off-gases to be treated
- Absence of combustion gases generated by conventional incinerators
- Smaller plants than for incinerators due to high energy densities, lower gas flows, and volume reduction
- Heat source is electricity rather than energy liberated from combustion and thus independent of the treated substances, providing flexibility, fast process control, and more options in process chemistry, including the possibility of generating valuable co-products
- Optimal control of the composition of the reaction gases
- As sufficiently high temperatures and homogeneous temperature distribution can be easily maintained in the whole reactor volume, the production of higher hydrocarbons, tars and other molecules is substantially reduced in comparison with incineration.

## 3. Gasification of biomass

The gasification of biomass to produce synthesis gas (*syngas*) offers an alternative to fossil fuels. Conventional biomass gasification technologies are based on the reaction between a solid or liquid carbonaceous material (containing mostly chemically bound carbon, hydrogen and oxygen) and limited amounts of air or oxygen. The exothermic reactions provide sufficient energy to produce a primary gaseous product containing mostly CO, $H_2$, $CO_2$ and $H_2O(g)$ and a small amount of higher hydrocarbons. Usually some heat is supplied into the reactor from external sources to control the process, but most of the heat comes from the calorific value of the biomass. The main problem of common biomass gasification is the production of tar.

Thermal plasma offers the possibility of decomposition of biomass by pure pyrolysis in the absence of oxygen. In this process all energy needed for gasification comes from the plasma, and no energy for decomposition is produced by combustion. The main advantage is a better control of the composition of the produced gas, higher heat capacity of the gas and reduction of unwanted contaminants like tar, $CO_2$ and higher hydrocarbons. Most of the experiments with plasma gasification/pyrolysis have been performed with electric-arc plasma torches with relatively high flow rates of plasma gas. The high flow rate of plasma ensures good mixing of plasma with the treated material and a uniform temperature in the reactor. However, the produced syngas contains plasma gas components, usually nitrogen and oxygen if air or nitrogen are used as plasma gases. The usage of mixtures of inert gas with hydrogen eliminates this disadvantage but it increases the costs. Therefore, steam is being used as plasma gas.

The present section presents experimental results obtained in the medium scale thermal plasma gasification reactor equipped with a hybrid (gas-water) plasma torch with arc power up to 160 kW at the Czech Academy of Sciences (IPP-CAS) in Prague. The DC plasma torch with gas/water stabilized arc, developed and patented by IPP-CAS and usually applied for plasma spraying, is used [2]. The plasma jet contains a mixture of steam with a small amount of argon. The hybrid torch is composed of two stabilizing chambers. Plasma produced in the cathode arc chamber with gas stabilization enters the second chamber with water stabilization. The gas flow in the cathode part of the torch protects the cathode tip and hence the consumable carbon cathode used in water torches can be replaced by a fixed tungsten cathode. Arc characteristics are dominated by processes in the water-stabilized part of the arc column which is much longer than the cathode section. Due to the principle of arc stabilization by a water vortex the flow rate of plasma gas (typically 0.3g/s $H_20$ plus 0.2g/s Ar ) is very low, the plasma enthalpy is higher than 200 MJ/kg, and the mean plasma temperature is higher than 15000 K. By changing the argon flow rate the hybrid torch provides the possibility of controlling the parameters of the plasma jet and the plasma composition (O-H-Ar) in a wide range from high enthalpy, low density plasmas typical for water stabilized torches to lower enthalpy, higher density plasmas generated in gas stabilized torches. An important characteristic feature of this hybrid torch is the very low mass flow rate of plasma. As a low amount of plasma carries high energy, the power needed for heating of plasma to reaction temperature is very low, and the efficiency of utilizing plasma power for waste destruction is extremely high. The clean synthesis gas consists mainly of hydrogen and CO (together minimum 90%). Several types of gas torches will also be used for biomass gasification on this reactor.

The experimental plasma-chemical reactor (see Figure 1) with water-cooling system has been designed to operate at a wall temperature up to 1700°C [3-4]. The inner volume of the reactor is 0.22 m$^3$. The inner lining of the reactor is made from special refractory ceramics. The insulating 400 mm thick ceramic liner reduces power losses to the walls. To prevent destruction of the ceramic insulation, prior to experiments the reactor is preheated for 24 hours to about 950 °C and further to the operational temperature by the plasma torch at a power level of 110 kW. All parts of the reactor chamber are water-cooled and calorimetric measurements on cooling circuits are made. The material container is equipped with a continuous material supply system with controlled flow rate. The reactor wall temperature (usually 1100-1400°C) can be regulated by the torch power and by the feeding rate of the material to be treated. The pressure in the supply conveyer is automatically kept higher then the pressure inside the reactor by controlled nitrogen flow to prevent the flow of reactor gases into the material supply system. Crushed wood with dimensions up to several mm has been injected into the plasma jet at a location of about 30cm downstream of the input plasma entrance nozzle at the reactor top. It is partially gasified during its flight within the jet. The non-gasified part of the wood falls to the bottom of the reactor where it is gasified in the hot gas flow. The exit tube for exhaust gas is located at the upper part of the reactor, forcing the produced gases to pass through the zone of high temperature within the plasma jet or close to it (mixed flow reactor). Furthermore, due to its high temperature this plasma jet is a very intensive source of UV radiation and hence has a strong gas cleaning effect. Additional gases (like $CO_2$ and oxygen) to control the reactor atmosphere can be injected at three positions in the upper part of the reactor. The gas produced in the reactor flows through a connecting tube to the cylindrical quenching chamber with length 2m. At the output of the quenching chamber the gas has a temperature of 300 °C, and flows into a combustion chamber where it is combusted in a flow of air.

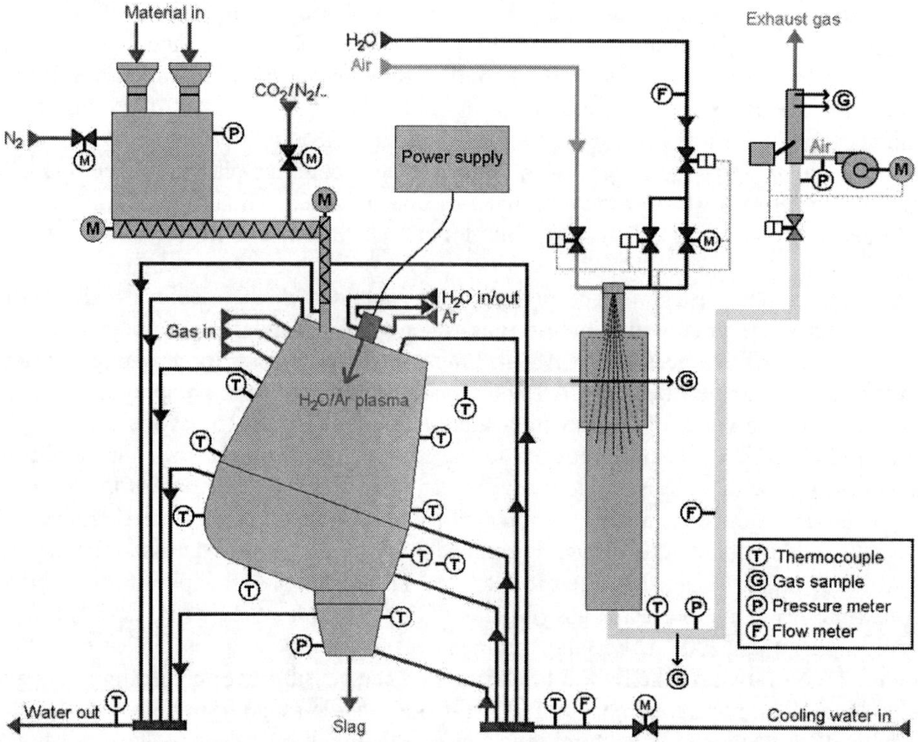


**Figure 1.** Schematic diagram of the experimental reactor for plasma pyrolysis & gasification.

The torch generates an oxygen-hydrogen-argon plasma jet with extremely high plasma enthalpy and temperature, and is attached at the top of the reactor. The anode of the torch consists of a rotating water-cooled disc, which is positioned outside the arc chamber downstream of the torch exit nozzle. The anode of the torch is thus located inside the reactor and separated from the reactor volume by a water-cooled wall. Additional argon has been supplied into the chamber at the location of the mounted torch to protect the anode from the flow of reaction gases. The free plasma jet containing a mixture of argon and steam enters the reactor volume through the nozzle with diameter of 40 mm in the wall. The measuring system includes monitoring of plasma torch operation parameters, temperatures in several positions inside the reactor and calorimetric measurements on cooling water loops. The temperature of the inner wall of the reactor is measured in six positions by thermocouples. The flow rate of reaction gas is measured using a Pitot tube flow meter. The gas temperature is measured at the input and output of the quenching chamber by thermocouples. The composition of the produced gas is measured at the output of reactor before the gas enters the quenching chamber. The main gas analysis is made by a quadrupole mass spectrometer Balzers QMS 200. Other analyses of the composition of the produced syngas and the content of tar are by means of mass spectroscopy with cryofocusing, gas and liquid chromatography and Fourier transform infrared spectroscopy (FTIR).

The results of experiments with *gasification of crushed wood* in argon/steam plasma demonstrate the complete gasification of wood with production of syngas with high content of hydrogen and carbon monoxide. In spite of the low plasma flow rate and the constricted plasma jet, homogeneous heating in the reactor as well as efficient mixing of the treated material with plasma and intensive energy transfer are observed. The flow within the reactor is almost completely controlled by gasification of the material as the flow rate of syngas coming from gasification is up to hundred times higher than the flow rate of plasma from the torch. The calorific value of the produced syngas is of the order of 2.5 times higher than the torch power. The process can act as an *energy storage* whereby electrical energy is converted into plasma energy and then stored as chemical energy of the produced syngas. Depending on operational conditions, the main components of produced syngas are hydrogen (28 - 46% vol.), CO (44 - 68%), $CO_2$ (2 - 8%) and Ar (0.2 - 8%). The content of complex hydrocarbons or tar is below 10 mg/Nm$^3$. The composition of the reaction products resulting from the pyrolysis of crushed wood is given in Figure 2. Furthermore, *wood pellets, polyethylene and plastic* waste have also been treated.

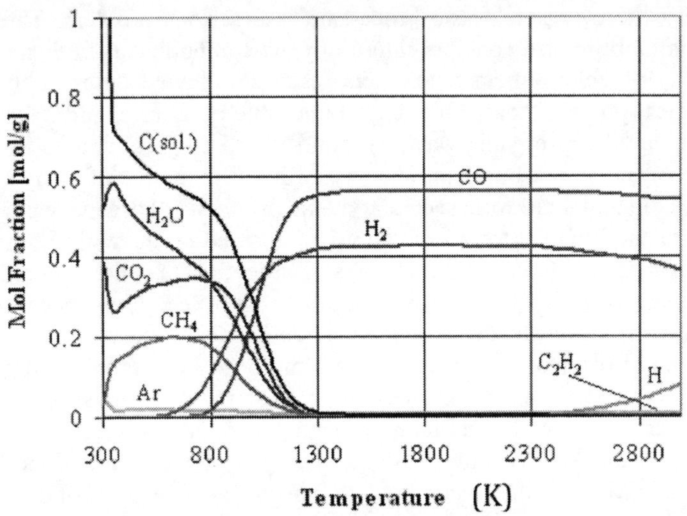


**Figure 2.** Gasification of wood in a steam/Ar plasma with addition of $CO_2$ and oxygen: composition of products of reaction as a function of the temperature in the reactor. Wood feeding rate 47 kg/h; humidity 6.5%; water plasma flow rate 18g/min; argon 13.6 slm (standard liter per minute); $CO_2$ 115 slm, $O_2$ 30 slm.

It can be concluded that homogeneous heating of the reactor volume and proper mixing of plasma with treated material occur in spite of the low plasma mass flow rate and constricted plasma jet. The mixing is more intensive at higher feeding rates. The conditions within the reactor ensure complete destruction of the tested substance. Furthermore, a novel and attractive procedure of *molecular dissociation of* $CO_2$ by utilizing it as oxidizing medium in the reactor has been demonstrated and will be optimized. No effect of arc power on gas composition and flow rate is observed for tested feeding rates up to 50 kg/h. It can be concluded that the maximum possible feeding rate at given power has not been reached.

The aim is also to demonstrate the economical viability, environmental performance and safety of *biofuels and hydrogen produced from syngas* resulting from the plasma-thermochemical gasification of a very broad range of second generation biomass feedstock (incl. the biodegradable fraction of waste) using several types of plasma torches and possible combinations of torches [4]. The plasma technology will be further optimized. Plasma-assisted combustion of biomass will also be investigated in existing (classical) biomass gasification installations. Apart from purely economic factors, other factors have to be taken into account such as greenhouse gas and mass & energy balances, the potential competition with food production and the impact of biomass production on the environment. The co-production of fuels, heat & power and co-products will enhance the overall economy and competitiveness of biofuels. The possibility of reforming of $CO_2$ is also very attractive.

## 4. Destruction of toxic organic compounds

The most widely used treatment of toxic organic waste is thermal processing. Common practice is e.g. direct combustion in industrial furnaces and boilers. This pertains mostly to liquid and solid waste with moderate and high calorific value and a minimum content of halogens. However, the combustion conditions in these furnaces and boilers do not always match the parameters that are necessary for complete combustion of organic waste, leading to large emissions of harmful substances into the atmosphere. As a result, exhaust gases may contain dangerous products of incomplete chemical burning. This is due to the fact that the process of neutralization of organic waste by thermal methods is carried out at temperatures sensitive to the formation of other harmful compounds. The combustion of chloro-organics at insufficiently high temperatures, in addition to nitrogen oxides and carbon monoxide, may lead to the formation of phosgene, dibenzofurans, dioxins, furans, polychlorinated benzopyrene, polyaromatic hydrocarbons, oxides of sulfur, soot and other very toxic products in quantities exceeding the maximum permissible concentration, in addition to nitrogen oxides and carbon monoxide. Chloro-organic waste processing is a serious problem. The amount of waste of production and consumption that contains chloro-organic compounds is millions of tons. They include end-of-life products of chloro-organic polymers, used chlorinated hydrocarbon solvents (including additives), dielectrics (polychlorinated biphenyls) unusable pesticides, and residues of chloro-organic production. Dumping chloro-organic waste at landfills and disposal should be completely excluded because of their poor biodegradability and the possibility of oxidation of many of them by air oxygen under the influence of sunlight to form secondary toxic products (phosgene, etc.).

An alternative method to treat toxic organic waste is based on thermal plasma technology [5]. The use of electric-arc plasma with mean temperatures of order 5000 K allows to effectively carry out the destruction of organic compounds into atoms and ions with very high speeds and a high degree of conversion. In addition, the destruction of complex compounds in the plasma is very efficient and can occur in the absence of oxygen, providing the opportunity to successfully carry out plasma pyrolysis reactions which in some cases has advantages in comparison with combustion.

A three-jet plasma reactor with electric-arc plasma generators with a total capacity of 200 kW was developed and built at the A.V. Luikov Heat and Mass Transfer Institute of the National Academy of Sciences of Belarus in Minsk. This reactor can use plasma-forming gases of different composition and is easily tunable for processing various types of toxic organic waste, including waste belonging to the group of Persistent Organic Pollutants (POPs). The heart of the plasma-chemical reactor is a three-jet mixing chamber which is equipped with three electric-arc plasma torches (power 50 kW; thermal

IOP Conf. Series: Journal of Physics: Conf. Series **941** (2017) 012014 doi:10.1088/1742-6596/941/1/012014

efficiency up to 70 %; plasma forming gas air; plasma forming gas consumption 3-6 g/s). The processing is characterized by highly turbulent plasma flow, which is formed in the three-jet mixing chamber, thus ensuring high-intensity mixing of the plasma flow and the substances to be treated. The high temperature in the reactor leads to a complete degassing of the inorganic ash residue. A shock cooling (quenching) module is used to avoid the formation of secondary toxic products. Acids are neutralized in the alkaline environment of a wet filter. An Ion Mobility Spectrometer RAID S2 is used for the detection of chemical agents in the air to control the exhaust gas composition and to prevent environmental disasters.

The tests showed that the thermal efficiency of the plasma torch in the temperature range 3000-5000 K is 60-70%. By using different composition of plasma forming gas the parameters of the medium inside the reactor can be controlled to achieve high level of toxic substance processing. Based on the results of the experiments, a pilot plant is being developed for the safe plasma-chemical processing of pesticides and toxic waste.

**References**

[1] Tendler M, P. Rutberg P and Van Oost G 2005 *Plasma Physics and Controlled Fusion* **47** A219
[2] Hrabovsky M et al 2006 *IEEE Trans. on Plasma Science* **34**, Issue 4, Part 3, 156
[3] Van Oost G et al 2006 *Vacuum* **80** 1132 and references therein
[4] Van Oost G et al 2009 *Vacuum* **83** 209 and references therein
[5] Van Oost G et al 2013 *Vacuum* **88** 165 and references therein

# On plasma sheath motion in coaxial electrode system

**V V Vikhrev**[1], **E I Dodulad**[2, *]

[1] National Research Center "Kurchatov Institute", Akademika Kurchatova pl. 1, Moscow 123182, Russia
[2] National Research Nuclear University MEPhI, Kashirskoe highway 31, Moscow 115409, Russia

E-mail: *dodulad@mail.ru

**Abstract.** Current-plasma sheath dynamics in coaxial pulsed plasma system was simulated in this work by means of one-fluid MHD code and two-dimensional "snow plough" model.

## 1. Introduction
The phenomenon of self-acceleration of current-plasma sheath with its own magnetic field during its motion between coaxial electrodes is used in various devices: plasma accelerators [1, 2], non-cylindrical z-pinches [3–5], plasma thrusters [6, 7] etc. Plasma sheath mass losses on the electrodes are often neglected in models that describe plasma motion in coaxial systems. It is correct for equilibrium flow. However, the phenomenon of electrode influence on plasma motion in pulsed plasma systems is much more complex compared with equilibrium flow theories, because the accumulation of matter takes place near the electrodes surface.

Such matter accumulation process is poorly studied experimentally [5, 8]. To carry out such researches one needs to develop special electrodes with diagnostic windows along them, get rid of spurious irradiation from plasma. Such electrode modification will lead to device operation changes. However such studies are essential for development and creation of workable equipment. Not all presently existing numerical models of plasma dynamics in coaxial electrode systems allow to simulate the interaction of current-plasma sheath with electrodes [9–11]. Consequently there is a challenge to create a code for numerical simulation of plasma motion and interaction in coaxial electrode system and to develop by its means optimal electrode system configurations for each specified task.

The problem of pulsed plasma dynamics in coaxial electrode system was considered. Plasma motion was simulated by one-fluid MHD code used earlier for modelling of Z-pinches [12] and two-dimensional model "Snow plough" which is often used for plasma focus sheath simulation [13].

## 2. Simulation results

### *2.1. MHD model*
For the sake of simplicity and clarity the dynamics of current-plasma sheath was simulated in electrode system with coaxial cylindrical electrodes.

Simulation of plasma motion between two coaxial electrodes by means of one-fluid MHD code shows that only a small fraction of working substance is carried away by current-plasma sheath, but substantial fraction is left behind.

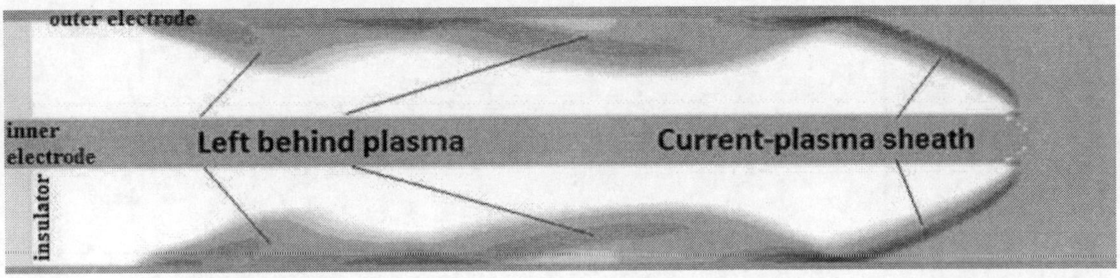


**Figure 1.** Plasma distribution between two coaxial electrodes, obtained by means of MHD simulation.

After consideration of working substance distribution changing in time (figure 2) one can notice that on the initial stage of movement plasma sheath completely pushes off the gas (figure 2a). During farther movement growth of plasma sheath's mass stops.Mass of gas that is carried away by the sheath in unit time becomes equal to the mass of leaking along the electrodes substance.

Simulation results show that the value of axial component of left behind plasma's velocity (~$10^5$–$10^6$ cm/s) is by 1–2 orders of magnitude less than plasma sheath's velocity (~$2 \cdot 10^7$ cm/s). The main mass of working substance is lost during plasma sheath interaction with outer electrode. Left behind plasma is distributed along the electrode in the shape of "wavy" structure, which period depends on the width of interelectrode gap (figure 3).

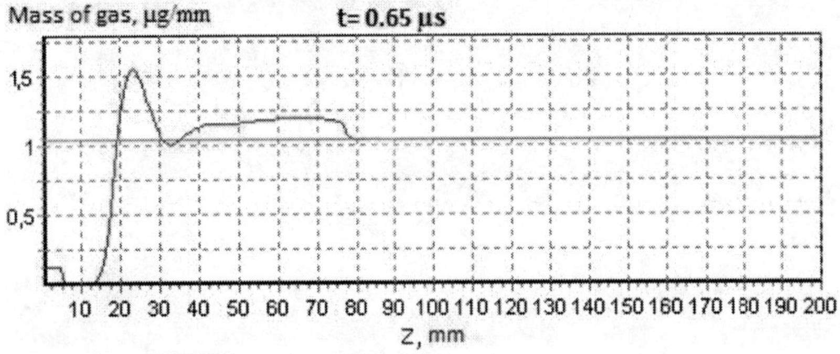


(a)

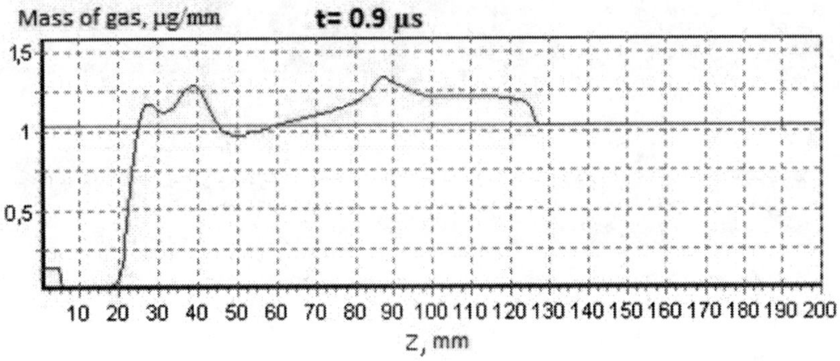


(b)

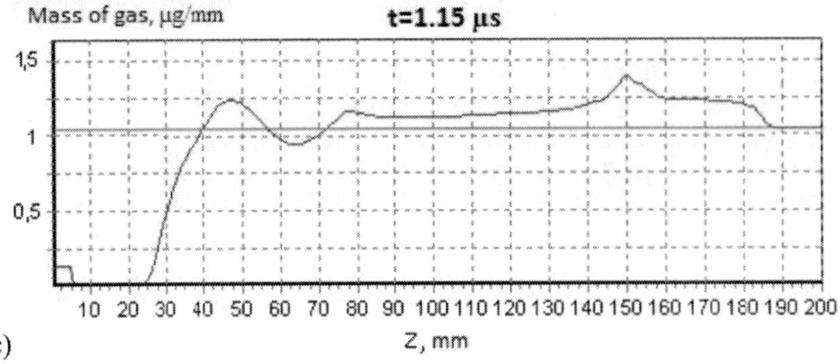


(c)

**Figure 2.** Working substance mass distribution along electrode system axis 0.65 µs, 0.9 µs, and 1.15 µs after discharge beginning.

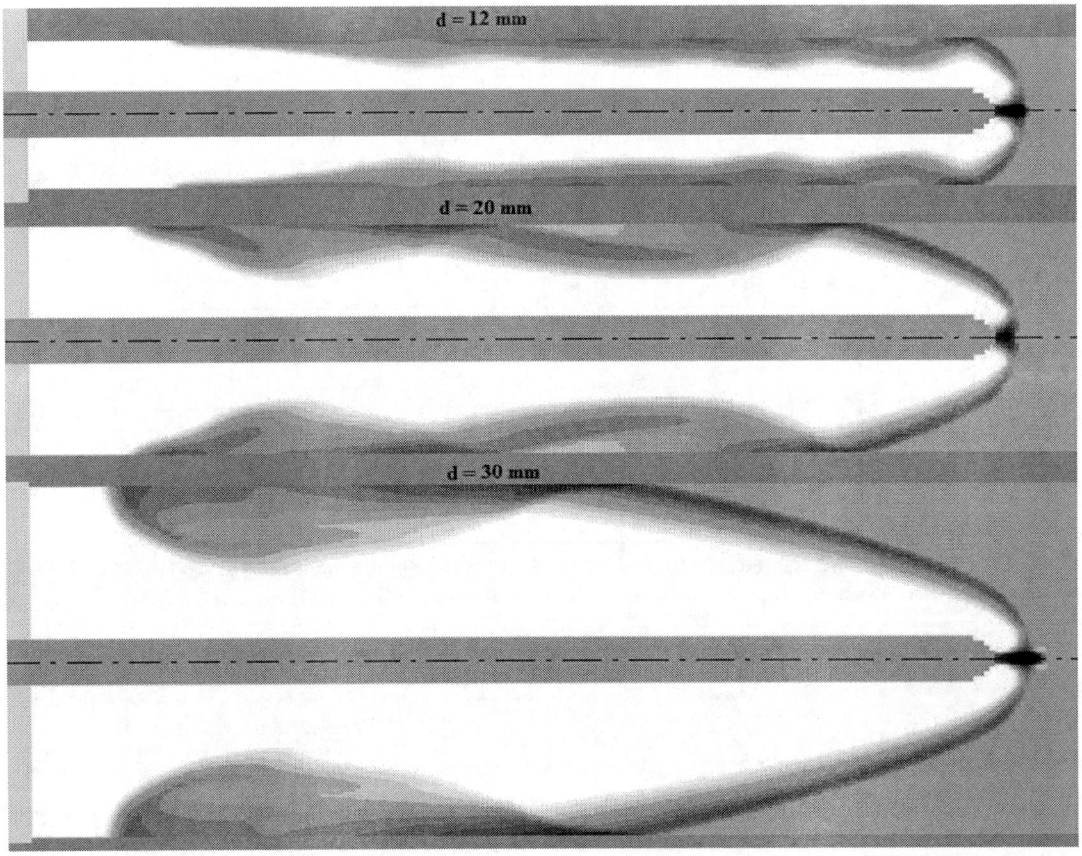


**Figure 3.** Working substance distribution dependence on the interelectrode gap width *d*.

## *2.2. "Snow Plough" model*

In addition to MHD modelling plasma sheath motion was simulated by means of two-dimensional "snow plough" model. Electrode system geometry and all initial parameters (pressure and working gas type, energy content, outer circuit parameters) were the same as in MHD model. Simulation of consecutive motion of current-plasma sheath is presented on figure 4. Contrary to MHD model where current-plasma sheath has finite thickness in "snow plough" model the sheath is infinitely thin.

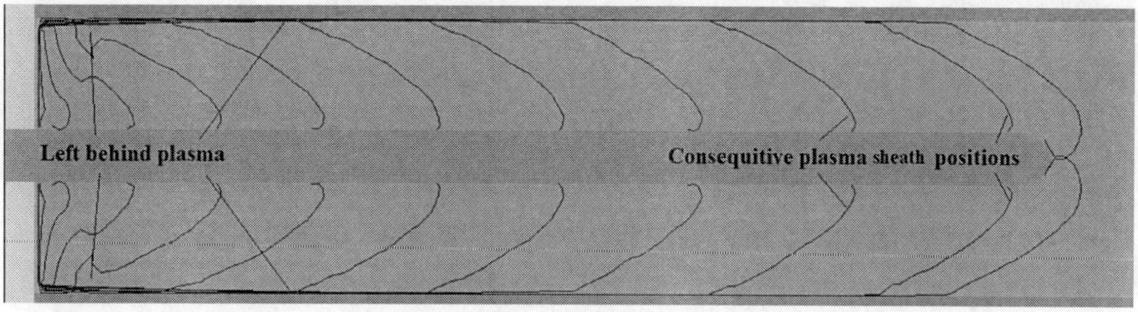


**Figure 4.** Consecutive motion of current-plasma sheath in coaxial cylindrical electrode system.

With the same initial parameters the value of current-plasma sheath velocity along the electrodes obtained by means of "snow plough" model differs from the one obtained by MHD code by less than 10%. Working substance distribution along the axis in "snow plough" simulation is also similar to MHD model (figure 5). The distribution has a sawtooth shape due to bigger step size in "snow plough" model.

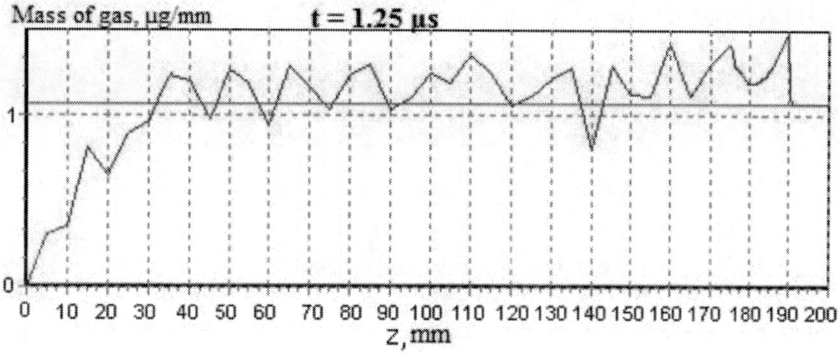


**Figure 5.** Working substance mass distribution along electrode system axis 1.25 μs after discharge beginning.

The main difference between current sheath dynamics in "snow plough" model and in MHD model involves lacking of "wavy" structure of substance flown from the sheath. Current-plasma sheath in "snow plough" model has curved inward sections only on the initial stages of its development. These sections will be considered below.

## 3. Results analysis

The results of MHD simulation show that particles from plasma sheath have a radial component of velocity besides of axial one (figure 6). In accordance with this data, one can suppose that carried away particles drift along plasma sheath towards outer electrode, collide with it, loose part of their kinetic energy and then travel in a "wavy path".

On the assumption of this theory, one can suppose that if the interelectrode gap is made widened then the particle losses on the electrode surface will be mitigated. For the sake of simplicity let us assume that radial and axial components of the velocities of particles in current-plasma sheath $V_r$ = const, $V_z$ = const (which is not technically correct) along all sheath. In this case if the electrode surface is at an angle $(\alpha) \geq {}^{V_r}/_{V_z}$, then the particles from plasma sheath will not reach it and the "wavy" structure will not develop. As a check on this assumption electrode system with widened outer electrode ($tg(\alpha) = {}^{V_r}/_{V_z}$) was simulated (figure 7). The results of simulation show that the "wavy" structure does not appear.

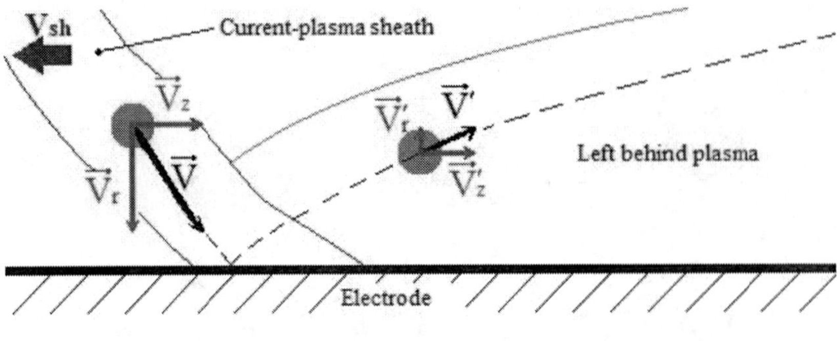


**Figure 6.** Particle motion pattern in relation to current-plasma shell.

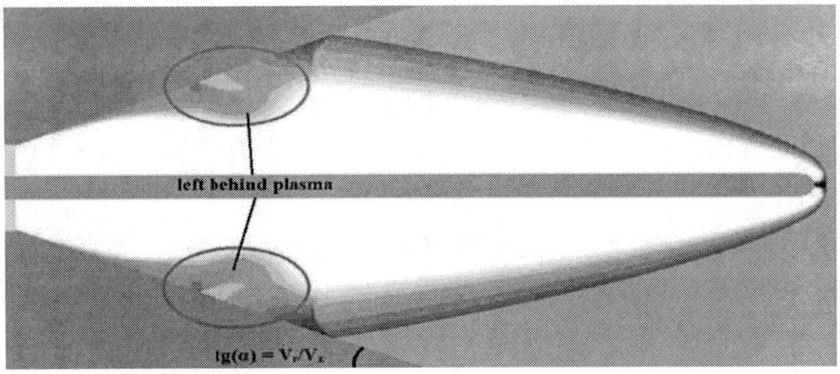


**Figure7.** Working substance distribution in electrode system with widened interelectrode gap.

Besides plasma that is flowing out of the current-plasma sheath along its path there is another slowly moving along the outer electrode plasma formation. This phenomenon is presented in MHD simulation as well as on early steps of plasma sheath development in “snow plough” simulation (figures 7 and 4 respectively).

Current-plasma sheath development was closely examined on the stage of its separation from the insulator in MHD model. It was noticed that current-plasma sheath consists of two components: slowly moving one and fast moving one drawn along the inner electrode (figure 8). The fast component outruns the slower one. The substance from their intersection area is pushed from the inner electrode towards the outer electrode and plasma is left behind fast moving sheath.

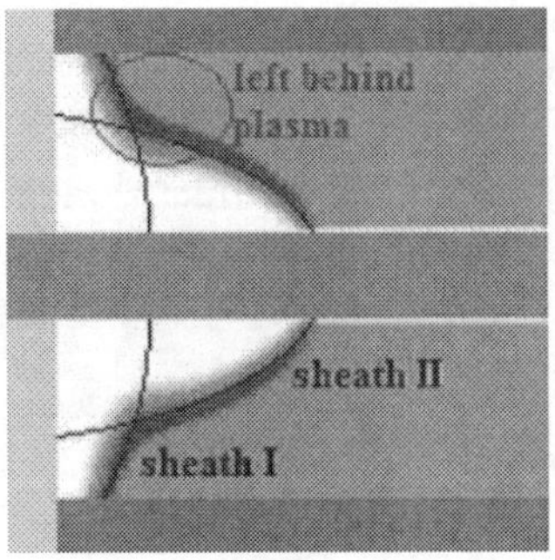


**Figure 8.** Current-plasma sheath development on the stage of its separation from the insulator.

## 4. Possible negative effects due to left behind plasma

Presence of left behind plasma can lead to appearance of parasitic bypassing currents in the interelectrode gap (figure 9). Such bypassing currents can lead to changes in electrical parameters of devices, their malfunction and failure in achievement of required parameters of operation.

For working substance losses mitigation it is recommended that the electrode system should have widening interelectrode gap. The law of gap widening for each device will depend on radial and axial plasma sheath particles velocities relation $V_r/V_z$. Velocity of particles depends on initial electrical parameters of the device, electrode system dimensions and used working substance.

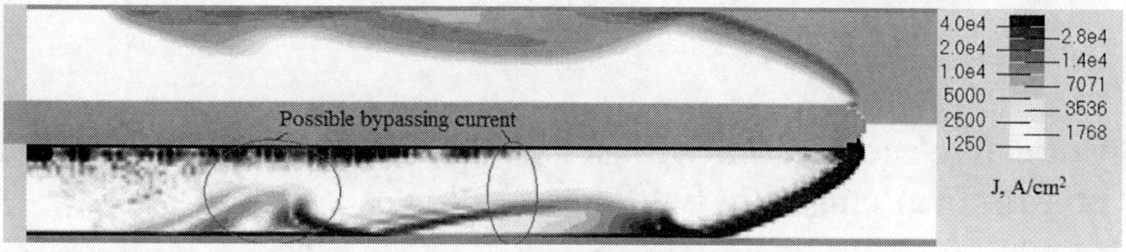


**Figure 9.** Distribution of working substance density (shades of blue) and current density (shades of gray) in discharge camera obtained by means of MHD simulation.

It is obvious that the widening interelectrode gap will result in increase of the discharge camera's basic dimensions and in increase of its inductance. But it is not necessary that the electrode system should have cylindrical shape. Selection of the optimal configuration of the discharge camera for each practical task will demand additional experimental and theoretical research.

## 5. Conclusions

In this work current-plasma sheath dynamics in coaxial electrode system was simulated by means of one-fluid MHD code and two dimensional "snow plough" model.

Unlike most of other, simulations used in current work allowed to obtain and analyze data about the influence of electrodes on effectiveness of working substance sweeping by current-plasma sheath. It is determined that only small fraction of gas is carried away by moving plasma sheath, but substantial fraction of substance is left behind the sheath.

Explanation of the observed phenomenon is given. It is supposed that the losses of working substance by the plasma sheath during its interaction with electrode occur due to special aspects of plasma motion inside the sheath and its dynamics during the stage of separation from the insulator.

In order to mitigate working substance losses it is recommended that the electrode system should have widening interelectrode gap.

**Acknowledgements**
This work has been partly supported by National Research Nuclear University MEPhI in the framework of the Russian Academic Excellence Project (contract No. 02.a03.21.0005, 27.08.2013).

## References

[1] Underwood T C, Loebner K T K and Kapelli M A 2017 *High energy density physics* **23** 73
[2] Putrik A B et al. 2015 *Physics of atomic nuclei* **78** 1174
[3] Klir D et al. 2012 *Plasma Phys. Control. Fusion* **54** 015001
[4] Pavez C, Pedreros J, Tarifeno-Saldivia A and Soto L 2015 *Physics of Plasmas* **22** 040705
[5] Piriaei D, YousefiH R, Mahabadi T D, Salar Elahi A and Ghoranneviss M 2017 *Physics of Plasmas* **24** 022509
[6] Kang B, Low K-S, Soon J J and Tran Q-V 2017*IEEE Transactions on Power Electronics* **32** 4503
[7] Christou A and Jugroot M 2017 *Vacuum* **141** 22
[8] Haines M G 2011 *Plasma Phys. Control. Fusion* **53** 093001
[9] Lee S 2014 *J. Fusion Energ.***33** 319
[10] Abd Al-Halim M A 2017 *Vacuum***138** 80
[11] Shmidt A, Tang V and Welch D 2012 *Phys. Rev. Lett.* **109** 205 003
[12] Vikhrev V V, Ivanov V V and Rozanova G A 1993 *Nuclear fusion* **33** 311
[13] Basque G, Jolas A and Watteau J P 1968 *Physics of Fluids* **11** 1384

IOP Conf. Series: Journal of Physics: Conf. Series **941** (2017) 012016 doi:10.1088/1742-6596/941/1/012016

# Collective acceleration of laser plasma in a nonstationary and nonuniform magnetic field

**A Isaev*, K Kozlovskiy, A Shikanov and E Vovchenko**

National Nuclear Research University MEPhI (Moscow Engineering Physics Institute), Kashirskoe highway str. 31, Moscow 115409, Russia

E-mail: *isaev@lenta.ru

**Abstract**. This paper presents the new experimental results concerning acceleration of deuterium ions extracted from laser plasma in the rapid-growing nonuniform magnetic field in order to initiate the nuclear reactions D(d, n)$^3$He and T(d, n)$^4$He. For obtaining of laser plasma a Nd: YAG laser ($\lambda$ = 1,06 µm) that generates in Q-switched mode the radiation pulses with the energy W $\leq$ 0,85 J and duration of $\tau \approx$ 10 ns was used. Rapid-growing magnetic field was created with the discharge of Arkadyev-Marx pulsed-voltage generator to conical coil with the inductance of 0,65 µH. At characteristic discharge time of 30 ns, the rate of magnetic field growth achieved $2{\cdot}10^7$ T/s. Ion velocity was determined with the time-of-flight technique. During the experiment on deuterium plasma an ion flux velocity of $\sim 3{\cdot}10^8$ cm/s was obtained, which corresponds to the deuteron energy of ~100 keV. Herewith, for target power density of $\sim 5{\cdot}10^{11}$ W/cm$^2$ obtaining of up to $10^{15}$ of accelerated deuterons and up to $10^8$ of neutrons per a pulse is expected.

## 1. Introduction

Currently, for generation of fast neutrons, as well as at implementation of a number of implantation and ecologic technologies, the devices based on deuterium ions electrostatic accelerators are often used. The drawbacks of these devices consist in restrictions for density of accelerated ions flux, which are related to impact of spatial charge, low efficiency at ions acceleration and to presence of associated X-ray emission induced with electrons of ion-electron, auto-electron or explosive emission. These drawbacks are to the great extent eliminated with implementation of effective acceleration of quasineutral plasma.

Another aspect proving work relevance is in active experimental research of magnetic-inertial thermonuclear synthesis that requires the description of physical processes occurring in high-temperature dense plasma within the strong magnetic field.

There exist known methods of plasma electrodynamic pulse acceleration consisting in introduction of working substance into interelectrode space and initiating of pulse high-current discharge between the electrodes via introduced working substance. At this discharge the substance is ionized and transformed into plasma, and the interaction between the discharge current through plasma and current magnetic field through the electrodes leads to plasma acceleration. However, plasma creation by means of the breakdown process in gas and the presence of a large portion of neutral component in these methods of plasma acceleration do not allow achieving of high efficiency and stable acceleration of dense homogeneous bunches with large number of ions up to high speeds.

Paper [1] considers the model of laser plasma acceleration in rapid-growing magnetic field formed at the discharge of a capacitor C charged up to voltage U to circuit in the form of conducting ring. Plasma is created at an instant of time t counted off from the beginning of capacitor discharge, as a result of pulse laser beam focusing on target of dielectric deuterated material. As the result of magnetic field growth, the eddy current in plasma is induced. At interaction of this current with the magnetic field a longitudinal ponderomotive force occurs that accelerates the plasma bunch.

In early theoretical works [2] the probability of acceleration of laser plasma spread from $10^7$ to $3\cdot10^7$ cm/s created with portable lasers emission was calculated. Experimentally the rate of magnetic field growth reached $10^5$–$10^6$ T/s.

This paper reports new experimental data on the research of plasma ion component by means of laser plasma bunch acceleration and generation of intense beams of accelerated ions under influence of rapid-growing magnetic field. In order to analyze the ion acceleration mechanism a model of laser plasma collective acceleration with nonstationary magnetic field [3] generated at the discharge of capacitor to circuit ring with current was discussed.

## 2. Experimental model

In order to resolve the set tasks an experimental model was developed that consists of: vacuum post and vacuum chamber with high-voltage and optical entries; the chamber contains the dielectric deuterium-containing laser target located within the helical line, small-size pulse laser, high-voltage generator of pulsed voltage (PVG) according to Arkadyev-Marx scheme and the system for time-of-flight recording of ion flux parameters.

Figure 1 presents the section of vacuum camera with the system for time-of-flight recording of ion flux parameters.

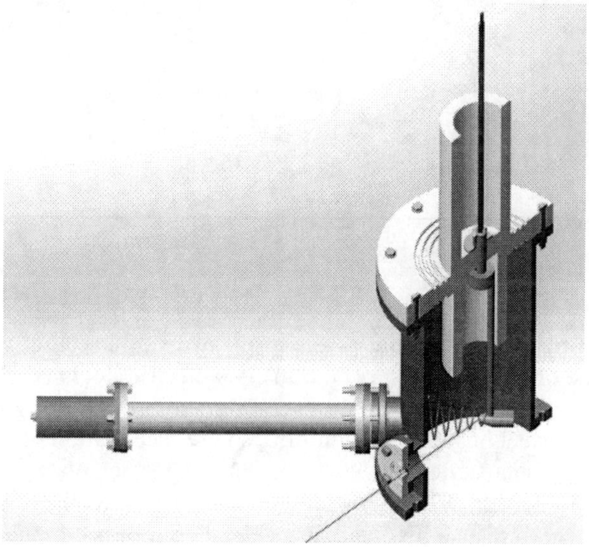

**Figure 1.** Illustration of test bench in section.

Rapid-growing magnetic field at laser target was created by means of PVG discharge to conical coil with the inductance of $L_1$ = 0,65 μH. Characteristic dimensions of conical coil amounted about 3 cm, cone opening was 20°. The discharge time was about 30 ns. Maximum and average velocity of ions was determined with the time-of-flight technique. Recording of ion bunch arrival time was performed by means of a collector – Faraday cylinder installed at the distance of 0,5 m from conical coil.

PVG allows obtaining of strong pulse of accelerating voltage with the amplitude of up to 400 kV at accumulated energy of up to 50 J, low-resistance load current of up to 1,5 kA with the discharge duration of up to 30 ns. Herewith, both PVG and laser can operate with the frequency of 1 Hz.

In order to obtain plasma a laser on yttrium-aluminum garnet activated with neodymium that generates in Q-switched mode the pulses of infrared radiation ($\lambda = 1.06$ μm) with the energy $W \leq 0.85$ J and duration of ≈10 ns was used. 10% of laser radiation power was focused on laser-controlled discharger commutating the first PVG cascade, thus providing a strict synchronization of pulse of rapid-growing magnetic field and creation of laser plasma bunches of required size.

Maximum and average velocity of ions was determined with the time-of-flight technique [4] by means of a collector in the form of a Faraday cylinder.

## 3. Schematic diagram of collective acceleration

Figure 2 presents the method schematic diagram. Laser radiation by means of focusing device was directed to dielectric laser target made of deuterated polyethylene. At this dielectric target a power density of about $5 \cdot 10^{15}$ $W \cdot m^{-2}$ was created.

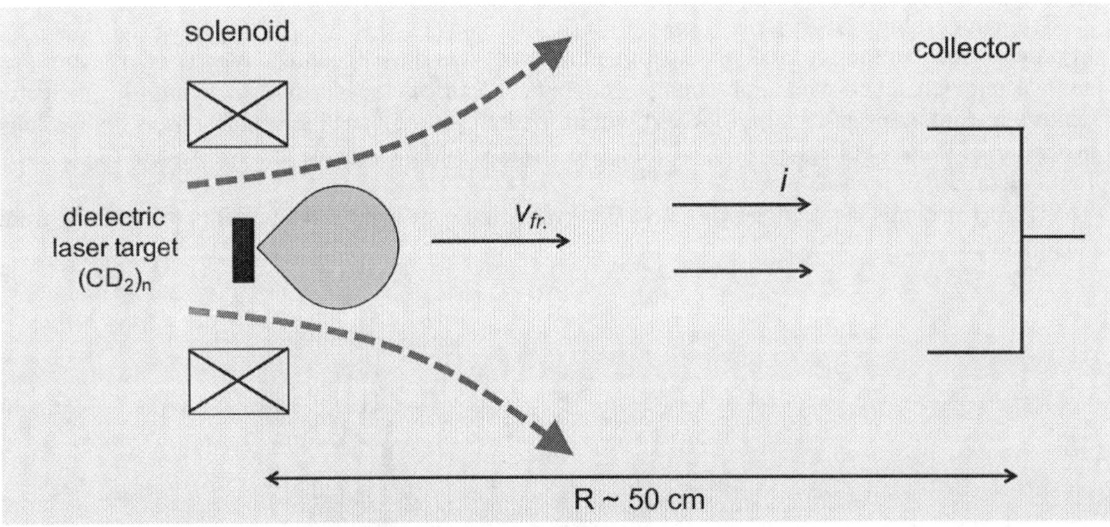


**Figure 2.** Method schematic diagram.

At the first stage of plasma expansion, as the result of its intersection with the magnetic lines of force, within the volume of plasma bunch occurs an eddy electric field and electric current creating the magnetic moment. Magnetic field at certain distance from the center of magnetic system in the form of cone-shaped helix with current becomes nonuniform, so the acceleration of part of ions in the center of plasma bunch occurs (gyromagnetic acceleration). During further plasma spread its periphery ions come into magnetic field domain with the radial component and, as they herewith possess azimuthal component of velocity, they are also involved in acceleration process. In the first case the acceleration energy is gained at the expense of plasma bunch kinetic energy, and in the second case it is gained at the expense of external sources providing field and the mode of its growth.

For plasma effective acceleration it is provided that the time of bunch creation and acceleration is less than the time of discharge current magnetic field penetration into this plasma. Within the scope of this work the rate of magnetic field growth of $2 \cdot 10^7$ T/s was achieved, which 1–2 orders of magnitude higher than in previous experiments [2–3].

## 4. Analysis of obtained results

Figure 3 presents the oscillogram of pulse of ions at collector with laser plasma, scanning is 1 μs, characteristic time was 3 μs. Figure 4 presents the oscillogram of focusing on the recording system in the absence of laser plasma, reference is the laser pulse recorded with coax photocell ФЭКом, scanning is 25 ns.

Figure 5 presents the oscillogram with scanning of 250 ns, in which are distinguished the two groups of ions: first group consists of accelerated deuterons, and the second one consists of carbon, as more heavy ions, but with greater energy, herewith, lag time between ion packages amounts not more than 300 ns.

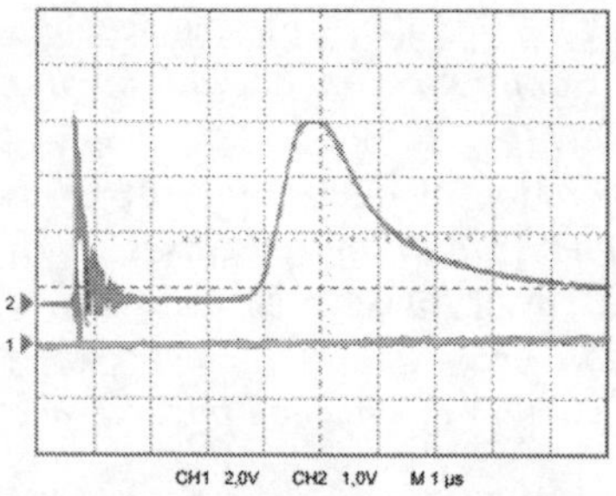


**Figure 3.** Oscillogram of signal from the collector with laser plasma, but without PVG.

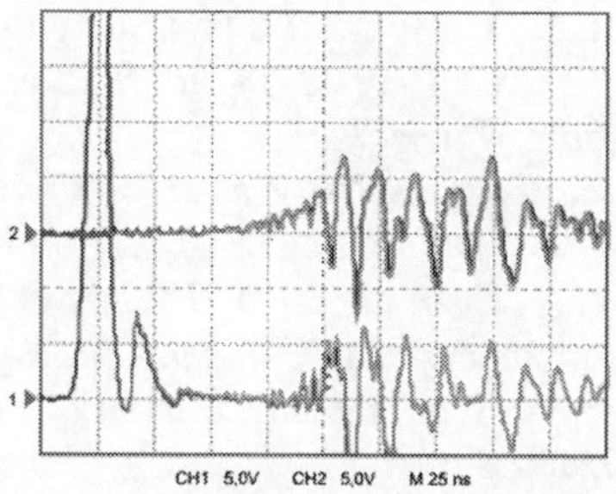


**Figure 4.** Oscillogram of signal from collector with PVG, without laser plasma.

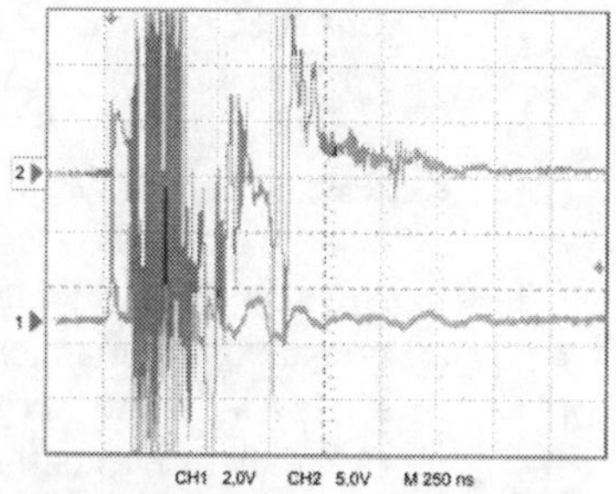


**Figure 5.** Oscillogram of signal from collector, with laser plasma and PVG.

Within the scope of work the method of plasma collective acceleration with rapid-growing magnetic field is implemented and the maximum speed of ion flux of $3 \cdot 10^8$ cm/s is achieved, which for deuterons corresponds to the energy of about 100 keV. The evaluation shows that the total number of accelerated deuterons amounts about $10^{15}$ (at collector, at the distance of 50 cm from coil, $10^{13}$ ions are recorded). Thus, at directing of accelerated deuterons beam to near-coil deuterium-containing target it is possible to achieve the emission of $10^8$ neutrons per a pulse. Herewith, the duration of neutron pulse is about 30 ns. Current density for accelerated deuterons reaches 1 kA/cm$^2$ at pulse duration of about 30 ns.

Consequently, by means of these neutron generators it is possible to implement the delayed neutrons method. Studied method of plasma acceleration with rapid-growing magnetic field may be used for development of neutron intense pulse generators, ion injectors for accelerators, plasma engines and other applications.

In order to optimize the system of laser plasma acceleration in rapid-growing magnetic field currently are performed works on modernization of test bench for provision of variation of synchronization of laser pulse initiation from start to maximum of magnetic field of coil with current.

## 5. Summary

In early works the increase of spread velocity of laser plasma created with radiation of portable lasers amounted from $10^7$ to $3 \cdot 10^7$ cm/s. Within the scope of the present paper a laser plasma deuteron accelerator to velocities higher than $10^8$ cm/s was created, which allows the initiation of nuclear reaction D+D, D+T by them. Within the scope of work the rate of magnetic field growth of $2 \cdot 10^7$ T/s was achieved, which is 1–2 orders of magnitude higher than in the previous experiments.

Calculations allow formulating of requirements to the experiment conditions and test plant design. They also allow compiling of recommendations on further experimental research. For example, it is necessary to provide a possibility of laser target motion along the magnetic system axis, as well as the possibility of laser and magnetic system independent startup. These activities allow selecting of optimal acceleration and neutron generation mode.

It is also worth noting that the developed system allows for mode of neutron generation (including thermonuclear ones) at counter fluxes during plasma generation and for locating of magnetic field generation systems in two different points of space on one axis. In this case the problem related to degradation of solid neutron-generating targets is resolved. There also occurs a possibility of fast accumulated running time of packed solid targets at using of deuteron-tritium laser targets.

**Acknowledgments**

This work has been supported by National Research Nuclear University MEPhI in the framework of the Russian Academic Excellence Project (contract No. 02.a03.21.0005, 27.08.2013).

**References**

[1] Tsybin A S, Kozlovsky K I, Kuznetsov A Yu and Shikanov A E 2003 Laser plasma accelerator for radiation-physical research *Proc. 12th International Conference on Radiation Physics and Chemistry of Inorganic Materials* 51

[2] Kozlovsky K I, Tsybin A S, Kozyrev Yu P et al. 1985 Generation of intense ion fluxes from laser plasma *Proc. of 17th International Conference on Phenomena in Ionized Gases* **2** 945

[3] Kozlovsky K I, Shikanov A E at al. 2003 The model of laser plasma acceleration with nonstationary magnetic fields *Proc. MEPhI scientific session* **4** 65

[4] Moskalev V A and Sergeyev G I 1991 *Measurement of parameters of charged particle beams* (Moscow: Energoatomizdat) 240

# Properties of plasma in case of capillary discharge interaction with metals

**D L Kirko[*], A S Savjolov**

National Research Nuclear University MEPhI (Moscow Engineering Physics Institute), Kashirskoe shosse 31, Moscow 115409, Russia

E-mail: [*]dlkirko@plasma.mephi.ru

**Abstract.** A consideration is given to the effect of capillary discharge torch on metal foils. The spectral characteristics of plasma have been measured in visible and ultraviolet ranges and plasma temperature has been determined in the area of interaction. The properties of autonomous plasma areas appearing in case of this interaction are being investigated. A study of plasma areas collision with the surface of liquid nitrogen has been presented. The motion of plasma areas in permanent electric field has been studied. A possible structure of plasma areas is being discussed.

## 1. Introduction

The formation of capillary discharge is accompanied by the appearance of torch of an extended plasma configuration [1, 2]. The plasma areas (PA) featuring size of 0.2–4 mm and lifetime of 0.1–1 s appear during different experiments with capillary and erosion discharges [3–5]. In order to initiate PA, the non-organic and organic substances, as, for instance, basalt wool and paraffin have been placed into discharges. After discharge discontinuation and PA extinction the microparticles have been studied, which supposedly could constitute a part of PA [5]. A torch of capillary discharge has been also used for experiments on interaction with metals [6, 7].

## 2. The experimental setup and spectral research

A capillary discharge has been used in interaction experiments (figure 1). A plate of plexiglas featuring thickness of 2–3 mm with capillary diameter equal to 1–3 mm has been used for a discharger, material of electrodes: graphite and copper. A power source with voltage of 250–300 V and energy content of 50–380 J was used for discharge functioning. The characteristic values in the experiments amounted to discharge current of 100–150 A, pulse duration of 5–10 ms. Metal foils of aluminum, tantalum and copper with thickness of 0.03–0.1 mm have been used for experiments.

The interaction of capillary discharge torch with metal foil results in its burning. An average quantity of evaporated material equals 0.1–5 mg. An emergence of plasma areas (PA) can be observed in the area of discharge and foil interaction (figure 1). The PA characteristic parameters are as follows: diameter 0.1–4 mm, lifetime 10 ms–0.5 s, quantity per shot – 1–20, radiation color – white, dark blue. PA motion speed is in the range of 0.1–5 m/s. In case of using aluminum and tantalum foil for PA maximum dimensions of 1–3 mm, maximum lifetime of 0.1-0.5 s and maximum quantity of 15–20 have been obtained. The photo and cine-shooting registration of luminescence has been performed by means of Nikon 1G camera (timing resolution of about 1 ms).

IOP Conf. Series: Journal of Physics: Conf. Series **941** (2017) 012017 doi:10.1088/1742-6596/941/1/012017

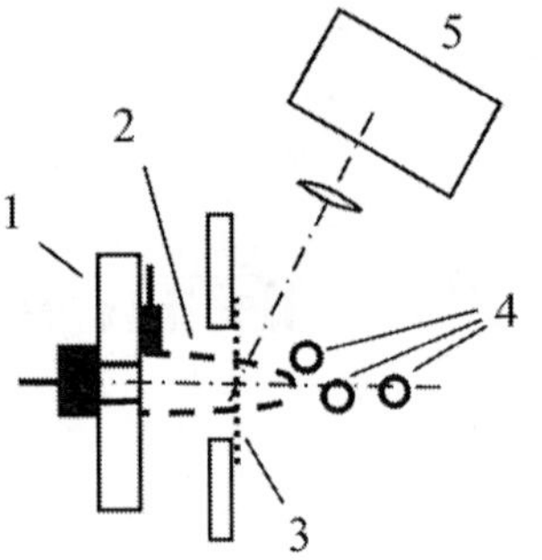


**Figure 1.** Experimental setup: 1 – discharger, 2 – capillary discharge torch, 3 – metallic foil, 4 – plasma areas, 5 – spectrometer.

The capillary discharge radiation in the visible and ultraviolet ranges of electromagnetic spectrum has been investigated during work. In order to obtain panoramic spectrum, Ava Spec 2048 spectrometer (range of 200–1000 nm, resolution of 0.3 nm) has been used. The radiation timing characteristics have been investigated by means of MUM monochromator (range timing resolution of 10 ns). Figure 2a shows a capillary discharge torch radiation spectrum in case of interaction with aluminum foil (thickness of 0.05 mm). A distance between the discharger and foil amounted to about 2.1 cm. The radiation comprises lines of hydrogen atoms $H_\alpha$ 656 nm, $H_\beta$ 486 nm and $H_\gamma$ 434 nm. The most intensive are the lines of: ion of nitrogen molecules $N_2^+$ 381…393 nm and copper atom Cu I 510.5 nm. The spectrum comprises lines of aluminum atoms Al I 309.2 nm. A characteristic time dependence of intensity of radiation of this aluminum line is shown in figure 2b. The existence of local thermal equilibrium for plasma has been supposed [8]. The temperature that amounted to value, $T = 2500 \pm 300$ K, has been calculated using relative intensities method along the lines of atomic hydrogen $H_\alpha$ 656 nm, $H_\beta$ 486 nm. The reviewed lines $H_\alpha$ and $H_\beta$ possess a widening in the range of 5–10 nm, which can be explained by the presence of the Stark effect. Therefore, the plasma concentration $n_e = (3.2 \pm 0.2) \cdot 10^{16}$ cm$^{-3}$ has been calculated by means of the Stark widening of these lines.

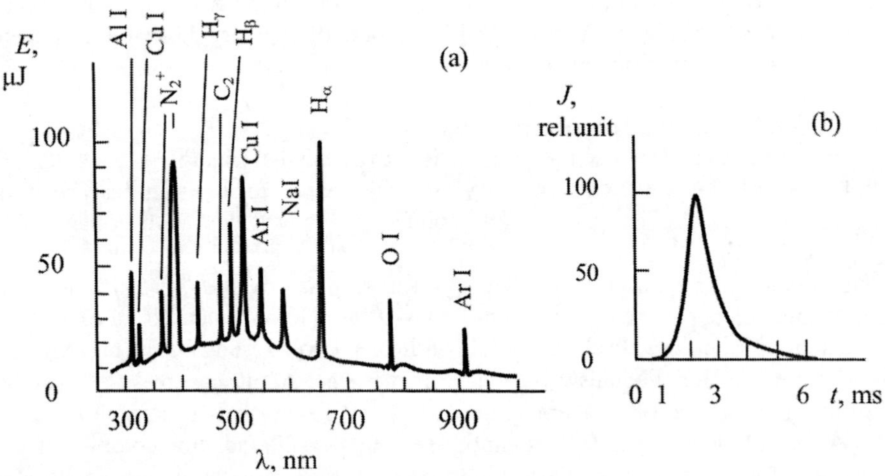


**Figure 2.** Spectral measurements of the torch radiation: (a) torch radiation spectrum at interaction with aluminum foil, (b) time dependence of intensity of the line radiation Al I 309.2 nm.

## 3. The interaction of plasma areas with liquid nitrogen

Let us give consideration to PA interaction with a surface of liquid nitrogen ($T_N = 77.4$ K) (figure 3). An axis of capillary discharge formed an angle $\alpha = 30$–$45°$ with the surface of liquid nitrogen. The aluminum and tantalum foils have been used in the experiments. Following the occurrence, PA began moving in the direction towards surface of liquid nitrogen. The approach of three PA to the surface of liquid nitrogen is shown in the frame of motion-picture photography (figure 4). The dimensions of PA are 2–3 mm. The motion has taken place from the area of discharger and interaction with foil (on the right) to the area of liquid nitrogen vessel (on thc lcft). The surface of liquid nitrogen is not registered in this picture due to high transparency of the liquid. Prior to touching the liquid nitrogen the PA featured, as a rule, nearly spherical shape and moved at the angle of about $\alpha \approx 45°$ to the liquid surface from right to left. The angular regularity of PA reflection from the liquid nitrogen surface with the initial angle of incidence $\alpha \approx 40$–$50°$ is shown in figure 3b. 40 interaction cases have been used. The maximum of this distribution is located close to the angle of reflection $\beta \approx 31$ . This specific reflection can be related to PA rotary motion. The PA extinction has, as a rule, taken place inadvertently in the process of motion, or in case of contact with liquid nitrogen.

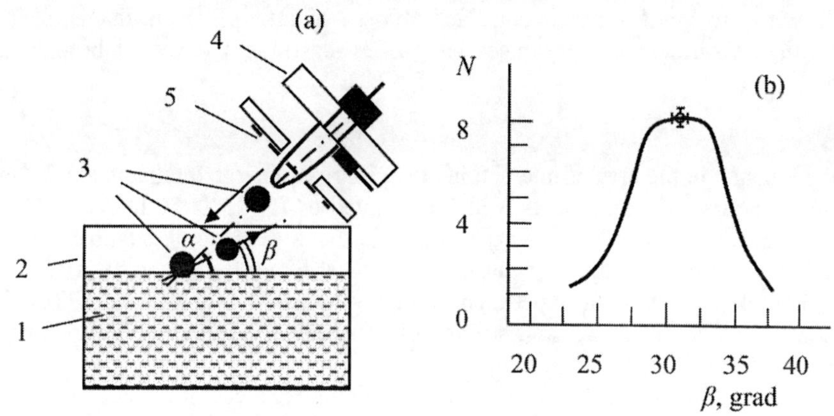


**Figure 3.** The interaction of plasma regions with liquid nitrogen: (a) experimental setup: 1 – liquid nitrogen, 2 – vessel, 3 – plasma areas, 4 – discharger, 5 – metallic foil, (b) angular distribution of plasma region refraction from liquid nitrogen surface.

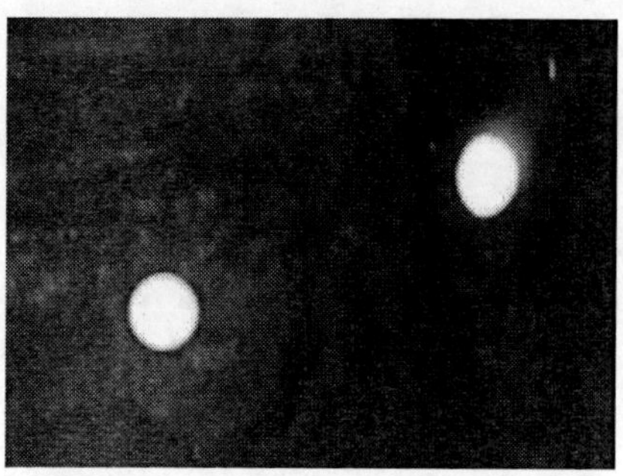

**Figure 4.** The motion of plasma areas near liquid nitrogen surface (time resolution $\Delta t \approx 1$ ms).

## 4. The studying characteristics of plasma areas

In order to study the content of electric charges in the substance of plasma areas, the experiments have been conducted with the presence of electric field in the space in the range of: $E = 10^3$–$10^4$ kV/cm. Tantalum foil has been used in the experiments. A bundle of PA has been preliminarily formed by means of two diaphragms. After that PA moved in the range of cone with the angle of about $\alpha \approx 6°$ and headed towards a uniform electric field (area dimensions: $l \approx 20$ cm, $a \approx 10$ cm). The PA speed has been in area of $v = 0.5–2.5$ m/s. A possible PA shift in the electric field has been studied. The PA path was registered by means of Nikon 1G camera. A maximum shift of PA has been observed within the limits of this angle $\alpha \approx 6°$ only. Great shifts as a result of electric field effect have not been registered. It testifies to an absence of positive or negative volume charge in PA volume. Therefore, plasma, apparently, forms a structure of PA matter.

The qualitative study of PA electromagnetic radiation has been carried out. These experiments have been complicated by low intensity of radiation and presence of PA motion. A spectral prism of crystal quartz as well as interference filters in the range of 400–700 nm was used for experiments. The PA images have been registered by means of Nikon 1G camera. The measurements have shown a presence of continuous electromagnetic radiation in the range under investigation.

The edges of holes in foil can become a possible area of PA emergence. The formation of toroidal vortexes at the edges of holes is observed in hydrodynamic vortexes in gas. In view of this fact one can suppose the formation of vortex plasma structure inside of PA volume. The density of PA substance proceeding from the analyses of data of uniformly accelerated motion in the field of Earth's gravity will amount to a value of $\rho \approx 1.5 \cdot 10^{-3}$ g/cm$^3$ and is close to air density of $\rho \approx 1.2 \cdot 10^{-3}$ g/cm$^3$ [6, 7]. In order to explain the existence of plasma configuration, including aluminum ions, a theoretical model has been built in works [9, 10]. The results of PA interaction with organic and non-organic materials testify to low PA temperature and about a specific effect mechanism [6, 7]. Some PA reflections from the surfaces of substances have been elastic and have not brought about PA destruction or extinction.

## 5. Conclusion

The temperature of plasma in the area of interaction of capillary discharge torch and aluminum foil has been measured in the course of experiments, which amounted to $T = 2500 \pm 300$ K. The PA motion in the permanent electric field demonstrates an absence of clearly expressed volume positive or negative charge in PA substance. In view of this fact plasma is the most likely PA structure. A reflection from the liquid surface has taken place in the experiments on PA interaction with the liquid nitrogen without vapor liberation in the place of contact and strong deformation of PA shape. These experiments of interaction of plasma areas testify to the presence of a resilient envelope preserving a shape of the area in case of external impact. In view of this fact one can admit an existence of molecular or cluster shell comprising aluminum atoms and possessing the stable interatomic links.

**Acknowledgements**
This work has been supported by National Research Nuclear University MEPhI in the framework of the Russian Academic Excellence Project (contract No. 02.a03.21.0005, 27.08.2013).

## References

[1] Avramenko R F, Bachtin B I, Nikolaeva V I and Poskacheeva L P 1990 *Techn. Phys. J.* **60** 57 (in Russian)
[2] Dimitrov S K, Zdanov S K, Kirko D L, Lutchko A S, Mihin S G, Smirnov V M and Telkovsky V G 1992 *Siberian Phys. Techn. J.* **2** 57 (in Russian)
[3] Emelin S E, Semenov V S, Bychkov V L, Belisheva N K and Kovshik A P 1997 *Tech. Phys.* **42** 269
[4] Bychkov V L, Bychkov A V and Timofeev I B 2004 *Techn. Phys. J.* **74** 128 (in Russian)
[5] Pirozerski A L, Mikhailovskii V Yu and Lebedeva E L, 2014 *Int. J. Appl. Engin. Res.* **9** 24047
[6] Kirko D L, Savjolov A S and Egorov I D 2014 *Proc. 4th Int. Conf. on Atmosphere, Ionosphere, Safety* (Kaliningrad: Imanuil Kant University) 179
[7] Kirko D L, Savjolov A S 2015 *High Temperature* **53** 921
[8] Huddlstone R, Leonard S 1965 *Plasma diagnostic techniques* (New York: Academic Press)
[9] Bychkov V L 2013 *Proc. 10th Int. Conf. on Wave Electrodynamics of a Conductive Liquid* (Yaroslavl, Yaroslavl State University) 27
[10] Bychkov V L 2010 *IEEE Transactions on Plasma Science* **38** 3289

# Investigation of plasma properties of discharge in electrolyte surface layer

**D L Kirko[*], A S Savjolov**

National Research Nuclear University MEPhI (Moscow Engineering Physics Institute), Kashirskoe shosse 31, Moscow 115409, Russia

E-mail: [*]dlkirko@plasma.mephi.ru

**Abstract.** The parameters of discharge close to electrolyte surface have been investigated in the work. The current-voltage discharge characteristics have been measured for the main discharge modes. The electric discharge oscillations have been studied and spectra determined in the range of 10 kHz–300 MHz for two electrolyte compositions. The study of discharge radiation in the visible and ultraviolet ranges of electromagnetic range has been carried out and plasma temperature determined. The influence of discharge on microstructure of electrodes surface has been investigated.

## 1. Introduction

The discharge formation in electrolyte is related to current flow through electrolyte environment and occurrence of luminosity near electrodes. In this case one of the electrodes can be placed over electrolyte surface [1, 2]. The emergence of discharge in electrolyte is accompanied by the appearance of radiating area and convective flows related to liquid heating. Various electric oscillations can take place in case of discharges in electrolytes [3–5]. These discharges are used for modifying surfaces of materials, polishing metal surfaces of intricate shape, more efficient hydrogen production.

## 2. The investigation of discharge electrical characteristics

The vessels featuring capacity of 150–350 $cm^3$ and electrolytes based on sodium carbonate and potassium hydroxide have been used for experiments in this work. The experimental setup is shown in figure 1. The rods of tungsten and titanium with diameter of 1–3 mm have been used for cathodes, and the plates of stainless steel and molybdenum with thickness of 0.1–0.5 mm have been used for anodes. The power source has corresponded to double half-period rectifier with voltage of 0–250 V and frequency of 100 Hz. The cathode has been placed in electrolyte surface layer. The discharge has been ignited from inside of electrolyte or at the boundary between electrolyte and air.

The dependences of influence of substance concentration on the shape of current-voltage characteristic (CVC) of discharge in electrolyte [4, 5] have been determined in the previous works. This work has been studying comparative characteristics in case of discharge in electrolyte based on sodium carbonate $Na_2CO_3$ and potassium hydroxide KOH. The investigations have been carried out in the range `of concentrations $C = 0.1$–1 M (concentration of $Na_2CO_3$ $C = 1$ M: 106 g per 1 l of distilled water). CVC for these two electrolytes for one and the same concentration $C = 0.4$ are presented in figure 2. A breakdown and ignition of discharge takes place in point 1. Section 2–3 is the working range of characteristic. The intensity of discharge grows weak at sector 3–4 and the discharge dies out in point 4. A shape resembling hysteresis is a common regularity of dependences. The values of current discharge density are in the range: $j = I/S = 10–50\ A/cm^2$. The basic differences of these CVC are as follows: different parameters for voltage and current in point 1 of discharge ignition, different run of working section (2–3) and different values for maximum voltage and current in point 3.

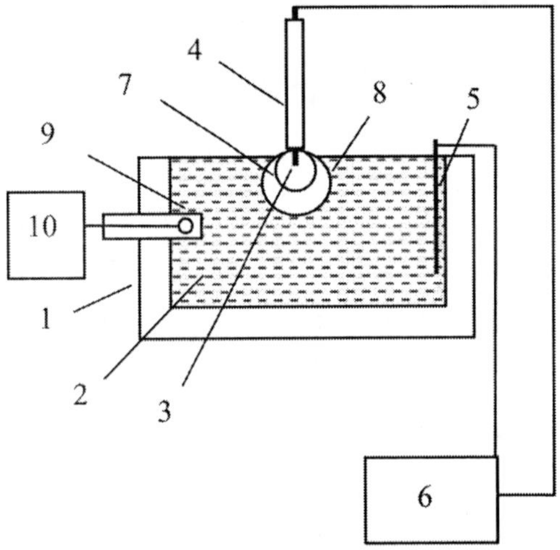


**Figure 1.** Experimental setup: 1 – chamber, 2 – electrolyte, 3 – cathode, 4 – ceramic tube, 5 – anode, 6 – power source, 7 – near cathode discharge region, 8 – middle region, 9 – magnetic probe, 10 – frequency analyzer.

A stage of discharge breakdown is related to phenomena in near cathode area of discharge with dimensions of 1–2 mm surrounding surface of cathode [4, 5]. Hydrogen bubbles measuring 0.1–0.5 mm are observed at cathode. It is supposed that hydrogen pressure in a separate bubble is close to atmospheric pressure. The evaluation of electric field intensity in near cathode area with thickness $d \sim 1$ mm equals $E \sim 1.1 \cdot 10^3$ V/cm. The formula for coefficient of ionization with electrons features the following view [6]: $\alpha = Ap\,\exp(-Bp/E)$. The evaluation of coefficient of ionization for characteristic discharge parameters gives the following value: $\alpha \approx 14\ cm^{-1}$. A model of spark discharge is more preferable for breakdown in the conductive liquid comprising gas bubbles [7]. The values of drift velocity of electrons calculated with the use of expression for mobility are available in the range: $v = 10^5–10^6$ cm/s. The assessment of time of cathode layer passage by electron avalanche gives the following values: $\Delta t = d/v = 10^{-7}–10^{-6}$ s.

## 3. The investigation of electric oscillations

The frequency distributions of electric oscillations in case of discharge in electrolyte have been obtained in the course of earlier works in case of using sodium carbonate [4, 5]. The sodium carbonate and potassium hydroxide have been used in this work as substances. The registration of electric fluctuations has been carried out by means of magnetic and electric probes and Tektronix TDS 2024B oscilloscope in the mode of spectrum analyzer. The individual pulses of discharge current had a shape of attenuating sine wave with a period of ~1.2 ms with periods number of 3–4. The high-frequency fluctuations have been registered at the beginning of every first current pulse period. The magnetic probes have been fabricated in the form of coils (diameter – 2–3 mm, quantity of turns – 80–150). The probes have been introduced into electrolyte in protective housing (figure 1), or have been arranged outside of a chamber. The electric oscillations have been investigated in the range of frequencies of 10 kHz–300 MHz.

The typical frequencies of oscillations are as follows for a discharge in electrolyte on the basis of sodium carbonate: (110±6) kHz, (350±18) kHz, (610±31) kHz, (820±41) kHz, (65±3) MHz, (102±5) MHz. The characteristic frequencies of oscillations are as follows in case of using potassium hydroxide for discharge in electrolyte: (350±18) kHz, (610±31) kHz, (41±2) MHz, (65±3) MHz, (102±5) MHz, (130±7) MHz. Figure 3a shows frequencies for potassium hydroxide in kilohertz range and figure 3b shows frequencies in Megahertz spectrum range. It should be noted that there exist frequency oscillations characteristic to particular electrolyte only. For sodium carbonate these are the following frequencies: (110±6) kHz, (820±41) kHz, and for potassium hydroxide: (41±2) MHz,

(130±7) MHz. In case of sodium carbonate the following number of frequencies in Megahertz range has been registered too: (1.2±0.1) MHz, (1.5±0.1) MHz, (1.8±0.1) MHz, (2.0±0.1) MHz, (2.2±0.1) MHz.

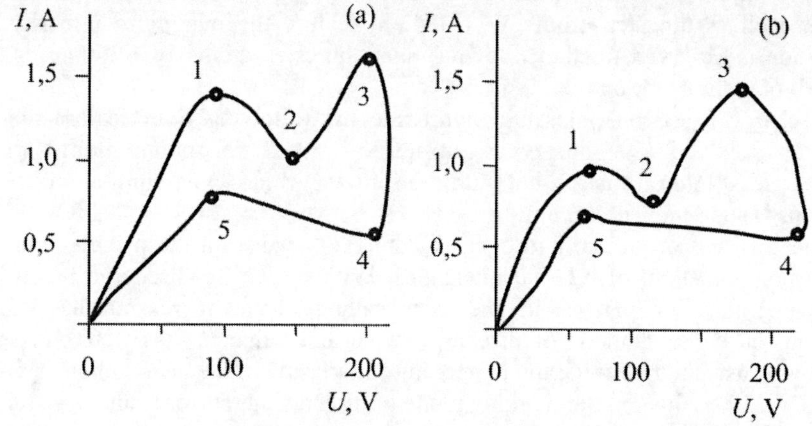


**Figure 2.** The current-voltage characteristic at the electrolyte discharge: (a) sodium carbonate $Na_2CO_3$ ($C$=0,4 M), (b) potassium hydroxide KOH ($C$ = 0,4 M).

The presence of frequencies of electric fluctuations has been interlinked earlier with shaping electronic-cyclotron and ionic-cyclotron plasma waves in the course of discharge in electrolyte [4, 5]. The frequency range has been determined: 20 kHz–600 MHz for characteristic discharge parameters as well as typical wave lengths have been calculated: $\lambda$ = 1–50 mm. Besides, the interrelation of frequencies oscillations with breakdown in the discharge cathode area is possible. The time of passing the near cathode area by electron avalanche obtained earlier equals $\Delta t = 10^{-7}$–$10^{-6}$ s. Using this method it is possible to generate electric fluctuations with characteristic frequencies: ν = 1–10 MHz. This frequency range is present in the registered sodium carbonate fluctuations. In view of this fact the emergence of electrical oscillations at the stage of breakdown of near cathode layer in electrolyte can be also assumed in line with the emergence of plasma waves.

In case of using a substance comprising sodium or potassium for the discharge in electrolyte the electric oscillations will be registered at the same frequencies, as well as there are frequencies characteristic for every electrolyte only. It shows a dependence of oscillations from the type of substance comprised by electrolyte and from the particular metal ions contained in electrolyte.

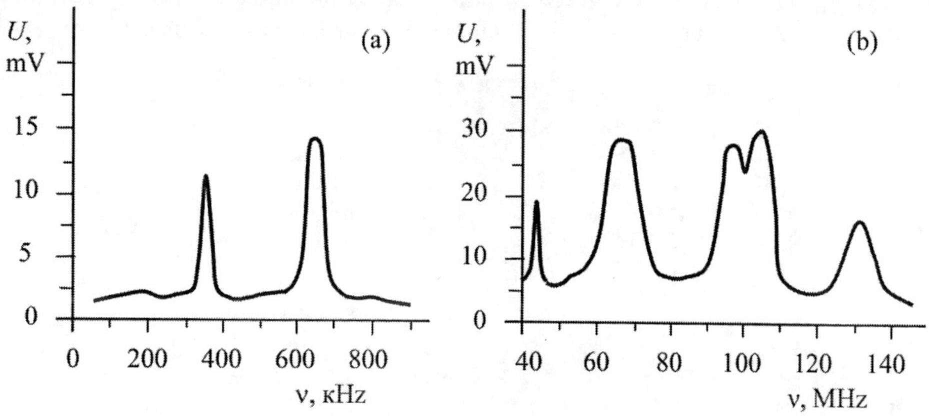


**Figure 3.** The electric oscillations spectra of the discharge at the potassium hydroxide using (concentration $C$ = 0,5 M): (a) region 50–900 kHz, (b) region 40–140 MHz.

## 4. The investigation of radiation characteristics of discharge

The radiation characteristics have been obtained in electromagnetic range of 200–1000 nm for the discharges in case of presence of sodium carbonate and potassium hydroxide in electrolyte. The lines of elements comprised by radiation have been determined in each of the above cases. In case of discharge based on sodium carbonate the lines of sodium atoms Na I 589 nm, 819 nm, and sodium ion Na II 309 nm, as well as tungsten atoms W I 384 nm feature the maximum intensity. The least intensity of radiation is observed for the following lines: tungsten atoms W I 360 nm, 420 nm, and hydrogen atoms $H_\alpha$ 656 nm, $H_\beta$ 486 nm.

In case of discharge based on potassium hydroxide the following lines feature the maximum intensity: potassium atom K I 769.9 nm, potassium ion K II 394.2 nm, sodium atom Na I 589.5 nm, Na I 588.9 nm. The least intensity is present with the following lines: potassium atom K I 404.4 nm, and hydrogen atoms $H_\alpha$ 656 nm, $H_\beta$ 486 nm.

Due to presence of lines of atomic hydrogen $H_\alpha$ and $H_\beta$ in radiation the plasma temperature has been calculated using a method of relative intensities. With respect to discharge based on sodium carbonate the temperature of plasma in the near cathode area of plasma has amounted to $T = 2100 \pm 200$ K in the working mode of discharge at current value: $I = (1.20 \pm 0.05)$ A. In case of discharge based on potassium hydroxide the temperature of plasma in the near cathode area of plasma has amounted to $T = 2300 \pm 200$ K in the working mode of discharge at current value $I = (1.10 \pm 0.05)$ A.

## 5. The investigation of surface of electrodes

The cathodes of tungsten and titanium and anodes of stainless steel and molybdenum have been used in experiments. A study of changes in structure of electrodes' surface has been carried out after effect of discharge in electrolyte. The surface of titanium cathode after affect of discharge in electrolyte based on sodium carbonate is shown in figure 4. The surface images have been obtained by means of VEGA 3 SEM and Hitachi TM1000 microscopes. The surface of cathode shows a formation of objects of irregular shape featuring size of 0.1–2 μm and bigger featuring length of 2–30 μm. At the same time the sections with micro porous structure featuring size of 0.2–1 μm are present at the surface (figure 4).

The analysis of element composition of this surface has been performed. In this case the following average elements distribution has been observed: oxygen (72%), titanium (16%), magnesium (7%), sodium (3%). High content of oxygen is, presumably, related to the presence of atomic oxygen close to cathode surface and possibility of oxidation processes. The sodium registration is caused by its deposition from electrolyte solution. Magnesium is included into original cathode alloy. Therefore, it may be supposed that electrolysis appeared close to cathode surface and atomic oxygen is present. Owing to the fact that temperature of plasma at cathode surface is higher than the temperature of titanium melting, a presence of titanium melt becomes possible in the form of seeds featuring size of 0.1–0.2 μm. At the same time a process of titanium oxidation at cathode surface is also possible.

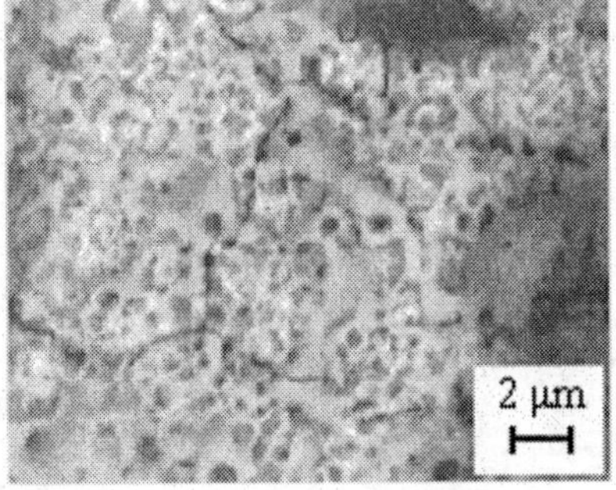


**Figure 4.** The image of micro porous structure on the cathode surface after the electrolyte discharge influence.

## 6. Conclusion

The current-voltage characteristics of discharge in electrolyte have been obtained in the course of this investigation in case of using sodium carbonate and potassium hydroxide as the working substance. The measured CVC possess definite characteristic features. A study of oscillation discharge processes has been carried out and a number of basic frequencies of electric fluctuations in the range of $\nu = 80$ kHz–200 MHz have been determined. The discriminative moments of spectra of sodium carbonate and potassium hydroxide are present. The emergence of electric oscillations at the stage of spark breakdown in case of shaping electronic avalanches is also possible alongside with the emergence of plasma waves. The temperature in near cathode area being in the range of $T = 2100$–2300 K is measured by means of spectral methods. The appearance of micro porous structure and presence of oxidation processes of titanium is observed at cathode surface.

## Acknowledgements

This work has been supported by National Research Nuclear University MEPhI in the framework of the Russian Academic Excellence Project (contract No. 02.a03.21.0005, 27.08.2013).

## References

[1] Gaisin Al F, Nasibullin R T 2011 *Plasma Phys. Rep.* **37** 896
[2] Kayumov R R, Gaysin Al F , Son E E, Gaysin Az F, and Gaysin F M 2010 *Physica Scripta* **T142** 014038
[3] Kanarev F M 2010 *Low temperature electrolysis of water* (Krasnodar: Krasnodarsk. Univ.) (in Russian)
[4] Kirko D L, Savjolov A S, Vizgalov I V 2013 *Russian Phys. J.* **55** 1243
[5] Kirko D L 2015 *Technical physics* **60** 505
[6] Raizer Yu P 2009 *Physics of gas discharge* (Dolgoprudny: Intellect) (in Russian)
[7] Korobeinikov S M, Melekhov A V, Besov A S 2002 *High temperature* **40** 652

# Extracting the differential inverse inelastic mean free path and differential surface excitation probability of Tungsten from X-ray photoelectron spectra and electron energy loss spectra

**V P Afanas'ev[1], A S Gryazev[1], D S Efremenko[2], P S Kaplya[3] and A V Kuznetcova[1]**

[1] National Research University "Moscow Power Engineering Institute", Department of General Physics and Nuclear Fusion, Krasnokazarmennaya 14, 111250 Moscow, Russia
[2] Deutsches Zentrum für Luft- und Raumfahrt (DLR), Institut für Methodik der Fernerkundung (IMF), 82234 Oberpfaffenhofen, Germany
[3] Raspletina St. 15, 123060 Moscow, Russia

E-mail: GryazevAS@gmail.com

**Abstract.** Precise knowledge of the differential inverse inelastic mean free path (DIIMFP) and differential surface excitation probability (DSEP) of Tungsten is essential for many fields of material science. In this paper, a fitting algorithm is applied for extracting DIIMFP and DSEP from X-ray photoelectron spectra and electron energy loss spectra. The algorithm uses the partial intensity approach as a forward model, in which a spectrum is given as a weighted sum of cross-convolved DIIMFPs and DSEPs. The weights are obtained as solutions of the Riccati and Lyapunov equations derived from the invariant imbedding principle. The inversion algorithm utilizes the parametrization of DIIMFPs and DSEPs on the base of a classical Lorentz oscillator. Unknown parameters of the model are found by using the fitting procedure, which minimizes the residual between measured spectra and forward simulations. It is found that the surface layer of Tungsten contains several sublayers with corresponding Langmuir resonances. The thicknesses of these sublayers are proportional to the periods of corresponding Langmuir oscillations, as predicted by the theory of R.H. Ritchie.

## 1. Introduction

For a long time, Tungsten has attracted interest in material science since it has the highest melting point of all metals, excellent corrosion resistance and low sputtering coefficient. For quantitative studies of energy loss processes of probing electrons, the information on differential inverse inelastic mean free path (DIIMFP) and differential surface excitation probability (DSEP) is required. Several techniques have been proposed for computing DIIMFP and DSEP involving linear response theory and density functional theory. However, such computations are computationally expensive for real atomic structures since a many-body quantum-mechanical problem has to be solved.

An alternative approach to get information on DIIMFP and DSEP is to extract them from electron energy loss spectra [1]. Important results using numerical deconvolution schemes have been given by W. Werner [2] and Afanas'ev et al [3]. However, the DIIMFP and DSEP extraction problem is severely ill-posed. Consequently, numerical deconvolution procedures can lead to unstable and noisy results containing a set of unphysical peaks in the retrieved functions. Afanas'ev et al [4] proposed a so-called fitting approach, in which the desired functions are parametrized on the base of a classical Lorentz oscillator. Unknown parameters are found by using the fitting procedure, which consists of computing the simulated spectra in a given energy loss range and matching them with the corresponding measurements. This method has been successfully applied for extracting inelastic scattering parameters of Be, Mg, Al, Si and Nb. The intent of this paper is to retrieve DIIMFP and DSEP for Tungsten from X-ray photoelectron spectra (PES) and transmission electron energy loss spectra (TEELS) using the fitting approach.

## 2. Methodology

### *2.1. Forward model: simulating PES and TEELS spectra*

The forward model is used for computations of PES and TEELS. It is based on the partial intensity approach [5], in which a spectrum is given as a weighted sum of the cross-convolved DIIMFP and DSEP. The weights can be found by using either the Monte-Carlo method, or the invariant imbedding method. In the latter case, a set of the matrix Riccati and Lyaponov equations is derived and solved numerically by using the backward differential formula (BDF). The computational details are described in [6] and references therein. Both PES and TEELS are computed within the same framework. Note that this procedure rigorously accounts for multiple scattering processes (unlike, e.g., $P_1$- and the transport approximations) and computationally efficient. For instance, it requires less than 0.01 seconds for computing one PES spectrum on Intel Xeon CPU E5-1620 3.60 GHz. Hence, the forward model performance is not an issue in the retrieval procedure.

### *2.2. Modeling of inelastic scattering properties*

The physics of inelastic energy losses in solids is well-known. It involves excitation of collective Langmuir oscillations of free electrons and local ionization processes. At the sample interface, there is a different loss mode due to a surface plasmon with frequency approximately by a factor $\sqrt{2}$ lower than that of the bulk plasmon. The surface excitations are taken into account by introducing an additional layer at the interface, in which the inelastic energy loss is described by DSEP. For transition metals, the situation is more complicated since two surface plasmons can be detected in high energy resolution spectra (e.g., see [4, 7] for Niobium studies). Consequently, a finer spatial discretization of inelastic scattering properties may be required, and the surface layer is divided into two sublayers. In this paper, the model with the following peculiarities is used:

- the sample is assumed to have three layers, namely, surface (S), transitional (G) and bulk (B); using this nomenclature, DIIMFP is related to the bulk, while DSEP is used for S- and G-layers;
- each layer contains a plasmon with a corresponding frequency $\omega = 2\pi/T$, where $T$ is the period of oscillations. In the energy loss spectrum, these plasmons form two peaks centered at energies $\hbar\omega_S$ and $\hbar\omega_G$ with $\hbar$ the reduced Planck constant;
- following Ritchie [8], the thicknesses of S- and G- sublayers are given as

$$d_S = T_S v_e,\ d_G = T_G v_e, \tag{1}$$

  where $T_S$ and $T_G$ are the periods of oscillations in S- and G-sublayers, respectively, and $v_e$ is the electron velocity.

Bearing this in mind, the unknown DIIMFP and DSEPs are sought using the following ansatz:

$$x_{\text{in}}(\Delta) = \left[\sum_{i=1}^{N_{\text{pl}}} \lambda_{\text{pl}\,i} x_{\text{pl}\,i}(\Delta) + \sum_{j=1}^{N_{\text{ion}}} \lambda_{\text{ion}\,j} x_{\text{ion}\,j}(\Delta)\right]. \tag{2}$$

Here $N_{pl}$ and $N_{ion}$ are the numbers of plasmon and ionization processes taken into account, respectively, $\lambda_{pl}$ and $\lambda_{ion}$ are the corresponding weights, while $x_{pl}(\Delta)$ and $x_{ion}(\Delta)$ are the basic functions for the plasmon and ionization processes, respectively. For $x_{pl}(\Delta)$ the following representation is used

$$x_{\mathrm{pl}\,i}(\Delta) = A_{\mathrm{pl}\,i}\frac{\Delta^{\beta}}{\left(\Delta^2-\varepsilon_{\mathrm{pl}\,i}^2\right)^2+\Delta^{\alpha}b_i^{4-\alpha}}. \tag{3}$$

Here, $A_{pl}$ is the normalization coefficient, $\varepsilon_{pl}$ is the effective plasmon energy related to the plasmon peak position, $b$ is the attenuation coefficient which governs the plasmon peak width, while $\alpha$ and $\beta$ are the tuning parameters which control the peak asymmetry and enhance the fitting. Essentially, Eq. (2) is a modification of the dispersion relation in solids and generalizes the excitation function of a classical Lorentz oscillator. The energy losses for ionization are described in the frameworks of classical Thompson theory [9]

$$x_{\mathrm{ion}\,j}(\Delta) = \frac{A_{\mathrm{ion}\,j}}{\Delta^{2+a_j}}\eta\left(\Delta - J_{\mathrm{ion}\,j}\right), \tag{4}$$

with $A_{ion\,j}$ the normalization coefficient, $\eta$ the Heaviside step function, $J_{ion\,j}$ the ionization potential, and $a$ the coefficient accounting for the electron screening of a Coulomb potential. Finally, for $\lambda_{pl}$ and $\lambda_{ion}$ the following normalization condition is used

$$\sum_{i=1}^{N_{\mathrm{pl}}}\lambda_{\mathrm{pl}\,i} + \sum_{j=1}^{N_{\mathrm{ion}}}\lambda_{\mathrm{ion}\,j} = 1. \tag{5}$$

Unknown parameters of the model are found by using the fitting procedure, which minimizes the residual between the forward model simulations and corresponding measurements. The constraints (1) and (5) are explicitly implied in the retrieval algorithm.

## 3. Results and discussion

For DIIMFP and DSEP retrieval we take PES and TEELS data from [10] and [11], respectively. Figure 1 shows the PES data. The photoelectron spectrum was induced by $K_\alpha$ line of Al corresponding to 1486.6 eV. The spectrum can be regarded as a superposition of several spectra induced by the photoelectron emission from levels $4f_{7/2}$, $4f_{5/2}$, $5p_{3/2}$, $5p_{1/2}$, $5s_{1/2}$. Such overlapping hinders the fine structure of the spectrum. Consequently, only one surface Plasmon can be observed in the PES. Therefore, here the two-layer model is used [3, 12]. The DIIMFP and DSEP are retrieved in the energy range 1360–1460 eV.

The high resolution TEELS data is shown in Figure 2. The measurements have been performed along the probe direction providing high intensity of the signal. Besides the bulk plasmon at energy $\hbar\omega_B = 24.0$ eV, two surface plasmons are observed at energies $\hbar\omega_S = 10.4$ eV, $\hbar\omega_G = 15.6$ eV. Hence, two DSEPs, namely, $x_{inS}(\Delta)$ and $x_{inG}(\Delta)$, should be used for the TEELS data interpretation.

Retrieved DIIMFPs and DSEPs within the two-layer model from PES data and the three-layer model from TEELS data are shown in Figure 3. The extracted functions are compared against data from W.S.M. Werner [13] obtained assuming homogeneous inelastic scattering properties throughout the sample. Table 1 lists the values of the fitted parameters. From Figure 3, it can be seen that the one-layer model can predict the energies of plasmons and ionization potentials. However, it fails to correctly reproduce the shape of DIIMFPs. Instead, it provides an effective DIIMFP being some kind of mixture of the actual DIIMFP, DSEP and their cross-convolutions. The functions retrieved from PES data agree well with those from TEELS.

Fitted PES and TEELS computed using retrieved DIIMFP/DSEPs are shown in Figures 1 and 2, respectively. The simulated spectra are in good agreement with experimental results. Note, that both EELS and PES data are interpreted within the same physical model for inelastic scattering energy losses without accounting intrinsic excitations [14].

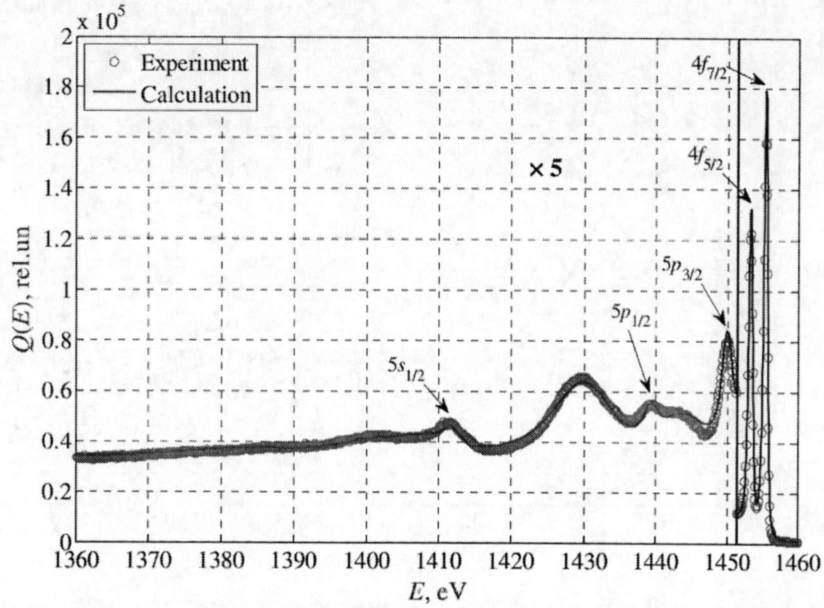


**Figure 1.** PES data of Tungsten. X-ray probing by monochromatic emission of line Al $K_\alpha$. Experimental data is taken from [10]. The circles refer to measurements, while the solid line corresponds to the computed spectrum using retrieved DIIMFP and DSEP functions.

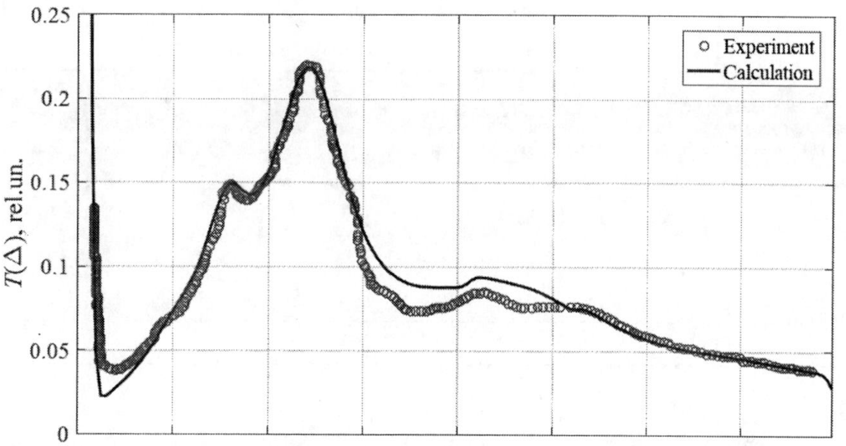


**Figure 2.** TEELS spectrum of the Tungsten foil. The thickness of the foil is about 30 nm. The incident electron energy is 25 keV. Sighting is in the probe direction. Measurements are taken from [11]. DIIMFP $x_{in}(\Delta)$ is extracted using the three-layer model.

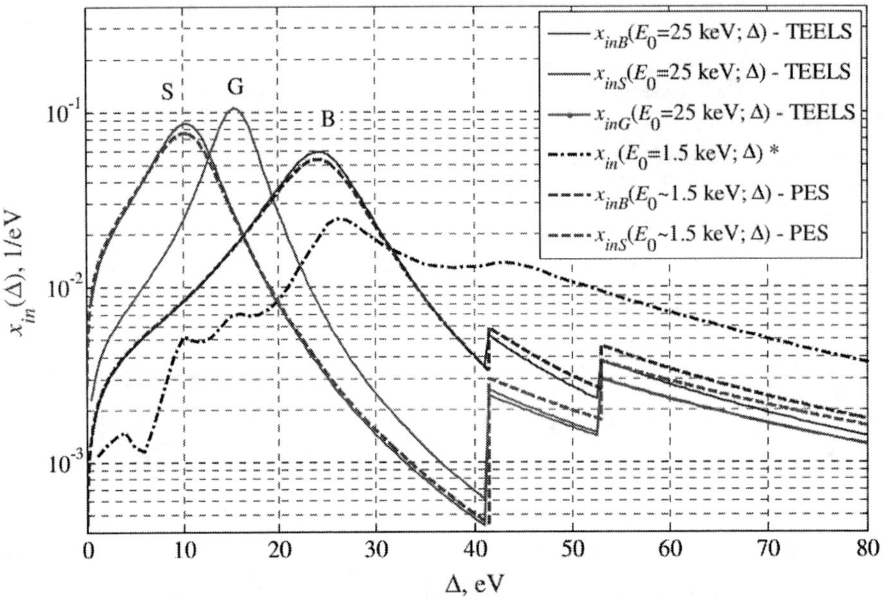


**Figure 3.** Extracted DIIMFP for bulk $x_{inB}(\Delta)$ and DSEPs for surface layers $x_{inG}(\Delta)$ and $x_{inS}(\Delta)$.
* Dash-and-dot line corresponds to the data from [13].

**Table 1.** Retrieved DIIMFP and DSEPs parameters for Tungsten and thicknesses of layers.

| Parameter | from TEELS spectrum, $E_0$=25 keV | | | from PES spectrum, $E_0$~1.5 keV | |
|---|---|---|---|---|---|
| | B-layer | G-layer | S-layer | B-layer | S-layer |
| $\varepsilon_{pl}$ | 24.6 | 15.8 | 11.0 | 24.6 | 11.0 |
| $b$ | 11.9 | 6.8 | 8 | 12.5 | 8.7 |
| $\alpha$ | | 1.5 | | 0.5 | |
| $\beta$ | | 0.5 | | 0.5 | |
| $J_{ion1}$ | | 41.5 | | 41.5 | |
| $\lambda_{ion1}$ | | 0.10 | | 0.13 | |
| $J_{ion2}$ | | 53 | | 53 | |
| $\lambda_{ion2}$ | | 0.10 | | 0.13 | |
| $d$, nm | 21.0 | 2·2.5 | 2·1.0 | ∞ | 0.68 |

## 4. Summary

The fitting approach has been applied to PES and TEELS data to retrieve DIIMFP and DSEP functions of Tungsten. It has been shown that for high-energy resolution TEELS data, it is mandatory to account for two surface plasmons at the interface, while for low resolution PES data it is sufficient to consider only one surface plasmon. The parametrization of DIIMFP/DSEP functions using the classical Lorentz oscillator expressions can be regarded as a regulation procedure, which guarantees a physically consistent result.

The fitting approach for extracting inelastic scattering properties is applicable to the hydrogen depth profiling [15, 16] using the elastic peak electron spectroscopy [17]. There, the DIIMFP overlaps the hydrogen peak and introduces a systematic error in the retrieved hydrogen concentration. The

IOP Conf. Series: Journal of Physics: Conf. Series **941** (2017) 012019 doi:10.1088/1742-6596/941/1/012019

method described in this paper is suitable for designing an inelastic scattering background subtraction tool. This is a topic for our future research.

**Acknowledgments**
This work has been supported by the grant of the Russian Science Foundation (project No. 16-19-10027).

**References**
[1] Tougaard S, Chorkendorff I 1987 *Phys. Rev.*B **35** 6570
[2] Werner W S M 2006 *Appl. Phys. Lett.* **89** 213106
[3] Afanas'ev V P, Efremenko D S, Lubenchenko A V 2011 *J. Surf. Invest.* **5**(2) 375
[4] Afanas'ev V P, Gryazev A S, Efremenko D S, Kaplya P S 2017 *Vacuum* **136** 146
[5] Werner W S M 1995 *Surface Interface Analysis* **23** 737
[6] Afanas'ev V P, Efremenko D S, Kaplya P S 2016 *J. Electron Spectrosc. Relat. Phenom.* **210** 16
[7] Afanasyev V P, Efremenko D S, Lubenchenko A V, Vos M, Went M R 2010 *Bull. Russian Acad. Sci.: Physics* **72** (2) 170
[8] Ritchie R H 1957 *Phys. Rev.***106** 874
[9] Radzig A A, Smirnov B M 1985 *Reference Data on Atoms, Molecules, and Ions* (Springer Series in Chemical Physics vol 31) ed J P Toennies (Berlin, Heidelberg: SpringerBerlin Heidelberg)
[10] Moulder J, Stickle W, Sobol P, Bomben K 1995 *Handbook of X-Ray Photoelectron Spectroscopy* (Eden Prairie, Minnesota: Perkin-Elmer Corporation, Physical Electronics Division)
[11] Zharnikov M V, Gorobchenko V D, Serpuchenko I L 1987 *JETP* **65** (1) 128
[12] Afanas'ev V P, Golovina O Y, Gryazev A S, Efremenko D S, Kaplya P S 2015 *J. Vac. Sci. Technol. B* **33** (3) 03D101
[13] Werner W S M, Glantschnig K, Ambrosch-Draxl C 2009 *J. Phys. Chem. Ref. Data.* **38** 1013
[14] Yubero F, Tougaard S 2005 *Phys. Rev.* B. **71** 045414
[15] Kostanovskiy I A, Afanas'ev V P, Naujoks D, Mayer M 2015 *J. Electron Spectrosc. Relat. Phenom.* **202** 22
[16] Afanas'ev V P, Gryazev A S, Efremenko D S, Kaplya P S, Ridzel O Y 2016 *J. Phys. Conf. Ser.* **748** 012005
[17] Vos M, Went M R 2009 *J. Phys.* B*: At. Mol. Opt. Phys.* **42** 065204

# Deuterium diffusion and retention in tungsten coated with barrier layer during ion irradiation

**L B Begrambekov, A S Kaplevsky, S S Dovganyuk, A E Evsin and I E Baryshnikova**

National Research Nuclear University MEPhI (Moscow Engineering Physics Institute), 115409 Kashirskoe shosse 31, Moscow, Russia Federation

E-mail: as.kaplevsky@plasma.mephi.ru

**Abstract.** The results of the comparative analysis of low-temperature desorption of deuterium from tungsten coated with aluminum and yttrium films under the irradiation by hydrogen plasma with oxygen impurity are presented. The irradiation of aluminum or yttrium coating by $H_2$+1%$O_2$ plasma leads to the desorption of implanted deuterium from the samples. It was shown that the number of atoms desorbed depends on the sign of enthalpy of hydrogen solution in the metal film.

## 1. Introduction

Beryllium and tungsten are chosen as the materials for the first wall and the diverter of the ITER tokamak, respectively. Beryllium and tungsten have oxide layers formed on their surfaces. It is known that the oxide layers function as barriers for hydrogen diffusion. The surface of tungsten and beryllium will be coated by oxides if the ITER plasma would have an oxygen impurity. Sputtered beryllium atoms will be deposited on the surfaces of tungsten tiles and form oxide layers under these conditions. The research on the influence of beryllium films on hydrogen isotope trapping and desorption from tungsten is of interest from the safety perspective as the limit for the accumulation of tritium in the walls cannot be exceeded.

The authors of paper [1] have discovered the effect of low-temperature hydrogen isotope desorption from stainless steel at $T \approx 320$ K under the irradiation by hydrogen ions in hydrogen atmosphere with oxygen impurity, as well as hydrogen plasma with oxygen impurity. This effect is explained by the processes initiated by inelastic collisions of deuterium and oxygen atoms/ions with the chromium oxide layer on the surface of stainless steel [2]. Trapping and retention of hydrogen in stainless steel and tungsten with oxide layers were studied in several works [3-5]. However, the influence of surface oxide layers on hydrogen trapping and desorption from metal during ion irradiation has not been properly investigated as of today.

In this work, a comparative research on the effect of low temperature desorption of hydrogen isotopes from tungsten samples with surface barrier coatings like aluminum, which is considered as a proxy for beryllium in laboratory experiments, and yttrium with oxidized surfaces. The task was to identify the characteristics of the process in cases when the metal (tungsten) and the barrier layer (aluminum) are metals with positive enthalpy of hydrogen solution (+1.1 eV/at [6] and +0.7 eV/at [6], respectively), and in case when the enthalpy of hydrogen solution in the barrier layer (yttrium) is negative (-0.85 eV/at [6]).

IOP Conf. Series: Journal of Physics: Conf. Series **941** (2017) 012020 doi:10.1088/1742-6596/941/1/012020

## 2. Experimental facility and experimental method

Aluminum and yttrium films 200±10 nm thick were deposited on one side of the 7×7×1 $mm^3$ tungsten samples. The Plansee-grade tungsten (99,96% wt. of W) samples were mechanically polished, cleaned with acetone and alcohol ultrasonic baths and then annealed in vacuum with the residual pressure of $\leq 5x10^{-6}$ Pa at a relatively low temperature (1600 K) to remove the "technological" hydrogen. The coating was formed by aluminum/yttrium atoms sputtered from the aluminum/yttrium target with argon plasma ions. The residual gas pressure during the deposition did not exceed $5\times10^{-4}$ Pa. The temperature of the samples during the deposition was 450-500 K, the duration of the deposition was 1 hour.

The X-ray Energy Dispersion Spectroscopy (EDS) and the Thermal Desorption Spectroscopy (TDS) analysis has shown that a significant number of hydrogen and oxygen atoms were trapped in aluminum and yttrium films during their respective depositions. In particular, concentrations of hydrogen and oxygen in the films were 30 at.% and 13 at.% for aluminum and 33at.% and 17 at.% for yttrium, respectively. Additionally, $Al_2O_3$ and $Y_2O_3$ oxide layers were formed on the surface of the films after being exposed to atmosphere.

The experiments were conducted in the "MIKMA" facility [7]. The pumping system allows for the residual pressure in both radiation chambers to not exceed $1\times10^{-4}$ Pa, as well as pump the TDS chamber down to $1\times10^{-5}$ Pa.

Three tungsten samples with aluminum and yttrium films were used to measure the average number of "technological" hydrogen atoms in the samples using TDS. Then deuterium was implanted into the coating-free side of the rest of the samples by irradiating them with 650 eV/at. energy deuterium plasma ions. The plasma consisted mostly of $D_2^+$ (82-85%) ions, with the rest being $D^+$ and $D_3^+$ ions. For this reason, the experimental results were analyzed assuming that the samples were irradiated by $D_2^+$ ions only. The parameters of implantation were as follows: ion flux density $j = 3.7\times10^{19}$ at/$m^2$s, fluence $F = 1.3\times10^{23}$ at/$m^2$, the temperature of the samples during the irradiation $T = 500$ K.

The first sample with implanted deuterium was used to measure the number of deuterium and hydrogen atoms after the irradiation by deuterium plasma. The second sample was kept in vacuum for 1 hour after the implantation at 500 K i.e. for the same duration at the same temperature as the one irradiated by $H_2$+1% $O_2$ plasma. It was made to see whether the exposure of the sample to vacuum would lead to the desorption of the implanted deuterium. The measurement of the number of deuterium atoms was then conducted using TDS. The third sample was irradiated by the ions with energy of 50 eV/at. (mostly $H_2^+$) of $H_2$+1% $O_2$ plasma. The parameters of irradiation were the following: ion flux density $j = 3.7\times10^{19}$ at/$m^2$s, fluence $F = 1.3\times10^{23}$ at/$m^2$, the temperature of the samples during the irradiation $T = 500$ K. The measurement of the number of deuterium and hydrogen atoms after the $H_2$+1% $O_2$ plasma irradiation was then conducted using TDS.

The heating rate of the samples during TDS was 5 K/s. The temperature of the samples during plasma irradiation and the TDS analysis was measured using W-Re thermocouple welded to the samples. Desorption signals of $H_2$, HD, $D_2$, $H_2O$, HDO and $D_2O$ molecules were measured during TDS. Desorption of another hydrogen isotopes-containing molecules was negligible.

## 3. Results and Discussion

The results of the experiments with tungsten samples coated withaluminum filmare shown in Table 1. It can be seen (Table 1.1) that during the coating deposition trapping of a large number of hydrogen molecules from the residual gas occurs. During the deuterium implantation about 14% of hydrogen is desorbed from the sample (Table 1.1, 1.2). The exposition to vacuum does not lead to the change in hydrogen or deuterium content in the sample (Table 1.2, 1.3). After the irradiation of the aluminum film by $H_2$+1% $O_2$ plasma ions at the energy of 50 eV/at., the concentration of deuterium in the sample decreased significantly. At least 48% of the implanted deuterium has passed through the bulk of the tungsten and aluminum oxide and was removed during the irradiation in hydrogen plasma with the oxygen impurity. At the same time, the number of hydrogen atoms in the sample was practically unchanged.

**Table 1**. The number of trapped hydrogen and deuterium atoms in tungsten samples with aluminium coating after deuterium implantation, exposition to vacuum and irradiation in $H_2$+1% $O_2$ plasma.

| № | Experiment | Deuterium trapping, $\times10^{20}$ at/m$^2$ | Hydrogen trapping, $\times10^{21}$at/m$^2$ | $D_{before}$/ $D_{after}$ | $H_{before}$/ $H_{after}$ |
|---|---|---|---|---|---|
| 1 | Before implantation in $D_2$ plasma | - | 3.6 | - | - |
| 2 | After implantation in $D_2$ plasma | 2.9 | 3.1 | - | ≈ 0.86 |
| 3 | After exposition to vacuum | 3.1 | 3.3 | ≈ 1.07 | ≈ 0.92 |
| 4 | After irradiation by $H_2$+1 at.% $O_2$ plasma | 1.5 | 3.2 | ≈ 0.52 | ≈ 0.89 |

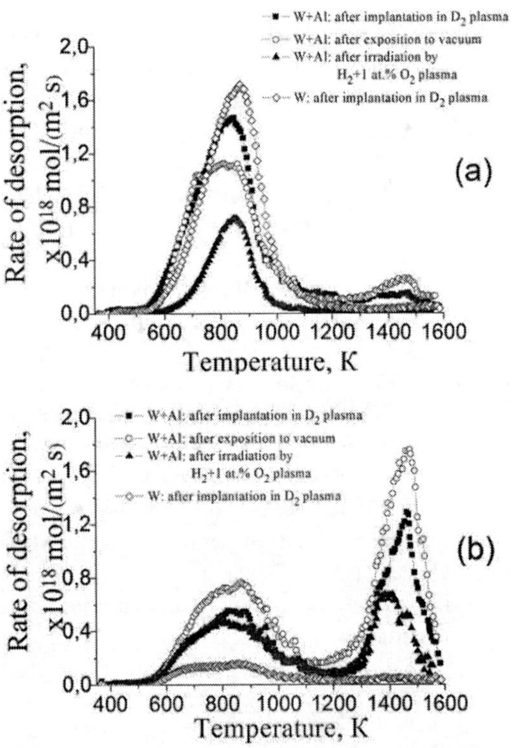


**Figure 1.** TDS spectra of $D_2$ (a) and HD (b) molecules from tungsten and tungsten with aluminum coating after deuterium implantation, exposition to vacuum and irradiation by $H_2$+1% $O_2$ plasma ions.

Thermal desorption spectra of $D_2$ and HD molecules, by which approx. 90% of deuterium is desorbed, from tungsten samples with aluminum coating shows two thermal desorption peaks at $T \approx 830$ K and 1450 K (Figure 1). Figure 1 also includes $D_2$ and HD molecules' spectra from a tungsten sample not coated with aluminum. It has only one peak at $T \approx 830$ K. The comparison of $D_2$ and HD molecules' spectra from both samples leads to the conclusion that the peak at 1450 K shows the desorption of deuterium from aluminum film. Deuterium implanted at $T$ = 500 K diffuses through tungsten and enters aluminum layer from the coating-free side of the sample. Deuterium can diffuse through tungsten for large distances (≈ 0.1 mm) at $T$ = 500K [8]. TDS analysis of the spectra in Figure 1, we can conclude that approx. 25% of deuterium implanted into tungsten passes into the aluminum film.

The results of the experiments with tungsten samples coated with yttrium film are shown in Table 2. It is shown (Table 2.1) that, as with aluminum coating, a significant number of hydrogen atoms is trapped from the residual gas during yttrium film deposition on tungsten. Deuterium implantation did not lead to hydrogen atoms desorption from the sample (Table 2.1, 2.2). Sample exposition to vacuum also did not lead to the change in deuterium and hydrogen content in the sample (Table 2.2, 2.3). After the irradiation of the yttrium coating surface by 50 eV/at $H_2$+1% $O_2$ plasma ions, the concentration of deuterium in the sample decreases. At least 21% of implanted hydrogen has passed through tungsten and yttrium coating and was removed by irradiation in $H_2$+1% $O_2$ plasma. At the same time, the number of hydrogen molecules has increased by a factor of 3.

Thermal desorption of $D_2$ and HD molecules from tungsten samples with yttrium coating have a peak at $T \approx 830$ K and two steep peaks at $T \approx 900$ K and 1100 K. It was shown before that the peak at 830 K is the desorption of deuterium from tungsten. It can be concluded, therefore, that the steep peaks at the range of $T$ = 900-1100K are caused by the desorption of deuterium from the yttrium coating. TDS analysis of the spectra (Figure 2) has shown that approx. 60% of deuterium implanted into tungsten passes into the yttrium film.

**Table 2**. The number of deuterium and hydrogen atoms in tungsten samples with yttrium coating after deuterium implantation, exposition to vacuum and irradiation in $H_2$+1% $O_2$ plasma.

| № | Experiment | Deuterium trapping, $\times 10^{20}$at/m$^2$ | Hydrogen trapping, $\times 10^{21}$at/m$^2$ | $D_{before}/D_{after}$ | $H_{before}/H_{after}$ |
|---|---|---|---|---|---|
| 1 | Before implantation in $D_2$ plasma | - | 2.0 | - | - |
| 2 | After implantation in $D_2$ plasma | 4.7 | 2.0 | - | 1.0 |
| 3 | After exposition to vacuum | 4.6 | 2.0 | ≈ 0.98 | 1.0 |
| 4 | After irradiation by $H_2$+1 at. % $O_2$ plasma | 3.7 | 5.8 | ≈ 0.79 | ≈ 2.9 |

As shown above, during deuterium implantation into the coating-free side of the sample, 25% and 60% of trapped atoms pass through into aluminum and yttrium layers, respectively. Enthalpy of hydrogen solution in yttrium is lower than that for aluminum; therefore, it is more energetically favorable for deuterium to pass into yttrium than into aluminum.

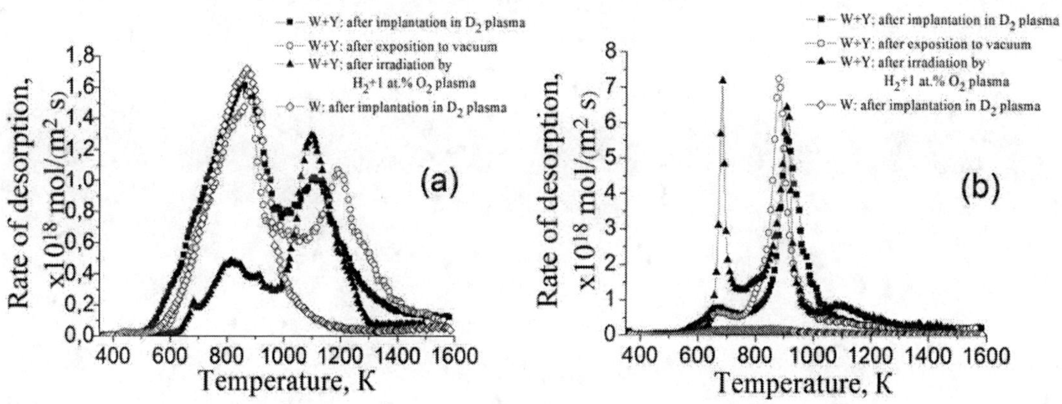


**Figure 2.** TDS spectra of $D_2$ (a) and HD (b) from tungsten and tungsten with yttrium coating after deuterium implantation, exposition to vacuum and irradiation with $H_2$+1%$O_2$ plasma ions.

The comparison (Table 1.3, 1.4 and 2.3, 2.4, respectively) shows that after the irradiation with $H_2$+1% $O_2$ plasma ions, there is twice as many desorbed deuterium atoms from metal with positive enthalpy of hydrogen solution (aluminum) than from the metal with negative enthalpy of hydrogen solution (yttrium). It can be explained by the fact that the energy levels of hydrogen in yttrium (negative enthalpy) and aluminum (positive enthalpy) are lower and higher than the vacuum level, respectively. Hydrogen needs to pass through the potential barrier in yttrium before it can desorb from its surface, which leads to the desorption of less deuterium atoms compared to the aluminum film.

IOP Conf. Series: Journal of Physics: Conf. Series **941** (2017) 012020 doi:10.1088/1742-6596/941/1/012020

## 4. Conclusion

A comparative study of low-temperature desorption of deuterium from tungsten with aluminum and yttrium coatings under the irradiation by $H_2$+1% $O_2$ plasma ions has been conducted. A dependence on the number of deuterium atoms desorbed from the value of enthalpy of hydrogen solution under the irradiation of $H_2$+1% $O_2$ plasma ions has been established.

**Acknowledgements**

The present work was supported by the Ministry of Education and Science of the Russian Federation (grant No 3.9750.2017/8.9).

## References

[1] Begrambekov L, Grunin A, Kaplevsky A, et al. 2015 *J. of Surf. Inv. Xray, Sunch. and Neut. Tech.* **9** 190
[2] Dvoichenkova O and Airapetov A 2015 *Physics Procedia* **71** 93
[3] Richards P 1988 *J. Nucl. Mater.* **152** 246
[4] Ogorodnikova O, Futterer M, Serra E, et al. 1999 *J. Nucl. Mater.* **273** 66
[5] Ogorodnikova O, Roth J, Mayer M, et al. 2003 *J. Nucl. Mater.* **313-316** 469
[6] Fukai Y 1993 *The metal-hydrogen system*, ed Hull R, Osgood R M, Parisi Jr J and WarlimontH (Berlin: Springer) p 16
[7] Airapetov A, BegrambekovL, Bremond S, et al. 2011 *J. Nucl. Mater.* **415** 1042
[8] Haasz A, Devis J, Poon M, et al. 1998 *J. Nucl. Mater.***258-263** 889

IOP Conf. Series: Journal of Physics: Conf. Series **941** (2017) 012021 doi:10.1088/1742-6596/941/1/012021

# Stochastic clustering of material surface under powerful load of high-temperature plasma

**V P Budaev**[1,2]

[1] National Research University "Moscow Power Engineering Institute", Moscow, Russia
[2] National Research Center "Kurchatov Institute", Moscow, Russia

**Abstract.** The results of a study of a surface formed by high-temperature plasma loads on refractory metals such as tungsten and stainless steel are presented. There is an inhomogeneous stochastic clustering of the surface with self-similarity properties of the granularity structure from nanoscale to macro scale, similar to the previously described fractal structure of carbon materials from fusion device. The statistical characteristics of hierarchical granularity and scale invariance are estimated that qualitatively differ from the properties of the roughness of the Brownian surface, which is possibly due to the universal mechanisms of stochastic clustering of materials under the influence of high-temperature plasma.

## 1. Introduction

Plasma-surface interaction in a fusion device leads to the change of the surface morphology influenced by several mechanisms, including erosion, melting and melt motion over the surface, redeposition of eroded materials, solidification and recrystallization (see [1]). Describing such a complex process requires consideration of the kinetic equations of the general form (e.g., the Smoluchowski kinetic equation). The solution of this equation is a complex theoretical problem, it depends on the symmetry of the problem including functional dependence of the kernel term in the kinetic equation and its property of self-similarity. Experimental data on the self-similarity property of stochastic surface (such as scalings of self-similarity - power laws of scale invariance) should be used to simplify the problem. In this way, it is important to find the most common power laws describing the surface clustering, which will be used for the data systematization and for the description of the material properties after irradiation by plasma in fusion device.

In the literature, it is discussed the formal analogy (see [2]) between the non-linear equation for the fragmentation - aggregation process and kinetic equation describing 3-wave turbulence, for which the power spectrum is proposed in the Kolmogorov-Zakharov approach. Distribution of clusters by mass in the fragmentation - aggregation process (aggregation / disintegration of clusters of different sizes) is similar to the cascade process of energy transfer in turbulence: the number $N$ of particles with mass $m$ is described by the power law $N(m)=Cm^{-(3+\eta)/2}$, where $C$ is a constant; the index $\eta$ is determined by the property of self-similarity. Agglomeration processes of different self-similarity have different spectra of surface relief that can be used for a classification and a systematization of experimental data.

## 2. Experiment and results

We investigated tungsten samples [3], manufactured from polycrystalline tungsten plates of originally smooth surface, after the high heat flux test in the QSPA-T facility by high-temperature plasma with loads expected in the ITER. After the tests the tungsten samples have a rough surface with stochastic structure [3] and different scales of granularity ranging from nanometers (Figure 1) similar to the previously detected on graphite samples from the T-10 tokamak [4] (Figure 2).

**Figure 1.** SEM micrograph of tungsten sample after the high heat plasma test in the QSPA-T [3].

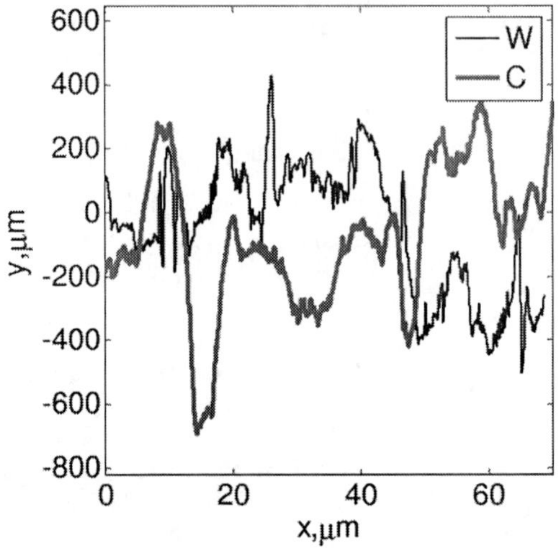


**Figure 2.** AFM height profiles of tungsten sample (W) from the QSPA-T [3] and the carbon film (C) from the T-10 tokamak [4].

Quantitative characteristics of the stochastic relief were found by spectral and statistical analysis of the experimental profiles. It allowed characterizing the qualitative difference between the stochastic clustering of samples after the treatment with high-temperature plasma from rough surfaces formed under other conditions. Fourier spectra of the relief profile (Figures 3, 4) characterize the heights (size of structures on the surface). The spectra are broadened without any resonances, indicating no dominant periodic structures in the relief. The spectra have the typical decaying shape observed usually in objects that have scale invariance and statistical self-similarity. The functional dependence of the Fourier spectrum on the wave number $k$ can be described by a power law $S(k) \sim k^p$, the power exponents $p$ are presented in Table 1. Note that for the tungsten samples [3], stainless steel [5] and graphite [4, 6] after exposure to high-temperature plasma in the T-10 tokamak, in the QSPA-T and in the NAGDIS-II plasma facilities the index $p$ has the value in the range of -2.4 to -2.8 or more. In contrast, for reference samples not treated by the high-temperature plasma (such as molybdenum after irradiation by low-temperature plasma in the magnetron discharge and industrial steel casting after solidification), the index $p$ ranges from -1.97 to -2.2. The statistical properties of stochastic relief self-similarity is described by the probability distribution function (PDF) of the sample relief heights. For the samples irradiated by high-temperature plasma the PDF typically has a "heavy" tail, and cannot be described by the Gaussian (normal) law (Figure 5). For comparison, a sample of the industrial steel casting, the rough surface of which was formed during solidification after melting, has the PDF of heights close to the Gaussian law (Figure 6) indicating the trivial statistical properties of clustering. Self-similarity of the stochastic relief is characterized by the Hurst exponent $H$ [4, 6] (Table 1); tungsten, graphite and stainless steel samples after exposure to high-temperature plasma have the rough surface relief characterized by the Hurst exponent in the range from 0.68 to 0.86, which corresponds to an irregular stochastic clustering with hierarchical granularity (fractality) [4, 6].

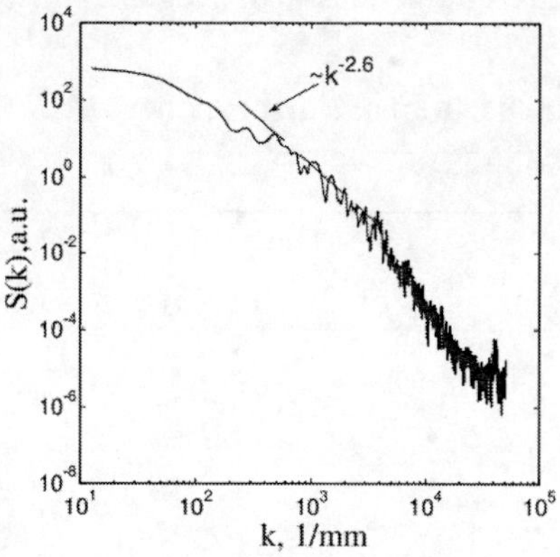


**Figure 3.** The Fourier spectrum of the stochastic relief of tungsten sample from QSPA-T [3] in Figure 2.

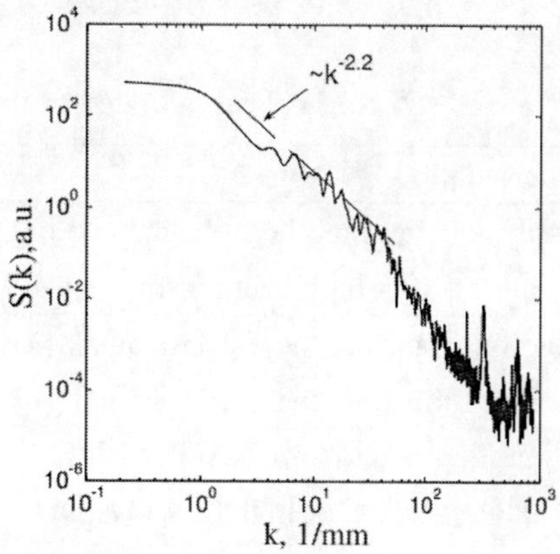


**Figure 4.** The Fourier spectrum of the stochastic relief of molybdenum surface after irradiation by plasma in the magnetron discharge.

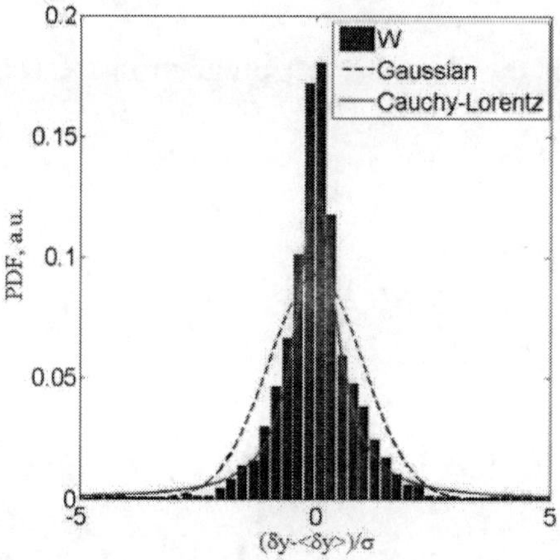


**Figure 5.** Probability distribution function for the surface height increments $\delta y = y(x+l) - y(x)$, tungsten samples in Figure 2, $l$ = 19.5 nm; Gaussian (dotted line) and the Cauchy-Lorentz (line) laws are shown for comparison.

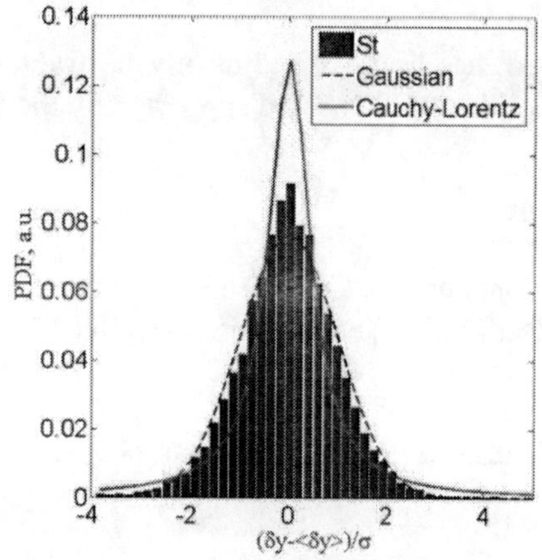


**Figure 6.** Probability distribution function for the surface height increments $\delta y = y(x+l) - y(x)$, industrial steel casting, $l$ = 0.5 μm; Gaussian (dotted line) and the Cauchy-Lorentz (line) laws are shown for comparison.

## 3. Conclusions

The paper presents the results of a study of the irregular surface structure formed after high-temperature plasma loads on samples of tungsten and stainless steel, the materials used for the manufacturing of the first wall components in fusion plasma devices. An inhomogeneous stochastic clustering of a surface with self-similarity properties of the granularity structure from nanoscale to macroscale is found, similar to that found earlier for carbon materials [1, 2]. It may indicate universal mechanisms of stochastic clustering of materials under the influence of high-temperature plasma. Such a clustering qualitatively differs from the trivial roughness of the Brownian surface; it was shown by analysis of reference samples - molybdenum after irradiation by low-temperature plasma in the magnetron discharge and a casting sample of industrial steel with the typical trivial roughness formed by solidification after melting. Revealed power laws of spectral and statistical scalings can be used to describe the symmetry of scale invariance of solids and agglomerates with irregular structure.

**Table 1.** Scaling index of the Fourier spectrum *p*, Hurst exponent *H*.

| N | The sample | *p* | *H* |
|---|---|---|---|
| 1 | Tungsten after high heat plasma test in QSPA-T, 2 pulses of 2 $MJm^{-2}$ [3] | -2.62 | 0.86 |
| 2 | Tungsten after high heat plasma test in QSPA-T, 50 pulses of 2 $MJm^{-2}$ [3] | -3.18 | 0.81 |
| 3 | Carbon film from the T-10 tokamak [4] | -2.49 | 0.78 |
| 4 | Carbon film from the T-10 tokamak [4] | -2.59 | 0.77 |
| 5 | Graphite limiter from NAGDIS-II [6] | -2.29 | 0.68 |
| 6 | Stainless steel after high heat plasma test in QSPA-T [5] | -2.85 | 0.72 |
| 7 | Stainless steel after high heat plasma test in QSPA-T [5] | -2.87 | 0.68 |
| 8 | Molybdenum after irradiation by plasma in the magnetron discharge | -2.20 | 0.60 |
| 9 | Industrial steel casting after solidification | -2.41 | 0.58 |

**Acknowledgments**
The work has been carried out in the framework of the materials test program in the ITER Project Centre of the ITER RA and supported by the Grant RSF № 16-19-10027.

**References**

[1] Budaev V P 2016 *Physics of Atomic Nuclei* **79** (7) 1137
[2] Connaughton C, Rajesh R, Zaboronski O 2005 *PRL* **94** (19) 194503
[3] Budaev V P, et. al. 2015 *JNM* **463** 237
[4] Budaev V P, et. al. 2012 *JETP Letters* **95** (2) 78
[5] Budaev V P, et. al. 2013 *Plasma Physics Reports* **39** (11) 910
[6] Budaev V P, et. al. 2007 *Physica A: Statistical Mechanics and its Applications* **382** 359

# On the possibility of thin layers thickness determination with low energy proton scattering

**D Bulgadaryan, V Kurnaev, D Sinelnikov and N Efimov**

National Research Nuclear University MEPhI (Moscow Engineering Physics Institute), Kashirskoe shosse 31, 115409 Moscow, Russia

E-mail: DGBulgadaryan@mephi.ru

**Abstract**. The analysis of erosion and redeposition processes plays an important role in the physics of fusion devices. In this work we present the results of computer simulation of plasma-facing materials surface analysis by use of the keV-energy proton scattering spectroscopy. It is shown that this technique can be used for the non-destructive analysis of thin surface layers. Energy spectra that correspond to different scattering and target parameters are presented.

## 1. Introduction

Plasma-surface interaction is an important topic in controlled nuclear fusion and, particularly, in tokamak physics. The interaction between plasma particles and plasma-facing materials leads to erosion and redeposition of these materials that further modifies the physical and chemical properties of the plasma-facing surfaces. Therefore, it is important to control erosion and redeposition processes in fusion facilities.

In modern fusion facilities, a combination of elements with small and big atomic number is usually used – in JET tokamak [1] the first wall is made from beryllium and tungsten is used as material of the divertor to simulate ITER conditions. Full-tungsten devices, such as ASDEX, use light materials [2] to enhance the plasma parameters. The average erosion rate, for example, of tungsten during JET ITER-like wall 2013-2014 campaign [3] was estimated as ~0.01 nm per second. Thus, the problem of erosion and redeposition processes control is reduced to the analysis of thin (nanometer scale) films of redeposited material on a surface of material with very different atomic mass.

Various analytical tools, such as Rutherford backscattering spectroscopy (RBS) [4], secondary ion mass spectrometry (SIMS) [5], X-ray photoelectron spectroscopy (XPS) [6], Auger electron spectroscopy (AES) [7], laser-induced breakdown spectroscopy (LIBS) [8] or other methods of surface analysis are used for this purpose; however, they have certain limitations – for example, SIMS destroys a sample while analyzing, XPS and AES do not have sufficient depth resolution for thin films analysis, RBS requires the use of expensive accelerators. Low energy helium ion scattering can be used for analysis of thin films [9], but the depth sensitivity of this method may be not sufficient because of high probability of helium ion neutralization during interaction with an analyzed surface.

It was shown earlier [10] that in this case the spectroscopy of keV-energy hydrogen ions scattering can be used as a nondestructive method of thin (1-10 nm) surface layers analysis with satisfactory sensitivity (0.3 nm for $B_4C$/Mo multilayer target [11]). In the presented study, we describe the optimized parameters of thin layers investigation procedure for different combinations of a surface layer and a substrate determined using computer simulation.

IOP Conf. Series: Journal of Physics: Conf. Series **941** (2017) 012022 doi:10.1088/1742-6596/941/1/012022

## 2. Computer simulation

The modelling of hydrogen ion beam scattering processes was performed using SCATTER computer code [12]. In the code the Monte-Carlo method and the binary collision approximation with Kr-C interatomic potential are used to simulate the energy and angular distributions of particles scattered from a target with the preset composition. In the simulation tungsten and beryllium were used as the analyzed materials. For every energy spectra of the scattered particles $10^8$ trajectories were calculated. The charge-state fraction of scattered particles was considered using approximation $\eta^+=0{,}08\ E^{1/2}$ [13]. The initial beam energies range was chosen from 1 to 20 keV as simply usable in laboratory conditions without the use of particle accelerators. The dependence of the energy spectra on the initial beam energy, scattering angle and surface layer thickness was investigated.

The use of hydrogen is due to its low sputtering yield and sufficient to analyze surface layers path in a target that can be regulated by the change of initial beam energy. If heavy element layer is present on surface of light material, this surface layer forms a high energy peak on the energy spectra of the scattered particles (figure 1), and the amplitude and width of this peak are dependent on the surface layer thickness. In the opposite case of the presence of layer of light material on surface of the heavy one (figure 2), the thickness of the light surface layer can be determined from the shift to the lower energy of the peak that is formed by scattering from the heavy substrate under the surface layer.

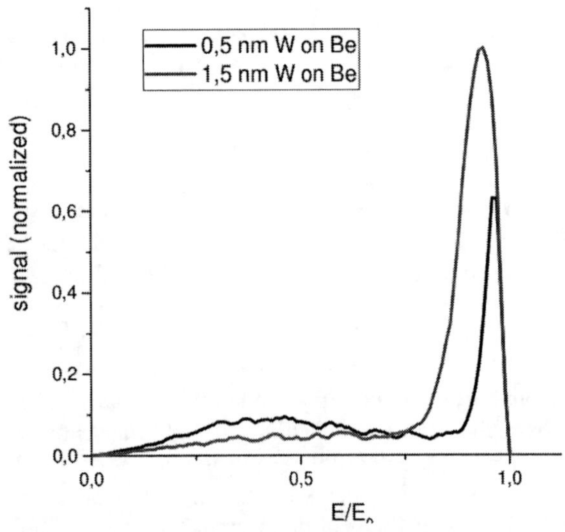


**Figure 1.** The energy spectra of $H^+$ with the initial energy $E_0$=6000 eV scattered at $\theta$=38° from the beryllium target with the surface layer of tungsten with the different thickness.

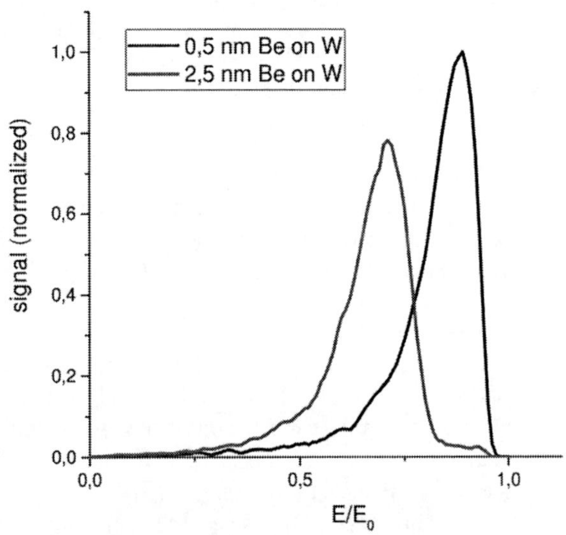


**Figure 2.** The energy spectra of $H^+$ with the initial energy $E_0$=4000 eV scattered at $\theta$=38° from the tungsten target with the surface layer of beryllium with the different thickness.

## 3. Results and discussion

The energy spectra of the proton beam with the initial energy $E_0$=9000 eV scattered at $\theta$=38° from the beryllium substrate with the tungsten surface layer of the different thickness are shown in figure 3. Figure 4, a) shows the dependence between the tungsten layer thickness and the full width at half maximum (FWHM) of the peak formed by scattering from that layer and linear approximation of the dependence. At the surface layer thicknesses up to 5 nm the dependence is quite linear, and the error of the thickness determination, assuming that the energy spectra are obtained with 1% resolution, is 0.3 nm. Increasing of the initial beam energy can reduce the thickness determination error and increase the range of thicknesses where their dependence with FWHM is linear, the dependence between tungsten thickness and FWHM when $E_0$=12000 eV is shown in figure 2, b). The use of $E_0$=15000 eV reduces the thickness determination error to 0.13 nm.

The increase of the scattering angle leads to appearance of a specific drop on the energy spectra (figure 5), that reduces the thickness determination error (0.25 nm for $E_0$=9000 eV), but at the same time the ion reflection coefficient, as well as while increasing the beam energy, is reduced, that can be undesirable in a laboratory experiment.

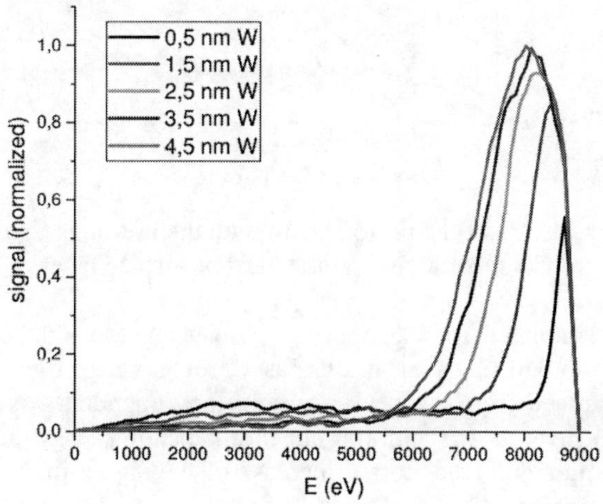


**Figure 3.** The energy spectra of the proton beam with the initial energy $E_0$=9000 eV scattered at $\theta$=38° from the beryllium substrate with the tungsten surface layer of the different thickness.

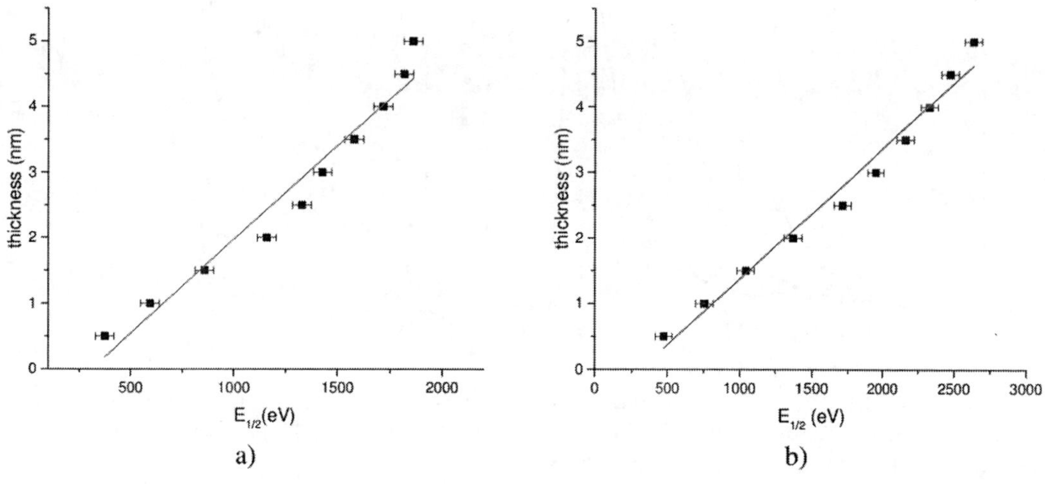


**Figure 4.** The dependence between the tungsten layer thickness and FWHM: a) $E_0$=9000 eV; b) $E_0$=12000 eV.

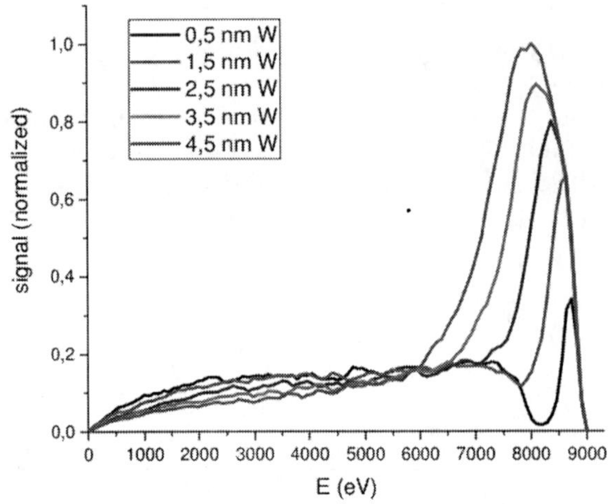


**Figure 5.** The energy spectra of the proton beam with the initial energy $E_0$=9000 eV scattered at $\theta$=60° from the beryllium substrate with the tungsten surface layer of the different thickness.

The energy spectra simulated for the opposite case when the light element is deposited on the heavy substrate (beryllium and tungsten, in our case) are shown in figure 6. In this case the surface layer thickness is determined from the peak position, the dependence is shown in figure 7, and the thickness determination error is 0.3 nm for the initial beam energy in the 4-10 keV range. The reduction of the scattering angle leads to the increase of the signal intensity depth resolution, but, at the same time, the real surface roughness plays a greater role, so in this case the optimal scattering angle should be determined experimentally.

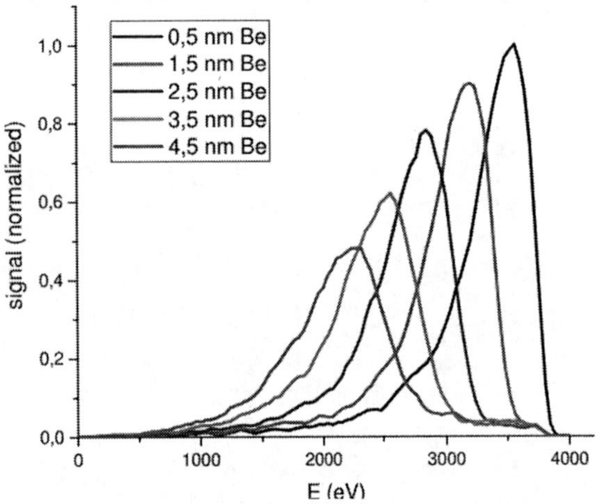


**Figure 6.** The energy spectra of H$^+$ with the initial energy $E_0$=4000 eV scattered at $\theta$=38° from the beryllium target with the surface layer of tungsten with the different thickness.

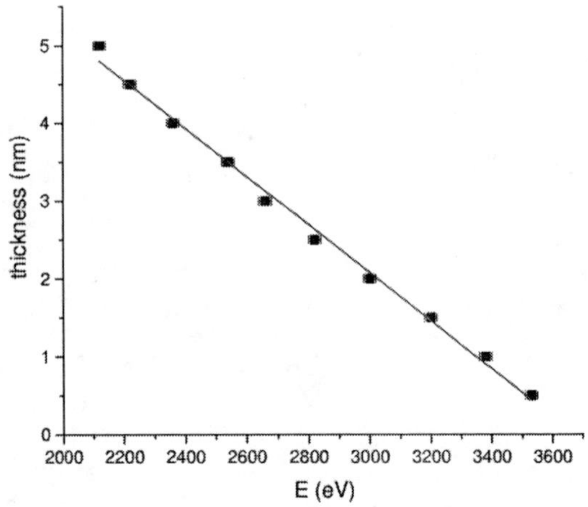


**Figure 7.** The dependence between the beryllium layer thickness and the peak position.

## 4. Conclusion

The results of the computer simulation show that scattering of hydrogen ions with keV energies at 30-60° can be used for the thin layer and, therefore, erosion and redeposition processes analysis. If a heavy element is present on a substrate with the low atomic mass, the thickness determination error decreases with the increase of the initial beam energy and scattering angle. In the opposite case when there is a surface layer of the light element on the heavy substrate, the thickness determination error does not significantly change for the initial beam energies in the 4-10 keV range, and smaller scattering angles are preferable. However, it should be noted that in laboratory experiments certain factors, such as the surface roughness or the ion reflection coefficient, should be taken into consideration.

## Acknowledgments

This work is supported by the Russian Science Foundation under grant № 17-12-01575.

## References

[1] Philipps V, Mertens P, Matthews G F and Maier H 2010 *Fusion Eng. Des.* **85** 1581
[2] Lang P T *et al* 2017 *Nucl. Fusion* **57**
[3] Krat S, Mayer M, Bykov I, Lungu C P, Aubin G de Saint, Widdowson A and Carvalho I S 2017 2013–2014 *Nucl. Mater. Energy* **11** 20
[4] Alves L C, Alves E, Barradas N P, Mateus R, Carvalho P, Coad J P, Widdowson A M, Likonen J and Koivuranta S 2010 *Nucl. Instruments Methods Phys. Res. Sect. B Beam Interact. with Mater. Atoms* **268**
[5] Likonen J, Coad J P, Vainonen-Ahlgren E, Renvall T, Hole D E, Rubel M and Widdowson A 2007 *J. Nucl. Mater.* **363–365**
[6] Lucia M, Kaita R, Majeski R, Bedoya F, Allain J P, Abrams T, Bell R E, Boyle D P, Jaworski M A and Schmitt J C 2015 *J. Nucl. Mater.***463** 907
[7] Skinner C H, Sullenberger R, Koel B E, Jaworski M A and Kugel H W 2013 *J. Nucl. Mater.***438** S647
[8] Almaviva S, Caneve L, Colao F, Fantoni R and Maddaluno G 2012 *J. Nucl. Mater.* **421**
[9] Brüner P, Grehl T, Brongersma H, Detlefs B, Nolot E, Grampeix H, Steinbauer E and Bauer P 2015 *J. Vac. Sci. Technol. A Vacuum, Surfaces, Film.* **33** 01A122
[10] Bulgadaryan D, Kolodko D, Kurnaev V and Sinelnikov D 2016 *J. Phys. Conf. Ser.* **748**
[11] Kurnaev V A, Trifonov N N, Drozdov M N and Salashchenko N N 2000 *Vacuum* **56** 253
[12] Koborov N N, Kuzovlev A I, Kurnaev V A, Remizovich V S and Trifonov N N 1997 *Nucl. Instruments Methods Phys. Res. Sect. B Beam Interact. with Mater. Atoms* **129** 5
[13] Mamedov N V, Kurnaev V A, Sinelnikov D N and Kolodko D V 2016 *J. Phys. Conf. Ser.* **666**

# Simulation of adsorbed hydrogen on tungsten surface

**N N Degtyarenko*[*] and A A Pisarev[**]**

National Research Nuclear University MEPhI (Moscow Engineering Physics Institute), 31 Kashirskoe sh., 115409 Moscow, Russia

E-mail: [*]NNDegtyarenko@.mephi.ru, [**]Pisarev@plasma.mephi.ru

**Abstract**. Calculations of the energy of the H-W system were performed using DFT method based on plane waves. Adsorption energies, equilibrium states, vibration spectra, saddle points, activation energies of jumps, and diffusion paths have been analyzed for H atom on W(100) and W(110). Diffusion coefficient for H on W(110) agrees very well with experimental data.

## 1. Introduction

Hydrogen adsorption on W was intensively studied both experimentally and theoretically [1-16]. Density functional theory (DFT) was previously used to calculate characteristic energies of H atom on the surface and analyze possible paths and activation energies for H diffusion on a reconstructed surface [10,11]. In this work, both energies and oscillation spectra were calculated for W(100) and W(110) surfaces. It was taken into account that the height of barriers for transitions along the surface are about the same as the energies of zero vibrations. Stability of hydrogen in some positions with minimal energies was also analyzed. Finally, paths of elementary jumps and the diffusion coefficients were obtained in the limit of non-reconstructed surface.

## 2. Features of calculations

Calculations were performed using DFT–RPBE–PW method. Usually, hydrogen atom is considered as a quantum oscillator vibrating in all directions in a deep potential well with infinite number of states. However, the well is high (1–2 eV) for vibrations perpendicular to the surface, but very shallow (fractions of eV) along the surface. In the latter case any state besides the zero one is over the well. Therefore, we take into account the whole range of oscillations perpendicular to the surface and only the zero oscillations along the surface. This feature was taken into account in calculations of the energy of the system both in equilibrium and saddle sites. Not only the energy, but also oscillation modes characterize these sites as well as possible paths of transitions of H-atom. The oscillation frequencies can be both real and imaginary. Imaginary frequency of oscillations in a direction indicates instability of the system in this direction. This allows selecting stable sites for H-atom and directions of possible jumps between equilibrium sites. Following this consideration, the diffusion coefficient was calculated as

$$D_i(T)=\frac{\lambda_i^2}{n_i}\cdot\left(\frac{Z^*_{as}}{Z_{gs}}\right)\cdot\langle\nu_{gs\leftrightarrow}\rangle\exp\left(-\frac{\varepsilon_{ad}}{kT}\right);\; Z_{gs}=\exp\left(-\frac{h\nu_{1\leftrightarrow}}{2kT}\right)\cdot\exp\left(-\frac{h\nu_{2\leftrightarrow}}{2kT}\right)\cdot\frac{\exp\left(-\frac{h\nu_{\updownarrow}}{2kT}\right)}{\left[1-\exp\left(-\frac{h\nu_{\updownarrow}}{kT}\right)\right]} \quad (1)$$

$$Z^*_{as}=\frac{\exp\left(-\frac{h\nu^*_{\updownarrow}}{2kT}\right)}{\left[1-\exp\left(-\frac{h\nu^*_{\updownarrow}}{kT}\right)\right]}\times\begin{cases}\exp\left(-\frac{h\nu^*_{\leftrightarrow}}{2kT}\right) & \text{- saddle point of the first order}\\ 1 & \text{- saddle point of the second order}\end{cases}$$

where $\lambda_i$ is the jump length; $n_i$ is the number of nearest sites to jump; $\varepsilon_i$ is the barrier in the adiabatic approximation; $\nu_{\updownarrow}$ and $\nu^*_{\updownarrow}$ – frequencies of normal vibrations of H in the equilibrium site and in the saddle point, respectively; $\nu_{\leftrightarrow}$ and $\nu^*_{\leftrightarrow}$ – frequencies of lateral oscillations of H in the equilibrium and saddle points; $\langle\nu_{gs\leftrightarrow}\rangle$ – the frequency of lateral ground state oscillations of H corresponding to the direction of the jump over the barrier. Normally, this vibration has the smallest frequency in the ground state and disappears in the saddle site.

## 3. Adsorption sites and jumps

Two optimized configurations of H atoms on W(100) are shown in Figures 1a,b: site **s1** (twofold bridge) and site **s2** (fourfold position). Adsorption energies were calculated as the difference between the total energy of the optimized system "W atoms plus an adsorbed H atom" and that for the system "W atoms plus a free H atom far from the surface". The heats of adsorption (the difference between the adsorption energy and the energy of one atom in $H_2$ molecule in gas phase (a half of the $H_2$ binding energy $1/2\varepsilon_{bond} \cong 2.3$ eV) for sites **s1** and **s2** on W(100) were equal to -2.21 and -1.91eV, respectively.

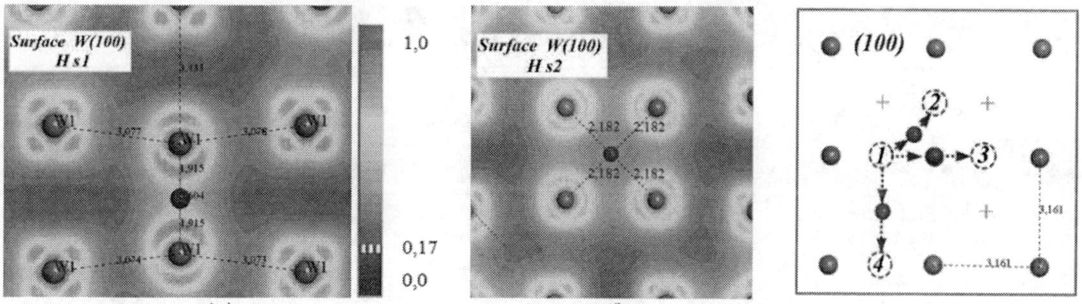


(*a*) (*b*) (*c*)

**Figure 1.** Hydrogen atom on W(100): *a, b)* – Optimized configuration in sites **s1** and **s2** (top view). Colors correspond to the electron density in electron/Å$^3$ in the plane of W surface atoms. *c)* – possible jumps of H atoms between sites **s1** from position 1 to positions 2, 3, and 4 through saddle points colored in red.

Sites **s1** and **s2** differ principally with respect to vibration spectra. H atom in site **s1** on the bridge has two real lateral oscillation modes ($h\nu_{2,3} \approx 50$, 130 meV) and one mode perpendicular to the surface ($h\nu_1 \approx 160$ meV). Site **s1** is dynamically stable, and H atom after small deviation from this site returns back to the original position. Contrary, site **s2** is unstable. H atom there has a real frequency only perpendicular to the surface ($h\nu_1 \approx 80$ meV), while frequencies in lateral directions are imaginary ($h\nu_{2,3} \approx i40$ meV). Besides, site **s2** is surrounded by four sites **s1**, which are energetically preferential. There are three kinds of jumps between sites **s1** on W(100) (Figure 1c): 1) curvilinear transition 1→2 to a perpendicular bridge through the threefold site, 2) jump 1→3 along the bridge over W atom, and 3) transition 1→4 to the opposite bridge through the fourfold site. The adiabatic barriers for these transitions are 0.22, 0.87, and 0.65 eV, respectively. The saddle points of transition 1→2 are of a simple type (one imaginary frequency – i75 meV, and two real frequencies 40 and 190 meV, while those for transitions 1→3 and 1→4 have two imaginary frequencies).

Four positions for H atom on W(110) are shown in Figure 2a: two normal bridge sites (long bridge **s1** and short bridge **s4**), as well as one diagonal bridge **s2** and a threefold site **s3**. The adsorption heats of sites **s1**, **s2**, **s3**, and **s4** on W(110) are -1.40, -1.55, -1.59 eV, and -1.45eV, respectively. Phonon spectra for **s1** (i94, 137, 143 meV) and **s4** (i98, 139, 150 meV) are similar and have one imaginary frequency. Contrary, all frequencies of sites **s2** (30, 145, 170 meV) and **s3** (90, 105, 162 meV) are real. That is, sites **s2** and **s3** are more stable from the point of view of both energy and vibration spectra. Possible transitions through sites **s2** and **s3** are shown in Figure 2b. Parameters of transitions between various positions are given in Table. Most barriers are of a simple saddle type (one imaginary frequency), and only saddle point for jump 2→3 over a W atom is of the second order (two imaginary frequencies).

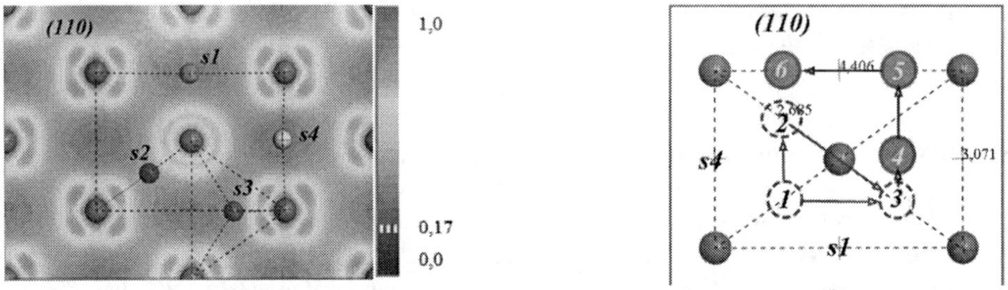


(*a*) (*b*)

**Figure 2.** *a)* Four sites for adsorption of hydrogen on W (110). Different colors correspond to different electron densities in electron/Å$^3$ calculated in the plane of W surface atoms; *b)* – possible jumps of H atoms between sites **s2** (positions 1,2,3) and **s3** (positions 4,5,6).

**Table.** Parameters of the H atom jumps on the surface W(110).

| Types of sites | Path | Adiabatic barrier, $\varepsilon_{ad}$, eV | Jump length, λ, Å | Number of neighbour sites | Oscillation frequencies at saddle points $h\upsilon$*, meV |
|---|---|---|---|---|---|
| s2→s2 | *1→2* | 0.20 | 1.56 | 2 | i 70, 137, 140 |
| s2→s2 | *1→3* | 0.58 | 2.23 | 2 | i 90, 139, 142 |
| s2→s2 | *2→3* | 0.72 | 3.13 | 2 | i 90, *i* 90, 145 |
| s2→s3 | *3→4* | 0.02-0.05 | 1.0 | 2 | i 80, 138, 140 |
| s3→s3 | *4→5* | 0.16 | 1.42 | 2 | i 80, 138, 140 |
| s3→s3 | *5→6* | 0.20 | 1.48 | 1 | i 82, 139, 142 |

## 4. Surface diffusion

Of the three jumps considered for W(100), the diagonal jumps 1→2 between bridge sites **s1** are responsible for H diffusion from the point of view both of energy and vibration spectra. Adiabatic activation energy for W(100) is $\varepsilon_{adiab} \approx$ 0.22 eV. Diffusion on W(110) is possible through various paths between sites **s2** and **s3**. Jumps between two **s2** sites (path 1→2) have the activation energy of 0.20 eV. Jumps between **s3** sites are sequence of two jumps (path 4→5→6); with activation energies of 0.16 eV for 4→5 and 0.2 eV for 5→6. The latter value can be considered as a diffusion barrier through **s3** sites. Interesting that the activation energies for diffusion through **s2** and **s3** sites on W(110) are the same.

Figure 3 shows the diffusion coefficient *D*(*T*) calculated according to (1) using barriers and frequencies obtained. Paths with higher barriers give small contribution. Also, two experimental points for W(110) from [3] are also shown. One can see a good agreement between the calculations and the experiment for W(110). Note that the slope of the curve for W(100) (1b) is smaller than that for W(110) (1a), while the adiabatic barriers $\varepsilon_{ad}$ on these surfaces are equal to 0.22 and 0.20 eV, respectively. This is due to the temperature dependence of the frequency factor (1) for these surfaces. Indeed, the difference between half the sum of energies of zero-point oscillations with real frequencies for the ground states and saddle points (energies ZPE) is $\Delta\varepsilon_{ZPE}=1/2[(h\nu_{1\leftrightarrow} + h\nu_{2\leftrightarrow} + h\nu_{\updownarrow})-(h\nu_{*\leftrightarrow}+ h\nu_{*\updownarrow})]$ and gives $\Delta\varepsilon_{ZPE}$ = 0.064 and 0.034 eV for the diffusion jumps on surfaces (100) and (110), respectively. Thus the values of effective barriers become 0.156 and 0.166 eV for W(100) and W(110), respectively.

One must mention that our activation energies 0.22(0.156) eV for W(100) agree well with the energies obtained in other DFT calculations 0.17(0.11) eV [10], 0.21 eV [9] but differ from the experimental values 0.47 eV [13]. Two numbers mean adiabatic barrier and barriers with ZPE taken into account. For W(110) our activation energies 0.20 (0.166) eV agree well with the experimental values 0.22 [14], while calculated values 0.31 eV [8] differ from both. One must mention that calculations [8-10] were performed for the reconstructed surface, while our calculations were performed for non-reconstructed surface and low concentration of hydrogen. Also there was a difference in the number of states taken into account in the potential well.

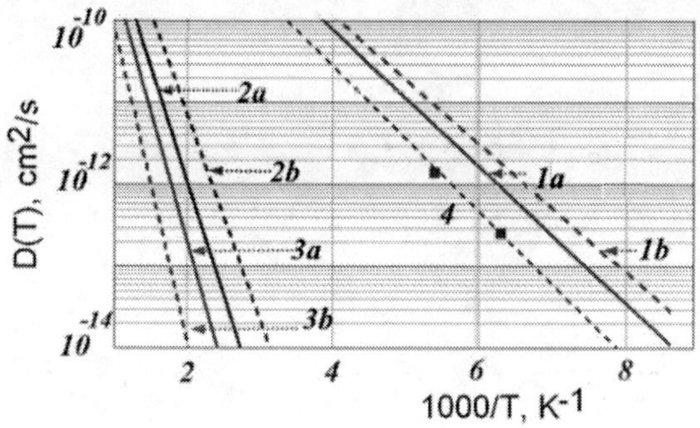


**Figure 3.** Diffusion coefficient of H on W surface: 1a) diffusion through sites **s2** on W (110); 1b) diffusion through sites **s1** on W (100); 2a, 2b, 3a, 3b) contribution of jumps with larger activation energies to $D(T)$; 4) two experimental points for W(110) are from [3].

## 5. Conclusions

Hydrogen adsorbed on W(100) and W(110) planes was analyzed by DFT methods for the case of non-reconstructed surface at the low H concentration. Calculations of the energy of the H-W system and oscillation spectra demonstrate that stable configuration of H atom on the W(100) plane is the short bridge, while the plane W(110) has two stable configurations: the diagonal bridge and threefold site with approximately similar energies and oscillation frequencies. Only these sites have only real frequencies of oscillations, while other sites have either one or two imaginary lateral frequencies, though they are associated with potential deeps. The heats of adsorption are –1.9 eV for W(100) and –(1.5÷1.6) eV for W(110). The number of H sites per W atom is 1:1 for W(100) and >2:1 for W(110). The minimal adiabatic barrier between neighbor equilibrium sites are ≈ 0.22 eV for W(100) surface and ≈ 0.20 eV for W(110). Vibrations of hydrogen atoms reduce the effective barrier to the value ≈ 0.156 and 0.166 eV, respectively. Diffusion coefficients were estimated using activation energies and frequencies obtained, and the values calculated for W(110) are in a good agreement with experimental data. For W(100) a good agreement with other DFT calculations was obtained though the later were performed for the reconstructed surface.

## 6. Acknowledgements

The work was supported by Grant № 14.Y26.31.0008 from the Ministry of Education and Science of the Russian Federation and by National Research Nuclear University MEPhI in the framework of the Russian Academic Excellence Project (contract No. 02.a03.21.0005, 27.08.2013).

IOP Conf. Series: Journal of Physics: Conf. Series **941** (2017) 012023 doi:10.1088/1742-6596/941/1/012023

**References**

[1] Grimleyt T B, Torrini M 1973 *J. Phys. C Solid State Phys.* **6** 868
[2] Zhdanov V I, Zamaraev K I 1986 *Physics-Uspekhi* (rus) **149** (4) 635
[3] Auerbach A, Freed K F and Gomer R 1987 *J. Chem. Phys.* **6** 2356
[4] Balden M, Lehwald S, Ibach H 1994 *Physical Review Letters* **73** (6) 854
[5] Meyerherm H L et al. 2001 *Surface Science* **475** 103
[6] Ammi D et al. 2004 *Surface Science* **554** 60
[7] Henriksson K O E et al. 2006 *Surface Science* **600** 3167
[8] Nojima A, Yamashita K 2007 *Surface Science* **601** 3003
[9] Heinola K, Ahlgren T 2010 *J. Appl. Phys.* **107** 113531
[10] Johnson D F, Carter E A 2010 *J. Mater. Res.* **25** (2) 315
[11] Lu G-H, Zhou H-B, Becquart C S 2014 *Nucl. Fusion* **54** 086001
[12] Kolasinski R D et al. 2015 *Journal of Nuclear Materials* **463** 1053
[13] Cai L et al. 2002 *Phys. Rev. Lett.* **88** 226105
[14] DiFoggio R and Gomer R 1982 *Phys. Rev. B* **25** 3490
[15] Degtyarenko N, Pisarev A 2015 *Physics Procedia* **71** 30
[16] Degtyarenko N, Pisarev A 2016 *Journal of Physics: Conference* Series **748** 012010

# The model of thermal field emission from tungsten fuzz

**D Sinelnikov, D Bulgadaryan, V Kurnaev and M Lobov**

National Research Nuclear University MEPhI (Moscow Engineering Physics Institute), Kashirskoe shosse 31, 115409 Moscow, Russia

E-mail: dnsinelnikov@mephi.ru

**Abstract.** Tungsten fuzz is a material formatted due to interaction of helium plasma with tungsten surface. This material consists of large number of thin nanostructures "growing" on the tungsten surface and having enhanced field emission properties. However, the calculation of field amplification factor from emission current-voltage characteristic gives large value which is not correlated with geometrical parameters of the nanostructures. Thermal field electron emission model that takes into account the Joule heating by emission current is presented in this work. This model predicts qualitatively similar current values corresponding to experiment and explains enhanced field amplification factor from the current-voltage characteristic.

## 1. Introduction

Tungsten is the main candidate for plasma facing material in the divertor zone of a fusion reactor due to high melting temperature, high thermal conductivity and low sputtering yield. However, its surface can be modified by "fuzz"– nanostructure, which can be formed in linear simulators and tokamaks under helium irradiation with fluence as high as $\sim 10^{25}$ $m^{-2}$, the sample temperature in the range of 1000-2000 K and energy > 20 eV. It was shown [1] that fuzzy tungsten surfaces can demonstrate enhanced frequency of arcing in linear simulator machines, but there are only few experimental works demonstrating the reason of such behaviour. Pre-breakdown current is often an initial step of the breakdown, but the mechanism of its emission is still unclear. Field emission from nanostructure filaments could be the reason of the pre-breakdown current, but in this case field amplification factor should be very high. As soon as the nanostructure consists of huge number of helium bubbles, its resistance could be high enough for intensive local heating and transition to thermal field emission (TFE). There are a lot of works devoted to calculation of the TFE current, but less of them take into account the heating of filaments by emission current.

## 2. Experimental results

Field emission from the flat cathode covered by tungsten fuzz was measured in the vacuum diode in [2]. In that work significant change of the current voltage characteristics (CVC) after microbreakdown was found. These CVCs were built in Fowler-Nordheim (F-N) plot (figure 1) and field amplification factors β were calculated from their slopes using the engineering formula:

$$\beta = -\frac{2.818 \cdot 10^{9} \cdot \phi^{3/2}}{slope}, \tag{1}$$

where $\varphi$ is work function in eV, $E$ is electric field in V/m and $I$ is emission current in A that are necessary for calculating the slope.

Photo of the fuzzy structure cross-section is shown in figure 2 and one can see a nanostructure protruding from fuzzy layer in it. The electric field strength may increase on such structure and the amplification factor $\beta$ can be roughly estimated as the height of filament divided by the radius [3]. However, there were no filaments found with $\beta>100$, while the factors $\beta$ obtained from CVC were 150 and 230 for the cases before and after breakdown, respectively. To explain this discrepancy one can suppose that the fuzzy nanostructure temperature could be increased due to Joule heating even by low field emission current and after that the main emission mechanism could be the TFE.

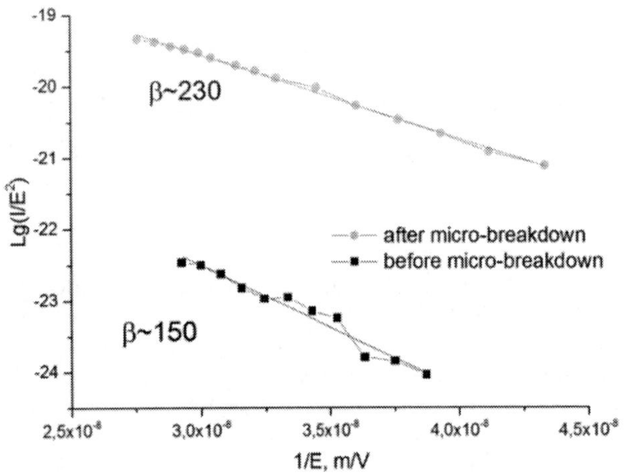


**Figure 1.** Experimental current-voltage characteristic of the tungsten fuzz sample before (blach squares) and after (grey circles) microbreakdown.

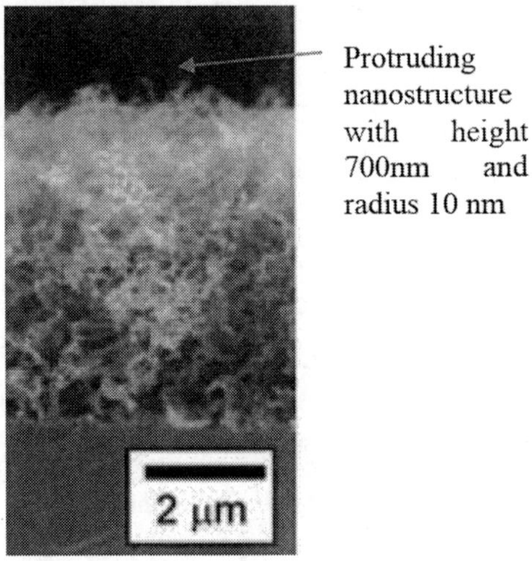


**Figure 2.** Photo of the tungsten fuzz cross-section.

## 3. Calculation model

For the TFE current calculation formula (1) from [4] was used:

$$j(E,T,\zeta)=\frac{4\pi mkTe}{h^3}\int_{W_a}^{W_l}\frac{\ln\left\{1+\exp\left[-\frac{W-\zeta}{kT}\right]\right\}dW}{1+\exp\left[\left(\frac{4}{3}\right)\sqrt{2}\left(\frac{E\hbar^4}{m^2e^5}\right)^{-\frac{1}{4}}y^{-\frac{3}{2}}v(y)\right]}+\frac{4\pi mkTe}{h^3}\int_{W_l}^{\infty}\ln\left\{1+\exp\left[-\frac{W-\zeta}{kT}\right]\right\}dW, \quad (2)$$

where $m$ – electron mass, $T$ – temperature, $k$ – Boltzmann constant, $h$ – Plank constant, $\zeta$ – Fermi energy level, $F$- electric field strength, $W$ – part of an electron kinetic energy corresponding to normal to the surface velocity. The first term in (2) strongly depends on the electric field strength while the second one depends on the temperature. The calculation model is based on the power balance (3) between Joule heating and radiation cooling of a filament.

$$I^2(E,T,S)R-\sigma T^4S_{surf.}=mc\Delta T/\Delta t, \quad (3)$$

where $I$ is emission current, $\sigma$ - the Stefan-Boltzmann constant, $T$ - temperature, $m$ - mass, $R$ - resistance, $S_{surf}$- surface area of a filament, $c$ - heat capacity of tungsten. Initially emission is only due to the electric field which leads to some heating. As soon as the TFE is significantly effective at $T>700$ K and current increase rate is rather high, energy losses due to the thermal conductivity were not taken into account.

If the temperature of the emitting filament exceeds some melting level, this will reduce its field amplification factor and, as a consequence, emitting current. In [5] it was shown that the melting level for tungsten fuzz is about 1200 K and in [6] density of the fuzz was estimated to be about 5% from bulk tungsten. Time dependence of the temperature (figure 3a) and the emission current (figure 3b) were calculated for two constant electric field strengths. Electric field strength larger than some critical value was sufficient for reaching the melting point and then emission sharply decreases, while at lower electric field the temperature and the current finally reach the constant value. The critical electric field depends on geometry of each filament because of the strong dependence of the field emission on field amplification factor.

The TFE current was calculated for 3 types of nanostructures with field amplification factors 50, 70, 120 and CVCs for them are shown in figure 4. The last point for each characteristic at maximum electric field strength corresponds to nanostructure melting and stopping of the emission. This could be the reason why the measured current is unstable and has the hysteresis effect with voltage between the electrodes increasing and decreasing. The dependences in *F-N* plot look linear, like a field emission current, and the $\beta$ factors calculated using formula (1) are 94, 133, and 233 respectively. Thus, the TFE could be one of the reasons leading to overestimation of $\beta$ by calculating a slope from CVC in *F-N* plot.

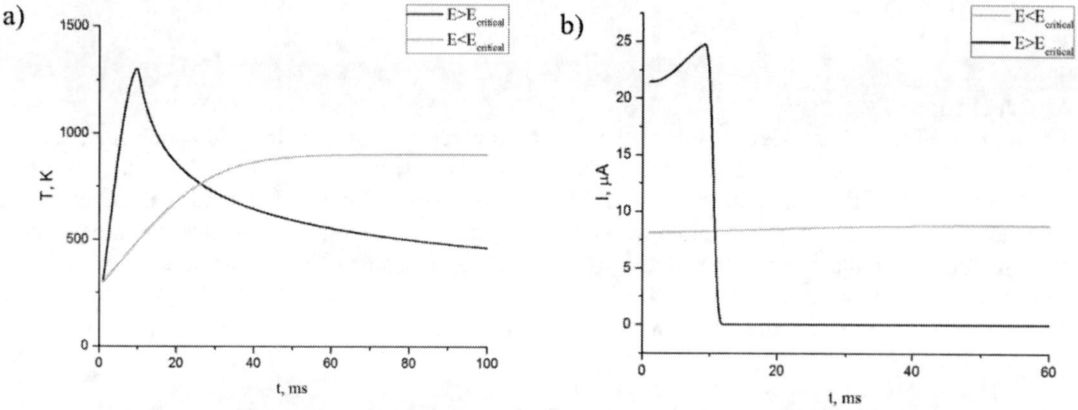


**Figure 3.** Evolution of the temperature (a) and emission current (b).

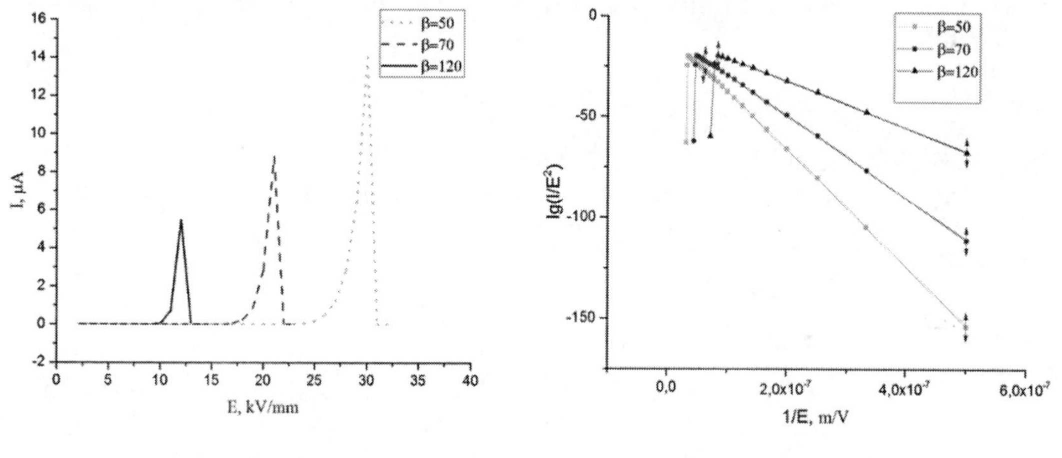


**Figure 4.** Current-voltage characteristics simulated in the model for field amplifications factors 50, 70 and 120.

In figure 5 experimental and simulated CVCs for the tungsten fuzz are shown. Each CVC was simulated for $10^6$ nanostructures that corresponds to 1% of nanostructures on the experimental emission area. Scanning electron microscope analysis detect similar number for protrude nanostructures on the average. Simulation was made for $\beta = 50$ and the currents in order of magnitude are rather close to experimental values. Typical current spread due to current instability is shown on the experimental curve.

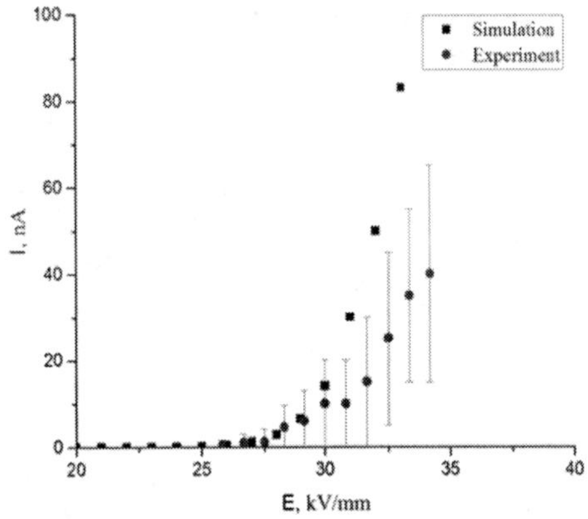


**Figure 5.** Experimental and simulated current-voltage characteristics for tungsten fuzz.

## 4. Conclusion

The numerical model was developed for thermal field electron emission from tungsten fuzz. Emission starts only due to electric field and then structures are heated according to the Joule law and cooled due to radiation. If the temperature reaches 1200K, then melting starts, and field amplification factor decreases, that usually stops the emission. Currents simulated using the model are comparable with experimental currents within the order of magnitude. Current-voltage characteristics in Fauler-Nordheim plot are linear as well as it should be for the field emission current. However, an attempt of calculation of the field amplification factor from the slope of them gives approximately twice overestimated value.

## Acknowledgments

Authors wish to acknowledge Russian President Grant for young scientist MK-3583.2017.8 for supporting this work and also the financial support from National Research Nuclear University MEPhI in the framework of the Russian Academic Excellence Project (contract No. 02.a03.21.0005, 27.08.2013).

## References

[1] Kajita S et. al. 2013 *Fusion Engineering and Design* **88** 2842
[2] Sinelnikov D et. al., *J. Phys.: Conf. Ser.* **748**(1) 012012
[3] Mesyats G *"Cathode phenomena in a vacuum discharge: the breakdown, the spark, and the arc"* (Nauka, 2000)
[4] Murphy E, Good R 1956 *Physical Review* **102** 1464
[5] Yajima M et. al. 2014 *J. Nucl. Mat.* **449** 9
[6] Baldwin M J, Doerner R P 2010 *J. Nucl. Mater.* **404** 165

# Modelling of lithium transport and its influence on the edge plasma parameters in T-15MD tokamak

**A A Pshenov[1,2] and A S Kukushkin[1,2]**

[1] National Research Center "Kurchatov Institute", Akademika Kurchatova pl. 1, 123182, Moscow, Russian Federation
[2] National Research Nuclear University MEPhI (Moscow Engineering Physics Institute), Kashirskoe sh. 31, 115409, Moscow, Russian Federation

E-mail: pshenov.andrey@gmail.com

**Abstract.** A possible application of a lithium emitter-collector scheme to the T-15MD tokamak is considered. 2D simulations with the SOLPS4.3 code package indicate that the outer midplane is the optimal position for the lithium emitter. In the magnetic configuration analysed, lithium deposition is primarily localized in the upper outer divertor. Lithium radiation enhancement due to non-coronal effects for the case studied is about a factor of 6. However, although much higher than normally observed for higher Z impurities (such as neon), this enhancement is clearly insufficient to provide significant energy dissipation, unless the lithium inventory in the edge becomes close to the deuterium one.

## 1. Introduction

Power exhaust, high erosion of the plasma facing components (PFC) caused by the extreme heat fluxes and probable PFC melting during the transient events, such as Type I ELMs and disruptions, are among the main challenges ITER and future tokamak-reactors will face [1]. One of the possible solutions for a steady-state tokamak-reactor or a fusion neutron source (FNS) is employing a liquid metal as the divertor material (for an overview of the different concepts proposed see [2] and references therein). The main advantages of the liquid metal concept are self-healing and virtual invulnerability to the neutron damage. One of the most mature and extensively used among the lithium technologies proposed is a capillary porous system (CPS) filled with liquid lithium [3]. In Russia, such systems have been successfully tested and routinely used on T-11M and T-10 tokamaks [4]. Now, with construction of new T-15MD tokamak at the "Kurchatov Institute" [5], a possible application of an emitter-collector lithium CPS limiter system similar to the one used on T-11M tokamak [6] should be analysed.

The first critical issue here is choosing the appropriate position for the lithium emitter that would provide as wide as possible distribution of lithium throughout the scrape-off layer (SOL). Uniform lithium distribution is important for both lithiation (wall covering with the lithium films) and maximisation of the lithium radiation power. Equally significant is determination of the location(s), where the lithium eroded from the emitter would migrate and where the most efficient position for the lithium collector would be. Enhancement of the lithium radiation due to non-coronal effects is often considered by the authors of lithium limiter/divertor concepts capable of making otherwise weak lithium radiation an efficient dissipation mechanism [7,8]. Therefore, evaluating the degree of the lithium radiation increase compared to its coronal level for the T-15MD plasma is another important objective. To answer these questions, we simulate lithium migration in the T-15MD geometry with the SOLPS4.3 code package [9].

## 2. Simulation setup

The SOLPS4.3 [9] code package consists of a 2D multi-fluid MHD code B2, which describes the transport of the energy and particles (both the main plasma and impurities) in the charged components of the plasma, coupled via the source terms to the 3D kinetic Monte-Carlo code EIRENE [10] that describes the neutral transport. In modelling, the effective rates of different atomic processes involving lithium impurity, such as electron impact ionization, recombination and electron impact excitation, must be taken into account. The data for these rates have been taken from the ADAS database [11] and approximated with the parametric fits used in the B2 and EIRENE codes. The maximum deviation of the resulting fits from the original ADAS data is within 20%.

The real geometry of the T15-MD chamber is used. The magnetic configuration considered for the first years of operation is an asymmetric double null with the upper X-point situated rather far from the inner separatrix. Therefore, the upper divertor is much less loaded compared to the "main", lower one. Up to 23 MW of the auxiliary heating power is planned for T-15MD in future [5]. In our modelling we simulate the deuterium edge plasma with the power coming from the core plasma across the separatrix, $Q_{\mathrm{SOL}} = 6\ \mathrm{MW}$, which is roughly equivalent to (for example) $Q_{\mathrm{SOL}} = 12\ \mathrm{MW}$ with 50% of this power reradiated by some seeded impurity. In simulations without lithium, the total amount of deuterium nuclei (both ions and neutrals), $N_{\mathrm{D}}^{\mathrm{edge}}$, is chosen as the second control parameter determining the edge plasma equilibrium [12]. $N_{\mathrm{D}}^{\mathrm{edge}}$ is taken equal to $6 \cdot 10^{20}$ particles, which corresponds to the onset of divertor detachment in pure deuterium plasma with the $Q_{\mathrm{SOL}}$ value selected.

Lithiation of the main chamber impedes the deuterium recycling. Following [13], we decrease the recycling coefficient from 1 to 0.93 for the cases with lithium injection. Changing the recycling coefficient affects the plasma solution significantly. The plasma density near the targets decreases noticeably. Consequently, the plasma temperature and the peak heat loads increase and the resulting divertor plasma state is now further away from detachment.

The lithium limiter is modelled with a toroidally-symmetric lithium atom source with the puffing rate of $\Gamma_{\mathrm{puff}}^{\mathrm{Li}} = 5 \cdot 10^{20}$ particles per second. Comparing the lithium influx chosen with the high performance FTU lithium limiter erosion rate calculated with the ERO code (see [14] for the details) one can see that $\Gamma_{\mathrm{puff}}^{\mathrm{Li}}$ used corresponds to the toroidal ring limiter of a roughly 1 cm width, which seems reasonable. For obvious reasons, it is much easier to insert any kind of movable limiter from the low field side (LFS) of the chamber. Therefore we limit our study to three possible lithium limiter positions – inside the lower outer divertor, inside the upper outer divertor and at the outer midplane (OMP). The recycling coefficient for lithium is set to 0 – i.e., the lithium ions and atoms stick to the wall as soon as they hit it.

## 3. Modelling results and discussion

The resulting lithium density distribution in the edge of T-15MD is shown in Figure 1 for the three lithium limiter positions examined. The low ionization potential of the neutral lithium (5.4 eV) results in a very short ionization length in the dense SOL and divertor plasmas. Hence, the lithium atoms can hardly penetrate to the main chamber SOL even through the cold plasma of the upper outer divertor, leave along the highly loaded "main", lower outer divertor. However, at the OMP, the distance the lithium neutrals need to fly to get inside the inner separatrix is very short compared to the divertor limiter positions. This results in a very wide lithium distribution, even if we do not allow for the re-erosion of lithium deposited on the main chamber walls. For this reason, in what follows, we limit ourselves to the most interesting case of the lithium limiter situated at the OMP.

To find the best possible position of the lithium collector, we compare the lithium deposition rate along the first wall with the rate of lithium erosion by the deuterium ions (we neglect lithium self-sputtering for now). For the modelled case of a rather high deuterium density, the energy of the impinging deuterium ions lies within the 10-50 eV range (allowing for the sheath acceleration). In this range, the lithium sputtering yield, $\gamma_{\mathrm{D+}}^{\mathrm{Li,coll}}$, calculated with the TRIM code [15], can be roughly approximated with a linear function. If we take into account that only 1/3 of the sputtered lithium leaves the surface as a neutral (and 2/3 leaving it as ions are immediately returned back to the surface by the sheath electric field), this function increases from 0.001 for the 10 eV deuterium ions to 0.01 for the 50 eV ions. Another factor that influences the lithium sputtering yield even more than the impinging deuterium ion energy is the surface temperature. To take the surface temperature into account we, following [16], calculate the total sputtering yield $\gamma_{\mathrm{D+}}^{\mathrm{Li,tot}} = \gamma_{\mathrm{D+}}^{\mathrm{Li,coll}} + \gamma_{\mathrm{D+}}^{\mathrm{Li,therm}}$, where

$$\gamma_{\mathrm{D+}}^{\mathrm{Li,therm}} = \frac{0.016}{\left\{k\left(T_s + 350\right)\right\}^{1/2}} \exp\left[-\frac{U}{k\left(T_s + 350\right)}\right] \tag{1}$$

Here $k$ is the Boltzmann constant, $U = 1.59$ eV is the lithium sublimation energy and $T_s$ is the surface temperature. What is left now is to estimate $T_s$. The T-15MD first wall and divertor will be made of the carbon fiber composite (CFC) and the pulse duration will be several (<10) seconds. Hence, for a crude estimate of the surface temperature we can use the well-known solution for the semi-infinite target:

$$T_s = \left(\pi \rho C_P \kappa\right)^{-1/2} \int_0^t q_w \left(t - t'\right)^{-1/2} dt' \tag{2}$$

Here $\rho$, $\kappa$ and $C_P$ are the CFC density, thermal conductivity and thermal capacity, respectively, and $q_w$ is the heat flux to the target surface. To obtain the zero-order estimate for $T_s$, we assume that the heat flux to the divertor targets equals the stationary heat load (calculated in our simulations) and lasts for $\tau_{\mathrm{pulse}} = 1$ s.

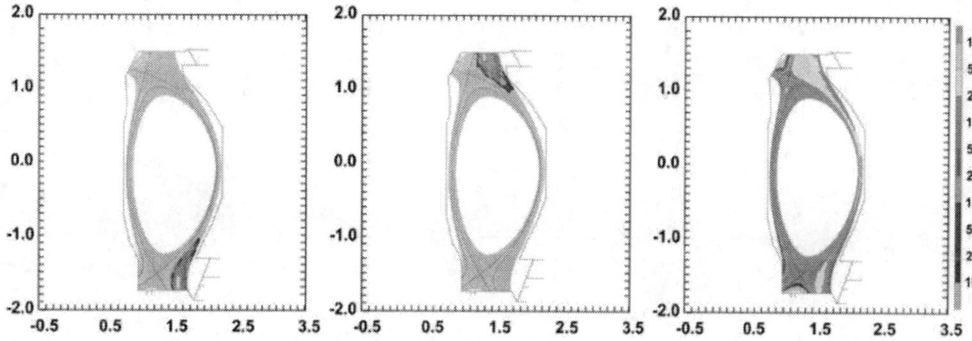


**Figure 1.** Density distribution of lithium (atoms and ions together) in the edge of T-15MD tokamak; the lithium limiter is situated (from left to right) at the lower outer divertor, at the upper outer divertor and at the outer midplane.

In Figure 2 the resulting lithium deposition profile is given. The boundary cells of the computational domain are stretched along the horizontal axis (we "go" along the main chamber wall from the lower private flux region (PFR) to the lower inner target (LIT), then to the upper inner target (UIT), etc., until the circle closes). The dashed red curve indicates $\gamma_{\mathrm{D+}}^{\mathrm{Li,tot}}$ cut off at the value of 10 roughly corresponding to the surface temperature of 500 °C. For given conditions, the sputtering yield above this value renders lithium deposition impossible. The solid blue curve shows the resulting lithium deposition rate (in the units of $10^{20}$ particles per second). It is clear from Figure 2 that the optimal position for the lithium collector is the outer (from the X-point to the shadow of the first wall) part of the upper outer divertor target. The reason for this is simple. Both the lower inner and lower outer divertors are highly loaded, therefore $\gamma_{\mathrm{D+}}^{\mathrm{Li,tot}}$ there is high and most of the lithium deposited there will be sputtered by the deuterium ions (note that we do not take into account evaporation and self sputtering that will make lithium deposition there even less probable). Because the lithium limiter is situated at the OMP, it is much easier for the lithium ions to get into the upper outer divertor compared to the upper inner one. Finally, the upper divertors are much cooler than the lower ones due to the asymmetry of the present magnetic configuration. This makes the upper outer divertor the ideal place for lithium to stick.

For given $\Gamma_{puff}^{Li}$ and the lithium limiter situated at the OMP, the resulting lithium inventory in the edge is about ~3% of the deuterium one. The lithium radiation is localized in a thin layer close to the inner separatrix, which agrees well with the existing experimental data [7] and 2D simulations [13]. Non-coronal effects for lithium are indeed significant. Close to the inner separatrix, the lithium radiation power is 10 to 30 times higher than the coronal one (see Figure 3). Overall, the increase of the radiation over the coronal one calculated for the same plasma profiles is approximately a factor of 6. This enhancement is much stronger than for the impurities conventionally used for the power exhaust, such as nitrogen or neon. For instance, the neon radiation enhancement due to the non-coronal effects rarely exceeds 20%. However, even the non-coronal effects cannot bring lithium on par with the conventional impurities with respect to the radiation power. It is negligible compared even to the deuterium radiation. For the particular case considered, the simulated lithium radiates approximately 170 kW, whereas the total deuterium radiation power is about 1.6 MW. Hence, significant dissipation through the lithium radiation channel would only be possible if the lithium inventory in the edge is of the same order as the deuterium one. This is hardly acceptable because of the core plasma contamination.

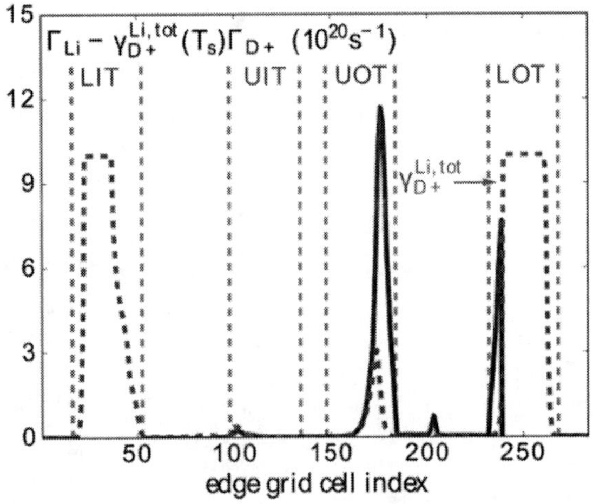


**Figure 2.** Lithium deposition rate (solid blue curve) and lithium sputtering yield by deuterium ions (red dashed curve) plotted along the computational domain boundary.

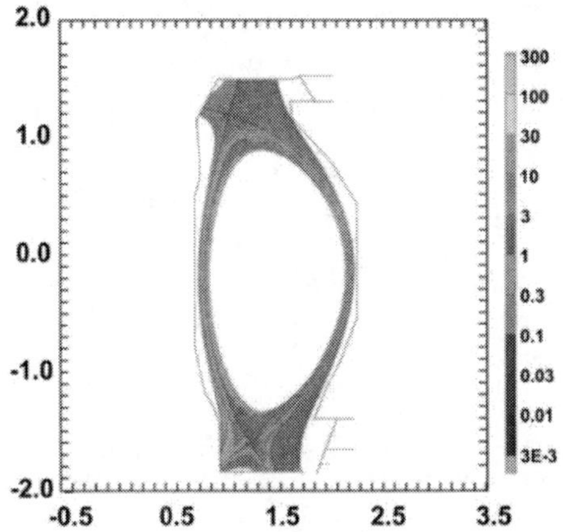


**Figure 3.** Ratio of calculated lithium radiation power to its coronal value, calculated based on the computed lithium distribution and background plasma profiles.

## 4. Conclusions

Simulations of lithium migration in the edge plasma of T-15MD tokamak have been performed with the SOLPS4.3 code package. For this purpose, the effective rates of different lithium atomic processes taken from the ADAS database were incorporated into the code. The simulations show that the lithium emitter positioned at the outer midplane provides a wide lithium distribution throughout the SOL and is optimal for wall lithiation and maximization of the lithium radiation. The emitted lithium is deposited mainly at the upper outer divertor and should be collected there. The non-coronal radiation enhances the lithium radiation significantly, by approximately a factor of 6. However, this enhancement is still insufficient for efficient energy dissipation with lithium radiation, unless the lithium inventory in the edge is of the same order as the deuterium one. This would result in a

IOP Conf. Series: Journal of Physics: Conf. Series **941** (2017) 012025 doi:10.1088/1742-6596/941/1/012025

significant core plasma contamination. Therefore, seeding the edge plasma with higher Z impurity is still required for the power exhaust during operation with the lithium emitter-collector system.

**Acknowledgements**
This work was supported by the Russian Science Foundation grant 15-12-30027.

**References**

[1] Loarte A *et al* 2007 *Nucl. Fusion* **47** S203
[2] Tabarés F L 2016 *Plasma Phys. Control. Fusion* **58** 014014
[3] Evtikhin V A *et al* 2002 *Plasma Phys. Control. Fusion* **44** 955
[4] Mirnov S V *et al* 2011 *Nucl. Fusion* **51** 073044
[5] Khvostenko P P *et al* 2015 *Fusion Eng. Des.* **98-99** 1090
[6] Mirnov S V *et al* 2015 *Nucl. Fusion* **55** 123015
[7] Mirnov S V *et al* 2006 *Plasma Phys. Control. Fusion* **48** 821
[8] Kuteev B V *et al* 2010 *Nucl. Fusion* **50** 075001
[9] Kukushkin A S *et al* 2011 *Fusion Eng. Des.* **86** 2865
[10] Reiter D, Baelmans M and Borner P 2005 *Fusion Sci. Technol.* **47** 172
[11] ADAS (http://open.adas.ac.uk/adf11)
[12] Krasheninnikov S I, Kukushkin A S and Pshenov A A 2016 *Phys. Plasmas* **23** 055602
[13] Sang C *et al* 2016 SOLPS *Nucl. Fusion* **56** 106018
[14] Ding R *et al* 2013 *J. Nucl. Mater.* **438** S690
[15] László J and Eckstein W 1991 *J. Nucl. Mater.* **184** 22
[16] Abrams T *et al* 2014 *Fusion Eng. Des.* **89** 2857–63

IOP Conf. Series: Journal of Physics: Conf. Series **941** (2017) 012026 doi:10.1088/1742-6596/941/1/012026

# The image of a nanosecond laser plasma in its own optical radiation

**A A. Fronya[1,2,*], N G Borisenko[1], V N Puzyrev[1], A T Sahakyan[1], A N Starodub[1] and O F Yakushev[1]**

[1] P.N. Lebedev Physical Institute of the Russian Academy of Sciences, Moscow, Russia
[2] National Research Nuclear University MEPhI (Moscow Engineering Physics Institute), Kashirskoe shosse 31, 115409 Moscow, Russia

E-mail: *nastya708@yandex.ru

**Abstract**. The results of experiments on the interaction of nanosecond laser radiation (wavelength of 1.06 μm and a radiation power density of $10^{12}$–$10^{13}$ W/cm$^2$) with targets from various materials (Cu, $(C_2H_4)_n$, TAC) are presented in the paper. In the experiments images of the plasma in own optical radiation in the wavelength range 0.4–1.1 μm were obtained. In one shot of laser images at wavelengths corresponding to the radiation of the harmonics $2\omega_0$, $3/2\omega_0$, $5/2\omega_0$, and at the frequency of laser radiation $\omega_0$ were recorded. Using the obtained images the spatial characteristics of the radiating regions of the plasma, as well as the radiated energy for each of the harmonics, were estimated.

## 1. Introduction

At laser-matter interaction it is essential problem to effectively expend such highly organized form of energy as laser radiation. The scattering processes lead to the loss of part of the laser radiation energy. In modern studies, there is a fundamental task of comprehensive study of the processes of harmonic generation, Raman scattering, stimulated Raman scattering (SRS), and stimulated Brillouin scattering (SBS) in a plasma [1–5].

In the plasma formation process the laser radiation energy is absorbed in a plasma region with a critical electron density $n_c$. Exactly in this region radiation scattering is occurring at the fundamental laser frequency $\omega_0$ and integer harmonics, for example, $2\omega_0$ are generating [6]. Another important region for the interaction of laser radiation with the plasma corona is the region with an electron density $n_c/4$. In its vicinity, the frequency of the laser is close to twice the plasma frequency, and it is possible to resonantly excite plasma waves, and as a consequence, an anomalous absorption of laser energy in plasma occurs. At the same time, due to the nonlinear interaction in this region, the generation of half-integer harmonics $1/2\omega_0$, $3/2\omega_0$, $5/2\omega_0$ are possible. In the region $n_c/4$ the main process responsible for the emission of harmonics is Raman scattering [7], wherein emission occurs at the frequencies $1/2\omega_0$, $3/2\omega_0$. In the next order in the pump wave amplitude, the Raman scattering process describes radiation emission at a frequency $5/2\omega_0$. Also in same region of the plasma the developing of the stimulated Raman scattering can lead to the direct generation of the harmonics $1/2\omega_0$, $3/2\omega_0$ in the form of the Stokes and anti-Stokes scattering components [7].

On the one hand, in studies on inertial confinement fusion the stimulated Raman scattering is potentially dangerous, and can lead to an acceleration of electrons, and premature heating of the irradiated target, on the other hand, it is potentially as a beneficial mechanism for optical pulse compression in plasma-based Raman amplifiers [7–9].

In case the wavelength laser irradiated the matter is 1.06 μm, the harmonics at frequencies $2\omega_0$, $3/2\omega_0$, $5/2\omega_0$ generated in the plasma fall into the optical range of the wavelengths of 0.4–1.1 μm. The most important characteristics of harmonics radiation generated in the plasma are spatial localization and spectral distribution. The study of the emission of harmonics emergence region makes it possible to diagnose the dynamics of the plasma region with a critical and a quarter of the critical electron densities $n_c$ and $n_c/4$.

## 2. Experimental data

Experiments on the interaction of laser radiation with matter were performed on the "Channel-2" facility [10]. The source of radiation was a neodymium-glass laser operating in a multimode regime. In experiments, the parameters of laser radiation were the following: pulse duration 2.5 ns, spectral width 20 Å, laser output energy 1–100 J, laser power density at target surface $10^{12}$–$10^{13}$ W/cm$^2$, focal spot size ~350 μm. The used targets were metal foils made of copper Cu, polyethylene $(C_2H_4)_n$ in the form of a film, a film of cellulose triacetate (TAC), and low density volume-structured targets from TAC. The density of the films from the TAC was ~1300 mg/cm$^3$, while the density of low-density target TAC varied from 2 to 10 mg/cm$^3$.

A four-frequency polarization microscope was used to record the image of laser plasma in the optical range 0.4–1.1 μm [11]. Microscope allows to record four images of laser plasma with the same spatial resolution the specified spectral range. The diagnostic channel with a microscope was located at an angle of 18° to the axis of the heating laser beam, measurements were made at a solid angle corresponding to the aperture of the microscope, and integrally over the entire time of the plasma emission. Laser radiation came along the normal to the plane of the target. For one shot of the laser, a plasma image was recorded at four wavelengths corresponding to the plasma radiation at the frequencies of the harmonics $2\omega_0$, $3/2\omega_0$, $5/2\omega_0$ and at the fundamental frequency $\omega_0$ (at wavelengths 0.53, 0.71, 0.42 and 1.06 μm, respectively). Also used microscope allows to estimate the size of the emission region, and the amount of energy radiated by the plasma at the harmonic frequency into the solid angle of the recording channel.

Figures 1–4 show the plasma images obtained by means a four-frequency polarization microscope for the investigated targets: A – low-density TAC target, B – film TAC target with density 1300 mg/cm$^3$, C – polyethylene $(C_2H_4)_n$ film, D – metal foils made of copper.

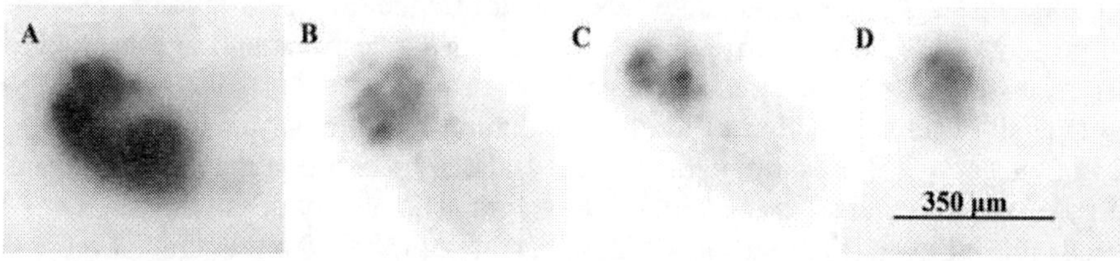


**Figure 1.** Images of plasma emission on fundamental frequency $\omega_0$.

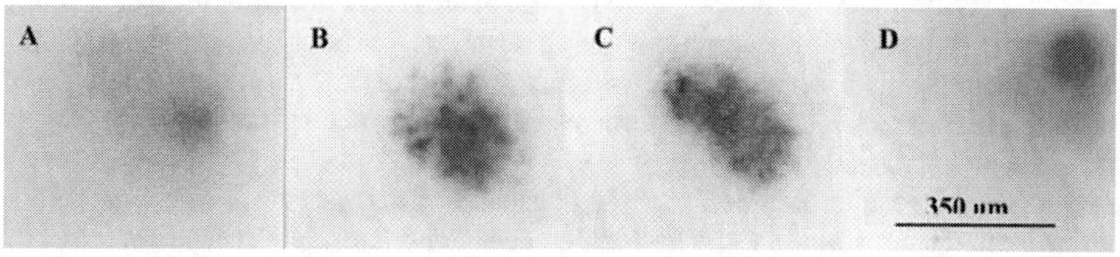


**Figure 2.** Images of plasma emission on harmonic frequency $2\omega_0$.

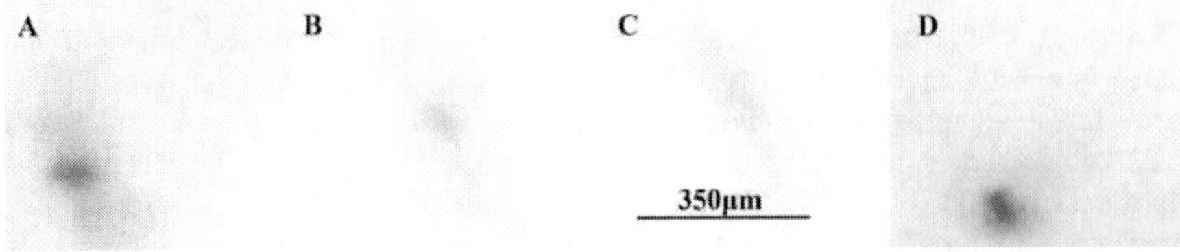


**Figure 3.** Images of plasma emission on harmonic frequency $3/2\omega_0$.

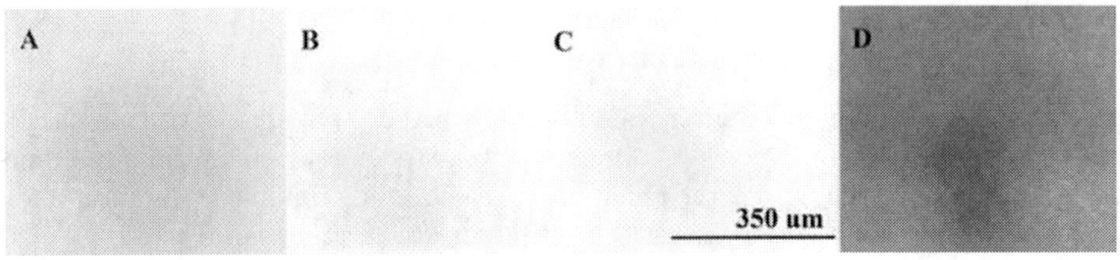


**Figure 4.** Images of plasma emission on harmonic frequency $5/2\omega_0$.

## 3. Discussion of experiments

For all types of targets, the radiation of the harmonics $2\omega_0$, $3/2\omega_0$, $5/2\omega_0$ and at the fundamental frequency $\omega_0$ was registered. A study of the plasma image with spatial resolution at the frequencies of the harmonics and the fundamental frequency showed that in the observation direction the emission distribution of the plasma has a complex character. According to plasma images, the localization size of the emitting regions was estimated. For emission area of the plasma the inhomogeneity intensity regions exist. And these areas can take different shapes and sizes with each laser shot.

In experiments, the plasma emission areas at frequencies $2\omega_0$ and $\omega_0$ in size correspond to the size of the focusing spot of laser radiation. As a rule, for all types of studied targets an inhomogeneous structure of the plasma emission in second-harmonic radiation was obtained in the experiments. This inhomogeneity indicates fluctuations in the electron density $n_e$ in the formed plasma, and is probably related to the formation in the plasma of regions corresponding to the most optimal second-harmonic generation conditions due to linear transformation process. At a frequency of $3/2\omega_0$, the characteristic size of the radiation localization is smaller than the size of the focusing spot of the laser radiation and depends on the target type and the laser radiation energy. For all types of studied targets, the emission area at the frequency $3/2\omega_0$ was the smallest region in comparison with the emission areas at the harmonic frequencies $2\omega_0$, $5/2\omega_0$ and at the fundamental frequency $\omega_0$.

Table 1 presents the results of processing plasma images by estimating the energy of the harmonics $2\omega_0$, $3/2\omega_0$, $5/2\omega_0$ and at the fundamental frequency $\omega_0$.

The coefficient of transformation of the laser radiation into the plasma harmonics was determined as the ratio of the energy of the harmonic recorded at the solid angle of diagnostic channel registration to the energy of the laser radiation incident on the target. The energy density emitted at a frequency of $3/2\omega_0$ exceeds this index for the frequency $5/2\omega_0$, in spite of the fact that the radiation region for $3/2\omega_0$ is much smaller than the radiation region for $5/2\omega_0$. This is because ehe process of harmonic generation $5/2\omega_0$ is of the next order of smallness in comparison with Raman scattering, and the harmonic intensity $5/2\omega_0$ should be less than the radiation intensity of the harmonic $3/2\omega_0$ [7].

Despite the small volume density of the volume-structured TAC target, the energy of the scattered radiation at frequencies $2\omega_0$, $3/2\omega_0$, $5/2\omega_0$, recorded in experiments, is comparable with the data for solid targets.

If we compare the results obtained for low-density TAC target and TAC films with a density of ~1300 mg/cm$^3$, then it turns out that the harmonic energy and harmonics transformation coefficient are large for targets with a density of 1300 mg/cm$^3$. According to the formulas for the transformation coefficient of the heating radiation into the second harmonic, the transformation coefficients are proportional to the scale of the density variation in the vicinity of the critical density [6], other parameters can be considered identical. Since for TAC target films the coefficient of transformation into the second harmonic is greater than for low-density TAC target, it can be concluded that the value of the characteristic size of density inhomogeneity for TAC target with a density of 1300 mg/cm³ is greater than for low-density TAC target.

**Table 1.** Results of energy estimating of the harmonics $2\omega_0$, $3/2\omega_0$, $5/2\omega_0$ and at the fundamental frequency $\omega_0$.

| Type of target → | Low-density TAC | TAC film | $(C_2H_4)_n$ | Cu |
|---|---|---|---|---|
| Target density, mg/cm³ → | 9 | 1300 | 950 | 8900 |
| Laser radiation energy, J → | 17.8 | 28.8 | 22.6 | 18.7 |
| Harmonics ↓ | Energy, scattered in solid angle of diagnostic channel, J | | | |
| $2\omega_0$ | $3.1\cdot10^{-11}$ | $5.7\cdot10^{-10}$ | $6.8\cdot10^{-10}$ | $7\cdot10^{-11}$ |
| $3/2\omega_0$ | $1.5\cdot10^{-9}$ | $10^{-9}$ | $5.4\cdot10^{-10}$ | $2.6\cdot10^{-9}$ |
| $5/2\omega_0$ | $2.6\cdot10^{-10}$ | $1.8\cdot10^{-10}$ | $1.5\cdot10^{-10}$ | $3.1\cdot10^{-10}$ |
| $\omega_0$ | $2.8\cdot10^{-6}$ | $1.9\cdot10^{-6}$ | $1.3\cdot10^{-6}$ | $2.3\cdot10^{-5}$ |
| Harmonics ↓ | Energy density, scattered in solid angle of diagnostic channel, J/cm² | | | |
| $2\omega_0$ | $4.2\cdot10^{-7}$ | $8.2\cdot10^{-7}$ | $9.7\cdot10^{-7}$ | $3.9\cdot10^{-7}$ |
| $3/2\omega_0$ | $1.8\cdot10^{-5}$ | $6.5\cdot10^{-6}$ | $3.4\cdot10^{-6}$ | $1.6\cdot10^{-5}$ |
| $5/2\omega_0$ | $2\cdot10^{-7}$ | $1.1\cdot10^{-7}$ | $1.1\cdot10^{-7}$ | $5.6\cdot10^{-8}$ |
| $\omega_0$ | $3.6\cdot10^{-3}$ | $3.9\cdot10^{-3}$ | $4.1\cdot10^{-3}$ | $7.8\cdot10^{-2}$ |
| Harmonics ↓ | Transformation coefficient of laser radiation into the harmonics | | | |
| $2\omega_0$ | $3.8\cdot10^{-12}$ | $2\cdot10^{-11}$ | $3\cdot10^{-11}$ | $3.7\cdot10^{-12}$ |
| $3/2\omega_0$ | $2.3\cdot10^{-10}$ | $3.6\cdot10^{-11}$ | $2.4\cdot10^{-11}$ | $1.4\cdot10^{-10}$ |
| $5/2\omega_0$ | $2.1\cdot10^{-11}$ | $6.3\cdot10^{-12}$ | $6.8\cdot10^{-12}$ | $1.7\cdot10^{-12}$ |
| $\omega_0$ | $1.8\cdot10^{-7}$ | $6.5\cdot10^{-8}$ | $5.6\cdot10^{-8}$ | $1.2\cdot10^{-6}$ |

For the TAC targets, a theoretical estimate of the value of the transformation coefficient of the heating radiation into the second harmonic was made, assuming that the $2\omega_0$ generation process is due to a linear transformation. This estimate gives a value of $\sim10^{-6}$, which is much higher than the value obtained experimentally. Here it should be noted that the harmonics radiation was registered in the limited solid angle of the diagnostic channel, and at an angle to the axis of the heating laser radiation.

## 4. Conclusion

As a result of the experiments on the nanosecond laser radiation interaction with metal foils of copper Cu, polyethylene $(C_2H_4)_n$ in the form of a film, films of cellulose triacetate (TAC) and low-density bulk-structured targets from TAC, Components corresponding to radiation at harmonic frequencies $2\omega_0$, $3/2\omega_0$, $5/2\omega_0$ for all types of targets studied. Radiation of the plasma at harmonic frequencies is localized in different spatial regions. An analysis of the experimental data indicates the emergence and development of parametric instabilities in the plasma regions, both with the critical density of electrons and a quarter of the critical one.

**Acknowledgments**
This work has been partly supported by National Research Nuclear University MEPhI in the framework of the Russian Academic Excellence Project (contract No. 02.a03.21.0005, 27.08.2013).

IOP Conf. Series: Journal of Physics: Conf. Series **941** (2017) 012026 doi:10.1088/1742-6596/941/1/012026

**References**

[1] Lindl J 1995 *Phys. Plasmas* **2** 3933
[2] Dunne M, Borghesi M, Iwase A, et al. 1995 *Phys. Rev. Lett.* **75** 3858
[3] Bugrov A E, Burdonskii I N, Gavrilov V V, et al. 2004 *Plasma Phys. Rep.* **30** 143
[4] Yaakobi O, Friedland L, Lindberg R R, et al. 2008 *Phys. Plasmas* **15** 032105
[5] Koester P, Antonelli L, Atzeni S, et al. 2013 *Plasma Physics and Controlled Fusion* **55 (12)** 124045
[6] Basov N G, Bychenkov V Yu, Krokhin O N, et al. 1979 *Kvantovaya Elektron.* **6** 1829
[7] Basov N G, Zakharenkov Yu A, Zorev, et al. 1982 *Itogi Nauki i Tekhniki. Ser. Radiotekhnika* **26**
[8] Bilak O M, Nikitin S Yu 1996 *Moscow University Bulletin, Physics and Astronomy* **4** 64
[9] Shvets G, Fisch N J, Pukhov A, and Meyer-ter-Vehn J 1998 *Phys. Rev. Lett.* **81** 4879
[10] Fedotov S I, Feoktistov L P, Osipov M V, Starodub A N, et al. 2004 *J. Russian Laser Research* **25** 79
[11] Vasin B L, Mal'kova S V, Osipov M V, et al. 2009 *Prikl. Fiz.* **6** 152

# Effect of the laser heat treatment on the formation of the gradient structures in alloys based on Fe – Cr – Ni system

**A O Andreev[1], D P Bykovskiy[1], A V Osintsev[1], V N Petrovskiy[1], I I Ryashko[1], E N Blinova[2], M A Libman[2] and A M Glezer[2]**

[1] National Research Nuclear University MEPhI (Moscow Engineering Physics Institute), Department of Laser Physics, Kashirskoe highway 31, Moscow 115409, Russia

[2] Federal State Unitary Enterprise (FSUE) I.P. Bardin Central Research Institute for Ferrous Metallurgy, Moscow, Radio st., 23/9, Russia

E-mail: andreev07@mail.ru

**Abstract.** The possibility of producing gradient materials, i.e. materials with pre-set distribution of areas having fundamentally different physical and mechanical characteristics, with the help of laser heat treatment was investigated. Using as an example austenitic-martensitic alloys of iron-chromium-nickel, subjected to cold plastic deformation led to formation of martensite, we show that using laser at the temperature higher than the temperature of reverse martensite transformation leads to the formation of areas of high-strength austenite having predetermined form inside the martensite matrix. Influence of austenite areas geometry on mechanical properties of gradient material was studied.

## 1. Introduction

Martensite transformation and reverse martensite transformation in iron-based alloys with considerable temperature hysteresis between them [1,2] provide opportunities for producing alloys with different content and combination of martensite and austenite, and therefore materials with various physical and mechanical properties. Using martensite transformation and reverse martensite transformation allows obtaining materials structure and properties of which are pre-set along the article – so-called gradient materials. For instance, there are materials with pre-set distribution of macroscopic areas with considerably different magnetic and mechanical properties. In [3,4] the possibility of producing such materials on the base of austenitic-martensitic alloys of iron-chromium-nickel was considered. In order to get ferromagnetic part of such materials cold plastic deformation was used, as a result of which martensite transformation took place and led to forming of deformation-induced martensite. Austenite areas formation was obtained by heating at the temperature higher than the temperature of reverse martensite transformation. In [3,4] laser irradiation was used as the heat source. Using lasers for heat treatment leading to austenite formation is necessary due to high heating rate which provides, according to [5], austenite stability in the wide range of temperatures including room temperature. Paper [6] investigates connection between mechanical properties and structure of austenite, formed by laser heating. Microhardness measurements of austenite reached 3000 MPa, which is lower than that of martensite (5500 MPa), but considerably higher than that of “ordinary” austenite, formed in alloy at hot deformation or long-term holding at temperatures higher than the temperature of reverse martensite transformation. The observed hardening of austenite is connected with the process of martensite-austenite transformation at ultra-fast heating, which occurs by a shearing mechanism. Electron-microscopic research [6] shows that transformations occurring by the same mechanism lead to keeping in austenite structure defects typical for deformation-induced martensite that provides its high strength. Using cyclic laser treatment (cycle of heating and cooling) leads to additional hardening of austenite that, according to [7], is connected to splitting its particles to nanocrystalline state.

This paper continues research of laser heat treatment influence on austenite formation in iron-chromium-nickel austenitic-martensitic alloys; in particular, direct mechanical tests were conducted by acquiring stress-deformation curves (stress-strain curves) for samples of gradient materials with various geometry of austenite areas.

## 2. Materials and methods of study

For our study we chose iron-chromium-nickel austenitic-martensitic alloys, their composition is given in table 1.

**Table 1**. The chemical composition of studied alloys.

| The content of alloying elements, masses (%) | | | | | | |
|---|---|---|---|---|---|---|
| Cr | Ni | V | Mn | Si | C | Fe |
| 16.5 | 7.5 | 0.04 | 1.0 | 0.47 | 0.07 | Balance |

Alloys of given composition were smelted in open induction furnace with using pure burden materials and applied on the billet 25 mm thick. The billet was firstly exposed to hot deformation in the temperature range of 1000 – 1200 °C on the sheet 4 mm thick and then to the cold deformation with reduction of 75% on the sheet 1 mm thick. After hot deformation the alloy had the structure of paramagnetic austenite with relatively low microhardness of 1500 MPa. During the cold deformation there was intensive formation of ferromagnetic austenite, and its microhardness increased to 5500 MPa, and there appeared saturation magnetic moment of 1.4 T. For obtaining austenite areas we used laser heating in the range of reverse martensite transformation temperatures, which were 850 – 900 °C according to magnetometrical measurements.

For the experiment on laser heat treatment we used YLS–1500–SM single-mode Ytterbium-doped continuous fiber laser (manufactured by IPG Photonics), its basic technical characteristics are given in table 2. For making local areas having pre-set geometry mask-plate 3 mm thick made of D16T alloy was put on the top of the sample. Focused laser radiation reached stack of the sample and mask-plate from above and in scanning mode it covered all the surface of the sample. Mask-plate protected the areas of the sample, which were not exposed to the radiation and therefore were not heated.

**Table 2.** Main technical characteristics of the laser.

| № | Characteristics | Type / Value |
|---|---|---|
| 1 | Mode of operation | continuous-wave |
| 2 | Maximum average power (W) | 1500 |
| 3 | Power tenability (%) | 10 – 100 |
| 4 | Polarization | random |
| 5 | Central wavelength range (nm) | 1070 ± 10 |
| 6 | Laser linewidth (nm) | 4.55 |
| 7 | Maximum power consumption (kW) | 4.2 |

In order to form austenite areas in martensite matrix of alloy the sample was placed under a mask-plate with a notch, geometrically corresponding to a processed zone. There were three types of notches: round-shaped area (G0), cross stripe (G1) and inclined stripe (G2). The notches in the mask-plates were made from the condition of equality of the zones areas of samples, which were laser heat treated. In the course of experiment studying the influence of laser heat treatment modes on mechanical properties of the gradient material samples the radiation treatment of the local zones was done one time or five times. Every cycle "heating-cooling" included laser treatment of samples local zones from two sides. That was necessary for phase transformation of martensite to austenite throughout the thickness of the sample with minimal tapering of the formed austenite area. The value of the austenite remained constant for all the samples regardless the geometry of the treated zone and it was 0.3 $cm^3$.

Table 3 shows main parameters of laser heat treatment mode, formed austenite areas of various geometry, and also the temperatures reached by the sample surface during the treatment.

**Table 3**. Parameters of laser heat treatment mode.

| Geometry of the processed area | G0 | G1 | G2 |
|---|---|---|---|
| Linear speed of sample surface scanning by laser beam (m/min) | | 60 | |
| Laser radiation power (W) | | 350 | |
| Density of the lines during scanning (line/mm) | | 20 | |
| Number of passes | | 6 | |
| Power density ($W/cm^2$) | | $7 \cdot 10^6$ | |
| Surface temperature (°C) | 875 | 850 | 900 |

To establish the fact of austenite formation during the laser heat treatment we conducted magnetooptic study, which allowed constructing spatial distribution of magnetization in the samples, exposed to laser treatment [8].

For uniaxial static tensile testing with using of the samples, exposed to plastic deformation and later laser heating, electromechanical tension testing machine Instron 5982 was used. In the course of experiment the values of loads applied to the samples and shifts of grip relatively to the initial position were recorded with high precision. These data allowed plotting the stress-strain curves for samples with various geometry of the treated areas.

Using method of measuring tangential displacements with the help of the Instron 5982 setup spatial distribution of longitudinal and transverse relative transformations on the surface of the samples containing various types of austenite areas was visualized.

Samples after 5 cycles of laser treatment were studied according to the method of transmission electron microscopy. Blanks for thin foils with thickness ~ 1 mm were cut so as to make the research zone correspond to the austenite zone. Samples preparation for transmission electron microscopy was carried out by electropolishing method with "window" technique at the temperature of 20 – 50 °C and voltage ~ 20 V. The study of fine structure was conducted at the accelerating voltage 120 kV. The analysis of microstructure was performed by comparing bright-field and dark-field electron microscope images and corresponding micro-diffraction images.

## 3. Experiments results

Figure 1 shows curves "Stress-Deformation" for samples of gradient materials with various geometry of austenite areas formed during laser heat treatment, and also stress-strain curve for cold deformation sample. The data shows that forming austenite areas leads to decreasing of ultimate strength compared to cold deformation sample. At the same time the maximum decrease is observed for round-shapes austenite samples, while the minimum decrease is observed for the transverse ones. Nevertheless, strength decrease that takes place during austenite formation is not so great compared to hot deformation induced austenite or austenite formed during heat treatment of cold-deformed samples at the temperature 850 °C during one hour.

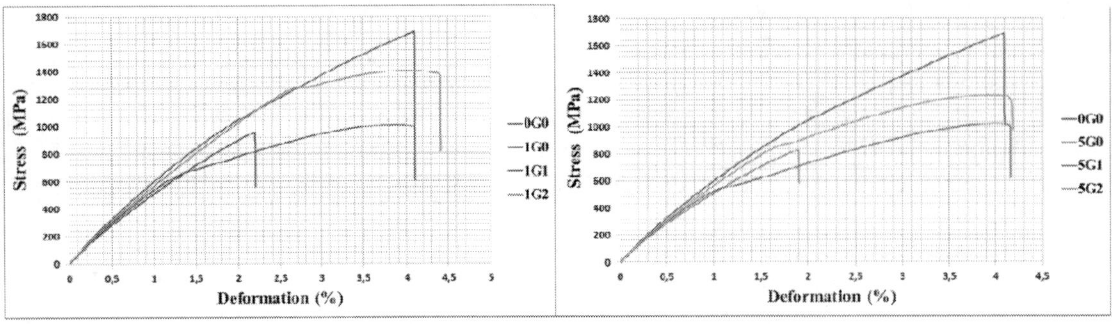


**Figure 1.** Graphical dependencies "Stress – Deformation" in the samples of different kinds: "left" is for 1 cycle "heating-cooling", "right" is for 5 cycles "heating-cooling".

Table 4 shows measurement results of mechanical properties of gradient materials with different austenite area geometry formed at one cycle "heating-cooling".

**Table 4**. Mechanical properties of gradient materials with different austenite area geometry.

| Sample structure | Austenite area geometry | Ultimate strength (MPa) | Microhardness (MPa) |
|---|---|---|---|
| Deformation martensite | – | 1700.95 | 5500 |
| Martensite with austenite areas | G0 | 956.01 | 3000 |
| | G1 | 995.14 | |
| | G2 | 1372.63 | |
| Ordinary austenite | – | 520.25 | 1500 |

Figure 2 shows electron microscope images of austenite microstructures in 5G0 and 5G1 samples after 5 cycles of laser treatment. Depending on geometry and location of irradiated zones in the samples, microstructure of austenite phase has different characteristics. 5G0 sample shows clearly expressed dislocation substructure with stacking faults, whereas 5G1 sample shows deformation twins.

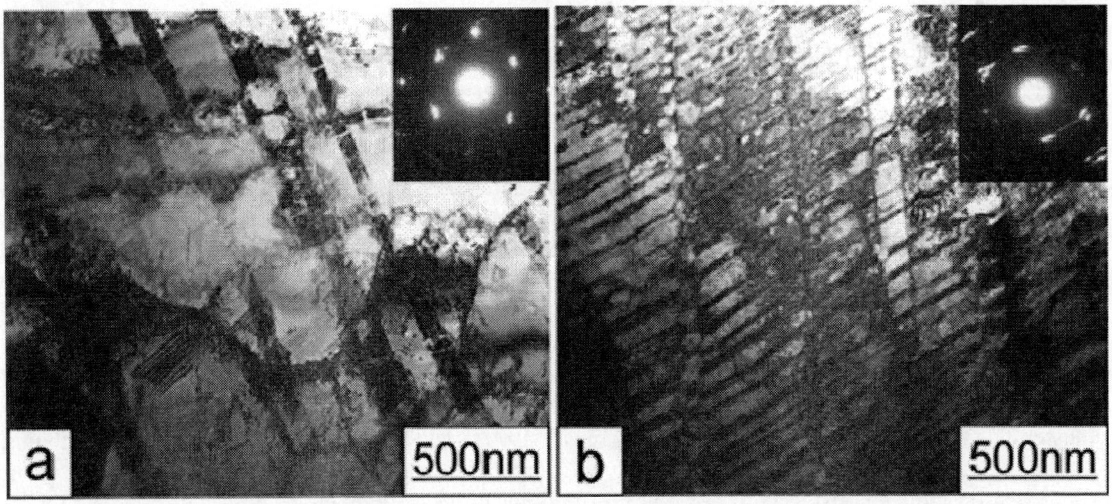


**Figure 2.** Electron microscope image of 5G0 and 5G1 samples microstructure after 5 cycles of laser treatment: a – austenite microstructure in 5G0 sample with corresponding micro-diffraction; b – austenite microstructure in 5G1 sample with corresponding micro-diffraction.

Data given in table 4 show considerable hardening of austenite obtained by laser heat treatment compared to ordinary austenite formed by hot deformation or annealing cold-deformed austenite sample at the temperature of 850 °C during 1 hour.

Differences in ultimate strength values of samples with different austenite geometry areas are connected with different structures of internal stresses, appearing during alloy deformation.

**Acknowledgments**

The study was carried out with partly financial support provided by the Russian Foundation for Basic Research (Grant № 14-08-00735).

This work was supported by National Research Nuclear University MEPhI in the framework of the Russian Academic Excellence Project (contract No. 02.a03.21.0005, 27.08.2013).

**References**

[1] Saragadze V V and Uvarova A M 2013 *Hardening and properties of austenitic steels* (Ekaterinburg: IFM) p 720
[2] Lobodyk V A and Estrin E I 2009 *Martensitic transformations* (Moscow: Physmatlit) p 351
[3] Galkin M P, Libman M A and Estrin E I 2014 *Inorganic Materials: Applied Research* **3** 25
[4] Andreev A O, Galkin M P, Libman M A, Mironov V D, Petrovskii V N and Estrin E I 2014 *Metal Science and Heat Treatment* **1** 50
[5] Libman M A and Estrin E I 2014 *Steel* **10** 67
[6] Blinova E N, Glezer A M, Libman M A and Estrin E I 2014 *News of universities. Physics* **4** 8
[7] Andreev A O, Blinova E N, Libman M A and Makushev S Y 2016 *Inorganic Materials: Applied Research* **10** 37
[8] Andreev A O, Brovin M A, Petrovskii V N and Protasov E A 2016 *Technical Physics. The Russian Journal of Applied Physics* **61** p 591

# Modification of polyacrylonitrile carbon fibers by high-fluence ion irradiation

**N N Andrianova[1], A M Borisov[1,5], V A Kazakov[1,2], E S Mashkova[3], M A Ovchinnikov[1,2], S V Savushkina[2] and N M Chernenko[4]**

[1] Moscow Aviation Institute (National Research University), Volokolamskoe sh. 4, 125993, Moscow, Russia
[2] Keldysh Research Center, Onezhskaya 8, 125438, Moscow, Russia
[3] Skobel'tsyn Institute of Nuclear Physics, Moscow State University, Leninskie gory 1(2), 119991, Moscow, Russia
[4] NIIgraphite, Electrodnaya 8, 111524, Moscow, Russia

E-mail: anatoly_borisov@mail.ru

**Abstract.** The Raman spectroscopy has been used to analyse ion-induced modifications of carbon PAN-fiber shell due to 10-30 keV $Ar^+$ high fluence ion irradiation at normal and oblique incidence in the temperature range from RT to 400 °C. It has shown that formed in ion-induced processes of amorphization, recrystallization and crimping the modifications of PAN-fiber shell are characterized by the presence of the amorphous phase with the A peak in the Raman spectra and the increased intensity of the D peak relative to the G peak in comparison with non-irradiated fiber. Amorphous phase in the PAN-fiber shell is the highest in case of amorphization and the least at the crimping. The increased intensity of the D peak in the Raman spectra and the G peak shift towards higher frequencies during recrystallization and crimping indicates ion-induced nanostructuring of the PAN- fiber shell.

## 1. Introduction

Carbon-carbon and carbon-ceramic composite materials reinforced by carbon polyacrylonitrile (PAN) fibers are considered as priority structural materials for aerospace engineering. PAN-fibers have a core with turbostratic structure and structurally perfect textured shell. High-fluence ion irradiation leads to significant changes in surface morphology and crystalline structure of PAN-fiber shell and this procedure is considered as the modification method necessary for obtaining high-temperature strength and high heat resistance carbon-ceramic composites reinforced by the fibers with optimal adhesive interaction with composite matrix [1]. Ion irradiation of analyzed in this work carbon-carbon composite KUP-VM reinforced by VMN-4 PAN-fibers can lead to such processes as amorphization, recrystallization and development of the surface topography, in particular, the fiber crimping [2,3]. In the present work for the characterization corresponding to these processes of ion-induced modification of carbon PAN fiber surface and also its structure changes during heat treatment along with electron-microscopy, laser Raman spectroscopy has been applied.

## 2. Experimental

The heat treatment process has been studied for samples of carbon fibers from PAN-raw materials used for the production of carbon fiber KULON. The Raman spectra of KULON PAN-fiber have been compared with the Raman spectra of PAN-fibers after carbonization at a temperature of 1000 °C and after graphitization at temperatures from 1450, 1550 and 1650 °C. The study of ion-beam modification has been performed for the carbon-carbon composite KUP-VM, reinforced carbon VMN-4 PAN-fibers. The samples of PAN-fibers have been analyzed using scanning electron microscope Quanta 600 and a Horiba JobinYvon T64000 Raman spectrometer at room temperature with the laser excitation wavelength $\lambda_L = 514.5$ nm. The ion irradiation of the samples of KUP-VM by 10, 25 and 30 keV $Ar^+$ ions at ion incidence angle $\theta = 0$ and $\theta = 70°$ relative to the fiber axis have been carried out on the mass-monochromator of the Skobel'tsyn Research Institute of Nuclear Physics Moscow State University [4]. A special target holder with the water cooling was used for controlling the target temperature from RT to 400 °C during ion irradiation, which was measured by means of an alumel-chromel thermocouple with the junction attached to the irradiated side of the target away from the irradiated area. Ion irradiation was monitored via the periodic recording of the ion current and secondary electrons to determine the irradiation fluence and the ion-induced electron yield. Note that studied ion-induced processes are dependent on irradiation temperature and radiation damage level v measured in the number of displacements per atom (*dpa*) as $v = \varphi t\, \sigma_{dam}$ ($\sigma_{dam}$ is the cross section of defect formation, $\varphi$ is the incident-ion flux density; $t$ is the time of irradiation). At high fluences radiation damage level is determined by ion energy and, according to the evaluation [2], $v$ is ranging from 30 *dpa* at the energy of the $Ar^+$ ions of 10 keV to 90 *dpa* at 30 keV ion energy.

## 3. Results and discussion

It is known that Raman spectra of graphite-like materials contain two main peaks: the G peak (graphite peak) at the Raman shift $\Delta k = \lambda_L^{-1} - \lambda_R^{-1}$ close to 1580 см$^{-1}$ and the D peak due to the lattice defects at $\Delta k \approx 1350$ см$^{-1}$, see, for example [5]. The Raman spectra reflect the processes of thermal treatment of the hydrocellulose and polyacrylonitrile fibers when processing respectively to these fibers [6,7]. It also displays the evolution of the Raman spectra for PAN-fibers with increasing treatment temperature in Figure 1. The trend of growth of the relative intensity of the G peak is clearly visible with increasing the treatment temperature of PAN-fiber. The most perfect graphite fiber shell is formed at the final treatment temperature 3000 °C, see carbon fiber KULON spectra. The degree of shell perfection is characterized by the size of $L_a$ graphite microcrystallites in the basic plane. Thus, the larger the $L_a$ of the graphite microcrystallites, the smaller the relative intensity of the D peak [8]. Figure 1 also shows that the relative intensity of the D peak for the Raman spectrum of VMH-4 PAN carbon fibers embedded to composite KUP-VM is several times less than the $I_D/I_G$ for carbon fiber KULON that indicates a more perfect shell of the VMN-4 carbon fiber.

Radiation exposure can significantly change the structure of graphite-like materials up to amorphization. Ion-induced amorphous carbon in the Raman spectra is manifested as band at $\Delta k \approx$ 1500 cm$^{-1}$ [5]. Radiation impact can also greatly reduce the size $L_a$ of the graphite crystallite. In this case the dependence of the intensities $I_D/I_G$ on $L_a$ in the Raman spectra will reverse if $L_a < 2$ nm namely, the greater $L_a$, then the greater relative intensity of the D peak [8]. The peculiarities of influence of radiation exposure on the Raman spectra of graphite-like materials are fully manifested in the present work for the carbon PAN-fibers.

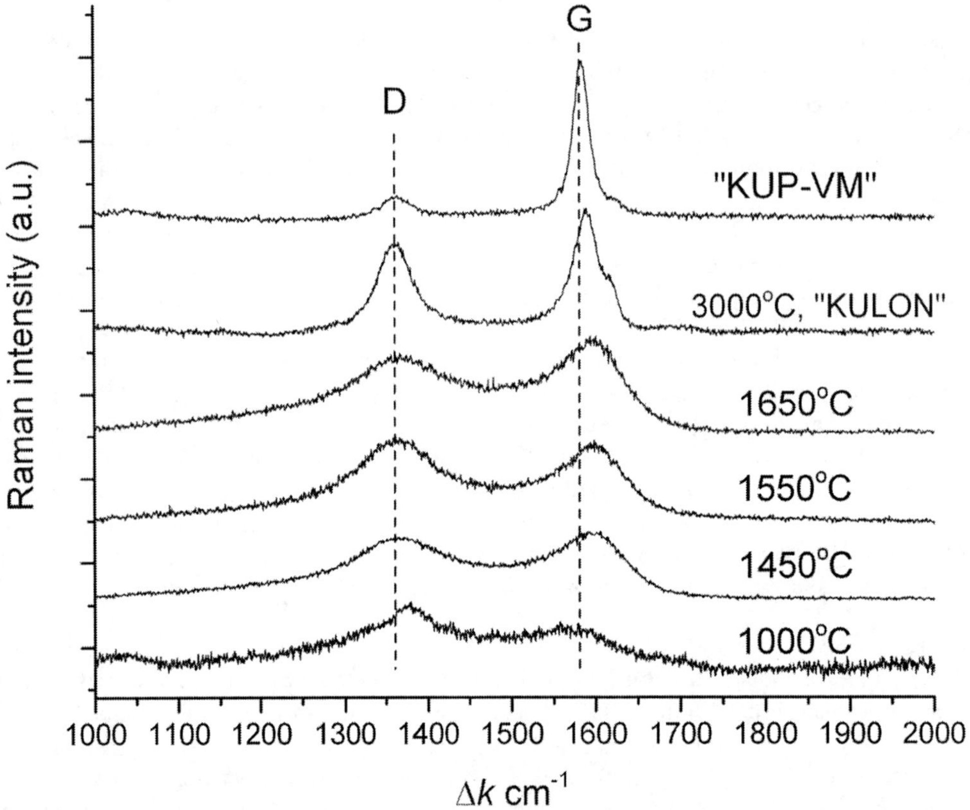


**Figure 1.** Raman spectra of PAN-fibers corresponding to different thermal treatment temperatures and VMN-4 PAN-fiber embedded in composite KUP-VM.

The Raman spectra and SEM images for the samples of the KUP-VM after irradiation, leading to different ion-induced surface modification of fibres, such as amorphization, recrystallization and recrystallization with surface crimping, are presented in Figure 2. The most typical cases of these surface modifications chosen according to the data of [2, 3], are given. Amorphization leads to a qualitative change in the Raman spectra. Namely, two peaked structure of the spectrum is not resolved, see Figure 2a. Two peaked structure of the Raman spectra is observed at target irradiation temperature $T$ larger than the temperature of dynamic annealing of radiation damage 150 – 200 °C [2,3]. The examples are the Raman spectra, corresponding to the ion energy of 10 and 30 keV at the irradiation temperature $T$ = 400 and 350 °C, respectively. Structural analysis shows that at 30 keV ion energy there occurs not only the recrystallization of the surface layer of a carbon fiber, like at the energy of 10 keV (figure 2b), but also crimping, see the SEM image in Figure 2c. To assess quantitative differences in the Raman spectra for different ion-induced modifications of PAN-fiber surface the decomposition of spectra into Gaussian components was carried. The decomposition for the irradiated samples in addition to the G and D peaks required the introduction of the A peak, taking into account radiation damage in the structure of short-range order upon amorphization of carbon materials.

The percentages of Gaussian components in the Raman spectra are shown in Table 1.

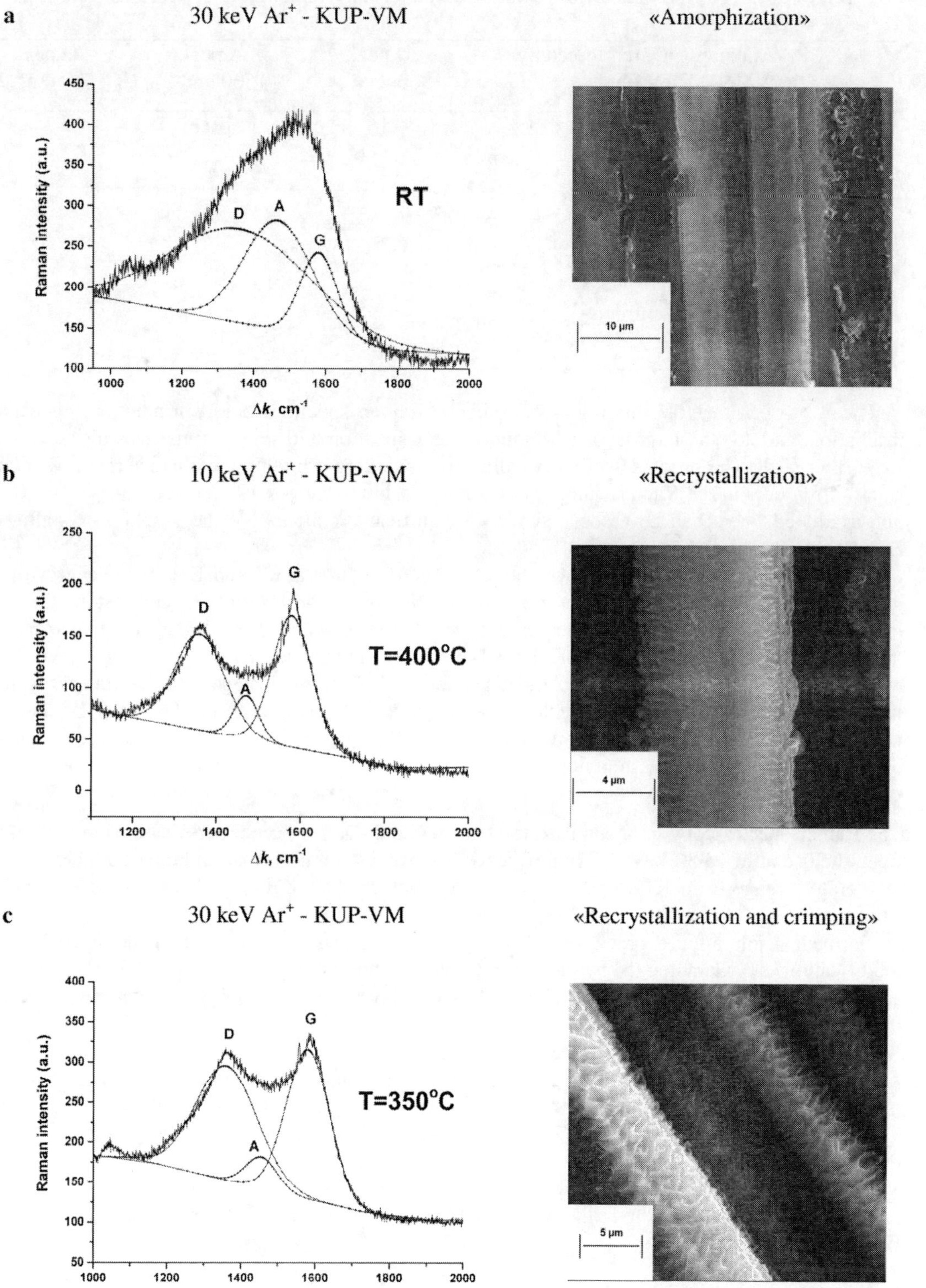


**Figure 2.** Raman spectra and SEM images of KUP-VM after ion irradiation.

**Table 1.** The percentages of Gaussian components in the Raman spectra of composite KUP-VM, reinforced by VMN-4 PAN fibers, before and after ion irradiation under various conditions.

| KUP-VM composite, reinforced by VMN-4 PAN fibers | D peak 1360, $cm^{-1}$, % | A peak ~1500 $cm^{-1}$, % | G peak 1580 $cm^{-1}$, % |
|---|---|---|---|
| Before ion irradiation | 16 | absent | 84 |
| Ion-induced amorphization 30 keV $Ar^+$, $T = 50$ °C | 56 | 31 | 13 |
| Ion-induced recrystallization 10 keV $Ar^+$, $T = 400$ °C | 44 | 10 | 46 |
| Ion-induced crimping 30 keV $Ar^+$, $T = 350$ °C | 49 | 7 | 44 |

It can be seen that the missing A peak in the Raman spectrum before irradiation, appears after irradiation, and it is the largest for ion-induced amorphization and several times less for recrystallized PAN-fiber shell. The processes of recrystallization including crimping are characterized by relatively small differences in the Raman spectra. Great relationship to $I_D/I_G \approx 1$ compared with $I_D/I_G \approx 0.14$ for non-irradiated fibers can be associated with a significant reduction of the graphite crystallites size $L_a < 10$ nm.

The Raman spectroscopy has shown that after ion irradiation at RT and 10 keV ion energy or at ion incidence angle 70° and 30 keV ion energy PAN-fiber amorphization in contrast to the normal incidence of 30 keV ions is not observed and the D and G peaks are separated. On the fiber surface there is ridge-like morphology, typical for oblique ion incidence on the surface of polycrystalline materials, see [9]. Taking into account the dependence of radiation damage level on ion energy and ion incidence angle [2] one may conclude about radiation damage threshold level of PAN-fiber amorphization. A similar conclusion was assumed in [2,3] by analyzing the temperature dependences of ion-induced electron emission yield at the normal ion incidence and different ion energies.

## 4. Conclusions

The Raman spectroscopy has been used to characterize the ion-induced modifications of carbon PAN-fiber shell due to 10-30 keV $Ar^+$ high fluence ion irradiation at normal and oblique incidence in the temperature range from RT to 400 °C and also structure change during thermal treatment of PAN-fiber.

Formed in ion-induced processes of amorphization, recrystallization and crimping of PAN-fiber modifications are characterized by the presence of the amorphous phase with the A peak in the Raman spectra and the increased intensity of the D peak relative to the G peak in comparison with non-irradiated fiber.

At high-fluence ion irradiation the decrease of radiation damage level with ion energy decreasing or with ion incidence angle increasing suppresses amorphization, the structure of the carbon fiber shell is recrystallized, which manifests in the Raman spectra in the form of separated the D and G-peaks.

Amorphous phase in the Raman spectra till irradiation is practically missed and it is the highest during the PAN-fiber amorphization and the least at crimping.

The increased intensity of the D peak and the G peak shift towards higher frequencies during recrystallization and crimping in the Raman spectra indicate ion-induced nanostructuring of the PAN-fiber shell.

## References

[1] Chernenko N M, Chernenko D N, Beylina N Yu, Elizarov P G, Borisov A M, Mashkova E S

and Andrianova N N *RU Patent* No. 25603 62, 23.04.2014
[2] Andrianova N N, Borisov A M, Mashkova E S, Parilis E S and Virgiliev Yu S 2013 *Horizons in World Physics* vol 280 (New-York, Nova Science Publishing) p 171
[3] Andrianova N N, Borisov A M, Mashkova E S, Virgiliev Yu S and Petrov D V 2014 *J.Surface Investigation. X-ray, Synchrotron and Neutron Techniques* **8** 513
[4] Mashkova E.S. and Molchanov V.A. 1985 *Medium-energy ion reflection from solids* (Amsterdam, North-Holland) p 444
[5] Niwase K 2012 *International Journal of Spectroscopy* **2012** 197609
[6] Kong K, Deng L, Kinloch I A, Young R J and Eichhorn S J 2012 *J. Mater Sci* **47** 5402
[7] Chieu T C, Dresselhaus M S and Endo M 1982 *Phys. Rev.* **B26** 10530
[8] Ferrari A C and Robertson J 2004 *Phil. Trans. R. Soc. Lond.* **A362** 2477
[9] Carter G 2001 *J. Phys. D: Appl. Phys.* **34** R1

IOP Conf. Series: Journal of Physics: Conf. Series **941** (2017) 012029 doi:10.1088/1742-6596/941/1/012029

# Physical and mechanical properties of high-modulus carbon fiber crimped by ion irradiation

**V A Anikin[1], N N Andrianova[1], A M Borisov[1], E S Mashkova[2], M A Ovchinnikov[1,2], S V Savushkina[3], D N Chernenko[4] and N M Chernenko[4]**

[1] Moscow Aviation Institute (National Research University), Volokolamskoe sh. 4, 125993 Moscow, Russia
[2] Skobel'tsyn Institute of Nuclear Physics, Moscow State University, Leninskie gory 1(2), 119991, Moscow, Russia
[3] Keldysh Research Center, Onezhskaya 8, 125438, Moscow, Russia
[4] NIIgraphite, Electrodnaya 8, 111524, Moscow, Russia

E-mail: anatoly_borisov@mail.ru

**Abstract.** The experimental results of ion-induced crimping of the high-modulus carbon PAN-fibers using nitrogen, neon and argon ions with energies of tens keV are presented. Ion-induced surface nanostructuring of the formed crimps provides more than a hundredfold increase in the specific surface area of the fiber and its shrinkage that retains in thermal processes in the production of composites.

Carbon-carbon composite materials (CCCM) reinforced with carbon polyacrylonitrile (PAN) fibers are considered as priority structural materials for nuclear reactors, thermonuclear devices and aerospace technique [1]. The carbon PAN-fibers are characterized by surface crimping at elevated temperatures $T$ of high-fluence irradiation by ions with energies of tens of keV and the level of radiation damage determined by the number of displacements per atom exceeding a certain threshold level. The crimping regularities of PAN-fibers were studied by irradiating of KUP-VM and Desna CCCM samples reinforced with PAN-fibers of VMN-4 and UKN 5000 grades, respectively [2-4].

In this paper a high-modulus carbon PAN-fibers Ravilon and Kulon were used as targets. Irradiation by nitrogen, neon and argon ions with energies of tens keV was carried out along the normal to the fiber axis using the mass-monochromator of the Skobel'tsyn Research Institute of Nuclear Physics Moscow State University [5] and the ion accelerator MIM-50 of Moscow Aviation Institute [6]. The ion current density on the irradiated target with the use of a mass-monochromator was ~ 0.4 mA/cm$^2$ for a beam cross section of 0.3 cm$^2$. The fluence irradiation was $10^{18}$-$10^{19}$ ion/cm$^2$. Irradiation using a non-separated beam of nitrogen ions was carried out at the MIM-50. The diameter of the homogeneous ion beam of MIM-50 was 36 mm. The ion current was 1-3 mA at the accelerating voltage of 30 kV. The 50x40 mm tape made of Kulon fibers were installed in a wire frame on the beam path. The irradiation was carried out along the normal to the plane of the carbon tape. Due to the small heat dissipation the target was heated up to the temperature at least 150 °C during the irradiation. The use of the non-separated wide-aperture beam increases irradiation area more than 40 times compared with a separated ion beam. The morphology of the surface of the fibers before and after irradiation was studied by Lyra 3 TESCAN and Quanta 600 scanning electron microscopes. Figure 1 (left) shows the SEM image of PAN-fibers prior to the irradiation. A typical example of the surface crimping effect of the Ravilon fiber by $Ar^+$ ions with an energy of 30 keV at a temperature of 250 °C is demonstrated in Figure 1 (right). The period and the height of the crimps are 0.5-1 μm.

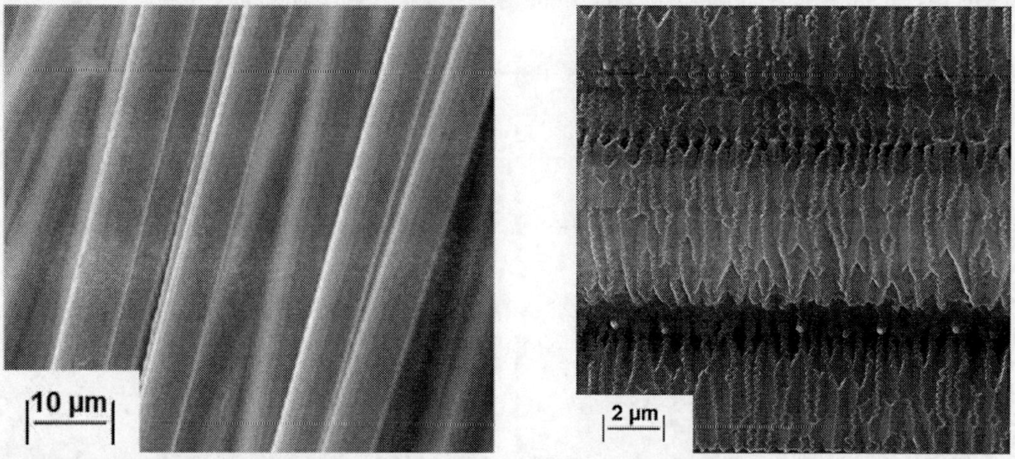


**Figure 1.** SEM images of Ravilon fiber before irradiation (left) and after irradiation (right) by 30 keV $Ar^+$ ions at $T = 250$ °C.

The similar crimping of the fiber surface is observed after nitrogen ion irradiation, see Figure 2. The magnified image on Figure 2 (right) demonstrates the prismatic shape of ion-induced crimps on the surface of the carbon fiber. The prisma-shaped crimps were observed earlier when the KUP-VM composite reinforced with the VMN-4 fibers was irradiated by 30 keV separated $N_2^+$ ion beam at the target temperature of 330 °C [2]. The estimations of the crimps microgeometry showed that the increase in the specific surface area of the Kulon and Ravilon PAN-fibers does not exceed a few times. It is also seen that the prismatic-shaped faces of the crimps have a nanosized roughness. Experimental measurements of the specific surface area were carried out by nitrogen adsorption on the surface area using porosimetry analyzer ASAP 2020 (Micrometrics, USA) and showed more than a hundredfold increase in the specific surface area of the Ravilon PAN-fiber, see Table. It can be assumed that the surface nanostructuring during ion irradiation provides the significant increase of the specific surface area.

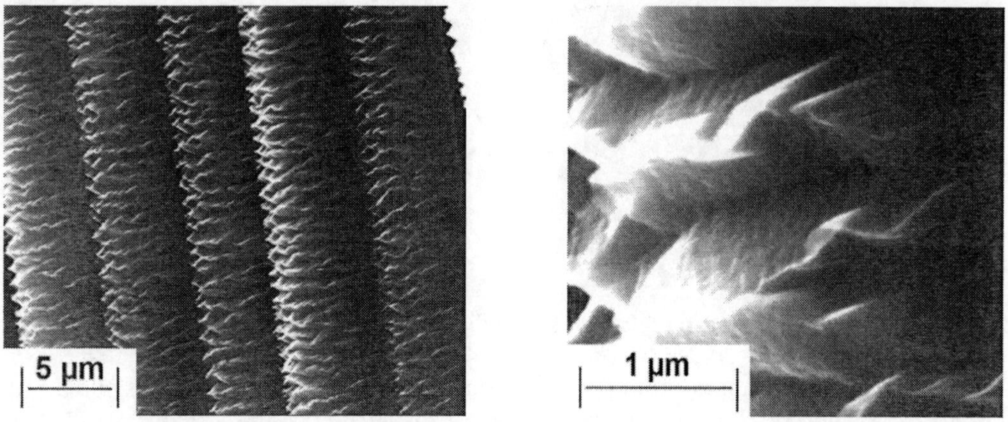


**Figure 2.** SEM image of Ravilon fiber after irradiation with non-separated 30 keV nitrogen ions.

IOP Conf. Series: Journal of Physics: Conf. Series **941** (2017) 012029 doi:10.1088/1742-6596/941/1/012029

Figure 3 shows that the ion-induced crimped structure remains practically unchanged during subsequent annealing. Thus, it can be assumed that in thermal processes in the production of CCCM on the basis of high-modulus PAN-fibers with a crimped surface structure will be remained. Due to the interweaving of fibers in carbon tape that leads to mutual masking during irradiation, SEM images have sections of carbon fibers with a sharp boundary of the irradiation zone. Analysis of such areas allows us to conclude that the ion-induced crimping results in fiber shrinkage.

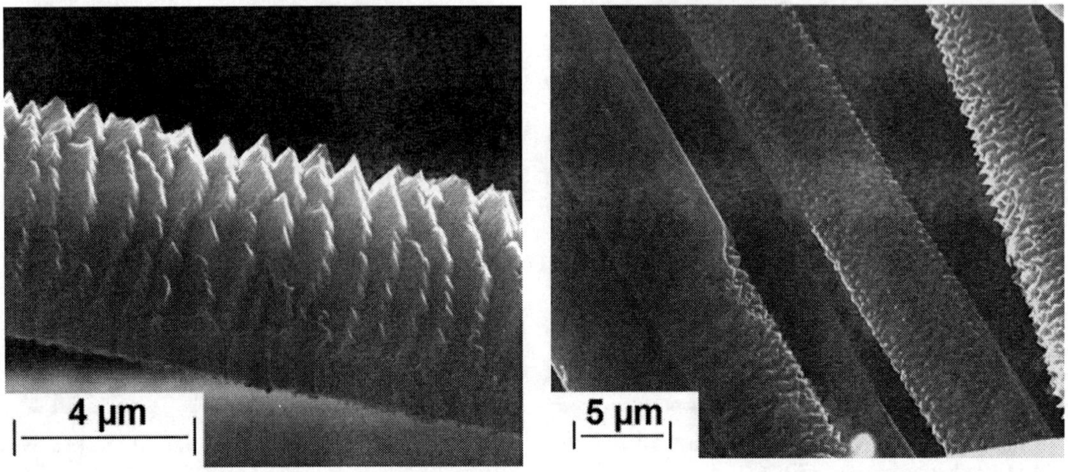


**Figure 3.** SEM images of Ravilon fiber after 30 keV $Ne^+$ ion irradiation at $T$ = 160 °C (left), and subsequent annealing at 2400 °C (right).

The shrinkage of the crimped Kulon fiber by non-separated beam of nitrogen ions is also clearly visible in Figure 4. On the SEM images the crimp tops are located no higher than the level of the fiber surface with a non irradiated smooth surface.

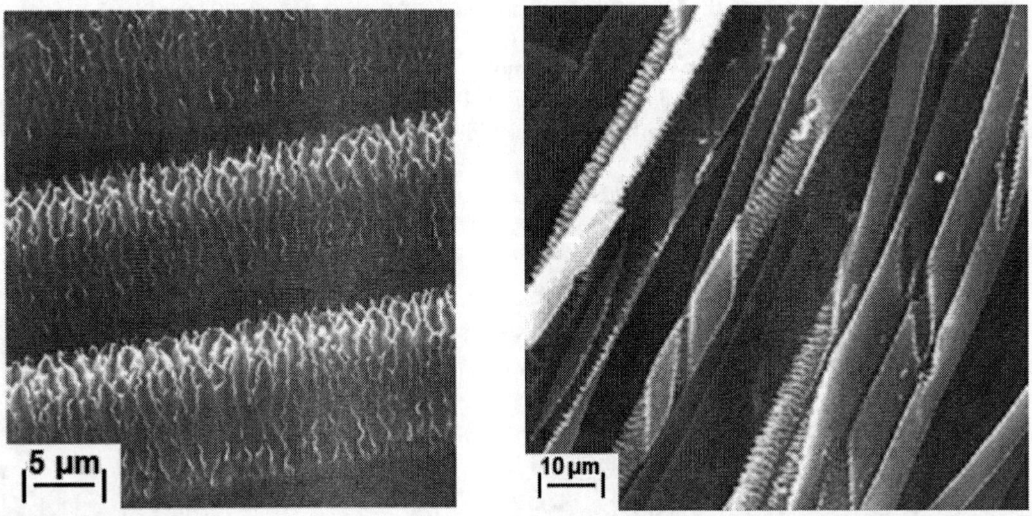


**Figure 4.** SEM images of a Kulon fiber after irradiation with non-separated 30 keV nitrogen ions.

For the irradiation by argon ions specific surface measurements and mechanical tests of non-irradiated and irradiated Ravilon PAN fibers were carried out. Mechanical properties were measured on a Zwick

IOP Conf. Series: Journal of Physics: Conf. Series **941** (2017) 012029 doi:10.1088/1742-6596/941/1/012029

test machine at a temperature of 20-22 °C and a relative humidity of 35-39%. The crosshead speed was 1 mm/min with a working length of 10 mm. The data of the mechanical properties in the table are averaged measurement results for 20-25 fiber samples.

**Table.** Diameter, tensile strength ($\sigma$), Young modulus ($E$), elongation ($\varepsilon$) and specific surface of Ravilon PAN-fiber till and after high-fluence irradiation with 30 keV $Ar^+$ ions at 250 °C.

| Carbon PAN-fiber | Diameter, μm | $\sigma$, GPa | $E$, GPa | $\varepsilon$, % | Specific surface area, $m^2/g$ |
|---|---|---|---|---|---|
| Before irradiation | 4.6±0.3 | 4.4±0.6 | $(4.6\pm0.5)\cdot10^2$ | 1.0±0.1 | 0.5-1.5 |
| After irradiation | 4.9±0.3 | 4.2±0.7 | $(4.5\pm0.4)\cdot10^2$ | 0.9±0.1 | 107 |

The data comparison for the non-irradiated and irradiated high-modulus PAN-fiber samples shows that mechanical properties of carbon fibers under ion irradiation practically do not change with the hundredfold increase in the specific surface.

Various methods are known for increasing of carbon fiber adhesion with the matrix in CCCM by modifying the carbon fiber surface [7, 8]. Among them, the high-fluence ion irradiation of high-modulus carbon PAN-fibers is the most suitable for reinforcing carbon-carbon and carbon-ceramic heat-resistant composite materials.

## References

[1] Virgil'ev Yu.S., Kalyagina I.P. 2004 *Inorganic Materials.* **40** Suppl.1. 33-49.

[2] Andrianova N N, Borisov A M, Virgil`ev Yu S, Mashkova E S, Nemov A S, Pitirimova E A, Timofeev M A 2008 *Journal of Surface Investigation. X-ray, Synchrotron and Neutron Techniques.* **2** 376-379

[3] Andrianova N N, Borisov A M, Mashkova E S, Parilis E S, Virgiliev Yu S 2013 *Horizons in World Physics* **280** 171 (Nova Science Publishing, New-York)

[4] Andrianova N N, Beylina N Yu, Borisov A M, Mashkova E S, Chernenko D N,Chernenko N M 2014 *Journal of Surface Investigation. X-ray, Synchrotron and Neutron Techniques* **8** 211

[5] Mashkova E.S. and Molchanov V.A. 1985 *Medium-energy ion reflection from* solids (North Holland, Amsterdam)

[6] Borisov A M, Goryaga N G, Romanovsky E A, Smagin M N, Dang L. *Izv. Akad. Nauk, Ser. Fiz.* 1992 **56** 194-197

[7] Blinov I V, Ostapovich A O 1994 Second Moscow conference of composite materials. *Abstract book.* 34-35 (In Russian)

[8] Chernenko N M, Chernenko D N, Beylina N Yu, Elizarov P G, Borisov A M, Mashkova E S and Andrianova N N *RU Patent* No.25603 62, 23.04.2014

# Calculation and experimental determination of the geometric parameters of the coatings by laser cladding

**V P Birukov and A A Fichkov**

Federal budget-funded research Institute for machine science named after A. A. Blagonravov of the Russian Academy of Sciences (IMASH PAN), Maly Kharitonievskiy Pereulok 4, 101990, Moscow, Russia

E-mail: laser-52@yandex.ru

**Abstract.** In the present work the experiments on laser cladding of powder Fe-B-Cr-6-2 on samples of steel 20. Metallographic studies of geometric parameters of deposited layers and the depth of the heat affected zone (HAZ). Using is the method of full factorial experiment (FFE) mathematical dependences of the geometrical sizes of the deposited layers of processing modes. Deviation of calculated values from experimental data is not more than 3%.

## 1. Introduction

Laser cladding is application on surface of the workpiece coating by melting basis and the filler material [1]. Unlike traditional methods of cladding for restoration of machine parts laser cladding has several advantages. Laser radiation is characterized by the local thermal influence on the material, and as a result, a relatively small thermal influence on the detail altogether. Owing to the fact that laser treatment allows to achieve ultra-high speeds of heating and cooling in the radiating zone, it is possible to reach unique mechanical and tribological properties of the machined surface, which cannot be achieved by traditional methods.

However, it should be noted that laser treatment is a quite expensive method. Equipment and supplies have high cost. Train to perfection of technological process of laser cladding demands large number of experiments. Of great interest is obtaining mathematical model of the process cladding of surfacing, which would allow to determine the parameters of the deposited layer without carrying out additional experiments.

The aim of this work is to obtain regression equations based on experimental data using the method of full factorial experiment to determine the geometrical parameters of the cladding rolls.

## 2. Materials and equipment

In our experiments the laser system "Kometa-M" equipped with a continuous $CO_2$-laser [2] was used. For surfacing a powder based on iron Fe-B-Cr-6-2 was used. Samples were made of steel 20 with dimensions 15×20×80 mm. Before cladding, the samples were cleaned off with acetone and dried in an oven at a temperature of 75 °C for 30 minutes. Filler powder material was applied on samples in the kind of a coating. The layer thickness was 0.9–1.0 mm. After application of the coating the samples were dried in the oven at the same modes.

Metallographic studies were performed on the hardness testing of the PMT-3, metallographic microscope Altami MET-1C.

## 3. Results of experiments and calculations

As factors in mathematical planning was chosen the radiation power *P*, speed of the beam *V* and beam diameter *d*. The responses of the system: the rolls height *H*, width *B*, depth of the zone of thermal influence of the HAZ – *Z*. As an additional discrete factor considered scanning the beam with a frequency $f = 220$ Hz. Surfacing was carried out at the maximum and minimum levels of the factors, which are designated respectively as $z+$ and $z-$. The results of preliminary experiments were chosen the upper and lower levels of factors in which the visually observed formation of deposited layers.

The experiments were carried out in two series – with and without scanning, it is for the accounting of discrete factors of scan. The levels of factors, the quantity of the center $z_0$ of the plan and the range of variation $\lambda$ and based on the coded variables from the natural quantities given in table. 1.

**Table 1**. The levels of factors of the experiment.

| Factor | $z+$ | $z-$ | $z_0$ | $\lambda$ | Dependence |
|---|---|---|---|---|---|
| *P*, W | 1000 | 700 | 850 | 150 | $x_i=(P-850)/150$ |
| *V*, mm/s | 9 | 5 | 7 | 2 | $x_i=0.5(V-7)$ |
| *d*, mm | 3 | 2 | 2.5 | 0.5 | $x_i=(d-2.5)/0.5$ |
| *f*, Hz | 220 | 0 | – | – | – |

According to the recommendations [3] as a mathematical model the algebraic linear polynomial of the form is adopted:

$$y = b_0 + b_1x_1 + \cdots + b_kx_k + b_{1,2}x_1x_2 \ldots + b_{k-1,k}x_{k-1}x_k \quad (1)$$

where is $x_i$ – levels of factors of the experiment, $b_i$ is linear regression coefficients.

According to the method of FFE, the resulting regression coefficients for all the feedback coefficients are substituted in equation (1), made identical to the conversion from coded variables to the natural variables and derived the resulting mathematical model of the form:

$$H, B, Z = f(P, V, d, f) \quad (2)$$

The obtained mathematical model was checked for according adequacy by Student and Fisher criterions. The regression coefficients are presented in table 2.

**Table 2.** The coefficients of regression equations.

| Parameter | $b_0$ | $b_1$ | $b_2$ | $b_3$ | $b_{12}$ | $b_{13}$ | $b_{23}$ | $b_{123}$ |
|---|---|---|---|---|---|---|---|---|
| H | 0.908 | 0.138 | –0.228 | 0.066 | –0.023 | 0.046 | 0.0038 | – |
| $H_{scan}$ | 0.917 | 0.035 | – | –0.020 | 0.035 | 0.017 | –0.07 | –0.027 |
| B | 1.712 | 0.185 | –0.217 | –0.107 | –0.040 | – | 0.027 | –0.04 |
| $B_{scan}$ | 3.170 | 0.540 | –0.255 | –0.065 | 0.200 | 0.155 | –0.100 | – |
| Z | 0.387 | 0.0920 | –0.006 | – | –0.003 | 0.028 | –0.052 | –0.024 |
| $Z_{scan}$ | 0.257 | 0.031 | –0.049 | –0.007 | 0.015 | –0.004 | 0.018 | – |

Removed coefficients are assumed to be zero as non-significant by Student criterion at the significance level $\alpha = 0.05$ [3].

Analysis of the obtained equations set is directly proportional to the dependence between power radiation and height rolls, and the inverse relationship between speed and height of roller. With the increase of the radiation power without scanning the height rolls increases faster than with scanning, due to the larger area of heat energy distribution of the laser when the high frequency oscillation of the beam. Scanning leads to an increase of the width with simultaneous decrease the height of the rolls and increases the productivity of cladding by 1.3–1.9 times as much. Analysis of the equations for the depth of HAZ shows that the greatest influence on depth of HAZ has the radiation power. The increase of the velocity $V$ leads to a decrease in the depth of the HAZ, due to the smaller momentary energy absorbed by a filler material per unit of time.

According to the obtained mathematical dependences the calculations and obtained data were compared with the experimental results. Built comparative surface defining the relationship between the response and factors the experiment. Comparative surface built for functions $B$ and $B_{scan}$= f (P, V), when $d$ = 2 mm, is shown (figure 1).

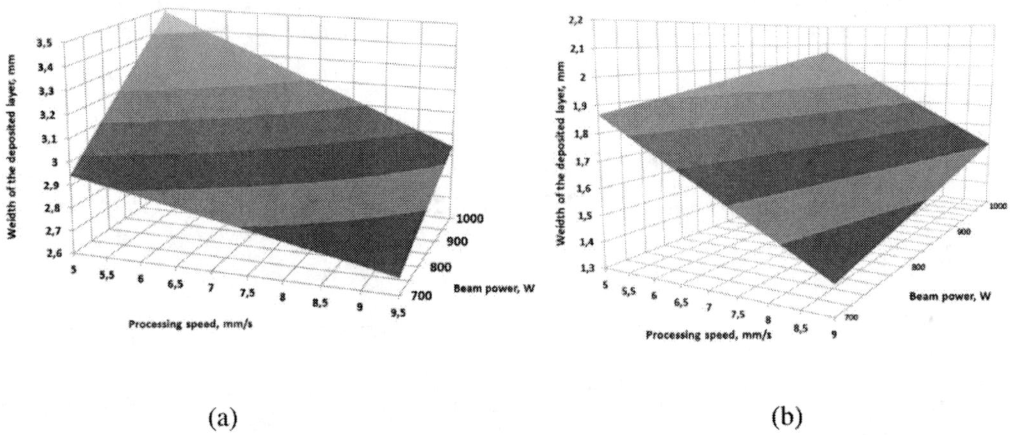


**Figure 1.** The dependence of the width of the cushion from the factors of the experiment (a - scanning, b – without scanning).

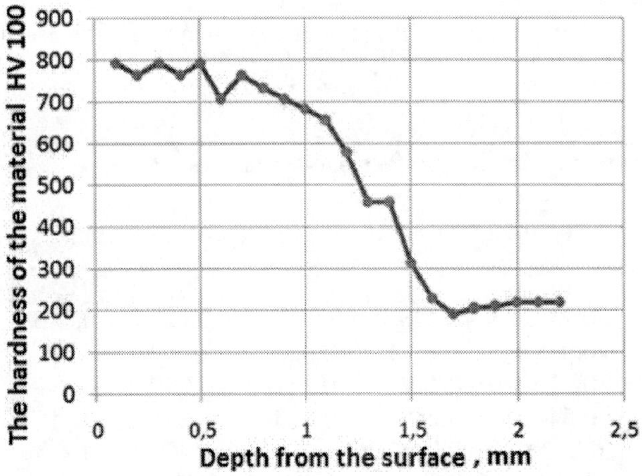


**Figure 2.** The dependence of micro-hardness on the depth.

According to the data obtained, the nature of the linear dependence obtained in the calculation is the same as that for the experimental values, but the estimated function is growing faster. The deviation of experimental data from calculated is not more than 3%.

Figure 2 shows a typical distribution of micro-hardness of the deposited layer from the surface to the base.

The deposited layer thickness of 0.5 mm has a high hardness of about 800 HV, under it is a HAZ with a hardness of HV 540-720, the thickness of this layer is 0.8 mm, below the follow zone - troostite and sorbitol. The base metal has the hardness 190–210 HV. Notably, the presence of the reflow zone the foundations to a depth of 50–100 μm, indicates high adhesion strength between the coating and the base metal.

## 4. Conclusion

On the basis of experimental data and mathematical models the surfaces, which allow to estimate the influence of the radiation power, processing speed, diameter of the beam on the geometrical parameters the deposited layers, are constructed. Estimate of the parameters of the deposited layers compared with experimental data the maximum differences do not exceed 3%.

It is established that under high-frequency beam scanning performance the coating process is 1.3–1.9 times as much higher than without scanning the beam.

## References

[1] Grygoryants A G, Shiganov I N, Misyurov A I 2008 Technological processes of laser processing (Moscow: MGTU) (in Russian)
[2] Biryukov V P and Dozorov A V 2006 *J. of Machinery Manufacture and Reliability* **1** 60–66
[3] Sokolovskaya I Y 2010 *Full factorial experiment* (Novosibirsk: SSUWT) (in Russian)

# Direct metal laser deposition of titanium powder Ti-6Al-4V

**D P Bykovskiy, V N Petrovskiy, K L Sergeev, A V Osintsev, P S Dzhumaev and V I Polskiy**

National Research Nuclear University MEPhI (Moscow Engineering Physics Institute), Kashirskoe sh. 31, Moscow, 115409, Russia

E-mail: d.bykofsky@gmail.com

**Abstract**. The paper presents the results of mechanical properties study of the material produced by direct metal laser deposition of VT6 titanium powder. The properties were determined by the results of stretching at tensile testing machine, as well as compared with the properties of the same rolled material. These results show that obtained samples have properties on the level or even higher than that ones of the samples obtained from the rolled material in a certain range of technological regimes.

## 1. Introduction

Titanium alloys have already well approved themselves in chemical, marine, aerospace industries. They are very strong, able to withstand heavy loads and have reduced mass. In addition, titanium alloys are used for manufacturing implants and prostheses. Tissues around such implants are not subjects to change. They are resistant to corrosion in aggressive environment of the human body, and oxide coating on their surface prevents the escape of the implant ions into the body. However, the manufacture of titanium parts and implants is strictly individual task, as it is a material, which is difficult to process. This fact significantly increases the cost of a product. There is a significant interest in using different additive technology for manufacturing titanium parts in recent times. It is necessary to provide wide complex of researches and tests. Titanium alloys are used in the following additive technology: melting layer by layer powder by electron beam (EBM) [1, 2] or laser (SLM, SLS methods) [3, 4], feeding melted powder or wire by plasma arc [5, 6] or laser (DMLD) [7, 8]. We suggest using laser metal deposition technology for manufacturing of titanium parts. This technology involves melting of metallic powder feeding through the nozzle coaxially with the radiation. When the nozzle moves along the surface, new metal layer will be formed after solidification. Thus, it is possible to produce parts of the desired shape. The research presents the results of the mechanical properties study of the samples obtained by laser metal deposition technology by different technological modes.

## 2. Materials and equipment

Experimental equipment is shown in figure 1. Cladding processing cell (1) is equipped with a control unit (2). Cell (1) consisting of 5-axis positioning system, optical head with water cooling and supply of protective gas and powder from the feeder (3) Sulzer Metco Twin 10 - C, as also industrial digital camera for positioning parts and monitoring the process. We use fiber laser (4) LS-3.5 power up to 3.5 kW.

The metal powder is fed into the treatment zone via a coaxial nozzle, under the impact of laser radiation it is heating and melting. Simultaneously, the previous layer is melting. The track is produced during the motion of the nozzle relative to the substrate after cooling. New layers are formed by sequential application of tracks with some overlapping.

**Figure 1.** Experimental equipment.

Tensile tests were performed on the machine INSTRON 5966, which has the following operational characteristics: allowable load is 10 kN, maximum speed is 1500 mm/min, minimum speed is 0.001 mm/min.

We used a titanium powder VT6, produced by "Normin". VT6 alloy powder consists of spherical particles with size distribution 40-110 microns, the chemical composition of which is shown in the table 1.

**Table 1**. The chemical composition of VT6 powder.

| The content of alloying elements, masses, % | | | | | | |
|---|---|---|---|---|---|---|
| Fe | Al | V | O | N | C | Ti |
| 0.11 | 6.1 | 3.9 | 0.08 | 0.01 | 0.03 | Balance |

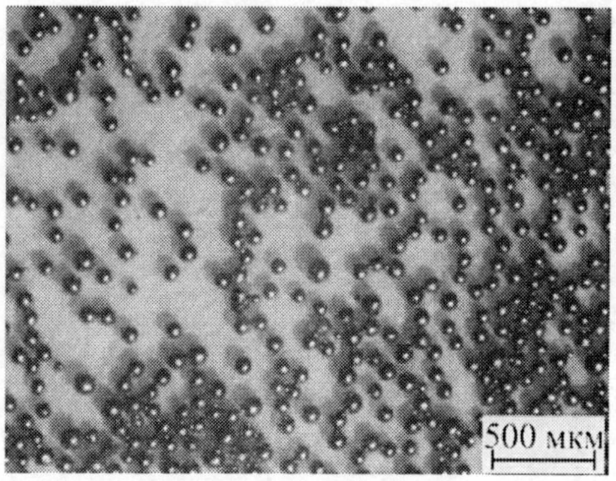


**Figure 2.** VT6 powder.

## 3. Results of experiments

Samples in the form of two-side blades of 316L stainless steel powder have been received and investigated in our previous works [9, 10]. Then the optimal strategy for the creation of such samples was selected. It has been shown that repeated melting of layers does not improve the mechanical properties. The optimal strategy for the creation of samples was the following. Two directions were chosen for double-sided cladding blades: parallel to the direction of tension (figure 3) and perpendicular (figure 4). If cladding direction is parallel, then contour of the blade 1 would be deposited first, and then it would be inner part, consisting of several regions, numbered 2-5.

Perpendicular application consists in drawing the contour, and then filling it with consistent cladding tracks. Also, alternating start peaks cladding were performed to improve thickness uniformity of the sample. After the deposition of two layers with different application areas rolls starting at point A, the nozzle moves to point B, where the cladding is also made of two layers. The operation was repeated further starting at point C and D.

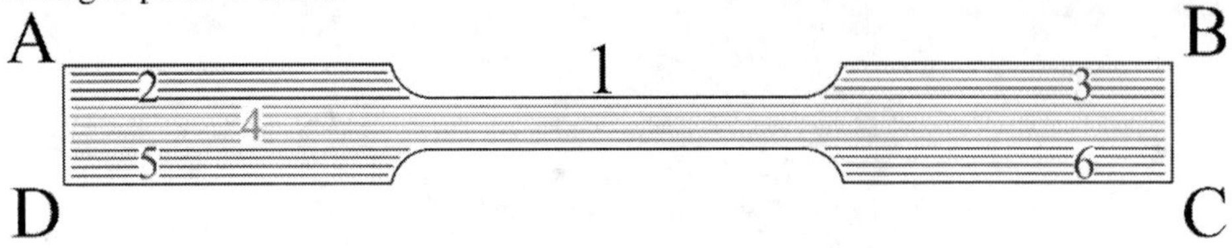


**Figure 3.** Scheme of cladding of rollers in a longitudinal direction.

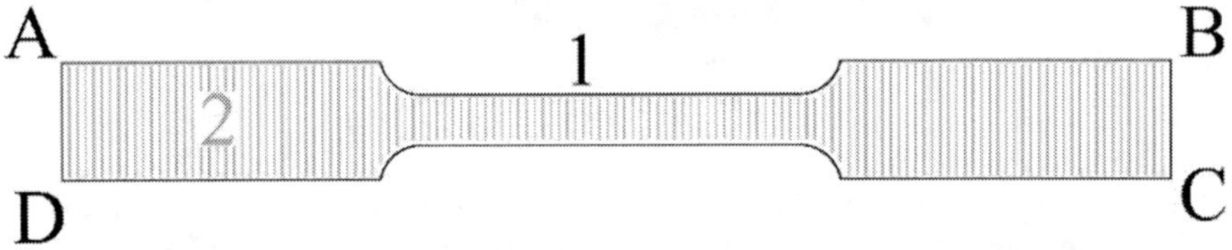


**Figure 4.** Scheme of cladding of rollers in a transverse direction.

The following process parameters were varied during operation in this process: laser power, scanning speed, powder feed rate. A scope of mechanical tensile tests were carried out for each technological regime at room temperature in accordance with GOST 1497-84. This standard specifies methods of static tensile tests of ferrous and non-ferrous metals and their products (samples) to determine at 20 °C the limit of strength $\sigma_{ts}$, yield strength $\sigma_{0,2}$ and others. Also, the samples of titanium rolled material were tested. The stress-strain diagrams were obtained during the test (figure 5) for each samples.

Yield strength $\sigma_{0,2}$ and tensile strength $\sigma_{ts}$ were determined according to the obtained stress-strain diagrams (figure 5) for the produced and rolled samples, and their dependences (figure 6-11) were built on the technological parameters of process: laser power, surface scanning speed, powder feed rate.

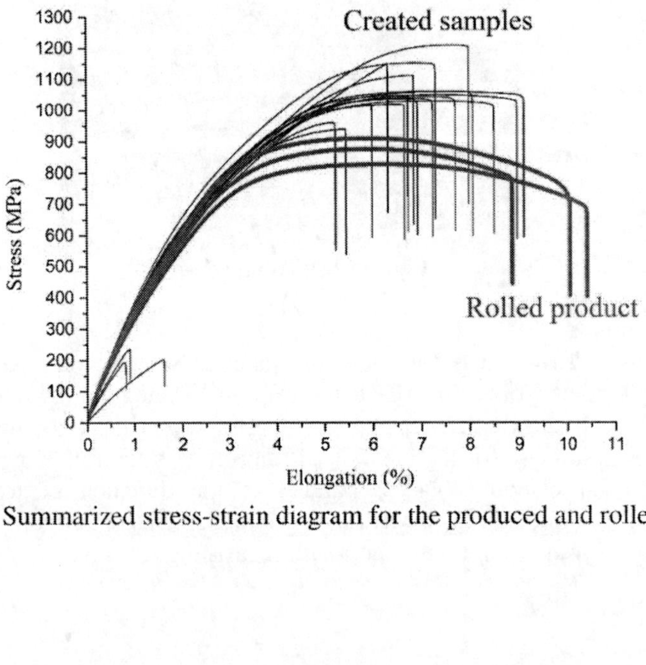


**Figure 5.** Summarized stress-strain diagram for the produced and rolled samples.

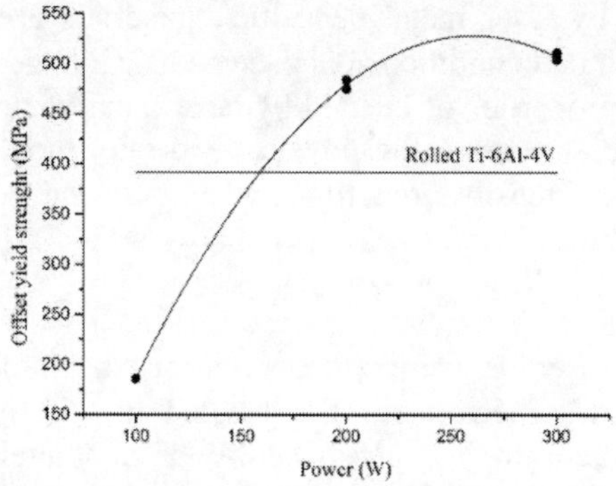


**Figure 6.** Dependence of yield strength $\sigma_{0,2}$ on the laser power.

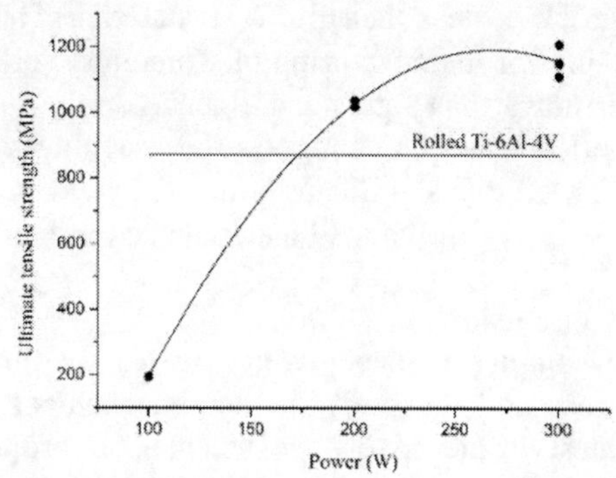


**Figure 7.** Dependence of tensile strength $\sigma_{ts}$ on the laser power.

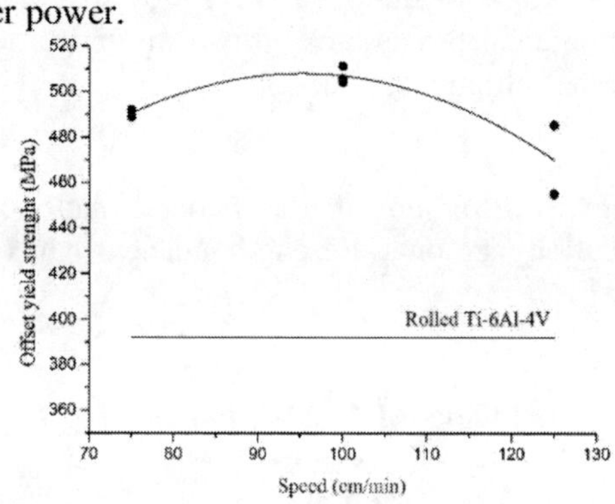


**Figure 8.** Dependence of yield strength $\sigma_{0,2}$ on scanning speed.

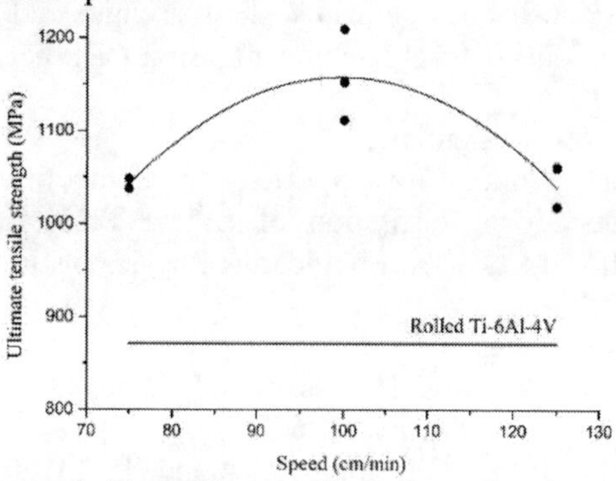


**Figure 9.** Dependence of tensile strength $\sigma_{ts}$ on scanning speed.

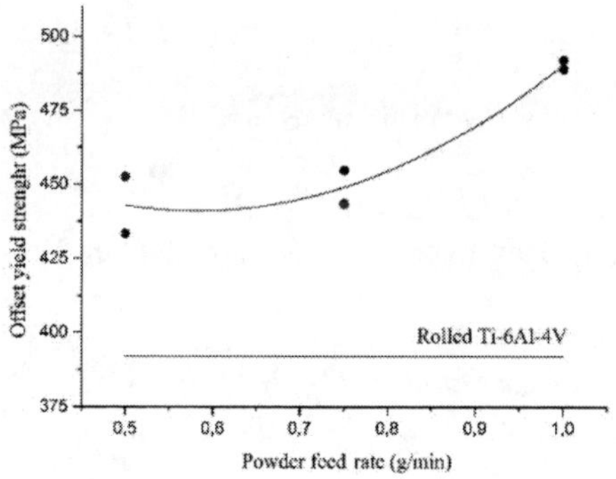


**Figure 10.** Dependence of yield strength $\sigma_{0,2}$ on powder feed rate.

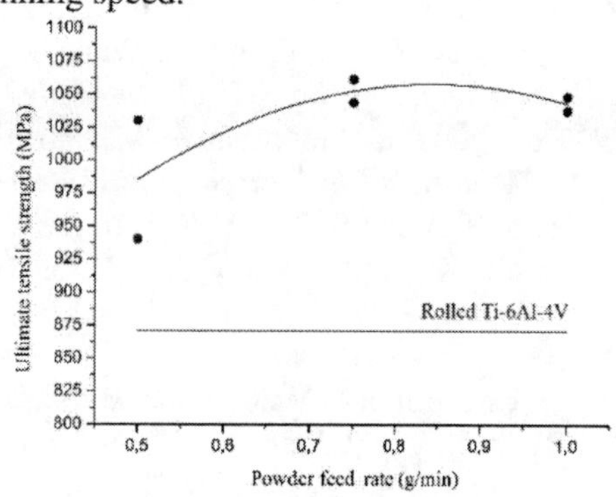


**Figure 11.** Dependence of tensile strength $\sigma_{ts}$ on powder feed rate.

The curve slope of the stress-strain is almost identical in the elastic region, it means equality of Young's modulus of created and cast materials. Yield strength $\sigma_{0,2}$ of created by laser metal deposition materials exceeds a similar value for the cast material by 10-25%. It leads to the fact that under equal stresses in structures made from cast material and the created by laser metal deposition material yield point (the occurrence of irreversible plastic deformation) occurs in the created material much later. The tensile strength $\sigma_{ts}$ of created by laser metal deposition materials exceeds the similar value of cast material by 20-30%. Therefore, the construction made from created by laser metal deposition materials is capable to withstand large external load with the same size of construction of the cast material, or it will allow making a smaller construction of the created material than from the cast material under equal external loads. The average value of the maximum relative deformation of the created by laser metal deposition material is ~ 7%, a similar value for the cast material is about 10%. In this regard, the plastic deformation of the structure, made of created by laser metal deposition materials is less than the cast material. Thus created by laser metal deposition materials are most suitable for manufacturing of structures working in cramped conditions displacements. Change in the laser power has a greater influence on the mechanical properties of created by laser metal deposition materials than the change in the scanning speed of the surface or the powder feed rate. Increase in power led to a significant growth of yield strength $\sigma_{0.2}$ and tensile strength $\sigma_{ts}$, which allowed to make properties of the created materials better than of cast ones.

## 4. Conclusion

The samples in paper were created by direct metal laser deposition under various technological regimes and the study of their mechanical properties was performed. The results showed that the greatest influence on the mechanical properties of the samples, produced by direct metal laser deposition, has a change of laser power. It also shows that the produced samples were not inferior to the characteristics of the rolled samples. Thus, the Young's modulus of the rolled material and produced ones coincide, yield strength and tensile strength materials have grown up and produced materials deform less under tension than the samples of the rolling material.

## Acknowledgments

This work was supported by Competitiveness Growth Program of the Federal Autonomous Educational Institution of Higher Professional Education National Research Nuclear University MEPhI (Moscow Engineering Physics Institute).

## References

[1] Galarraga H, Warren R J, Lados D A, Dehoff R R and Kirka M M 2017 *Engineering Fracture Mechanics* **176** 263
[2] Cao J, Gharghouri M A and Nash P 2016 *J. Mat. Proc. Tech.* **237** 409
[3] Yang J, Yu H, Wang Z and Zeng X 2017 *Materials Characterization* **127** 137
[4] Yang Z, Yu Y, Wei Y and Huang C 2017 *Thin-Walled Structures* **111** 1
[5] Lin J, Lv Y, Liu Y, Sun Z, Wang K, Li Z, Wu Y and Xu B 2017 *Journal of the Mechanical Behavior of Biomedical Materials* **69** 19
[6] Donoghue J, Antonysamy A, Martina F, Colegrove P, Williams S and Prangnell P 2016 *Materials Characterization* **114** 103
[7] Book T A and Sangid M D 2016 *Materials Characterization* **122** 104
[8] Li W, Karnati S, Kriewall C, Liou F, Newkirk J, Taminger K M B and Seufzer W J 2017 *Additive Manufacturing* **14** 95
[9] Bykovskiy D P, Andreev A O, Mironov V D, Petrovskiy V N, Popkova I S, Solonin A N and Cheverikin V V 2016 *International Conference Laser Optics (LO)* pp 33
[10] Bykovskiy D P, Petrovskiy V N, Mironov V D, Osintsev A V and Ochkov K Y 2016 *Journal of Physics: Conference Series* **747** 012057

# Laser annealing of porous defects in metals

**M S Grigoryeva[1,2], I N Zavestovskaya[1,2] and A P Kanavin[1,2]**

[1] National Research Nuclear University MEPhI (Moscow Engineering Physics Institute), 31, Kashirskoe shosse, 115409 Moscow, Russia
[2] P.N. Lebedev Physical Institute of the Russian Academy of Sciences, 53, Leninskiy Prospekt, 119991 Moscow, Russia

E-mail: MSGrigoryeva@mephi.ru

**Abstract.** In present paper a self-consistent model of laser annealing of porous defects in metals is presented. In the proposed model, the defects represent empty isolated pores whose dimensions are much less than thickness of the metal coating. Reduction of porosity occurs due to pore collapse in the molten pool under surface tension forces. In this case the research showed that pore collapse dynamics depends considerably on the pressure in the molten material. From the equality of the negative pressure in the melt and the capillary pressure the applicability bound of the model is determined. It is the critical melting rate of porous material at which pore collapse is still possible.

## 1. Introduction

The porosity is one of the common defects of metal coatings produced by plasma or magnetron deposition. The use of laser annealing makes it possible to remove these defects, to change the surface relief, microstructure, and roughness [1]. Depending on the coating deposition method, pores can be both empty and gas-filled. The nature of the change in the porosity of the material under laser annealing depends significantly on the gas content in individual pores. For example, in the case of gas-filled pores, the reduction of porosity may occur due to thermocapillary drift of bubble in an inhomogeneously heated melt [2]. For empty pores, a probable mechanism for reducing porosity will be their collapse in the molten material under the action of surface tension forces. In this study, we developed a self-consistent model of pore collapse dynamics during laser annealing of porous defects in metals. For the case of constant velocity of the melting front motion, the thickness of the melt layer in which complete pore closure occurs is determined. The critical melting rate of porous material is found, which is the criterion for the applicability of the proposed model.

## 2. Theoretical model

Consider a porous metal material exposed to the action of a pulsed laser radiation. Laser pulse irradiation of the coating surface results in material heating and melting. The porosity of the material is small, and isolated spherical pores with dimensions equal in order of magnitude to the size of the original defects are formed on the melting front moving to the depth of the solid phase.

### *2.1. Thermocapillary drift*

In the case of gas-filled pores, the main mechanism for the movement of bubbles in the molten pool may be of thermocapillary drift (the effect of Marangoni-Benard). This mechanism is due to the surface tension temperature dependence and the inhomogeneity of heating of the bubble surface during the melting of metallic materials by laser radiation. Due to the presence of high values of temperature gradients in the melt, the surface of the bubble turns out to be heated non-uniformly, which results in the appearance of tangential thermocapillary force on it, causing the liquid to move in the direction of decreasing temperature, and the bubble relative to the fluid in the opposite direction. Thus thermocapillary drift can lead to a significant reduction of porosity in the near-surface layers of the material.

We estimate the velocity of the bubble motion in a vertical temperature gradient. To calculate, we use formula [3]:

$$V = -\frac{R_0}{3\mu}\left(\frac{\partial\sigma}{\partial T}\right)\frac{\partial T}{\partial z}, \tag{1}$$

where $R_0$ is the bubble radius, $\mu$ is the dynamic viscosity coefficient, $\sigma$ is the surface tension coefficient, $\frac{\partial T}{\partial z}$ is the temperature gradient.

Then for the case of indium with initial porosity $\Phi_0 = 0.01$, pore radius $R_0 = 10^{-4}$ cm, $\mu = 1.9\cdot10^{-2}$ g/cm·s, the surface tension gradient $\left(\frac{\partial\sigma}{\partial T}\right) = -0.09$ erg/cm$^2$·°C, and the temperature gradient $\left(\frac{\partial T}{\partial z}\right)$=65·10$^4$ °C/cm, corresponding to laser action of nanosecond duration with the flux density $q = 0.1$ J/cm$^2$, the bubble velocity is $V$ = 100 cm/s. It should be noted that this estimate is valid only for a small porosity of the material, when the pores can be considered single.

*2.2. Pore collapse under surface tension forces*

If the pores are considered empty, then fluid motion in the melt can be characterized by two spatial scales: the local motion on the pore size scale and the averaged homogeneous melt motion. Moreover, these movements can be separated. The local fluid motion causes the molten material to flow into pores and their collapse. The velocity of such flow depends on the relation between the capillary and the near-pore pressures. It is obvious that flowing of material into the pores should be accompanied by melt motion from the irradiated surface to the collapse region and hence the appearance of the corresponding averaged macroscopic motion, which is defined by the pressure difference on the free surface and at a given melt depth. Thus, the dynamics of the melt flow into pores and its translational averaged motion from the surface to the collapse region appear self-consistently related by the pressure at a given depth.

The equation of motion and the continuity equation which describe the macroscopic averaged melt motion, neglecting the viscosity, are given by

$$\frac{\partial V}{\partial t} + V\frac{\partial V}{\partial z} + \frac{1}{\rho}\frac{\partial P}{\partial z} = 0, \tag{2}$$

$$\frac{\partial \rho}{\partial t} + \frac{\partial}{\partial z}\rho V = 0. \tag{3}$$

The spherical cavity boundary motion in the moving liquid element under the action of surface tension and external pressure is described by the Rayleigh equation [4]

$$R(t')\frac{d^2R}{dt'^2} + \frac{3}{2}\frac{dR}{dt'}^2 + \frac{2\sigma}{\rho R(t')} + \frac{4\mu}{\rho R(t')}\frac{dR}{dt'} = -\frac{P}{\rho} \tag{4}$$

with initial conditions

$$R(0) = R_0,\ \dot{R}(0) = 0, \tag{5}$$

where $R$ is the pore radius, $t'$ is the time after the passage of the melting front through the pore, $\sigma$ is the surface tension coefficient, $\mu$ is the dynamic viscosity coefficient, $\rho$ is the material density, and $P$ is the external pressure.

**3. Results and discussion**

We consider the case of constant melting front velocity, and hence, the melt motion velocity. Then the equations (2) and (3) take on form:

$$\frac{P}{\rho} + \frac{V^2}{2} = const, \tag{6}$$

$$\rho V = const. \tag{7}$$

From the solution of the system of equations (6) - (7), taking into account the relation between the average density of the material and the porosity:

$$\Phi = 1 - \frac{\rho}{\rho_0}, \tag{8}$$

where $\rho_0$ is the material density in the absence of pores, in the first-order smallness approximation, we obtain the pressure in the melt near the melting front:

$$P = P_0 - \rho_0 \cdot V_m^2 \cdot \Phi, \tag{9}$$

where $P_0$ is the pressure on the free surface, $V_m$ is the melting front velocity.

We can see that this pressure is lower than that on the free surface by $\rho_0 V_m^2 \Phi_0$, which provides molten material flow from the surface to the melting front. The pressure (9) is external for the moving boundary of the spherical cavity described by equation (4).

To solve equation (4) it is convenient to pass to the coordinate system associated with the melting front motion, that is, we pass from the dependence $R(t')$ to the dependence on the melting front coordinate z, then $\frac{d}{dt} = V_m \frac{d}{dz}$.

We introduce the following dimensionless variables: the radius $y = \frac{R}{R_0}$ and coordinate $u = \frac{z}{L_0}$, where $L_0 = \left(\frac{\rho_0 V_m^2 \cdot R_0{}^3}{2\sigma}\right)^{1/2}$ determines the spatial scale of the region in which the pore radius decreases.

In the new variables, equation (4) can be reduced to the form

$$y\frac{d^2y}{du^2} + \frac{3}{2}\left(\frac{dy}{du}\right)^2 + \frac{1}{y}\left(1 + \alpha\frac{dy}{du}\right) + \beta(1 - \delta y^3) = 0 \tag{10}$$

with the initial conditions

$$y(0) = 1, \quad \left.\frac{dy}{du}\right|_{u=0} = 0, \tag{11}$$

where

$$\alpha = \sqrt{\frac{8 \cdot \mu^2}{\sigma \cdot \rho_0 \cdot R_0}}, \quad \beta = \frac{P_0 \cdot R_0}{2 \cdot \sigma}, \quad \delta = \frac{\rho_0 \cdot V_m{}^2 \cdot \Phi_0}{P_0}. \tag{12}$$

The coefficient $\alpha$ is defined by the liquid viscosity, $\beta$ is the ratio of external and capillary pressures, $\delta$ defines the pressure difference between the free surface and at the melting front.

If viscosity is neglected in equation (10), that is, set $\alpha = 0$, then the solution can be represented in the implicit form:

$$u = \int_y^1 \frac{y^{3/2}dy}{\sqrt{1-y^2+\frac{2\beta}{3}(1-y^3)(1-\frac{\delta}{2}(1+y^3))}} . \tag{13}$$

In the case of the external pressure smallness in comparison with the capillary pressure, when $\beta \ll 1$, integral (13) is

$$u = \int_y^1 \frac{y^{3/2}dy}{\sqrt{1-y^2}} = \sqrt{y-y^3} + \frac{1}{\sqrt{2}}K\left(\frac{1}{\sqrt{2}}\right) - \frac{1}{\sqrt{2}}F\left(\left(\frac{2\cdot y}{y+1}\right)^{1/2}, \frac{1}{\sqrt{2}}\right), \tag{14}$$

where $K\left(\frac{1}{\sqrt{2}}\right)$ is the complete elliptic integral of the first kind, $F\left(\left(\frac{2\cdot y}{y+1}\right)^{1/2}, \frac{1}{\sqrt{2}}\right)$ is the incomplete elliptic integral of the first kind.

Then the region in which complete pore collapse occurs, that is radius $y \to 0$, is $u_0 = \frac{1}{\sqrt{2}}K\left(\frac{1}{\sqrt{2}}\right) \cong 1.31$.

In the case of nonzero parameters (12) equation (10) can be solved only numerically. In the figure 1 the pore radius variation with the distance to the melting front, normalized to the full width of the pore collapse region is presented. This solution is obtained for the case of indium with initial porosity $\Phi_0 = 0.01$, the pore radius $R_0 = 10^{-4}$ cm, the melting rate $V_m = 7\cdot 10^3$ cm/s, and the pressure at infinity $P_0 = 1$ atm. The calculation used the following material parameters: the compact material density is $\rho_0 = 7.31$ g/cm$^3$, the dynamic viscosity coefficient is $\mu = 1.9\cdot 10^{-2}$ g/cm·c, and the surface tension coefficient is $\sigma = 556$ erg/cm$^2$. The parameter $L_0$ turned out to be equal to $L_0 = 5.7\cdot 10^{-4}$ cm.

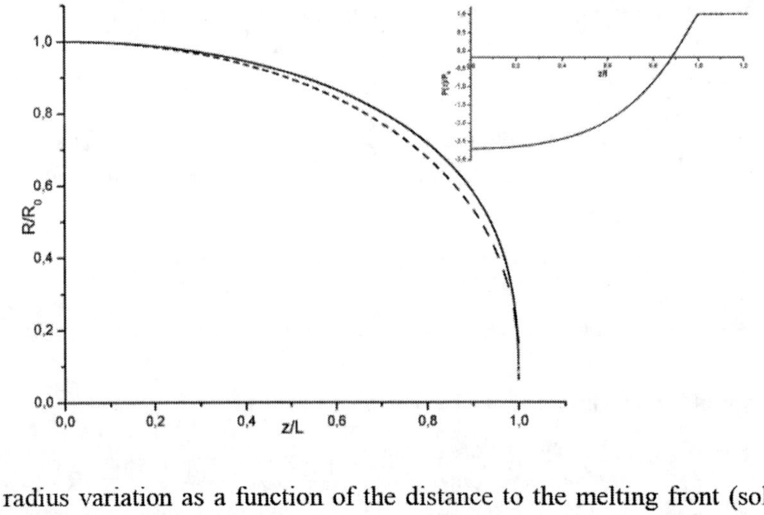


**Figure 1.** Pore radius variation as a function of the distance to the melting front (solid curve is the numerical solution; dashed curve is the analytical approximation). The inset shows the dependence of the pressure variation in the pore collapse region on the distance from the melting front.

For the given problem parameters, the width of the region in which the pores collapses was calculated. It was $L = u_0 L_0 = 7.5\cdot 10^{-4}$ cm. In addition, in figure 1, an analytical approximation of the numerical solution (10) and solution (13) is given by a function of the form

$$\tilde{y} = (1 - u^2)^{7/18}. \tag{15}$$

Also on the inset of figure 1 the dependence of the pressure variation in the pore collapse region on the distance to the melting front is represented. According to (9), the pressure difference in the collapse region is proportional to $V_m^2 \Phi_0$. At the considered melting rate and initial porosity, the negative values of the pressure are implemented in the collapse region. In order to molten material flow into empty pores during melting and pore collapse occur under surface tension forces, the condition

$$\frac{2\sigma}{R_0} > |P_0 - \rho_0 \cdot V_m^2 \cdot \Phi_0| \tag{16}$$

should be satisfied.

When the pressure in the melt and capillary pressure are equal, the process of pore collapse stops, which imposes constraints on the applicability of the proposed model. In order that pore collapse would occur during porous material melting by laser radiation in the quasisteady mode, the melting rate should be lower than a certain critical melting rate $V_{cr}$. The critical melting rate is determined from the equality of the right- and left-hand sides of inequality (16). At the pressure on the free surface $P_0 \ll \rho_0 \cdot V_m^2 \cdot \Phi_0$ critical melting rate is

$$V_{cr} = \sqrt{\frac{2\sigma}{\Phi_0 \rho_0 R_0}} \tag{17}$$

In our case the critical melting rate is $V_{cr}$=1,2·10$^4$ cm/s. An analysis of the numerical solution of equation (10) at variations of $\Phi_0$, $R_0$, $V_m$ shows that the width of the region in which pore collapse occurs is a universal function depending on the ratio $\frac{V_m}{V_{cr}}$; as the melting rate $V_m$ approaches critical value $V_{cr}$, the collapse region size increases. The asymptotics of the increase in the region in which collapse occurs, can be obtained at $\frac{V_m}{V_{cr}} \to 1$; in this case $L \sim L_0 \cdot \left(1 - \frac{V_m^2}{V_{cr}^2}\right)^{-1/2}$.

The estimate of the pore collapse time showed that for indium with a pore radius $R_0$ =10$^{-4}$ cm, it is about 10 ns. While for the same parameters of the problem, the time of defect removal to the surface due to the thermocapillary drift does not exceed 10 μs.

## 4. Conclusions

The present paper considers pore removing mechanisms during laser annealing of porous defects such as the thermocapillary drift of a bubble caused by the surface tension temperature dependence and the pore collapse under surface tension forces. The analysis of the pore collapse dynamics shows that the principal role is played by the pressure in the molten material. On the one hand, it defines the velocity of melt homogeneous motion from the surface to the collapse region; on the other hand, the velocity of the melt flow into empty pores. In addition, the equality of pressure in the melt and capillary one leads to stop the pore collapse, which imposes a limitation on the applicability of the proposed model. The criterion of model applicability is the critical melting rate of the porous material, in which the pore collapse is still possible.

## Acknowledgments

This work has been supported by National Research Nuclear University MEPhI in the framework of the Russian Academic Excellence Project (contract No. 02.a03.21.0005, 27.08.2013).

## References

[1] Gladush G G and Smurov I 2011 *Physics of Laser Materials Processing. Theory and Experiment* (Berlin: Springer)
[2] Young N O, Goldstein J S and Block M J 1959 *J. Fluid Mech.* **6** (3) 350
[3] Kuznetsov V M, Lugovtsov B A and Sher E I 1966 *Zhurnal Prikladnoi Mekhaniki I Tekhnicheskoi fiziki* **1** 88
[4] Brenner C E 1995 *Cavitation and Bubble Dynamics* (New York: Oxford University Press) p 50

IOP Conf. Series: Journal of Physics: Conf. Series **941** (2017) 012033 doi:10.1088/1742-6596/941/1/012033

# Attenuation of laser radiation by the flame of burning hydrocarbons and efficiency of remote cutting of metals

**S V Gvozdev[1], A F Glova[1,2], V Yu Dubrovskii[1], S T Durmanov[1], A G Krasyukov[1], A Yu Lysikov[1], G V Smirnov[1] and V M Pleshkov[1]**

[1] State Research Center of Russian Federation Troitsk Institute for Innovation and Fusion Research (SRC RF TRINITI), Moscow, Troitsk, 142190, Russia
[2] National Research Nuclear University MEPhI (Moscow Engineering Physics Institute), Moscow, 115409, Russia

E-mail: afglova@triniti.ru, gsv@triniti.ru

**Abstract.** Mobile laser technological complex MLTC-20 with radiation power 20 kW and radiation wavelength 1.07 μm created in SRC RF TRINITI on the base of a three cw fiber Yb lasers is used successfully at remote cutting of the metalworks at carrying out of the emergency-reduction works on the out of control gas wells. In this work the results of the investigation of the possibility and the efficiency of laser radiation application for remote cutting of metals on the emergency oil wells have been presented. Measurements of the mean absorption coefficient of the radiation of a cw fiber Yb laser under its propagation in a flame of burning oil in dependence on radiation intensity have been carried out. It was shown that at the intensity ~$10^4$ W/cm$^2$ the absorption coefficient traverses the maximum where its value is equal to ~0.1 cm$^{-1}$, and at the intensity increasing to the values $10^5$ – $10^6$ W/cm$^2$ it stabilizes on a small level ~$5 \cdot 10^{-3}$ – $10^{-2}$ cm$^{-1}$. It is established that the maximal velocity and the efficiency of remote cutting of the steel plates with a thickness up to 10 mm by the radiation with the intensity $10^6$ W/cm$^2$ exceed these factors at the intensity $10^4$ W/cm$^2$. The possibility of the efficient remote cutting of steel plate with a thickness of 60 mm by laser radiation having the power 7.5 kW and the intensity $10^5$ W/cm$^2$ has been demonstrated.

## 1. Introduction

Dissection of the metalworks at the accident on the out of control oil or gas well is the actual problem since it allows the works to perform upon the accident elimination and to replace the damaged equipment. Modern and safety method of the performing of emergency-reduction works in these conditions is remote separating cutting of the constructions by laser radiation [1]. The efficiency of this method at elimination of the consequences of real accidents on the gas wells is demonstrated successfully beginning from 2011 by the mobile laser technological complex MLTC-20 with radiation power of 20 kW [2]. Three cw fiber Yb lasers with radiation wavelength of 1.07 μm serve as the basis of the complex. Lasers are being housed in the same container; in a separate container three-channel lens forming telescope with a systems of guidance and control for the complex operation is housed (see figure 1). The range of the telescope radiation focusing is equal to 20 – 75 m, the cutting process is carried out at the laser radiation intensity on the metal surface $I_0 \approx 10^4$ W/cm$^2$. Figure 2*a* shows the typical view of the accident on gas well and operation of the MLTC-20 complex at the accident: the bright point on the dark background indicates the position of laser beam incidence. The fragment of the gas fittings sheared on the accident site by laser beam of the complex MLTC-20 in figure 2*b* is presented.

**Figure 1.** Total view (*a*) and telescope view (*b*) of the complex MLTC-20.

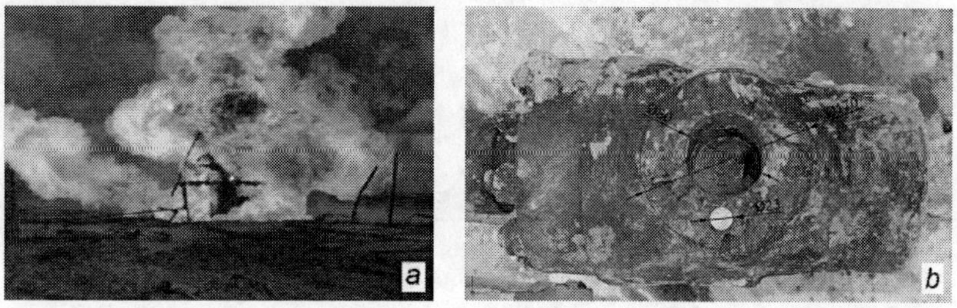


**Figure 2.** Operation of the complex MLTC-20 at the accident (*a*) and the fragment of the gas fittings sheared on the accident site by laser beam (*b*).

The success of the laser radiation application for these goals is largely dependent on the efficient and lossless delivery of the radiation to the operation site. Obviously, in the case of the radiation beam propagating through the flame, the beam is attenuated, depending on the composition of the burning hydrocarbons and the length of the burning zone [3]. The conditions for radiation transmission through the flame of burning natural gas are relatively favourable. However, these conditions may change in the case of the radiation passing through another flame, for example, the flame of burning oil.

The goal of the given paper is to study experimentally the dependence of the attenuation of high-power radiation from cw fiber Yb laser on the radiation intensity in the course of radiation propagation through a diffusion flame of burning oil and determination of the efficiency of remote cutting of metals. These studies were performed in lab conditions.

## 2. Attenuation of laser radiation by the diffusion flame of burning oil

The flame was initiated in a 50 mm × 50 mm metallic cuvette of depth 5 mm by the ignition of the oil vapours filling the cuvette under normal conditions. The flame was stabilized by the side walls of the cuvette. The radiation source was a cw fibre Yb laser with a radiation wavelength 1.07 μm and a maximal power 1.5 kW or 2.7 kW when single-mode or multimode laser was used. The laser beam propagates in the yellow part of the flame, the separation between the beam axis and the oil surface being equal to $h = 10$ mm. The length of the flame along the beam trace was $l \approx 30$ mm. The diameter of the beam waist $d_0$ amounted to 1 or 0.4 mm for single-mode laser and to 0.9 mm for multimode laser. The caustic length for both lasers greatly exceeds the flame length therefore the diameter of the beam at the distance $l$ was only slightly different from $d_0$. The input intensity $I_{in}$ of the radiation entering the flame was varied by changing the power and $d_0$ in the case of single-mode laser.

The value of the radiation absorption coefficient is a main criterion of the power attenuation of laser radiation under its propagation through the flame. Really, the power of the radiation output from the flame can be presented in the form $P_{out} = P_{in}\exp[-(\alpha + \mu)l]$, where $\alpha$, $\mu$ are the radiation absorption and scattering coefficients averaged over the length $l$, respectively, and $P_{in}$ is the input power at the flame entrance. Based on the results of measuring the fraction of the radiation power, scattered by the similarly-looking flame of the aviation kerosene [3], we assumed that $\mu << \alpha$ and for mean absorption coefficient obtain the following expression which was used in an experiments for its value determination: $\alpha = (-1/l)\ln(P_{out}/P_{in})$.

The dependence of $\alpha$ on $I_{in}$ for single-mode laser in figure 3 is presented. The measurements of the flame temperature under the conditions of laser beam propagation have shown that in the range of intensities $3.5 \cdot 10^4 - 10^5$ W/cm$^2$ the temperature linearly grows from 2000 to 3000 K. With a further increase in intensity, the temperature rise becomes slower, and at the maximal value of $I_{in} = 1.2 \cdot 10^6$ W/cm$^2$ the temperature becomes equal to 3400 K.

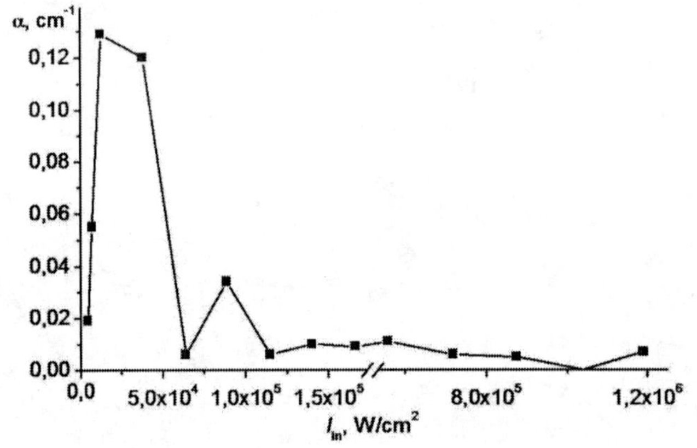


**Figure 3.** Dependence of $\alpha$ on $I_{in}$ at propagation of the radiation of single-mode laser in oil flame.

For qualitative interpretation of the data presented in figure 3 let us restrict ourselves to considering the influence of solid carbon particles on the radiation transmission [4]. These particles are always present in the hydrocarbon flame in the form of soot [5]. On the growing branch of the dependence $\alpha(I_{in})$, the radiation intensity increases, the temperature also increases and the heating power losses of laser radiation grow, leading to the grows of the absorption coefficient. Due to relatively low temperature of the flame the evaporation of the particles in the laser radiation field does not occur. However, with a rise of temperature, the rate of particle combustion dramatically increases [6]. Therefore, the section in figure 3, corresponding to the band of the intensity change $\sim 10^4 - 5 \cdot 10^4$ W/cm$^2$, can be considered as the region of unstable dynamic equilibrium between combustion of the particles and their convection supply to the interaction zone. At further increase in the intensity, strong, by more than an order of magnitude, decrease in the absorption coefficient takes place, and then its value practically stabilizes at a very low level of $5 \cdot 10^{-3} - 10^{-2}$ cm$^{-1}$. One can say that at $I_{in} \geq 10^5$ W/cm$^2$ the brightening of the channel of the radiation propagation occurs. The reason of this can be the following. At $I_{in} \geq 10^5$ W/cm$^2$ the radiation intensity is so high, that the absorbing particles are getting burned already in the peripheral zone of the beam. As a result, the major central part of the beam appears to be free of particles and propagates in the flame with small power losses due to the weak absorption of the laser radiation with the wavelength 1.07 μm by gaseous burning products.

Similar measurements were carried out using radiation of a multimode laser. Results of the measurements for oil flame and aviation kerosene flame in figure 4 are presented. The dependences $\alpha(I_{in})$ have the same features as the one presented in figure 3, and the absolute values of the absorption coefficient for the fixed values of $I_{in}$ for the oil flame are virtually the same for the single-mode and multimode lasers. This may evidence in favour of the conclusion that the main role in the cleaning of the light propagation channel is played by the intensity of radiation, rather than by the focusing spot size $d_0$ and the radiation quality.

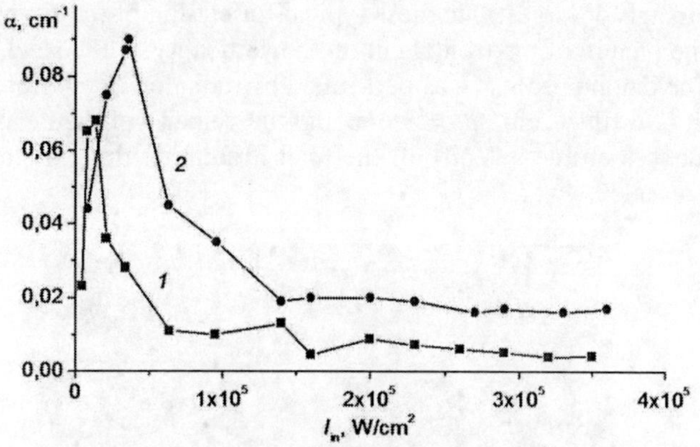


**Figure 4.** Dependences of $\alpha$ on $I_{in}$ at propagation of the radiation of multimode laser in oil flame (*1*) and aviation kerosene flame (*2*).

## 3. Remote cutting of metals by laser radiation

Reduction of the radiation absorption coefficient in oil flame at radiation intensities $I_{in} > 10^5$ W/cm$^2$ can evidence about the basic possibility of the remote cutting of metalworks on the emergency oil well. But the question arises for the cutting efficiency, because due to decreasing of the focal spot size at the intensity increasing and absence of the auxiliary gas jet for melt removal, their removal can be hampered along of the viscosity of the melt.

To answer this question, the comparative experiments were carried out upon remote cutting of the verticality St.3 steel plates with a thickness $H$ = 3, 4, 6 and 10 mm by the horizontal beam of a cw fiber Yb laser with the radiation power $P_0$ = 2.7 kW at and two values of the radiation intensity on the metal surface $I_0 = 10^4$ W/cm$^2$ (the size of the focal spot $d_0$ = 6 mm) and $I_0 = 10^6$ W/cm$^2$ ($d_0$ = 0.6 mm). The results of the experiments are shown in figure 5.

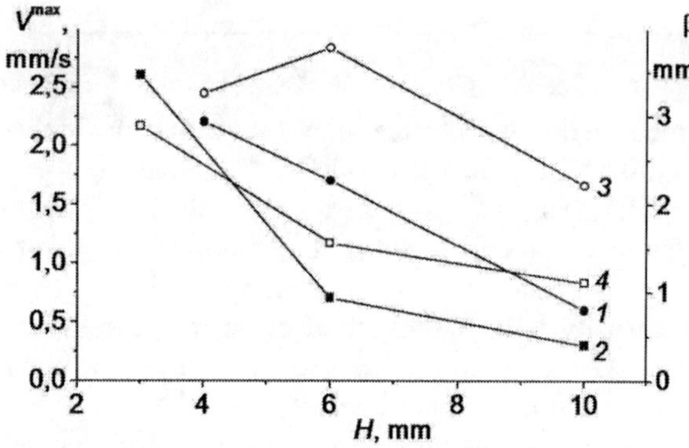


**Figure 5.** Dependences on $H$ of the maximal velocity $V^{max}$ (*1,2*) and cutting efficiency $\beta$ (*3,4*) at $P_0$ = 2.7 kW and $I_0 = 10^6$ (*1,3*) and $10^4$ W/cm$^2$ (*2,4*).

In figure 5 $V^{max}$ is the maximal velocity of the beam moving at which the solid cut still exists, $\beta = V^{max}H/P_0$ is the cutting efficiency determined as the area of the cut related to the unity of the expended energy [7]. It is seen, that at $I_0 = 10^6$ W/cm$^2$ as maximal velocity as well as the efficiency exceed these factors at $I_0 = 10^4$ W/cm$^2$ for all values of the plate thickness. The reason of this is increase of the role of the vapour recoil momentum at melt removal [8], the influence of which can neglect at removal under action of the gravity force [9,10] at $I_0 \leq 10^4$ W/cm$^2$.

The efficient remote cutting by laser radiation with an increased intensity ~$10^5$ W/cm$^2$ is possible also at cutting of the metals of largest thickness typical for cutting of the fittings at the accident. As an example in figure 6 the picture of the frontal cut on the verticality A40G steel plate with a thickness $H$ = 60 mm is shown. The cutting process was performed by radiation of cw fibre Yb laser with radiation power $P_0$ = 7.5 kW at $I_0 = 10^5$ W/cm$^2$, $d_0$ = 3 mm and the velocity of the beam moving $V$ = 0.1 mm/s. In table 1 the dependences on the velocity of the total irradiation time, the cut depth and the frontal width of the cut are presented.

**Figure 6.** View of the frontal cut on the verticality A40G steel plate with a thickness $H$ = 60 mm at $P_0$ = 7.5 kW, $I_0 = 10^5$ W/cm$^2$ and $V$ = 0.1 mm/s.

**Table 1.** Parameters of cutting of the verticality steel plate with a thickness of 60 mm at $P_0$ = 7.5 kW and $I_0 = 10^5$ W/cm$^2$.

| Velocity (mms$^{-1}$) | Time (min) | Depth (mm) | Width (mm) |
|---|---|---|---|
| 0.1 | 12 | 50 | 10 |
| 0.2 | 6 | 34 | 10 |
| 0.3 | 4 | 27 | 10 |
| 0.4 | 3 | 23 | 10 |

## 4. Conclusions

The investigations performed in this work have shown that the mean absorption coefficient of the radiation of a cw fiber Yb laser by the diffusion oil flame changes non-monotonic in dependence on the intensity of incident radiation and at the intensity more than $10^5$ W/cm$^2$ stabilizes on a level of $5\cdot10^{-3} - 10^{-2}$ cm$^{-1}$. The possibility of the efficient remote cutting of metals by radiation of these lasers with the intensity $\geq 10^5$ W/cm$^2$ has been demonstrated exceeding 1.5 – 2 times the efficiency at cutting by the radiation with the intensity $\leq 10^4$ W/cm$^2$. It became possible due to the change of the main mechanism of the melt removal under action of the gravity force at low intensity on the removal under action of the vapour recoil momentum at higher intensity.

The obtained results should be taken into account in the development and application of new-generation mobile laser technological complexes providing the radiation intensity on the level of $10^5$ W/cm$^2$ in the active zone and aimed at remote cutting of metalworks of gas-condensate and oil wells, spouting as a result of an accident.

## References

[1] Blokhin O, Vostrikov V, Gavrilyuk V, Krasyukov A, Naumov V, Svotin P, Shashkov V and Shachkin L 2001 *Chemical and Petroleum Engineering* **5** 52

[2] Krasyukov A, Durmanov S and Cherkovets V 2012 *6-th Int. Symp. on High-Power Fiber Lasers and their Applications (St. Petersburg)* p 72

IOP Conf. Series: Journal of Physics: Conf. Series **941** (2017) 012033 doi:10.1088/1742-6596/941/1/012033


[3] Gvozdev S, Glova A, Dubrovskii V, Durmanov S, Krasyukov A, Lysikov A. Smirnov G and Solomakhin G 2012 *Quant. Electron.* **42** 350

[4] Gvozdev S, Glova A, Dubrovskii V, Durmanov S, Krasyukov A, Lysikov A. Smirnov G and Pleshkov V 2015 *Quant. Electron.* **45** 582

[5] Lee C, Lee W, Oh K, Shin H and Yoon J 2006 *Combust., Explos., and Shock Waves* **42** 688

[6] Khitrin L 1957 *Physics of Combustion and Explosion* (Moscow: Moscow State University Press)

[7] Gladush G, Drobyazko S, Rodionov N, Antonova L and Senatorov Yu 2000 *Quant. Electron.* **30** 1072

[8] Gvozdev S, Glova A, Gladush G, Dubrovskii V, Durmanov S, Krasyukov A, Lysikov A, Smirnov G, and Pleshkov V 2017 *J. of Adv. Mater.* **4** 44

[9] Likhanskii V, Loboiko A, Antonova G, Krasyukov A and Sayapin V 1999 *Quant. Electron.* **26** 139

[10] Antonova G, Gladush G, Krasyukov A, Kosyrev F and Sayapin V 2000 *High Temperature* **38** 477

IOP Conf. Series: Journal of Physics: Conf. Series **941** (2017) 012034 doi:10.1088/1742-6596/941/1/012034

# Synthesis of oxide and nitride ceramics in high-power gyrotron discharge

**N S Akhmadullina[1,*], N N Skvortsova[2,3,**], E A Obraztsova[2,4], V D Stepakhin[2,5], E M Konchekov[2,5], Yu F Kargin[1], and O N Shishilov[6,7]**

[1] A.A. Baikov Institute of Metallurgy and Material Science, Russian Academy of Sciences, Lenisky prospect 49, Moscow, 119991, Russia
[2] Prokhorov General Physics Institute, Russian Academy of Sciences, ul. Vavilova 38, Moscow, 119991 Russia
[3] National Research Nuclear University MEPhI (Moscow Engineering Physics Institute), Kashirskoe shosse, 31, Moscow, 115409, Russia
[4] M.M. Shemyakin and Yu.A. Ovchinnikov Institute of Bioorganic Chemistry, Russian Academy of Sciences, Miklukho-Maklaya st. 16/10, Moscow, 117997, Russia
[5] N.I. Pirogov Russian National Research Medical University, Ostrovitianov st. 1, Moscow, 117997, Russia
[6] Institute of Fine Chemical Technologies, Moscow Technological University, Vernadskogo prospect 86, Moscow, 119571, Russia
[7] O3-Innovation, LLC, Nobel str. 1, Skolkovo Innovation Center, Moscow, Russia

E-mail: [*]nakhmadullina@mail.ru, [**]mukudori@mail.ru

**Abstract**. Synthesis of oxides, nitrides, and oxynitrides of silicon and aluminium by a pulsed microwave discharge in the mixtures of metal and dielectric powders is described. The microwave pulses were generated by high-power gyrotron (frequency 75 GHz, power up to 550 kW, pulse duration from 0.1 to 15ms). $SiO_2$ + β-$Si_3N_4$ (1:1 by molar) and α-$Al_2O_3$ + AlN (2:1 by molar) mixtures with Mg (1 and 5wt%) were treated in air with microwave pulses with power of 250÷400 kW and duration of 2÷8 ms. It was found that the discharge cannot be initiated for both mixtures in absence of Mg at any pulse power and duration. When 1% of Mg was added, the discharge was observed for both mixtures under 8 ms pulses of 400 kW; however, the amounts of materials produced were not enough for analysis. With 5% of Mg the discharge was observed for both mixtures under 8 ms pulses of 350 kW, and products of the plasma-chemical processes in the $Al_2O_3$ + AlN mixture were analyzed.

## 1. Introduction

Ceramic materials are widely used in modern technology due to their unique mechanical and physical properties. Plasma methods are among the most promising techniques for the synthesis, modification, and processing of micro- and nanostructured ceramic materials[1–3]. There have been many examples of nanostructured material syntheses using various discharges: spark, arc, glow, high-frequency, microwave, and so on. Gyrotrons, being one of the most powerful sources of microwave discharges, are also used for the technologies mentioned, however, their application is rather limited to the experiments on high-density ceramics sintering [4] and deposition of thin films [5,6].

Recently we have shown that gyrotron discharge in mixtures of metal and dielectric powders can be used for synthesis of micro- and nanostructured materials. The synthetic approach was developed on a special stand incorporated into the standard design of the MIG-3 gyrotron complex [7]. The complex consists of two gyrotrons with a total power up to 1.5 MW. It was constructed in 2012 for plasma generation and heating in the L-2M stellarator. The plasma-chemical experiments were carried out in a specially constructed reactor [8] using gyrotron with a power up to 0.55 MW, frequency of 75 GHz, and microwave pulse duration between 0.1 and 15 ms. An actual gyrotron power was measured with a calorimeter [9]. The discharge was initiated in Ti + B, Mo + B, W + B, Mo + W + B, Mo + BN, Ti + $KBF_4$, and Mg + $CB_4$ mixtures in the air and nitrogen atmospheres [8,10,11]. It was found that the surface temperature of the powder mixture in the breakdown phase for these discharges was 2000 to 5000 K, while the temperature in the plasma–gas layer reached 10000 K [11]. The main products of the process were metal and boron oxides and metal borides [10,11]. Also formation of BN was found in some cases.

Here we describe our results of studies of plasma-chemical reactions under high-power gyrotron-initiated discharge in mixtures of metal and dielectric powders. We consider the possibility of obtaining new oxide, nitride and oxynitride materials from $SiO_2$ + $\beta$-$Si_3N_4$ and $\alpha$-$Al_2O_3$ + AlN mixtures.

## 2. Experimental

The initial materials used were commercial powders of AlN (chemical grade, AlN powders for sealants), $\alpha$-$Al_2O_3$ (chemical grade), $SiO_2$ (pyrogenic, Cabot), and Mg (chemical grade, Sigma-Aldrich). Initial powder of $\beta$-$Si_3N_4$ was prepared by self-propagating high-temperature synthesis at the Institute of Structural Macrokinetics and Materials Science RAS (Chernogolovka) [12].

The materials produced in plasma were analyzed using X-ray diffraction analysis (Shimadzu XRD 6000 diffractometer, CuK$\alpha$ radiation, graphite monochromator, $\lambda$ = 1.54178 Å) and energy-dispersive microanalysis using the EDMA, SEM JSM5910-LV with the analytical system INCA.ENERGY. Scanning electron microscopy (JEOL JSM5910-LV, Zeiss Merlin) was used to get images of produced structures.

### *2.1. Synthesis in gyrotron-initiated discharge in metal-dielectric powder mixtures*

Experiments were carried out with stand-mounted plasma-chemical complex of the MIG-3 installation described above. A specially designed plasma-chemical reactor (Figure 1, left) [8] was placed between two mirrors of the quasi-optical tract of the complex. Microwave breakdown in the reactor was initiated in air in the mixture of metal and dielectric powders placed on a quartz substrate with free upper surface (Figure 1, right). A quartz tube was set inside the reactor to collect the synthesized substances for further analysis.The gyrotron power was measured by a flow calorimeter. The evolution of direct and reflected gyrotron power during the microwave discharge was monitored using a system of microwave detectors, which were calibrated using the calorimeter [9].

The samples for plasma treatment (Figure 1, right) were prepared in the following way: 1-mm-thick layer of boron powder (2) followed by a 0.5-0.7 mm layer of metal-dielectric powder mixture (3) was put on the quartz plate (1). The layers were slightly squeezed with a flat quartz plate, and a quartz cylinder was installed over the plate. The plate with cylinder was set into the reactor. The upper surface of the powder remained open to ensure free plasma-gas-dynamic expansion of reaction products into the quartz cylinder volume.

The gyrotron microwave radiation was supplied through the bottom quartz plate. Irradiation of the samples was carried out with 5÷100 single 2÷12-ms pulses with intervals no less than 20 s. The gyrotron radiation power was 250÷400 kW which corresponds to intensity of ~6÷10 kW·cm$^{-2}$ inside the powder mixture. The Gaussian microwave beam with a characteristic radius of 40 mm penetrated the reactor volume and was absorbed in ceramic sinks located at a distance of 60 cm above the upper reactor exit window. The synthesized substances were deposited on both the walls of the quartz cylinder (6) and the remaining surfaces of the reactor including the upper quartz window.

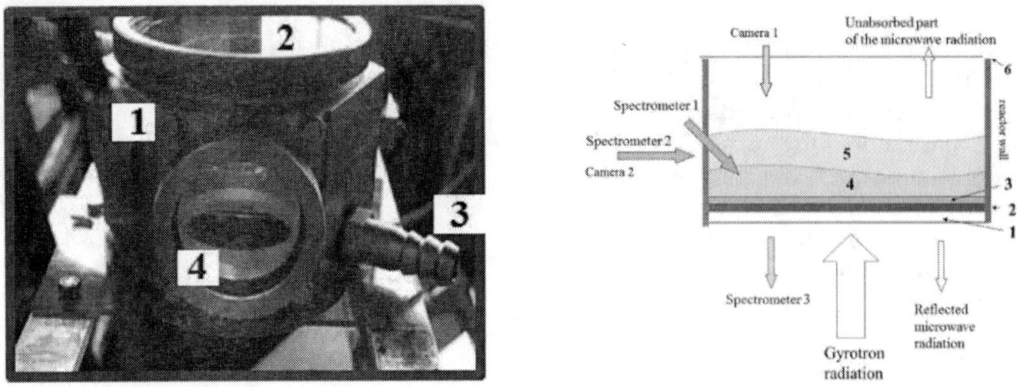


**Figure 1.** Left: Plasma chemical reactor: 1 – hermetic chamber, 2 – exit quartz window, 3 – vacuum nozzle, 4 – diagnostic window. Right: General scheme of the reactor and measurements: 1 – quartz plate, 2 – boron powder layer, 3 – mixture of metal and dielectricpowders, 4 – plasma, 5 – working gas, 6 – quartz cylinder ($d$ = 70 mm, $L$ = 350 mm). The arrows on the left-hand side indicate directions along which the microwave radiation was received by two cameras (#1, 2) and three spectrometers (#1, 2, 3).

### *2.2. Discharge parameters and evolution*

The microwave glow discharge, powder and synthetic substances were observed through the viewing window in the plasma-chemical reactor. The evolution of integrated microwave glow discharge and afterglow phase was recorded by a high-speed camera «Fastec Imaging IN250M512» (camera 1 in Figure 1b) synchronized with measurements of three spectrometers. Figure 1b shows the locations of the three spectrometers. The time evolution of discharge radiation in the optical wavelength range was recorded with the help of three spectrometers «Ava-Spec» in the range 370 ÷ 920 nm with a resolution of 0.7 and 0.3 nm and in the range of 250 ÷ 800 nm with a resolution of 0.8 nm. From the recorded spectra we analysed evolution of characteristic atomic and molecular wavelengths of reaction components in the different sites of the reactor. The receiving system of spectrometer no. 1 was directed on the powder and allowed to record the glow of the powder surface as well as that of the plasma and gas phase (layers 2–4 in Figure 1b). The receiving systems of spectrometers no. 2, 3 were directed parallel to the powder mixture surface allowing to record radiation mainly from the plasma and gas regions of the discharge (layers 3, 4 in Figure 1b). In some experiments, one of the spectrometers was used to record radiation from the lower surface of the powder mixture. During a single microwave discharge, spectrometers recorded up to 100 optical spectra at intervals of about 4 ms.

## 3. Result and discussion

Previously it was shown that in order to transform the initial mixture of powders into a new material the gyrotron discharge parameters should be chosen so that the microwave discharge was followed by an afterglow phase [11]. Only under this condition almost complete absorption of microwave radiation of gyrotron in the powder mixture and initiation of chemical reactions of synthesis of substances was possible [13]. So we have treated the systems listed in table 1 with 2÷8 ms microwave pulses generated by gyrotron at power levels from 250 to 400 kW. For all the systems, no discharge was observed if the pulse duration was below 2÷6 ms at any power level. When the mixtures were treated with 8 ms pulses, the discharges were observed at the power level 400 kW both with Al- and Si-containing mixtures with 1% of Mg. However, it was not possible to collect significant amounts of the products (table 1, entries 3 and 4). When 5% of Mg was added to the mixtures, the discharges could be initiated at 400 kW power level (table 1, entries 5 and 6); but only in the case of Al-containing mixture we collected an amount of new product materials appropriate for analysis with SEM and XRD. Only one pulse lead to the discharge for the systems with 1% of Mg, while up to three pulses were successful when 5% of Mg was added. This means that Mg is consumed during the discharge taking part in plasma-chemical processes.

**Table 1.** Results of microwave discharge initiation in the AlN + $Al_2O_3$ and $Si_3N_4$ + $SiO_2$ systems with 8 ms pulses of 250÷400 kW.

| Entry | System | 250 kW | 300 kW | 350 kW | 400 kW |
|---|---|---|---|---|---|
| 1 | $SiO_2$ + $Si_3N_4$ | no/no* | no/no | no/no | no/no |
| 2 | $Al_2O_3$ + AlN | no/no | no/no | no/no | no/no |
| 3 | $SiO_2$ + $Si_3N_4$ + 1% Mg | no/no | no/no | no/no | **yes**/no |
| 4 | $Al_2O_3$ + AlN + 1% Mg | no/no | no/no | no/no | **yes**/no |
| 5 | $SiO_2$ + $Si_3N_4$ + 5% Mg | no/no | no/no | **yes**/no | – |
| 6 | $Al_2O_3$ + AlN + 5% Mg | no/no | no/no | **yes/yes** | – |

*Discharge initiated/significant amount of products formed in plasma-chemical process.

SEM images of the obtained materials are given in Figure 2 (top and bottom left). One can see that the material basically consists of amorphous particles 2-10 microns in size and well-shaped spherical crystals. Elemental analyses showed that spherical crystals mostly consist of Mg, Al, and O, so one can assume that the crystals are $MgAl_2O_4$. XRD-analysis confirmed the suggestion (Figure 2, bottom right). Characteristic diffraction peaks of $MgAl_2O_4$ spinel were identified in XRD pattern, although according to XRD data $MgAl_2O_4$ is not a major phase, which is actually aluminum nitride AlN coming most likely from starting material. Thus, we can conclude that plasma-chemical processes under microwave discharge in the mixtures of metal and dielectric powders can lead to formation of new materials. These materials can be highly dispersive and show high crystallinity.

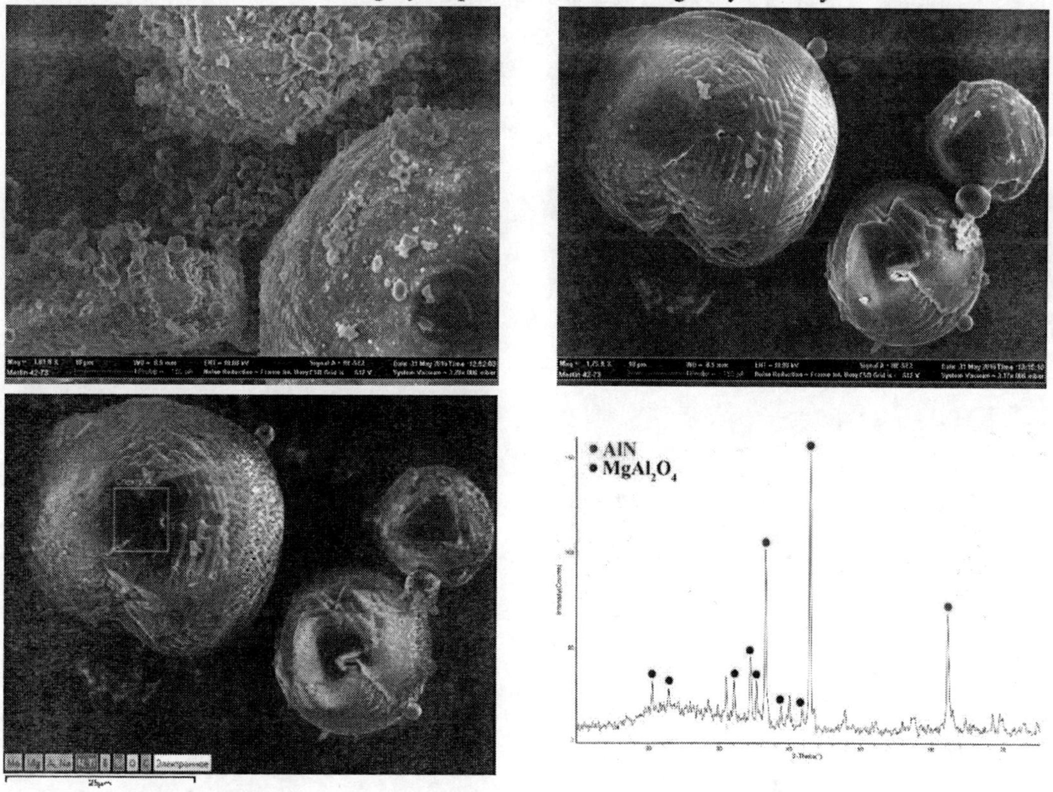


**Figure 2.** Top: SEM-images of products from the system **6**. Bottom left: Element mapping of products from the system **6**. Bottom right: XRD pattern of products from the system **6**.

## 4. Conclusions

Usability of the microwave discharge generated by high-power gyrotron in mixtures of metal and dielectric powders for synthesis of oxide and nitride ceramic materials containing Al and Si was demonstrated. It was shown that the discharge cannot be initiated in mixtures of dielectric powders $Al_2O_3$ + AlN and $SiO_2$ + $Si_3N_4$ even with 8 ms pulses of 450 kW. Addition of 1wt% of Mg allowed, in principle, to initiate the discharge with 8 ms pulses of 400 kW, and 5% is enough for obtaining significant amounts of new synthesized materials including micro dispersed magnesium aluminate $MgAl_2O_4$. SEM images showed that microcrystals of $MgAl_2O_4$ are well shaped. XRD analysis confirmed spinel structure for them.

## Acknowledgements

We are grateful to Russian Foundation for Basic Research for financial support (projects #15-32-70014-mol_a_mos, 17-03-00630-a). One of the authors acknowledges the financial support from National Research Nuclear University MEPhI in the framework of the Russian Academic Excellence Project (contract No. 02.a03.21.0005, 27.08.2013).

## References

[1] Clark D E and Sutton W H 1996 *Annu. Rev. Mater. Sci.* **26** 299
[2] Thostenson E T and Chou T W 1999 *Compos. Part A: Appl. Sci. Manuf.* **30** (9) 1055
[3] Ivanovskii A L 2002 *Russ. Chem. Rev.* **71** (3) 175
[4] Bykov Y V, Rybakov K I and Semenov V E 2001 *J. Phys. D: Appl. Phys.* **34** R55
[5] Nakano H, Ozono K, Saji T, Miyake S, Hayashi H 2013 *Optical Materials* **35** 2045
[6] Mansfeld D A, Vodopyanov A V, Golubev S V et al. 2014 *Thin Solid Films* **562** 114
[7] Batanov G M etal. 2013 *PlasmaPhys. Rep.* **39** (13) 1088
[8] Batanov G M et al. 2013 *Plasma Phys. Rep.* **39** (10) 843
[9] Kharchev N K et al. 2013 *Rev. Sci. Instrum.* **84** 013507
[10] Batanov G M et al. 2013 *J. Nanoelectron. Optoelectron.* **8** 58
[11] Batanov G M et al. 2016 *J. Nanophotonics* **10** (1) 012520
[12] Zakorzhevsky V V, Borovinskaya I P, Sachkova N V 2002 *Inorg.Mater.* **38** 1131
[13] Ukryukov G V et al. 2017 *Engineering Physics* **2** 27

# Influence of the modes of laser cladding on bond strength and wear resistance of coatings

**V P Birukov[1,*], D Yu Tatarkin[2], E V Chriptovish[2] and A A Fichkov[1]**

[1] Federal budget-funded research Institute for machine science named after A. A. Blagonravov of the Russian Academy of Sciences (IMASH RAN), Maly Kharitonievskiy Pereulok 4, Moscow101990, Russia
[2] IPG IRE-Polus, Vvedenskogo Sq. 1, Fryazino, Moscow141190, Russia

E-mail: *laser-52@yandex.ru

**Abstract**. The paper presents the results of metallographic studies and laboratory comparative tests on the adhesion strength of the coating to the substrate and abrasion on the scheme Brinell-Haworth cladding powder coatings on Nickel-based and samples of steel 40X. Strength of adhesion of the first coating layer with a hardness of HRC 38–42 was 400–480 MPa. It is shown that when the hardness of the deposited layer HRC 58–61 wear resistance of the coatings is higher than 40X steel in the normalized and improved in 10 and 4.6 times, respectively.

## 1. Introduction

In industry there are different methods for cladding of metals: electric arc, plasma, laser, gas, high frequency cladding and others. Each of existing methods has its advantages and disadvantages, which are characterized by specific technical and economic indicators in determining the effective areas of their application. For arc cladding in conditions of hydro-abrasive wear should be preferred deposited coating carbide phase having a higher wear resistance compared to martensite-austenitic structures [1]. After electric arc cladding the allowance on subsequent machining can reach 2 mm, which increases the complexity of manufacturing products. Modern plasma and gas-dynamic methods allow obtain coatings with porosity 1–10% [2, 3]. The allowance on machining after laser cladding should not exceed 0.5–0.7 mm. In the deposited layer do not contain defects like pores, cavities and cracks. Methods of feeding the filler materials can be divided into two main groups: pre-location of the powder materials on the surface and flow directly into the treatment area simultaneously with the laser radiation [4]. A method for pre-coating by using the slip coating is characterized by high utilization of powder material. However, a more common method of feeding the powder material is forced feed directly into the molten bath of base material. Investigations of the coatings based on Nickel with a hardness of 48–54 HRC, obtained by laser cladding with a gas and fiber lasers show improved wear resistance of friction sliding 2 times as much in comparison with steel 38CrNi3Mo [5, 6].

The aim of this work was to increase the utilization rate of the powder material, the determination of the strength of adhesion coatings which uses as intermediate layer and optimization laser cladding for increase abrasive wear resistance, the deposited layers.

## 2. Materials and equipment

In the experiments used the universal equipment: IPG IRE-Polus with fiber laser LS-5, the robot of the company KUKA, optical head company PRECITEC. As a finishing material was chosen powders are Nickel-based firm Höganäs, Sweden and OJSC "POLEMA", Tula. The size of particle the powder was 40 to 150 μm. The laser power was varied in the range of 1800–4800 W. The speed is of movement of the optical head abaut 0.5–2 m/s in the transverse direction and from 1 to 15 mm/s in the longitudinal direction. In laser cladding of pre-poured the powder layer thickness 3–10 mm. The samples were cut from steel plate with dimensions of 140×140×20 mm with electric-spark method with dimensions 25×70×20 mm. Metallographic studies were performed with use of hardness testing of the PHT-3M, metallographic microscope Altami met-1C. The abrasion testing was carried out on the machine friction of BH-4 upgraded to IMASH RAN scheme Brinell-Haworth [7]. Samples for testing the strength of adhesion of the coating to the substrate was carried out on the testing machine "Instron" 1115 in accordance with GM250-87 "Determination of the adhesion strength of thermal spray coating with the base metal. Guidance materials ", revised taking into account the specifics of laser cladding [8].

## 3. Results and experiment

Sample of the investigated material deposited on it a coating with a hardness of HRC 38–42, symmetrically on both sides mechanically processed to the width of the deposited layer in the direction of compression is 2 mm. Then the sample was mounted in a fixture and pressed through a die. Thus, under the action of shearing stresses was the cut of the deposited layer, and the shear stress determined the adhesion strength of the deposited layer to the substrate.

The sample made of the base metal in the form of a plate thickness of 10 mm, a width and a length of 25 mm. Thickness of the deposited layer was 1.5–3 mm. The results of the tests obtained values of the strength adhesion of coatings in the range of 400 to 480 MPa, depending from the processing mode.

For testing technological regimes used powder firms Höganäs brand 1360 (claimed hardness HRC 58) with different thickness of the poured layer of 3.5–10 mm. Obtained deposited layers of thickness 3–9 mm, width 20–80 mm in one pass of hand robot. Figure 1 is the metallographic of a single claddings tracks. By change power, speed and diameter of the laser beam on the cladding surface had been bring to light conformity to natural laws of change micro-hardness in the deposited layer. The overlapping cladding paths is shown in figure 2. In it there are no defects like pores, cracks and cavities, as and track cladding under optimum processing conditions.

**Figure 1.** Metallographic of a cladding a single track in a single pass×7.

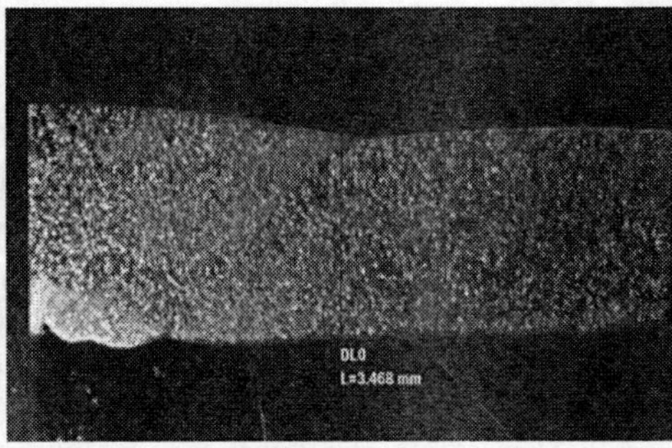


**Figure 2.** Metallographic of the overlap area of the deposited layers ×10.

When change mode cladding surfacing micro-hardness in the deposited layer varies widely 6120–12000 MPa. At high energy density laser beam of alloying elements and carbon burn and micro-hardness in the deposited layer decreases, defects appear in the form of pores. At low values of the energy density of the laser radiation is not a complete fusion in the powder material, and reduced micro-hardness.

Table 1 shows the results of testing abrasion wear according to the scheme Brinell-Haworth. Rotating rubber disk pressed on flat sample with the deposited coating. In the friction zone was fed silica sand with a size of particle 200–600 μm. The test duration was 10 minutes. By results the tests of three samples was determined average value of the mass loss of the deposited coating at each mode of processing.

**Table 1.** Results of the Brinell-Havorth abrasion test.

| | Powder grade | Micro-hardness, MPa | Loss of mass, g |
|---|---|---|---|
| 1 | 1360 | 7180–8300 | 0.0071 |
| 2 | 1360 | 7180–10800 | 0.0035 |
| 3 | 1360 | 6120–9180 | 0.0052 |
| 4 | 1360 | 7040–9540 | 0.0048 |
| 5 | 1360 | 7180–8340 | 0.0062 |

As a result of the carry out studies on abrasive wear of the first batch of samples were found, optimal regimes of laser cladding of powder coatings. The second batch of samples was melted with powders of different hardness and both manufacturers of powder materials. Figure 3 shows the dependence of Brinell-Havort's abrasive wear from the hardness of the deposited layer and abrasive wear of the samples 40X steel. From the given data , it follows that the cladding layers with a hardness of HRC 58–61 (powders of the company Höganäs-1 and OJSC POLEMA-2) are 10 times as much than wear resistance the normalized steel 40X (HB180) and 4.6 times as much higher, than the improved steel 40X (HB 250–270).

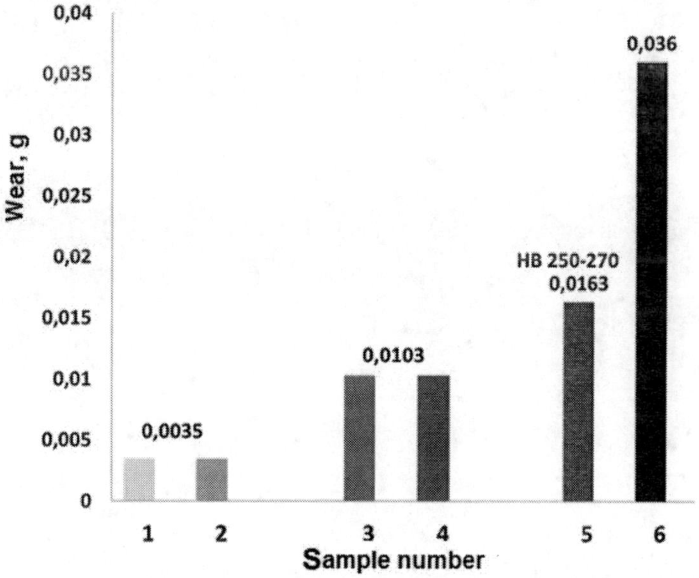


**Figure 3.** Dependence of abrasive wear from the hardness of the deposited layer powder materials and a sample of steel 40X: 1 – Höganäs, HRC 58–61; 2 – OJSC POLEMA, HRC 58–61; 3 – Höganäs, HRC 38–42; 4 – OJSC POLEMA , HRC 38–42; 5 – steel 40X, HB250–270; 6 – steel 40X, HB 180.

## 4. Conclusion

The technology was developed of laser cladding of powder materials on the basis of nickel with a width of the deposited layer of 20–80 mm and a roller height of 3–9 mm at one pass hand of robot. The carried out tests for abrasive wear according to the Brinell-Havorth scheme showed an increase in

IOP Conf. Series: Journal of Physics: Conf. Series **941** (2017) 012035 doi:10.1088/1742-6596/941/1/012035

wear resistance of the cladding coatings by 10 times as much in comparison with the normalized steel 40X. The adhesion strength of a coating powder based on nickel with hardness of HRC 38–42 was 400–480 MPa.

## References


[1] Pogodaev L I and Ezhov Y E 2014 *J. of Machinery Manufacture and Reliability* **43** 528
[2] Zabelin A M, Shiganov I and Chirkov A M 2007 *Hybrid Laser Surfacing Technology* (Moscow: MGOU) in Russian
[3] da Costa Alves Borges B M F 2008 *Laser Cladding Using Filler Powder and Wire. Productivity and Quality* Ph.D. thesis Technical University of Lisbon
[4] Grigoryants A G and Misyurov A I 2005 *Tekhnologiya Mashinostroeniya* **10** 32
[5] Biryukov V P and Dozorov A V 2006 *J. of Machinery Manufacture and Reliability* **1** 60
[6] Biryukov V P, Lapteva V G and Khrennikova I A 2013 *Problems of mechanical engineering and reliability of machines* **5** 1
[7] Khrushchov M M and Babichev M M 1970 *Abrasion wear* (Moscow: Nauka) in Russian
[8] Biryukov V P 2014 *Proc. of GOSNITI* **117** 196

# Pulsed abnormal glow discharge with hollow cathode for nitriding of internal cylindrical surfaces

**Yu V Borisyuk*, V V Kozlova, D V Mozgrin, P S Nenashev, N M Oreshnikova, M M Berdnikova and A A Pisarev**

National Research Nuclear University MEPhI (Moscow Engineering Physics Institute), Kashirskoe shosse 31, Moscow 115409, Russia

E-mail: *yvb65@mail.ru

**Abstract.** Possibility of nitriding of internal cylindrical surfaces in plasma of abnormal glow discharge with hollow cathode in the pulse-periodic mode in $N_2$+$H_2$ was investigated. Tubes with internal diameters of 6, 8, and 9 mm and with the aspect ratio from 12 to 50 were made a low carbon steel were used. Regimes of the discharge were investigated. Nitrided layers in different parts of the tubes were analyzed.

## 1. Introduction

Plasma nitriding is a method of thermochemical treatment, which is widely used to enhance surface hardness, fatigue strength, and corrosion resistance of steels [1, 2]. In industry, abnormal glow discharge is commonly used for this purpose. The specific technological task is treating of cavities, which are usually inaccessible for plasma. Still, the need to modify cavities and tubes of small diameters and large aspect ratios exists. In [3], it was proposed to use the abnormal glow discharge (AGD) with hollow cathode in the pulse-periodic mode for this purpose. The present work is devoted to investigation of ignition and properties of the abnormal glow discharge in tubes and demonstration of nitriding of internal surfaces of the tubes.

## 2. Discharge features

Experiments were performed in the experimental set up described in details in [4]. The treated tube served as the cathode, and vacuum wall served as the anode. Nitriding was performed in plasma of the abnormal hollow cathode discharge. Tubes with the internal diameter of 6, 8, and 9 mm and with the aspect ratio from 12 to 50 were used as the cathode. Tube dimensions are denoted in the text below by three numbers in mm (external diameter × internal diameter × length). The tubes were made of Russian low-carbon steel 30ХН2МФА (its analogues in Poland and Bulgaria are 30HN2MFA and 30ChN2MFA, respectively). Before nitriding, the samples were cleaned in an ultrasonic bath and then in a glow-discharge plasma to remove oxides (discharge in argon, pressure 0.5–0.8 mbar, voltage 450–500 V, duration 15–30 minutes). Plasma nitriding was performed in the 50–50% mixture of nitrogen and hydrogen. Plasma parameters during nitriding were the following: voltage 450–650 V, frequency 1–100 kHz, duty cycle 25–85%, pressure 0.6-6.6 mbar, duration 4–4.5 hours. Voltage and discharge current as well as the temperature of samples were monitored.

The discharge was initiated on the outer surface of the tube at the voltage of 600 V, and then it propagated into the tube. Penetration into the tube was accompanied by sharp increase in the discharge current and decrease in the voltage to 450 V (mention that the voltage is negative). The current and voltage oscillograms of the discharge are shown in figure 1. Discharge transition into the tube is shown on a larger scale in the circular inset. The small initial step of the current of the order of 0.1 A is connected with the discharge on the outer surface, while the following sharp increase of the current up to 1.7 A is due to transition of the discharge into the tube. Typical final volt-ampere characteristics of the discharge are shown in figure 2. The rising dependence is typical for the abnormal discharge [5].

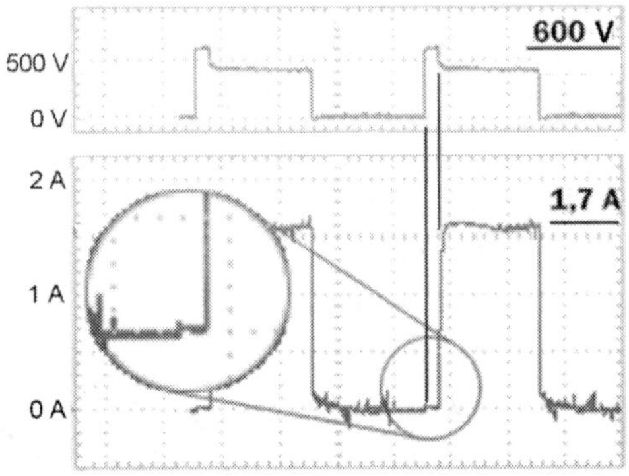


**Figure 1.** Current and voltage oscillograms of the discharge (gas argon, pressure 4 mbar, frequency 1 kHz, duty cycle 50%, the tube 20×8×100 mm). The time of ignition and transition inside the tube are shown by vertical lines. Transition of the discharge into the tube is shown in the circle insert.

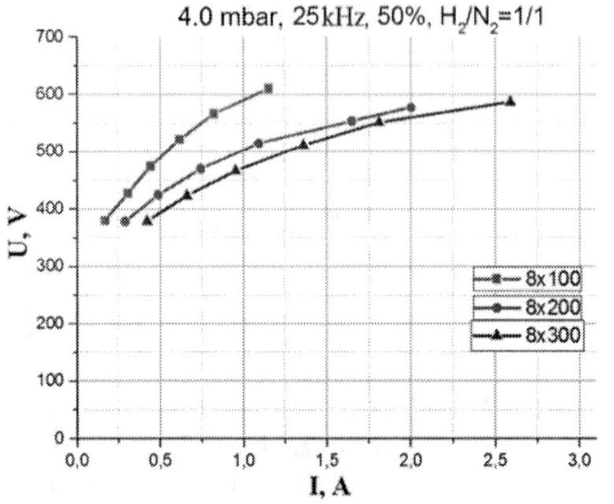


**Figure 2.** Volt-ampere characteristics of the discharge in mixture of nitrogen and hydrogen 50–50% at a pressure of 4.0 mbar, frequency of 25 kHz, and duty cycle of 50%

## 3. Nitriding results

After nitriding, the samples were cut perpendicular to the axis in several parts, and the cross sections were polished and investigated under the electron microscope. Microhardness over the cross-section was measured using a FM-ARS 9000 hardness tester. The load on the indenter was 10 g, and it was applied perpendicular to the cross section at different distances from the tube surface. The respective depth profiles of three nitrided and one not nitrided sample are shown in figure 3 for the cross section near the central part of the tube between its edges (figure 3a) and in the cross section near the tube edge (figure 3b). After nitriding, the microhardness close to the surface increased 1.5–2.5 times. The thickness of the hardened layer was 300–400 µm. Tube 14×9×150 (green line) was overheated, and therefore the hardness of its core decreased compared to that of the virgin sample.

Microstructure of samples was analyzed using a TESCAN VEGA3 scanning electron microscope. Polished cross sections were chemically etched in the 16:3:1 solution of hydrogen peroxide, water, and hydrofluoric acid ($H_2O_2$+$H_2O$+HF) prior to the analyses. Elemental composition in the cross section was measured by energy dispersive analyses (EDS) as it was made in [6].

Figure 4 shows typical SEM images of the cross-sections of treated samples. The samples were nitrided at a temperature of 450 °C. After plasma nitriding at temperatures below 500 °C the samples still have their original ferrite-pearlite structure with distinct grain boundaries. The compound layer was about 2–5 µm thick. One can see that both the white layer and the structure of the diffusion zone are similar for the external and internal surfaces of the tube, though currents and therefore fluxes of active nitrogen were very different outside and inside the tube.

IOP Conf. Series: Journal of Physics: Conf. Series **941** (2017) 012036 doi:10.1088/1742-6596/941/1/012036

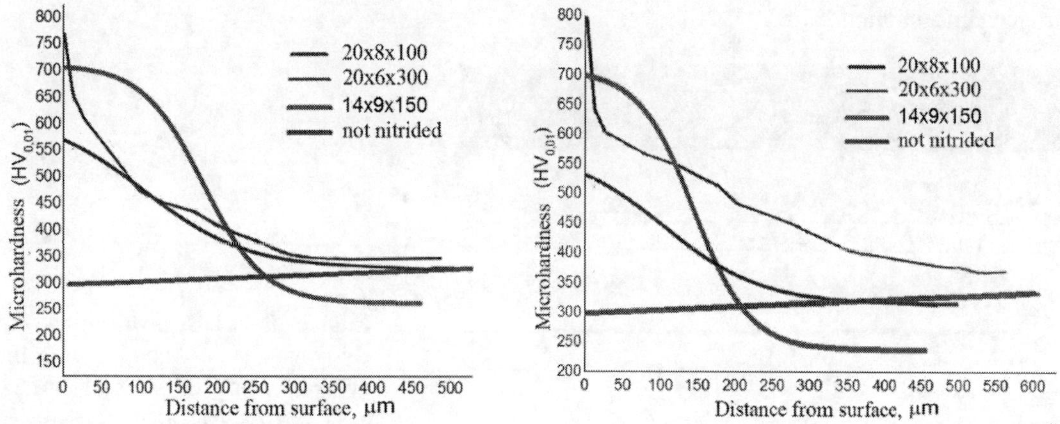


(a) (b)

**Figure 3.** Hardness as a function of the distance from the internal surface of the tubes nitrided at 450–500 °C measured in two cross sections: (a) in the middle area of the tube between its edges, (b) near the edge of the tube.

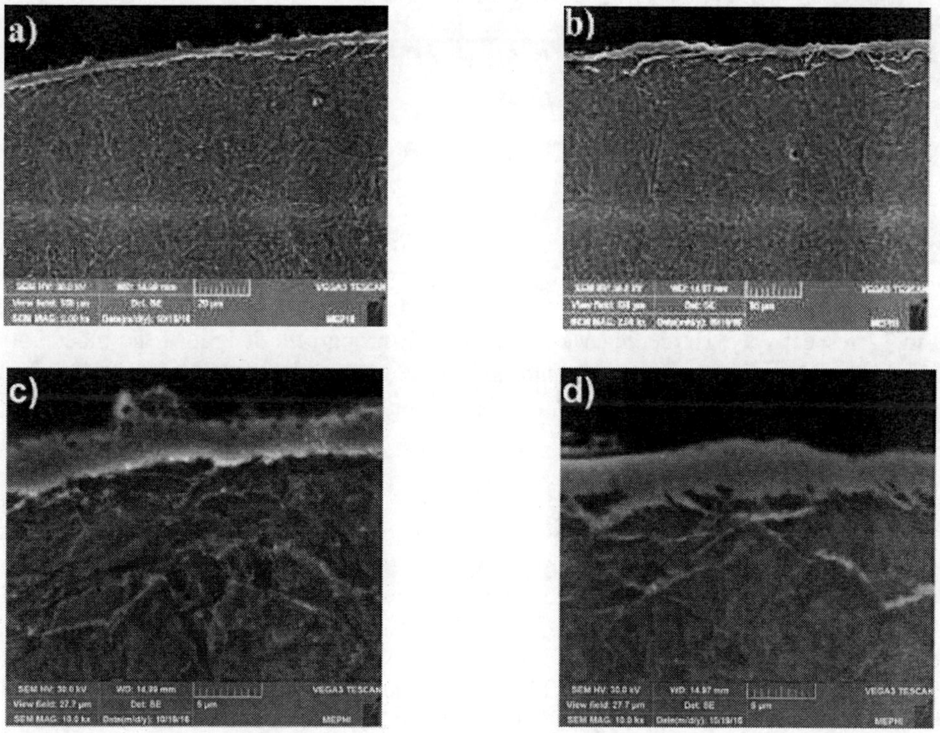


**Figure 4.** The microstructure of a cross-section in the middle area between the edges of the tube nitrided at 450 °C. The cross sections are shown near: (a), (c) internal surface, (b), (d) external surface.

Figure 5 shows the distribution of elements through the depth near the internal surface of samples after the plasma modification. The nitrogen concentration is high within 5 µm, which correlates with the thickness of the compound layer. Although the hardness is high within 150–200 µm, the nitrogen concentration is small behind the compound layer and only slightly exceeds the background at these depths. It is usually suggested that hardening at large depths is due to nitrogen in solid solution where nitrogen concentration is low.

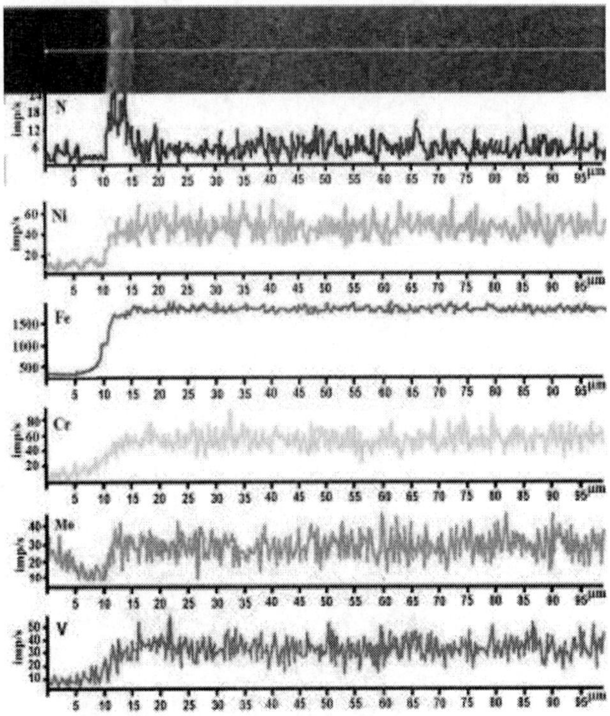


**Figure 5.** EDS distributions of elements as a function of the distance from the internal surface of the tube nitrided at 450 °C. Measurements were made in the middle part between the edges of the tube 20×8×100 mm.

## 4. Conclusion

Hardening of the internal surfaces of tubes by nitriding at 400-550°C in abnormal glow discharge with hollow cathode was investigated for tubes with internal diameter from 6 to 9 mm and the aspect ratio from 50 to 12 respectively. Hardening was observed all through the length of the tubes both on the external and internal surfaces. The compound layer was about 2-5 μm thick, and the hardened zone was about 300–400 μm thick. The hardness of close to the internal surface below the compound layer increased 1.5–2.5 times after nitriding.

## Acknowledgments

This work was partially supported by Ministry of Education and Science of the Russian Federation (Contract № 14.Y26.31.0008) and by byNational Research Nuclear University MEPhI in the framework of the Russian Academic Excellence Project (contract No. 02.a03.21.0005,27.08.2013).

## References

[1] Kashaev N, Stock H-R and Mayr P 2005 *J. Surf. Coat. Technol.* **200** 502
[2] Ramazanov K N, Ishmukhametov D Z and Sadkova N S 2011 *Vestnik UGATU* **15 3 (43)** 67
[3] Pokorný Z, Hrubý V, Stránský K and Kadlec J 2011 *Metal,* Materials of Conference 18–20 May, Brno, Czech Republic
[4] Borisyuk Yu V, Oreshnikova N M, Berdnikova M A, Tumarkin A V, Khodachenko G V and Pisarev A A 2015 *Phys. Procedia* **71** 105
[5] Borisyuk Yu V, Oreshnikova N M, Mozgrin D V, Stepanova T V and Pisarev A A 2016 *Journal of Physics: Conference Series* **747** 012019
[6] Borisyuk Yu V, Oreshnikova N M and Pisarev A A 2017 *Journal of Surface Investigation: X-ray, Synchrotron and Neutron Techniques* **11 (1)** 259

IOP Conf. Series: Journal of Physics: Conf. Series **941** (2017) 012037 doi:10.1088/1742-6596/941/1/012037

# Influence of technological parameters on the geometry of single-track laser clad nickel based alloy on grey cast iron substrate

**D S Gusev**[1,2,*], **A B Lyukhter**[1,**]

[1] Vladimir State University named after Alexander and Nikolay Stoletovs, Vladimir, Russia
[2] LLC "Engineering Center at VlSU", Vladimir, Russia

E-mail: [*]gusev@laser33.ru; [**]3699137@mail.ru

**Abstract.**On the example of glass forming equipment, the surfaces of which must have high wear resistance during repeated contacts with molten glass, a study was made of laser cladding of nickel based alloys on a substrate of gray cast iron. In study the shapes of individual tracks are investigated with varying laser radiation power, processing speed and powder feed rate. The influence of technological parameters on the width and height of the clad is shown. A similarity is found between the two principles of measuring the dilution through linear dimensions and the areas of track in cross section. A high correlation between dilution and laser radiation power over a wide range of speeds has been established, which has made it possible to develop a scheme of control laser cladding process with achieving a low level of dilution in order to minimize the heat effect zone (HAZ).

## 1. Introduction

In the manufacture of shaped glass products in large batches (millions of pieces), glass container manufacturers are faced with a problem associated with the permanent failure of the critical components of equipment. In the process of using the glass forming equipment, the details of glass molds operating under conditions of cyclic alternating thermomechanical loads and produced, in most cases, from cast iron, are subject to the greatest defectiveness. A significant influence on the stability of glass forming equipment is the desire of container manufacturers to reduce the mass of produced glass products (reducing the wall thickness), increase the rate of release of packaging, while improving the quality of melted glass [1, 2].

One of the most promising technologies to improve the service life of glass forming tooling is to restore the working surfaces of wear-resistant, corrosion resistant and other coatings obtainable laser cladding a nickel-based powders. [3, 4]

The wide application of laser cladding is limited to the use of automated process control systems that do not have in their arsenal of algorithms for adjusting operation windows with changing geometric parameters of the deposited layer. This is caused by an insignificant share of research in the field of dependencies and regularities of the formation of the clad on the technological parameters of the processing time.

The aim of this work is to identify patterns of geometric characteristics of the tracks on the technological processing parameters for the creation of an automated process control system of laser cladding.

In studies [5–8] on the assessment of the influence of technological parameters, the main characteristics of the clad were the height, width and shape of the clad, the depth of penetration, and the percentage of defects. Parameters of control were the laser radiation power, processing speed and powder feed rate.

## 2. Experimental

The experiments were carried out on an installation consisting of: a 6-axis Kuka HR 60LH robot, a Sulzer Metco Twin 10C powder feeder, ytterbium fiber laser YLS-6 (max. average power 6 kW), COAX powder nozzle.

The chemical composition used in the experiments of cast iron and surfacing powder Eutroloy 16233 (particle size 40–120 μm, initial hardness 31–36 HRC) was determined using a Bruker Q-8 Magellan spectrometer and is presented in tables 1 and 2.

**Table 1**. Chemical composition of the experimental cast iron.

| Element (wt %) | | | | | | | | | | | |
|---|---|---|---|---|---|---|---|---|---|---|---|
| Fe | C | Si | Cr | Mn | Ni | Cu | Na | Al | P | S | Mo |
| Bal | 3.09 | 1.35 | 0.54 | 0.5 | 0.25 | 0.09 | 0.08 | 0.07 | 0.05 | 0.04 | 0.03 |

**Table 2.** Chemical composition of the surfacing powder Eutroloy 16233.

| Element (wt %) | | | | | |
|---|---|---|---|---|---|
| Ni | Cr | Si | Fe | B | C |
| 85 | 7 | <3.1 | <3 | 1.3 | 0.3 |

Deposition process was performed using a continuous laser radiation in inert gas (argon). The height of the deposited layer was significantly lower than the thickness of the substrate to ensure sufficient heat removal from the treatment zone.

Review papers [5–8] allowed to choose the following processing parameters: laser power $P$ of 700 to 2800 W, the processing speed $S$ from 9 to 18 mm/s, powder flow rate $F$ of 12 to 25 g/min.

The clad height $H$ [mm] is defined as the height of laser track over surrounding substrate, the clad width $W$ [mm] is a distance between side points where the laser track touches and $b$ is the clad depth representing the thickness of substrate melted during the cladding and added to the clad region. The clad area $A_c$ [mm$^2$] is the cross-section of the clad area above the surrounded substrate surface and the molten area $A_m$ [mm$^2$] is the cross-section molten area under the surrounded substrate surface. The area of defects were measered in the clad and melted areas and was determined as $(A_{def}/(A_c+A_m))\cdot 100\%$.

The measurements were carried out with the help of JMicroVision software on cross sections. Photographing of the microstructure was performed on an inverted microscope with magnification x25 on the Nikon Epiphot TME200.

Regression analysis and processing of experimental results are implemented in STATISTICS software [9].

## 3. Results and discussion

To evaluate the effect of laser radiation power $P$, processing speed $S$ and powder flow $F$ on geometric parameters, 32 clad tracks were manufactured. The results of measuring are shown in table 3.

With constant powder flow ($F$ = 12 g/min), the height and width of the deposited layer were plotted against laser radiation power and processing speed (figures 1 ... 4).

**Table 3.** Results of measuring the geometric dimensions of the clad.

| track | $P$, W | $S$, mm/s | $F$, g/min | $W$, μm | $H$, mm | $b$, μm | $A_c$, mm$^2$ | $A_m$, mm$^2$ | $A_{def}$, mm$^2$ |
|---|---|---|---|---|---|---|---|---|---|
| 1 | 2100 | 9 | 15 | 3.052 | 0.479 | 0.549 | 1.117 | 1.089 | 0.00 |
| 2 | 2100 | 12 | 15 | 2.902 | 0.531 | 0.687 | 1.188 | 1.131 | 3.62 |
| 3 | 2100 | 15 | 15 | 3.392 | 0.651 | 0.496 | 1.540 | 1.068 | 1.15 |
| 4 | 2100 | 18 | 15 | 3.691 | 0.737 | 0.721 | 1.895 | 1.641 | 0.00 |
| 5 | 1750 | 9 | 12 | 3.260 | 0.745 | 0.543 | 1.585 | 1.197 | 2.37 |
| 6 | 1750 | 12 | 12 | 2.937 | 0.525 | 0.626 | 1.002 | 1.144 | 0.28 |
| 7 | 1750 | 15 | 12 | 2.801 | 0.475 | 0.415 | 0.909 | 0.658 | 6.70 |
| 8 | 1750 | 18 | 12 | 2.741 | 0.449 | 0.478 | 0.849 | 0.813 | 0.00 |
| 9 | 1400 | 18 | 12 | 2.283 | 0.449 | 0.342 | 0.603 | 0.505 | 1.44 |
| 10 | 1400 | 15 | 12 | 2.640 | 0.457 | 0.367 | 0.779 | 0.531 | 1.07 |
| 11 | 1400 | 12 | 12 | 2.725 | 0.614 | 0.291 | 1.165 | 0.481 | 2.43 |
| 12 | 1400 | 9 | 12 | 2.983 | 0.684 | 0.342 | 1.364 | 0.609 | 0.00 |
| 13 | 700 | 9 | 12 | 1.946 | 0.572 | 0.164 | 0.700 | 0.350 | 0.00 |
| 14 | 700 | 12 | 12 | 1.797 | 0.515 | 0.131 | 0.600 | 0.290 | 0.00 |
| 15 | 700 | 15 | 12 | 1.725 | 0.401 | 0.176 | 0.516 | 0.233 | 0.00 |
| 16 | 700 | 18 | 12 | 1.598 | 0.296 | 0.172 | 0.340 | 0.171 | 0.78 |
| 17 | 2800 | 18 | 12 | 3.064 | 0.380 | 0.737 | 0.813 | 1.636 | 1.59 |
| 18 | 2800 | 15 | 12 | 3.235 | 0.449 | 0.766 | 0.958 | 1.644 | 8.72 |
| 19 | 2800 | 12 | 12 | 3.829 | 0.471 | 0.784 | 1.120 | 2.038 | 6.33 |
| 20 | 2800 | 9 | 12 | 3.895 | 0.546 | 0.905 | 1.423 | 2.444 | 3.26 |
| 21 | 2800 | 9 | 20 | 4.445 | 1.034 | 0.669 | 3.200 | 2.084 | 3.97 |
| 22 | 2800 | 12 | 20 | 4.143 | 0.823 | 0.589 | 2.453 | 1.594 | 3.58 |
| 23 | 2800 | 15 | 20 | 3.927 | 0.722 | 0.599 | 1.982 | 1.487 | 2.71 |
| 24 | 2800 | 18 | 20 | 3.614 | 0.631 | 0.584 | 1.657 | 1.307 | 0.98 |
| 25 | 2100 | 18 | 12 | 3.018 | 0.369 | 0.708 | 0.819 | 1.473 | 1.92 |
| 26 | 2100 | 15 | 12 | 3.087 | 0.451 | 0.540 | 0.832 | 1.083 | 3.71 |
| 27 | 2100 | 12 | 12 | 3.621 | 0.728 | 0.511 | 1.594 | 1.346 | 1.56 |
| 28 | 2100 | 9 | 12 | 3.869 | 0.902 | 0.573 | 2.298 | 1.162 | 3.93 |
| 29 | 2800 | 9 | 25 | 4.764 | 1.119 | 0.415 | 4.030 | 0.940 | 0.70 |
| 30 | 2800 | 12 | 25 | 4.331 | 0.997 | 0.476 | 3.210 | 0.979 | 1.46 |
| 31 | 2800 | 15 | 25 | 4.081 | 0.802 | 0.521 | 2.377 | 1.157 | 0.71 |
| 32 | 2800 | 18 | 25 | 3.827 | 0.650 | 0.466 | 1.795 | 1.232 | 1.45 |

IOP Conf. Series: Journal of Physics: Conf. Series **941** (2017) 012037 doi:10.1088/1742-6596/941/1/012037

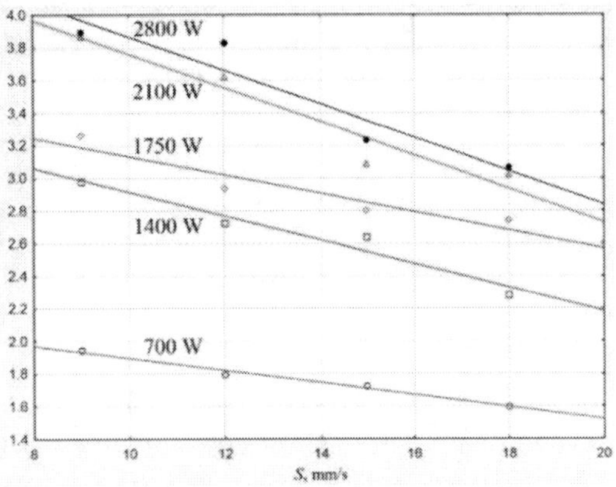


**Figure 1.** The dependence of the clad width on the processing speed at different laser radiation powers:

$W_{700} = 2.2687–0.0372·S$ ($R = -0.99$)
$W_{1400} = 3.6410–0.0728·S$ ($R = -0.97$)
$W_{1750} = 3.6966–0.0564·S$ ($R = -0.94$)
$W_{2100} = 4.7879–0.1029·S$ ($R = -0.96$)
$W_{2800} = 4.8949–0.1029·S$ ($R = -0.95$).

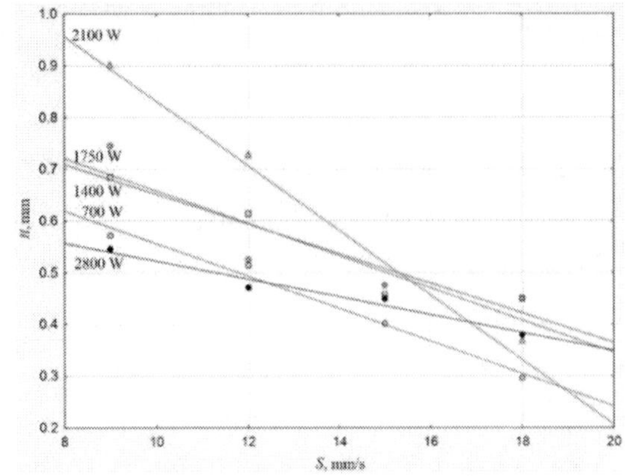


**Figure 2.** The dependence of the clad height on the processing speed at different laser radiation powers:

$H_{700} = 0.8699–0.0314·S$ ($R = -0.99$)
$H_{1400} = 0.9389–0.0287·S$ ($R = -0.95$)
$H_{1750} = 0.9706–0.0313·S$ ($R = -0.89$)
$H_{2100} = 1.4567–0.0625·S$ ($R = -0.98$)
$H_{2800} = 0.6955–0.0173·S$ ($R = -0.98$).

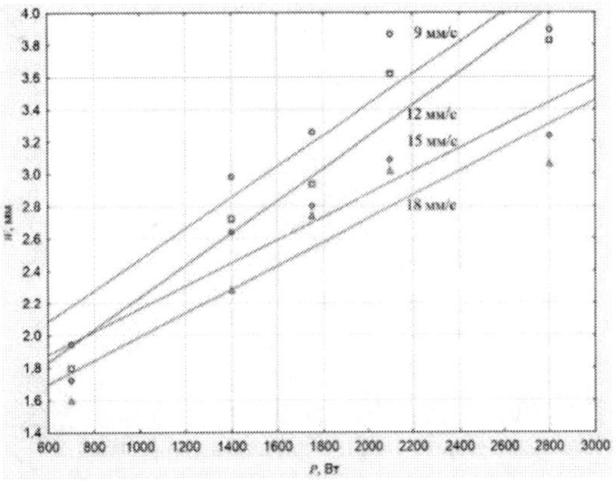


**Figure 3.** The dependence of the clad width on the laser radiation power at different laser radiation powers:

$W_9 = 1.5073+0.00096\,P$ ($R = 0.94$)
$W_{12} = 1.2338+0.00100\,P$ ($R = 0.97$)
$W_{15} = 1.4534+0.00071\,P$ ($R = 0.94$)
$W_{18} = 1.2576+0.00073\,P$ ($R = 0.94$).

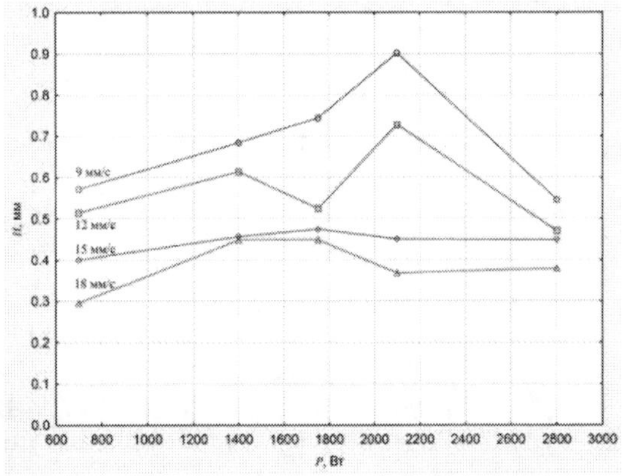


**Figure 4.** The dependence of the clad height on the laser radiation power at different laser radiation powers.

As the processing speed increases (Figures 1 and 2). the width $W$ and the height $H$ of the clad decrease according to a linear relationship: the correlation coefficient $R$ for $W$ is from -0.94 to -0.99. and for $H$ – from -0.89 to - 0.99. but at a laser radiation power $P = 2800$ W there is a drop in height. especially noticeable at lower processing speeds (9 and 12 mm/s). It is caused by excess heat. which is expressed in a significant depth of penetration (from 0.737 to 0.905 mm) and lack of powder: the base metal is melted to great depths and feed powder fills the melt pool instead of forming effective deposited layer.

Increasing the power of the laser radiation causes linear increase ($R$ from 0.94 to 0.97) of the track width (figure 3). and the maximum track height for each speed is at a power level of 1400–2100 W with a peak at 2100 W (figure 4).

Figure 5 represents the laser track cross-section map for *P* and *S* variations at a constant consumption of the powder feed (12 g/min).

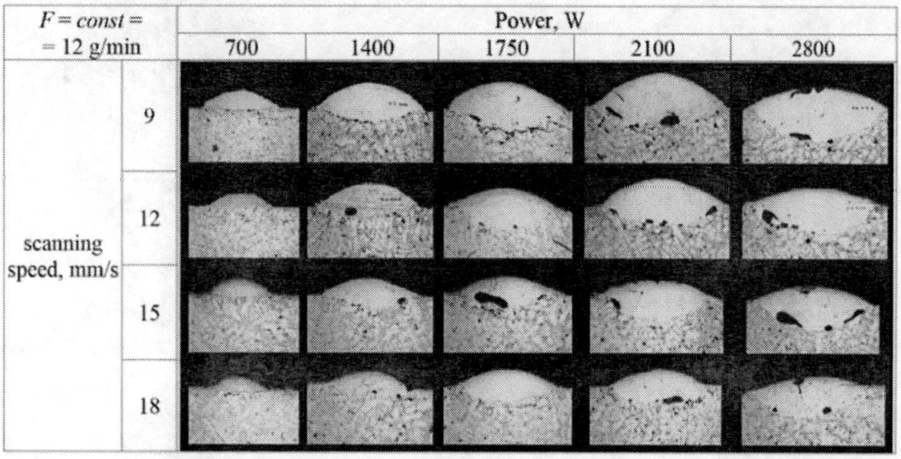


**Figure 5.** Laser tracks cross-section map for side cladding of Ni-based alloy on gray cast iron substrate with different scanning speed and laser power.

To evaluate the effect of powder flow on geometric parameters of clads prepared at different scanning speeds (9–18 mm/s) with a different powder flow (12–25 g/min) at constant power $P = 2800$ W for maximum deposition rate process. The cross-sections map of these clads are shown in figure 6.

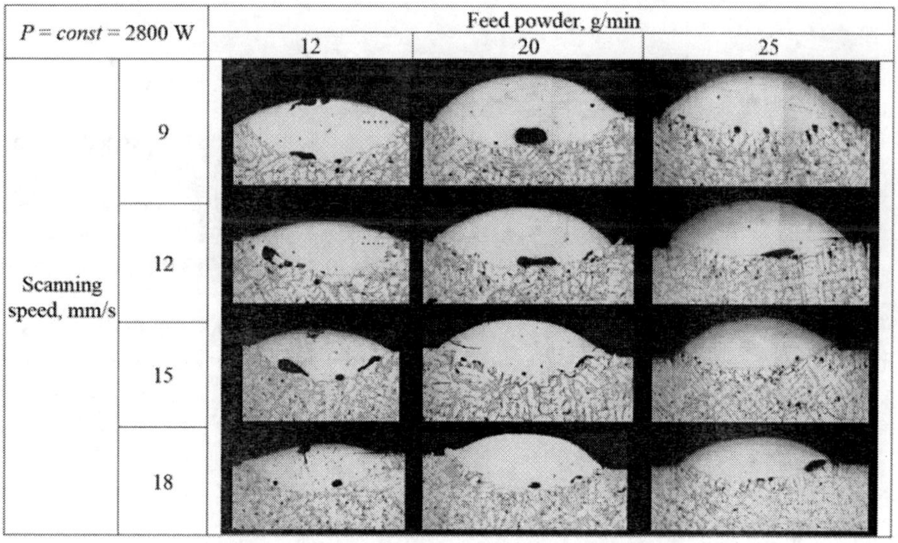


**Figure 6.** Laser tracks cross-section map for side cladding of Ni-based alloy on gray cast iron substrate with different scanning speed and feed powder.

Increasing the powder feed allows to increase the width (figure 7a) and the height (figure 7b) of the clad layer. thereby forming tracks with a lower dilution (figure 7c). obtaining up to 5 mm$^2$ of effective area of clad in a single track in a cross section (figure 7d).

In this case. the effect of the scanning speed on the geometric parameters of clads is preserved: with its growth. the width. height of the track. dilution and effective area of clad decrease as well as at constant feed powder.

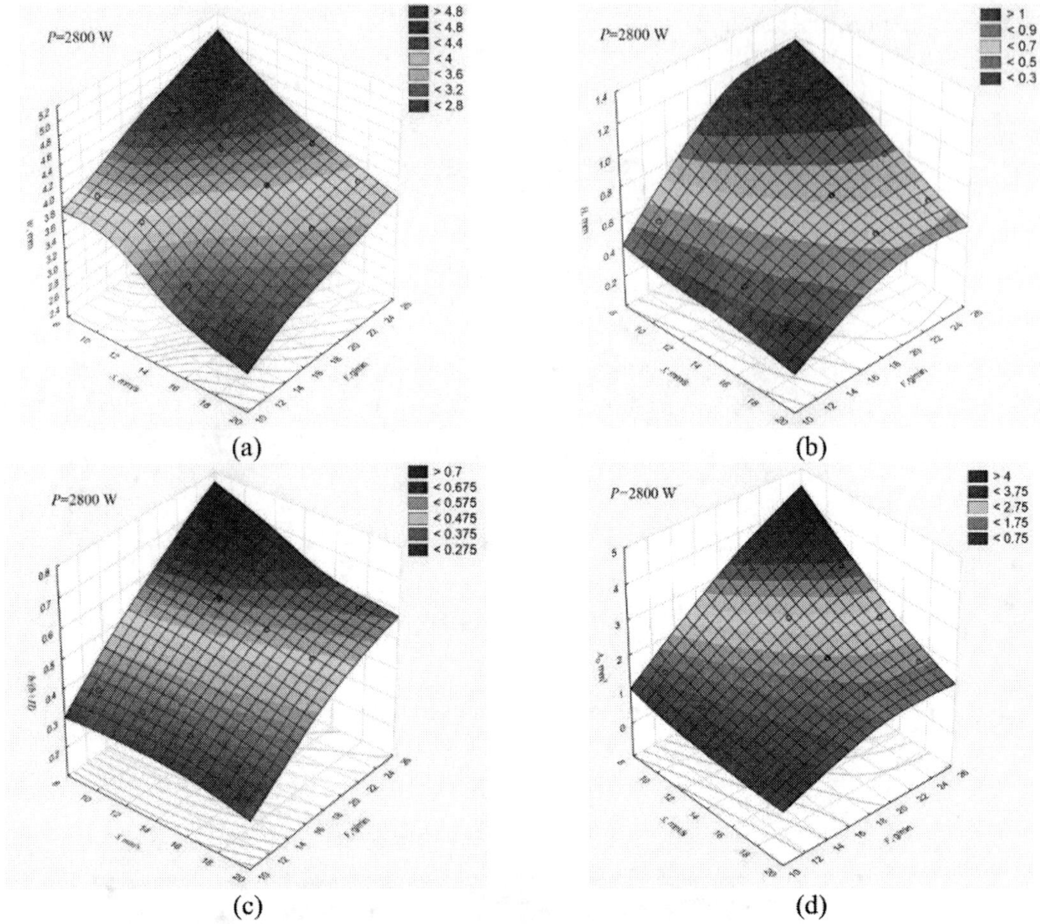


**Figure 7.** Dependence of the geometrical parameters of the clad on the processing speed and feed powder:
(a) – width track $W = 4.1543\text{-}0.0984 \cdot S\text{+}0.0581 \cdot F$;
(b) – height track $H = 0.5875\text{-}0.0381 \cdot S\text{+}0.034 \cdot F$;
(c) – dilution $b/(H\text{+}b) = 0.2453\text{-}0.0103 \cdot S\text{+}0.0221 \cdot F$;
(d) – effective area of clad $A_c = 1.6521\text{-}0.1626 \cdot S\text{+}0.1383 \cdot F$.

The obtained dependences (figures 1–4. 7a. 7b) made it possible to confirm the existing regularities of the influence of technological parameters on the width and height of clad. but not enough to create a control system for the laser cladding process. To solve this problem. the effect of the technological parameters of laser cladding on the dilution. which is one of the indicators of the quality of the deposited layer. was analyzed.

Dilution $b/(H+b)$. reflecting the ratio of the thickness of substrate that was melted during the cladding process $h$ to the total melt height $H+b$.

This calculation is similar to the area ratio $A_m/(A_c+A_m)$ of the melted base metal Am to the sum of the areas of the clad layer and the melted base metal ($A_c+A_m$). since these ratios are almost identical for all tracks (figure 8). This can be explained by the use source of a laser radiation with a Gaussian energy distribution and the use of a coaxial feed powder which results in a uniform melt pool relative to the laser beam axis (the laser beam in all experiments was directed along the normal to the substrate). The low dilution indicates a small penetration depth of the substrate. a decrease in the HAZ and the probability of defects in the clad. so it is more preferable.

IOP Conf. Series: Journal of Physics: Conf. Series **941** (2017) 012037 doi:10.1088/1742-6596/941/1/012037

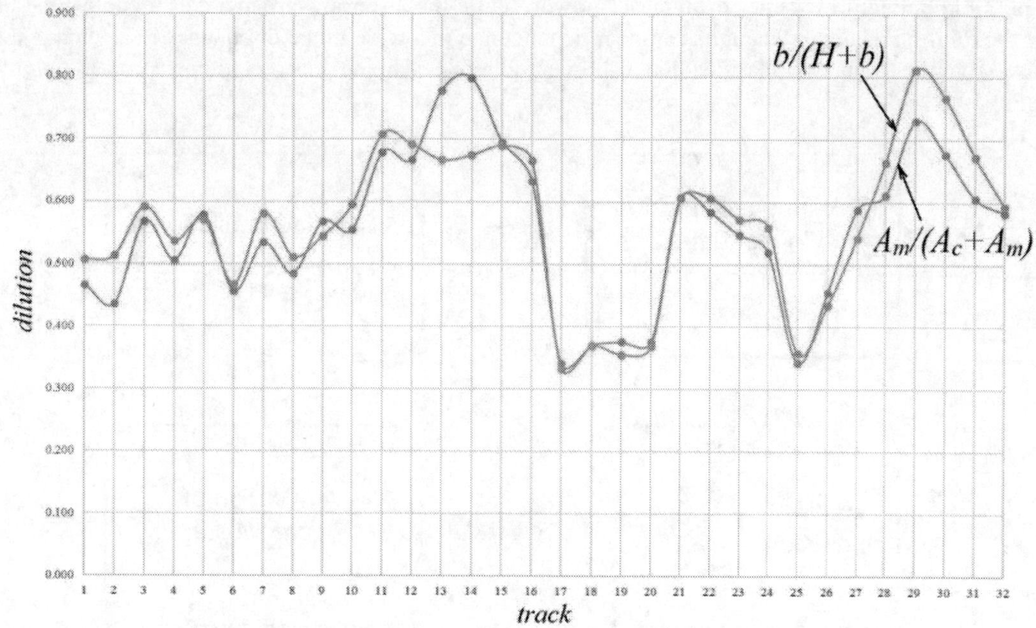


**Figure 8.** Dilution obtained by formulas $b/(H+b)$ and $A_c/(A_c+A_m)$.

Dilution $b/(H+b)$ has a high level of correlation with the laser radiation power at different processing speeds. dependence is shown in figure 9.

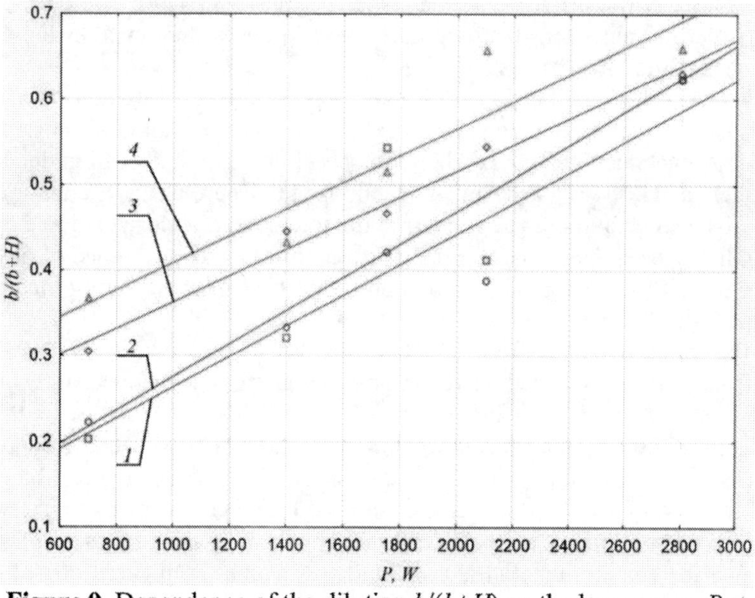


**Figure 9.** Dependence of the dilution $b/(b+H)$ on the laser power $P$ at different speeds:

1 – 9 mm/s; $b/(b+H) = 0.08354 + 0.00018 \cdot P$ ($R = 0.95$);

2 – 12 mm/s; $b/(b+H) = 0.08191 + 0.00019 \cdot P$ ($R = 0.90$);

3 – 15 mm/s; $b/(b+H) = 0.20946 + 0.00015 \cdot P$ ($R = 0.99$);

4 – 18 mm/s; $b/(b+H) = 0.25108 + 0.00016 \cdot P$ ($R = 0.93$).

It can be seen from the figure 9 that for coaxial laser cladding the tendency to obtain a low dilution is possible only at low values of the laser power. while the laser power can act as a control parameter for the dilution quality. since it has a linear dependence in a wide range of changes in scanning speeds. Based on the found regularities the following control scheme laser cladding process. shown in figure 10.

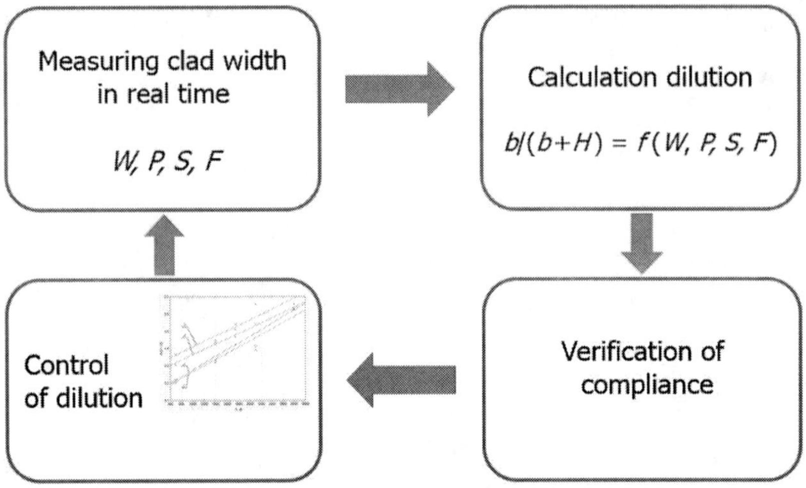


**Figure 10.** Control scheme of laser cladding process.

The principle of this system is monitor the width clad with a laser scanner in real time; the control system calculates the dilution and compares it with the value set by the operator. introducing a correction for the laser power through the established regularities (figure 9). Repeated measurement of the width clad is carried out after obtaining correction in the system or after the set time (for cases when the dilution is within the permissible range).

Another indicator of the quality of the deposited layer are the coating defects. in particular cracks. pores and sinks. Defects detected in cross sections of experimental samples occurred at the boundaries of the remelted base metal and the HAZ. The presence of pores in this location due to the solubility of gases in the molten metal. as well as chemical reactions that occur during its cooling.

Changes cause the generation of gas clusters at the interface. If a closed volume of gas doesn't have time to float to the surface. then closed pores are formed. Thus. the presence of gas and liquid in the powder mixture leads to increased pore formation in the clad metal if during the coating process the gas didn't escape [10].

Defects in the samples ranges from 0–9% while. for 20 processing modes. this value doesn't exceed 2%. defects were not detected on 7 samples. Obtaining defect-free coatings is one of the most important tasks. connected with the increase of the strength characteristics of the part and solves additional difficulties in the fight against corrosion of cast iron. This is because the nickel coating with respect to the cast iron is cathodic. while cathodic coatings can serve as a reliable protection against corrosion only in the absence of pores. cracks and other defects in them. In case of damage to the coating or the presence of pores. a corrosion element appears in which the cast iron serves as an anode and dissolves. and the coating material is a cathode. on which the process of hydrogen reduction or ionization of oxygen [11].

High performance correlation between the process parameters (including complex) and defect area (or percentage defect of total area) were not detected. Thus. it is necessary to continue searching for alternative approaches. which can have a direct impact on the formation of defect-free clad tracks. An example is preheating. carried out by various methods and considered in [12]. including. by means of laser radiation.

## 4. Conclusions

1. Investigated 32 single track with a variation in laser radiation power *P* from 700 W to 2800 W. processing speed *S* from 9 mm/s to 18 mm/s. powder feed rate *F* from 12 g/min to 25 g/min.

2. Confirmed existing patterns of influence of process parameters on the geometry of clad. The greatest influence on the width of the clad is the laser power. the effective height of clad is the feed powder. when the excess energy is reached. and the effective height of the clad is falls.

3. The desire to obtain clad layers with a low dilution limits the use of high laser power.

4. A similarity is found between the two principles of measuring the dilution through linear dimensions and the areas of the track in the cross section.

5. The laser radiation power can serve as a control parameter for the dilution. An algorithm for controlling the laser cladding process is proposed.

**Acknowledgments**
The authors thank D.Yu. Tatarkin for the experiments in the territory of NTO IRE-Polus in Fryazino. The work was carried out with the financial support of the Foundation for Assistance to Small Innovative Enterprises under the program "UMNIK".

## References

[1] Chistyakov D G 2014 The development of manufacturing technologies mould of iron castings with increased operational resource: Diss. of masterdegree in mechanical engineering: 05.16.04 (Nizhny Novgorod) 245
[2] Leushin I O and Chistyakov D G 2014 Defect in the details of cast iron mould and measures against premature of their failure *Vestnik UGATU* (Ufa: UGATU) **18** 27
[3] Birykov V P. Petrova I M and Gadolina I V 2013 Laser cladding influence on characteristics of fatigue resistance*Mechanical engineering and engineering education* **2** 54
[4] Ocelík V. Oliveira U. Boer M and Hosson J Th M 2007 Thick Co-based coating on cast iron by side laser cladding: Analysis of processing conditions and coating properties *Surface & Coatings Technology* **201** 5875
[5] A V Dolgovechniy. Demidova L A and Morozov E A 2012 Technologies of alloy steel building-up welding on the basis from carbon steel *Proceedings of the Samara Scientific Center of the Russian Academy of Sciences*(Samara: Samara Scientific Center of the Russian Academy of Sciences) **14** 550
[6] Grigor'yanc A G. Misyurov A I and Tret'yakov R S 2011 Analysis of the influence of the coaxial laser cladding parameters on the formation of clad *Tekhnologiya Mashinostroeniya* (Moscow: Technology of Mechanical Engineering) **11** 19
[7] Oliveira U. Ocelík V and Hosson J Th M 2005 Analysis of coaxial laser cladding processing conditions *Surface & Coatings Technology* **197** 127
[8] Egunov A I. Artemenko Yu A and Rodionova I N 2012 Formation of pores in composite alloy WC+NiCrMoFe during laser cladding *Zagotovitel'nieproizvodstva v mashinostrornii* **4** 8
[9] Borovikov V 2008 Neural networks. STATISTICA Neural Networks. Methodology and technology of modern data analysis. 2-nd edition (Moscow.GoryachayaLiniya – Telecom) 392
[10] Grigor'yanc A G. Misyurov A I. Shiganov I N.Tret'yakov R S and Stavertii A Ya 2012 Porosity investigation of the cobalt- and nickel-based laser coatings *Vestnik MGTU im. N.E. Baumana* (Moscow: MGTU im. N.E. Baumana) **6** 165
[11] Maltseva G N 2000 Corrosion and protection of the equipment against corrosion ed Vinogradov S N (Penza: Penza State University)p 211
[12] Yi P. Xu P. Fan C. Li C and Shi Y 2015The effect of dynamic local self-preheating in laser cladding on grey cast iron *StrojniškiVestnik - Journal of Mechanical Engineering* **61** 43

# Laser-ablative fabrication of nanoparticle inks for 3D inkjet-printing of multifunctional coatings

**A A Ionin[1], A K Ivanova[1,2], R A Khmel'nitskii[1], Yu V Klevkov[1], S I Kudryashov[1,2], N N Mel'nik[1], A A Nastulyavichus[1], A A Rudenko[1], I N Saraeva[1*], N A Smirnov[1] and D A Zayarny[1]**

[1] Lebedev Physical Institute, Leninsky prospect 53, Moscow 119991, Russia
[2] National Research Nuclear University MEPhI (Moscow Engineering Physics Institute), Kashirskoe shosse 31, Moscow 115409, Russia

E-mail: *heddwch@mail.ru

**Abstract.** We report the fabrication of multifunctional coatings via inkjet printing using water-based nanoinks in the form of selenium (Se) and gold (Au) nanoparticle (NP) colloids, prepared by laser ablation of solid targets in deionized water or 50%-isopropyl alcohol solution. Nanoparticles and NP-based coatings were deposited onto silver films, magnetron-sputtered to silica-glass substrates, and characterized by means of scanning and transmission electron microscopy (SEM, TEM), UV-vis-IR, Raman and energy-dispersive X-ray spectroscopies.

## 1. Introduction

Many techniques, widely used for the nanostructured coatings manufacturing, such as lithography, chemical etching or vacuum sputtering are efficient, although complex and time-consuming, also reducing the spectrum of usable substrate materials due to high temperature or pressure during the machining process. The emerging field of 3D inkjet printing allows overcoming the imposed restrains, making it possible to create the required nanostructured areas with necessary NPs concentration on the surface of any possible material, including elastic and soft plastics, dielectrics, paper etc.

3D inkjet printing with the use of nanoparticle (NP) colloids as an active ink-pigment components is considered an eligible and simple-to-implement technique for manufacturing of thin-film transistors, SERS-substrates, accumulators, solar batteries, memory basic components etc. [1–4], based on the multi-elemental sandwich structure. Two problems emerge while constructing functional coatings with nanoinks: (1) the fabrication of NPs with the required morphology without any contaminations and (2) their precise deposition on given substrates. Chemical and mechanical methods of NPs manufacturing are well known for high impurity, caused by side reaction products or machining process, and are hard to implement. Laser ablation of solid targets in liquid media seem an eligible candidate for NPs preparation, as it ensures the purity of the resulting ablation products as well as their abundant yield and rapid fabrication [5–8].

In this work, we present a technique of mono- and multilayer nanostructure coatings' inkjet printing with NPs colloidal solutions, prepared by laser ablation of solid targets in liquid media, on the magnetron-sputtered silica-supported Ag films. The resulting coatings were characterized by means of scanning electron and transmission microscopy (SEM, TEM), UV-vis-IR, Raman and energy-dispersive X-ray spectroscopy modules.

## 2. Experimental details

### *2.1. Fabrication of nanoinks and inkjet printing*

The ablative fabrication of NP-based inks was implemented using an $Yb^{+}$-doped fiber laser HTFMark, Bulat (central wavelength $\lambda = 1040$ nm, pulsewidth FWHM $\tau = 200$ ns, average pulse energy up to 1 mJ, repetition rate up to 80 kHz), equipped with a galvanometric scanner and an f-theta objective (focal length F = 160 mm) (figure 1). The 2×2 $mm^2$ areas of solid targets (golden plate, purity 99.99%, Savings Bank of Russian Federation, and selenium plate, purity 99%, hexagonal γ-phase, homemade) were processed via multipass (10 passes) laser beam scanning with the constant speed of 40 m/s under the 2-mm layer of liquid (deionized water or isopropyl alcohol (IPA) solutions of different concentration). The inkjet printing of nanocoatings on silica glass-supported silver films was implemented via multicartrige printer Epson L800.

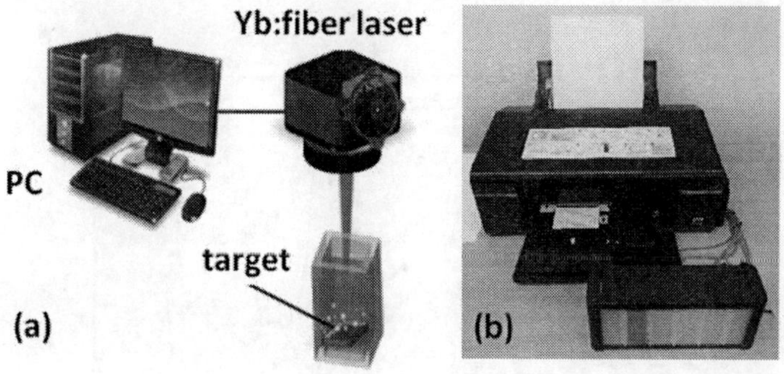


**Figure 1.** (a) – experimental setup for nanoinks fabrication; (b) – inkjet printer Epson L800.

### *2.2. Characterization of nanoinks*

The resulting ablation products in form of Au and Se colloidal solutions were analyzed in order to define their morphology and properties. The NPs size distribution was estimated with the use of dynamic light scattering (DLS) (PhotocorFC), scanning electron microscopy (SEM, JSM 7001F, JEOL) and transmission electron microscopy (TEM, LEO912 ABOMEGA). NPs for SEM characterisation were deposited onto atomically-smooth silicon wafers and dried at room temperature. Plasmon-related absorption band (for Au nanoinks) and interband absorption peak (Au, Se) positions were detected via UV-vis transmission spectroscopy (V-70, Brucker).

Se colloids, prepared in a 50%-IPA solution (opaque red), exhibited an interband absorption band <400 nm and the strong Mie-resonance absorption band at >400 nm (figure 2b). The position of the main $E_g \approx 233$ $cm^{-1}$ in bulk selenium corresponds to its γ-crystalline structure, while in NPs deposits a new band appears $A_{1g} \approx 250$ $cm^{-1}$ (figure 3), illustrating the amorphization of target material during the laser-induced ejection [9]. The mean Se NPs radii equalled to 75 and 167 nm, which is supported by the SEM and DLS data (figure 2a).

Au hydrosols of reddish-violet colour had an absorption band at ~530 nm (figure 2d), which is red-shifted comparing to the typical Au localised plasmon resonance ($\lambda_{LPR} \approx 520$ nm) [10] and can be explained by the formation of spheroids and overall NPs size increase [11]. The average Au NPs size equalled to ~20 nm, but with a broad polydispersity range (figure 2c).

Se NPs, fabricated in 99% IPA solution (opaque, dark brown colour), were nucleated in form of trigonal selenium (t-Se) nanorods, which was supported by the TEM photographs (figure 4) and Raman spectroscopy (figure 3, blue curve).

**Figure 2.** SEM-images of Se (left) and Au (right) NPs, printed on silica-supported silver films. Insets: (a), (c) – the corresponding DLS spectra of NPs size distribution; absorbance spectra of Se NPs (b) and Au NPs (d).

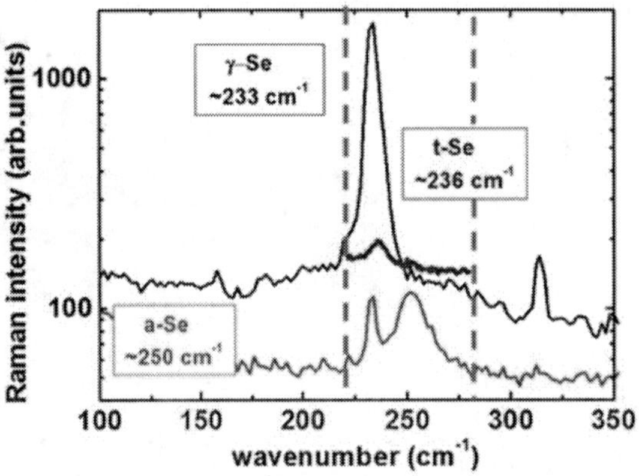


**Figure 3.** Raman spectra of solid γ-Se (black curve), amorphous a-Se NPs (red curve) and trigonal t-Se nanorods (blue curve).

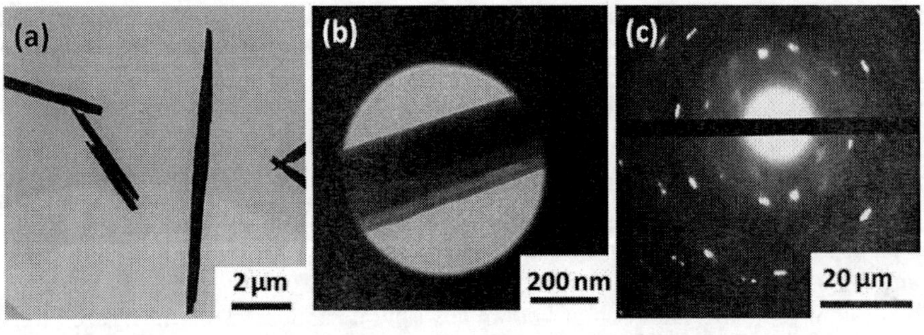


**Figure 4.** TEM-photographs of Se nanorods (a), (b) and the corresponding electron diffraction pattern (c).

## 3. Fabrication of 2D and 3D functional nanocoatings

### *3.1. Inkjet printing*

The as-prepared nanoinks (Au NPs hydrosols and Se NPs in a 50%-IPA solution) were used for inkjet printing of nanocoatings on silica-supported silver films via multicartrige printer Epson L800. The resulted coatings were analyzed with the use of optical microscopy (biological microscope Altami-6), SEM and the built-in module for energy-dispersive X-ray spectroscopy (EDS, module INCA, Oxford Instruments), allowing characterization of elemental composition of the nanocoatings.

During the printing nanoinks were ejected in form of 1-pL drops and rapidly dried on the Ag film surface, creating quasiperiodical structure of discrete spots with mean diameter of about tens of μm (figure 5). The percent of spot area filling with NPs expectedly increased during multipass printing, accompanied by the layer height change from hundreds of nm to several μm.

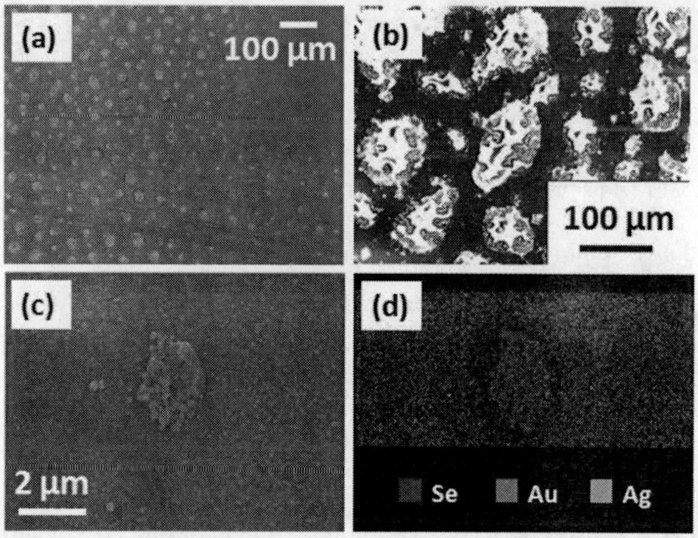


**Figure 5.** SEM-images of inkjet printed multilayers of Se and Au NPs on the silica-supported Ag film (a), (b), (c) and the corresponding EDS-map (d): Se 0.79 at. %; Au 0.38 at. %; Ag 7.69 at. %.

Au nanionks inkjet printing over the Se-layer resulted in a second inhomogenously deposited layer, which seems to be isolated from the supported metal film, as it exhibits a strong surface charge, detected during its SEM-visualization as the corresponding over-saturated bright areas (figure 5b). The resulted 3D-structure can be considered as a simple semiconductor-based photoresistor.

### *3.2. Sedimentation of colloids*

Se NPs, prepared in the 99% IPA solution, were deposited on silica-supported Ag films and dried at room temperature. During the evaporation of liquid medium amorphous NPs organized in form of t-Se nanorods (vide supra in Section 2: figures 3, 4; 6). This phenomenon can be explained in terms of Ostwald ripening process, which involves the dissolution of unstable a-Se NPs and their aggregation into the c-axis oriented hexagonal structure of the Se helical chains [12]. The resulting nanorods have the mean diameter of 600 nm and the average length of 10 μm. The increase of laser exposition time from 5 to 15 min resulted in a more rapid NRs formation, which lead to the colloid colour change from opaque brick-red to dark gray and was detected during SEM-analysis (figure 6).

## 4. Conclusion

In this work we report the preparation of nanionks with high pigment (Au, Se NPs) concentration, provided by efficient laser ablation of solid targets in liquid media, and the following manufacturing of multifunctional nanocoatings via inkjet printing and sedimentation of colloids. The resulting coatings were composed of spherical NPs (Au, Se, prepared in a 50% IPA solution) and nanorods (Se colloids @ 99% IPA, air-dried on a substrate). Monolayer nanocoatings of plasmonic NPs can be used as highly-sensitive elements for SERS chemo- and biosensors, and a sandwich structure of Ag/Se/Au NPs can be basically considered as a simple microscale photoresistor.

**Figure 6.** SEM-images of dried Se NRs deposits on silica-supported Ag films with 5 min (a) and 15 min (b) laser exposition of Se target; (c) – a photograph of the corresponding Se-colloids in medical syringes.

**Acknowledgments**
This work was supported in part of ink-jet printing by the Russian Science Foundation (Grant No. 15-19-00208), and in the characterization part of laser-ablative colloids by the by National Research Nuclear University MEPhI in the framework of the Russian Academic Excellence Project (contract No. 02.a03.21.0005, 27.08.2013).

**References**

[1] Sirringhaus H, Kawase T, Friend R H, Shimoda T, Inbasekaran M, Wu W and Woo E P 2000 *Science* **290** 5499
[2] Singh M, Haverinen H M, Dhagat P and Jabbour Gh E 2010 *Adv. Mater.* **22** 673
[3] Janoschka T, Teichler A, Häupler B, Jähnert T, Hager M D and Schubert U S 2013 *Adv. Energy Mater.* **3** 1025
[4] Secor E B, Prabhumirashi P L, Puntambekar K, Geier M L and Hersam M C 2013 *J. Phys. Chem. Lett.* **4** 8
[5] Streubel R, Barcikowski S and Gökce B 2016 *Optics Letters* **41** 7
[6] Scaramuzza S, Zerbetto M and Amendola V 2016 J. Phys. Chem. C. **120** 17
[7] Amendola V, Scaramuzza S, Carraro F and Cattaruzza E 2017 *Journal of Colloid and Interface Science* **489** 18
[8] Dell'Aglio M, Mangini V, Valenza G, De Pascale O, De Stradis A, Natile G, Arnesano F and De Giacomo A 2016 *Applied Surface Science* **374** 297
[9] Gates B, Mayers B, Cattle B and Xia Y 2002 *Adv. Funct. Mater.* **12** 3
[10] Kreibig U, Vollmer M 1995 *Optical Properties of Metal Clusters* (Berlin: Springer)
[11] Khlebtsov N G, Bogatyrev V A, Dykman L A and Melnikov A G 1996 *Journal of colloid and interface science* **180** 0323
[12] Jeong U, Camargo P H C, Leeb Y H and Xia Y 2006 *J. Mater. Chem.* **16** 3893

IOP Conf. Series: Journal of Physics: Conf. Series **941** (2017) 012039 doi:10.1088/1742-6596/941/1/012039

# Dimensional processing of composite materials by picosecond pulsed ytterbium fiber laser

**S A Kotov**[*]

IRE-Polus Co, Moscow region, Fryazino, Russia

E-mail: [*]skotov@ntoire-polus.ru

**Abstract.** In this paper, an experimental study of laser dimensional processing of thermoset carbon fiber reinforced plastics with a thickness of 2 and 3 mm was performed. In the process of work test rig setup based on picosecond pulsed fiber laser with 1.06 microns wavelength and 30 W average power was developed. Experimental tests were carried out at the maximum average power, with laser beam moved by a galvanometric mirrors system. Cutting tests were executed with different scanning velocity, using different laser modes, number of repetitions, hatching distance and focal plane position without process gas. As a result of the research recommendations for the selection processing mode parameters, providing minimal heat affected zone, good kerf geometry and high cutting speed were produced.

## 1. Introduction

The widespread introduction of composite materials due to their high performance properties raises the problem of their processing. The carbon fiber reinforced plastics (CFRP) with 2 and 3 mm thickness based on thermosetting binders are widely used in various constructions [1]. At the final stage of the manufacture of items from CFRP, there is a need for additional processing for forming windows, through holes and other elements that are impossible or impractical to produce in the preform stages. Traditionally, mechanical and abrasive water jet machining is used to perform these operations [2]. They have some significant shortcomings – high tool wear, high cost of consumables, delamination due to vibration and shock loadings, some limitations on the cutting contour and treatment of complex shape details. Laser machining by industrial ytterbium fiber lasers is a contact-free, flexible, stable processing method requiring no consumables. A comparative analysis of the characteristics of these treatment methods is presented in the article [3]. These sources reveal that laser processing is one of the promising directions for eliminating lacks of conventional machining methods.

The main technological problems of laser treatment are formation of heat affected zone (HAZ) and defects of cutting kerf geometry (taper and roughness). The HAZ is formed due to high heterogeneity and anisotropy of thermophysical properties of CFRP. In HAZ area adhesion strength between carbon fibers and matrix is broken, this dramatically reduces material strength properties. Taper and roughness of the cutting kerf determine detail accuracy grade and grade of finish. So to achieve high quality of CFRP laser machining, it is necessary to carry out complex experimental research to optimize technological parameters and processing logic. The use of pulsed-periodic laser sources can almost completely eliminate the HAZ and substantially improve the quality. For this reason, machining was carried out on a modern industrial pulsed laser with 1.06 microns wavelength.

## 2. Experimental setup

Experimental investigations of laser machining CFRP were carried out on the system consisting of the following keycomponents:

1. Picosecond industrial ytterbium fiber laser YLPP-1-150-30M (IPG) with $\lambda = 1.06$ microns wavelength and 30 W average power. Typical operating modes of this laser are shown in table 1.;
2. A two-axis high-speed and high-accuracy scan head with a clear mirror aperture of 12 mm;
3. F-Theta lens with effective focal length EFL =330 mm and processing field size 220×220 mm;

**Table 1.** Parameters of typical operation modes of ytterbium fiber laser model YLPP-1-150-30-M.

| Mode number | Average power (W) | Frequency (kHz) | Pulse duration (ns) | Pulse energy (mJ) |
|---|---|---|---|---|
| T1 | 30 | 600 | 0.15 | 0.05 |
| T2 | | 100 | 1 | 0.3 |
| T3 | | 60 | 2 | 0.5 |
| T4 | | 30 | 5 | 1 |

The choice of the laser and the scanning optical system was due to necessity to provide high power density in the focused spot for material removal due to mechanism of evaporation and short interaction time between laser and CFRP specimen because the reduction ofheat accumulation effect. The evaluation of the main characteristics of the experimental setup (focus spot diameter, depth of focus, pulse power density) were carried out according to the known equations (1–3) [4]:

$$D = \frac{4\cdot\lambda}{\pi\cdot D_0}\cdot M^2 \cdot F \simeq 100\ \mu\text{m}, \quad (1)$$

$$2 \cdot z_0 = \frac{\pi\cdot D^2}{4\cdot\lambda} \simeq 14\ \text{mm}, \quad (2)$$

$$W_{pulse} = \frac{P_{avg}}{f\cdot\tau\cdot S} \simeq 10^9\,\frac{\text{W}}{\text{cm}^2}, \quad (3)$$

where $\lambda = 1.06$ µm – laser wavelength; $D_0$ – diameter of the collimated beam; $M^2$ – beam quality parameter; $F$ – focal length of the F-Theta lens; $P_{avg}$ – average laser power; $f$ – pulse repetition rate; $\tau$ – pulse duration; $S$ – area of the focused laser spot. The block diagram of the experimental setup is shown in figure 1.

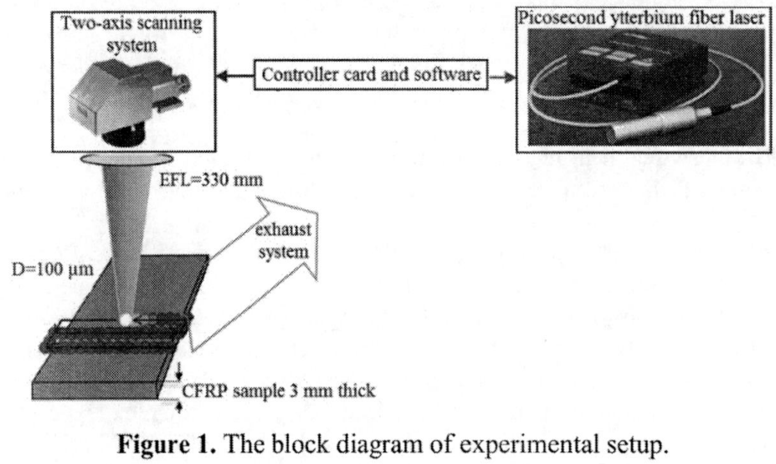


**Figure 1.** The block diagram of experimental setup.

In the investigation, we use CFRP with 2 and 3 mm thicknessproduced by direct compression molding.Reinforcing fiber – Toray T700, binding epoxy matrix – Hexcel M21. The mass content of reinforcing in the material is 65%. Orthotropic reinforcement scheme. The structure of the investigated material is schematically shown in figure 2.

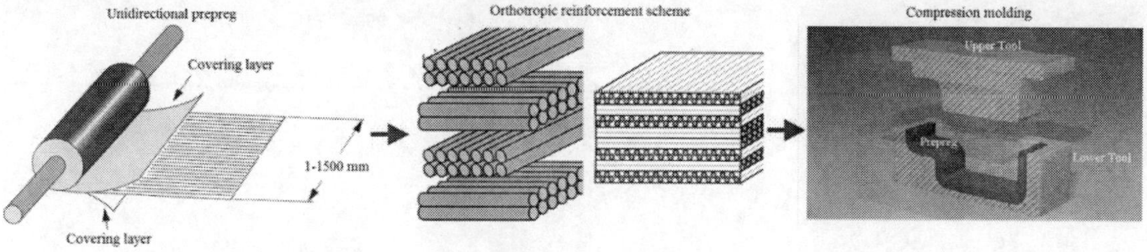


**Figure 2.** The structure of the investigated CFRP.

## 3. Method of evaluation of processing quality

The main criteria to evaluate the quality of laser processing CFRP parts are shown in figure 3.These criteria are discussed in detail in the articles [5, 6].

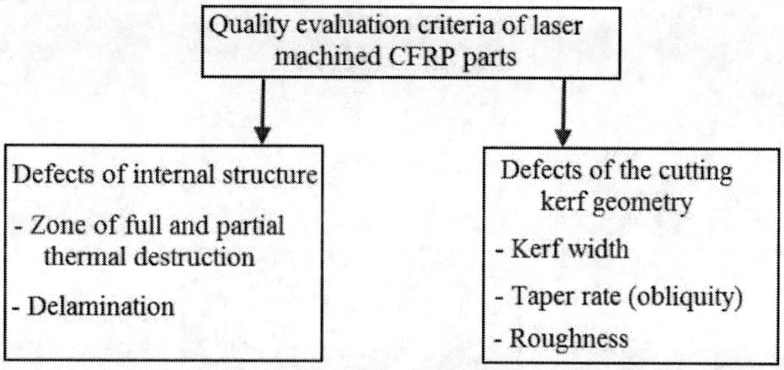


**Figure 3.** The main quality evaluation criteria of laser machined CFRP parts.

Estimation of the kerf width and HAZ on the sample surface was carried out with the optical microscope Olympus GX-51. Defects of internal structure were investigated on microsections of the cutting kerf cross-section with the same optical microscope. Estimation of the kerf taper was carried out using a contour measuring station Mitutoyo Contracer CV-2100. The roughness was measured with a surface analyzer (profilometer) Proton Model 130. The investigation of the kerf surface was also carried out using electronic scanning microscope Zeiss Gemini Merlin Compact VP-60-13.

## 4. Experimental results

In the research, the cutting of CFRP specimens was carried out in a straight line 30 mm length due to mechanism of evaporation in a multipass mode by the hatching algorithm without the assist gas in the air atmosphere. The algorithm for single pass processing is schematically represented in figure 4.

As a result of a large number of experiments, the recommended technological parameters and the processing algorithm were determined: average laser power 30 W; pulse repetition rate 30 kHz; pulse width 5 ns; scanning speed 0.5–1.5 m/s; hatching distance 25–50 μm. These settings provide high-quality CFRP parts on the chosen criteria. The results of machining 2 and 3 mm thick CFRP are shown in figure 5 and figure 6.

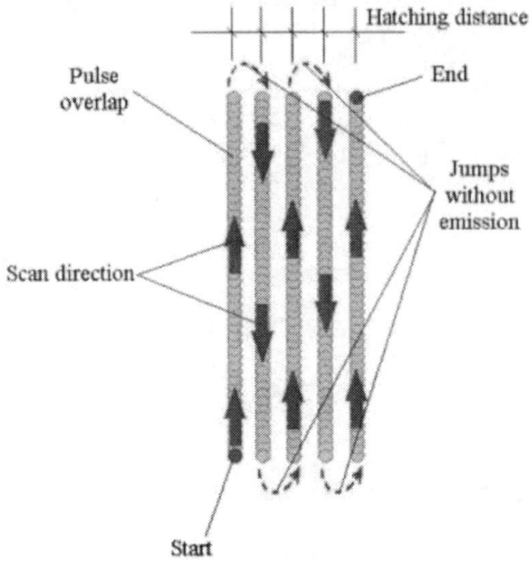


**Figure 4.** Machining method.

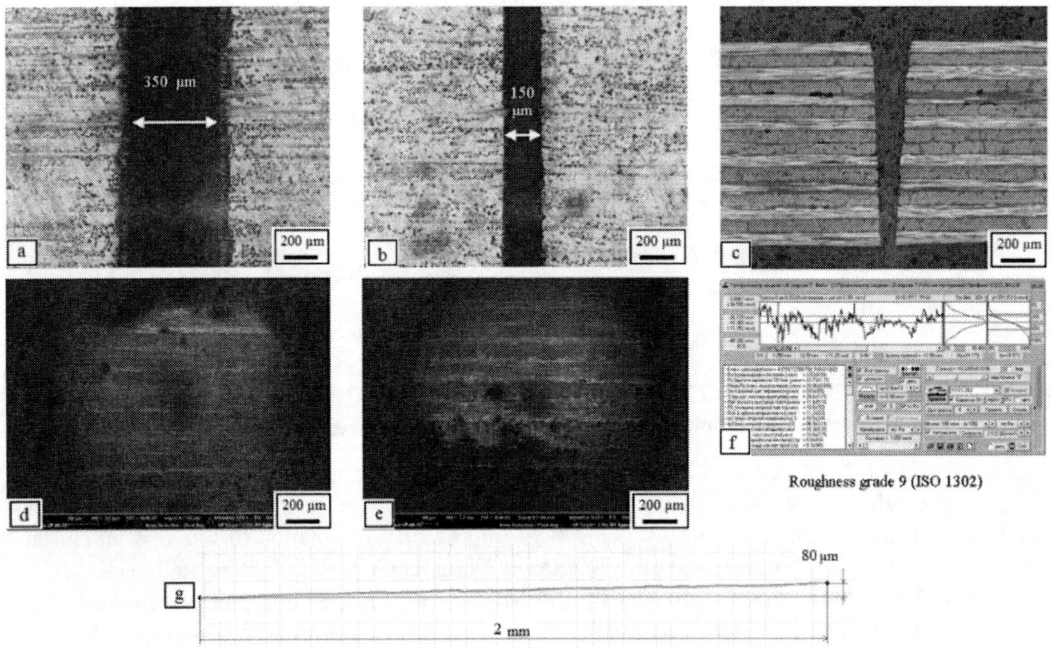


**Figure 5.** The surface of the CFRP sample from the beam entrance (a) and exit (b); microsection of the cutting kerf cross-section (c); kerf surface on an electronic scanning microscope (d), (e); the result of roughness measuring (f); the result of taper measuring (e).

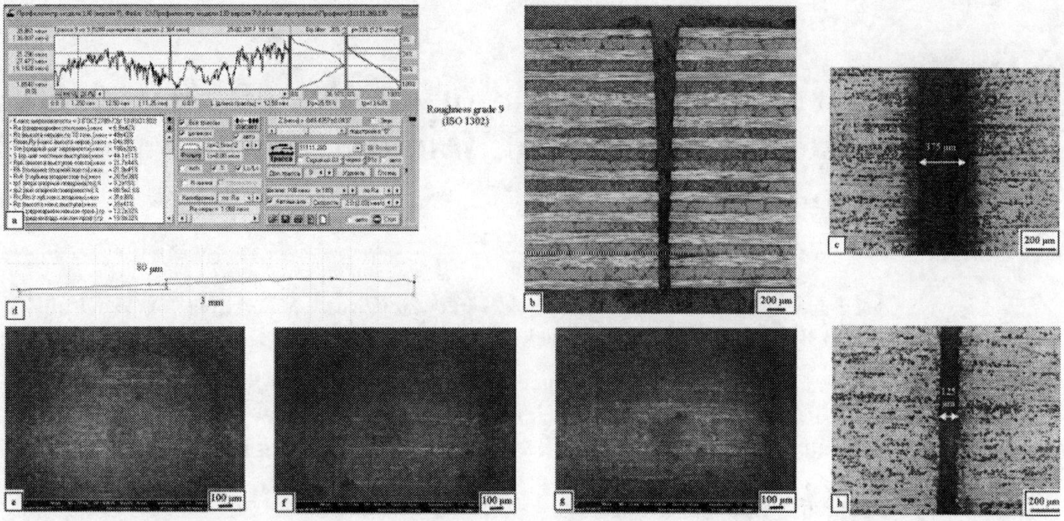

**Figure 6.** The result of roughness measuring (a); microsection of the cutting kerf cross-section (b); the surface of the CFRP sample from the beam entrance (c); the result of taper measuring (d); kerf surface on an electronic scanning microscope (e, f, g); the surface of the CFRP sample from the beam exit (h).

## 5. Conclusions

In the course of this research work, a test rig setup based on picosecond fiber laser model YLPP-1-150-30-M (IPG) with wavelength $\lambda = 1.06$ µm and average power 30 W was developed. Using this equipment the experimental evaluation of laser dimensional processing modes CFRP with 2 and 3 mm thickness was carried out. The high-quality of CFRP was provided at the recommended technological parameters and laser processing algorithm. The HAZ (zone of full and partial matrix thermal destruction) was less than 50 µm. The taper of the cutting kerf wall was less than 100 µm. The laser machining provided roughness grade 9 in accordance with ISO 1302. The cutting speed was about 10 mm/min. On microsections of the cutting kerf cross-section and its surface, there are no visible signs of delamination. These results testify to the prospects of high-quality machining composite materials by short pulse industrial fiber lasers.

## References

[1] Michailin U A 2015 *The fibrous polymer composite material in technics* (Saint-Petersburg: Science basis and technologies) 720
[2] Sheikh-Ahmad J Y 2009 *Machining of polymer composites* (Berlin: Springer) 315
[3] Negarestani R and Li L 2012 *Laser machining of fiber-reinforced polymeric composite materials* (Cambridge: Woodhead Publishing Limited) **11** 288
[4] Vaks E D, Milenkiy M N and Saprykin L G 2013 *Practice of precision laser processing* (Moscow: Technosphere) 696
[5] Goeke A and Emmelmann C 2010 *Physics Procedia* B **5** 253
[6] Cenna A and Mathew P 1997 *Int. J. Mach. Tools Manufact.* **37** 723

IOP Conf. Series: Journal of Physics: Conf. Series **941** (2017) 012040 doi:10.1088/1742-6596/941/1/012040

# The structure and mechanical properties of parts elaborated by direct laser deposition 316L stainless steel powder obtained in various ways

**I S Loginova[1,*], A N Solonin[1], A S Prosviryakov[1], S B Adisa[1], A M Khalil[1], D P Bykovskiy[2], V N Petrovskiy[2]**

[1] National University of Science and Technology "MISiS", Moscow, Russia
[2] National Research Nuclear University MEPhI (Moscow Engineering Physics Institute), Kashirskoe shosse 31, Moscow 115409, Russia

E-mail: *i-popkova@list.ru

**Abstract.** In this work the morphology, the size and the chemical composition of the powders of steel 316L received by the two methods was studied: fusion dispersion by a gas stream and reduction of metal chlorides with the subsequent plasma atomization of the received powder particles. The powder particles received by the first method have a spherical shape (aspect ratio 1,0–1,2) with an average size of 77 μm and are characterized by the absence of internal porosity. Particles of the powder received by the second method also have a spherical shape and faultless structure, however, their chemical composition may vary in different particles. The average size of particles is 32 μm. Though the obtained powders had different properties, the experimental samples received by DLD technology demonstrated by equally high durability (Ultimate strength is 623±5 and of 623±18 MPa respectively) and plasticity (38 and 41% respectively). It is established that mechanical properties of DLD samples increase for 7-10% after treatment of the surface.

## 1. Introduction

Additive technologies are a new type of production of the prototypes and the functional parts of complex shape from powders of aluminum, nickel, titanium, cobalt alloys and alloyed steels [1–10]. The products received by additive technologies can be characterized with high dispersion of structure and enhanced mechanical properties (in comparison with the standard products received by casting, rolling, forging and stamping) [2,5]. The DLD technology is one of the types of additive production, it implies direct supply of powder material by means of a nozzle into the zone of influence of a continuous wave laser beam. By means of DLD technologies, it is possible to produce prototypes and finished products of any shape for arms, medical, aviation and other types of industry.

The initial material in additive manufacturing is a metal powder with special properties: a spherical shape of particles with a narrow particle size distribution ranging from 30 to 70 μm, a homogeneous chemical composition without internal defects (pores). Such powders have better flow rate and compaction than powders with a dendritic particle shape, which ensures continuous supply to the treatment zone through a narrow coaxial nozzle [9]. Formation of a product with a high density and chemical uniformity is possible due to the homogeneous chemical composition of powders and the absence of internal defects. Therefore, development of the fundamentals of technology for the production of powders, which would fully satisfy all the requirements of DLD-process, is a promising research trend. The most common methods of producing powders for DLD technologies are pulveri of a liquid alloy with a gas jet and plasma atomization of powder particles with an irregular (angular) shape. In the present work, a comparative analysis of the morphology, size, and composition of 316L grade austenitic steel powders, obtained by two methods, was made. The first powder (P1) was obtained by spraying with a gas melt jet, the second (P2) – a reduction of the metal chloride and the subsequent plasma atomization of the resulting solid powder particles. A comparative analysis of the structure and properties of the DLD-manufactured samples, obtained using these powders, was carried out.

## 2. The objects of research and the methodology of experiments

P1 and P2 powders were used for deposition of the experimental samples. Powder P1 has a spherical particle shape (aspect ratio of 1.0–1.2) with an average size of 77±13.5 μm and can be characterized by dense structure (figure 1a) and a uniform chemical composition (table 1). Powder particles P2 also have a spherical particle shape (aspect ratio of 1.0-1.2) and a defect-free structure (figure 1b). The particle size of the powder P2 is finer than P1 – 32±8.5 μm. Separate powder particles have a non-uniform chemical composition (table 2). However, the average composition of all powder particles P2 corresponds to the 316L steel grade.

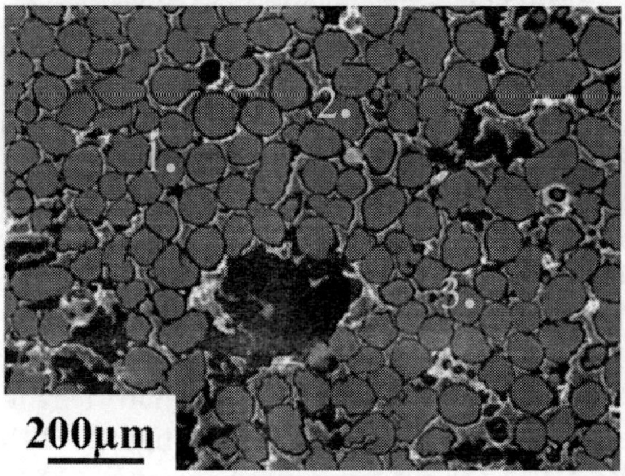

(a)

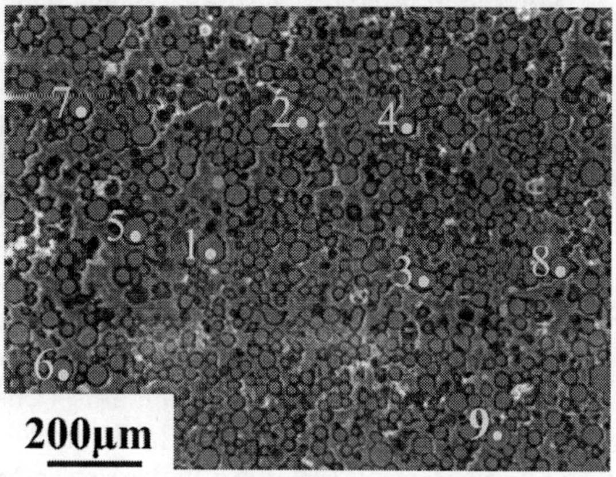

(b)

**Figure 1.** Powder particle structure P1 (a) and P2 (b).

The aspect ratio of powders was determined using SEM images of particles. Image magnification was chosen so as to observe no more than 150 particles. The aspect ratio represents the relation of the maximum linear size of a projection of a particle lmax to its minimum lmin size ($l_{max}/l_{min}$). The form of particles is defined as follows:

$l_{max}/l_{min}$ = 1,0–1,2, the form is defined as spherical;
$l_{max}/l_{min}$ = 1,2–2,0 – rounded shape;
$l_{max}/l_{min}$ = 2,0–5,0 – the form is angular;
$l_{max}/l_{min}$ = 2,0–25,0 – the form is rod;
$l_{max}/l_{min}$ = 25 – needle-shaped.

**Table 1.** Chemical composition of powder P1.

| Spectrum | The content of alloying elements, mass. % | | | | | | |
|---|---|---|---|---|---|---|---|
| | Cr | Ni | Mo | Mn | Si | C | Fe |
| 1 | 17.5 | 11.8 | 2.4 | 1.8 | 0.7 | | |
| 2 | 17.9 | 11.1 | 2.7 | 1.5 | 0.8 | <0.03 | bal. |
| 3 | 16.8 | 11.5 | 2.6 | 1.7 | 0.85 | | |
| Average composition | 17.3±1.0 | 11.7±0.6 | 2.5±0.3 | 1.7±0.2 | 0.9±0.25 | <0.03 | bal. |

**Table 2.** Chemical composition of powder P2.

| Spectrum | The content of alloying elements, mass. % | | | | | | |
|---|---|---|---|---|---|---|---|
| 1 | Cr | Ni | Mo | Mn | Si | C | Fe |
| 2 | 17.2 | 12.5 | 2.8 | 1.5 | 0.6 | | |
| 3 | 17.9 | 12.6 | - | 1.7 | 0.9 | | |
| 4 | 17.1 | 12.3 | 2.1 | 1.5 | 0.7 | | |
| 5 | 19.0 | 8.6 | - | - | - | <0.03 | bal. |
| 6 | 24.4 | 1.2 | - | - | - | | |
| 7 | 13.1 | 11.5 | - | - | 0.3 | | |
| 8 | 21.4 | 2.6 | - | - | 0.2 | | |
| 9 | 19.4 | 8.6 | - | - | 0.1 | | |
| Average composition | 18.6±2.7 | 8.7±3.7 | 2.4±0.4 | 1.6±0.2 | 0.5±0.3 | <0.03 | bal. |

Multilayer three-dimensional objects were fabricated on a substrate of A284 Steel.

DLD process was carried out on a HUFFMAN HC-205 industrial plant equipped with a fiber laser LS-3.5 with a maximum power of 3.5 kW with a radiation wavelength $\lambda$ 1069 nm. The powder mixture was transported by an argon stream and focused on the treatment zone with a coaxial nozzle of diameter 1.9 mm.

In the design of the coaxial head. an axisymmetric beam was formed. It made possible to obtain a thin gas-powder flow resulting in track deposition with a width more than 200 µm and a thickness of 200 µm. In operation. the width of the track was 500 µm. The overlap of the adjacent tracks of one layer was 10–15%. and the diameter of the laser beam was 370 µm.

Laser surfacing of samples in the form of a brick (width – 12 mm. length – 70 mm. height – 10 mm. figure 2a) and dumb-bells (width of the working part – 4 mm. length – 67 mm. height – 10 mm. figure 2b) using a crosswise strategy of the scanning (figure 3). The laser power was 250 W. the powder flow rate was 0.013 g/s; scanning speed 16 mm/s.

In order to study the structure. the surface of the samples was polished on a Struers Labopol-5. The structure of the alloys was studied using electron scanning microscope TESCAN VEGA LMH with a cathode LaB6 and an X-ray energy dispersive microanalysis system of Oxford Instruments Advanced AZTec Energy X-max 50. The backscattered electron mode (BSE) was used.

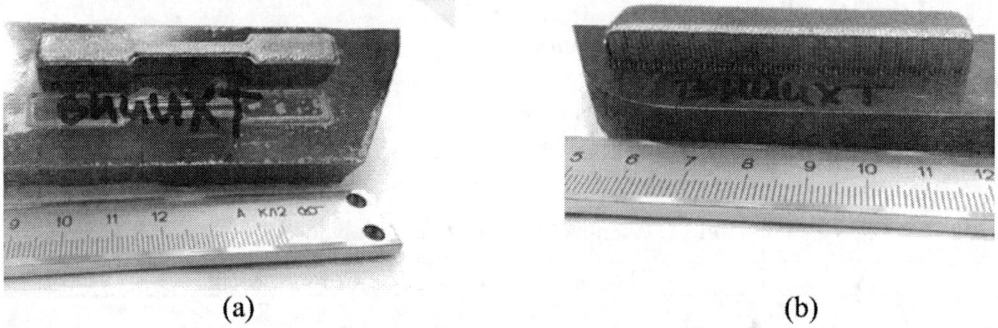

(a) (b)

**Figure 2.** Form of samples: dumb-bell (a) and brick (b).

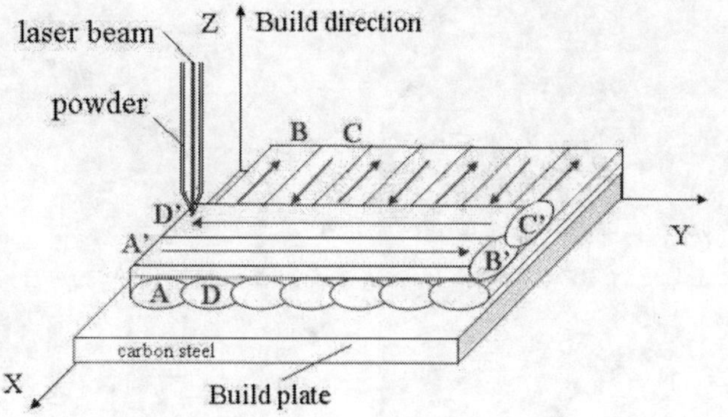


**Figure 3.** Scheme of scanning strategy.

The mechanical properties of the samples were evaluated by the yield strength (YS). and the ultimate strength (UTS) values. which were determined by the uniaxial tensile test on a Zwick Z250 with a test speed of 4 mm/min

## 3. Results and discussion

The structure of samples made from P1 and P2 powders was the same. It is shown in figure 4 at different magnifications. On the general image of the structure. tracks distinguish in parallel layers. It was found that crystals of elongated shape were located in these layers facing away from the arc-shaped interface.

The reason for this can be a direct heat removing during crystallization of the melt. The track has the shape of a semicircle. During cooling. the heat is removed in all directions from its surface. so that the crystals grow inward parallel to this direction. (figure 4a). Samples were characterized by high continuity. absence of gas pores and cracks (figure 4a). At high magnifications. it was found that an ultrafine austenite structure with an average crystallite size of 2.5 μm was formed. This crystallite size corresponds to high cooling rates of $10^5$-$10^6$ K/s [8].

There are two types of crystals: fine columnar and fine equiaxed grains (figure 4b). It was established by means of EDX that samples made of powder P2 (which have low chemical homogeneity of the particles) do not have a segregation of the chemical composition at their cross section. Apparently powder particles of different composition were melted and form a drop of melt with an average chemical composition during the DLD process. Thus. the chemical composition was leveled after the solidification of layers.

To assess the mechanical properties of the experimental samples. uniaxial tension tests were carried out. The results are represented in table 3. In both samples made from powders P1 and P2. the strength and ductility were high. It should be noted that the strength of the samples was higher than 30–40 MPa in comparison with samples of the same steel obtained by other researchers [8]. Surface treatment of the test samples (EDM of cutting edges) leads to an increase of mechanical properties by 7–10% (table 3).

IOP Conf. Series: Journal of Physics: Conf. Series **941** (2017) 012040 doi:10.1088/1742-6596/941/1/012040

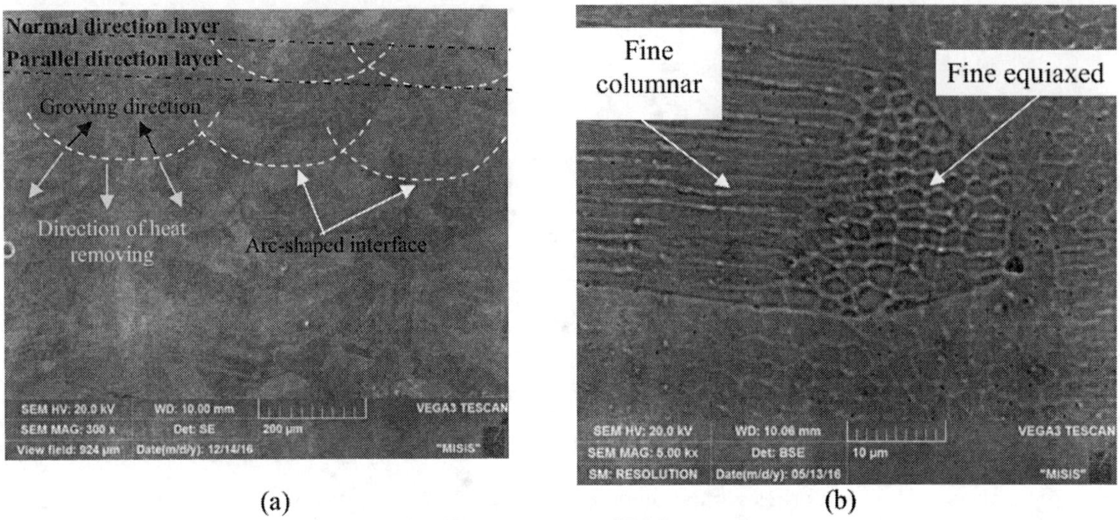


**Figure 4.** Microstructure of DLD-samples.

**Table 3.** Average values of properties.

| Conditions/powder | YS. MPa | UTS. MPa | Elongation. % |
|---|---|---|---|
| DLD/P1 | 489±20 | 624±4 | 38±14 |
| DLD/P1 (surface treatment) | 495±7 | 685±21 | 31±2 |
| DLD/P2 | 400±25 | 632±18 | 41±14 |
| DLD/P2 (surface treatment) | 447±3 | 675±10 | 46±10 |
| Wrought (cold finished) [8] | 255–310 | 525–623 | 30 |
| DLD/other studies [8] | 330–395 | 540–625 | 35–85 |

The fractures were mainly ductile. as evidenced by the presence of pits at the surface of the fractures (figure 5a. b).

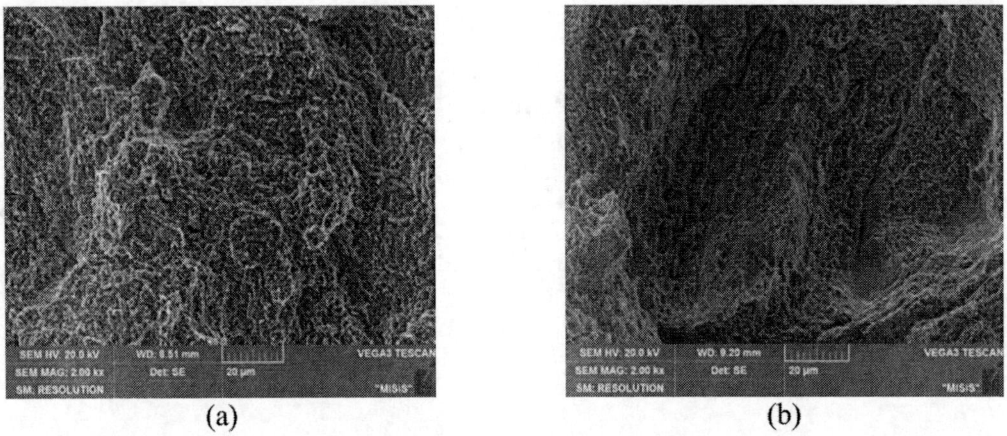


**Figure 5.** Typical fracture image of DLD-samples in the different zones.

## 4. Conclusions

The morphology. size. and chemical composition of 316L steel powders obtained by fusion dispersion by a gas stream and reduction of metal chlorides with the subsequent plasma atomization were studied. Particles of powder P1 have a spherical shape (aspect ratio of 1.0–1.2) with an average size of 77 μm and have a dense structure. Powder particles P2 also have a spherical shape and a defect-free structure but the chemical composition of the individual particles may vary. The average particle size is 32 µm.

It was shown that. the experimental samples P1 and P2 obtained by DLD process demonstrated by equally high strength (ultimate strength is 623±5 and 623±18 MPa. respectively) and plasticity (38 and 41%. respectively). despite the different properties of 316L steel powders.

## 5. Acknowledgments

This work was supported by the Ministry of Education and Science of the Russian Federation in the framework of Increase Competitiveness Program of NUST"MISiS"(No. K1-2014-026) and by National Research Nuclear University MEPhI in the framework of the Russian Academic Excellence Project (contract No. 02.a03.21.0005,27.08.2013).

## References

[1] Zaitsev A A. Sentyurina Zh A. Levashov E A. Pogozhev Yu S. Sanin V N. Loginov P A and Petrzhi M I 2017 Structure and properties of NiAl-Cr(Co.Hf) alloys prepared by centrifugal SHS casting. Part 1 – Room temperature investigations *Materials Science and Engineering: A* **690** (6) 463

[2] Shamsaei N. Yadollahi A. Bian L and Thompson S M 2015 An overview of Direct Laser Deposition for additive manufacturing Part II: Mechanical behavior. process parameter optimization and control *Additive Manufacturing* **8** 12

[3] Zhang B. Dembinski L and Coddet C 2013 The study of the laser parameters and environment variables effect on mechanical properties of high compact parts elaborated by selective laser melting316L powder *Materials Science&Engineering* **A 584** 21

[4] Liu Y. Li A. Cheng X. Zhang S Q and Wang H M 2016 Effects of heat treatment on microstructure and tensile properties of laser melting deposited AISI 431 martensitic stainless steel *Materials Science&Engineering* **A 666** 27

[5] Yadollahi A. Shamsaei N. Thompson St M and Seely D W 2015 Effects of process time interval and heat treatment on the mechanical and microstructural properties of direct laser deposited 316L stainless steel *Materials Science&Engineering* **A 644** 171

[6] Chlebus E. Kuźnicka B. Kurzynowski T and Dybała B 2011 Microstructure and mechanical behaviour of Ti—6Al—7Nb alloy produced by selective laser melting *Materials Characterization* **62** 488

[7] Louvis E. Fox P and Sutcliffe Christopher J 2011 Selective laser melting of aluminium components *Journal of Materials Processing Technology* **211** 275

[8] Ma M. Wang Z and Zeng X 2017 A comparison on metallurgical behaviors of 316L stainless steel by selective laser melting and laser cladding deposition *Materials Science&Engineering* **A 685** 265

[9] Cheikh H E. Courant B. Branchu S. Huang X. Hasco J Y and Guillen G 2012 Direct Laser Fabrication process with coaxial powder projection of 316L steel. Geometrical characteristics and microstructure characterization of wall structures. *Optics and Lasers in Engineering* **50** 1779

IOP Conf. Series: Journal of Physics: Conf. Series **941** (2017) 012041 doi:10.1088/1742-6596/941/1/012041

# The influence of laser hardening on the performance of steel 30KHGSA by the finite elements method

**Y A Lyutikov[1, 2, *], A B Lyuchter[2, 3] M A Korobov[2, 3]**

[1] Bauman Moscow State Technical University, 2-ya Baumanskaya Street, 5, Moscow
[2] Vladimir engineering center of using laser technologies in mechanical engineering at VlSU, 3b Belokonskaya street, Vladimir
[3] Vladimir State University named after Alexander and Nikolay Stoletovs, Gorky Street 87, Vladimir

E-mail: *bakela01@yandex.ru

**Abstract.** The article presents the results of the computer simulation in the Solid Works using the method of finite elements. In the research described a computer model that was tested on experimental sample that was made of steel 30KHGSA. Data of thermal processes and residual stresses in a sample after laser processing were received during the research. Calculations of the depth of the hardened zones and the strength of the prototype were obtained during the simulation. The process of laser hardening a "Water jet engine sleeve" was simulated in the research. The results and conclusions obtained during this work are given in the conclusion.

## 1. Introduction

Requirements to the quality of manufactured machine parts are continuously increasing. They stipulate using highly effective methods of processing and improvement of technological processes. Applying of laser hardening is one of the most promising and productive methods. The main advantages are: reducing the duration of the thermal cycle of hardening, locality impact, no deformation and high processing speed and efficiency [1,2].

Modeling processes of heat treatment will allow to solve the problem of structurization and modernization surfaces in real products [4]. Modeling of laser hardening will allow considering a large number of factors with different nature (thermal, physic-chemical, etc.). The analysis of these factors will ensure adequate forecasting of the processes of hardening the structure and improving the qualities of the surface.

The computer model of a laser hardening process was developed for steel 30KHGSA [5]. The steel is pre-tempering and annealing which increases its abrasion resistance, strength, plasticity. The advantages of steel are the possibility to temper and anneal further. This steel is cheap enough, as it doesn't have expensive additives. The steel is highly weldable, resistant to corrosion and impact, it also has a moderate toughness. The steel is mostly applied in aviation industry, but due to high consumer properties it is widespread in other branches of engineering [5].

This paper discusses examples of the application of the computer simulation for the selection of conditions and modes of laser thermal hardening and study of its influence on performance characteristics of parts, enable optimizing the technological process of laser hardening.

IOP Conf. Series: Journal of Physics: Conf. Series **941** (2017) 012041 doi:10.1088/1742-6596/941/1/012041

Computer simulation will allow to reduce the cost of research, to get the automatic calculation module of processing modes in the production, and to reduce the time to design and field research on selection of modes.

The aim of this work is to obtain a computer model of laser thermal strengthening of steel 30KHGSA which is able automatically to select the processing modes for different parts, as well as identifying improvements to operational characteristics from the use of this method of processing.

## 2. Model description

Computer modeling of laser hardening was worked out on a plate with a thickness of 5 mm made of steel 30KHGSA.

This research was conducted to check the truthfulness computer model and assessment of its performance. Computer simulation was carried out by the finite element method in Solid Works to solve three tasks:

• Calculation of the temperature fields in the material of steel 30KHGSA, taking into account the spatial and temporal characteristics of laser radiation.

• Determination of the effect of laser hardening on the performance of the material and the influence of residual stresses in the sample.

• Determination of the optimal conditions of processing for a specific technological process.

Temperature field's formation research with the continuous laser radiation in three dimensional statements was carried out in the program Solid Works Simulation. The type of analysis was thermal, with the selection of the transition process. Formulas of the temperature fields were used for the process of propagation of heat instantly a localized source emitted at the surface of a semi-infinite body (1).

$$T(R,t) = \frac{2Q}{c\gamma(4\pi at)^{3/2}} e^{-\frac{R^4}{4at}} + T_0 \tag{1}$$

and the basic amount of heat distributed in the body in a certain period of time (2) [1].

$$dT(R,t-t') = \frac{2dQ}{c\gamma[4\pi a(t-t')]^{3/2}} e^{-\frac{R^2}{4a(t-t')}} \tag{2}$$

Where in formulas (1) and (2),
$T$ – the temperature in a considered point;
$t$ – time counted from the moment of introduction of heat;
$R$ – the distance from the heat source Q to the points of the body;
$a$ – thermal diffusivity;
$T_0$ – the ambient temperature;
$c\gamma$ – heat capacity material;
$Q$ – quantity of heat;
$t$–$t'$ – a period of time [1].

The material properties of the sample corresponded to the steel 30KHGSA. Plastic properties of steel 30HGSA such as modulus of elasticity E, Poisson's ratio v, tensile strength σ and yield strength σ were used for the prototype (plate). The prototype is divided into 10 zones 1 mm. For each zone in a certain time used laser radiation acting on the surface. Simulation of heating continuous laser radiation applied on the plate with a total duration of 0.5 s with a time increment of 0.05 seconds. The cooling time of the workpiece was – 1 s. The total spot size of the laser radiation was 1×1 mm. On each plot the distribution of the laser energy density was taken evenly. The convection coefficient was used in simulation with room temperature of 300 K. Laser radiation power need to be programmed in Solid Works during the simulation. The optimum power range of laser hardening is 800 to 2500 Wt for the required structure of the material [3]. For the model was created standard high density grid which is allowed to simulate the fast process of laser hardening with sufficient accuracy.

## 3. The results of the experiment

According to the results of modeling were obtained the plot figure1, which shows the movement of the laser beam, heat distribution in the underlying layers of material, the depth of thermal influence and the heating temperature of a material by laser radiation. It should be noted that in the simulation of thermal processes were carried out iteration for selecting the optimal heat range. It is necessary to pay more attention on the iterations in Solid Works environment because of model does not reflect the true result without it.

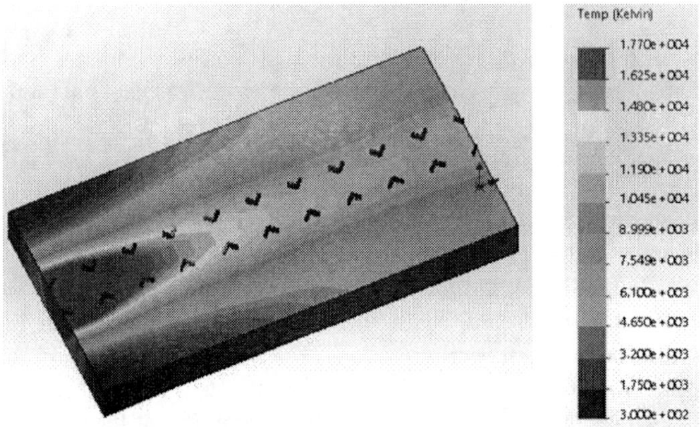


**Figure 1.** Heat plot, with visible movement of the laser beam and the heat distribution around the prototype.

Then, the residual stress of a test sample determined after exposure to laser radiation. For this, in simulations, the thermal effect of laser radiation programmed on each heat affected zone at a particular time and incorporated the type of fastening. The fixation was carried out on the sides of the plate as seen as in figure 2. In the simulation on the prototype can be seen compressive and tensile stresses, which can show the strength of the material in zone of laser irradiation. Similarly, it is necessary to pay more attention the iteration of the selection of an optimum range of residual stresses with the aim of obtaining accurate results.

**Figure 2.** Plot of residual stresses after laser irradiation on the prototype.

According to the results of computer modeling in Solid Works environment was measured the depth of the hardened zones and calculated the strength of the prototype under different conditions of laser treatment (table 1).

**Table 1.** Characteristics of heat-affected zones (depth, strength).

| Cipher zone | 1 | 2 | 3 | 4 | 5 | 6 | 7 | 8 | 9 | 10 |
|---|---|---|---|---|---|---|---|---|---|---|
| Laser radiation power *P*, W | 2500 | 2000 | 1500 | 1200 | 1100 | 1050 | 1000 | 900 | 850 | 800 |
| Depth heat affected zone, mm | 3.56 | 3.22 | 3.170 | 2.766 | 2.554 | 2.243 | 1.994 | 1.98 | 1.87 | 1.83 |
| Strength N/mm$^2$ | 2125 | 1922 | 2228 | 2318 | 2234 | 2155 | 2147 | 2098 | 1970 | 1889 |

Data obtained by computer simulation, used as a base for conducting practical confirmation of the truthfulness of the model. Practical confirmation consists of 3 stages:

- Testing technological modes of laser thermal strengthening on a plate made of steel 30KHGSA.
- Measure the depth of the hardened layer and the hardness of zones with different modes of processed.
- Study the microstructure on the different modes of processed.

To confirm the truthfulness of the model and to estimate the selection of the modes of laser treatment, the results obtained during the practical part of the experiment were compared with the simulation results.

## 4. Verification

The experimental research was carried out on laser robotized complex of the Russian development which includes:

- Ytterbium fiber laser power 3 kW production of company IRE Polus (Fryazino Moscow region);
- FLW D50 laser head with lens Ø 50 mm production of company IRE Polus (Fryazino Moscow region);
- Interface, the cockpit door, automatic disabling the laser radiation production of Engineering center (Vladimir);
- A six-axis robot and two-axis positioner, production of company Fanuc.

For the identification deviation between theoretical and experimental data, and identification of the error model were compared the results of computer simulation and experiment. Table 2 shows the comparison of the depths of heat-affected zones and the strength of the surface depending on the modes of laser hardening.

The experimental results practically coincide with results of computer simulation as you see from table 2. The average deviation in depth of heat-affected zones is 3.9% and the average deviation for strength is 4.6%. It should be noted that the results of verification while using laser radiation power 2600 and 2000 W their average deviation higher than the others. The reason for this was overheating of the part and on the surface appeared melt. Also it can be concluded that the temperature plots and plots of residual stresses of the prototype show the correct values consistent with the values obtained as a result of laser hardening. This proves the adequacy of the model and makes it workable, what enables to use this model in the calculation of the modes of laser hardening of the real details.

For example, was selected the "Water jet engine sleeve" detail made of steel 30KHGSA. New modes and conditions of processing were used to simulate the process of laser hardening on Sleeve. Plots of distribution of temperature fields and residual stresses of a 3D computer model of the part the Sleeve was obtained in Solid Works environment (figure 3).

**Table 2.** Comparing computer models and experimental data.

| Cipher zone | 1 | 2 | 3 | 4 | 5 | 6 | 7 | 8 | 9 | 10 |
|---|---|---|---|---|---|---|---|---|---|---|
| Laser radiation power *P*, W | 2600 | 2000 | 1500 | 1200 | 1100 | 1050 | 1000 | 900 | 850 | 800 |
| Depth heat affected zone in modeling, mm | 3.56 | 3.22 | 3.17 | 2.77 | 2.55 | 2.24 | 2.0 | 1.98 | 1.87 | 1.83 |
| Depth heat affected zone in experiment, mm | 3.34 | 3.18 | 3.07 | 2.67 | 2.46 | 2.13 | 1.92 | 1.89 | 1.79 | 1.78 |
| Deviation, % | 6.2 | 2.3 | 3.2 | 3.5 | 3.7 | 5.1 | 3.9 | 4.8 | 4.3 | 2.6 |
| Strength in modeling, N/mm$^2$ | 2125 | 1922 | 2228 | 2318 | 2234 | 2155 | 2147 | 2098 | 1970 | 1889 |
| Strength in experiment N/mm$^2$ | 1920 | 1775 | 2145 | 2345 | 2310 | 2310 | 2225 | 2200 | 2100 | 1920 |
| Deviation, % | 9.0 | 7.0 | 3.4 | 1.2 | 3.3 | 6.8 | 3.6 | 4.7 | 6.2 | 1.7 |

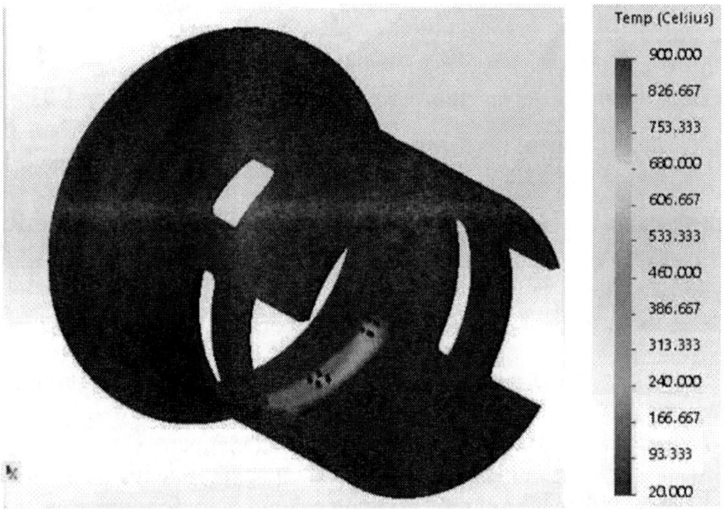


**Figure 3.** Heat plot, with visible movement of the laser beam and the heat distribution in detail Sleeve.

According to simulation results, the depth of heat-affected zones and their strength obtained in (table 3).

**Table 3.** The depth of heat-affected zones and tensile strength with the optimum modes of laser hardening while modeling.

| Cipher zone | 5 | 6 | 7 |
|---|---|---|---|
| Laser radiation power *P*, W | 1100 | 1050 | 1000 |
| Depth heat affected zone, mm | 2,554 | 2,343 | 2,240 |
| Strength, N/mm$^2$ | 2234 | 2155 | 2047 |

IOP Conf. Series: Journal of Physics: Conf. Series **941** (2017) 012041 doi:10.1088/1742-6596/941/1/012041

## 5. Conclusion

In the conducted research were obtained:

1. Models of thermal plots, which shows: the depth of thermal influence and heat temperature of a material by laser radiation.
2. Model plots of residual stress that can determine the strength characteristics of material in zone of laser irradiation.
3. The good tests results of computer models were obtained by the verification in experiment.
4. The computer model of laser thermal hardening of steel 30HGSA which can obtain optimal processing modes and strength characteristics of the part by 3D simulation.
5. Results that can help to develop a module for selection of technological modes of laser handling in the laser robotic complex.

## References

[1] Grigoryants A G and Safonov A N 1987 *Methods of surface laser processing*, ed. by Grigoryants A G (Moscow: publishing house of the Higher school)

[2] Grigoryants A G, Shiganov I N and Misyurov A I 2006 *Technological processes of laser processing*, ed. by Grigoryants A G (Moscow: publishing by Bauman Moscow State Technical University)

[3] 1985 *Laser and electron beam processing of materials* Ed. by Uglova A A, Rikaline N N (Moscow: Mashinostroenie)

[4] Yaresko S I and Goryainov D S 2013 Development of technologies for pulsed laser processing of cutting tool using finite element method. ed. by Bryansk state engineering-technological Academy

[5] Lyutikov Y A, Luchter A B and Kravchenko I I 2017 Researching the technology of laser thermal harfening of water jet engine cylinder parts from steel 30KHGSA, *Reference book. Engeneering Journal* (ed. by Publishing house Spektr)

IOP Conf. Series: Journal of Physics: Conf. Series **941** (2017) 012042 doi:10.1088/1742-6596/941/1/012042

# Application of low-temperature plasma for the synthesis of hydrogenated graphene (graphane)

**M B Shavelkina*, R H Amirov, V A Katarzhis and V I Kiselev**

Joint Institute for High Temperatures of Russian Academy of Sciences, Izhorskaya Street 13/2, 125412 Moscow, Russia

E-mail: *mshavelkina@gmail.com

**Abstract**. The possibility of a direct synthesis of hydrogenated graphene in decomposition of methane by means of low-temperature plasma was investigated. A DC plasma torch with an expanding channel-anode, a vortex gas supply and a self-setting arc length was used as a generator of low-temperature plasma. Argon was used as the plasma-forming gas. The temperatures of argon plasma and with methane addition to it were determined on the basis of spectral measurements. The synthesis products were characterized by electron microscopy and thermogravimetry. The effect of hydrogenated graphene as a nanomodifier on the properties of the cubic boron nitride based functional ceramics was investigated.

## 1. Introduction

Low-temperature plasma physics is a very active area of research located on the boundaries between physics, chemistry and materials science. Recent technological developments, e.g. in plasma etching or plasma deposition, have led to a revived interest in plasma physics and technology.

Plasma torches provide a stationary temperature at the level of several tens of thousands of degrees in a gas of arbitrary composition. This makes it possible to use them for synthesis of nanoscale particles with widely varying set of properties [1].

At present, low-temperature plasma is used not only for synthesis of carbon nanostructures, but also for their functionalization. In the work [2], with the help of a hydrogen plasma, the surface of carbon nanotubes was doped with hydrogen, for hydrogen doping of graphene obtained by micromechanical exfoliation the authors [3] applied hydrogen afterglow plasma.

The present work is devoted to the possibility of using a low-temperature plasma for direct synthesis of hydrogenated graphene (graphane).

## 2. Methods

To synthesize graphane we used a plasma jet reactor, which was used early for the synthesis of carbon nanotubes and graphene. A detailed description of the experimental setup was given in the studies [4, 5]. The key element of the installation is a direct current plasma torch with an expanding channel-anode and a self-setting arc length. The use of the expanding channel of the output electrode in the plasma torch allows to significantly increase the velocity of the cold gas at the inlet to the channel and to intensify the heat exchange between the arc and plasma-forming gas, which contributes to a decrease in arc length and more uniform distribution of heat fluxes to the electrode walls [6].

The experiment involved a simultaneous input of carbon precursor (methane) with a plasma forming gas (argon) into the plasma torch, wherein heating and decompositions occurred in the plasma jet and in the region of the arc discharge, followed by condensation of the synthesis product on the metal surface of the target in a vacuum chamber. The rate of carbon precursor, plasma gas flow and the power of the plasma torch varied independently of each other. For the experimental conditions the electric power of plasma torch was set up to 35 kW.

Figure 1 shows the experimental setup.

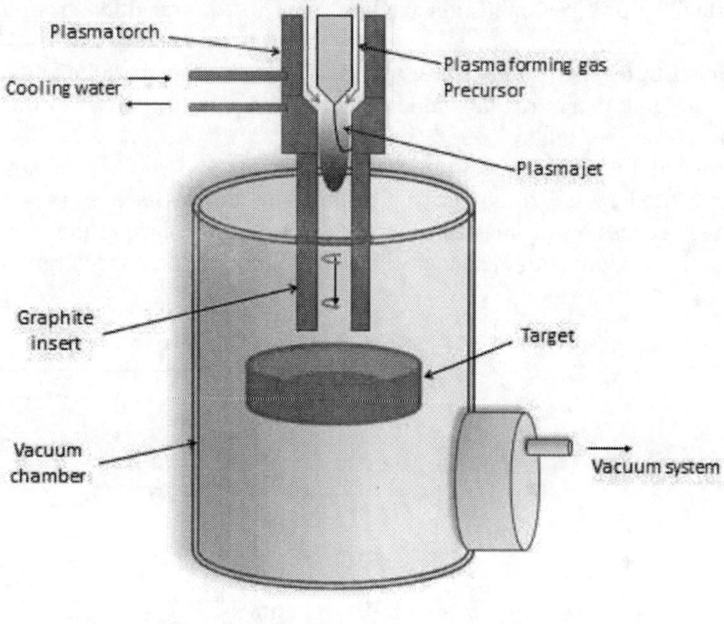


**Figure 1.** The experimental setup.

The experimental conditions are presented in table 1.

**Table 1.** Technological conditions.

| Power (kW) | Current (A) | Voltage (V) | Gas pressure (Torr) | Argon flow rate (g/s) | Methane flow rate (g/s) |
|---|---|---|---|---|---|
| 22–35 | 300–350 | 60–90 | 150–740 | 3.0–3.75 | 0.05–0.37 |

The synthesis products formed in the volume of the plasma jet reactor were investigated by the standard methods for diagnostics of carbon nanomaterials structure. The method of electron microscopy (scanning electron microscope Hitachi S5500) was used in three modes: SE (displays the surface morphology), BF-STEM (provides a transmission mode), DF-STEM (displays regions that efficiently scatter electrons). To evaluate the efficiency of the synthesis, thermogravimetry was used - (STA 449 Netzsch) with linear sample heating in argon at the rate of 5 K/min at temperatures up to 1000 °C.

The spectra of argon plasma were taken on a three-channel fiber-optic spectrometer AvaSpec 2048 with spectral resolution of 0.1–0.4 nm and spectral range 200–1100 nm.

The electron temperature was found, using the method of relative intensities of the same lines of ionization particles (method of Boltzmann exhibitors).

Stark broadening of the hydrogen lines at 468 and 656 nm for a mixture of argon and methane was used to determine the electron concentration [8].

Volt-ampere characteristics of argon plasma generator were investigated at pressures in the planetary jet reactor of 350 and 710 Torr.

IOP Conf. Series: Journal of Physics: Conf. Series **941** (2017) 012042 doi:10.1088/1742-6596/941/1/012042

## 3. Results and discussions

It is known that the efficiency of technological processes increases significantly when using plasma torches with an increasing volt-ampere characteristic. Figure 2 presents the results of experimental studies of volt-ampere characteristics of argon plasma generator for argon flow rate of 3.0 g/s. They have clearly expressed growing character. Whereas, a larger inclination angle is observed at a pressure of 710 Torr. The increasing character of the volt-ampere characteristic is due to the peculiarity of the used plasma torch, in which the electric discharge burns completely in the expanding channel, and therefore the stability of arc combustion persists even in the transition from argon to a mixture of argon and methane.

Figure 3 demonstrates that when adding methane to argon, the arc voltage jump and, consequently, power jump occur, and afterwards the values do not change until the end of the experiment (duration of the experiments was 5–12 min).

Figure 3 shows the emission spectrum of argon plasma and mixture of argon and methane. The spectra were recorded in the discharge region near anode on its axis at a pressure of 710 Torr. According to the processing of the emission spectra of argon plasma, the electron temperature was 11700 K, and in the mixture with methane – 11300 K. The electron concentration was $6.75 \cdot 10^{16}$ cm$^{-3}$.

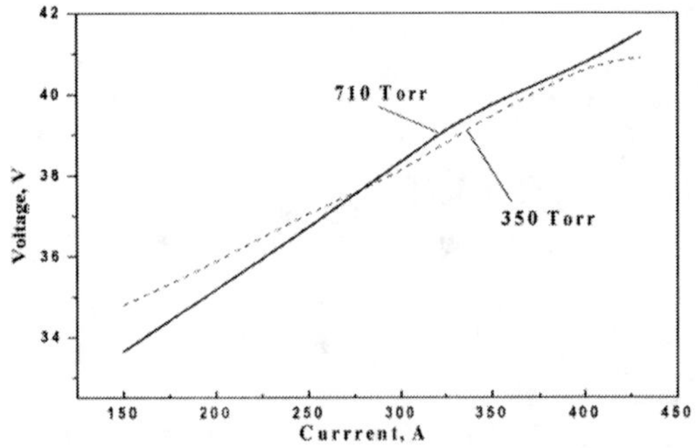


**Figure 2.** Volt-ampere dependence for argon plasma. Argon flowrate is 3.0 g/s.

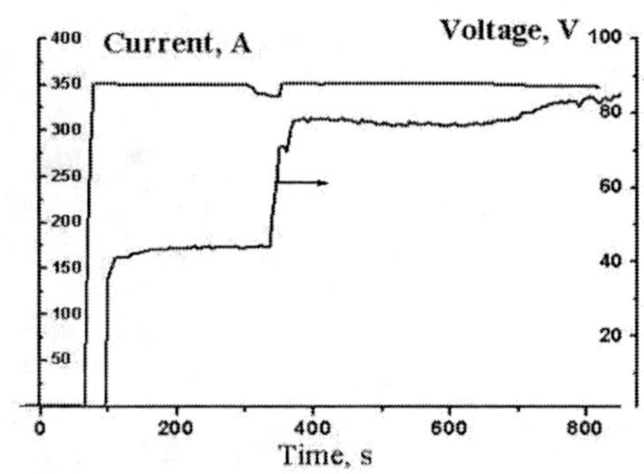


**Figure 3.** The change in current and voltage during the experiment. Argon flow rate is 3.0 g/s, methane flow rate is 0.3 g/s. Pressure of the medium is 710 Torr.

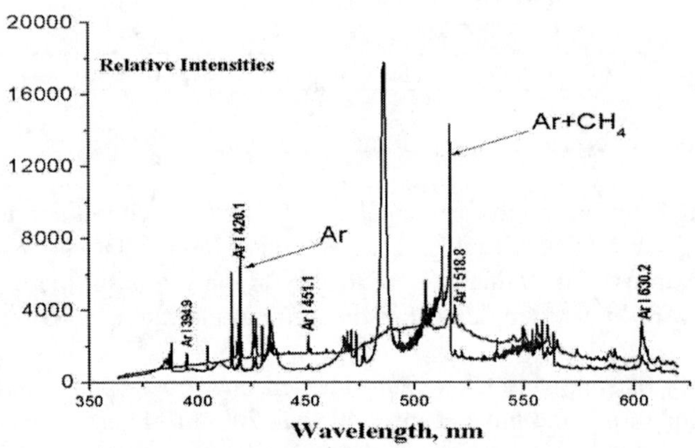


**Figure 4.** Emission spectra of plasma Ar and Ar-$CH_4$.

IOP Conf. Series: Journal of Physics: Conf. Series **941** (2017) 012042 doi:10.1088/1742-6596/941/1/012042

Figures 5–6 present the morphology of graphene materials obtained by a plasma jet. The synthesis products are flakes, which are usually observed in the synthesis of graphene with the use of thermal plasma [8–10]. Inflation is noticeable on the surface of the cup-shaped graphene. Comparison of the two images obtained from different detectors gives the main result - the inflation is represented by gas-filled bubbles. These inflations in their morphology are the same as in [8], where graphite and graphene were treated with hydrogen-argon plasma. The authors associate the structural changes with surface doping by hydrogen.

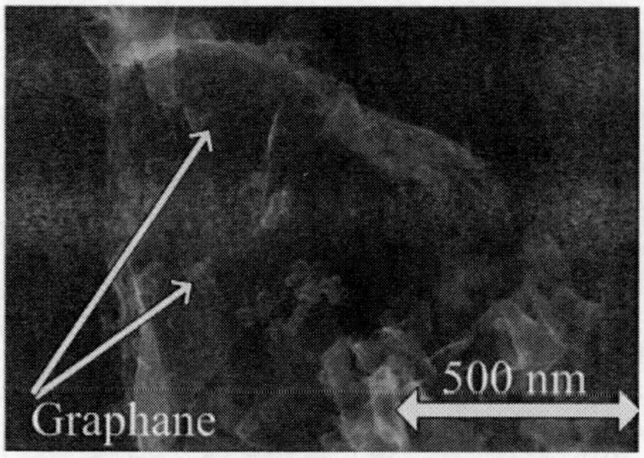


**Figure 5.** SEM image of graphane flakes.

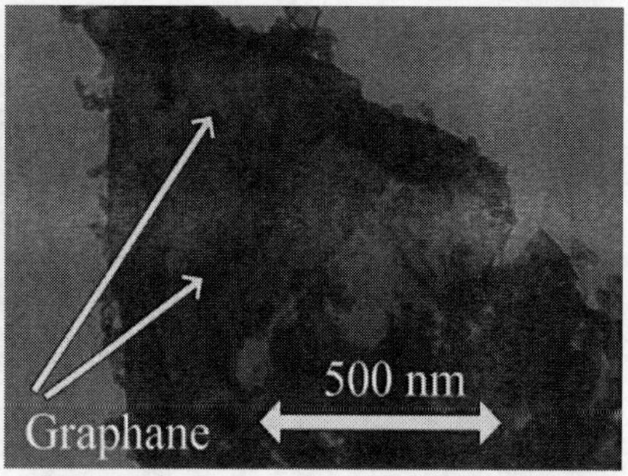


**Figure 6.** STEM image of graphane flakes.

The results of thermogravimetric analysis confirmed the presence of hydrogen in the synthesized samples. Figure 7 shows the mass loss curves when samples are heated, the rate of its change, and desorption of hydrogen. Emission of hydrogen is observed throughout the entire temperature increase region. The peak is observed near 300 °C as in [3].

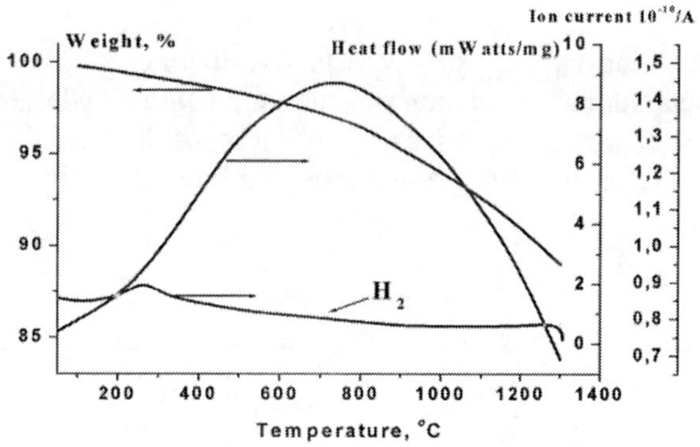


**Figure 7.** Thermogravimetry of graphane flakes.

The obtained hydrogenated graphene were introduced into the composition of ceramic bond of a high-strength material based on cubic boron nitride (its hardness is close to diamond) fabricated under hot pressing conditions. According to [11], in these conditions, the use of fine-dispersed highly active composite powders, which ensure the formation of a dense fine-grained matrix in the process of hot pressing, is especially promising. At 1450 °C and pressure of 5.5 GPa a modified ceramic composition with a content of up to 70 mass % of cubic boron nitride was obtained.

Microstructure of the obtained samples, average particle size, nature and value of the closed porosity were studied by scanning electron microscopy. Figure 6 shows the surface of a sample destruction with a matrix containing from 0.05 to 1 mass % of graphane. Analysis of the ceramic samples fractures microstructure was carried out on a scanning electron microscope Nova NanoSem 650 using the standard procedure after testing the samples for bending. The study showed that the microstructures of fracture surfaces of these three samples, both in the locations of localization of nanofibers and in zones free from them were similar to each other in the basic features. On the bends, boron nitride grains with characteristic dimensions of ~1 to 2.5 μm and surrounding shells of bond composed of more dispersed formations based on Al and Ti are observed. Both - nanotube groups and individual nanotubes with a diameter of 20–40 nm are protruded from those shells, (i.e. formation of a framework in a ceramic matrix is seen). Thus, under the conditions of hot pressing due to the presence of hydrogen in graphene materials, graphene flakes are folded into filamentary nanostructures and self-reinforcement of the ceramic matrix takes place.

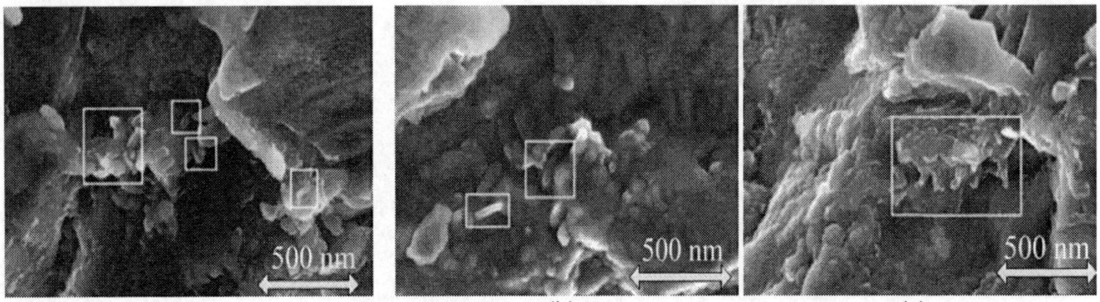


(a) (b) (c)

**Figure 8.** Microphotographs of fractures of ceramic samples based on cubic BN with addition of CNS in the ratio: a – 0.05 mass %, b – 0.5 mass %, c – 1 mass %. Separate formed zones of carbon nanotubes localization are marked by rectangles.

Physical and mechanical properties (apparent density, open porosity and average tensile strength) of the samples were studied. The obtained data are presented in table 2. Determination of the average density and open porosity of the items was carried out by kerosene saturation with preliminary weighing of dry samples and subsequent hydrostatic weighing of saturated samples on precision scales. Density of kerosene for the calculation was measured with a hydrometer. While at the graphene content of 0.05 and 1.0 mass % the values of the composite strength are practically equal, the composite with 0.5 mass % has the strength properties by 1.12 times higher. The other parameters are approximately similar.

**Table 2.** Physical and mechanical properties of ceramics samples.

| Content of carbon nanostructures (mass %) | Apparent density ($g/cm^3$) | Open porosity (%) | Average tensile strength (MPa) |
|---|---|---|---|
| 0 | 3.47–3.50 | 0.5–1.5 | 750 |
| 0.05 | 3.491 | 0.631 | 730 |
| 0.5 | 3.495 | 0.892 | 820 |
| 1.0 | 3.435 | 1.535 | 730 |

## 4. Conclusion

In this paper, the possibility of hydrogen doping of graphene structures during their synthesis with the use of low-temperature plasma was demonstrated. To generate plasma with stable characteristics, we used a constant-current plasma torch with an expanding channel-anode, a vortex gas supply and a self-setting arc length. Studies of the plasma torch characteristics have shown the stability of its operation

IOP Conf. Series: Journal of Physics: Conf. Series **941** (2017) 012042 doi:10.1088/1742-6596/941/1/012042

when introducing hydrocarbon into the plasma jet. The synthesized samples are flakes with a lateral dimension of 50 to 700 nm. The use of synthesized hydrogenated graphene samples in the composition of a binder in high-temperature hot-pressed nitride ceramics has led to the increase of its average strength.

**Acknowledgements**
The authors gratefully acknowledge Russian Foundation for Basic Research for the support by grants No 15-08-00165, 16-08-00145, 16-08-00057 and 16-08-00081.

**References**
[1] Rempel A 2007 *J. Russian Chemical Reviews* **76** (5) 435
[2] Khare, Bishun and Meyyappan M 2004 *NASA Tech Briefs* **01** 01
[3] Elias D, Nair R, Mohiuddin T, Morozov S, Blake P, Halsall M, Ferrari A, Boukhvalov D, Katsnelson M, Geim A and Novoselov K 2009 *Science.* **323** 610
[4] Amirov R, Isakaev E, Shavelkina M and Shatalova T 2014 *J. of Phys.: Conf. Ser.* **550** 0120
[5] Amirov R, Shavelkina M, Alihanov N, Shkolnikov E, Tyuftyaev A and Vorob'eva N 2015 *J. of Nanomaterials* **2015**
[6] Isakaev E, Sinkevich O, Tyuftyaev A and Chinnov V 2010 *High Temperature.* **48** (1) 97
[7] Belevtsev A, Chinnov V and Isakaev E 2006 *Plasma Sources Science and Technology.* 15 (3) 450
[8] Castellanos-Gomez A, Wojtaszek M, Arramel, Tombros N and van Wees B 2012 *Small* **8** (10) 1607
[9] Legrand U, Mendoza Gonzalez N-Y, Pascone P, Meunier J-L and Berk D 2016 *Carbon* **102** 216
[10] Kim J, Heo S, Gu G and Suh 2010 *J. Nanotechnology* **21** 095601
[11] Kurakevych O and Solozhenko V 2016 *Molecules* **21** 1399

# Propagation of Circularly Polarized Light Through a Two-Dimensional Random Medium

**E E Gorodnichev**

National Research Nuclear University MEPhI, Kashirskoe Shosse 31, 115409 Moscow, Russia

E-mail: gorodn@theor.mephi.ru

**Abstract.** The problem of small-angle multiple-scattering of circularly polarized light in a two-dimensional medium with large fiberlike inhomogeneities is studied. The attenuation lengths for elements the density matrix are calculated. It is found that with increasing the sample thickness the intensity of waves polarized along the fibers decays faster than the other density matrix elements. With further increase in the thickness, the off-diagonal element which is responsible for correlation between the cross-polarized waves dissapears. In the case of very thick samples the scattered field proves to be polarized perpendicular to the fibers. It is shown that the difference in the attenuation lengths of the density matrix elements results in a non-monotonic depth dependence of the degree of polarization.

## 1. Introduction

Studies of multiple scattering of light in disordered two-dimensional media have attracted much attention (see, e.g., [1–12]). This problem is of interest due to practical applications in optics of artificial two-dimensional nanomedia [6, 10, 12], biotechnology [7] etc.

The multiple scattering of electromagnetic waves polarized perpendicular or along the fiberlike inhomogeneties is described within the scalar theory [1–9, 11]. However, the scalar field approximation fails to account for correlations between cross-polarized waves and therefore does not allow one to describe the propagation of circularly polarized waves.

Below the problem of the small-angle multiple scattering of circularly polarized light in a medium composed of thick (the radius larger than the wavelength) parallel fibers is studied within the Fokker-Planck approximation. It is assumed that the medium is highly absorbing and the angular distribution of the scattered light remains sharply anisotropic at all depths [2, 11, 13, 14]. The Stokes parameters of scattered waves are calculated. It is found that with increasing the sample thickness the intensity of waves polarized along the fibers decays faster than the other density matrix elements. With further increase in the thickness the off-diagonal element which is responsible for correlation between the cross-polarized waves vanishes. In the case of very thick samples the scattered wave field proves to be polarized perpendicular to the fibers. It is shown that the difference in the attenuation lengths of the density matrix elements results in a non-monotonic dependence of the degree of polarization on the sample thickness. In propagation through a relatively thin layer, the degree of polarization decreases due to decay of correlations between the cross-polarized waves. For a thick sample the polarization of scattered wave field is independent on the initial polarization of the incident beam, and the degree of polarization tends to unity.

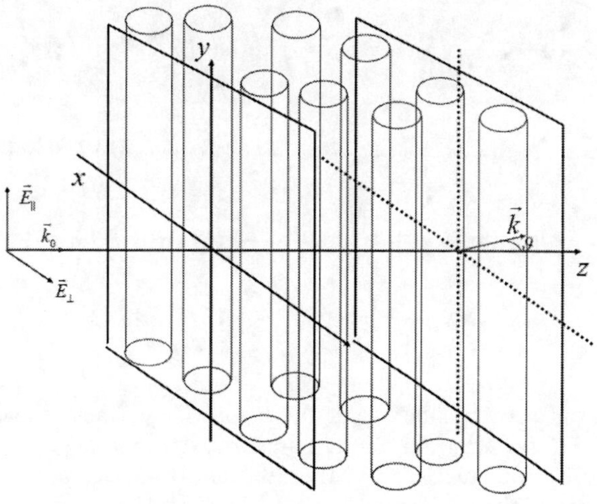


**Figure 1.** Scattering geometry.

The results obtained in this paper can be used in studies of optical properties of inhomogeneous media containing a fiberlike microstructure [6, 7, 10, 12].

## 2. General relation

Consider a plane wave of circularly polarized light incident normally to the plane boundary of a two-dimensional medium. The medium is composed of parallel fiberlike inhomogeneities, the $y$-axis is directed along the fibers and the beam is incident along the $z$-axis. The direction of light propagation is specified by angle $\theta$ ($\theta$ is deviation of the wave-vector from the $y0z$ plane, $-\pi < \vartheta < \pi$ (see Fig. 1). Transverse size $a$ of the scatterers is supposed to be much greater than the wavelength of light $\lambda$. We assume that the refraction index $n$ of fibers is close to unity and for single scattering the phase shift is small: $n - 1 \ll \lambda/a \ll 1$.

For the considered geometry, multiple scattering is conveniently described by a density matrix in the linear representation [1]

$$\hat{\rho} = \begin{pmatrix} |E_{\parallel}|^2 & E_{\parallel}E_{\perp}^* \\ E_{\parallel}^*E_{\perp} & |E_{\perp}|^2 \end{pmatrix} = \frac{1}{2}\begin{pmatrix} I+Q & U-\mathrm{i}V \\ U+\mathrm{i}V & I-Q \end{pmatrix} \tag{1}$$

where $E_{\parallel}$ and $E_{\perp}$ are the electric-field components that are parallel and perpendicular to the fibers (see Fig. 1). The Stokes parameters $I$, $Q$, $U$ and $V$ of the scattered radiation are conveniently written in the form

$$\begin{pmatrix} I \\ Q \\ U \\ V \end{pmatrix} = \frac{1}{2}\begin{pmatrix} I_{\parallel}+I_{\perp} & I_{\parallel}-I_{\perp} & 0 & 0 \\ I_{\parallel}-I_{\perp} & I_{\parallel}+I_{\perp} & 0 & 0 \\ 0 & 0 & 2\mathrm{Re}\, I_{\parallel,\perp} & 2\mathrm{Im}\, I_{\parallel,\perp} \\ 0 & 0 & -2\mathrm{Im}\, I_{\parallel,\perp} & 2\mathrm{Re}\, I_{\parallel,\perp} \end{pmatrix}\begin{pmatrix} I_0 \\ Q_0 \\ U_0 \\ V_0 \end{pmatrix} \tag{2}$$

where $I_0$, $Q_0$, $U_0$ and $V_0$ are the Stokes parameters of the incident beam.

The elements $I_{\parallel}$, $I_{\perp}$ and $I_{\parallel,\perp}$ appearing in the linear representation of the density matrix (1) is convenient in that obey independent transport equations

$$\left\{\cos\vartheta\frac{\partial}{\partial z}+\sigma_{\parallel}+\sigma_a\right\}I_{\parallel}(z,\vartheta)=\int\limits_{-\pi}^{\pi}d\vartheta'\;a_{\parallel}(\vartheta-\vartheta')\,I_{\parallel}(z,\vartheta'),$$

$$\left\{\cos\vartheta\frac{\partial}{\partial z}+\sigma_{\perp}+\sigma_a\right\}I_{\perp}(z,\vartheta)=\int\limits_{-\pi}^{\pi}d\vartheta'\;a_{\perp}(\vartheta-\vartheta')\,I_{\perp}(z,\vartheta'),$$

$$\left\{\cos\vartheta\frac{\partial}{\partial z}+\frac{1}{2}\left[\sigma_{\parallel}+\sigma_{\perp}\right]+\left[\sigma_a-\frac{4in_0}{k_0}\mathrm{Im}\;(T_1(0)-T_2(0))\right]\right\}I_{\parallel,\perp}(z,\vartheta)=$$
$$\int\limits_{-\pi}^{\pi}d\vartheta'\;a_{\parallel,\perp}(\vartheta-\vartheta')\,I_{\parallel,\perp}(z,\vartheta'), \tag{3}$$

where $k_0=2\pi/\lambda$ is the wave number, $\sigma_a$ is the absorption coefficient and $T_{1,2}(\psi)$ $(\psi=\vartheta-\vartheta')$ are the amplitudes of scattering for the cross-polarized waves ($T_{1,2}(0)$ are the amplitudes of forward scattering). The differential scattering coefficients $a_{\parallel}$, $a_{\perp}$, and $a_{\parallel,\perp}$ entering into Eq.(3) are expressed in terms of the amplitudes $T_{1,2}$ as follows [1, 15]:

$$a_{\parallel}(\psi)=\frac{2n_0}{\pi k_0}\,|T_1(\psi)|^2,\;a_{\perp}(\psi)=\frac{2n_0}{\pi k_0}\,|T_2(\psi)|^2,\;a_{\parallel,\perp}(\psi)=\frac{2n_0}{\pi k_0}\,T_1(\psi)\,T_2^*(\psi) \tag{4}$$

where $n_0$ is the number of fibers per unit area in the plane normal to their axes. The scattering coefficients $\sigma_{\parallel}$ and $\sigma_{\perp}$, are written as

$$\sigma_{\parallel}=\int\limits_{-\pi}^{\pi}d\psi\;a_{\parallel}(\cos\psi),\;\sigma_{\perp}=\int\limits_{-\pi}^{\pi}d\psi\;a_{\perp}(\cos\psi) \tag{5}$$

The boundary conditions for Eq.(3) have the form

$$I_{\parallel}(z=0,\vartheta)=I_{\perp}(z=0,\vartheta)=I_{\parallel,\perp}(z=0,\vartheta)=\delta(\vartheta) \tag{6}$$

The first two equations entering into Eqs.(3) were derived in Ref. [1]. Equation for the off-diagonal element $I_{\parallel,\perp}$ can be obtained from the equations for the third $U$ and fourth $V$ Stokes parameters, that were also given in Ref. [1].

In the Born approximation (i.e., for weakly refractive fibers), the scattering amplitudes of the cross-polarized waves are related by the relation

$$T_2(\psi)=T_1(\psi)\cos\psi, \tag{7}$$

where the amplitude $T_1$ coincides with the inhomogeneity form-factor [16]. For three-dimensional Born inhomogeneities, the scattering amplitudes of the cross-polarized waves are related to each other by a similar equation (7) [15, 16].

The equation for the intensity of the co-polarized waves $I_{\parallel}$ (see the first equation in the system (3)) coincides with the scalar transfer equation. At the same time, propagation of the cross-polarized field is sensitive to the vector nature of light. In particular, within the Born approximation the differential scattering cross section for the intensity $I_{\perp}$ contains the additional factor $\cos^2\psi$ (see Eqs.(3) and (7)) [4] .

For weakly refractive fibers the third equation entering into the system (3) takes the form:

$$\left\{\cos\vartheta\frac{\partial}{\partial z}+\frac{1}{2}\left[\sigma_{\|}+\sigma_{\perp}\right]+\sigma_a\right\}I_{\|,\perp}(z,\vartheta)=\frac{1}{2}\int\limits_{-\pi}^{\pi}d\vartheta'\ \left[a_{\|}(\vartheta-\vartheta')+a_{\perp}(\vartheta-\vartheta')\right]I_{\|,\perp}(z,\vartheta')-$$
$$\frac{1}{2}\left(\frac{2n_0}{\pi k_0}\right)\int\limits_{-\pi}^{\pi}d\vartheta'\ \left|T_1(\vartheta-\vartheta')-T_2(\vartheta-\vartheta')\right|^2 I_{\|,\perp}(z,\vartheta') \tag{8}$$

The second term in the right-hand-side of Eq.(8) is responsible for decrease of correlations between the cross-polarized fields.

For weakly refractive fibers, as follows from Eqs.(6) and (8), the off-diagonal elements of the density matrix are real, Im $I_{\|,\perp}=0$.

The system of equations (3) and Eq.(8) under boundary conditions (6) completely determines the Stokes parameters of multiply scattered light in a medium with two-dimensional weakly refractive inhomogeneities.

## 3. Fokker-Planck approximation

Propagation of polarized light through a medium with large imhomogeneities can be best understood within the exactly solvable model, which is based on the small-angle diffusion (or Fokker-Planck) approximation [11, 13, 14]. The latter assumes that the single scattering angle is small. Apart from that, we consider the case of high absorption: the absorption length $l_a = (\sigma_a)^{-1}$ is assumed to be less than the transport mean free path $l_{tr}^{\|} = \left(\int_{-\pi}^{\pi}d\psi(1-\cos\psi)a_{\|}(\psi)\right)^{-1}$. Then, for multiple scattering, the deflection angle remains also small at arbitrary depth $z$ and the contribution of backscattered radiation is negligibly small [13, 14]. Therefore, the angular variable $\vartheta$ can vary in infinite limits, $-\infty<\vartheta<\infty$.

Within the Fokker-Planck approximation the transport equation for modified intensity $\tilde{I}_{\|}(z,\vartheta)=\exp(\sigma_a z)I_{\|}(z,\vartheta)$ has the form (see, e.g., [2, 11])

$$\left\{\frac{\partial}{\partial z}+\sigma_a\frac{\vartheta^2}{2}\right\}\tilde{I}_{\|}(z,\vartheta)=D_{\|}\frac{\partial^2\tilde{I}_{\|}(z,\vartheta)}{\partial\vartheta^2} \tag{9}$$

where $D_{\|}=\sigma_{\|}\left\langle\psi^2\right\rangle/2$ is the diffusion coefficient for waves polarized along the fibers and $\left\langle\psi^2\right\rangle = \int\limits_{-\infty}^{\infty}d\psi\ a_{\|}(\psi)\,\psi^2/\sigma_{\|}$ is the variance of single scattering angle. In the diffusion approximation the coefficient $D_{\|}$ is equal to the scattering transport coefficient of the co-polarized waves $\sigma_{tr}^{\|}$ [11]

$$\sigma_{tr}^{\|}=\int_{-\pi}^{\pi}d\psi(1-\cos\psi)a_{\|}(\psi)\approx\frac{1}{2}\int\limits_{-\infty}^{\infty}d\psi\ a_{\|}(\psi)\,\psi^2 \tag{10}$$

The equation for $\tilde{I}_{\perp}(z,\vartheta)=\exp(\sigma_a z)I_{\perp}(z,\vartheta)$ differs from Eq.(9) by the substitution of the coefficient diffusion $D_{\|}$ by

$$D_{\perp}=\int_{-\pi}^{\pi}d\psi(1-\cos\psi)a_{\perp}(\psi)\approx\frac{1}{2}\int\limits_{-\infty}^{\infty}d\psi\ a_{\perp}(\psi)\,\psi^2=D_{\|}(1-\eta),\ \eta=\frac{13}{12}\frac{\left\langle\psi^4\right\rangle}{\left\langle\psi^2\right\rangle}\ll 1. \tag{11}$$

As the analysis shows, the effects associated with changes in light polarization are noticeable in the asymptotic regime at depths $z > l_d = \left(\sqrt{D_{\parallel}\sigma_a/2}\right)^{-1}$. Using the results [2, 11], for the modified intensities at depths $z > l_d$ we obtain

$$\tilde{I}_{\parallel}(z,\vartheta) = \frac{1}{\sqrt{\pi\langle\vartheta^2\rangle_\infty}} \exp\left(-\frac{z}{l_d} - \frac{\vartheta^2}{2\langle\vartheta^2\rangle_\infty}\right) \tag{12}$$

$$\tilde{I}_{\perp}(z,\vartheta) = \frac{1}{\sqrt{\pi\langle\vartheta^2\rangle_\infty}} \exp\left(-\frac{z}{l_d}\left(1-\frac{\eta}{2}\right) - \frac{\vartheta^2}{2\langle\vartheta^2\rangle_\infty}\right) \tag{13}$$

where $\langle\vartheta^2\rangle_\infty = \sqrt{2D_{\parallel}/\sigma_a} \approx \sqrt{2D_{\perp}/\sigma_a}$ is the variance of multiple scattering angle in the asymptotic regime.

The small-angle approximation is applied to Eq.(8) in the following way. The expansion of the angle-dependent coefficients on the left- and right-hand sides of Eq.(8) in terms of small angle $\vartheta$ with allowance for only the first non-vanishing terms yields the equation

$$\left\{\frac{\partial}{\partial z} + \sigma_a\frac{\vartheta^2}{2}\right\}\tilde{I}_{\parallel,\perp}(z,\vartheta) = \left(\frac{D_{\parallel}+D_{\perp}}{2}\right)\frac{\partial^2}{\partial\vartheta^2}\tilde{I}_{\parallel,\perp}(z,\vartheta) - \frac{\sigma_{dep}}{2\pi}\int_{-\infty}^{\infty} d\vartheta'\tilde{I}_{\parallel,\perp}(z,\vartheta'), \tag{14}$$

where

$$\sigma_{dep} = \frac{1}{2}\left(\frac{2n_0}{\pi k_0}\right)\int_{-\pi}^{\pi} d\psi\ \left|T_1(\psi) - T_2(\psi)\right|^2 \tag{15}$$

is the effective "absorption" coefficient for correlations between cross-polarized waves.

The last term in right-hand-side of Eq.(14) results from the approximation

$$\frac{n_0}{\pi k_0}\int_{-\pi}^{\pi} d\vartheta'\ \left|T_1(\psi) - T_2(\psi)\right|^2 \tilde{I}_{\parallel,\perp}(z,\vartheta') \approx \frac{1}{4\pi}\int_{-\pi}^{\pi} d\vartheta'\ a_{\parallel}(\psi)\left(1-\cos(\psi)\right)^2 \int_{-\infty}^{\infty} d\vartheta'\tilde{I}_{\parallel,\perp}(z,\vartheta') \tag{16}$$

Equality (16) is based on the assumption that the angle dependence of $\left|T_1(\vartheta-\vartheta') - T_2(\vartheta-\vartheta')\right|^2$ is practically isotropic in comparison with $\tilde{I}_{\parallel,\perp}(z,\vartheta')$ (see [17]).

In the case of multiple scattering of light in a medium with three-dimensional inhomogeneities the effective "absorption" coefficient $\sigma_{dep} = (n_0/2)\int d\mathbf{n}'\left|A_{\parallel}(\mathbf{nn}') - A_{\perp}(\mathbf{nn}')\right|^2$ ($A_{\parallel}$ and $A_{\perp}$ are the amplitudes of the waves polarized parallel and perpendicularly to the scattering plane) is responsible for the "dynamical" mechanism of depolarization [18, 19].

The off-diagonal element $\tilde{I}_{\parallel,\perp}$ in the asymptotic regime can be easily calculated by the perturbation theory using as the first approximation the solution of Eq.(14) for $\sigma_{dep} = 0$ (see, e.g., [18]). As a result, for the modified parameter $\tilde{I}_{\parallel,\perp}$ we obtain

$$\tilde{I}_{\parallel,\perp}(z,\vartheta) = \frac{1}{\sqrt{\pi\langle\vartheta^2\rangle_\infty}} \exp\left(-\frac{z}{l_d}\left(1-\frac{\eta}{4}\right) - \frac{z}{l_{dep}}\sqrt{\frac{\langle\vartheta^2\rangle_\infty}{\pi}} - \frac{\vartheta^2}{2\langle\vartheta^2\rangle_\infty}\right) \tag{17}$$

where $l_{dep} = (\sigma_{dep})^{-1}$ is the mean free path with respect to random shift of phase difference between cross-polarized waves.

Relying on Eqs.(12), (13) and (17), we proceed to analyze the effects of multiple scattering on the polarization state of light for the incident circularly polarized waves.

## 4. Discussion

According to Eqs.(12), (13), (17) in propagation of light through a two-dimensional medium with thick fibers, the hierarchy of attenuation lengths of the density matrix elements can be distinguished. The different attenuation lengths result from the difference in the scattering amplitudes of the cross-polarized waves. The diagonal element which describes the scattering of waves polarized along the fibers decays rapidly. The attenuation length $l_d^{\parallel}$ coincides with the asymptotic length $l_d^{\parallel} = l_d = (D_{\parallel}\sigma_a/2)^{-1/2}$. The off-diagonal element which is responsible for correlations of the cross-polarized waves decays at a greater length, $z \geq l_d^{\parallel,\perp} = \left(((D_{\parallel} + D_{\perp})\sigma_a/4)^{1/2} + \sigma_{dep}(\langle\vartheta\rangle_{\infty}/\pi)^{1/2}\right)^{-1} \sim l_d(1 + l_d/(2l_{dep}))$. The diagonal element that describes propagation of the waves polarized perpendicular to the fibers is the most long-lived one. The length of attenuation of this element proves to be of the order of $l_d^{\perp} = (D_{\perp}\sigma_a/2)^{-1/2} \approx l_d(1 + l_d/l_{dep})$. As known [18, 19], in a medium with three-dimensional inhomogeneities the differences in the scattering amplitudes of the cross-polarized waves are responsible for the "dynamical" mechanism of the depolarization. Thus, in the case of two-dimensional scatterers, any changes in polarization of light are due to the "dynamical" mechanism of depolarization.

The depth dependence of the density matrix elements is illustrated in Fig. 4, where polarization ratios $I_{\parallel}/I_{\perp}$ and $I_{\parallel,\perp}/I_{\perp}$ as functions of $z$ are shown.

For the forward scattered light with the circular initial polarization ($I_0 = V_0 = 1$, $Q_0 = U_0 = 0$), the degree of polarization $P = \sqrt{Q^2 + V^2}/I$ and the ellipticity $\beta = (1/2)\arctan(V/|Q|)$ in the asymptotic regime are equal to

$$P = \sqrt{1 - \cosh^{-2}\left(\frac{\eta}{4}\frac{z}{l_d}\right)\left(1 - \exp\left(-\frac{2z}{l_{dep}}\sqrt{\frac{\langle\vartheta^2\rangle_{\infty}}{\pi}}\right)\right)} \tag{18}$$

$$\beta = \frac{1}{2}\arctan\left[\sinh^{-1}\left(\frac{\eta}{4}\frac{z}{l_d}\right)\exp\left(-\frac{z}{l_{dep}}\sqrt{\frac{\langle\vartheta^2\rangle_{\infty}}{\pi}}\right)\right] \tag{19}$$

At relatively small depths, $l_d < z < l_d/\eta \ll l_{dep}$, the decrease of $P$ is due to decay of correlations between the cross-polarized fields, $P = 1 - (z/l_{dep})\sqrt{\langle\vartheta^2\rangle_{\infty}/\pi}$. At large depths, $z > l_{dep}\sqrt{\pi/\langle\vartheta^2\rangle_{\infty}}$, the correlations vanish and the value of $P$ tends to unity. In such a limit the scattered light is polarized perpendicularly to the fibers. This is due to the fact that $I_{\parallel}$ decreases rapidly compared to $I_{\perp}$ and $I_{\perp,\parallel}$. The degree of polarization $P$ reaches its minimum at an intermediate depth, $z \sim (4l_d/\eta)$ (see Fig. 4).

The ellipticity $\beta$ at relatively small depths, $z < l_d/\eta$, proves to be close to the initial value, $\beta = \pi/4$. With increasing $z$, $z > l_d/\eta$, the ellipticity $\beta$ tends exponentially to zero. As a result, the polarization plane of scattered waves becomes to be orthogonal with respect to the fibers. As follows from Eq.(19), the decrease of $\beta$ with increasing depth $z$ is governed by the differences between the diffusion coefficients $D_{\parallel}$ and $D_{\perp}$, and, in contrast to the degree of polarization $P$, is virtually independent of the presence of the phase coherence between the cross-polarized waves.

## 5. Conclusions

Small-angle scattering of circularly polarized light in a medium with large two-dimensional inhomogeneities has been studied. The attenuation lengths of the density matrix elements have been calculated within the Fokker-Planck (or diffusion) approximation. It has been found that, at a relatively small depth, the degree of polarization changes owing to decay of correlations between the cross-polarized waves. As the depth increases, the intensity of waves polarized along

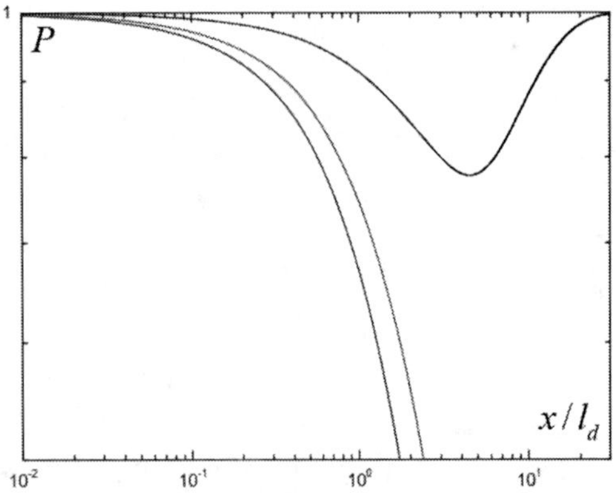


**Figure 2.** Degree of polarization $P$ and ratios $I_{\|,\perp}/I_\perp$, $I_\|/I_\perp$ (black, red and blue curves, respectively) as a functions of depth $z$ ($\eta = 0.8$, $l_d/l_{dep} = 0.12$, $\langle\vartheta^2\rangle_\infty = 0.6$).

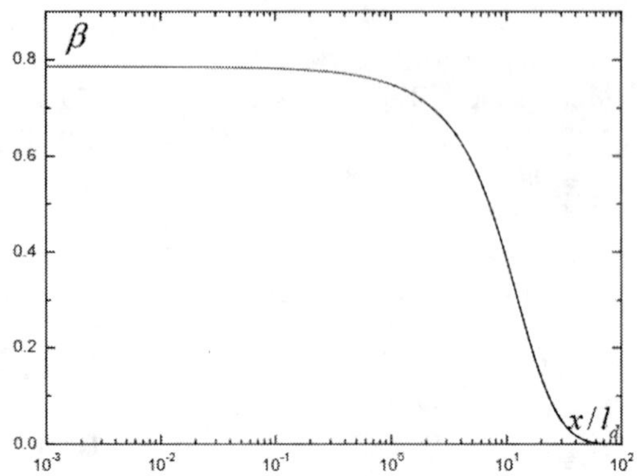


**Figure 3.** Degree of ellipticity $\beta$ as a function of depth $z$ ($\eta = 0.8$, $l_d/l_{dep} = 0.12$, $\langle\vartheta^2\rangle_\infty = 0.6$).

the fibers decays faster than the other density matrix elements. The off-diagonal element that is responsible for correlations between the cross-polarized waves disappears at greater depths. In the asymptotic regime the scattered light proves to be polarized perpendicular to the fibers, and the degree of polarization tends to unity. The difference in the attenuation rate between the density matrix elements results in a non-monotonic dependence of the degree of polarization on the depth.

**Acknowledgment.** The author thanks V.V. Marinyuk and D.B. Rogozkin for their interest in this work and useful discussions. This work was supported by the MEPhI Academic Excellence Project (Contract No. 02.a03.21.0005).

## References

[1] Mishchenko M I, Dlugach J M and Yanovitskij E G 1992 *J. Quant. Spectrosc. Radiat. Transfer* **47** 401
[2] Alexandrov M D and Remizovich V S 1995 *J. Opt. Soc. Am.* A **12** 2726
[3] McLean A S and Pendry J B 1995 *J. Mod. Opt.* **42** 339
[4] Kirchner A, Busch K and Soukoulis C M 1998 *Phys. Rev.* E **57** 277
[5] Roux L, Mareschal P, Vukadinovic N, Thibaud J-B and Greffet J-J 2001 *J. Opt. Soc. Am.* A **18** 374
[6] Ito T and Tomita M 2002 *Phys. Rev.* E **66** 027601
[7] Kienle A, Forster F K and Hibst R 2004 *Opt. Letters* **29** 2617
[8] Volkov S N, Saarinen J J, and Sipe J E 2012 *J. Mod. Opt.* **59** 954
[9] Frezza F, Pajewski L, Ponti C and Schettini G 2010 *J. Opt. Soc. Am.* A **27** 687
[10] Liu W, Miroshnichenko A E, Oulton R F, Neshev D N, Hess O and Kivshar Y S 2013 *Opt. Lett.* **38** 2621
[11] Marinyuk V V 2015 *J. Opt. Soc. Am.* A **32** 1330
[12] Arruda T J, Martinez A S and Pinheiro F A 2016 *Phys. Rev.* A **94** 033825
[13] Dolin L S 1981 *Sov. Phys. Dokl.* **26** 976
[14] Remizovich V S, Rogozkin D B and Ryazanov M I 1982 *Izv. Acad. Sci. USSR, Atmos. Ocean. Phys.* **18** 480
[15] Bohren C F and Huffman D R 1983 *Absorption and Scattering of Light by Small Particles* (New York: Wiley-Interscience)
[16] Newton R G 1982 *Scattering Theory of Waves and Particles* (New York: Springer, 2nd ed.)
[17] Gorodnichev E E, Ivliev S V, Kuzovlev A I and Rogozkin D B 2010 *Laser Physics* **20** 2021
[18] Gorodnichev E E, Kuzovlev A I and Rogozkin D B 2007 *JETP* **104** 319
[19] Gorodnichev E E, Kuzovlev A I and Rogozkin D B 2014 *Phys. Rev.* E **90** 043205

IOP Conf. Series: Journal of Physics: Conf. Series **941** (2017) 012044 doi:10.1088/1742-6596/941/1/012044

# Depolarization of an Ultrashort Pulse in a Disordered Ensemble of Mie Particles

**E E Gorodnichev, S V Ivliev, A I Kuzovlev and D B Rogozkin**

National Research Nuclear University MEPhI, Kashirskoe Shosse 31, 115409 Moscow, Russia

E-mail: gorodn@theor.mephi.ru

**Abstract.** We study propagation of an ultrashort pulse of polarized light through a turbid medium with the Reynolds-McCormick phase function. Within the basic mode approach to the vector radiative transfer equation, the temporal profile of the degree of polarization is calculated analytically with the use of the small-angle approximation. The degree of polarization is shown to be described by the self-similar dependence on some combination of the transport scattering coefficient, the temporal delay and the sample thickness. Our results are in excellent agreement with the data of numerical simulations carried out previously for aqueous suspension of polystyrene microspheres.

## 1. Introduction

Over past two decades, the propagation of light through highly scattering media has attracted great attention due to its potential applications to optical tomography [1].

An ultrashort pulse of light propagating in a scattering medium experiences multiple scattering events, resulting in broadening of its temporal profile and depolarization. Early arrival photons of the transmitted pulse propagate along nearly straight lines. These photons are the least depolarized. Therefore the polarization can be used for gating the early arrival component of the transmitted pulse with the gate of the order of a polarization decay time (in accordance with Refs. [1–3] the degree of polarization decays over the initial $100ps$ after the arrival of the ballistic photons).

Depolarization of polarized pulses transmitted through a thick turbid slab was studied both experimentally [2–5] and numerically with Monte Carlo [6] and DISORT [7] codes. The results [2–7] pertain to turbid media composed of Mie-particles (polystyrene microspheres suspended in water). As follows from Refs. [2–7], the transmitted light retains its initial polarization over a small temporal interval following the arrival of the ballistic photons, $\Delta = ct - z \ll z/2$ ($c$ is the velocity of light in the medium, $z$ is the slab thickness). The photons arriving with delays $\Delta > z$ are completely depolarized.

An analytical treatment of the depolarization process was carried out within the small-angle diffusion (or Fokker-Planck) approximation [8] which holds provided that the phase function falls off rapidly as the scattering angle increases. Such an approximation fails for many realistic situations where the phase function decreases rather slowly.

It what follows the depolarization of an ultrashort pulse of polarized light propagating through a turbid medium is considered. Within the basic mode approximation [9–11], we develop a theoretical approach to calculating the temporal profile of the degree of polarization. The

IOP Conf. Series: Journal of Physics: Conf. Series **941** (2017) 012044 doi:10.1088/1742-6596/941/1/012044

calculations are carried out within the small-angle approximation in application to the Reynolds-McCormick phase function [12] which is frequently used for modeling the single-scattering by Mie-particles. The degree of polarization is shown to depend only on a certain combination of the delay $\Delta$, the slab thickness $z$ and the transport scattering coefficient $\sigma_{tr}$. Our results are in excellent agreement with the data of numerical simulations [6].

## 2. Basic mode approximation

To calculate the temporal profile of the degree of polarization, we take advantage of the nonstationary vector radiative transfer equation. The normal incidence of a $\delta$-pulse of polarized light on the surface is considered. The beam width is assumed to be greater than the beam transverse spread in the medium. For highly forward single-scattering, we can neglect the off-diagonal elements of the scattering matrix [9, 10]. Then, the vector radiative transfer equation decomposes into three independent equations for the basic modes [9–11, 13].

The intensity satisfies the scalar transfer equation

$$\left\{\frac{1}{c}\frac{\partial}{\partial t}+\mathbf{\Omega}\frac{\partial}{\partial \mathbf{r}}+\sigma_{tot}\right\} I\left(\mathbf{r},\mathbf{\Omega},t\right)=\sigma\int d\mathbf{\Omega}' a_1(\mathbf{\Omega}\mathbf{\Omega}') I\left(\mathbf{r},\mathbf{\Omega}',t\right) \quad (1)$$

where $\sigma_{tot}=\sigma+\sigma_a$ is the total extinction coefficient; $\sigma$ and $\sigma_a$ are the scattering and absorption coefficients of the medium, respectively, the vectors $\mathbf{\Omega}$ and $\mathbf{\Omega}'$ denote the directions of photon propagation, $a_1(\mathbf{\Omega}\mathbf{\Omega}')$ is the phase function of single-scattering. For the unit incident flux, the boundary condition for Eq.(1) is written as

$$I\left(z=0,\mathbf{\Omega},t\right)=\delta\left(t\right)\delta\left(\mathbf{\Omega}-\mathbf{\Omega}_0\right) \quad (2)$$

where $\mathbf{\Omega}_0$ is the internal normal to the surface. The basic modes of linear and circular polarizations, $W\left(\mathbf{r},\mathbf{\Omega},t\right)$ and $V\left(\mathbf{r},\mathbf{\Omega},t\right)$ , are subject to the following transfer equations:

$$\left\{\frac{1}{c}\frac{\partial}{\partial t}+\mathbf{\Omega}\frac{\partial}{\partial \mathbf{r}}+\sigma_{tot}\right\} W\left(\mathbf{r},\mathbf{\Omega},t\right)=\sigma\int d\mathbf{\Omega}'\left(\frac{a_1(\mathbf{\Omega}\mathbf{\Omega}')+a_2(\mathbf{\Omega}\mathbf{\Omega}')}{2}\right) \mathrm{e}^{2\mathrm{i}(\chi_+-\psi)} W\left(\mathbf{r},\mathbf{\Omega}',t\right) \quad (3)$$

$$\left\{\frac{1}{c}\frac{\partial}{\partial t}+\mathbf{\Omega}\frac{\partial}{\partial \mathbf{r}}+\sigma_{tot}\right\} V\left(\mathbf{r},\mathbf{\Omega},t\right)=\sigma\int d\mathbf{\Omega}' a_2(\mathbf{\Omega}\mathbf{\Omega}') V\left(\mathbf{r},\mathbf{\Omega}',t\right) \quad (4)$$

where $a_2(\mathbf{\Omega}\mathbf{\Omega}')$ is the second diagonal element of the scattering matrix [10, 11], the definition of the angle $\chi_+$ is given in Refs. [10, 11], $\psi=\varphi-\varphi'$ is the difference between the azimuth angles of the vectors $\mathbf{\Omega}$ and $\mathbf{\Omega}'$. For polarized light, the boundary conditions to Eqs.(3) and (4) are similar to Eq.(2).

## 3. Small-angle multiple scattering

The small-angle approximation is valid provided that the mean square of the multiple-scattering angle $\theta$ at depth $z$ and time instant $t$ satisfies inequality $\langle\theta^2\rangle_{z,t}\ll 1$. According to Refs. [14, 15] $\langle\theta^2\rangle_{z,t}\sim\Delta/z$ and, therefore, the early arrival component of the transmitted pulse, $\Delta\ll z$ can universally be treated within the small-angle approximation.

To describe the scattering of light by large particles, we take advantage of the two-parameter Reynolds-McCormick phase function [12]

$$a_1(\cos\gamma)=\frac{\alpha-2}{2\pi}\frac{g\left(1-g^2\right)^{\alpha-2}}{\left(1+g\right)^{\alpha-2}-\left(1-g\right)^{\alpha-2}}\frac{1}{\left(1+g^2-2g\cos\gamma\right)^{\alpha/2}}, \quad (5)$$

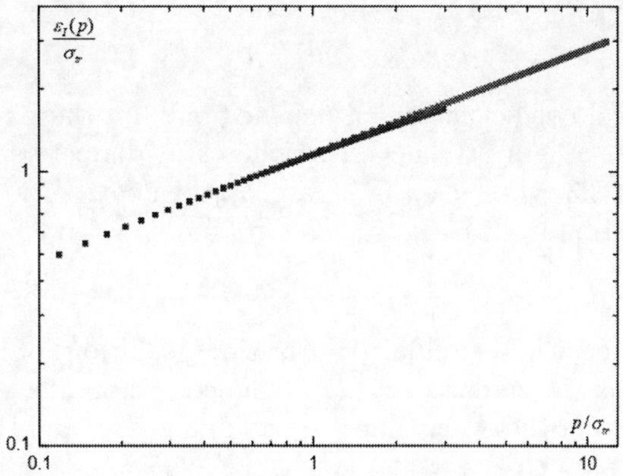


**Figure 1.** Eigenvalue $\varepsilon_I(p)$ as a function of variable $p/\sigma_{tr}$. Results of numerical calculations with the Mie theory for aqueous suspension of polystyrene microspheres (red and blue symbols correspond to diameters of 0.993 $\mu m$ and 0.300 $\mu m$, respectively).

where $\alpha$ and $g$ are the parameters. For $\alpha = 3$, Eq.(5) coincides with the well-known Henyey-Greenstein phase function. In scattering through small angles the Reynolds-McCorrmick phase function falls off with increasing angle $\gamma$ in accordance with a power law $a_1 \sim 1/\gamma^\alpha$, $2 \leq \alpha \leq 4$.

Within the small-angle approximation, Eq.(1) takes the form [15]

$$\left\{\frac{\partial}{\partial z} + \frac{\theta^2}{2}\frac{\partial}{\partial \Delta}\right\} \tilde{I}(z,\theta,\Delta) = \frac{\sigma_{tr}}{4\pi}2^{\alpha-2}(4-\alpha)\int d\boldsymbol{\theta}' \frac{\tilde{I}(z,\theta,\Delta) - \tilde{I}(z,\theta',\Delta)}{|\boldsymbol{\theta}-\boldsymbol{\theta}'|^\alpha} \quad (6)$$

where $\tilde{I}(z,\theta,\Delta) = c^{-1}\exp(-\sigma_a ct)I(z,\theta,\Delta)$, $\theta$ is the angle between vectors $\mathbf{\Omega}$ and $\mathbf{\Omega}_0$, and $\Delta = ct - z$ is the difference between path length $ct$ and depth $z$, $\sigma_{tr} = \sigma(1 - \langle\cos\gamma\rangle)$ is the transport scattering coefficient, and $\langle\cos\gamma\rangle$ is the mean cosine of single scattering.

Using the Bessel transform with respect to the angular variable $\theta$ and the Laplace transform with respect to $\Delta$, we can reduce Eq.(6) to an eigenvalue problem for the equation that resembles the stationary Schrödinger equation [15]. For relatively small delays, $\Delta \ll z$, only the minimum eigenvalue and the corresponding eigenfunction contributes to the intensity of the transmitted pulse. Therefore $\tilde{I}(z,\theta,\Delta)$ can be presented in the form [15]

$$\tilde{I}(z,\theta,\Delta) = \int_{-\mathrm{i}\infty}^{\mathrm{i}\infty} \frac{dp}{2\pi\mathrm{i}} \exp(p\Delta - \varepsilon_I(p)z) \int_0^{\mathrm{i}\infty} \frac{ldl}{2\pi} J_0(l\theta)\Phi_I(l,p) \int_0^{\mathrm{i}\infty} \frac{l'dl'}{2\pi}\Phi_I(l',p) \quad (7)$$

where $\varepsilon_I(p)$ and $\Phi_I(l,p)$ are determined from the equation

$$\left(-\frac{p}{2}\Delta + \sigma(1 - a_1(l))\right)\Phi_I(l,p) = \varepsilon_I(p)\Phi_I(l,p) \quad (8)$$

The function $a_1(l)$ appearing in Eq.(8) is the Bessel transform of the small-angle phase function $a_1(\mathbf{\Omega}\mathbf{\Omega}') \approx a_1(|\boldsymbol{\theta}-\boldsymbol{\theta}'|)$. For the Reynolds-McCormick phase function

$$\sigma(1 - a_1(l)) = \frac{\sigma_{tr}}{2}\left(\frac{4-\alpha}{\alpha-2}\right)\frac{\Gamma(2-\alpha/2)}{\Gamma(\alpha/2)} \cdot l^{\alpha-2} \quad (9)$$

In the case of $\alpha = 3$ which corresponds to the Henyey-Greenstein phase function, $\sigma(1 - a_1(l)) = \sigma_{tr}l$.

From Eqs.(8), (9) it follows that the eigenvalue $\varepsilon_I(p)$ can be written as [15]

$$\varepsilon_I(p) = c_\alpha \sigma_{tr}^{2/\alpha} p^{1-2/\alpha} \tag{10}$$

where $c_\alpha$ is the numerical coefficient which depends only on the parameter $\alpha$. Our numerical calculations carried out for polystyrene microspheres of diameters 0.3 $\mu m$ and 0.993 $\mu m$ at the wavelength $\lambda = 0.633$ $\mu m$ (see Fig.1) are approximated by the relations $\epsilon_I(p)/\sigma_{tr} = 1.15(p/\sigma_{tr})^{0.357}$, $\alpha = 3.11$ and $\epsilon_I(p)/\sigma_{tr} = 1.62(p/\sigma_{tr})^{0.381}$, $\alpha = 3.23$, respectively.

## 4. Degree of polarization

For the basic modes of linear and circular polarizations, solutions of Eqs.(3) and (4) can also be presented in the form that is similar to Eq.(7). The corresponding solutions differ from Eq.(7) only by the specific values of the eigenvalues, $\varepsilon_W(p)$ and $\varepsilon_V(p)$, and the eigenfunctions, $\Phi_W(l,p)$ and $\Phi_V(l,p)$.

In the case of highly forward multiple scattering, the values of $\varepsilon_W(p)$ and $\varepsilon_V(p)$ turn out to be close to $\varepsilon_I(p)$ provided that the value of $p$ is rather great, $p > \sigma_{tr}$. This follows directly from our numerical calculations of $\varepsilon_I(p)$ , $\varepsilon_W(p)$ and $\varepsilon_V(p)$ with the use of the corresponding characteristic equations [11] both for the Reynolds-McCormick phase function and for large Mie-particles (polystyrene and silica microspheres in water). Therefore we can calculate the values of $\varepsilon_W(p)$ and $\varepsilon_V(p)$ analytically with a perturbation theory taking the solution of the scalar transfer equation (8) as the first approximation. Within such an approach, the difference between the values of $\varepsilon_I(p)$, $\varepsilon_W(p)$ and $\varepsilon_V(p)$ can be written in the form

$$\delta\varepsilon_{W,V}(p) = \sigma \int_0^\infty \frac{ldl}{2\pi} \Phi_I(l,p) \Delta a_{W,V}(l) \Phi_I(l,p) \tag{11}$$

where $\Delta a_{W,V}(l)$ is the difference between $a_1(l)$ and the Bessel transforms of the "effective" phase functions appearing in Eqs.(3) and (4).

The small-angle transfer equation for the basic mode of linear polarization differs from Eq.(8) only by the "perturbation" term $\sigma \Delta a_{W,V}(l)$ which has the form [9, 10]

$$\sigma \Delta a_W \approx \sigma \Delta a_W^{geom} + \sigma \Delta a_W^{dyn} = \frac{\sigma}{2} \left[ \frac{1}{l} \frac{\partial}{\partial l} \left[ \frac{\partial a_1(l)}{\partial l} \frac{\partial}{\partial l} \right] \right] + \frac{\sigma}{2} (a_1(l) - a_2(l)) \tag{12}$$

The first term in Eq.(12) is responsible for the "geometrical" depolarization due to the Rytov effect. The second term describes the "dynamical" depolarization. With the standard perturbation theory [16], the value of $\varepsilon_W(p)$ can be written as $\varepsilon_W = \varepsilon_I + \delta\varepsilon_W^{geom} + \delta\varepsilon_W^{dyn}$. The "geometrical" contribution to $\varepsilon_W$ is determined by

$$\delta\varepsilon_W^{geom}(p) = \frac{\sigma}{2} \int_0^\infty \frac{ldl}{2\pi} \Phi_I(l,p) \left[ \frac{1}{l} \frac{\partial}{\partial l} \left[ \frac{\partial a_1(l)}{\partial l} \frac{\partial}{\partial l} \right] \right] \Phi_I(l,p) = \frac{\Gamma(2-\alpha/2)}{2} \sigma_{tr} \left( \frac{\varepsilon_I(p)}{p} \right)^{3-\alpha/2} \tag{13}$$

When deriving Eq.(13), we use the eigenfunction $\Phi_I(l,p)$ approximated by [15]

$$\Phi_I(l,p) = \sqrt{\frac{4\pi\varepsilon_I(p)}{p}} \exp\left( -\frac{l^2 \varepsilon_I(p)}{2p} \right) \tag{14}$$

The "dynamical" contribution to $\varepsilon_W$ depends on ratio $\varepsilon_I(p)/p$ in a similar way,

$$\delta\varepsilon_W^{dyn}(p) = \frac{\sigma}{2} \int_0^\infty \frac{ldl}{2\pi} \Phi_I(l,p) \left[ a_1(l) - a_2(l) \right] \Phi_I(l,p) = \frac{\sigma_{tr}(4-\alpha)}{2^\alpha} \cdot \left( \frac{\varepsilon_I(p)}{p} \right)^{3-\alpha/2} \tag{15}$$

IOP Conf. Series: Journal of Physics: Conf. Series **941** (2017) 012044 doi:10.1088/1742-6596/941/1/012044

For the basic mode of circular polarization, the difference $\Delta a_V$ is only due to the "dynamical" mechanism of depolarization. The value of $\delta\varepsilon_V$ turns out to be two times greater than $\delta\varepsilon_W^{dyn}$.

Evaluating the integral in Eq.(7) and the integrals in the corresponding representations for the modes $W$ and $V$ by the saddle-point method, we obtain the following expressions for the degree of polarization of linearly and circularly polarized pulses

$$P_L = \frac{W}{I} = \exp\left(-\left(\beta(\alpha)+\eta(\alpha)\right)\cdot\sigma_{tr}z\cdot\langle\theta^2\rangle_{z,t}^{3-\alpha/2}\right),$$
$$P_C = \frac{V}{I} = \exp\left(-2\eta(\alpha)\cdot\sigma_{tr}z\cdot\langle\theta^2\rangle_{z,t}^{3-\alpha/2}\right) \quad (16)$$

where $\beta(\alpha) = \Gamma\left(3-\alpha/2\right)/2^{4-\alpha/2}$, $\eta(\alpha) = (4-\alpha)/2^{3+\alpha/2}$ and

$$\langle\theta^2\rangle_{z,t} = \frac{2\alpha}{\alpha-2}\left(\frac{\Delta}{z}\right)$$

is the mean square of the multiple-scattering angle $\theta$ for the Reynolds-McCormick phase function [15]. From Eq.(16) it follows that the degree of polarization depends on one dimensionless variable

$$\xi = \sigma_{tr}z\left(\frac{\Delta}{z}\right)^{\alpha-3/2}. \quad (17)$$

This conclusion is confirmed by comparison of our result (16) with data of Monte Carlo simulations for aqueous suspension of polystyrene microspheres [6] (see Fig. 2). Numerical simulation [6] was carried out for linearly polarized light and for various values of the transport optical thickness $\sigma_{tr}z$. Depolarization ratio $D_L = (1-P_L)/(1+P_L)$ was presented in [6] as a function of the normalized delay $\Delta/z$. When going from $\Delta/z$ to the variable $\xi$ we obtain a virtually universal pattern of data [6]. As follows from Fig. 2 the results of our calculations are in excellent agreement with the data of numerical simulations.

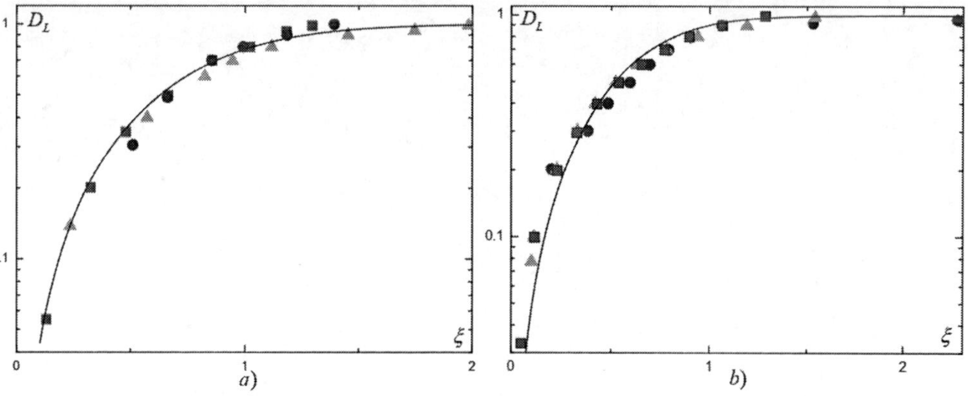


**Figure 2.** Depolarization ratio as a function of variable $\xi$. Symbols (■ – $\sigma_{tr}z = 2$, ▲ – $\sigma_{tr}z = 4$, ● – $\sigma_{tr}z = 10$) are the results of Monte Carlo simulations [6] for aqueous suspension of polystyrene microspheres of diameter 0.3 $\mu m$ ($a$) and 0.993 $\mu m$ ($b$). Solid lines are the results of our calculations ($\sigma_{tr}/\sigma = 0.339$, $\alpha = 3.11$ ($a$) and $\sigma_{tr}/\sigma = 0.085$, $\alpha = 3.23$ ($b$)).

## 5. Conclusions

We have studied the transmission of an ultrashort polarized pulse through highly scattering media. A theoretical model based on the Reynolds-McCormick phase function has been proposed for the analysis of depolarization of light in the medium with large inhomogeneities. In accordance with the model, the polarization state is determined by the intensity and the basic modes of linear and circular polarizations. The values of $W$ and $V$ differ from the intensity $I$ only by the factors that are responsible for additional attenuation of $W$ and $V$ with increasing the delay $\Delta = ct - z$. The results of our calculations are in excellent agreement with the data of numerical simulations [6] and make it possible to estimate the dependence of the degree of polarization on time and characteristics of the medium.

**Acknowledgment.** This work was supported by the MEPhI Academic Excellence Project (Contract No.02.a03.21.0005).

## References

[1] *Photonics: Biomedical photonics, spectroscopy, and microscopy* 2015 ed. D.L. Andrews **4** (Wiley)
[2] Horinaka H, Hashimoto K, Wada K and Cho Y 1995 *Opt. Lett.* **20** 1501
[3] Demos S G and Alfano R R 1996 *Opt.Lett.* **21** 161
[4] Wang X, Wang L V, Sun C W and Yang C C 2003 *J.Biomed.Opt.* **8** 608
[5] Ni X, Xing Q, Cai W and Alfano R R 2003 *Opt. Lett.* **28** 343
[6] Bruscaglioni P, Zaccanti G and Wei Q 1993 *Appl. Opt.* **32** 6142
[7] Ishimaru A, Jaruwatanadilok S and Kuga Y 2001 *Appl.Opt.* **40** 5495
[8] Gorodnichev E E, Kuzovlev A I and Rogozkin D B 2010 *Laser Phys.* **20** 2021
[9] Gorodnichev E E, Kuzovlev A I and Rogozkin D B 1999 *Laser Phys.* **9** 1210
[10] Gorodnichev E E, Kuzovlev A I and Rogozkin D B 2007 *JETP* **104** 319
[11] Gorodnichev E E, Kuzovlev A I and Rogozkin D B 2014 *Phys.Rev.* E **90**, 043205
[12] Reynolds L and McCormick N J 1980 *JOSA* A **70** 1206
[13] Zege E P and Chaikovskaya L I 2000 *JQSRT* **66** 413
[14] Remizovich V S, Rogozkin D B and Ryazanov M I 1983 *Izv. Acad. Sci. USSR, Atmos. Ocean. Phys.* **19** 796
[15] Rogozkin D B 1987 *Izv.Acad.Sci. USSR. Atmos.Oceanic Phys.* **23** 366
[16] Landau L D and Lifshitz E M 1977 *Quantum Mechanics (NonrelativisticTheory)* (3rd ed., Pergamon, New York)

# Long-Range Correlations Between Transmitted and Reflected Fluxes of Electromagnetic Waves

**E E Gorodnichev, A I Kuzovlev and D B Rogozkin**

National Research Nuclear University MEPhI, Kashirskoe Shosse 31, 115409 Moscow, Russia

E-mail: gorodn@theor.mephi.ru

**Abstract.** We study the long-range spatial correlations between intensity fluctuations in speckles formed by multiply scattered light. The correlation function between intensity fluctuations at the opposite boundaries of the slab are analyzed under the conditions of circular polarization memory. It shown that, until the scattered light is depolarized completely, the polarization and scalar contributions to the correlation function are of the same order of magnitude. As the slab thickness increases, their ratio falls off in inverse proportion to the thickness.

## 1. Introduction

Long-range correlations between the wave fields scattered in a disordered medium are responsible for numerous mesoscopic effects [1], in particular, for correlations between transmitted and reflected speckle patterns (see, e.g., [2, 3]). The corresponding spatial correlation function was calculated recently [4] within the scalar wave approximation without regard to the polarization state of scattered waves. Such an approach is not valid under the conditions of the polarization memory effect [5–11] where the depolarization of light occurs deep in the diffusive regime, and the scalar and polarization contributions to the correlation function are of the same order of magnitude.

The pronounced effect of circular polarization memory has been discovered recently [9–11] for resonant Mie particles with high refractive index. In this case the polarization contribution to intensity correlations through a disordered slab should be observed up to the optical thickness of the order of a few tens.

Below we study the intensity correlation function for two points lying on the input and output boundaries of diffusive-scattering medium. For the slab of great optical thickness $L \gg l_{tr}$, where $l_{tr}$ is the transport mean free path, the intensity fluctuations at the opposite boundaries of the slab are negatively correlated at spatial difference $\rho < L$. As $\rho$ increases, the correlation function changes its sign. The polarization contribution to the correlation function, as compared to the scalar one, changes its sign at smaller values of $\rho$ and is characterized by a greater amplitude of positive correlations. This contribution remains noticeable even though the transmitted light is completely depolarized.

IOP Conf. Series: Journal of Physics: Conf. Series **941** (2017) 012045 doi:10.1088/1742-6596/941/1/012045

## 2. General relations

The correlation function between the local values of irradiance (or, for brevity, the correlation function of intensity fluctuations) is defined as

$$C_{RT}\left(z_f^{(1)}, z_f^{(2)} | \boldsymbol{\rho} = \mathbf{r}_{\|} - \mathbf{r}_{1\|}\right) = \int\limits_{(\mathbf{n}\mathbf{n}_{ext}>0)} d\mathbf{n} \int\limits_{(\mathbf{n}_1\mathbf{n}_{ext}>0)} d\mathbf{n}_1 \left(\mathbf{n}\mathbf{n}_0\right)\left(\mathbf{n}_1\mathbf{n}_0\right)$$
$$\langle I(\mathbf{r}_1, \mathbf{n}) I(\mathbf{r}_2, \mathbf{n}_1)\rangle - I(\mathbf{r}_1, \mathbf{n})\rangle\langle I(\mathbf{r}_2, \mathbf{n}_1)\rangle \tag{1}$$

where $z_f$ denotes position of the output boundary of the slab ($z_f = 0$ and $z_f = L$ in reflection and transmission, respectively), $\mathbf{n}_{ext}$ is the external normal to the output boundary, $I(\mathbf{r}, \mathbf{n})$ is the intensity of radiation propagating in direction $\mathbf{n} = \mathbf{k}/k_0$ ($\mathbf{k}$ is the wave vector, $k_0 = 2\pi/\lambda$) at point $\mathbf{r}$ [12]. Brackets $\langle \ldots \rangle$ appearing in Eq.(1) denote averaging over disorder.

In addition to the correlation function, it is convenient to introduce the spatial spectrum of intensity fluctuations [12]

$$M_{RT}(z_f^{(1)}, z_f^{(2)}, \mathbf{q}) = \int d\boldsymbol{\rho} \exp(-i\mathbf{q}\boldsymbol{\rho}) C(z_f, \boldsymbol{\rho}) \tag{2}$$

Behavior of the spectrum at low spatial frequencies $q$ is related directly to the long-range asymptotics of the correlation function and is sensitive to the regime of wave propagation in the medium. The value of the spectrum at $q = 0$ determines the variance of total transmission $T$ or reflection $R$ coefficients [13, 14]

$$\langle(\delta T)^2\rangle = \frac{1}{A} M(z_f^{(1)} = z_f^{(2)} = L, \mathbf{q} = 0), \ \ \langle(\delta R)^2\rangle = \frac{1}{A} M(z_f^{(1)} = z_f^{(2)} = 0, \mathbf{q} = 0) \tag{3}$$

and also the covariance between these coefficients

$$\langle(\delta T)(\delta R)\rangle = \frac{1}{A} M(z_f^{(1)} = L, z_f^{(2)} = 0, \mathbf{q} = 0) \tag{4}$$

where $A$ is the area of the slab surface, $T$ is the ratio of the total transmitted flux to the incident one ($R$ is defined analogously).

In the scalar wave approximation the spectrum of intensity fluctuations was calculated in Refs. [13, 14]. Extension of results [13, 14] with allowance for the polarization state of the incident light was performed in Refs. [15, 16]. Within the circular representation [17, 18], the spectrum of transmitted (or reflected) intensity can be written as [16]

$$M(z_f^{(1)} = z_f^{(2)}, \mathbf{q}) = \left(\frac{2\pi}{k_0}\right)^2 n_0 \int\limits_0^L dz \int d\mathbf{n} \int d\mathbf{n}'$$
$$|I_{\mathbf{q}}(z_f|z, \mathbf{n}) - I_{\mathbf{q}}(z_f|z, \mathbf{n}')|^2 I_m^*(z, \mathbf{n}) d_{mn}(\mathbf{n}, \mathbf{n}') I_n(z, \mathbf{n}') \tag{5}$$

Quantities $I_m$ ($m = \pm 0, \pm 2$) are expressed in terms of the Stokes parameters $I$, $Q$, $U$ and $V$

$$\hat{I} = \frac{1}{\sqrt{2}} \left(I_2, I_0, I_{-0}, I_{-2}\right) = \frac{1}{2} \left(Q - \mathrm{i}U, I - V, I + V, Q + \mathrm{i}U\right), \tag{6}$$

and are subject to the vector radiative transfer equation [17, 18]

$$\left\{\mathbf{n}\frac{\partial}{\partial \mathbf{r}} + n_0\sigma_{tot}\right\} I_m(\mathbf{r}, \mathbf{n}) = n_0 \int d\mathbf{n}' d_{mn}(\mathbf{n}, \mathbf{n}') I_n(\mathbf{r}, \mathbf{n}') \tag{7}$$

where $\sigma_{tot} = \sigma + \sigma_a$ is the total cross section, $\sigma$ and $\sigma_a$ are the cross sections of elastic scattering and absorption, respectively. Matrix $d_{mn}(\mathbf{n}, \mathbf{n}')$ entering into Eqs.(5) and (7) is the single-scattering phase matrix in the circular representation [18, 19]

$$d_{mn} = \begin{pmatrix} a_+ e^{2i\chi_+} & \frac{b_1}{2} e^{-2i\beta} & \frac{b_1}{2} e^{-2i\beta} & a_- e^{2i\chi_-} \\ \frac{b_1}{2} e^{-2i\beta'} & a_+ & a_- & \frac{b_1}{2} e^{2i\beta'} \\ \frac{b_1}{2} e^{-2i\beta'} & a_- & a_+ & \frac{b_1}{2} e^{2i\beta'} \\ a_- e^{-2i\chi_-} & \frac{b_1}{2} e^{2i\beta} & \frac{b_1}{2} e^{2i\beta} & a_+ e^{-2i\chi_+} \end{pmatrix} \tag{8}$$

where angles $\beta$, $\beta'$ and $\chi_\pm = \pi - (\beta \pm \beta')$ are defined in [19], $a_\pm = (a_1 \pm a_2)/2$, quantities $a_1$, $a_2$ and $b_1$ are the elements of single scattering matrix (see, e.g., [18]).

Expression (5) can easily be generalized to the case of correlations between the reflected and transmitted fluxes (in this case the values of $z_f^{(1)}$ and $z_f^{(2)}$ are different from each other). To do this, the product of "outgoing" propagators which enters into Eq.(5) should be modified as follows

$$M_{RT}(\mathbf{q}) = \left(\frac{2\pi}{k_0}\right)^2 n_0 \int_0^L dz \int d\mathbf{n} \int d\mathbf{n}' \left( I_{\mathbf{q}}\left(z_f^{(1)} = 0 \middle| z, \mathbf{n}\right) - I_{\mathbf{q}}\left(z_f^{(1)} = 0 \middle| z, \mathbf{n}'\right)\right)$$
$$\left( I_{-\mathbf{q}}\left(z_f^{(2)} = L \middle| z, \mathbf{n}\right) - I_{-\mathbf{q}}\left(z_f^{(2)} = L \middle| z, \mathbf{n}'\right)\right) I_m^*(z, \mathbf{n}) d_{mn}(\mathbf{n}, \mathbf{n}') I_n(z, \mathbf{n}') \tag{9}$$

Equations (5) and (9) can be combined, to give the spectrum of total (transmitted plus reflected) intensity fluctuations $M_{tot}(\mathbf{q}) = M_R(\mathbf{q}) + M_T(\mathbf{q}) + 2M_{RT}(\mathbf{q})$.

As has been shown in Refs. [15, 16] the polarization of light can reveal itself in the intensity correlations only under the conditions of the polarization memory effect [6–11, 19]. This effect is observable for circularly polarized light provided that the scattering matrix elements $a_1$ and $a_2$ are very close to each other (e.g., for microspheres with high refractive index in the vicinity of the first Kerker point [9–11, 20] or for large (compared to the wavelength of light) weakly refracting particles [6–8, 19]).

For circularly polarized light Eq.(5) is simplified, as only two components of the Stokes vector make a contribution to that equation,

$$I_m^*(z, \mathbf{n}) d_{mn}(\mathbf{n}, \mathbf{n}') I_n(z, \mathbf{n}') = a_1 (\mathbf{n}\mathbf{n}')\, I(z, \mu) I(z, \mu') + a_2 (\mathbf{n}\mathbf{n}')\, V(z, \mu) V(z, \mu'). \tag{10}$$

According to Eq.(10) the spatial spectra (5) and (9) can be presented as the sum of the scalar contribution and the polarization one.

## 3. Intensity correlations in diffusive transport of circularly polarized light

In the diffusive approximation the intensity propagators entering into (9) and (10) can be expanded in directions $\mathbf{n}$ [12]. Then the spectrum (9) takes the form

$$M_{RT}(\mathbf{q}) = \frac{l_{tr}}{48\pi k_0^2} \int_0^L dz \left[ \left( \frac{\partial}{\partial z} I_{\mathbf{q}}\left(z_f^{(1)} = 0 \middle| z\right)\right) \left( \frac{\partial}{\partial z} I_{\mathbf{q}}\left(z_f^{(2)} = L \middle| z\right)\right) + \right.$$
$$\left. \mathbf{q}^2 I_{\mathbf{q}}\left(z_f^{(1)} = 0 \middle| z\right) I_{\mathbf{q}}\left(z_f^{(2)} = L \middle| z\right)\right] \left[I^2(z) + V^2(z)\right] \tag{11}$$

Substituting the intensity propagators $I_{\mathbf{q}}(z_f | z)$, $I(z)$ and $V(z)$ calculated in the diffusive approximation [16] to Eq.(11) we obtain the spectrum in the form

$$M_{RT} = M_{RT}^{(I)} + M_{RT}^{(V)} = \frac{9\pi}{k_0^2} \frac{l_{tr}}{L} \left( F_{RT}(qL, 0) + F_{RT}\left(qL, \frac{L}{l_{circ}}\right)\right) \tag{12}$$

where $l_{tr}$ and $l_{circ}$ are the transport mean free path and the depolarization length for circularly polarized light (see, e.g., [7, 10]), and

$$F_{RT}(x,y) = -\left(\frac{x}{\sinh x}\right)^2 \left[\frac{\coth^2 y}{x^2 - y^2}(x \sinh x - y \tanh y \cosh x) - \frac{\sinh x}{x}\frac{1}{\sinh^2 y}\right] \quad (13)$$

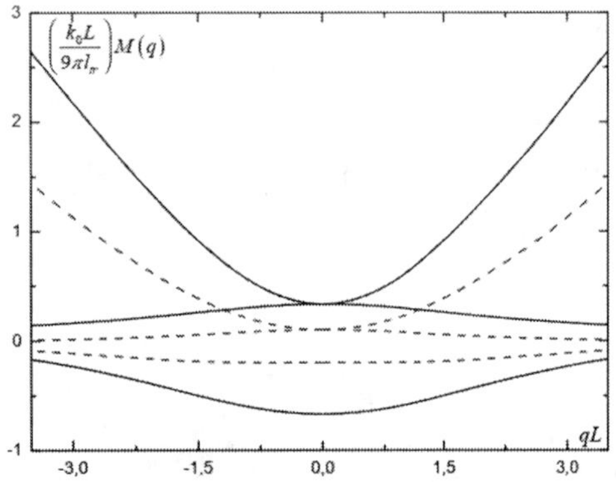

**Figure 1.** Spatial spectrum of intensity fluctuations. Scalar and polarization contributions are shown by solid and dashed lines, respectively. The upper and middle curves correspond to the reflection and transmission geometries, the lower curves describe correlations between the reflected and transmitted fluxes. The polarization contribution is calculated for $L = 5\ l_{circ}$.

The $q$-dependence of the scalar and polarization contributions to the spectrum $M_{RT}$ is illustrated in Fig. 1. The both contributions are negative, which means anti-correlation between the transmitted and reflected fluxes. The spectra $M_T$ and $M_R$ are also shown in Fig. 1 for comparison. Due to the total flux conservation the equality $M_R = M_T = -M_{RT}$ is fulfilled at $\mathbf{q} = 0$.

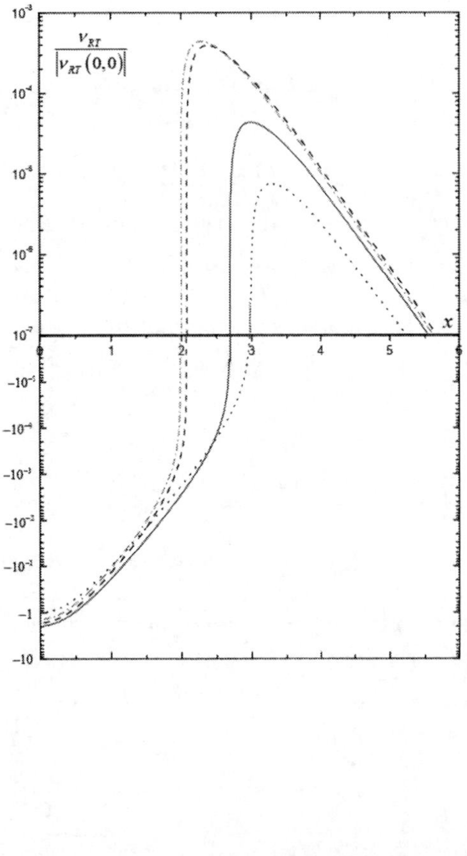

**Figure 2.** Correlation function between transmitted and reflected intensity fluctuations $\nu_{RT} = \nu_{RT}(x,0) + \nu_{RT}(x,y)$ ($y = 1$ solid red line, $y = 3$ dashed blue line, $y = 7$ dash-dotted green line). The scalar contribution $\nu_{RT}(x,0)$ is shown by dotted black line. The curves are normalized to the value of $|\nu_{RT}(0,0)| = 0.5104$.

From Eq.(12) it follows that the intensity-intensity correlation function can be written as

$$C_{RT} = C_{RT}^{(I)} + C_{RT}^{(V)} = -\frac{9\pi l_{tr}}{k_0^2 L^3}\left[\nu_{RT}\left(\frac{\rho}{L}, 0\right) + \nu_{RT}\left(\frac{\rho}{L}, \frac{L}{l_{circ}}\right)\right] \tag{14}$$

where

$$\nu_{RT}(x, y) = -\frac{\Phi_1(x)}{\sinh^2 y} + \coth^2 y\left[\Phi_2(x, y) - y\ \tanh y\ \Phi_3(x, y)\right]\ , \tag{15}$$

$$\Phi_1(x) = \frac{1}{\pi}\sum_{n=1}^{\infty}(-1)^{n+1}\pi^2 n^2 K_0(\pi n x)\,,\ \Phi_2(x, y) = \frac{1}{\pi}\sum_{n=1}^{\infty}\frac{(-1)^{n+1}\pi^4 n^4}{\pi^2 n^2 + y^2}K_0(\pi n x)\,,$$

$$\Phi_3(x, y)\,(x, y) = \frac{1}{\pi}\sum_{n=1}^{\infty}\frac{(-1)^{n+1}\pi^2 n^2}{(\pi^2 n^2 + y^2)}\left[\pi n x K_1(\pi n x) - \frac{(\pi^2 n^2 + 3y^2)}{(\pi^2 n^2 + y^2)}K_0(\pi n x)\right] \tag{16}$$

Functions $K_n(z)$ appearing in Eq.(16) are the modified Bessel functions.

The $\rho$-dependence of the correlation function (14) is illustrated in Fig. 2. From Fig. 2 it follows that the intensity fluctuations at the opposite boundaries of the slab are negatively correlated at $\rho < L$. As $\rho$ increases, the correlation function changes its sign. As compared to the scalar contribution, the resulting correlation function changes its sign at smaller values of $\rho$ and is characterized by a greater amplitude of positive correlations. The asymptotic behavior of the correlation function $C_{RT}$ at $\rho \gg L$ is identical to that of $C_T$, $C_R$.

Our conclusions regarding the scalar contribution are in qualitative agreement with those of Ref. [4] where the density-density type of correlations were studied (contrary to the current-current one considered here).

## 4. Discussion

The results presented above give insight into the spatial correlations between the transmitted flux and the reflected one, including the impact of the polarization memory effect on the correlations.

Intensity fluctuations at the opposite boundaries of a disordered slab are always negatively correlated. This is explained by the total flux conservation $\langle(\delta R + \delta T)^2\rangle = 0$ [13, 14], resulting in the anti-correlation effect for the covariance $\langle\delta T \delta R\rangle = -(\langle(\delta R)^2\rangle + \langle(\delta T)^2\rangle)/2$ and, as a consequence, both for the angular-resolved intensity correlations (see, e.g., [2, 3]) and for fluctuations of the local values of irradiance considered above.

The polarization contribution to the intensity correlation function is rather noticeable provided that the circular polarization decreases slowly as compared to the flux randomization over directions of propagation, $l_{tr} < l_{circ}$. In this case the relative value of the covariance $\langle\delta T \delta R\rangle$ can be written in the form

$$\frac{\langle\delta T \delta R\rangle}{\langle T\rangle\langle R\rangle} = -\frac{3^{3/2}}{2N}\left(1 + \frac{\sinh 2y - 2y}{4\ y\ \sinh^2 y}\right) \tag{17}$$

where $N = Ak_0^2/(2\pi)$ is the number of propagating modes and $y = L/l_{circ}$. The scalar contribution to Eq.(17) is $L$-independent, while the polarization one decreases with increasing $L$. As long as the scattered light remains polarized, $L < l_{circ}$, the scalar and polarization contributions to Eq.(17) are of the same order of magnitude. As the slab thickness $L$ increases, the polarization term in Eq.(17) decreases as $l_{circ}/L$.

For a thick slab, the correlations between intensity fluctuations at the opposite points of the slab boundaries fall off as $1/L^3$ with $L$ (see Eq.(14) at $\rho = 0$). In the limit $L > l_{circ}$

$$C_{RT}(\rho = 0) = -\frac{9\pi l_{tr}}{k_0^2 L^3}\left(0.51 + 1.94\,\frac{l_{circ}}{L}\right) \tag{18}$$

Ratio between the polarization contribution and the scalar one decreases with $L$ as $l_{circ}/L$ while the degree of circular polarization $V/I$ proves to be exponentially small, $V/I \sim \exp(-L/l_{circ})$.

## 5. Conclusions

In conclusion, we have developed an analytical theory of long-range intensity correlations in speckles arising in multiple scattering of polarized light. Under the conditions of circular polarization memory, the correlation function has been presented as a sum of the contributions from the intensity and from the basic mode of circular polarization. We have shown that the polarization contribution to mesoscopic reflection-transmission fluctuations remains noticeable even for thick samples where the polarization of transmitted light vanishes. The results obtained above give a rather complete insight into the manifestation of the vector nature of light in mesoscopic fluctuations and can be useful for studies of optical properties of disordered media.

**Acknowledgment.** This work was supported by the MEPhI Academic Excellence Project (Contract No.02.a03.21.0005).

## References

[1] Akkermans E and Montambaux G 2011 *Mesoscopic Physics of Electrons and Photons* (Cambrige University Press)
[2] Rogozkin D B 1999 *JETP Lett.* **69** 117
[3] Froufe-Perez L S, García-Martin A, Cwilich G and Sáenz J. J 2007*Physica* A **386** 625
[4] Fayard N, Caze A, PierratR and Carminati R 2015 *Phys.Rev.* A **92** 033827
[5] MacKintosh F C, Zhu J X, Pine D J and D. A. Weitz 1989 *Phys. Rev.* B **40** 9342
[6] Bicout D, Brosseau C, Martinez A S and Schmitt J M 1994 *Phys. Rev.* E **49** 1767
[7] Gorodnichev E E, Kuzovlev A I and Rogozkin D B 1998 *JETP Lett.* **68** 22
[8] Sankaran V, Everett M J, Maitland D J and Walsh J T 1999 *Opt.Lett.* **24** 1044
[9] Schmidt M K, Aizpurua J, Zambrana-Puyalto X, Vidal X, Molina-Terriza G and Saenz J J 2015 *Phys. Rev. Lett.* **114** 113902
[10] Gorodnichev E E, Kuzovlev A I and Rogozkin D B 2016 *JETP Lett.* **104** 157
[11] Gorodnichev E E, Kuzovlev A I and Rogozkin D B 2017 *J. Physics: Conf. Series* **788** 012039
[12] Ishimaru A 1997 *Wave Propagation and Scattering in Random Media* (Institute of Electrical and Electronics Engineers, New York, 1997).
[13] Rogozkin D B and Cherkasov M Yu 1993 *JETP Lett.* **58** 585
[14] Rogozkin D B and Cherkasov M Yu 1995 *Phys. Rev.* B **51** 12256
[15] Gorodnichev E E, Kuzovlev A I and Rogozkin D B 2009 *JETP Lett.* **89** 547
[16] Gorodnichev E E, Kuzovlev A I and Rogozkin D B 2016 *JOSA* A **33** 95
[17] Kuščer I and Ribarič M 1959 *Opt. Acta* **6** 42
[18] Mishchenko M I, Travis L D and Lacis A A 2006 *Multiple Scattering of Light by Particles* (Cambridge Univ. Press, Cambridge)
[19] Gorodnichev E E, Kuzovlev A I and Rogozkin D B 2014 *Phys.Rev.* E **90** 043205
[20] Gorodnichev E E, Kuzovlev A I and Rogozkin D B 2016 *Quantum Electronics* **46** 947

IOP Conf. Series: Journal of Physics: Conf. Series **941** (2017) 012046 doi:10.1088/1742-6596/941/1/012046

# High energy electrons accelerated in the field of tightly focused relativistic laser pulse for peak intensity evaluation

**K A Ivanov[1], I N Tsymbalov[1], A V Rusakov[2], R V Volkov[1], A B Savel'ev[1]**

[1]Physics faculty and International Laser Center of M.V. Lomonosov Moscow State University, Moscow, Leninskie gory 1/62, Russia
[2]Institute for Nuclear Research of Russian Academy of Sciences, Moscow, Prosp. 60-letia Oktyabrya 7A, Russia

E-mail: akvonavi@gmail.com

**Abstract**. The action of relativistic femtosecond laser pulse with peak intensity over $10^{19}$ W/cm$^2$ onto low density plasma is studied experimentally and numerically. It is demonstrated that electrons may be accelerated up to the relativistic energy (>500 keV) directly by the strong field of tightly focused laser radiation. In dependence on the peak laser intensity a dominant angle of high energy electrons is observed, which is around 45 degree for experimental intensity. On the basis of particles energy and angular distribution the laser pulse peak intensity may be evaluated.

## 1. Introduction

The progress of the laser technologies within the last decade allowed to reach the tera- and petawatt level of peak power of the ultrashort pulses opening for researchers the experimental investigation of a promising field of high intensity laser interaction with matter and plasma [1,2]. Among various scientific tasks and applications, the area of laser acceleration of electrons is one that has been intensively studied both theoretically and experimentally. With the access to the relativistic regime of laser interaction (when the peak intensity exceeds $10^{18}$ W/cm$^2$), a myriad of new concepts of particle acceleration to relativistic energy in the low density plasma have been demonstrated: wakefield acceleration [3], direct laser acceleration [4,5], in vacuum energy gain [6], electron acceleration in the presence of magnetic field [7] etc. Together with the precise control of key beam parameters, such as phase, polarization, transverse intensity distribution in the beam waist the efficient energy gain up to a GeV level may be achieved [8,9]. At the same time, the interaction of strong laser field with low density plasma may be also utilized for studying the complex particles dynamics, non-linear physics, atoms ionization [10-12]. One of the challenging tasks is the laser beam parameters evaluation, especially for ultrahigh peak power machines. The measurements of particles momentum and spatial distribution, ionization state may serve as a foundation for intensity estimation and focal energy distribution of the beam [13-16]. Since the collective plasma effects may be neglected (due to low plasma density), the energy and angle of scattering of the outgoing particles are the result of the net motion in the laser field.

In this paper we experimentally demonstrate, that electrons may be accelerated to the energy over 500 keV in the waist of tightly focused linearly polarized femtosecond laser pulse with an estimated peak intensity around $10^{19}$ W/cm$^2$. The particles are born via field ionization of noble gas (Argon) at a pressure of ~$10^{-1}$ Torr. Measurement of angular distribution in the polarization plane of outgoing electrons revealed the dominant direction of about 45 degree from the laser wavevector $k$ for the high energy electrons. At the same time, the larger number of slow particles (<200 keV) is detected along the polarization direction. Numerical 2D3V PIC simulations show that the electrons are accelerated in the high intensity region and mainly scattered from the caustic at an angle between the polarization direction and the $k$-vector in accordance with the theory of single electron motion in a planar wave. On the basis of angular-energy electrons distribution a simple model for peak intensity evaluation may be developed.

## 2. Experimental results

The experiment was made on the Ti:Sa laser system of the International Laser Center of M.V. Lomonosov Moscow State University, capable to deliver pulses of 50±5 fs duration and energy after the compressor 50±5 mJ at 10 Hz repetition rate. The central wavelength is 805 nm. The horizontally polarized beam was focused by a 90° off-axis parabola (F/D=3) mounted inside a gas filled vacuum chamber, see Fig.1(a). The residual pressure of the air was $10^{-5}$ torr (achieved by a turbopump). A gas leaker was used to add pure Argon into the chamber and maintain its pressure on the level of $10^{-2}$-$10^{-1}$ Torr, enough to make the collective plasma effects negligible [17].

Using a microobjective (NA=0.3) with optical resolution of ~1 μm we measured the diameter of the focal spot, which is found to be 2.2±0.1 μm (FWHM). Approximately 30% of laser pulse energy is contained in the main peak, see Fig.1(b). Hence the estimated peak vacuum intensity in our experiment was ~$1.5\times10^{19}$ W/cm$^2$.

To accumulate electron spectra for a long time (at 10 Hz), we used a simple scintillator based detector (3 cm thick plastic scintillator coupled to PMT), absolutely calibrated by Sr90 beta-source, with a linear response in the range 100-4000 keV and energy resolution of 10%. However due to ADC limitations (the signal was digitized by 8-bit digital oscilloscope with 500MHz bandwidth) the maximal detectable energy in our case was ~600 keV. The device was installed outside the chamber at a distance of 30 cm from the beam waist. For measurements four angles of observation were available in the plane of beam polarization: along $k$-vector, at 22.5 and 45 degrees and almost in the polarization direction (~85 degrees). The entrance aperture of the detector was 11 mm in diameter. Between the interaction region and the sensitive area of the detector there was a set of filters: a 20 μm mylar as vacuum chamber window, ~3 cm of air at atmospheric pressure and 12 μm Al foil to protect the scintillator against visible light. Altogether it gives a 100 keV low energy cut off and high attenuation of electrons within this energy range. By Monte-Carlo simulation of electrons transport through matte using CASINO code [18] we processed the accumulated data to take into account the spectral distortions.

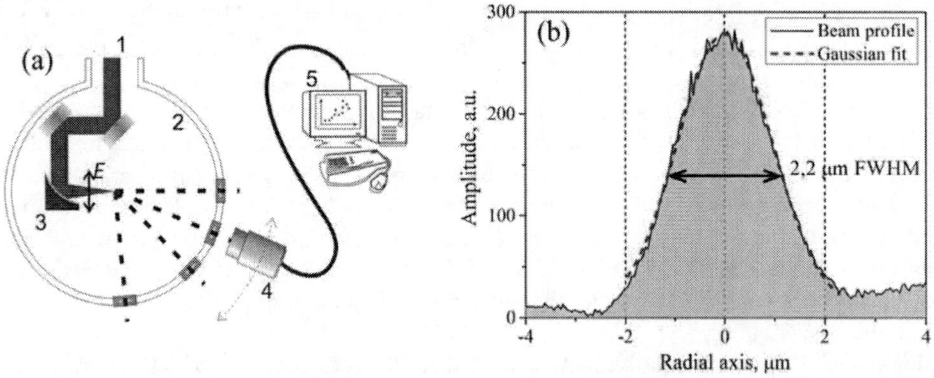


**Figure 1.** (a) Experimental setup: 1 – laser pulse, 2 – vacuum chamber, 3 – off-axis parabolic mirror (OAP), 4 – scintillation electron detector, 5 – PC with digital oscilloscope. (b) Transvers profile of the beam in the focus of the OAP. The red dashed line shows the Gaussian fit of the experimental curve.

We also made the "null" experiment. The placement of a 1 cm thick piece of plastic in front of the detector (stopping all the electrons below 2 MeV and transparent for X-ray above 10 keV) lead to the decrease of signal down to the zero pedestal level. Hence we were convinced that the signal on the detector without the shielding is caused by the electrons and not by the X-rays, which may be generated via attenuation of electrons on the chamber walls etc.

The Figure 2 illustrates the Monte-Carlo processed electron spectra, measured at different angles, and also the integrated over all angles spectrum (dashed curve). One may notice the exponentially decreasing behavior of each curve with energy growth. At the same time two peculiarities are observed. At first, a greater number of low energy electrons (below 200 keV) is detected along the polarization direction (~85 degrees). The particle, trapped by the laser pulse at its periphery, oscillates dominantly by the action of the electric field of the pulse and leaves the caustic with a finite residual energy due to permanent drift from high to low intensity region. Second, the particles with maximal energy (above 500 keV), are measured at 45 degrees from the laser *k*-vector. The quantity of high energy electrons (above 200 keV) is also higher in this specific direction, compared to the others. Because of the limited angular resolution in our experimental setup, we cannot state definitely, that electrons with higher energy are not lying in some intermediate angle. However such a behavior may be roughly described by the model of isolated electron movement in a planar electromagnetic wave, which gives the escape angle $\varphi = arctg(2/a)$, where a is the classical normalized momentum of the electron quivering in the electric field [16]. For intensity $I$=$10^{19}$ W/cm$^2$ the angle $\varphi$ is around $\pi/4$. The rather unexpected result is the relatively high flux of particles along the beam axis. According to the simulations with test particle the electrons are expelled along the *k*-vector at much higher peak intensities [19]. It is worth mention, that all the measurements were carried out in the polarization plane of the radiation. The studies in the perpendicular plane may reveal other features.

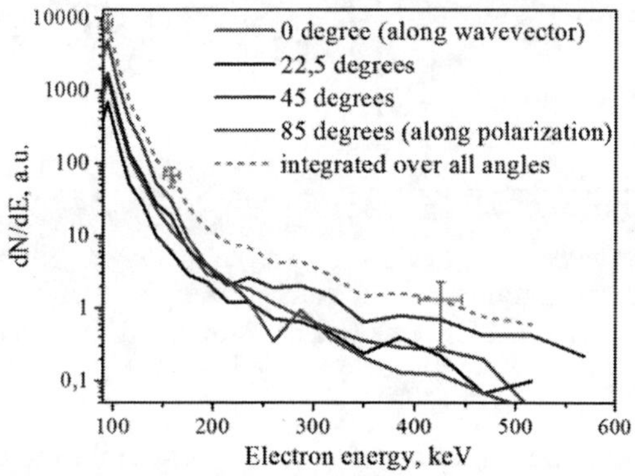


**Figure 2.** Electron energy spectra, registered at different angles of observation in the polarization plane. The dashed curve depicts the integrated over all angles spectrum.

## 3. Numerical simulations and discussion

To make an insight on the experimental data we carried out the numerical PIC simulations using the fully relativistic code MANDOR [20]. Here 2D3V simulation was made with box size 30x60 microns, grid resolution 0.01 microns and temporal resolution 0.003 fs. The Gaussian pulse with a duration of 50 fs (FWHM) was focused into a 2.5 μm (FWHM as well) spot. The polarization was in the plane of the simulation box. We varied the peak intensity from 2.5x to 25x$10^{18}$ W/cm$^2$. To make the effects of interaction and collision between the charged particles negligible the initial electron density was taken $10^{14}$ cm$^{-3}$, which corresponds to the gas pressure $10^{-2}$ Torr, similar to the experimental value. In order to follow the particles in the region with the maximal intensity we were also able to take into account the ionization on the basis of Ammosov, Delone, Krainov ADK model [21]. Otherwise, the ponderomotive force already at the front of the pulse will expels the particles from the caustic area. The ionization potentials for Argon were introduced.

The ionization process was exhaustively regarded, for instance, by N. Ekanayake et al. [10], so here we pay attention on the particles distribution. Figure 3 illustrates angular-energy diagrams of particles for different laser intensities. Basically, we observe the similar tendency of the appearance of the dominant direction of high energy electrons as it was in the experiment and as it follows from the theory: the angle of maximal emission coincides well with the theoretical angle $\varphi$, mentioned above. This angle decreases (if measured from the *k*-vector) with the intensity growth. Similar angular-energy diagrams were observed in [10]. The maximal ionization state of Argon in the simulation was +16.

However, the direct comparison of the numerical and experimental data (Fig.2 and Fig.3(b)), reveals some discrepancy, which is manifested the most in the high energy part of spectrum. It may be partially explained by the fact, that the experimental spectra are averaged over a few thousands shots. Since the energy of the pulse is distributed around the mean value with an error of 10-15% the measured spectra are slightly blurred in comparison to the calculated. The experimentally observed acceleration of electrons along the beam axis also cannot be explained by the model to the moment. It should be noted, that the non-uniformities in the transvers energy distribution over the focal spot may be responsible for the divergence. However, some rings or hot spots play a role at much higher peak intensity (above $10^{20}$ W/cm$^2$). Whereas for the discussed intensity range (~$10^{19}$ W/cm$^2$) only the central focal spot must be considered, in our opinion.

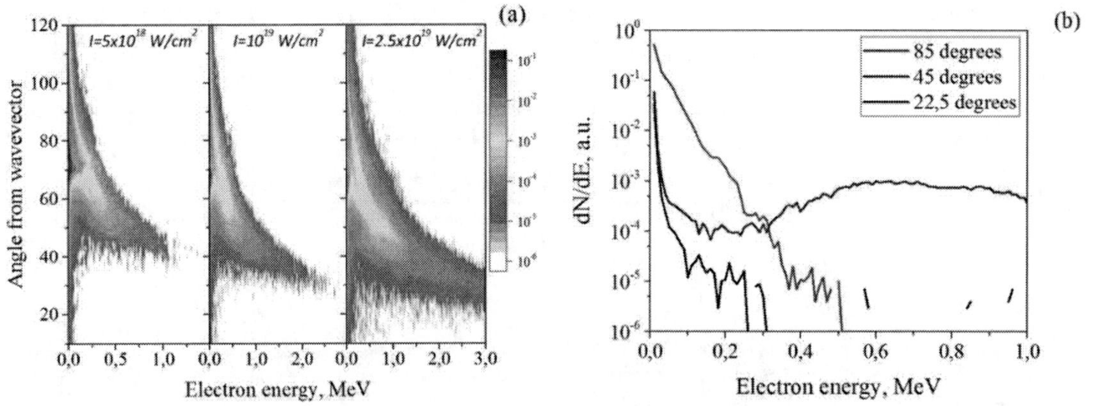


**Figure 3.** (a) Angular energy diagrams of electrons in the plane of beam polarization accelerated by the field of laser pulse with varied peak intensity. (b) Calculated spectra of electrons at different angles for intensity $I=10^{19}$ W/cm$^2$.

As for the intensity determination, two techniques in principle may be proposed. The first, more simple but requiring a sufficient number of laser shots, is based on the measurement of electron energy spectra in a specific direction: ~45 degrees from the laser wavevector for intensity ~$10^{19}$ W/cm$^2$ and closer to the beam axis (~15-25 degrees) at ~$10^{20}$ W/cm$^2$ and higher. For this a single scintillator coupled PMT may be utilized. The main problem is that the total number of electrons, encountered in the caustic region, must be low enough to avoid collective effects. Hence, the single shot methods, such as magnetic separation of particles by energy and registration on an imaging plate, are complicated as requiring high dose to trig the detector. Afterwards, the peak intensity may be calculated through the comparison with the numerical modeling. The second method is based on the measurement of the angular position of high energy peak, which is strictly connected with the intensity, see Fig.3(a). A position-sensitive detector (sensitive to single particles though) could afford such task. The main advantage is the ability to evaluate the intensity in each laser shot.

## 4. Conclusions

In summary, the acceleration of electrons directly in the field of tightly focused femtosecond laser pulse with peak intensity around $10^{18}$ W/cm$^2$ was studied experimentally. The electrons were born via field ionization or noble gas (Argon) with a pressure around $10^{-2}$-$10^{-1}$ Torr in the focal region. It is shown that the electron may gain residual energy exceeding 500 keV with an exponential decaying spectrum. At the same time the angular distribution of accelerated particles reveals the dominant direction of escape for low energy particles along the polarization direction of the radiation, whereas the high energy electrons are expelled mainly at an angle of 45 degrees from the laser *k*-vector in the polarization plane, which is in a reasonable agreement with the theory of single electron motion in a plane electromagnetic wave. Numerical simulations of laser interaction with low density plasma have shown, that electrons may gain energy up to a few MeV in a specific direction, also in accordance with the theoretical model, under the influence of a laser pulse with intensity around $10^{19}$ W/cm$^2$. At the same time the angular-energy distribution of electrons is very sensitive to the peak laser intensity, which gives the opportunity to build a simple model of peak laser intensity evaluation, based on detection and spectral measurements of the outgoing from the beam waist particles.

The work was supported by Russian Foundation for Basic Research, grant #16-32-60174-mol_a_dk.

**References**

[1] Esarey E, Schroeder C B and Leemans W P 2009 *Rev. Mod. Phys.* **81** 1229
[2] Ledingham K W D and Galster W 2010 *New J. Phys.* **12** 45005
[3] Malka V, Fritzler S, Lefebre E, Aleonard M M, Burgy F, Cham-Baret J P, Chemin J F, Krushelnick K, Malka G, Mangles S P D, Naj-Mudin Z, Pittman M, Rousseaul J P, Scheurer J N, Walton B and Dangor A E 2002 *Science* **298** 1596
[4] Li Y Y, Gu Y J, Zhu Z, Li X F, Ban H Y, Kong Q and Kawata S 2011 *Phys. Plasmas* **18** 1
[5] Arefiev A V., Khudik V N, Robinson A P L, Shvets G, Willingale L and Schollmeier M 2016 *Phys. Plasmas* **23** 056704
[6] Cline D, Shao L, Ding X, Ho Y, Kong Q and Wang P 2013 *J. Mod. Phys.* **4** 1
[7] Liu H, He X T and Chen S G 2004 *Phys. Rev. E - Stat. Nonlinear, Soft Matter Phys.* **69** 066409
[8] Payeur S, Fourmaux S, Schmidt B E, MacLean J P, Tchervenkov C, Légaré F, Piché M and Kieffer J C 2012 *Appl. Phys. Lett.* **101** 1
[9] Ghotra H S and Kant N 2015 *Opt. Commun.* **356** 118
[10] Ekanayake N, Luo S, Grugan P D, Crosby W B, Camilo A D, McCowan C V., Scalzi R, Tramontozzi A, Howard L E, Wells S J, Mancuso C, Stanev T, Decamp M F and Walker B C 2013 *Phys. Rev. Lett.* **110** 1
[11] Eloy M, Guerreiro A, Mendonça J T and Bingham R 2007 *J. Plasma Phys.* **73** 635
[12] Hua J F, Lin Y Z, Tang C X, Ho Y K and Kong Q 2007 *J. Appl. Phys.* **101** 023102
[13] Gao J 2006 *Appl. Phys. Lett.* **88** 091105
[14] Galkin A L, Kalashnikov M P, Klinkov V K, Korobkin V V, Romanovsky M Yu and Shiryaev O B 2010 *Phys. Plasmas* **17** 053105
[15] Link A, Chowdhury E A, Morrison J T, Ovchinnikov V M, Offermann D, Van Woerkom L, Freeman R R, Pasley J, Shipton E, Beg F, Rambo P, Schwarz J, Geissel M, Edens A and Porter J L 2006 *Rev. Sci. Instrum.* **77** 10E723
[16] Kalashnikov M, Andreev A, Ivanov K, Galkin A, Korobkin V, Romanovsky M, Shiryaev O, Schnuerer M, Braenzel J and Trofimov V 2015 *Laser Part. Beams* **33** 361
[17] Vais O E, Bochkarev S G and Bychenkov V Yu 2016 *Plasma Phys. Rep.* **42** 818
[18] Drouin D, Couture A R, Joly D, Tastet X, Aimez V and Gauvin R 2007 *Scanning* **29** 92
[19] Vais O E, Bochkarev S G, Ter-Avetisyan S and Bychenkov V Yu 2017 *Quantum Electron.* **47** 38
[20] Romanov D V, Bychenkov V Yu, Rozmus W, Capjack C E, Fedosejevs R 2004 *Phys. Rev Lett.* **93** 215004
[21] Ammosov M V, Delone N B, Krainov V P 1986 *Sov. Phys. JETP* **64** 1191

# The Kubo-Greenwood formula as a result of the random phase approximation for the electrons of the metal

**S V Ivliev**

National Research Nuclear University MEPhI, Kashirskoe Shosse 31, Moscow 115409, Russia

E-mail: ivliev@theor.mephi.ru

**Abstract.** For calculation of short laser pulse absorption in metal the imaginary part of permittivity, which is simply related to the conductivity, is required. Currently to find the static and dynamic conductivity the Kubo-Greenwood formula is most commonly used. It describes the electromagnetic energy absorption in the one-electron approach . In the present study, this formula is derived directly from the expression for the permittivity expression in the random phase approximation, which in fact is equivalent to the method of the mean field. The detailed analysis of the role of electron-electron interaction in the calculation of the matrix elements of the velocity operator is given. It is shown that in the one-electron random phase approximation the single-particle conductive electron wave functions in the field of fixed ions should be used. The possibility of considering the exchange and correlation effects by means of an amendment to a local function field is discussed.

## 1. Introduction

At present, a large number of fundamental and applied problems are related to the physics of extreme states of matter (see, e.g., [1]). One way of obtaining a substance with a high energy density is the use of femtosecond laser radiation [2, 3]. At laser radiation intensities $I \sim 10^{15}\ W/cm^2$ , the behavior of a metal sample under the action of a femtosecond pulse can be qualitatively represented as follows. Laser radiation is absorbed by electrons in a thin near-surface layer. The electron temperature $T_e$ increases sharply, the temperature of the $T_i$ ions remains relatively small. The volume occupied by the substance is practically unchanged, that is, it can be assumed that the heating occurs isochorically. Later, by means of electronic thermal conductivity, the deeper layers of the sample are heated. As a result of electron-ion exchange, the ion temperature also increases. These phenomena are accompanied by the flow of complex hydrodynamic processes and phase transformations, in particular, shock-wave compression of matter to increased densities is possible. Depending on the design of the experiment, ablation or high-speed splitting is possible, due to a significant decrease in the density of matter. A two-temperature model is widely used for quantitative description of the phenomena described [3]. To describe propagation in matter and absorption of laser radiation, information is needed on the complex dielectric constant or dynamic conductivity, that is, the optical properties of metals. When analyzing the absorption of laser radiation, it is important to understand the optical properties of the metal, as well as the static conductivity or conductivity. Optical properties include complex dynamic conductivity, complex dielectric permeability, complex refractive index,

IOP Conf. Series: Journal of Physics: Conf. Series **941** (2017) 012047 doi:10.1088/1742-6596/941/1/012047

reflectivity and absorptivity A free electron does not absorb the energy of an alternating electric field either under classical or quantum considerations. In order to obtain a nonzero one, one must consider the interaction of an electron with other particles. The properties of matter in this state are significantly influenced by Coulomb interactions: between electrons and ions, electrons and electrons, ions and ions. One of the main methods for studying the optical properties of metals is the theory of linear response. An external electric field $E(t)$ is superimposed on a system of interacting electrons and ions, which leads to a change in the dynamics of the electrons. This change can be calculated with the help of perturbation theory and obtain the current $j(t)$, which then arises, and then the dynamic current. Information about the system in the theory of linear response is contained in the dynamics of the system in the absence of an electric field. This approach is based on the Kubo formalism [3], but the direct application of the Kubo formula is practically impossible

$$\sigma_{\mu\nu} = \lim_{\alpha \to 0} \frac{1}{i\hbar} \int\limits_{-\infty}^{0} dt \exp(-i\omega t + \alpha t) \int\limits_{0}^{1/T} d\lambda \langle j_\mu(0) j_\nu(t - i\hbar\lambda) \rangle. \quad (1)$$

Based on the approach of Kubo, Greenwood and other authors [6, 7], simpler formulas for the conductivity were obtained, based on the one-electron approximation. Spatial parts are calculated for electrons in the external ion potential. Wave functions that depend on time are obtained by multiplying the spatial parts by the corresponding time multipliers. Then, in the first order of the temporal perturbation theory, the current arising in the system under the action of an external electric field can be calculated. An alternative approach is to calculate the total energy absorbed by the electrons; From this energy can be calculated. With this consideration, the quantum Kubo-Greenwood formula is obtained. The Kubo-Greenwood formula includes only the spatial parts of the wave functions, and they contain all information about the system under consideration. It is the calculation by the Kubo-Greenwood formula that is used most often. In many works (see, for example, [7,8]) (in the same papers some experimental data are also given), an even simpler Kubo-Greenwood formula for conductivity, obtained in [7], is used:

$$\sigma_1(\omega) = \frac{2\pi e^2 \hbar^2}{3m^2 \omega \Omega} \sum_{\mathbf{k},n,m} (f_0(E_{n,\mathbf{k}}) - f_0(E_{m,\mathbf{k}})) \times |\langle \Psi_{n,\mathbf{k}} | \nabla_\alpha | \Psi_{m,\mathbf{k}} \rangle|^2 \delta(E_{n,\mathbf{k}} - E_{m,\mathbf{k}} - \hbar\omega). \quad (2)$$

The calculation by this formula is usually carried out as follows [9,10]. First, ion positions are set at a given ion temperature. Their position is usually determined by the methods of quantum or classical molecular-dynamic modeling. A computational cell can contain a lot of atoms, and then it repeats itself in space. Then the self-consistent Kohn-Shem equations for finding the electron wave functions are solved numerically. At the same time, the effective potential includes the Hartree potential and the exchange-correlation potential Further, the matrix elements from the gradients are calculated and finally the sum or integral over. The delta function, as a rule, is replaced by some sharp analytic function. Often, further averaging is performed on the positions of the ions, which greatly increases the amount of computation. We will analyze such calculations. The most complexity of such an approach is the finding of one-electron wave functions. According to the very structure of the formula for the conductivity in this form, the connection with the imaginary part of the dielectric permittivity function in the random phase approximation (RPA) is clearly visible. Actually, the very derivation of these expressions is almost identical. The imaginary part of dielectric permittivity function $\varepsilon(\omega)$ is related to the conductivity $\sigma(\omega)$ by the known relation

$$\varepsilon(\omega) = 1 + \frac{4\pi i \sigma(\omega)}{\omega}.$$

## 2. Dielectric Permittivity of Electrons in the Random Phase Approximation

We introduce the exact susceptibility $\chi$

$$\rho_{ind} = \chi \cdot V_{ext}$$

$$\frac{1}{\varepsilon} = 1 + \frac{4\pi e^2}{q^2}\frac{\rho_{ind}}{V_{ext}}.$$

This function can be written in terms of the matrix elements from the density operator in terms of the exact many-particle wave functions of electrons and the exact many-particle energy values of $E_\xi$, where $\xi$ is the set of quantum numbers. Unfortunately, in this approach exact multiparticle energy values and wave functions are required, which can not really be obtained. In this case, one can write down the formula for the inverse permittivity, for example, for the ground state of a system of electrons. From the expression obtained, one can obtain a rather cumbersome expression for the conductivity, which is unsuitable for further calculations.

$$\frac{1}{\varepsilon(q,\omega)} = 1 - \frac{4\pi e^2}{\Omega q^2}\sum_{\xi}|\langle 0|\hat{\rho}(q)|\xi\rangle|^2\left(\frac{1}{E_\xi - E_0 - \hbar\omega - i\delta} + \frac{1}{E_\xi - E_0 + \hbar\omega + i\delta}\right). \quad (3)$$

If, on the other hand, the susceptibility of noninteracting electrons is applied to $\chi_0$,

$$\rho_{ind} = \chi_0 \cdot V_{ext}$$

this leads to fundamentally incorrect results for the electrons of the metal.

However, one can still use the susceptibility of noninteracting electrons to $\chi_0$, but the response of the electron system should be sought not on the external but on the total potential. In fact, it is the random phase approximation (RPA).

$$\rho_{ind} = \chi_0 \cdot (V_{ext} + \frac{4\pi e^2}{q^2}\rho_{ind}). \quad (4)$$

Going beyond RPA (accounting for exchange-correlation effects) is possible by introducing the local field correction function [12]

$$\rho_{ind} = \chi_0 \cdot (V_{ext} + (1-G)\frac{4\pi e^2}{q^2}\rho_{ind}). \quad (5)$$

The random phase approximation for free electrons imaginary part of $\varepsilon(q,\omega)$ is determined by a known equation [9]

$$Im\varepsilon(q,\omega) = -\pi\frac{4\pi e^2}{\Omega q^2}\sum_{\mathbf{k}}\left(f_0(\varepsilon_{\mathbf{k}}^{(0)}) - f_0(\varepsilon_{\mathbf{k+q}}^{(0)})\right)\delta(\varepsilon_{\mathbf{k}}^{(0)} - \varepsilon_{\mathbf{k+q}}^{(0)} + \hbar\omega). \quad (6)$$

When calculating the dielectric constant of an electron gas in the ion lattice potential [12], it suffices to introduce the square of the matrix element of the plane wave $\exp(i\mathbf{qr})$

$$\varepsilon(q,\omega) = 1 + \frac{4\pi e^2}{\Omega q^2}\sum_{\mathbf{k},n,m}\frac{(f_0(E_{n,\mathbf{k}}) - f_0(E_{m,\mathbf{k+q}}))|\langle n,\mathbf{k}|e^{i\mathbf{qr}}|m,\mathbf{k+q}\rangle|^2}{\hbar\omega + E_{n,\mathbf{k}} - E_{m,\mathbf{k+q}} + i\delta}, \quad (7)$$

where $n, m$ - the numbers of Bloch functions. In the limit of small $q$, the expression for the imaginary part $\varepsilon(\omega)$ turns into

$$Im\varepsilon(\omega) = \frac{4\pi^2 e^2}{3\Omega} \sum_{\mathbf{k},n,m} (f_0(E_{n,\mathbf{k}}) - f_0(E_{m,\mathbf{k}}))|\langle n,\mathbf{k}|r_\alpha|m,\mathbf{k}\rangle|^2 \delta(\hbar\omega + E_{n,\mathbf{k}} - E_{m,\mathbf{k}}). \quad (8)$$

By the very structure of the record of this formula, in this form, the apparent part of the permittivity function in the random phase approximation with the Kubo-Greenwood formula for the conductivity. The transition from matrix elements $r_\mu$ to matrix elements $\nabla_\mu$ is carried out with the aid of relation

$$v_\mu = \frac{i}{\hbar}[H, r_\mu] = -\frac{i\hbar\nabla_\mu}{m}.$$

Then the matrix elements $r_\mu$ can be expressed in terms of matrix elements $\nabla_\mu$ as follows

$$\langle\Psi_i|r_\mu|\Psi_j\rangle = \frac{\hbar^2}{2m}\frac{\langle\Psi_i|\nabla_\mu|\Psi_j\rangle}{(E_i - E_j)},$$

below, because of the presence of the $\delta$-function, the energy difference in the denominator is replaced by $\hbar\omega$. As a result, taking into account the connection between the imaginary part of $\varepsilon(\omega)$ and the conductivity $\sigma(\omega)$, we get

$$\sigma_1(\omega) = \frac{2\pi e^2\hbar^2}{3m^2\omega\Omega} \sum_{\mathbf{k},n,m} (f_0(E_{n,\mathbf{k}}) - f_0(E_{m,\mathbf{k}}))|\langle n,\mathbf{k}|\nabla_\alpha|m,\mathbf{k}\rangle|^2 \delta(E_{m,\mathbf{k}} - E_{n,\mathbf{k}} - \hbar\omega), \quad (9)$$

which exactly coincides with the Kubo-Greenwood formula.

## 3. Random phases approximation as a mean-field method

But the basic idea of the RPA method (in fact it is the mean-field method) is to find the response of a system of noninteracting electrons to a common (external plus induced) potential or field. When calculating conductivity, the Kubo-Greenwood method also looks for a response to a common (total) field. The interelectronic Coulomb interaction is taken into account both in the mean-field method and in the mean-field method. The response is sought not for the external, but for the general field.

$$H = H_{kin} + H_{e-ion} + H_{e-e} + H_{ext}$$

$$H_{ext} = \sum_{\mathbf{q}} \hat{\rho}_e V_{ext}(\mathbf{q}, \omega)$$

Further in the last formula, one of the electronic densities is averaged and a response is obtained to the common field

$$H_{e-e} = \sum_{\mathbf{q}} \frac{4\pi e^2}{q^2} \hat{\rho}_e(\mathbf{q})\hat{\rho}_e(-\mathbf{q})$$

$$\tilde{H}_{e-ext} = \sum_{\mathbf{q}} \hat{\rho}_e(\mathbf{q}) \left( V_{ext}(\mathbf{q}, \omega) + \frac{4\pi e^2}{q^2}\langle\rho_e(\mathbf{q})\rangle \right). \quad (10)$$

IOP Conf. Series: Journal of Physics: Conf. Series **941** (2017) 012047 doi:10.1088/1742-6596/941/1/012047

## 4. Conclusions

Thus, in the pseudopotential approach, when conduction electrons are considered in the ion lattice, the wave functions of electrons and band energies and the matrix elements should apparently be calculated without allowance for the electron-electron interaction A more accurate account of the interelectron interaction is possible through a correction to the local field.

$$Im\varepsilon(\omega) = \frac{v_q Im\chi_0}{(1 + Gv_q Re\chi_0)^2 + (Gv_q Im\chi_0)^2}, \tag{11}$$

where $v_q = 4\pi e^2/q^2$. In general, it is obvious that taking into account the electron-electron interaction should somewhat decrease the conductivity.

And, finally, we make a remark about the possible nonequilibrium of the electron subsystem itself in the process of absorption of a short laser pulse. In the Kubo-Greenwood formula, we mean the local equilibrium of the electrons at each instant of time. This means that the electronic processes of relaxation (the establishment of the electron temperature) occur much faster than the time of the action of the pulse. However, for femtosecond pulses this co-relation can be violated, and then the approach with energy absorption through conductivity of the Kubo-Greenwood type can turn out to be illegal.

## 5. References

[1] Fortov V E 2009 *Extreme states of matter* (Moscow: Fizmatlit)
[2] Gamaly E G 2011 *Phys. Rep.* **508** 91
[3] Anisimov S I and Rethfeld 1997 *Proc. SPIE.* **3093** 192
[4] Lin Zh, Zhigilei L V and Celli V 2008 *Phys. Rev.* B **77** 075133
[5] Kubo R 1957 *Journ. Phys.Soc.Japan* **12** N6 570
[6] Greenwood D A 1958 *Proc. Phys. Soc.* **A71** 585
[7] Moseley L L and Lukes T 1978 *Am. J. Phys.* **46** 676
[8] Pozzo M, Desjarlais M P and Alfe D 2011 *Phys. Rev.* B **84** 054203
[9] Povarnitsyn M E, Knyazev D V and Levashov P R 2012 *Contrib. Plasma Phys.* **52** N2 145
[10] Knyazev D V and Levashov P R 2014 *Phys.Plasmas.* **21** 073302
[11] Ichimaru S 1982 *Rev. Mod. Phys.* **54** 1017
[12] Ziman J M 1972 *Principles of the theory of solids* (Cambridge: At the University Press )

IOP Conf. Series: Journal of Physics: Conf. Series **941** (2017) 012048 doi:10.1088/1742-6596/941/1/012048

# The Scattering of Massive Holes by Supercritical Impurity

**V M Kuleshov**[1], **V D Mur**[1], **A M Fedotov**[1], **Yu E Lozovik**[1,2]

[1] National Research Nuclear University MEPhI, Kashirskoe Shosse 31, 115409 Moscow, Russia

[2] Institute for Spectroscopy of the Russian Academy of Sciences, Fizicheskaya Str. 5, 108840 Troitsk, Russia

E-mail: `kuleshov.val@gmail.com`

**Abstract.** The properties of charge carriers in doped graphene are considered. The closed set of explicit equations determining the spectrum and wavefunctions of charge carriers for total angular momentum $J = M + 1/2 = 0, \pm 1/2, \pm 1 \ldots$ is obtained for the case of the Coulomb potential modified at small distances. The critical values $Z_{cr}$ of the dopant charge at which the energy level with the given quantum numbers crosses the lower continuum boundary are determined. For $Z < Z_{cr}$ for several values of the orbital angular momentum dependence of the position of the energy level as a function of charge $Z$ is obtained. For $Z > Z_{cr}$, the position $\varepsilon_0$ and width $\gamma$ of the lowest quasidiscrete state, which may manifest itself as a resonance in hole-dopant scattering, are calculated. It's shown that there is no electron-hole pairs creation because of unitarity of the partial scattering matrix.

## 1. Introduction

According to [1–3], the electronic properties of graphene doped by atomic nuclei or ions with charge $Z$ can be described by the effective two-dimensional Dirac equation with the gap in electronic spectrum and with the Coulomb potential

$$\hbar v_F \left( -i\boldsymbol{\sigma}\frac{\partial}{\partial \mathbf{x}} - \frac{q}{\mathrm{x}} + \frac{m_* v_F}{\hbar}\sigma_z \right) \Psi_E(\mathbf{x}) = E\Psi_E(\mathbf{x}) \tag{1}$$

Here $v_F$ is the velocity at the Fermi surface, $\mathbf{x} = (x_1, x_2)$, $\boldsymbol{\sigma} = (\sigma_x, \sigma_y, \sigma_z)$ are the Pauli matrices, $q = Z\alpha_F$, $\alpha_F = e^2/\hbar v_F$ is the effective fine structure constant for graphene, $\Psi_E(\mathbf{x})$ - is the two-component wavefunction for an electron with the effective mass $m_*$ and energy $E$.

In the case of doped graphene the energy spectrum of the two-dimensional heterostructure is absolutely similar to one for the three-dimensional Coulomb problem, including the case with the nuclear charge $Z > 137$, e.g. [4–6]. In [3] the authors discuss the screening by electrons created together with holes from the Dirac sea (by analogy with the three-dimensional case [8, 9]). In the present work, we demonstrate that such a mechanism of screening cannot be put into effect, see also [4–6], but resonances in the scattering of the holes by overcritical impurities occur.

Accounting for [3, 7] the gap width $\Delta = 2m_* v_F^2 = 0.26\, eV$, the effective constant $\alpha_F = 0.4$ and the distance between carbon nuclei $a_{cc} = 1.42$ Å for graphene deposited onto SiC substrate we have

$$v_F = 5.5 \cdot 10^8\, cm/s, \quad m_* = 7.6 \cdot 10^{-4} m_e, \quad a_{CC}/l_F = 5.5 \cdot 10^{-3}, \tag{2}$$

where $m_e$ is the mass of the electron, $l_F = \hbar/m_* v_F$ is the "Compton length" in graphene.

Owing to the axial symmetry, the conserving quantum number [10] is the total angular momentum $J = M + 1/2$ (in $\hbar$ units), i.e., the eigenvalue of the generator for the two-dimensional rotations $-i\partial/\partial\varphi + 1/2\sigma_3$. The orbital angular momentum $M$ is the eigenvalue of the self-adjoint operator $-i\partial/\partial\varphi$ acting in the Hilbert space of square integrable functions on the circle $0 < \varphi < 2\pi$, which eigenfunctions are [11,12][1]:

$$\Phi_M(\varphi) = \frac{1}{2\pi} e^{\imath M\varphi}, \quad M = \delta + m; \ \delta = 0, \frac{1}{2}, \quad m = 0, \pm 1, \pm 2, \ldots \tag{3}$$

These functions realize irreducible single-valued ($\delta = 0$) or double-valued ($\delta = 1/2$) representations of $SO(2)$ group. In terms of

$$E = m_* v_F^2 \varepsilon, \ \mathbf{x} = l_F \boldsymbol{\rho}, \ \boldsymbol{\rho} = (\rho\cos\varphi, \rho\sin\varphi),$$

the wavefunction with fixed total angular momentum is represented as:

$$\Psi_{\varepsilon,J}(\boldsymbol{\rho}) = \frac{1}{\sqrt{2\pi\rho}} e^{iJ\varphi} \begin{pmatrix} e^{-i\varphi/2} F(\rho) \\ i e^{i\varphi/2} G(\rho) \end{pmatrix} \tag{4}$$

## 2. Boundary conditions and short-range Coulomb problem

The functions $F(\rho)$ and $G(\rho)$ satisfy the two-dimensional Dirac equation:

$$H_D \Psi_{\varepsilon,J}(\rho) = \varepsilon \Psi_{\varepsilon,J}(\rho), \ H_D = \begin{pmatrix} 1 - \frac{q}{\rho} & \frac{J}{\rho} + \frac{d}{d\rho} \\ \frac{J}{\rho} - \frac{d}{d\rho} & -1 - \frac{q}{\rho} \end{pmatrix}, \ \Psi_{\varepsilon,J} = \begin{pmatrix} F(\rho) \\ G(\rho) \end{pmatrix}, \ J = M + \frac{1}{2}, \tag{5}$$

which up to the notation coincides with the set of equations for the radial functions for the three-dimensional problem [8]. In the two-dimensional case, however, half-integer values of $J = m + 1/2, \ m = 0, \pm 1, \pm 2, \ldots$ are possible.

The equation for the critical charge $q_{cr}^{(n)} = Z_{cr}\alpha_F$, at which the $n$-th level with the given quantum number $J$ reaches the boundary of the lower continuum of the solutions of Eq. (5), was obtained in [4,6]:

$$\arg\Gamma\left(2i\sqrt{\left(q_{cr}^{(n)}\right)^2 - J^2}\right) = \sqrt{\left(q_{cr}^{(n)}\right)^2 - J^2} \cdot \ln\left(2q_{cr}^{(n)}\right) - \theta_\tau(J) + \pi n, \ n = 0, 1, 2, \ldots \tag{6}$$

In these papers it was also shown that the parameters $\theta_\sigma(J)$ and $\theta_\tau(J)$ defininig one-parameter families of self-adjoint radial Dirac hamiltonians, are defined by the following conditions:

$$\begin{aligned} \frac{u_\sigma}{u_{-\sigma}} &= \left(\frac{u_\sigma}{u_{-\sigma}}\right)^* = \tan\theta_\sigma(J), \ -\frac{\pi}{2} \le \theta_\sigma(J) \le \frac{\pi}{2}, \quad 0 < \sigma < 1/2 \\ \frac{u_\tau}{u_{-\tau}} &= \left(\frac{u_{-\tau}}{u_\tau}\right)^* = e^{2i\theta_\tau(J)}, \ \operatorname{Im}\theta_\tau(J) = 0, \ \tau = \sqrt{q^2 - J^2} > 0, \ q > |J|, \\ u_{\pm\sigma} &= \frac{\Gamma(\mp 2\sigma)(2\lambda)^{\pm\sigma}}{\Gamma\left(1 \mp \sigma - \frac{\varepsilon}{\lambda} q\right)} [q\sqrt{1-\varepsilon} - (J \pm \sigma)\sqrt{1+\varepsilon}] \end{aligned} \tag{7}$$

[1] The half-integer quantization of the orbital angular momentum comes into effect for two-dimensional quantum dots with an odd number of electrons [13]

**Table 1.** Critical charge values $q_{cr}^{(n)}(J) = Z_{cr}^{(n)}(J)\alpha_F$ at which the lowest $n = 0$ and first excited $n = 1$ levels with the given $J$ reach the boundary of the lower continuum for $R = 1/20$.

| $R = 1/20$ | $J = 1/2$ | $J = -1/2$ | $J = 1$ | $J = -1$ |
|---|---|---|---|---|
| $q_{cr}^{(0)}(J)$ | 0.90 | 1.14 | 1.36 | 1.51 |
| $q_{cr}^{(1)}(J)$ | 1.61 | 1.91 | 1.97 | 2.19 |

To fix the parameters $\theta_\sigma(J)$ and $\theta_\tau(J)$ we need, according to [14], to consider a Coulomb potential modified at small distances $\rho \le R_0 \ll 1$:

$$V_R(\rho) = -\frac{q}{R}\begin{cases} R/\rho, & \rho \ge R \\ f(\rho/R), & \rho \le R \end{cases} \tag{8}$$

Equation (5) with the replacement $V_C(\rho) \to V_R(\rho)$ at $J \neq 0$ has an analytical solution for $f(\rho/R) \equiv 1$. Within the framework of this model we have

$$\tan\theta_\sigma(J;R) = -\frac{(J+\sigma)[qJ_{\mp 1/2 \pm J}(q) \mp (J-\sigma)J_{\pm 1/2 \pm J}(q)]R^{-\sigma}}{(J-\sigma)[qJ_{\mp 1/2 \pm J}(q) \mp (J+\sigma)J_{\pm 1/2 \pm J}(q)]R^{\sigma}}, \tag{9}$$

where $J_\nu(q)$ is the Bessel functions with upper (lower) signs corresponding to $J > 0$ ($J < 0$) respectively.

If $q > |J| > 0$, assuming in (9) $\sigma = i\tau$, we obtain the equality

$$\exp(2i\theta_\tau(J;R)) = \tan\theta_{i\tau}(J;R), \ \ \tau = \sqrt{q^2 - J^2} > 0, \ \ J \neq 0 \tag{10}$$

The case of $J = 0$ may be solved analytically. This case was considered in [6], so we will consider cases with $J = \pm 1/2$ and $J = \pm 1$.

Equation (6) determines the critical charge $q_{cr}(J;R)$ as a function of the quantum number $J$ and the cutoff radius $R$, see Table 1, whereas conditions (7) with (9) and (10) give the charge dependence of the energy $\varepsilon_J(q,R)$ at the fixed cutoff radius (see Fig. 1).

The parameters $\theta_\sigma(J;R)$ and $\theta_\tau(J;R)$ completely determine the energy spectrum and stationary wavefunctions of (5) at any values of the charge $q = Z\alpha_F$.

In [4, 6] we obtained the equation for the spectrum of complex energies of quasistationary states of holes corresponding to the poles of the scattering matrix $S_J(k;q) = \exp(2i\delta_J(k;q))$:

$$\frac{(-2ik)^{i\tau}[q\sqrt{-\varepsilon+1} + (iJ-\tau)\sqrt{-\varepsilon-1}]\Gamma(-2i\tau)\Gamma\left(1+i\tau-i\frac{\varepsilon}{k}q\right)}{(-2ik)^{-i\tau}[q\sqrt{-\varepsilon+1} + (iJ+\tau)\sqrt{-\varepsilon-1}]\Gamma(2i\tau)\Gamma\left(1-i\tau-i\frac{\varepsilon}{k}q\right)} = e^{2i\theta_J(q;R)}. \tag{11}$$

The solutions of this equation determine both the position of a quasidiscrete level $\varepsilon_0$ and its width $\gamma$:

$$\varepsilon_{qs} = -\varepsilon_p = \varepsilon_0 - \frac{i}{2}\gamma, \ \ \varepsilon_0 > 0, \ \ \gamma > 0, \tag{12}$$

Due to the Coulomb barrier in the lower continuum the width $\gamma \ll 1$

$$k = k_0' - ik_0'', \ \ k_0' = \sqrt{\varepsilon_0^2 - 1} > 0, \ \ k_0'' = \frac{\gamma}{2k_0'} > 0,$$

IOP Conf. Series: Journal of Physics: Conf. Series **941** (2017) 012048 doi:10.1088/1742-6596/941/1/012048

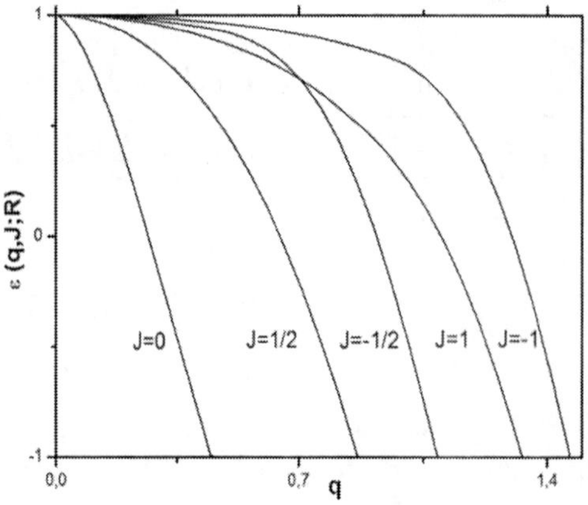


**Figure 1.** Energy of the ground state $\varepsilon(q, J; R)$ versus the charge $q = Z\alpha_F$ for the Coulomb potential cutoff radius $R = 1/20$ for different values of quantum number $J$.

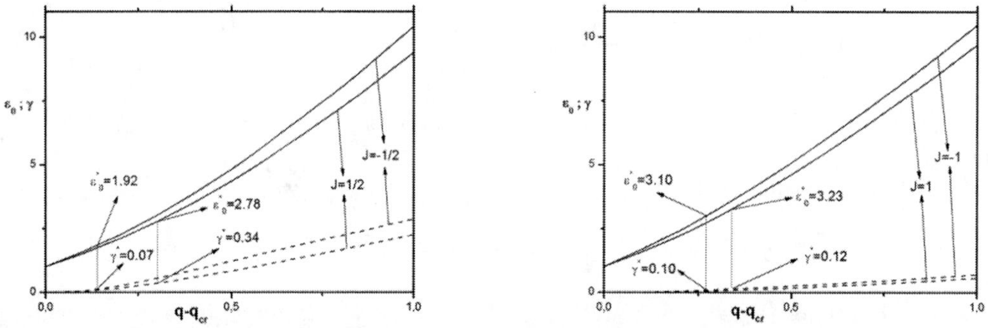


**Figure 2.** Positions of the quasidiscrete level $\varepsilon_0$ and its widths $\gamma$ versus the difference $q - q_{cr}$ at the cutoff radius $R = 1/20$ for $J = \pm 1/2$ and $J = \pm 1$. Values $\varepsilon^*$ and $\gamma^*$ are the position and width of the pole of $S$-matrix corresponding to the resonance, see Fig. 3:
a) for $J = 1/2$ at $q - q_{cr} = 0.3$ and for $J = -1/2$ at $q - q_{cr} = 0.16$;
b) for $J = 1$ at $q - q_{cr} = 0.34$ and for $J = -1$ at $q - q_{cr} = 0.29$.

and these poles are located on the second (unphyisical) sheet [15], in accordance with the unitarity of the partial scattering matrix.

Figure 2 shows the position of the lowest quasidiscrete level $\varepsilon_0$ and its width $\gamma$ as a function of $q - q_{cr}$ for $J = \pm 1/2$ and $J = \pm 1$. If the hole energy $\overline{\varepsilon}$ is in the range of dramatic change of scattering phase, such a states are evident as resonances in the hole-impurity scattering.

The scattering phases $\delta_J(\overline{\varepsilon}, q; R)$ for several $q = Z\alpha_F$ as functions of hole energy $\overline{\varepsilon} = -\varepsilon > 1$ for $J = \pm 1/2$ and $J = \pm 1$ when $q > q_{cr}$ are shown in Fig. 3. The values of $q$ are chosen so that the integer values of $Z$ were close to $Z_{cr}$ from Table 1 for $SiC$ graphene with $\alpha_F = 0.4$. If the hole energy $\overline{\varepsilon}$ is in the range of dramatic change of scattering phase, such a states are evident as resonances in the hole-impurity scattering and the partial cross-section $\sigma_J(\overline{\varepsilon}) = \sin^2 \delta_J$.

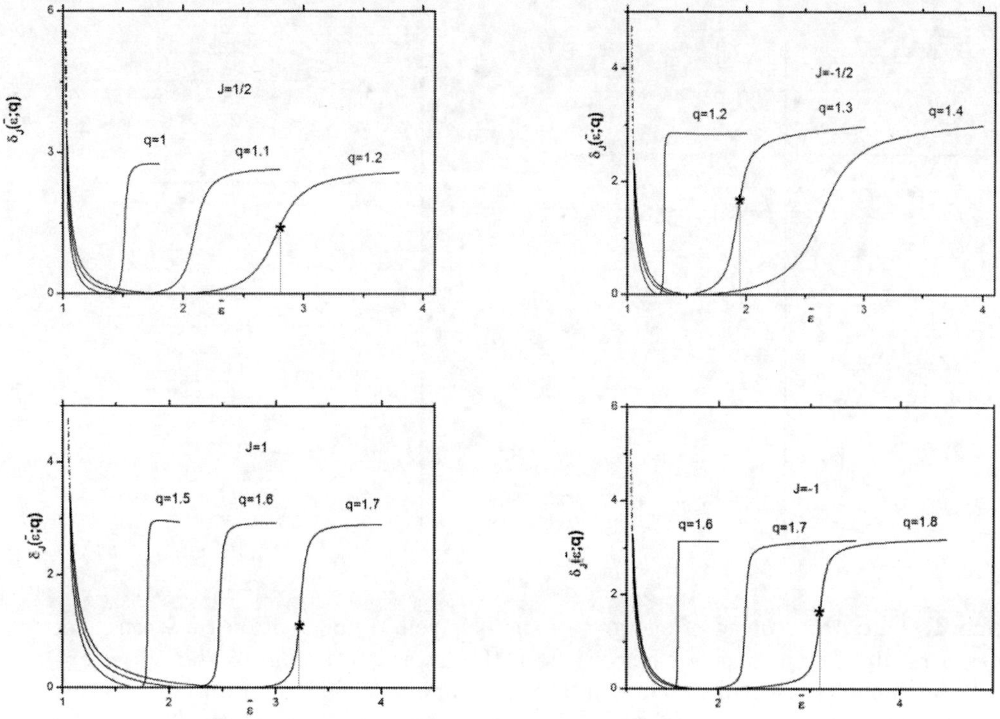


**Figure 3.** Scattering phases $\delta_J(\bar{\varepsilon}, q; R)$ as functions of hole energy for total momenta $J = 1/2$, $J = -1/2$, $J = 1$, $J = -1$. Resonances in the scattering of holes on impurity are marked with gears. Positions and widths of these resonances are given on Fig. 2. Dot-dashed line correspond to the asymptotic (13).

For controlling of numerical calculations asymptotic behaviour of scattering phases at small wavevectors $k = \sqrt{\varepsilon^2 - 1} \ll 1$ was used

$$\delta_J(k; q) \simeq \frac{q}{k}\left(\ln\frac{q}{k} - 1\right) + \frac{\pi}{4} + O(k \ln k), \ k \ll 1. \tag{13}$$

## 3. Conclusions

Thus, the "cutoff" i.e., the regularization of the Coulomb potential at small distances, ensures that the Dirac Hamiltonian is self-adjoint both for two-dimensional and for three-dimensional problems. This implies that the one-particle approximation for the Dirac equiation is consistent at any values of the Coulomb charge, including the overcritical range. It means that there is no spontaneous creation of electron-hole pairs. At the same time, owing to the presence of low permeable effective Coulomb barrier, the continuum states near the boundary of lower continuum form quasistationary states which may appear as resonances in the scattering of holes on supercritical impurity.

Since incident and diverging waves with total momentum $J$ are unitary equivalent, overcritical radial Coulomb problem differs markedly from "Klein paradox" [16]. Let's consider the example with $J = 0$.

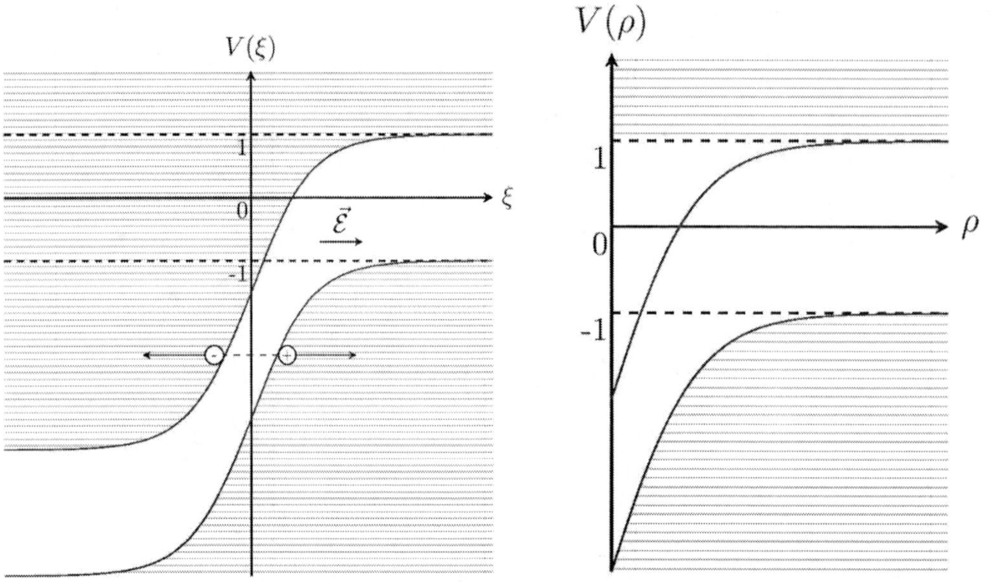


**Figure 4.** Left: the scheme of energies of the electron. Pair creation by strong electrostatic field as a result of tunneling of electrons from the lower continuum to the upper one.; Rigth: there is no spontaneous pair creation in strong Coulomb field because lower continuum does not overlap wit the upper one although well depth is more than gap width. The hatching marks the continuum.

Along with radial equation (1) with $J = 0$ with regularized potential determined in (8), we consider one-dimensional (along $z$ axis) Dirac equation

$$\begin{pmatrix} 1+V(\xi) & \frac{d}{d\xi} \\ -\frac{d}{d\xi} & -1+V(\xi) \end{pmatrix} \begin{pmatrix} F(\xi) \\ G(\xi) \end{pmatrix} = \varepsilon \begin{pmatrix} F(\xi) \\ G(\xi) \end{pmatrix}, \quad z = l_F \xi, \; -\infty < \xi < \infty. \tag{14}$$

For the energy of the electron from upper (lower) continuum within the quasiclassical approximation we get $\varepsilon = \pm\sqrt{P^2(\xi)+1} + V(\xi)$. So, for the upper continuum bound $P(\xi) = 0$ we have $\varepsilon_+ = 1 + V(\xi)$, and for lower one – $\varepsilon_- = -1 + V(\xi)$, see Fig. 4a in the case of potential $V(\xi) = \mathcal{E} \tanh \xi$, where $\mathcal{E} > 1$ is the reduced electric field.

The electron-hole pair creation may be interpreted as tunneling of the electron from the Dirac sea to the upper continuum. The motion of the electron with $\varepsilon < -1$ should be considered as the "time-opposite" motion, see Fig. 4a. Such a tunneling is possible if the energetic spectra of these continua are overlapping. At the same time, in the radial problem electronic spectra does not overlap with hole one, see Fig. 4b.

This difference is related to the fact that in one-dimensional case the electron-hole pair is pulled apart by the electric field, whereas in the radial case both electron and hole are attracted to the impurity. So, spontaneous creation of electron-hole pairs do not realize in the relativistic Coulomb problem and it is an additional argument for the validity of one-particle approximation for the Dirac equation at $Z > Z_{cr}$.

IOP Conf. Series: Journal of Physics: Conf. Series **941** (2017) 012048 doi:10.1088/1742-6596/941/1/012048

## References

[1] Biswas R R, Sachdev S and Son D T 2007 *Phys. Rev.* B **76** 205122
[2] Novikov D S 2007 *Phys. Rev.* B **76** 245435
[3] Pereira V M, Kotov V N and Castro Neto A H 2008 *Phys. Rev.* B **78** 085101
[4] Kuleshov V M, Mur V D, Narozhny N B, Fedotov A M, Lozovik Yu E 2015 *JETP Letters* **101** 264
[5] Kuleshov V M, Mur V D, Narozhny N B, Fedotov A M, Lozovik Yu E, Popov V S 2015 *Physics-Uspekhi* **58** 785
[6] Kuleshov V M, Mur V D and Narozhny N B 2017 *Journal of Physics: Conf. Series* **788** 012044
[7] Zhou S Y, Gweon G-H, Fedorov A V, First P N, de Heer W A, Lee D-H, Guinea F, Castro Neto A H and Lanzara A 2007 *Nature Materials* **6** 770.
[8] Akhiezer A I, Berestetskii 1965 *Quantum electrodynamics* (New York: Wiley)
[9] Zel'dovich Ya B, Popov V S 1971 *Sov. Phys. Usp.* **14** 673
[10] DiVincenzo D P and Mele E J 1984 *Phys. Rev.* B **29** 1685
[11] Pauli W 1939 *Helv. Phys. Acta* **12** 147
[12] van Vinter C 1968 *Ann. Phys.* **47** 232
[13] Mur V D, Narozhny N B, Petrosyan A N, Lozovik Yu E 2008 *JETP Letters* **88** 688
[14] Pomeranchuk I, Smorodinsky Ya 1945 *J. Phys. USSR* **9** 97
[15] Mur V D, Popov V S 1976 *Teor. Math. Phys.* **27** 204
[16] Klein O 1929 *Z. Physik* **53** 157

# Crossover in mesoscopic conductance fluctuations in a quasi-one-dimensional system

**V V Marinyuk, K A Kondratiev, D B Rogozkin**

National Research Nuclear University MEPhI, Kashirskoe Shosse 31, Moscow 115409, Russia

E-mail: vmarinyuk@mail.ru

**Abstract.** We study the conductance fluctuations in the crossover between the ballistic and diffusive regimes of phase coherent transport. For a quasi-1D disordered system, the correlation function of the conductance at different frequencies is calculated beyond the diffusion approximation. The result obtained establishes the interrelation between conductance fluctuations in the crossover regime and the characteristics of the disordered system. The frequency dependence of different contributions to the correlation function is analyzed at subdiffusion length scales.

## 1. Introduction

Previous studies of mesoscopic conductance fluctuations (see, e.g., Refs. [1–11]) were mostly focused on the diffusive regime of wave transport where length $L$ of the sample is much larger than transport mean free path $l_{tr}$. In this regime, the effect of "universal" conductance fluctuations (UCF) is observed [1,4]. The variance of the dimensionless conductance appears to be of the order of unity and, to a large extent, does not depend on the size of the sample. Not much is known as regards the crossover from ballistic to diffusive regime and, correspondingly, the fluctuations at subdiffusion length scales, $L \leq l_{tr}$. The currently available results concerned with the crossover regime are based mostly on direct numerical modelling [12–15]. Analytical results within the random matrix theory (RMT) approach [4, 16–18] were obtained only for waves identical in frequency.

In this paper we present a result of diagrammatic calculations for the conductance correlation function of a quasi-1D system (a waveguide with bulk disorder). The correlation function is expressed explicitly in terms of the cross-section of scattering by inhomogeneities of the medium and the propagators that obey the standard transport equation. In the large-length limit, our result transforms with no divergencies to the well-known diffusion formula [6,8]. The crossover between the quasi-ballistic and diffusive regimes is studied for a system of point-like centers. Our calculations within the two-stream version of the discrete-ordinate method generalize the well-known RMT formula for the conductance variance [16–18] to the case of waves different in frequency.

## 2. Results of diagrammatic calculations

We consider transmission of classical waves through a disordered waveguide of length $L$. The dimensionless conductance (or transmittance) of the waveguide can be defined as the sum of transmission coefficients $T_{ab}$ connecting incoming and outgoing modes $a$ and $b$, respectively (see, e.g., Refs. [4–6])

$$G = \sum_{a,b} T_{ab}. \tag{1}$$

The transmission coefficients $T_{ab}$ depend on spatial configuration of the scattering centers and vary from sample to sample. Under conditions of weak localization ($G \gg 1$), the value of $\langle G \rangle$ averaged over an ensemble of disordered samples is governed, within the standard impurity technique, by the sum of ladder diagrams. The correlation function of conductance fluctuations at different frequencies $C(\Delta\omega) = \langle \delta G(\omega_0 + \Delta\omega/2)\delta G(\omega_0 - \Delta\omega/2)\rangle$ ($\Delta\omega$ is the shift of incident waves in frequency, $\omega_0$ is the carrying frequency) can be expressed in terms of the ensemble-average fourth moment of a wave field and represented as expansion in orders of interference between ladders. Each interference event between the ladders contains the Hikami vertex [19]. The correlation function is governed by diagrams containing two vertices (see Fig.1). In the presence of time-reversal symmetry, the diagrams shown explicitly in Fig.1 should be supplemented by those that are obtained by interchanging initial and final states in one pair of conjugated wave fields. These diagrams contain the maximally crossed internal graphs (or cooperons) instead of the ladders.

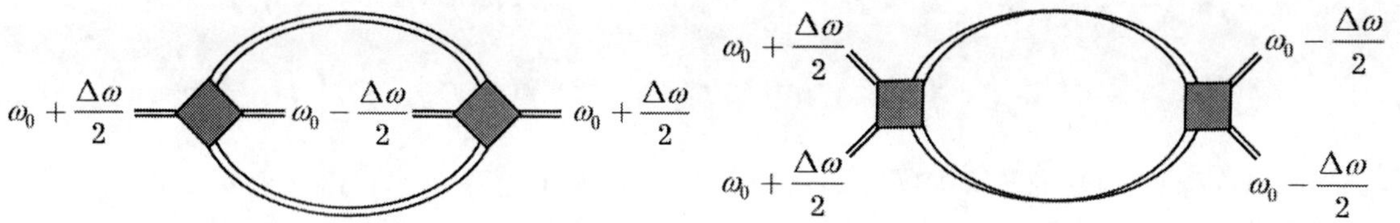


**Figure 1.** Diagrams contributing to the conductance correlation function. The paired lines correspond to ladder graphs. The shaded boxes are the Hikami vertex [19].

For great number $N$ of propagating modes ($N = k_0^2 A/4\pi$, $k_0$ is the wavenumber, $A$ is the area of the waveguide cross-section) the summation over modes can be replaced by integration over directions $\mathbf{\Omega}$ of wave propagation (see, e.g., Ref. [6]),

$$\sum_a \ldots = \int \frac{A d\mathbf{q}_a}{(2\pi)^2} \ldots = \frac{k_0^2 A}{(2\pi)^2} \int d\mathbf{\Omega}_a |\mu_a| \ldots \quad , \tag{2}$$

where $\mathbf{q}_a$ is the transverse momentum ($q_a < k_0$), $\mu_a = \Omega_{az}$, the $z$-axis is directed along the waveguide. Hemispheres $\Omega_{az} > 0$ and $\Omega_{az} < 0$ correspond to the waves that propagate in the forward and backward directions, respectively. The average conductance is expressed in terms of the average intensity,

$$\langle G \rangle = \frac{N}{\pi} \int d\mathbf{\Omega}_a d\mathbf{\Omega}_b |\mu_a||\mu_b| I_{ab}(z_f = L|z_i = 0), \tag{3}$$

where propagator $I_{ab}(z|z') = I(z, \mathbf{\Omega}_a|z', \mathbf{\Omega}_b)$ denotes the intensity at depth $z$ in direction $\mathbf{\Omega}_a$ from a source placed at depth $z'$ and emitting waves in direction $\mathbf{\Omega}_b$. Intensity $I_{ab}(z|z')$ is subject to the transport equation [20]

$$\left(\mu_a \frac{\partial}{\partial z} + n\sigma_{\text{tot}}\right) I_{ab}(z|z') = \delta(z - z')\delta(\mathbf{\Omega}_a - \mathbf{\Omega}_b) + \int d\mathbf{\Omega}_c \sigma_{ac} I_{cb}(z|z'), \tag{4}$$

where $\sigma_{ac} = n\mathrm{d}\sigma(\mathbf{\Omega}_a\mathbf{\Omega}_c)/\mathrm{d}\Omega$, $n$ is the number of scattering centres per unit volume, $\mathrm{d}\sigma/\mathrm{d}\Omega$ is the differential scattering cross-section, $\sigma_{\mathrm{tot}} = \sigma + \sigma_{\mathrm{a}}$ is the total cross-section of interaction, $\sigma$ and $\sigma_{\mathrm{a}}$ are the cross-sections of elastic scattering and absorption, respectively.

The diagrams shown in Fig.1 can be evaluated by a straightforward manner (see, e.g., Refs. [21, 22]). In what follows, we take into account the ladders incorporating an arbitrary number of scattering events, among them the graphs without any scattering. These latter describe nonscattered waves. Contrary to calculations, performed within the diffusion approximation [6, 8], we need not introduce particular diagrams containing only one internal ladder propagator and the six-point Hikami vertex. Such diagrams are already contained among the diagrams depicted in Fig.1. They correspond to the pair of nonscattered waves in either of two internal ladders.

Our diagrammatic calculations are similar to those of Ref. [22] where the conductance variance was found. Extending results [22] to the case of the waves differing in frequency, we derive the following expression for the correlation function:

$$C(\Delta\omega) = C^A(\Delta\omega) + C^B(\Delta\omega) + C^C(\Delta\omega), \tag{5}$$

$$C^A(\Delta\omega) = \int_0^L \mathrm{d}z\,\mathrm{d}z' \int \mathrm{d}\mathbf{\Omega}_a \mathrm{d}\mathbf{\Omega}_b \mathrm{d}\mathbf{\Omega}_c \mathrm{d}\mathbf{\Omega}_d\, \sigma_{ab}\sigma_{cd} \times \tag{6}$$

$$\left[(I_a^f - I_b^f)^2\, I_{ac}(\Delta\omega) I_{bd}(-\Delta\omega)\, (I_c^i - I_d^i)^2 + (I_a^f - I_b^f)(I_{-a}^i - I_{-b}^i)\, I_{ac}(\Delta\omega) I_{bd}(-\Delta\omega)\, (I_{-c}^f - I_{-d}^f)(I_c^i - I_d^i)\right]$$

$$C^B(\Delta\omega) = \int_0^L \mathrm{d}z\,\mathrm{d}z' \int \mathrm{d}\mathbf{\Omega}_a \mathrm{d}\mathbf{\Omega}_b \mathrm{d}\mathbf{\Omega}_c \mathrm{d}\mathbf{\Omega}_d\, \sigma_{ab}\sigma_{cd} \left[(I_a^f - I_b^f) I_b^i + I_{-b}^f (I_{-a}^i - I_{-b}^i)\right] \times \tag{7}$$

$$\mathrm{Re}\left\{(I_{ac}(\Delta\omega) - I_{ad}(\Delta\omega))(I_{-a-c}(\Delta\omega) - I_{-b-c}(\Delta\omega))\right\} \left[(I_c^f - I_d^f) I_d^i + I_{-d}^f (I_{-c}^i - I_{-d}^i)\right]$$

$$C^C(\Delta\omega) = \int_0^L \mathrm{d}z \int \mathrm{d}\mathbf{\Omega}_a \mathrm{d}\mathbf{\Omega}_b\, \sigma_{ab} \left[(I_a^f - I_b^f) I_b^i + I_{-b}^f (I_{-a}^i - I_{-b}^i)\right]^2 \mathrm{Re}\left\{I_{ab}(z|z, \Delta\omega)\right\} \tag{8}$$

where internal propagator $I_{ab}(\Delta\omega) = I_{ab}(z|z', \Delta\omega)$ obeys the transport equation that is obtained from Eq.(4) by substitution of complex absorption coefficient $(n\sigma_a + \mathrm{i}\Delta\omega/c)$ for $n\sigma_a$ ($c$ is the velocity of waves). The incoming and outgoing propagators are defined as:

$$I_a^i(z) = \int \mathrm{d}\mathbf{\Omega}_b\, |\mu_b|\, I_{ab}(z|z_i = 0, \Delta\omega = 0), \quad I_a^f(z) = \int \mathrm{d}\mathbf{\Omega}_b\, |\mu_b|\, I_{ba}(z_f = L|z, \Delta\omega = 0). \tag{9}$$

The change of sign in a subscript of any propagator entering into Eq.(5) implies the reverse of the direction (i.e. substitution of $-\mathbf{\Omega}_a$ for $\mathbf{\Omega}_a$ into the propagator). The incoming and outgoing propagators appearing in Eq.(5) with subscripts $\pm a$, $\pm b$ are functions of $z$, while those with subscripts $\pm c$, $\pm d$ are functions of $z'$.

The terms that contain products of the incoming and outgoing propagators with subscripts of opposite sign (e.g., $[(I_a^f - I_b^f) I_b^i][\ldots][I_{-d}^f (I_{-c}^i - I_{-d}^i)]$) correspond to the cooperon contribution which includes the maximally crossed internal graphs and, in the low-order scattering limit, the graphs that describe the waves propagating in opposite directions. Equations (6)-(8) has the presented form provided that the time-reversal symmetry is not violated.

Note that the third term in Eq.(5), $C^C(\Delta\omega)$, corresponds to the contribution which, within the diffusion approach [6, 8], is given by the six-point Hikami vertex.

Equation (5) generalizes the result of the diffusion approximation [6, 8], much as the transport equation generalizes the equation of diffusion. Equation (5) establishes the interrelation between the conductance fluctuations and the characteristics of scattering centers of the disordered system, and enables us to study evolution of the fluctuations in going from the quasi-ballistic propagation to the diffusive regime.

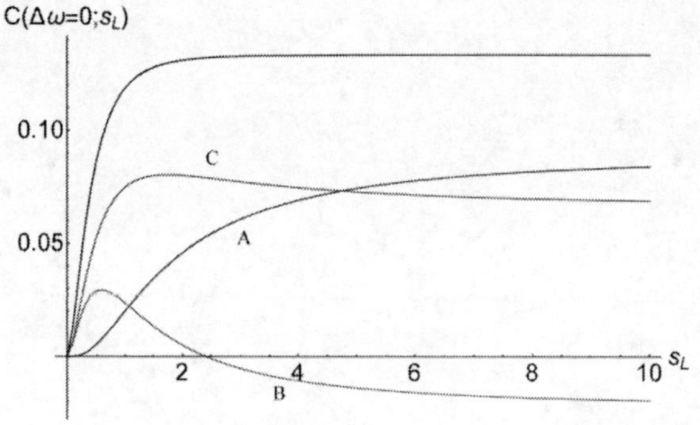


**Figure 2.** Different contributions to $C(\Delta\omega = 0, s_L)$ as functions of the waveguide length. The upper curve is the sum of all contributions.

## 3. Crossover between the quasi-ballistic and diffusive regimes

As an illustration of application of Eq.(5) to calculating the correlation function of conductance fluctuations beyond the diffusion approximation, we take advantage of the two-stream version of the discrete-ordinate method [20]. This simplest model enables us to perform integration in Eq.(5) explicitly and to derive analytical results for $C(\Delta\omega)$ which describe the crossover between the quasi-ballistic and diffusive regimes.

Within this approach, each integral over $\mathbf{\Omega}$ is supposed to be equal to the sum of the values of an integrand quantity at $\Omega_z = \pm\mu_0$,

$$\int d\mathbf{\Omega}_a I(z, \mathbf{\Omega}_a) = 2\pi I_+(z) + 2\pi I_-(z), \tag{10}$$

where $I_\pm(z) = I(z, \Omega_z = \pm\mu_0)$, and $\pm\mu_0$ are the discrete ordinates [20].

For a widely used model of pointlike centers, quantity $\sigma_{ab}$ in Eqs. (4) and (5) is independent of the directions, and the disorder is characterized only by mean free path $l$, $\sigma_{ab} = (4\pi l)^{-1}$ (the medium with no absorption is considered, $\sigma_a = 0$). In this case, the $C$-functions entering into Eq.(5) are expressed in terms of the values of intensity propagators at $\pm\mu_0$ as follows

$$C^A(\Delta\omega) = 2\left(\frac{\pi}{l}\right)^2 \mathrm{Re}\int_0^L dz\, dz'[I_{++}(\Delta\omega)I_{--}(-\Delta\omega) + I_{+-}(\Delta\omega)I_{-+}(-\Delta\omega)]\times \tag{11}$$

$$[(I_+^f - I_-^f)^2(I_+^i - I_-^i)'^2 + (I_+^f - I_-^f)(I_+^i - I_-^i)(I_+^f - I_-^f)'(I_+^i - I_-^i)']$$

$$C^B(\Delta\omega) = \left(\frac{\pi}{l}\right)^2 \mathrm{Re}\int_0^L dz\, dz'[h_+(I_{++}(\Delta\omega) - I_{+-}(\Delta\omega)) + h_-(I_{--}(\Delta\omega) - I_{-+}(\Delta\omega))]\times \tag{12}$$

$$[h'_+(I_{--}(\Delta\omega) - I_{+-}(\Delta\omega)) + h'_-(I_{++}(\Delta\omega) - I_{-+}(\Delta\omega))]$$

$$C^C(\Delta\omega) = \frac{\pi}{l}\mathrm{Re}\int_0^L dz[h_+^2 I_{+-}(z|z, \Delta\omega) + h_-^2 I_{-+}(z|z, \Delta\omega)], \tag{13}$$

where $h_\pm = [\pm(I_+^f - I_-^f)I_\mp^i \mp (I_+^i - I_-^i)I_\pm^f]$. The propagators $I_{\pm\pm}(\Delta\omega) = I_{\pm\pm}(z|z', \Delta\omega)$ and $I_\pm^{i,f} = I_\pm^{i,f}(z)$ entering into Eqs. (11)-(13) are determined analytically from the transport equation and given by

$$I_{++}(z|z', \Delta\omega) = \frac{1}{2\pi\mu_0 \sinh(\gamma)\sinh(\xi_L + \gamma)}\begin{cases} \sinh(\xi)\sinh(\xi_L - \xi'), & z < z' \\ \sinh(\xi' + \gamma)\sinh(\xi_L - \xi + \gamma), & z > z', \end{cases} \tag{14}$$

$$I_{-+}(z|z', \Delta\omega) = \frac{1}{2\pi\mu_0 \sinh(\gamma)\sinh(\xi_L + \gamma)}\begin{cases} \sinh(\xi + \gamma)\sinh(\xi_L - \xi'), & z < z' \\ \sinh(\xi' + \gamma)\sinh(\xi_L - \xi), & z > z', \end{cases} \tag{15}$$

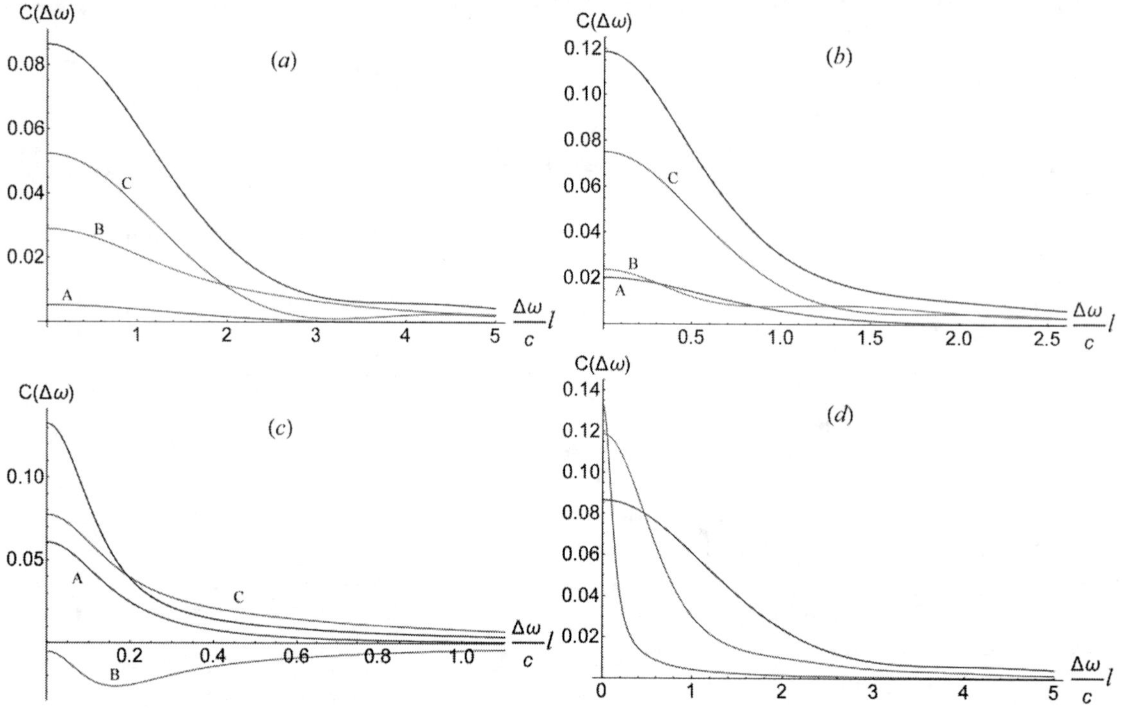


**Figure 3.** Frequency dependence of the different contributions to the correlation function for various waveguide lengths. (a) $s_L = 0.5$, (b) $s_L = 1$, (c) $s_L = 3$. The upper curves correspond to the sum of all contributions. (d) Correlation function $C(\Delta\omega)$, from upper to lower curves, $s_L = 0.5$, 1 and 3.

where $\xi = s\sinh(\gamma)$, $\xi_L = s_L\sinh(\gamma)$ and $\sinh(\gamma) = 2\sqrt{\mathrm{i}\Delta\omega l(1+\mathrm{i}\Delta\omega l/c)/c}$, $s = z/2\mu_0 l$, $s_L = L/2\mu_0 l$. The propagators $I_{\pm,-}(z|z',\Delta\omega)$ are obtained from Eqs.(14) and (15) with the relations $I_{--}(z|z',\Delta\omega) = I_{++}(z'|z,\Delta\omega)$, $I_{+-}(z|z',\Delta\omega) = I_{-+}(L-z|L-z',\Delta\omega)$. The incoming and outgoing propagators are equal to

$$I^i_{\pm}(z) = \frac{s_L - s + (1\pm 1)/2}{s_L+1}, \quad I^f_{\pm}(z) = \frac{s+(1\pm 1)/2}{s_L+1}. \tag{16}$$

In Eqs.(11)-(13), the incoming and outgoing propagators marked by prime are functions of $z'$, otherwise they are functions of $z$.

The length dependence of the different contributions to the correlation function at $\Delta\omega = 0$ is shown in Fig.2. Their sum is equal to

$$C(\Delta\omega = 0, s_L) = \frac{2}{15}\left(1 - \frac{1+6s_L}{(1+s_L)^6}\right), \tag{17}$$

The obtained value of $C(\Delta\omega = 0, s_L)$ (i.e., the conductance variance $\langle(\delta G)^2\rangle$) coincides with the well-known result of RMT calculations [16–18].

An analytical formula that extends Eq.(17) to a nonzero value of the frequency shift $\Delta\omega$ is too cumbersome to be presented here and, therefore, we illustrate the frequency-dependence of the conductance correlation function in the graphical form. Evolution of the frequency dependence both of the different contributions to the correlation function and of their sum with the waveguide length is illustrated in Fig.3. As follows from Fig.3(d), the frequency profile of $C(\Delta\omega)$ becomes

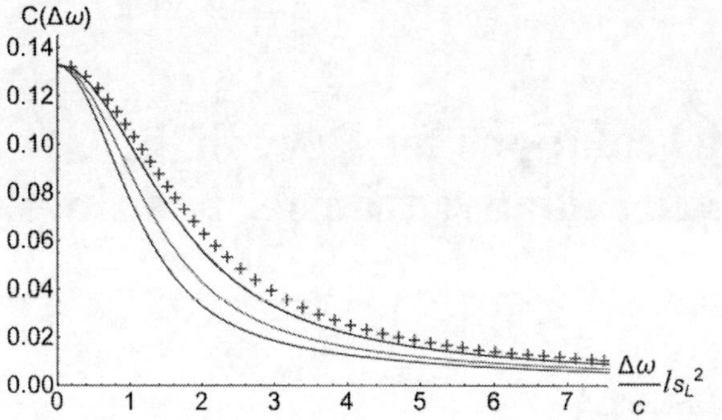


**Figure 4.** Correlation function of conductance fluctuations versus variable $\Delta\omega l s_L^2/c$ (from upper to lower curve $s_L = 5, 10, 15$). Crosses are the results of the diffusion approximation [6]. The discrete ordinate is chosen to be equal to $\mu_0 = 1/\sqrt{3}$.

narrower as the waveguide length increases. In the diffusive limit ($L \gg l$), the correlation function as a function of the variable $\Delta\omega l s_L^2/c$ tends to the universal law which coincides with the corresponding result of the diffusion approximation [6, 8] (see Fig.4).

## 4. Conclusions

In conclusion, we have developed a theoretical approach to calculating the correlation function of conductance fluctuations in a quasi-1D system, paying special attention to the subdiffusion length scales. We have presented an analytical result which relates the correlation function to the characteristics of the disordered system. The correlation function has been calculated within the two-stream model. It has been shown that $C(\Delta\omega)$ becomes narrower with increasing the waveguide length $L$ and tends to the diffusion result [6, 8] in the large $L$ limit. The results obtained generalize RMT calculations [4, 16–18] to the case of a nonzero frequency shift, and can be useful for experimental studies of the conductance correlation function [10].

## 5. Acknowledgments

We acknowledge support from the MEPhI Academic Excellence Project (contract 02.a03.21.0005, 27.08.2013).

## 6. References

[1] Altshuler B L 1985 *JETP Lett.* **41** 648
[2] Lee P A and Stone A D 1985 *Phys. Rev. Lett.* **55** 1622
[3] Lee P A, Stone A D and Fukuyama H, 1987 *Phys. Rev.* B **35** 1039
[4] Mello P A, Akkermans E and Shapiro B 1988 *Phys. Rev. Lett.* **61** 459
[5] Feng S, Kane C, Lee P A, and Stone A D 1988 *Phys. Rev. Lett.* **61** 834
[6] van Rossum M C W, Nieuwenhuizen Th M, and Vlaming R 1995 *Phys. Rev.* E **51** 6158
[7] Scheffold F and Maret G 1998 *Phys. Rev. Lett.* **81** 5800
[8] van Rossum M C W and Nieuwenhuizen Th M 1999 *Rev. Mod. Phys.* **71** 313
[9] Ludwig T, Blanter Ya M and Mirlin A D 2004 *Phys. Rev.* B **70** 235315
[10] Chabanov A A, Tregoures N P, van Tiggelen B A and Genack A Z 2004 *Phys. Rev. Lett.* **92** 173901
[11] Cherroret N, Pena A, Chabanov A A and Skipetrov S E 2009 *Phys. Rev.* B **80** 045118
[12] Nikolic K and MacKinnon A 1994 *Phys. Rev.* B **50** 11008
[13] Ando T 2004 *J. Phys. Soc. Jpn.* **73** 1895
[14] Markussen T, Rurali R, Jauho A-P and Brandbyge M 2007 *Phys. Rev. Lett.* **99** 076803
[15] Iori F, Ossicini S and Rurali R 2014 *J. Appl. Phys.* **116** 074303
[16] Beenakker C W J 1997 *Rev. Mod. Phys.* **69** 731
[17] Garcia-Martin A, Scheffold F, Nieto-Vesperinas M and Saenz J J 2002 *Phys. Rev. Lett.* **88** 143901
[18] Froufe-Perez L S, Garcia-Martin A, Cwilich G and Saenz J J 2007 *Physica* A **386** 625
[19] Hikami S 1981 *Phys. Rev.* B **24** 2671
[20] Ishimaru A 1999 *Wave Propagation and Scattering in Random Media*(New York: Wiley)
[21] Rogozkin D B and Cherkasov M Y 1995 *Phys. Rev.* B **51** 12256
[22] Marinyuk V V and Rogozkin D B 2015 *Phys. Rev.* B **91** 125125

# Accuracy analysis of automodel solutions for Lévy flight-based transport: from resonance radiative transfer to a simple general model

**A B Kukushkin[1,2] and P A Sdvizhenskii[1]**

[1] National Research Centre 'Kurchatov Institute', Moscow, 123182, Russian Federation

[2] National Research Nuclear University MEPhI, Moscow, 115409, Russian Federation

E-mail: kukushkin_ab@nrcki.ru

**Abstract.** The results of accuracy analysis of automodel solutions for Lévy flight-based transport on a uniform background are presented. These approximate solutions have been obtained for Green's function of the following equations: the non-stationary Biberman-Holstein equation for three-dimensional (3D) radiative transfer in plasma and gases, for various (Doppler, Lorentz, Voigt and Holtsmark) spectral line shapes, and the 1D transport equation with a simple long-tailed step-length probability distribution function with various power-law exponents. The results suggest the possibility of substantial extension of the developed method of automodel solution to other fields far beyond physics.

## 1. Introduction

The analysis of the Green's function of the non-stationary Biberman-Holstein equation for radiative transfer in plasma and gases has shown [1] that there is an approximate automodel solution based on three scaling laws: for the propagation front (i.e. relevant-to-superdiffusion average displacement of perturbation's carrier from an instant point source) and asymptotic solutions far beyond and far in advance of the propagation front. All these scaling laws are determined essentially by the long-free-path carriers (named, by Mandelbrot [2], Lévy flights, cf. page IX in Ref. [3]). The validity of the suggested automodel solution was proved by its comparison with analytical solutions by Veklenko [4] in the 3D case of the Biberman–Holstein equation of the resonance radiation transfer for various (Doppler, Lorentz, Voigt and Holtsmark) spectral line shapes with complete redistribution over frequency (within spectral line width) in the elementary act of the resonance scattering (i.e. absorption and subsequent emission) of the photon by an atom/ion. The approach [1] was extended in [5] on the wide class of non-stationary superdiffusive transport on a uniform background with a simple long-tailed step-length probability distribution function (PDF) with various power-law exponents.

In this paper, we present the results of accuracy analysis of automodel solutions for Lévy flight-based transport, including the resonance radiative transfer (section 2) and a simple general model (section 3). Possible applications of the method [1, 5] to other problems, where the role of the processes with dominating role of Lévy flights is identified and widely used, is discussed in section 4.

## 2. Self-similarity accuracy estimations for the Biberman-Holstein equation

The Biberman–Holstein equation for radiative transfer in a uniform medium of two-level atoms/ions is obtained from a system of equations for spatial density of excited atoms, $F(\mathbf{r}, t)$, and spectral intensity of resonance radiation. This system is reduced to a single equation for $F(\mathbf{r}, t)$, which appears to be an integral equation, non-reducible to a differential diffusion-type equation:

$$\frac{\partial F(\mathbf{r},t)}{\partial t}=\frac{1}{\tau}\int_V G(|\mathbf{r}-\mathbf{r}_1|)F(\mathbf{r}_1,t)\mathrm{d}V_1-\left(\frac{1}{\tau}+\sigma\right)F(\mathbf{r},t)+q(\mathbf{r},t), \quad (1)$$

where τ is the lifetime of the excited atomic state with respect to spontaneous radiative decay; σ is the rate of the collisional quenching of excitation; $q$ is the source of excited atoms, different from populating by the absorption of resonant photons (e.g. collisional excitation). Hereafter we use the dimensionless time coordinate, assuming the normalization of time by τ. The kernel $G$ is determined by the (normalized) emission spectral line shape $\varepsilon_\omega$ and the absorption coefficient $k_\omega$. In homogeneous media, $G$ depends on the distance between the points of emission and the absorption of the photon:

$$G(r)=-\frac{1}{4\pi r^2}\frac{dT(r)}{dr},\quad T(r)=\int_0^\infty \varepsilon_\omega \exp(-k_\omega r)\mathrm{d}\omega. \quad (2)$$

The automodel solution of equation (1) with a point instant source $q(\mathbf{r}, t) = \delta(\mathbf{r} - \mathbf{r}_0)\delta(t - t_0)$ (i.e. Green's function) was suggested in [5], which in the 3D case and for arbitrary space-time coordinates of the instant point source takes the form:

$$F_{\text{auto}}(\mathbf{r},t;\mathbf{r}_0,t_0)=(t-t_0)\cdot G\left(|\mathbf{r}-\mathbf{r}_0|\cdot g\left(\frac{r_{\text{fr}}(t-t_0)}{|\mathbf{r}-\mathbf{r}_0|}\right)\right). \quad (3)$$

where $g$ is a function of a single variable, and its asymptotic behavior is known: $g(s) = 1$, $s \ll 1$; $g(s) \propto s$, $s \gg 1$. The relation between $g$ and the exact solution of equation (1), $f_{\text{exact}}$, is described by the following equations:

$$Q_G(\rho,t)\equiv\frac{1}{\rho}\hat{G}^{-1}\left(\frac{f_{\text{exact}}(\rho,t)}{t}\right), \quad (4)$$

where $\hat{G}^{-1}$ is the function reciprocal to the $G$ function, $\rho \equiv k_0|\mathbf{r} - \mathbf{r}_0|$, $k_0$ is the absorption coefficient for photons with frequency $\omega_0$, corresponding to the line shape center,

$$Q_G(\rho,t(\rho,s))\equiv Q_{G1}(s,\rho)=g(s), \quad (5)$$

$$Q_G(\rho(t,s),t)\equiv Q_{G2}(s,t)=g(s), \quad (6)$$

where the functions $t(\rho, s)$ and $\rho(t, s)$ are determined by the relation

$$s=\rho_{\text{fr}}(t)/\rho. \quad (7)$$

To prove the automodel solution one has to show weak dependence (independence) of $Q_{W1}$ and $Q_{W2}$ functions on, respectively, space coordinate and time. The results of the validation of the automodel solution and the reconstruction of function $g$ from comparison of function (3) with computations of the Green's function [4] for the Lorentz, Doppler, Voigt, and Holtsmark line shapes are shown in figure 1.

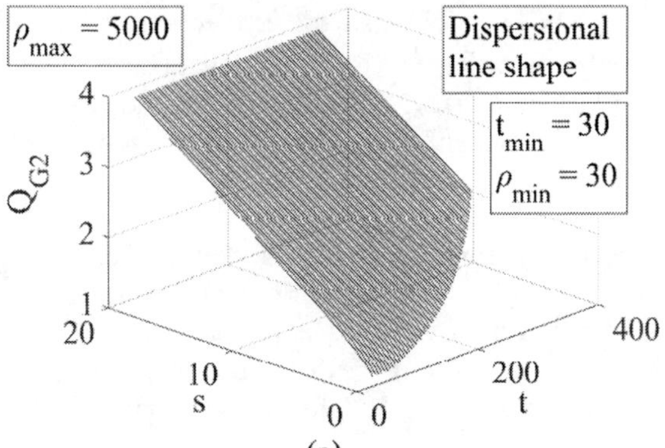


(a)

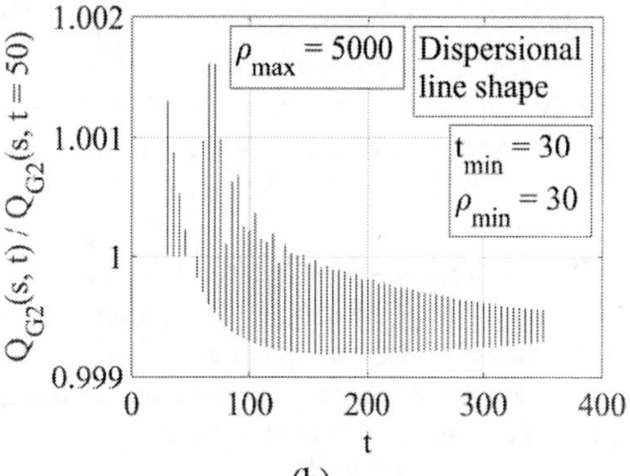


(b)

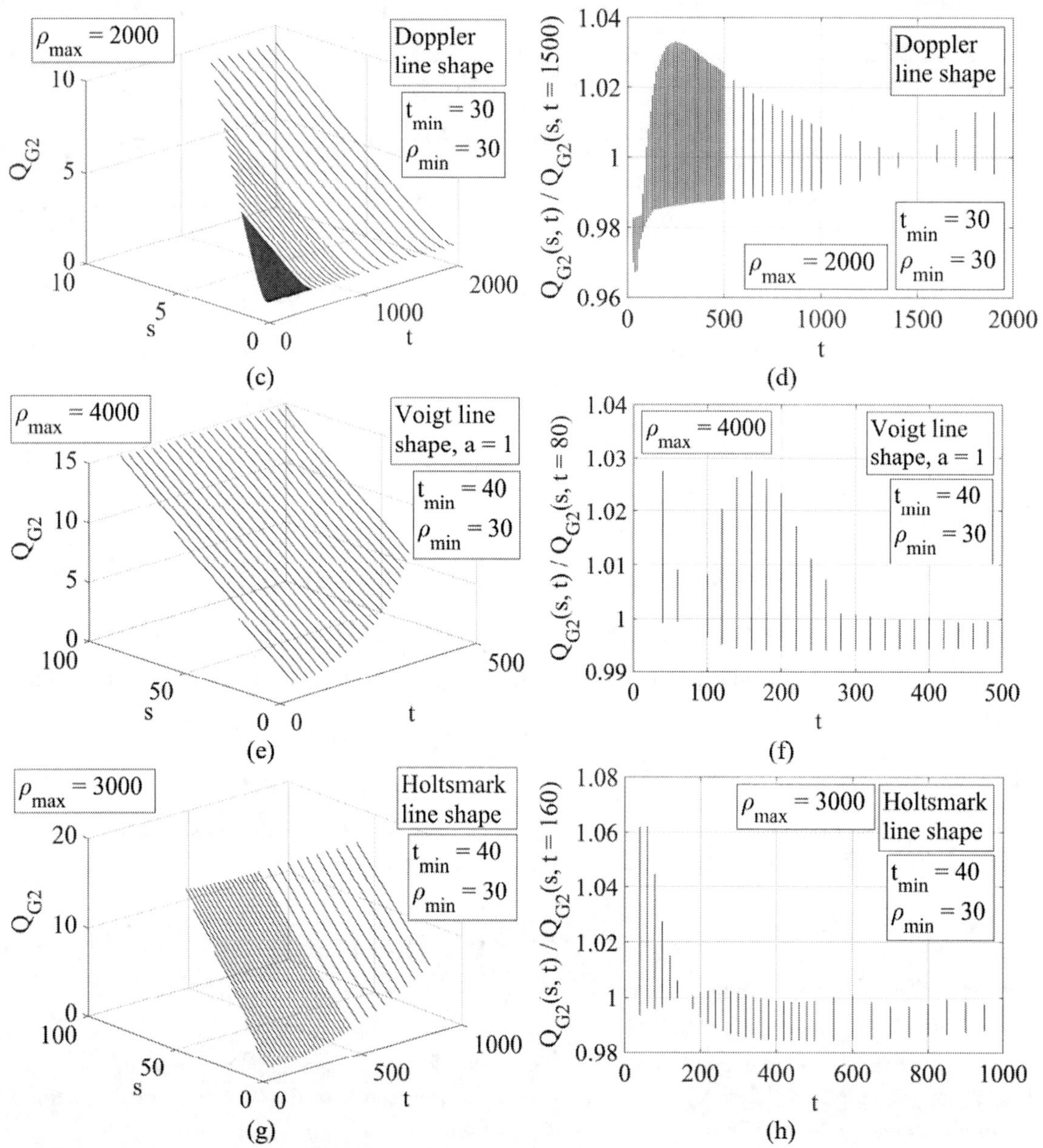


**Figure 1.** Function $Q_{G2}$ for $\rho > \rho_{min}$, $t > t_{min}$ for different line shapes: (a) dispersional (Lorentz), (c) – Doppler, (e) Voigt ($a = \sqrt{ln\,2}\ \Delta\omega_{\text{Lorentz}}/\Delta\omega_{\text{Doppler}} = 1$), (g) Holtsmark. Projection of the ratio $Q_{G2}(s, t)/Q_{G2}(s, t^*)$ onto the $\{Q_{G2}, t\}$ plane for different line shapes: (b) dispersional (Lorentz), $t^*$ = 50, (d) Doppler, $t^*$ = 1500, (f) Voigt, a = 1, $t^*$ = 80, (h) Holtsmark, $t^*$ = 160. The dependence of the error, which is given by the deviation from unity, on the $t_{min}$ value is shown in the right-hand-side figures. The range of variable $s$ is limited by the conditions $\rho < \rho_{max}$.

It is seen from figure 1 that the function (3) is indeed an automodel Green's function of equation (1) with an accuracy defined by the coincidence of the curves $Q_{G2}(s, t)$), for various values of the time $t$. Maximum deviation of $Q_{G1}$ and $Q_{G2}$ functions from a function of $s$ only is as follows. For $\{\rho > 30, t > 40\}$ we have 0.2% for the Lorentz line shape, and 4%, 3%, and 8% for, respectively, Doppler, Voigt ($a$ = 1), and Holtsmark line shapes. The maximum deviation amounts to 1% for the last-named three line shapes if we, for $\rho > 30$, restrict the time to $t > 950, 280, 700$, respectively. The largest deviations take place at $s \sim 1$.

## 3. Self-similarity accuracy estimations for general model (simple PDF, 1D case)

We consider the 1D transport on a uniform background, described by the equation for spatial density f(x, t) of an excitation of the background medium, which may evolve due to the exchange of excitation between various points of the medium via emission and absorption of the carriers (here the retardation caused by the finite velocity of carriers is neglected; the derivation of this equation for a possible mechanism of interaction between the medium and the carriers of medium's perturbation is given in Appendix in [5]):

$$\frac{\partial f(x,t)}{\partial t} = \frac{1}{\tau}\int_{-\infty}^{\infty} W(|x-x_1|)f(x_1,t)\mathrm{d}x_1 - \left(\frac{1}{\tau}+\sigma\right)f(x,t)+q(x,t) \tag{8}$$

where $W(x)$ is a step-length PDF (i.e. the probability that the carrier, emitted at some point, is absorbed at a distance $x$ from that point), $1/\tau$ is the absolute value of the emission rate (i.e. $\tau$ is the average waiting time between the absorption and reemission of the carrier), $q$ is the source function, which is the rate of production of excitation by an external source (i.e. a source which differs from the excitation of the medium due to absorption described by the $W$ function), and $\sigma$ is the rate of the quenching of excitation. The uniformity of the background assumes that, first, the $W$ is a function of only one variable — the distance between the points of emission and absorption — and, second, $\tau$ and $\sigma$ are the constants. The latter makes the role of quenching simply described by the time exponent $\exp(-\sigma\tau)$, therefore in what follows we omit this process. Hereafter we use the dimensionless time and space coordinate, assuming the normalization of time by $\tau$ and using a dimensionless PDF. We will seek for the Green's function, taking, respectively, the source function as a point instant source, $q(x, t) = \delta(x)\delta(t)$.

We take the PDF in the following simple form which possesses a long tail and the infinite value of the mean square displacement:

$$W(\rho) = \frac{\gamma}{2(1+\rho)^{\gamma+1}}, \quad 0<\gamma<2, \quad \rho = |x-x'|, \quad \int_{-\infty}^{\infty} W(|x-x'|)\mathrm{d}x' = 1. \tag{9}$$

It was shown in [5] that one can derive automodel Green's function in the form

$$f(x,t) = t\,W\left(\rho g\left(\frac{\rho_{\mathrm{fr}}(t)}{\rho}\right)\right), \quad \rho = |x|, \tag{10}$$

or, for the PDF (9),

$$f(x,t) = t\frac{\gamma}{2\left[1+\rho g\left(\rho_{\mathrm{fr}}(t)/\rho\right)\right]^{\gamma+1}}, \quad \rho = |x|, \tag{11}$$

where $\rho_{\mathrm{fr}}(t) = (t+1)^{1/\gamma} - 1$. The procedure of the reconstruction of function $g$ is quite similar to that of equations (4)–(7):

$$Q_W(x,t) \equiv \frac{1}{\rho}\hat{W}^{-1}\left(\frac{f_{\mathrm{exact}}(x,t)}{t}\right) = \frac{1}{\rho}\left[\left(\frac{\gamma t}{2 f_{\mathrm{exact}}(x,t)}\right)^{\frac{1}{\gamma+1}} - 1\right], \quad \rho \equiv |x|, \tag{12}$$

where $\hat{W}^{-1}$ is the function reciprocal to $W$ function,

$$Q_W(\rho, t(\rho,s)) \equiv Q_{W1}(s,\rho) = g(s), \tag{13}$$

$$Q_W(\rho(t,s),t) \equiv Q_{W2}(s,t) = g(s), \tag{14}$$

where the functions $t(\rho, s)$ and $\rho(t, s)$ are determined by the relation

$$s = ((t+1)^{1/\gamma} - 1)/\rho \tag{15}$$

(the propagation front $\rho_{\mathrm{fr}}(t) = \rho(t, s = 1)$).

The $Q_{W2}$ function for various values of time coordinates is shown in figure 2 for three values of the exponent $\gamma$.

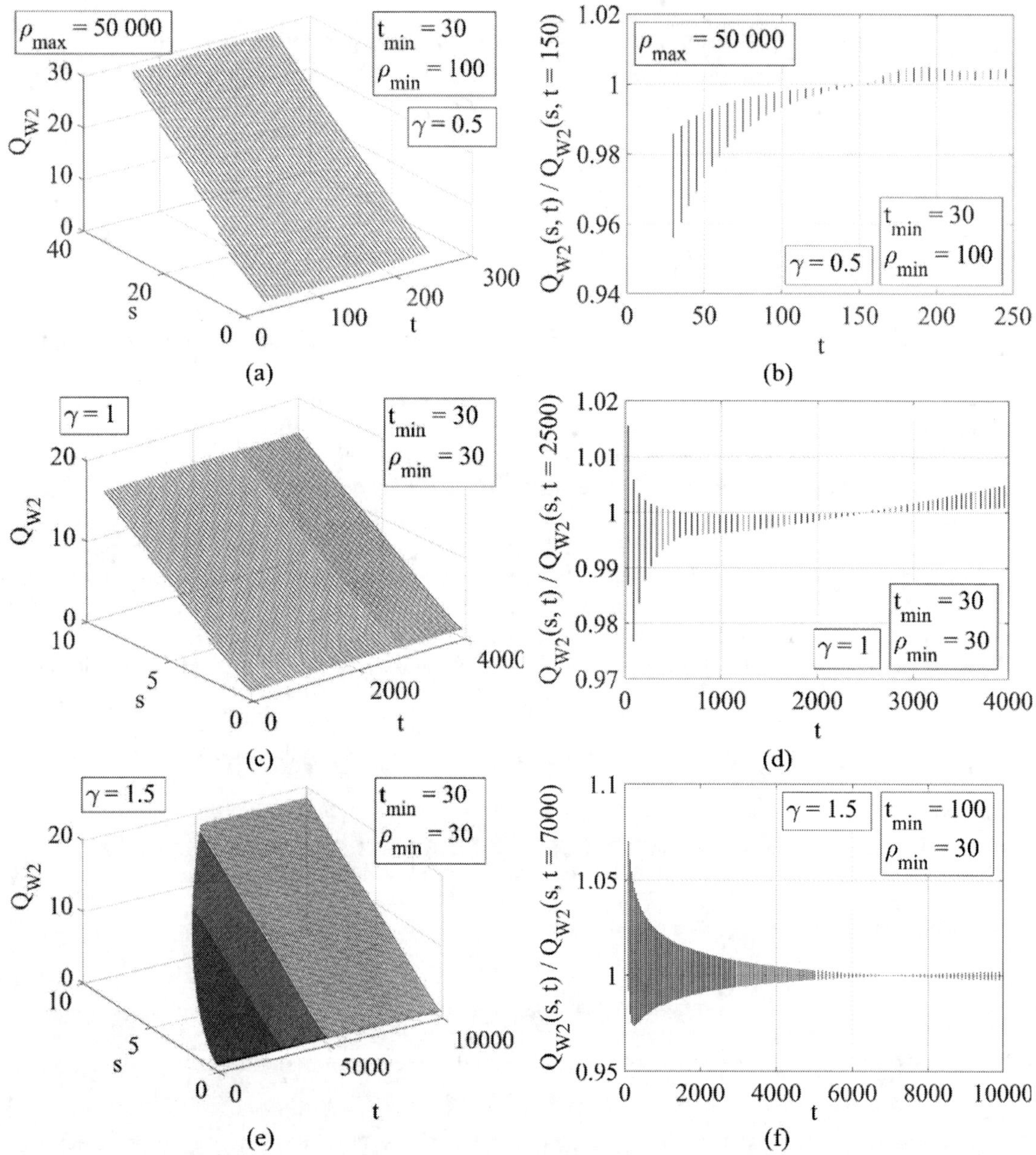


**Figure 2.** Function $Q_{W2}$ for $\rho > \rho_{min}$, $t > t_{min}$ for different $\gamma$ values: (a) $\gamma = 0.5$; (c) $\gamma = 1$; (e) $\gamma = 1.5$. Projection of the relation $Q_{W2}(s, t)/Q_{W2}(s, t^*)$ in the plane $\{Q_{W2}, t\}$ for different $\gamma$ values: (b) $\gamma = 0.5$ for $\rho > 100$, $t > 30$, $t^* = 150$; (d) $\gamma = 1$ for $\rho > 30$, $t > 30$, $t^* = 2500$; (f) $\gamma = 1.5$ for $\rho > 30$, $t > 100$, $t^* = 7000$. The dependence of the error, which is given by the deviation from unity, on the $t_{min}$ value is shown in the right-hand-side figures. The range of variable $s$ is limited by the conditions $\rho < \rho_{max}$.

It is seen from figure 2 that for given values of $\gamma$ the function (11) is indeed an automodel solution of equation (8) with the PDF of equation (9) with an accuracy defined by the coincidence of the curves $Q_{W2}(s, t)$, for various values of the time $t$. According to exact asymptotic solutions, used for constructing solution (10), one may expect high accuracy of the respective solution (11) for large enough values of dimensionless time and distance from the source. Indeed, we have the following numbers for the maximum deviation of $Q_{W1}$ and $Q_{W2}$ functions from a function of $s$ only. For instance, for $\gamma = 1$ this deviation amounts to 3% for $\{\rho > 30, t > 30\}$ and 1% for $\{\rho > 30, t > 270\}$ and $\{\rho > 100, t > 30\}$. The accuracy for a weaker tail appears to be worse: for $\gamma = 1.5$ and $\rho > 30$ the deviation amounts to 12%, 7% and1% for, respectively, $t > 30$, t > 100, and $t > 2400$. For a stronger tail, $\gamma = 0.5$, the propagation front moves substantially faster, and high accuracy is achieved at larger distances: 5% for $\rho > 100$, $t > 30$ and 1% for $\rho > 100$, $t > 100$. The largest deviations take place at $s \sim 1$.

## 4. Conclusions

The results of Sec. 2 and 3 shows that the identification of main features of nonlocal (superdiffusive) mechanism of transport of resonance radiation may be fruitful for extending the method of approximate automodel solution to other problems. There is an example of physics model which gives precisely the power-law PDF of Eq. (9): see Eq. (1) in [6] for photon-assisted transport of minority carriers in semiconductors (photo-excited holes in n-type InP [7]). The simple long-tailed step-length probability distribution function (9) with various power-law exponents may serve a bridge between physics and other fields. The general features of superdiffusion based on Lévy flights are recognized and applied in many fields (see, e.g., [1], [8]). Derivation of scaling laws, and especially of approximate automodel solutions, may be of practical interest.

**Acknowledgments**
This work is partially supported by the Russian Foundation for Basic Research (project RFBR-15-07-07850-a).

## References

[1] Kukushkin A B, Sdvizhenskii P A 2014 *Proc. 41st EPS Conference on Plasma Physics (Berlin)* P4.133 http://ocs.ciemat.es/EPS2014PAP/pdf/P4.133.pdf
[2] Mandelbrot B B 1982 *The Fractal Geometry of Nature* (New York: Freeman)
[3] Shlesinger M, Zaslavsky G M and Frisch U. (ed) 1995 *Lévy Flights and Related Topics in Physics* (New York: Springer)
[4] Veklenko B A 1959 *Soviet Phys. JETP* **9** 138
[5] Kukushkin A B and Sdvizhenskii P A 2016 *J. Phys. A: Math. Theor.* **49** 255002
[6] Subashiev A V, Semyonov O, Chen Z, Luryi S 2014 *Phys. Lett. A* **378** pp 266–269
[7] Luryi S, Semyonov O, Subashiev A V, Chen Z 2012 *Phys. Rev. B* **86** 201201(R)
[8] Klafter J, Sokolov I M 2005 Anomalous diffusion spreads its wings *Physics world* **18** p 29

# Hydrodynamic instability at the interface between colliding inhomogeneous mediums

**Kirill Gorodnichev[1], Sergey Kuratov[1], Pavel Zakharov[1] and Evgenii Gorodnichev[2]**

[1] The Centre for Fundamental and Applied Research, Dukhov Research Institute of Automatics, ul. Suschchevskaya, 22, 127055, Moscow, Russia
[2] National Research Nuclear University MEPhI, Kashirskoe sh., 31, 115409, Moscow, Russia

E-mail: cyrgo85@gmail.com

**Abstract.** This paper deals with the analysis of the perturbation growth at the interface between the colliding mediums, one of them has an initial perturbation field of density. The main difference of this hydrodynamic instability from the classical Richtmyer–Meshkov instability is the nature of the initial disturbance. The analyzed hydrodynamic instability is caused by the initial perturbation of the medium density.

The time evolution of interface between two mediums has been studied analytically in nonlinear approximation. The problem has been solved for the third-order corrections to hydrodynamic quantities. Nonlinear interactions cause the appearance of surface sound wave near the contact discontinuity. This wave decreases the growth of the spike velocity. Spike and bubble evolutions have a saturation stage.

The numerical calculations confirm the results of the theoretical analysis.

## 1. Introduction

The perturbation growth at the interface between mediums was studied in the last half century in numerous theoretical and experimental works [1–11]. Most of them deal with the analysis of Richtmyer–Meshkov instability (RMI) in the linear approximation [3–5, 9]. The analysis in the nonlinear approximation is submitted in [6, 7, 10]. The time evolution of the interface is considered, providing the interface has the curved form at the initial moment.

In the article [11] the interface instability between the mediums called Richtmyer–Meshkov-like (RMLI) is analyzed. The main difference of this hydrodynamic instability from the classical RMI is the nature of the initial disturbance. The RMLI is caused by the initial perturbation of the medium density. The authors of this paper argue that they can describe the development of this hydrodynamic instability at the interface between the mediums by their theory [12]. But the description of this development has not been obtained yet.

The problem of the collision of two mediums, one of which has the initial perturbation field of density, was studied in [13] in the linear approximation. The evolution of hydrodynamic instabilities in the mediums has been analyzed.

In this paper we deal with the development of the hydrodynamic instability at the interface between the mediums. This hydrodynamic instability is similar to RMLI. The common origin of these instabilities is caused by the following. During the impact of two mediums two shock waves begin to propagate in different directions within them. The interaction of shock wave

IOP Conf. Series: Journal of Physics: Conf. Series **941** (2017) 012051 doi:10.1088/1742-6596/941/1/012051

with the perturbation of the density is the cause of the entropy-vortex perturbations. These entropy-vortex perturbations cause the perturbation growth at the interface.

In this paper we analyze the perturbation growth at the interface between the mediums in the nonlinear approximation. Further this problem is solved for the first-order, the second-order and the third-order corrections to hydrodynamic quantities. We use the solution of this problem in the linear approximation obtained in [13].

Carrying out nonlinear analyses is caused by the following. The authors of the paper [14, 15] suggest that the perturbation growth at the interface between the mediums is valid till a certain time moment. At the last stage the velocity growth reaches saturation and the velocity stops changing. This conclusion is based on experimental and numerical results. In this paper we confirm this conclusion and analytically demonstrate that the velocity saturation has been obtained only if we take into account the third-order correction to hydrodynamic quantities.

## 2. Theoretical analysis

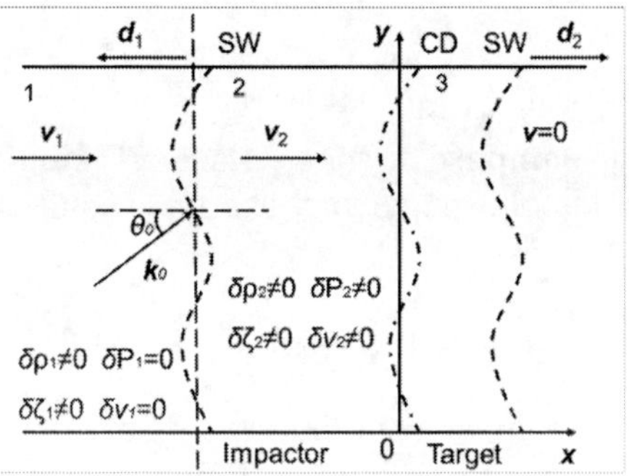


**Figure 1.** Statement of the problem.

The collision of two semi-infinite plates one of which is at rest (see Fig. 1) is studied in the paper. A spatially anisotropic density perturbation field initially exists in the incident plate (the impactor).

It is assumed that the velocity of the impactor is rather high (about 5 km/s). For this reason, the elastoplastic properties of the colliding plates can be neglected and the interaction of the shock wave with the density perturbation field in the impactor can be described by the equations of hydrodynamic [16]. The system of the equations of hydrodynamics should be supplemented by the Rankine–Hugoniot conditions at the shock front and equation of the medium state. Equation of the medium state is used in the form of the Mie–Gruneisen equation [17].

As a result of the colliding event we obtain two shock waves diverging in both plates from the contact discontinuity see Fig. 1. At the interaction of the density perturbation field with the shock wave, acoustic and entropy–vortex waves are formed behind the shock front [16]. Our analysis is carried out in the contact discontinuity system of coordinates, $x$ axis being directed towards the movement of the impactor (in the right direction) and orthogonal $y$ axis being directed upwards (see Fig. 1). Let the initial density perturbation field in the impactor be described by the expression:

$$\delta\rho_1(x,y) = A_{\delta\rho_1} \exp\left(ik_{0x}x + ik_{0y}y\right). \tag{1}$$

The axis of anisotropy for such perturbation is directed along the vector $\mathbf{k}_0 = (k_{0x}, k_{0y})$ (see Fig. 1). The shock wave moves from the right at the velocity $(d_1 - v_1)$ in the negative direction of the $x$ axis $(d_1 - v_1) < 0$. The angle between the vector $\mathbf{k_0}$ and the $x$ axis is $\theta_0$ ($\tan\theta_0 = k_{0y}/k_{0x}$).

It was shown [13] that the regime of perturbation propagation in mediums depends on the angle $\theta_0$. In this paper we consider the angle $\theta_0$ in the entire range $(\theta_0^*, \pi/2)$, where $\theta_0^* = \operatorname{atan}\left\{(v_1 - d_1)/\sqrt{c_2^2 - (v_2 - d_1)^2}\right\}$. In this case only the entropy-vortex wave is formed near the contact discontinuity.

To solve the system of the hydrodynamic equations for perturbation of hydrodynamic values we need to use the condition of the velocity equality on the contact discontinuity:

$$\begin{aligned} \delta v_{2x} &+ \delta'' v_{2x} - ik_{0y}\xi\delta v_{2y} + \delta''' v_{2x} - ik_{0y}\xi\delta'' v_{2y} - 2ik_{0y}\delta v_{2y}\delta''\xi = \\ &= \delta v_{3x} + \delta'' v_{3x} - ik_{0y}\xi\delta v_{3y} + \delta''' v_{3x} - ik_{0y}\xi\delta'' v_{3y} - 2ik_{0y}\delta v_{3y}\delta''\xi, \end{aligned} \tag{2}$$

where $\xi$ – the contact discontinuity perturbation. We use the following expression for the third-order correction to $\mathbf{n} = \left(1 + 0.5k_{0y}^2\xi^2 - 2k_{0y}^2\xi\delta''\xi; -ik_{0y}\xi - 2ik_{0y}\delta''\xi - i(2+\varepsilon)k_{0y}\delta'''\xi\right)$. The expression being used $\delta''\xi \sim \exp\{2ik_{0y}y\}$, $\delta'''\xi \sim \exp\{i(2+\varepsilon)k_{0y}y\}$. Constant $\varepsilon = \pm 1$. Using $d\xi/dt = \delta v_{2x} + \delta'' v_{2x} + \dots$ we have the following expressions $\xi = \delta v_{2x}t$, $\delta''\xi = 0.5\left(\delta'' v_{2x} + ik_{2x}\left(\delta v_{2x}\right)^2\right)t$.

### *2.1. Linear analysis*

It follows from the linear analysis [13] that the $x$ projection of the perturbation velocity of contact discontinuity is given by the following expression:

$$\delta v_{2x} = A_{\delta v_{2x}} \exp\left\{ik_{0y}y\right\}, \; A_{\delta v_{2x}} = \text{const}. \tag{3}$$

### *2.2. Nonlinear Analysis*

The area of influence of contact discontinuity deformation has finite width. So the deformation can only cause the sound wave appearance. The phase of the sound wave remains on the contact discontinuity ($x = \xi$), therefore we obtain the following $\Omega_{2,3} = \kappa_{2,3\,x}\delta v_{2x}$, where $\Omega$, $\kappa$ are the frequency and the wave vector of the sound wave in the target and the impactor, accordingly. The expression being used $x_{CD} = \delta v_{2x}t$.

It follows from dispersion relation $\Omega_{2,3} = \sqrt{\kappa_{2,3\,x}^2 + 4k_{0y}^2}c_{2,3}$ that $\kappa_{2,3\,x} = \mp 2ik_{0y}$ and $\Omega_{2,3} = \mp 2ik_{0y}\delta v_{2x}$ . We take into account the spatial damping of sound wave. Taking into account time increasing quadratic terms in (2) we have the following expression for second-order correction to the velocity:

$$\delta'' v_{2x} = -i\rho_3\left(k_{2x} - k_{3x}\right)\left(A_{\delta v_{2x}}\right)^2 \exp\left(2ik_{0y}y\right)t/\left(\rho_2 + \rho_3\right). \tag{4}$$

We applied $x \to x + v_{2x}t$ for contact discontinuity system of coordinates. The $k_{3x}$ value is obtained using the solution of the equations of hydrodynamics and the Rankine–Hugoniot conditions at the shock front in the target.

Taking time increasing ($\sim t^2$) terms in (2) into account we have the following for the third-order correction to the velocity:

$$\delta''' v_{2x} = \frac{\rho_2}{\rho_1 + \rho_2}\left[-i\left(\frac{\rho_2 - \rho_3}{\rho_2 + \rho_3}k_{0y} - i\left(\frac{\rho_3\left(k_{2x} - k_{3x}\right)}{\rho_2 + \rho_3} - k_{2x}\right)\right)\left(k_{2x} - k_{3x}\right)\left(\delta v_{2x}\right)^3 t^2 + \right.$$
$$\left. + \delta''' v_{3\,x}^{(se)} - \delta''' v_{2\,x}^{(se)}\right] \exp\left\{i\left(2 + \epsilon\right)k_{0y}y\right\}, \tag{5}$$

where $\kappa_{2,3x}^{(3)} = \kappa_{2,3x} + \varepsilon k_{2,3x}$, $\kappa_{2,3\,y}^{(3)} = \left(2 + \epsilon\right)k_{0y}$, $\gamma_{2,3\,x} = \left(\varepsilon - 2\right)k_{2,3\,x} + 2k_{0y}^2\epsilon/\kappa_{2,3x}$, $\gamma_{2,3\,y} = -\left(\epsilon k_{2,3\,x}^2/k_{0y} + 2k_{2,3\,x}k_{0x}/\kappa_{2,3\,x}\left(2 + \epsilon\right)\right)$, $\delta''' v_{2,3\,x}^{(se)} = 0.5\left(\gamma_{2,3\,y}\kappa_{2,3\,x}^{(3)}\kappa_{2,3\,y}^{(3)} - \gamma_{2,3\,x}\left(\kappa_{2,3\,y}^{(3)}\right)^2\right)/\left(\left(\kappa_{2,3\,x}^{(3)}\right)^2 + \left(\kappa_{2,3\,y}^{(3)}\right)^2\right)\delta v_{2x}\delta'' v_{2,3x}$.

### *2.3. Discussing results*

As it follows from the analytical analysis nonlinear interactions cause the appearance of the surface sound wave near the contact discontinuity. This wave has the following dispersive ratio $\Omega_2 = \kappa_2 c_2$ and causes to take into account the second-order and the third-order corrections to hydrodynamic quantities. The $x$ projection of the velocity perturbation of contact discontinuity

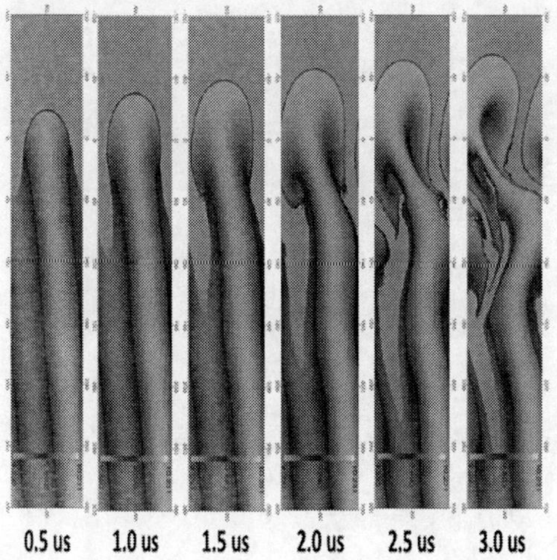


**Figure 2.** Profile of ferrum density with $\theta_0 = 1.471$ at different time moments. The color of aluminium is grey.

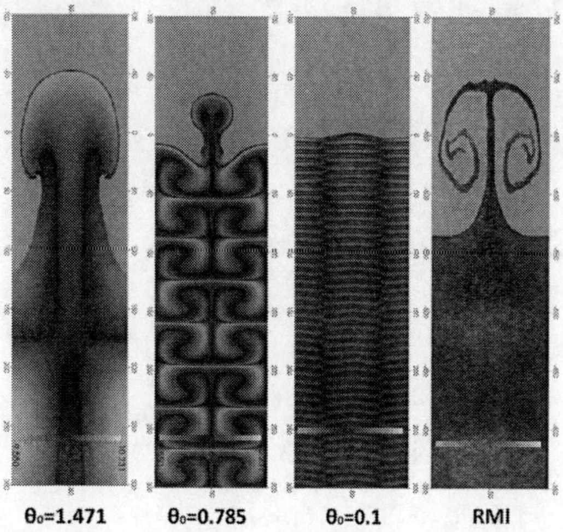


**Figure 3.** Density profile of ferrum for different angles $\theta_0$. The color of aluminium is grey.

has the following final expression: $v_{CD} = \delta v_{2x} + \delta'' \tilde{v}_{2x} t + \delta''' \tilde{v}_{2x} t^2$, where $v_x$, $\delta'' \tilde{v}_{2x}$, $\delta''' \tilde{v}_{2x}$ from (3), (4), (5), accordingly.

As we take into account the second-order and the third-order corrections to the hydrodynamic quantities the bubble velocity growth is a monotonically increasing function of time till a certain time moment. This time moment coincides with the maximum value of the bubble velocity, $V_{b,max}$. Following [14, 15], we consider that this time moment is the bubble saturation time, $t_b$. It is suggested that the bubble velocity is equal $V_{b,max}$ for $t > t_b$.

The spike velocity growth is decreasing function of time. So, we can not obtain the value of spike saturation time, $t_s$. In this paper we consider two different values of saturation time.

## 3. Numerical calculations

We analyze the collision of two semi-infinite aluminium and ferrum plates. The ferrum plate has the initial perturbation field of density (1) with $\theta_0 = 1.471$, $A_{\delta\rho_1}/\rho_1 = 0.05$. The numerical analysis has been carried out in the contact discontinuity system of coordinates. The colliding velocity value is 5 km/s. The initial position of contact discontinuity is $x = 0$. This statement of the problem is equivalent to the task in the previous section. The calculations have been performed with the method from [18].

Fig. 2 demonstrates the numerically calculated profile of ferrum density near the interface at different time moments. The color of aluminium is grey. According to Fig. 7 the value of the bubble velocity is twice as many as the value of the spike velocity, approximately.

Fig. 6 shows the analytically calculated time profile of the bubble and the spike velocity obtained for the first-order, the second-order and the third-order corrections to hydrodynamic quantities. Dashed segment of the red curve corresponds to time interval $t > t_b$. The results of the theoretical analysis of the bubble velocity are in good agreement with the results obtained from the numerical calculation (see Fig. 7).

Fig. 7 demonstrates the spike and bubble amplitude as a function of time. The spike development has two stages. At the first stage ($t < t_s$) the spike velocity is decreasing function of time. At the second stage ($t \geq t_s$) spike velocity is slowly decreasing function of time. Assuming

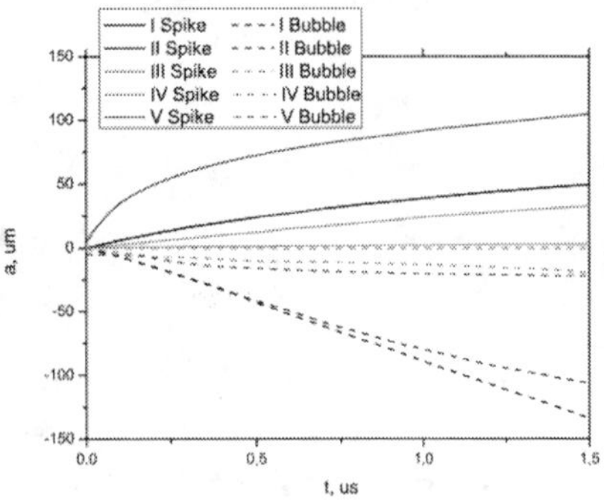


**Figure 4.** The bubble and the spike amplitude as a function of time for I – $\theta_0 = 1.471$, $\delta\rho_1$ – (1); II – $\theta_0 = 1.471$, $\delta\rho_1$ – (6); III – $\theta_0 = 0.785$, $\delta\rho_1$ – (6); IV – $\theta_0 = 0.1$, $\delta\rho_1$ – (6); V – RMI.

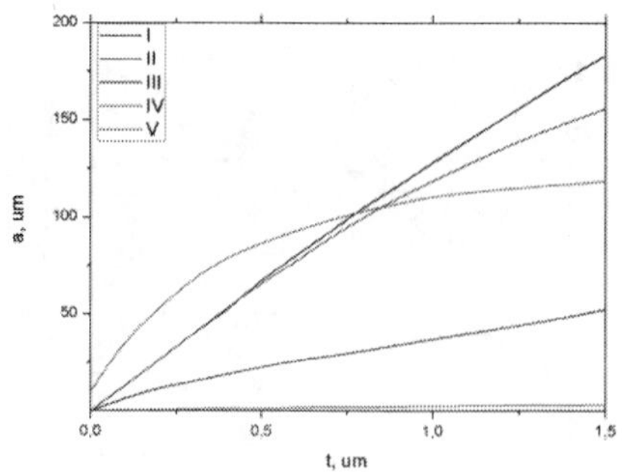


**Figure 5.** Mixing width as a function of time for I – $\theta_0 = 1.471$, $\delta\rho_1$ – (1); II – $\theta_0 = 1.471$, $\delta\rho_1$ – (6); III – $\theta_0 = 0.785$, $\delta\rho_1$ – (6); IV – $\theta_0 = 0.1$, $\delta\rho_1$ – (6); V – RMI.

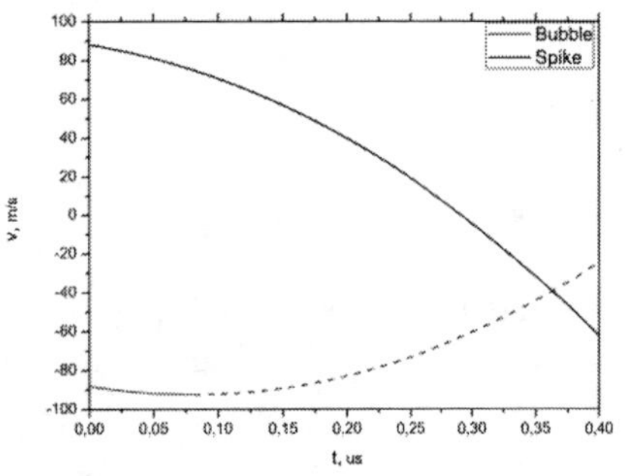


**Figure 6.** The analytically calculated time profile of the bubble velocity and the spike velocity obtained for the first-order, the second-order and the third-order corrections to hydrodynamic quantities.

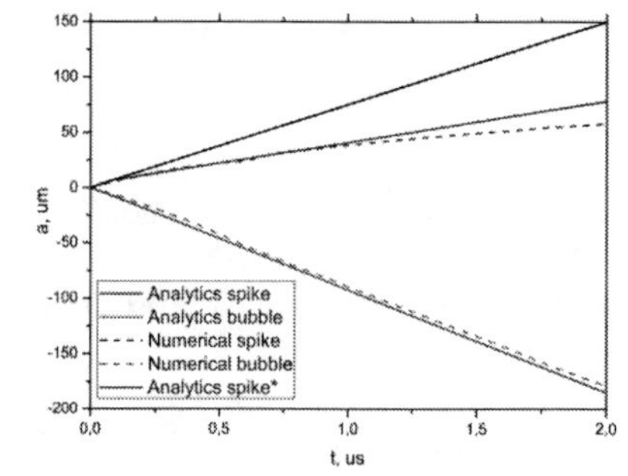


**Figure 7.** Time dependence of spike and bubble amplitude. Black line – $t_s = t_b$; blue line – $t_s = 2.5t_b$.

$t_s = 2.5t_b$ we obtain a good agreement between analytical and numerical results.

The analysis of these numerical calculations shows that the development of instability at the interface differs from the classical case of RMI. According to Fig. 2 the direction of the perturbation development is orthogonally to the wave vector of the initial perturbation (1). In the case of the classical RMI the direction of the perturbation development is orthogonally to the interface (see Fig. 3).

To compare RMLI and RMI we numerically analyze the case of RMLI with the following initial density perturbation of ferrum plate:

$$\delta\rho_1(x, y) = A_{\delta\rho_1}(\cos(k_{0x}x + k_{0y}y) + \cos(k_{0x}x - k_{0y}y)), \qquad (6)$$

with $A_{\delta\rho_1}/\rho_1 = 0.05$. In this case the density field has a staggered structure. For the case of the RMI we take $a/\lambda = 0.05$, $a$ is the perturbation amplitude, $\lambda$ is the perturbation wavelength of contact discontinuity. Interface velocity for RMI and RMLI cases is equal. Fig. 3 shows the numerically calculated profile of density near the interface at the time moment $t = 1.7$ us for the different angles $\theta_0$. Fig. 4 demonstrates the bubble and the spike amplitude as a function of time. Fig. 5 demonstrates the mixing width as a function of time.

Let we compare results of the initial density distribution for two cases, (1) and (6), ($\theta_0 = 1.471$). The mixing width development is identical at the initial stage. At the later stage in case of (6) the bubble velocity decreases due to interaction between the bubble and the neighbouring entropy cell formed behind the shock wave (see Fig. 4). The spike growth is equivalent in both cases, provided perturbations are absent in aluminium plate ($\theta_0 = 1.471$). We note that the mixing width depends on value of $\theta_0$.

## 4. Conclusions

The agreement between the analytical result and numerical calculation has been obtained only due to the fact that the third-order correction to hydrodynamic quantities is taken into account. As it follows from the obtained results the second-order and the third-order corrections to the velocity are proportional to $t$ and $t^2$, accordingly. We describe the anisotropy of the process using nonlinear corrections to hydrodynamic quantities. It is shown that bubble amplitude grows with time faster than spike amplitude. The bubble velocity is a monotonous function of time till a certain time moment. This time moment coincides with the maximum value of the bubble velocity. At the last stage the bubble amplitude growth reaches saturation and the bubble velocity stops changing. Taking into account the the second and the third-order correction we can not obtain the value of spike saturation time. It is numerically shown that the spike saturation time is larger than bubble saturation time. Appearance of vortex cells depends on initial conditions, (1) and (6). These cells influence the perturbation growth of contact discontinuity.

## Acknowledgments

The work one of the authors (E.G.) was supported by the Center of Applied Mathematics and Theoretical Physics, National Research Nuclear University MEPhI (project no. 02.a03.21.0005, August 27, 2013).

## References

[1] Richtmyer R 1960 *Comm. on Pure and Appl. Math.* **XIII** 297–319
[2] Meshkov E 1970 *NASA Tech. Trans.* **NASA TT F-13**
[3] Yang Y, Zhang Q and Sharp D 1994 *Phys. Fluids* **6** 1856–1873
[4] Wouchuk J and Lopez Cavada J 2004 *Phys.Rev.E* **70** 046303
[5] Velikovich A, Wouchuk J, Huete Ruiz de Lira C, Metzler N, Zalesak S and Schmitt A 2007 *Phys. Plasmas* **14** 072706
[6] Nishihara K, Wouchuk J, Matsuoka C, Ishizaki R and Zhakhovsky V 2010 *Phil. Trans. R. Soc. A* **368** 1769–1807
[7] Matsuoka C, Nishihara K and Fukuda Y 2003 *Phys. Rev. E* **67** 036301
[8] Prestridge K, Orlicz G, Balasubramanian S and Balakumar B 2013 *Phil. Trans. R. Soc. A* **371** 20120165
[9] Griffond J 2006 *Phys. Fluids* **18** 054106
[10] Brouillette M 2002 *Annu. Rev. Fluid Mech.* **34** 445–468
[11] Velikovich A, Dahlburg J, Schmitt A, Gardner J, Phillips L, Cochran F, Chong Y, Dimonte G and Metzler N 2000 *Comm. on Pure and Appl. Math.* **7** 1662–1671
[12] Wouchuk J, Huete Ruiz de Lira C and Velikovich A 2009 *Phys. Rev. E* **79** 066315
[13] Gorodnichev K, Kuratov S and Gorodnichev E 2017 *J. of Phys.: Conf. Ser.* submitted
[14] Berning M and Rubenchik A 1998 *Phys. of Fluids* **10** 1564–1587
[15] Jacobs J and Catton I 1988 *J. Fluid Mech.* 329
[16] Landau L D and Lifshitz E M 1987 *Fluid Mechanics* (New York: Pergamon)
[17] Zababakhin E 1997 *Nekotorye voprosy gazodinamiki vzryva* (Snezhinsk [in Russian])
[18] Menshov I and Zakharov P 2014 *Int. J. Numer. Meth. Fluids* **76** 109

IOP Conf. Series: Journal of Physics: Conf. Series **941** (2017) 012052 doi:10.1088/1742-6596/941/1/012052

# Response of explosive HMX to low-velocity impact: modeling by the crystal plasticity finite element method

**Denis Ilnitsky**[1,2], **Nail Inogamov**[1,2], **Vasily Zhakhovsky**[1,2]

[1] Dukhov Research Institute of Automatics (VNIIA), 22, ul. Sushchevskaya, Moscow 127055, Russia

[2] L.D. Landau Institute for Theoretical Physics, 1a, ul. Akademika Semenova, Chernogolovka, Moscow Region, 142432, Russia

E-mail: denisilnitskiy@gmail.com

**Abstract.** Crystal plasticity finite element method (CPFEM) is a powerful tool for modeling the various deformation problems, which takes into account the different plasticity mechanisms at microscale of grain sizes and contribution of anisotropic behavior of each grain to macroscopic deformation pattern. Using this method we simulated deformation and plasticity of high explosive HMX produced by relatively low velocity impact. It was found that such plastic deformations of grains cause local heating which is sufficient to induce chemical reactions.

## 1. Introduction

Safety of different devices, which contain high explosives (HE) like octahydro-1,3,5,7-tetranitro-1,3,5,7-tetrazocine (HMX), is determined in many aspects by chemical reactions initiated by impact on HE with different velocities. The impact loadings may lead to high pressure (more than 10 GPa) or to low pressure (less then 1 GPa). Many phenomenological models were developed for high-pressure impacts [1, 2]. Those models are widely used for HE reactions predictions in engineering practice. But the commonly accepted models for low-pressure impacts are lacking. This fact is connected with complex and multiscale processes involved to studied phenomena. Complexity of determination of the leading process stems from many reasons for generation of detonation centers such as HE heterogeneity, HE deformation process instability, and friction of material grains.

Some researchers consider adiabatic shear bands formation during shear loading of HE as the main mechanism of the non-homogeneous high-temperature heating which leads to detonation [3, 4]. There are several phenomenological models for describing this process [3–5]. The main weakness of those models is the usage of a number of artificial parameters which are not based on clear physical picture. For clarifying physical processes we have to obtain the non-homogeneous pattern of heating HE explicitly. Several authors made attempts to receive the non-homogeneous pattern inside the separate grains of material with using a model for metals only [6]. Here we present the CPFEM simulations of impact loading of HE with taking into account anisotropic orientation of HE grains and their mutual interaction. As far as we know this approach is used for the first time.

## 2. Mathematical model of HE deformation

The elasto-plastic deformation of material grains conglomerate is known to be sufficiently anisotropic, and thus its behavior depends on loading direction. The anisotropic mechanical properties of the material is defined by its bulk anisotropic elastic characteristics and various plasticity deformation mechanisms (dislocations, twins and martensite transformations) for different spatial orientations of material grains. Complicated picture of elasto-plastic deformation is formed on account of anisotropic properties of material grains and its interaction.

For accurate calculation of mutual interaction between HE grains, HE grains and plasticizer, and to account the various plastic responses of individual grains (in particular dislocation dynamics inside) in modeling we have to choose a method which able to account those deformation processes. Three decades ago the finite element method of crystal plasticity [7] was proposed with the aim to model deformation processes in metals and alloys, where anisotropic properties of material grains play a significant role in material response.

Here we apply this method to consider deformation process in HE. It allows us to account such processes as mutual interactions between grains as well as interactions between HE grains and plasticizer. The method makes feasible the calculation of plastic response of each grain including dynamics of dislocations and other processes involved in plasticity. The last processes are twinning, non crystallographic adiabatic shear, interaction of grain boundary with dislocations, twin-dislocation interactions, interaction between dislocations of opposite signs and etc. All of them can be implemented in finite-element model of crystal plasticity using a set of kinetic equations. Lets consider an implementation of dislocation plasticity, adiabatic shear and twinning, proposed in [8].

In this method a decomposition of deformation gradient $\mathbf{F}$ to factors is given as:

$$\mathbf{F} = \mathbf{F_e}\mathbf{F_p}, \tag{1}$$

where $\mathbf{F_e}$ is an elastic part and $\mathbf{F_p}$ is a plastic part.

Plastic deformation varies as:

$$\dot{\mathbf{F_p}} = \mathbf{L_p}\mathbf{F_p}. \tag{2}$$

In the case of plasticity produced by dislocation motion only:

$$\mathbf{L_p} = \sum_{\alpha=1}^{N} \dot{\gamma^{\alpha}} \mathbf{m}^{\alpha} \otimes \mathbf{n}^{\alpha}, \tag{3}$$

where $\mathbf{m}^{\alpha}$ and $\mathbf{n}^{\alpha}$ are unit vectors, which describe a dislocation slip direction and a normal to slip plane for slip system $\alpha$, respectively; $N$ is a number of slip systems, and $\dot{\gamma}^{\alpha}$ - a shear rate for this system.

Equation (3) is expanded by additional terms:

$$\mathbf{L_p} = \sum_{\alpha=1}^{N} \dot{\gamma^{\alpha}} \mathbf{m}^{\alpha} \otimes \mathbf{n}^{\alpha} + \sum_{\beta=1}^{M} \dot{\gamma^{\beta}} \mathbf{m}^{\beta}_{twin} \otimes \mathbf{n}^{\beta}_{twin} + \sum_{\chi=1}^{P} \dot{\gamma^{\chi}} \mathbf{m}^{\chi}_{sb} \otimes \mathbf{n}^{\chi}_{sb}, \tag{4}$$

which account for twinning and formation of non-crystallographic shear bands. For describing crystal microstructure, it is necessary to define expressions for shear rates of different plasticity mechanisms. For this purpose the small scale calculations can be used. For example in [8], to account the non-crystallographic deformation due to adiabatic shear banding the kinetic formulae discussed belowere used. At first the next parameters introduced for microstructure description are $\rho_{sgl}$, $\rho_{dip}$ - dislocation density and dislocation dipole density, respectively, and $f$ - twins volumetric density. Shear rate of shear system $\alpha$ with taking into account the thermoactivated dislocation motion through "forest" of dislocations from other slip planes is given by:

$$\dot{\gamma^\alpha} = \rho^\alpha_{sgl} b v_0 exp\left[\left(-\frac{Q_0}{k_B T}\left(1-\frac{|\tau^\alpha|}{\hat{\tau^\alpha}}\right)^p\right)^q\right] sign(\tau^\alpha), \tag{5}$$

where $\tau^\alpha$ - shear stress, $\hat{\tau}^\alpha$ - resistance of "forest" dislocation, $b$ - the Burgers vector, $v_0$ - dislocation velocity of current slip system at shear stress equal to $\hat{\tau^\alpha}$, $Q_0$ - activation energy of dislocation slip, $k_B$ and $T$ - the Boltzmann constant and temperature, $p$ and $q$ - numerical parameters, which describe an obstacle profile for dislocation motion. Slip resistance stress $\hat{\tau}^\alpha$ depends on local dislocation density as:

$$\hat{\tau^\alpha} = \tau_{solute} + Gb\left(\sum_{\alpha'=1}^{N_{slip}} \xi_{\alpha\alpha'}\left(\rho^{\alpha'}_{sgl} + \rho^{\alpha'}_{dip}\right)\right)^{1/2}, \tag{6}$$

where $\tau_{solute}$ - a resistance constant, $G$ - shear modulus, $\xi_{\alpha\alpha'}$ - parameter characterizing interaction stress between all possible dislocation types (mobile, immobile, coplanar, collinear,orthogonal etc.). Twin system volumetric share $f^\beta$ change is based on a twin nucleation theory [8]. It is proposed that twin nucleation occurs at three partial dislocations bowing between three immobile points, where the distance between points is $L_0$. Twin formation critical stress is:

$$\hat{\tau}_{twin} = \frac{\gamma_{sf}}{3b_{twin}} + \frac{3Gb_{twin}}{L_0}, \tag{7}$$

where $b_{twin}$ - moving partial dislocations Burgers vector and $\gamma_{sf}$

$$\dot{N}^\beta = \dot{N}_0 exp\left[-\left(\frac{\hat{\tau}_{twin}}{\tau^\beta}\right)\right], \tag{8}$$

where $\tau^\beta$ - reduced shear stress in twin system, $N_0$ and $r$ - numerical parameters. We assume that any twin grows until contact with an obstacle such as grain boundary or other twin boundary. Volume of new twin:

$$V^\beta = \frac{\pi}{6} s \lambda^{\beta 2}, \tag{9}$$

here $s$ - constant twin thickness, $\lambda^\beta$ - effective distance between obstacles for twin growth:

$$\frac{1}{\lambda^\beta} = \frac{1}{d_{grain}} + \sum_{\beta'=1}^{M} \xi_{\beta\beta'} \frac{1}{d^{\beta'}_{twin}}, \tag{10}$$

$d_{grain}$ - constant grain size, $d_{twin}$ - twin size changes with volumetric twin fraction as:

$$d^\beta_{twin} = s(1-f)/f^{beta}, \tag{11}$$

where $f$ - a total volumetric twin fraction. Twin-twin interaction parameter $\xi_{\beta\beta'}$ equals zero for coplanar twin systems $\beta$ and $\beta'$ and unity for non coplanar. Thus, result deformation rate for each twin system calculated as:

$$\dot{\gamma}^\beta = (1-f)\gamma_{twin} V^\beta \dot{N}^\beta, \tag{12}$$

where $\gamma_{twin}$ - characteristic twin stress.

IOP Conf. Series: Journal of Physics: Conf. Series **941** (2017) 012052 doi:10.1088/1742-6596/941/1/012052

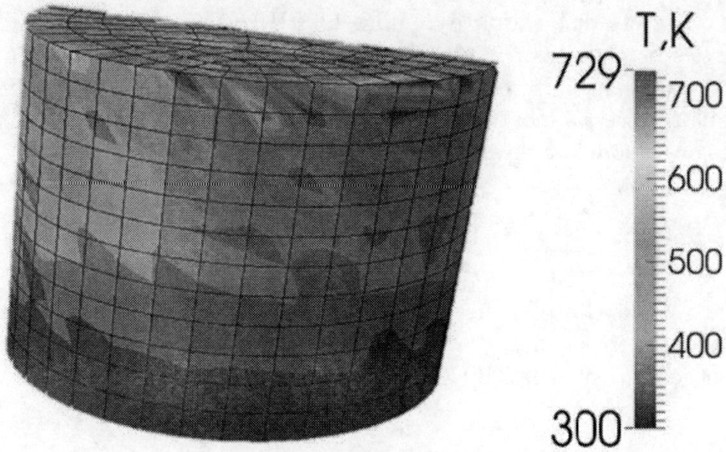


**Figure 1.** Temperature in Kelvin degrees at 4 microseconds.

**Figure 2.** Burn fraction at 4 microseconds.

A term describing the adiabatic shear bands is added because of non-crystallographic defects formation in alloys with low stacking fault, where shear dislocation motion is almost completely blocked by dense twin boundaries. For this processes shear rate is:

$$\dot{\gamma^{\chi}} = \dot{\gamma}_0^{\chi} exp\left[\left(-\frac{Q_0}{k_B T}\left(1-\frac{|\tau^{\chi}|}{\hat{\tau_{sb}}}\right)^p\right)^q\right] sign(\tau^{\chi}), \tag{13}$$

where $\dot{\gamma}_0^{\chi}$ - characteristic shear rate, $\tau^{\chi}$ - reduced stress at system $\chi$ and $\hat{\tau}_{sb}$ - a threshold stress for shear band formation.

The method described above is applied for HE deformation problem, in particular, for HMX. Such application is difficult due uncertainty of HE anisotropic elastic characteristics – they may differ in several times for HMX [9]. Dislocation motion parameters are completely

IOP Conf. Series: Journal of Physics: Conf. Series **941** (2017) 012052 doi:10.1088/1742-6596/941/1/012052

lacking in literature. For proper HE deformation modeling the many calculations at different spatial-temporal scales are required, such as quantum mechanics modeling for adjustment of HE anisotropic elastic properties, classical molecular dynamics for single dislocation motion parameters, and discrete dislocation dynamics for HE plastic behavior inside material grains. Material properties of plasticizer is also needed, because CPFEM allows to consider different material grain response in finite element node. In this work, the phenomenological plasticity model is used due to reasons mentioned above.

Phenomenological model of shear rate is:

$$\dot{\gamma}^{\alpha} = \dot{\gamma}_0 \left| \frac{\tau^{\alpha}}{\tau_c^{\alpha}} \right|^n sign(\tau^{\alpha}), \tag{14}$$

where $\dot{\gamma^{\alpha}}$ - shear rate at slip plane $\alpha$ under reduced stress $\tau^{\alpha}$ at slip resistance $\tau_c^{\alpha}$, $\dot{\gamma}_0$ and $n$ - material parameters of initial shear rate and shear sensitivity from shear rate, respectively.

Influence of other slip system $\beta$ on $\alpha$ system hardening is given by:

$$\dot{\tau}_c^{\alpha} = h_{\alpha\beta}|\dot{\gamma}^{\beta}|, \tag{15}$$

where $h_{\alpha\beta}$ - hardening matrix:

$$h_{\alpha\beta} = q_{\alpha\beta} \left[ h_0 \left( 1 - \frac{\tau_c^{\beta}}{\tau_s} \right) \right]. \tag{16}$$

For burn rate the Arrhenius-type relation is applied:

$$\frac{d\lambda}{dt} = Z(1-\lambda)exp\left(-\frac{E_a}{RT}\right), \tag{17}$$

where $T$ - temperature, $Z$ - kinetic rate constant, $E$ - activation energy per mole, $R$ - the universal gas constant, $\lambda$ - burn rate ($\lambda = 0$ means no reaction, $\lambda = 1$ corresponds to fully burned). These parameters for HMX were taken from [5].

All kinetic models mentioned above have been implemented in the LS-DYNA finite element code using user-defined subroutines. For finite element modeling the eight-node solid elements are used, ten grains randomly orientated were assigned in each computational point of element.

In order to obtain macroscopic deformation we need to average plastic response of all material grains in each finite element node. In our case isostrain method is used for averaging, which assumes equal deformation of each grain in finite element nodes [7].

Let's consider HE cylinder with height 1 cm deformed under motion of its upper plane with velocity 100 m/s. We apply the phenomenological model described above. With parameters $\gamma_0 = 10^8$ 1/s and $\tau_c = 6.3 * 10^7$ Pa at 4 microseconds the local heating above 400 K (Fig.1) and HE burning (somewhere completely) are obtained as shown on Fig.2. This result corresponds to the experimental data [5], where a steel ball with velocity 100 m/s impacts HE cylinder, which explodes at 4 microseconds after impact beginning. This result shows that proposed method is reasonable and applicable for description of low velocity impacts on HE.

## 3. Conclusions

We demonstrated that the process of HE explosive transformation is initiated by plastic deformation of HMX simulated by the crystal plasticity finite element method without the usage of artificial phenomenological parameters.

**Acknowledgments**
This work was supported by the Russian Science Foundation (project no. 14-19-01599). D.I. thanks Dr. Martin Diehl and Dr. Sergey Kuratov for fruitful discussions.

## References

[1] Lee E and Tarver C 1980 *Physics of Fluids* **23** 2362–2372
[2] Forest C 1981 *LA-8790 LANL report*
[3] Zerilli F, Guirguis R and Coffey C 2002 *12 International Detonation Symposium*
[4] Kerrisk J 1996 *LA-13127 LANL Report*
[5] Gushanov A, Volodina N, Belov G, Khvorostin N and Isheev D 2010 *Journal of Energetic Materials* 50–65
[6] Kubin L, Fressengeas C and Ananthakrishna G 2002 *Collective Behaviour of Dislocations in Plasticity in book: Dislocations in Solids, vol. 11* (Elsevier Science B.V.)
[7] Roters F, Eisenlohr P, Bieler T and Raabe D 2010 *Crystal Plasticity Finite Element Methods in Materials Science and Engineering* (Wiley-VCH,Weincheim)
[8] Jia N, Roters F, Eisenlohr P, Kords C and Raabe D 2012 *Acta Materialia* 1099–1115
[9] Li M, Tan W J, Kang B, Xu R J and Tang W 2010 *Propellants Explos. Pyrotech.* 81–99

IOP Conf. Series: Journal of Physics: Conf. Series **941** (2017) 012053 doi:10.1088/1742-6596/941/1/012053

# Hyperons, $\Delta$ resonances and condensate of charged $\rho$ mesons within relativistic mean-field models with scaled hadron masses and couplings

**K A Maslov**[1], **E E Kolomeitsev**[2], **D N Voskresensky**[1]

[1]National Research Nuclear University MEPhI (Moscow Engineering Physics Institute), Kashirskoe highway 31, 115409 Moscow, Russia
[2]Matej Bel University, Tajovského 40, 97401, Banská Bystrica, Slovakia

E-mail: maslov@theor.mephi.ru

**Abstract.** We study equation of state of cold and dense baryon matter within the relativistic mean-field framework with hadron masses and coupling constants dependent on the mean scalar field. Previously constructed models with included hyperons and $\Delta$ isobars are extended taking into account possibility of the condensation of charged $\rho$ mesons. We demonstrate that the results obtained in the models with the charged $\rho$-meson condensation taken into account exhibit a strong model dependence. In some of our (so-called KVORcut-based) models the charged $\rho$ condensation does not significantly affect the value of the neutron star maximum mass. In other (so called MKVOR-based) models the neutron star maximum mass decreases substantially. All thus constructed models pass the observational constraint on the minimal value of the maximum neutron star mass.

## 1. Introduction

Knowledge of the equation of state (EoS) of the cold and dense hadronic matter is required for the description of the neutron star (NS) matter. Nowadays there exists a vast number of EoSs and a large set of experimental and observational constraints which an appropriate EoS should satisfy [1]. None of the EoSs satisfies all the existing constraints. Recent measurement of two solar mass ($2M_\odot$) NS [2] rules out many soft EoSs. Possibility of the appearance of additional degrees of freedom in dense NS interiors, such as hyperons and $\Delta$ isobars or meson condensates, leads to a softening of the EoS and a decrease of the maximum NS mass. In our previous works [3, 4, 5] we constructed relativistic mean-field (RMF) models with effective hadron masses and coupling constants dependent on the scalar field with hyperons and $\Delta$ resonances taken into account, which successfully pass the majority of the constraints. However, it has been shown in [6, 7] that with increasing density in the isospin asymmetric matter the condensation of charged $\rho$ mesons becomes possible, provided the model includes the non-Abelian coupling of the $\rho$ fields and the decrease of the $\rho$ meson effective mass with increasing density. In this work we focus on the possibility of the charged $\rho$ meson condensation within our models and its effect on the EoS of the NS matter.

## 2. RMF models with charged $\rho$ condensation

We use the framework proposed in [7]. The model is a generalization of the non-linear Walecka model with effective coupling constants $g^*_{mb} = g_{mb}\chi_{mb}(\sigma)$ and hadron masses $m^*_i = m_i\Phi_i(\sigma)$ dependent on the scalar field $\sigma$. Here $m = \{\sigma, \omega, \rho, \phi\}$ denotes mesons, $b = (N, H, \Delta)$ lists baryon species, where $N$ denotes nucleons $\{p, n\}$, H stands for hyperons $\{\Lambda, \Sigma, \Xi\}$ and $\Delta$ denotes the $\Delta$-isobars, and index $i$ runs through all hadrons $(m, b)$, $\chi_{mb}(\sigma)$ and $\Phi_i(\sigma)$ are some scaling functions. With the charged $\rho$ meson condensation switched off, the standard solutions for mean meson fields can be obtained, which lead to the following energy density:

$$E[\{n_b\},\{n_l\},f] = \sum_b E_{\rm kin}(p_{{\rm F},b}, m_b\,\Phi_b(f), s_b) + \sum_{l=e,\mu} E_{\rm kin}(p_{{\rm F},l}, m_l, s_l) + \frac{m_N^4 f^2}{2C_\sigma^2}\eta_\sigma(f)$$

$$+\frac{1}{2m_N^2}\left[\frac{C_\omega^2 n^2}{\eta_\omega(f)} + \frac{C_\rho^2 n_I^2}{\eta_\rho(f)} + \frac{C_\phi^2 n_S^2}{\eta_\phi(f)}\right], \quad E_{\rm kin}(p_{\rm F}, m, s) = (2s+1)\int_0^{p_{\rm F}} \frac{p^2 dp}{2\pi^2}\sqrt{p^2+m^2}, \qquad (1)$$

$$n = \sum_b x_{\omega b} n_b\,, \quad n_I = \sum_b x_{\rho b} t_{3b} n_b\,, \quad n_S = \sum_H x_{\phi H} n_H\,,$$

where we use the dimensionless scalar field variable $f = g_{\sigma N}\chi_{\sigma N}(\sigma)\sigma/m_N$, the coupling constant ratios $x_{mb} = g_{mb}/g_{mN}$. The $t_{3b}$ is the isospin projection of baryon $b$, and the Fermi momentum of a fermion $j$ is $p_{{\rm F},j} = (6\pi^2 n_j/(2s_j+1))^{1/3}$, where $s_j$ is the fermion spin, $j = (b, l)$. In case of infinite matter meson coupling constants, masses and scaling functions enter the energy density only in combinations $C_M = g_{MN} m_N/m_M$, $M = \sigma, \omega, \rho$, $C_\phi = g_{\omega N}\, m_N/m_\phi$,

$$\eta_\omega(f) = \Phi_m^2(f)/\chi_{\omega N}^2(f)\,, \eta_\rho(f) = \Phi_m^2(f)/\chi_{\rho N}^2(f)\,, \eta_\sigma(f) = \frac{\Phi_\sigma^2[\sigma(f)]}{\chi_{\sigma N}^2[\sigma(f)]} + \frac{2\,C_\sigma^2}{m_N^4 f^2}U[\sigma(f)]\,,$$

where the scaling function $\eta_\sigma(f)$ is expressed in terms of the self-interaction potential $U(\sigma)$ entering the Lagrangian of the model, and we simplifying use $\Phi_N = \Phi_m = 1 - f$ (cf. Brown-Rho scaling), $\chi_{\phi H}(f) = 1$, $\chi_{\phi N}(f) = \chi_{\phi\Delta}(f) = 0$. The scaling function of the baryon mass is $\Phi_b(f) = 1 - x_{\sigma b}\xi_{\sigma b} m_N f/m_b$, where $\xi_{\sigma b} = \chi_{\sigma b}/\chi_{\sigma N}$, and we suppose that $\chi_{\omega b}(f) = \chi_{\omega N}(f)$, $\chi_{\rho b}(f) = \chi_{\rho N}(f)$. Explicit expressions for the scaling functions $\eta_M(f)$ are given in [3, 4, 5].

The vector-meson coupling constants to hyperons are chosen following the quark SU(6) symmetry:

$$x_{\omega\Lambda} = x_{\omega\Sigma} = 2x_{\omega\Xi} = \tfrac{2}{3}\,, \ x_{\rho\Sigma} = 2x_{\rho\Xi} = 2\,, \ x_{\phi\Lambda} = x_{\phi\Sigma} = x_{\phi\Xi} = -\tfrac{\sqrt{2}}{3}\,, \ x_{\rho\Lambda} = x_{\phi N} = 0. \qquad (2)$$

The hyperon coupling constants with the scalar field are deduced from the hyperon binding energies $\mathcal{E}^H_{\rm bind}$ in isospin-symmetric matter (ISM) at the saturation density $n = n_0$ given by $\mathcal{E}^H_{\rm bind}(n_0) = C_\omega^2 m_N^{-2} x_{\omega H} n_0 - m_N + m_N^*(n_0)$, and the empirical values $\mathcal{E}^\Lambda_{\rm bind}(n_0) = -28\,{\rm MeV}$, $\mathcal{E}^\Sigma_{\rm bind}(n_0) = 30\,{\rm MeV}$ and $\mathcal{E}^\Xi_{\rm bind}(n_0) = -15\,{\rm MeV}$. The value of the $\Delta$ potential $U_\Delta(n_0)$ is poorly constrained by the data. Here we use $U_\Delta(n_0) = -50\,{\rm MeV}$ as the most realistic estimate. We incorporate the $\phi$ meson field assuming the universal mass scaling with other hadrons, $\Phi_\phi = 1-f$, and taking the vacuum coupling constant $\chi_{\phi N} = 1$, that leads to $\eta_\phi(f) = (1-f)^2$. We denote models with hyperons included with this choice of $\eta_\phi$ by "H$\phi$" suffix.

The free parameters of the model are fitted to reproduce properties of nuclear matter near the saturation point. These properties are defined as the coefficients of the Taylor expansion of the energy per particle in the ISM in terms of $\epsilon = (n - n_0)/3n_0$ and $\beta = (n_n - n_p)/n$, $\mathcal{E} = \mathcal{E}_0 + \frac{K}{2}\epsilon^2 - \frac{K'}{6}\epsilon^3 + \beta^2\widetilde{J}(n) + \ldots$ and $\widetilde{J}(n) = \widetilde{J} + L\epsilon + \frac{K_{\rm sym}}{2}\epsilon^2 + \ldots$.

The $\rho$ meson part of the Lagrangian is as follows [6, 7],

$$\mathcal{L}_\rho = -\frac{1}{4}\vec{R}_{\mu\nu}\vec{R}^{\mu\nu} + \frac{1}{2}m_\rho^2\Phi_\rho^2\vec{\rho}_\mu\vec{\rho}^\mu - g_\rho\chi_\rho\vec{\rho}_\mu\vec{j}^\mu_I\,, \quad \vec{j}^\mu_I = \delta^{a3}\delta^{\mu0}\bar{\psi}\gamma^0\tau\psi\,, \qquad (3)$$

$$\vec{R}_{\mu\nu} = \partial_\mu\vec{\rho}_\nu - \partial_\nu\vec{\rho}_\mu + g'_\rho\chi'_\rho[\vec{\rho}_\mu\times\vec{\rho}_\nu] + \mu_{{\rm ch},\rho}\delta_{\nu0}[\vec{n}_3\times\vec{\rho}_\mu] - \mu_{{\rm ch},\rho}\delta_{\mu0}[\vec{n}_3\times\vec{\rho}_\nu]\,,$$

where $(\vec{n}_3)^a = \delta^{a3}$ is the unit vector in the isospin space and $\mu_{\mathrm{ch},\rho}$ is the chemical potential for charged mesons. From the hidden local symmetry arguments it follows that $g'_\rho = g_\rho$, which we adopt here. The new ansatz for $\rho$ meson field includes non-zero $\rho_0^{(3)}$ and $\rho_i^{\pm} = (\rho_i^{(1)} + \rho_i^{(2)})/\sqrt{2}$, $i = 1,2,3$ components, with the condition $\rho_i^{(+)}\rho_j^{(-)} - \rho_i^{(-)}\rho_j^{(+)} = 0$, which implies that the ratio $\rho_i^{(+)}/\rho_i^{(-)}$ is constant, independent of the spatial index $i$. Thus we assume $\rho_i^{(-)} = a_i\,\rho_c$ and $\rho_i^{(+)} = a_i\,\rho_c^{\dagger}$, where $\vec{a} = \{a_i\}$ is the spatial unit vector, and $\rho_c$ is a complex amplitude of the charged $\rho$ meson field. With this ansatz the $\rho$ field contribution to the thermodynamic potential $\Omega[\mu]$ is as follows

$$\Omega_\rho[\{n_b\},\mu_{\mathrm{ch},\rho},f,\rho_0^{(3)},\rho_c] = g_\rho\,\chi_\rho n_I\,\rho_0^{(3)} - \frac{1}{2}(\rho_0^{(3)})^2\,m_\rho^2\,\Phi_\rho^2 - \left[\left(g_\rho\chi'_\rho\,\rho_0^{(3)} - \mu_{\mathrm{ch},\rho}\right)^2 - m_\rho^2\,\Phi_\rho^2\right]|\rho_c|^2\,. \quad (4)$$

Equations of motion for the $\rho_0^{(3)}$ and $\rho_c$ fields given by minimization of the thermodynamic potential have two solutions. One of them is the traditional one

$$\rho_0^{(3)} = \frac{g_\rho}{m_\rho^2}\frac{\chi_\rho}{\Phi_\rho^2}n_I,\ \rho_c = 0\,,\quad \Omega_\rho^{(1)} = \frac{C_\rho^2 n_I^2}{2m_N^2\eta_\rho(f)}. \quad (5)$$

The second solution is

$$\rho_0^{(3)} = \frac{\mu_{\mathrm{ch},\rho} - m_\rho\Phi_\rho}{g_\rho\chi'_\rho}\,,\quad |\rho_c|^2 = \frac{(-n_I - n_\rho)\theta(-n_I - n_\rho)}{2\,m_\rho\,\eta_\rho^{1/2}\,\chi'_\rho}\,, \quad (6)$$

where

$$n_\rho = a\,(m_\rho\,\Phi_\rho - \mu_{\mathrm{ch},\rho})\,,\ a = \frac{m_N^2\eta_\rho^{1/2}\,\Phi_\rho}{C_\rho^2\chi'_\rho}\,, \quad (7)$$

resulting in

$$\Omega_\rho^{(2)} = \Omega_\rho^{(1)} - \frac{C_\rho^2}{2\,m_N^2\,\eta_\rho}(n_I + n_\rho)^2\theta(-n_I - n_\rho)\,. \quad (8)$$

The $\rho$ charge density is given by $n_{\mathrm{ch},\rho} = \partial\Omega/\partial\mu_{\mathrm{ch},\rho} = -2m_\rho\Phi_\rho(f)|\rho_c|^2 < 0$. The contribution to the energy density from the charged $\rho$ meson condensate (1) is then given by

$$\Delta E_{\mathrm{ch},\rho}[\{n_b\};f] = -\frac{C_\rho^2}{2\,m_N^2\,\eta_\rho}(n_I + n_\rho)^2\theta(-n_I - n_\rho) - \mu_{\mathrm{ch},\rho}n_{\mathrm{ch},\rho}\,, \quad (9)$$

$\theta(-n_I - n_\rho) = 1$ for $n_I + n_\rho < 0$ and zero otherwise.

The charge neutrality condition is now $\sum_b Q_b n_b - n_e - n_\mu + n_{\mathrm{ch},\rho} = 0$ and the relations between chemical potentials in beta-equilibrium matter (BEM) $\mu_e = \mu_\mu = \mu_{\mathrm{ch},\rho}$, $\mu_b = \mu_n - Q_b\mu_l$.

All equations are solved self-consistently with the equation of motion for the scalar field $\partial E/\partial f = 0$. The pressure is given by $P = \sum_j \mu_j n_j - E$, $j = b \cup l \cup \{\mathrm{ch},\rho\}$. To be specific in our numerical calculations we use $\chi' = 1$.

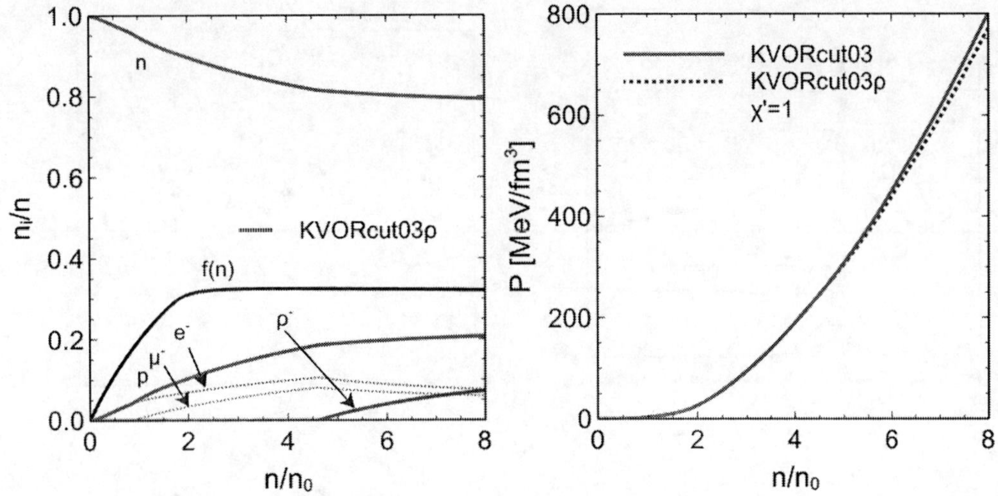


**Figure 1.** Left panel: particle concentrations and the scalar field $f$ in BEM as functions of the total baryon density, $n$, for the KVORcut03$\rho$ model. Right panel: pressure in BEM as a function of the baryon density for KVORcut03 and KVORcut03$\rho$ models.

## 3. Numerical results

First, we study two models constructed in [4, 5], which use the "cut-mechanism" suggested in [8] to make the EoS stiffer at high densities without altering its properties near the saturation density. The KVORcut03 model is a modification of the KVOR model (labeled as MW(nu) model in [7]). The models labeled as KVORcut exploit a sharp decrease of the $\eta_\omega(f)$ for $f > f_c^{\rm KVORcut}$, $f_c^{\rm KVORcut} > f(n_0)$, thus making the EoS stiffer in both ISM and BEM at high densities. This allows to pass the maximum NS mass constraint even provided hyperons and $\Delta$s are included. The saturation properties of the model are the same as in the original KVOR model. Below we choose the KVORcut03-based models, cf. [7]. Then we consider extensions of the MKVOR* model, cf. [5]. In the MKVOR* model a sharp decrease in the $\eta_\rho(f)$ with increasing $f$ for $f > f_c^{\rm MKVOR*} > f(n_0)$ is used to make the EoS stiffer in the BEM keeping it unchanged in the ISM. The input parameters for the MKVOR* model are: $n_0 = 0.16\,{\rm fm}^{-3}$, the binding energy $\mathcal{E}_0 = -16\,{\rm MeV}$, the incompressibility $K = 250\,{\rm MeV}$, the symmetry energy $\widetilde{J} = 32\,{\rm MeV}$ and the nucleon effective mass $m_N^*(n_0) = 0.73\,m_N$. The models satisfy many constraints from NS observations, HICs analyses and nuclear phenomenology.

### 3.1. *KVORcut03-based models*

On the left panel in Fig. 1 we show particle fractions together with the scalar field for the KVORcut03$\rho$ model (KVORcut03 model with included possibility of the charged $\rho$ condensation) in the BEM. The charged $\rho$ condensate appears with a second order phase transition with the critical density $n_{c,\rho} \simeq 4.62\,n_0$, which is larger than that found in [7] for the original KVOR model. The reason of this shift of the charged $\rho$ condensation threshold to a higher density is the limiting of the hadron effective mass decrease, which is a general feature of models with the "cut-mechanism" included in the scaling functions $\eta_\sigma$ or $\eta_\omega$. On the right panel in Fig. 1 we show the pressure in the BEM as a function of the density for the KVORcut03$\rho$ model. The effect of the charged $\rho$ condensation on the EoS is minor, and the maximum NS mass decreases only by $0.01\,M_\odot$. In the KVORcut03 model with the hyperons and/or $\Delta$s included into calculation the charged $\rho$ condensation does not occur at densities relevant for NSs.

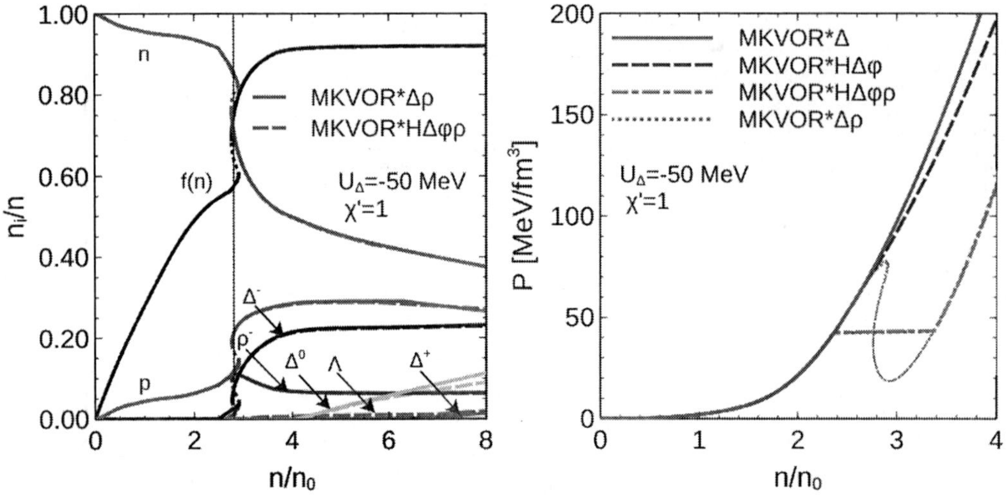


**Figure 2.** Left panel: particle concentrations and the scalar field $f$ in BEM as functions of the total baryon density for MKVOR*$\Delta\rho$ and MKVOR*H$\Delta\rho\phi$ models. Dashed line regions denote the unstable branch of solutions for the equilibrium concentrations. The dotted vertical line denotes the density $n^* \simeq 2.81\, n_0$, at which the second stable branch becomes energetically favorable for both models. Right panel: pressure in BEM as a function of the baryon density for MKVOR*$\Delta$, MKVOR*H$\Delta\phi$, MKVOR*$\Delta\rho$, MKVOR*H$\Delta\rho\phi$ models. For models with the charged $\rho$ condensation thin lines show the initial multi-valued solution and horizontal lines show the Maxwell construction.

### *3.2. MKVOR*-based models*

On the left panel in Fig. 2 we show the particle concentrations and the scalar field $f$ in the BEM for the MKVOR*$\Delta\rho$ (MKVOR* model with included $\Delta$ and charged $\rho$ condensation) and MKVOR*H$\Delta\phi\rho$ models. In the presence of the charged $\rho$ condensation the hyperon concentrations are suppressed, so the lines for MKVOR*$\Delta\rho$, MKVOR*H$\Delta\phi\rho$ visually coincide. There exists a region with three solutions for equilibrium particle concentrations for a given total baryon density, which is a feature of the first order phase transition. One of these three solutions corresponds to the maximum of the energy density and the configuration is unstable (shown by the dashed line regions), while two other solutions are the local minima of the energy density. The stable solution with greater $n_{\mathrm{ch},\rho}$ becomes energetically favorable at $n^* = 2.81\, n_0$ for both models, which is shown by the dotted vertical line. The pressure for the MKVOR*$\Delta$, MKVOR*H$\Delta\phi$, MKVOR*H$\Delta\phi\rho$ and MKVOR*$\Delta\rho$ models in the BEM is shown on the right panel of Fig. 2. The result of calculation with self-consistent particle concentrations is shown by thin lines, and horizontal lines indicate the Maxwell construction, which spans from $2.36\, n_0$ to $3.37\, n_0$.

In Fig. 3 we show the resulting NS mass as a function of the central density (left panel), mass-radius relation (central panel), and NS radius as a function of the central density (right panel) for the MKVOR*[$\Delta$, H$\Delta\phi$, $\Delta\rho$, H$\Delta\phi\rho$] models. Due to the description of the phase transition with the Maxwell construction, the region of central densities corresponding to the Maxwell construction (horizontal dotted lines between bars in left and right panels) cannot be realized in stable NS configurations. Contrary to the KVORcut03 model, in the MKVOR*-based models the appearance of the charged $\rho$ condensate leads to a substantial decrease of the maximum NS mass from 2.3 $M_\odot$ for MKVOR*$\Delta$ model (2.21$M_\odot$ for MKVOR*H$\Delta\phi$) to 2.03 $M_\odot$ for MKVOR*$\Delta\rho$ and MKVOR*H$\Delta\phi\rho$ models (for latter two models the curves visually

IOP Conf. Series: Journal of Physics: Conf. Series **941** (2017) 012053 doi:10.1088/1742-6596/941/1/012053

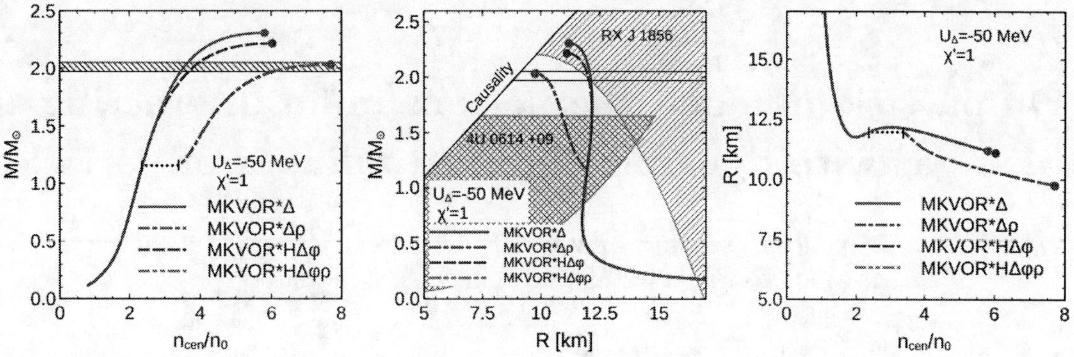


**Figure 3.** NS masses versus the central density (left panel) and NS radii (central panel), and NS radius as a function of the central density (right panel) for MKVOR*$\Delta$, MKVOR*H$\Delta\phi$, MKVOR*$\Delta\rho$, MKVOR*H$\Delta\phi\rho$ models. Dotted lines between bars indicate the central density interval corresponding to the Maxwell construction line in Fig. 2 (right), which is not realized in NSs.

coincide). However, the resulting models still pass the maximum NS mass constraint with $M_{\rm max} = 2.03\,M_\odot$. On the mass-radius relationship the phase transition leads to a connected stable branch. The radius, corresponding to the NS with the maximum mass, decreases from 12 km to 9.75 km.

## 4. Conclusion

In this contribution we studied the possibility of the charged $\rho$ meson condensation in the KVORcut03 and MKVOR* relativistic mean-field models and their extensions with scaled hadron masses and coupling constants including hyperons and $\Delta$ isobars, have being constructed in our previous works. Results are model dependent. In the KVORcut03-based models the charged $\rho$ meson condensate appears by a second order phase transition, and the effect on the EoS and on the maximum neutron star mass is minor. In the MKVOR*-based models the charged $\rho$ meson condensate appears by the first order phase transition, that results in a substantial decrease of the maximum neutron star mass.

## Acknowledgements

The reported study was funded by the Russian Foundation for Basic Research (RFBR) according to the research project No 16-02-00023-A. The work was also supported by the Slovak Grant No. VEGA-1/0469/15, and "NewCompStar", COST Action MP1304 and the Ministry of Education and Science of the Russian Federation within the state assignment, project No 3.6062.2017/BY.

## References

[1] Klähn T *et al.* 2006 *Phys. Rev.* C **74** 035802.
[2] Antoniadis J *et al.* 2013 *Science* **340** 6131.
[3] Maslov K A, Kolomeitsev E E and Voskresensky D N 2015 *Phys. Lett.* B **748** 369.
[4] Maslov K A, Kolomeitsev E E and Voskresensky D N 2016 *Nucl. Phys.* A **950** 64.
[5] Kolomeitsev E E, Maslov K A and Voskresensky D N 2017 *Nucl. Phys.* A **961** 106.
[6] Voskresensky D N 1997 *Phys. Lett.* B **392** 262.
[7] Kolomeitsev E E and Voskresensky D N 2005 *Nucl. Phys.* A **759** 373.
[8] Maslov K A, Kolomeitsev E E and Voskresensky D N 2015 *Phys. Rev.* C **92** 052801.
[9] Danielewicz P, Lacey R and Lynch W G 2002 *Science* **298** 1592.
[10] Vidana I 2016 *J. Phys. Conf. Ser.* **668** 012031.
[11] van Dalen E N E, Colucci G and Sedrakian A 2014 *Phys. Lett.* B **734** 383.

# On particle number fluctuations in an interacting pion gas with dynamically fixed number of particles

**D N Voskresensky**[1] **and E E Kolomeitsev**[2]

[1]National Research Nuclear University (MEPhI), Moscow, Russia
[2]Matej Bel University, Banska Bystrica, Slovakia

E-mail: d.voskresensky@gsi.de

**Abstract.** We consider a hot isospin-symmetric pion gas with the dynamically fixed number of particles in the model with a $\lambda\phi^4$ interaction. In the thermodynamic limit, for temperature above the critical value for the Bose-Einstein condensation we calculate the effective pion mass, the chemical potential and the normalized variance. In contrast to the ideal gas, the normalized variance remains finite in the critical point of the Bose-Einstein condensation.

## 1. Introduction

More than a thousand of pions can be produced in central heavy-ion collisions at RHIC and LHC energies [1]. The measured pion spectra show an enhancement at low transverse momenta, $p_T \lesssim m_\pi$, where $m_\pi$ is the pion mass, whereas spectra of harder pions are approximately exponential, cf. [2]. The latter circumstance allows to assume that pions form a hot expanding fireball characterized by the temperature $T(\vec{r},t)$ and density $n(\vec{r},t)$. The kinetic freeze-out temperature is estimated to be $T_{\rm kin} \sim 100-120$ MeV. An estimate [3] shows that at $T \sim 120-140$ MeV processes of the pion absorption are rather suppressed, whereas re-scattering processes are still important. Thus, one may assume that the total pion number is dynamically fixed on a time scale from a chemical freeze out, when the total pion number becomes frozen, till the thermal freeze-out, when the momentum distributions stop changing [4]. The existing estimates for the typical pion density in the pion fireball are contradictory [5, 6] yielding values varying in a broad interval, for LHC from $n \sim 0.8$ to $2.5 n_0$, where $n_0$ is the nuclear saturation density, whereas from measured volume in HBT correlations and pion yield [1] follow much higher values.

Already the first estimates [7, 8] have demonstrated that experimental low-$p_T$ pion yields in collisions at 200 $A$GeV can be fitted if the pion chemical potential $\sim 120-130$ MeV is introduced. Recent work [5] evaluated the non-equilibrium chemical potential of pions as $\mu \simeq 134.9$ MeV, being very close to the critical one ($\mu = m_\pi \simeq 135-140$ MeV) for a Bose-Einstein condensation (BEC) in an ideal pion gas. Non-equilibrium supercooling effects [9, 10, 11] as well as decays of resonances [12] may provide effective mechanisms to drive the pion system to the Bose-Einstein condensation (BEC). Reference [9] studied a possibility of the pion BEC in the interacting relativistic pion gas with a $\lambda\varphi^4$ interaction under the assumption that the number of pions created in the course of ultrarelativistic heavy-ion collisions is (dynamically) fixed in the time interval between the chemical and thermal freeze-outs. The most central collisions with high pion multiplicity were proposed as the most preferable to observe effects of the pion BEC. Utilizing the Weinberg Lagrangian, the investigation of the pion BEC was continued in [13, 14].

Reference [14] incorporated a possibility of a change of the initial isotopic composition due to $\pi^+\pi^- \leftrightarrow \pi^0\pi^0$ reactions. Reference [11] demonstrated that before the formation of the BEC the interacting pion gas in ultrarelativistic heavy-ion collisions may pass several stages, including a wave-turbulence stage, effects studied in [15] in relation to the BEC in application to non-relativistic gaseous systems.

A growth of fluctuations near the critical point of the second-order and first-order phase transitions is a general property manifested in various critical opalescence phenomena, cf. [16]. References [17, 18] returned to consideration of the ideal pion gas and argued for the divergence of the normalized variance in the critical point of the BEC in the thermodynamic limit. The result was supported by the analysis in the micro-canonical approach. An enhancement of the normalized variance was observed in the high pion multiplicity events in *pp* collisions in the energy range 50–70 GeV [19]. Here, a care should be taken, when one compares naive theoretical expectations for thermal fluctuation characteristics with actual measurements, which incorporate background contributions, the dependence on center-of-mass energy, other dynamical effects, collision centrality, kinematic cuts, etc.

In this contribution we will compute normalized variance in the non-ideal hot pion gas with the dynamically fixed number of particles considered in thermodynamical limit.

## 2. Effective Lagrangian for pions with dynamically fixed particle number

As a model for the description of the pion system we use the Lagrangian [20]

$$\mathcal{L} = \tfrac{1}{2}(\partial\vec{\varphi})^2 - \tfrac{1}{2}m_\pi^2\vec{\varphi}^2 - \tfrac{1}{4}\lambda\vec{\varphi}^4. \quad (1)$$

Simplifying we exploit the $\lambda\varphi^4$ model with $\lambda \sim 1$. Indeed, comparing with the leading terms of the Weinberg Lagrangian we find $\lambda = m_\pi^2/2f_\pi^2 \simeq 1.13$, where $f_\pi = 93\,\mathrm{MeV}$ is the weak pion decay constant. The isospin vector $\vec{\varphi} = (\varphi_1, \varphi_2, \varphi_3)$ is associated with the fields of the positive ($\pi^+$), negative ($\pi^-$) and neutral ($\pi^0$) pions $\pi^\pm = \frac{1}{\sqrt{2}}(\varphi_1 \pm i\varphi_2)$, $\pi^0 = \varphi_3$ . The positive and negative pions, being introduced as particles and anti-particles, are described by the one complex field, $(\pi^-)^\dagger = \pi^+$, where $(...)^\dagger$ means the Hermitian conjugation. The neutral pions, as the self-conjugated particles, are represented by the real field $(\pi^0)^\dagger = \pi^0$.

The 4-point interaction among the fields $\vec{\varphi}$ included in the Lagrangian (1) modifies the pion properties and the pion-pion interaction in medium. The resulting in-medium retarded pion propagator is determined by the Dyson equation

$$G(\omega, k) = G_0(\omega, k)/(1 - G_0(\omega, k)\Pi(\omega, k)) = \quad (2)$$

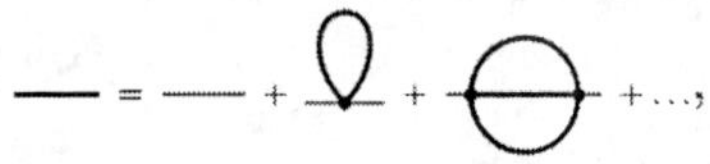

where the thin line is the free pion retarded propagator $G_0(\omega, k) = 1/(\omega^2 - k^2 - m_\pi^2 + i0)$, and $\omega$ and $k$ stand for the frequency and momentum of the pion, and $\Pi$ is the full pion retarded polarization operator determined by the diagrams shown in (2). The first diagram in (2) is the "tadpole" diagram and the second one is the "sandwich" diagram.

Now we will consider the pion system with the dynamically fixed particle number. To do this we turn in the Lagrangian (1) to the new complex fields $\varphi_-$, $\varphi_+$, $\varphi_0$ corresponding to pions with the positive frequencies $\pi^- = \varphi_- + \varphi_+{}^\dagger$, $\pi^+ = \varphi_+ + \varphi_-{}^\dagger$, and $\pi^0 = \varphi_0 + \varphi_0{}^\dagger$. In a system with the fixed and, in general, different numbers of pions of each species the particle–anti-particle symmetry is lost, and therefore, it is possible that $\varphi_- \neq \varphi_+^\dagger$, and neutral pions are described by a complex field $\varphi_0$. In the second quantization new fields of $\pi^-$, $\pi^+$, $\pi^0$ mesons possess the following operator representations in the quasi-particle approximation $\hat{\varphi}_i = \sum_{\vec{k}} \hat{a}_{\vec{k}}^{(i)}\, Z_k^{(i)}\, e^{-i\omega_i(k)t + i\vec{k}\vec{r}}$, $\quad i = +, -, 0$. We choose positive solutions of the dispersion relations, $\omega_{\pm 0}(k) > 0$, and, in general case, $\omega_-(k) \neq \omega_+(k) \neq \omega_0(k)$ with

$Z^-_k \neq Z^+_k \neq Z^0_k$. The operators $\hat{a}$, $\hat{b}$ and $\hat{c}$ define now the numbers of pions of the corresponding sorts $\sum_{\vec{k}} 2\omega_i Z^{(i)}_k \overline{\hat{a}^{(i)\dagger}_{\vec{k}} \hat{a}^{(i)}_{\vec{k}}} = N_i$ , where $\overline{(...)}$ denotes averaging over the vector of state of the quantum system with the Gibbs weight factor. The mean value $\overline{\hat{a}^{(i)\dagger}_{\vec{k}} \hat{a}^{(i)}_{\vec{k}}}$, is determined by the Bose distribution with frequencies $\omega_i$ and chemical potentials $\mu_i$ fixing the particle numbers $N_i$.

Replacing new fields in the Lagrangian (1), we are able to separate terms, which correspond to the particle number conservation. In terms of the new fields the Lagrangian density (1) is

$$\mathcal{L} = \mathcal{L}_{\rm fix} + \mathcal{L}^{'} \,, \tag{3}$$

$$\mathcal{L}_{\rm fix} = \mathcal{L}_{\pi^-} + \mathcal{L}_{\pi^+} + \mathcal{L}_{\pi^0} + \mathcal{L}_{\pi^+\pi^-\pi^0} \,, \quad \mathcal{L}^{'} = \mathcal{L}_{2\leftrightarrow 2} + \mathcal{L}_{3\leftrightarrow 1}. \tag{4}$$

Here the terms $\mathcal{L}_{\pi^i} = |\partial\varphi_i|^2 - m_\pi^2|\varphi_i|^2 - \lambda|\varphi_i|^4$ describe the interaction of pions of a certain sort and we neglected a difference in pion masses for various species, the term

$$\mathcal{L}_{\pi^+\pi^-\pi^0} = -4\lambda|\varphi_-|^2|\varphi_+|^2 - 2\lambda\left(|\varphi_+|^2 + |\varphi_-|^2\right)|\varphi_0|^2$$

describes the interaction of the different pion species with each other via reactions $\pi^+\pi^- \leftrightarrow \pi^+\pi^-$ and $\pi^\pm\pi^0 \leftrightarrow \pi^\pm\pi^0$. These processes conserve the total number of pions and also the number of pions of each sort. Note that the pion self-interaction contained in the terms $\mathcal{L}_{\rm fix}$ gives a contribution to the pionic polarization operator already in a first-order in the coupling constant $\lambda$. Such a contribution is depicted by the tadpole diagram in (2). The part of the Lagrangian density

$$\mathcal{L}_{2\leftrightarrow 2} = -\lambda\left(\varphi_+\varphi_-\varphi_0^{\dagger 2} + \varphi_0^2\varphi_+^\dagger\varphi_-^\dagger\right)$$

corresponds to the processes $\pi^0\pi^0 \leftrightarrow \pi^+\pi^-$, which, while keeping the total number of pions fixed, change the relative fractions of the pion species so that the chemical potentials obey the relation $2\mu_0 = \mu_+ + \mu_-$. For the case of the isospin-symmetric pion gas, that we will focus on in this contribution, $\mu_0 = \mu_+ = \mu_- \equiv \mu$. The term containing non-equal numbers of the pion creation and annihilation operators, $\mathcal{L}_{3\leftrightarrow 1} = \mathcal{L}(\varphi_i^\dagger\varphi_j\varphi_k\varphi_l; \varphi_i\varphi_j^\dagger\varphi_k^\dagger\varphi_l^\dagger)$, $\;i,j,k,l = +,-,0,$ corresponds to the processes with a change of the number of pions in the system, $\pi \leftrightarrow \pi\pi\pi$. These processes bring the system to the chemical equilibrium. The kinetic equation for the pion gas allowing for these reactions has the Bose distribution with vanishing chemical potential as an equilibrium solution.

The terms $\mathcal{L}_{2\leftrightarrow 2}$ and $\mathcal{L}_{3\leftrightarrow 1}$ give contributions to the pion polarization operator only in the second order of the coupling constant. They are represented by a sandwich diagram in (2) with different pion species in the internal lines allowed by charge conservation. The direction of the internal lines takes also into account the distinct processes: $\pi\pi \leftrightarrow \pi\pi$ (two lines are directed to the one side and one line to the other side) and $\pi \leftrightarrow \pi\pi\pi$ (all lines go from the left to the right). For $T \lesssim m_\pi$ of our interest the real part of the sandwich diagram proves to be small in comparison with the tadpole contribution. Thereby we omit the former contribution. The imaginary part of the sandwich diagram determines the rates of rescattering and absorption/production reactions. As we have discussed, the processes $\pi\pi \leftrightarrow \pi\pi$ responsible for the thermal equilibration occur essentially faster then processes $\pi \leftrightarrow \pi\pi\pi$ responsible for the chemical equilibration at $T \lesssim m_\pi$. Therefore, we drop furthermore the term $\mathcal{L}_{3\leftrightarrow 1}$ of the Lagrangian, introducing instead the non-vanishing pionic chemical potentials fixing the pion numbers determined by reactions $\pi^+\pi^- \leftrightarrow \pi^0\pi^0$. Doing so we assume that these reactions are operative on the time scale of the pion fireball expansion up to its break up. Being armed by the effective Lagrangian (3) describing the system with the fixed total particle number and an arbitrary electric charge, we can study its properties.

## 3. Pion spectrum in self-consistent Hartree approximation

Varying the Lagrangian (3) with respect to $\varphi_+^\dagger$, $\varphi_-^\dagger$, $\varphi_0^\dagger$ fields we obtain a system of coupled non-linear equations of motion. For isospin-symmetric system under consideration $\varphi_+^\dagger = \varphi_-^\dagger = \varphi_0^\dagger \equiv \varphi$ and we have only one equation of motion. We solve this equation within the self-consistent Hartree approximation, which is rather appropriate at temperatures of our interest, $T \lesssim m_\pi$. Within this approximation the behaviour of a certain particle is determined by an averaged interaction with surrounding particles, which form a thermal bath. The properties of particles in the bath are, in turn, determined by the same equation of motion as that for the given pion (cf. bold line in (2)).

Formally, we represent the field $\varphi$ as a superposition of some picked-out field $\tilde{\varphi}$ and an environmental field $\xi$, $\varphi \to \tilde{\varphi} + \xi$. Then we keep in the equation of motion only the terms that are linear in the $\tilde{\varphi}$ and quadratic in $\xi$ (other terms vanish after averaging). As the result we find equation of motion for the field $\tilde{\varphi}$ in the Hartree approximation

$$\partial^2 \tilde{\varphi} + m_\pi^2 \tilde{\varphi} + \Pi \tilde{\varphi} = 0 \tag{5}$$

with the spectrum

$$\omega(k) = \sqrt{k^2 + m^{*2}}, \quad m^{*2} = m_\pi^2 + \Pi\,, \tag{6}$$

where the polarization operator

$$\Pi = 10\lambda\,\overline{|\xi^2|}\,, \quad \overline{|\xi|^2} = \Big\langle \frac{f_k}{2\omega(k)} \Big\rangle \equiv \int \frac{d^3k}{(2\pi)^3} \frac{f_k}{2\omega(k)}\,, \quad f_k = \frac{1}{e^{(\omega(k)-\mu)/T} - 1}\,, \tag{7}$$

and the total pion density $n = 3\langle f_k \rangle$. The non-commutativity of the creation-annihilation operators produces divergent contributions of quantum fluctuations to the macroscopic characteristics of the system, e.g. in (7), which have to be renormalized by subtraction of the corresponding vacuum values. The remaining finite contributions from quantum fluctuations are proved to be rather small and we omit them.

When the temperature decreases up to the value $T_c^{\rm ind}$ there appears an "induced" BEC. It occurs for $\mu = m^*$. Following [14], we say about induced BEC, since the BEC could also occur in a first-order phase transition at a higher temperature, when $\mu$ reaches the free pion mass, $m_\pi$, if it were energetically favorable. Any first-order phase transition requires an extra time and may occur only in rare events. Therefore, we will ignore this possibility in our study here.

The dependence of the effective pion mass $m^*$ and the chemical potential $\mu$ on the temperature in the interacting pion gas for the isospin-symmetric matter is shown on the left panel in figure 1. On the middle panel we demonstrate the critical temperature of the induced BEC as a function of the pion gas density. We see that the effective pion mass is larger than the free pion mass and increases with a decrease of the temperature and an increase of $\lambda$. Also the mass increases with an increase of the density. The critical temperature $T_c^{\rm ind}$ for the interacting pion gas is smaller than that for the ideal gas.

## 4. Normalized variance of the particle number

In the case of the uniform system the normalized variance of the particle number $N$ is given by [21]

$$\frac{\overline{(\Delta N)^2}}{N} = \frac{T}{n} \frac{\partial n}{\partial \mu}\Big|_T\,. \tag{8}$$

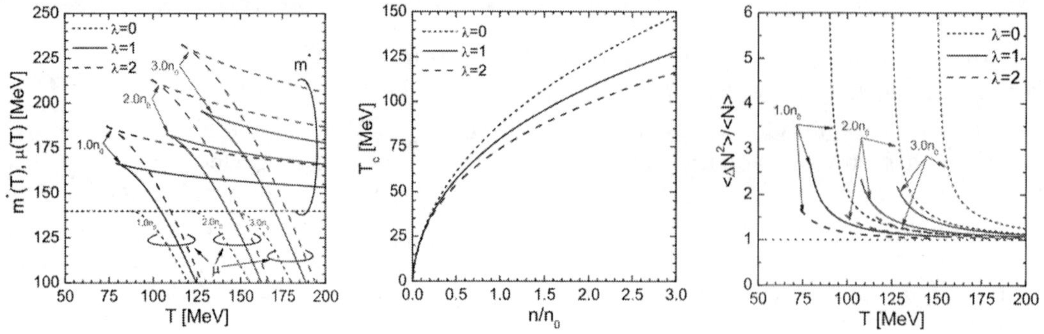


**Figure 1.** Left panel: effective pion mass and chemical potential in the isospin-symmetric pion gas as functions of temperature for three values of the pion density $n = n_0$, $2n_0$, and $3n_0$ and for the interaction parameter $\lambda = 0$ (short-dash lines), $\lambda = 1$ (solid lines) and $\lambda = 2$ (long-dash lines). Horizontal short-dash line depicts the free pion mass. Middle panel: the critical temperature of the BEC for the isospin-symmetric pion gas as a function of the pion density for $\lambda = 0, 1, 2$. Right panel: normalized variance of the pion number for the interacting isospin-symmetric pion gas (with $\lambda = 1$ and 2) and for ideal gas ($\lambda = 0$) as a function of the temperature for $n = n_0$, $2n_0$, and $3n_0$, here $n_0 = 0.15\,\mathrm{fm}^{-3}$.

Using parial integrations for the isospin-symmetric pion gas we get

$$\frac{\overline{(\Delta N)^2}}{N} = 3\frac{m^{*2}T}{n}\frac{I_1 + 10\lambda(I_1 I_2 - I_3^2)}{1 + 10\lambda I_2}, \tag{9}$$

where dimensionless integrals are

$$I_1 = \frac{\langle f_k(1+f_k)\rangle_k}{m^{*2}T}, \quad 4I_2 = \Big\langle \frac{f_k}{k^2\omega}\Big\rangle + \Big(\frac{f_k k}{2\pi^2\omega}\Big)\Big|_{k\to 0}, \quad I_3 = \Big\langle \frac{f_k(1+f_k)}{2m^* T\omega}\Big\rangle_k . \tag{10}$$

In the critical point of the induced pion BEC, i.e. for $T = T_c^{\rm ind}$, we have $\mu = m^*$, for $\lambda \neq 0$, and the distribution behaves as $f_k \to 2m^*T/k^2$ for $k \to 0$. Using these limiting expressions we obtain that $I_{1,2,3}$ diverge. Separating the divergent parts we have

$$I_1 = \frac{2T}{\pi^2 k_0} + \delta I_1, \quad I_2 = \frac{T}{2\pi^2 k_0} + \delta I_2\,, \quad I_3 = \frac{T}{\pi^2 k_0} + \delta I_3 \tag{11}$$

for $k_0 \to 0$, where $\delta I_{1,2,3}$ are already convergent terms.

For the non-ideal gas perturbation expansion in $\lambda$ is not valid for any finite $\lambda \neq 0$, since the integrals in denominator in (9) diverge. However the self-consistent account of the interaction in the Hartree approximation utilized here leads to the finite result

$$\frac{\overline{(\Delta N)^2}}{N}\Big|_{T=T_c^{\rm ind}} = \Big(\frac{6m^{*2}T}{5n}\Big[\frac{1}{\lambda} + \frac{5}{2}(\delta I_1 + 4\delta I_2 - 4\delta I_3)\Big]\Big)\Big|_{T=T_c^{\rm ind}} . \tag{12}$$

For the ideal pion gas, $\lambda = 0$, at $T = T_c^{\rm id}$ one has $\mu = m_\pi$ and $f_k \to 2m_\pi T/k^2$ for $k \to 0$. Thus,

$$\frac{\overline{(\Delta N)^2}}{N}\Big|_{\rm id} \to \frac{6m_\pi^2 (T_c^{\rm id})^2}{\pi^2 k_0}$$

diverges as $1/k_0$ for $k_0 \to 0$.

On the right panel in figure 1 we show the normalized variance of the pion number (9) for the interacting pion gas (for $\lambda = 1$ and 2) and for the ideal gas ($\lambda = 0$) as functions of a temperature for three values of the density. For the ideal gas the normalized variance diverges at the critical temperature of the BEC in agreement with the statement of [18]. Therefore, the authors [18] suggested that the strong enhancement of the normalized variance can be considered as a clear signature of the BEC although effects of the finite volume smear the singularity. On the contrary, we see that for the interacting gas even in the infinite volume limit the singularity disappears provided the interaction is taken into account self-consistently (within the self-consistent Hartree approximation in our current study).

Finally, we would like to remind that the comparison of the idealized thermal fluctuations with those appeared in real experiments is in any case quite indirect because of many additional effects not taken into account, like a finite volume, realistic pion-pion interaction, the dependence on the collision energy, collision centrality, kinematic cuts, and other dynamical effects. Thus, we may only say that, if a significant growth of various pion number fluctuation characteristics were observed, it could be associated with a proximity of the system to the pion BEC either at the chemical freeze-out or at the thermal freeze-out, depending on the specifics of the measurements. Measurements of the normalized variance and higher moments like sckewness and kurtosis in pre-selecting high multiplicity events are desirable in order to observe a manifestation of the signatures of the Bose-Einstein pion condensation in heavy-ion collisions.

**Acknowlegments.** We thank M.E. Borisov for interest to this work and help. The reported results were obtained within the state assignment by Ministry of Education and Science of the Russian Federation, project No 3.6062.2017/BY. The study was supported by the Russian Foundation for Basic Research (RFBR) according to the research project No 16-02-00023-A, by Slovak grant VEGA-1/0469/15 and by THOR the COST Action CA15213.

## References

[1] Nayak T K 2012 *Pramana* **79** 719.
[2] Abelev B et al. [ALICE Collaboration] 2013 *Phys. Rev. C* **88** 044910.
[3] Goity J L and Leutwyler H 1989 *Phys. Lett. B* **228** 517.
[4] Gerber P, Leutwyler H and Goity J L 1990 *Phys. Lett. B* **246** 513.
[5] Begun V, Florkowski W and Rybczynski M 2014 *Phys. Rev. C* **90** 054912.
[6] Teaney D 2002 *Preprint* nucl-th/0204023.
[7] Kataja M and Ruuskanen P V 1990 *Phys. Lett. B* **243** 181.
[8] Mishustin I N, Satarov L N, Maruhn J, Stöcker H and Greiner W 1992 *Phys. Lett. B* **276** 403.
[9] Voskresensky D N 1994 *J. Exp. Theor. Phys.* **78** 793 [*Zh. Eksp. Teor. Fiz.* **105** 1473].
[10] Voskresensky D N, Blaschke D, Röpke G and Schulz H 1995 *Int. J. Mod. Phys. E* **4** 1.
[11] Voskresensky D N 1996 *Phys. Atom. Nucl.* **59** 2015 [*Yad. Fiz.* **59** 2090].
[12] Ornik U, Plümer M and Strottmann D 1993 *Phys. Lett. B* **314** 401.
[13] Kolomeitsev E E and Voskresensky D N, 1995 *Phys. Atom. Nucl.* **58** 2082 [*Yad. Fiz.* **58** 2195].
[14] Kolomeitsev E E, Kämpfer B and Voskresensky D N 1996 Acta Phys. Polon. B **27** 3263.
[15] Semikoz D V and Tkachev I I 1995 *Phys. Rev. Lett.* **74** 3093;
Semikoz D V and Tkachev I I 1997 *Phys. Rev. D* **55** 489.
[16] Landau L D and E. M. Lifshitz E M 1984 *Electrodynamics of Continuous Media* vol 8 (Oxford: Pergamon press).
[17] Begun V V and Gorenstein M I 2007 *Phys. Lett. B* **653** 190.
[18] Begun V V and Gorenstein M I 2008 *Phys. Rev. C* **77** 064903.
[19] Kokoulina E 2012 *Prog. Theor. Phys. Suppl.* **193**, 306 (2012);
Kokoulina E [SVD-2 Collaboration] 2013 *PoS ICHEP* **2012** 259.
[20] Itzykson C and Zuber J B 1980 *Quantum Field Theory* (New York: McGraw-Hill).
[21] Landau L D and Lifshiz E M 1980 *Statistical Physics, Part I* (Oxford: Pergamon Press).

# Controlling the light shift of the CPT resonance by modulation technique

**E A Tsygankov[1], S V Petropavlovsky[2], M I Vaskovskaya[3], S A Zibrov[3], V L Velichansky[1,3] and V P Yakovlev[1]**

[1]National Research Nuclear University MEPhI, Kashirskoe sh. 31, Moscow, 115409 Russia
[2]National Research University Higher School of Economics, 20 Myasnitskaya Ulitsa, Moscow 101000, Russia
[3]Lebedev Physical Institute, Russian Academy of Sciences, Leninskii pr. 53, Moscow, 119991 Russia

E-mail: selentinthebright@gmail.com

**Abstract.** Motivated by recent developments in atomic frequency standards employing the effect of coherent population trapping (CPT), we propose a theoretical framework for the frequency modulation spectroscopy of the CPT resonances. Under realistic assumptions we provide simple yet non-trivial analytical formulae for the major spectroscopic signals such as the CPT resonance line and the in-phase/quadrature responses. We discuss the influence of the light shift and, in particular, derive a simple expression for the displacement of the resonance as a function of modulation index. The performance of the model is checked against numerical simulations, the agreement is good to perfect. The obtained results can be used in more general models accounting for light absorption in the thick optical medium.

## 1. Introduction

The effect of coherent population trapping (CPT) allows to realize narrow resonances [1, 2, 3] and therefore finds promising applications in frequency standards and quantum metrology. In a conventional setup, a local oscillator is locked to the energy gap between the lower states of the three-level atomic system, see Fig. 1: at an exact resonance the atoms reside in the so-called dark state not absorbing the probing light. Deviations from the resonance are detected by the frequency modulation technique by means of the in-phase response of the slowly modulated probing field. A challenging issue in the design of such atomic standards is the dynamic Stark shift of the atomic levels causing drift and noise in the clock output. There are different remedies for that adverse effect among which we point out the technique [4] where the authors propose to use a combination of both in-phase and quadrature responses to control the deviations from the resonance and simultaneously suppress the light shift. To the best of our knowledge, this technique has never been applied to the CPT-based atomic clocks. Therefore, it makes practical sense to investigate the behavior of the three-level atomic system under conditions used in the FM spectroscopy and, in particular, in the context of the proposal [4]. To be more specific, in the model we introduce below we consider a laser wave interacting with atomic $^{87}Rb$ vapor and resonant to $D_1$-line transition that is quite conventional in the real CPT-experiments.

## 2. Model

The internal structure of the Rb atoms interacting with light under the CPT resonance conditions is modeled by a standard $\Lambda$-scheme of levels, see Fig. 1. The laser field $\mathcal{E}_0 \cos[\omega_0 t + a \sin\Phi(t)]$ propagating through the atomic cell is tuned to the optical frequency $\omega_0$ such that $\hbar\omega_0 = E_c - (E_a + E_b)/2$ where $E_a$, $E_b$, $E_c$ are the energies of the respective atomic states. The laser field is frequency modulated due to the term $a \sin\Phi(t)$ where $a$ is a modulation index and $\Phi(t) = \Omega t + b \sin\omega_m t$ is the time dependent phase. The instantaneous frequency of modulation,

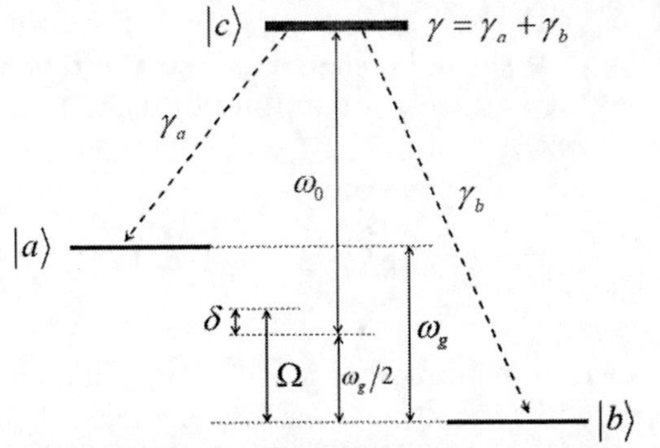


**Figure 1.** $\Lambda$-scheme of levels. The splitting between the lower levels $\omega_g/2\pi \approx 6.8$ GHz, the natural width of the upper level $\gamma/2\pi \approx 5$ MHz.

$\Phi'(t) = \Omega + b\omega_m \cos\omega_m t$, suggests that the modulation is not one-tone at the frequency $\Omega$. Rather, the microwave frequency $\Omega \equiv \omega_g/2 + \delta$ is modulated as well at a much lower rate $\omega_m$ with the modulation index $b$ and frequency deviation $b\omega_m$. The rationale behind such nested modulation is simple: the modulation at $\Omega$ creates the sideband frequencies $\omega_0 \mp \Omega$ resonant to the transitions $|a\rangle, |b\rangle \leftrightarrow |c\rangle$ and thus creates the CPT resonance *per se* whereas the modulation at $\omega_m$ provides additional spectroscopic signals measured over time against the detuning $\delta$ from the exact resonance and used for stabilizing frequency of the atomic clocks. We emphasize that the microwave frequency $\Omega/2\pi \approx 3.4$ GHz differs from the slow modulation rate $\omega_m/2\pi \sim 0.1 \div 1$ kHz by 6 to 7 orders of magnitude that makes the direct numerical simulations somewhat expensive but, fortunately, facilitates the analytical treatment of the problem.

## 3. Approximate solution

To describe the model outlined above, we use the optical Bloch equations for the atomic density matrix $\hat{\rho}(t) = \sum_{\alpha,\beta=\{a,b,c\}} \rho^{\alpha\beta}(t) |\alpha\rangle\langle\beta|$:

$$i\left(\frac{d}{dt} + \hat{\Gamma}\right)\hat{\rho} = \left[\hat{H}, \hat{\rho}\right], \tag{1}$$

where the Hamiltonian $\hat{H} = \hat{H}_0 + \hat{H}_{\rm int}$ reads

$$\hat{H}_0 = \frac{\omega_g}{2}|a\rangle\langle a| - \frac{\omega_g}{2}|b\rangle\langle b|, \quad \hat{H}_{\rm int} = -V(t)|c\rangle\langle a| - V(t)|c\rangle\langle b| + \text{h.c.}$$

The interaction Hamiltonian, $\hat{H}_{\rm int}$, is written under the resonant approximation with respect to optical transitions $|a\rangle, |b\rangle \leftrightarrow |c\rangle$ with the time dependent interaction term $V(t) = d\mathcal{E}_0/2\hbar \times \exp(-ia\sin\Phi(t))$ where $d = d^* \equiv \langle c|\hat{d}|a\rangle = \langle c|\hat{d}|b\rangle$ is a dipole matrix element which is supposed to be real and equal for both transitions.

The operator $\hat{\Gamma}$ in (1)

$$\hat{\Gamma}\hat{\rho} = -\rho^{cc}\{\gamma_a|a\rangle\langle a| + \gamma_b|b\rangle\langle b|\} + (\gamma_a + \gamma_b)\rho^{cc}|c\rangle\langle c| + \Gamma\left\{\rho^{ca}|c\rangle\langle a| + \rho^{cb}|c\rangle\langle b| + \text{h.c.}\right\} \tag{2}$$

accounts for the incoherent phenomena of two different kinds: (i) pumping of the lower levels due to radiative decay of the upper state at a rate of $\gamma = \gamma_a + \gamma_b$ (the first and second terms in (2)) and (ii) decay of optical coherences $\rho_{ca}$, $\rho_{cb}$ at a rate of $\Gamma \gg \gamma$ due to other incoherent processes such as collisions of Rb atoms with atoms of the buffer gas (the third term).

The interaction term $V(t)$ admits the following decomposition:

$$V(t) = \frac{d\mathcal{E}_0}{2\hbar} e^{-ia\sin\Phi(t)} = \sum_{k=-\infty}^{\infty} \tilde{V}_k(t) e^{-ik\Omega t}, \quad \tilde{V}_k(t) = V_k e^{-ik\varphi(t)}, \quad V_k = \frac{d\mathcal{E}_0}{2\hbar} J_k(a), \tag{3}$$

where $\varphi(t) = b\sin\omega_m t$ and $J_k(\cdot)$ is the Bessel function of the first kind. As $\omega_m/\Omega$ is of the order of $10^{-7} \div 10^{-6}$, representation (3) means that the interaction can be viewed as a sum of rapidly oscillating harmonics with slowly varying amplitudes. It is therefore tempting to separate the time dependence of different scales for the density matrix as well,

$$\rho^{\alpha\beta}(t) = \sum_{k=-\infty}^{\infty} \rho_k^{\alpha\beta}(t) \exp(-ik\Omega t), \quad \alpha, \beta = \{a, b, c\}, \tag{4}$$

where the amplitudes $\rho_k^{\alpha\beta}(t)$ are supposed to change at multiples of the characteristic rate $\omega_m$. Note also that the spectroscopic signals measured in the experiment are the mean light absorption in the atomic cell, i.e., the mean population of the upper level, $\langle\rho^{ee}(t)\rangle$, and its Fourier amplitudes at the frequency $\omega_m$, i.e., $\rho^{ee}(t) \propto M_S \cos\omega_m t + M_Q \sin\omega_m t$. Obviously, both signals are recorded over times of the order of $\omega_m^{-1}$ for which the fast oscillations at multiples of $\Omega$ will be averaged out due to inertia of the experimental setup. As such, it suffices to calculate only the zeroth amplitude, $\rho_0^{ee}(t)$, i.e., the time average *over fast oscillations* and use it further to obtain the above mentioned spectroscopic signals.

Projecting equations (1) onto the basis $\exp(ik\Omega t)$ we get the equations for amplitudes in (4). In doing so, we treat the slow quantities $\rho_k^{\alpha\beta}(t)$, $\tilde{V}_k(t)$ as constants neglecting thus small corrections of the order of $\omega_m/\Omega \sim 10^{-7} \div 10^{-6}$. Below we provide a sketch of the solution to the resulting equations.

First, by nature of the experiment, we seek for a steady-state solution at large times $\sim \omega_m^{-1}$ neglecting the time derivatives $d/dt \sim \omega_m \ll \gamma \ll \Gamma \ll \Omega$ wherever possible. Second, the light field is assumed weak, $d\mathcal{E}_0/2\hbar \ll \gamma \ll \Gamma \ll \Omega$, that holds in the real experiments (the low saturation regime). This implies that there are only three leading amplitudes of the order of unity in series (4), specifically, $\rho_0^{aa}, \rho_0^{bb}, \rho_{-2}^{ba}$. To demonstrate this, let us consider the equation for $\rho_k^{ba}(t)$ as an example:

$$i\left(\frac{d}{dt} + k\Omega + \omega_g\right)\rho_k^{ba} = \sum_s (\tilde{V}_{s-k}^*(t)\rho_k^{ca} - \tilde{V}_{s+k}(t)(\rho_k^{cb})^*).$$

For any $k \neq -2$ the term $k\Omega + \omega_g$ is non-zero and one can neglect the time derivative obtaining thus a steady-state $\rho_k^{ba}(t)$. This solution is obviously suppressed by the magnitude of the parameter $|V_k/\Omega| \ll 1$ and can be neglected. Conversely, if $k = -2$, the sum $-2\Omega + \omega_g = \delta$ can be arbitrarily small, the derivative cannot be omitted, and $\rho_{-2}^{ba}(t)$ is not necessarily small. The similar reasoning applies to amplitudes $\rho_k^{aa}$, $\rho_k^{bb}$ with a leading term at $k = 0$, but not to $\rho_k^{ca}$ and $\rho_k^{ca}$. The respective equations in the latter case contain an additional imaginary term $i\Gamma$ on the left-hand side that has no counterpart to cancel out for any $k$. As such, all of the amplitudes $\rho_k^{ca}$, $\rho_k^{cb}$ are small.

Note also that though the population of the upper level is small under low saturation, $\rho_0^{ee}(t) \ll 1$, it is a quantity of primary interest responsible for the spectroscopic signals mentioned above. The equation for $\rho_0^{ee}(t)$ is obtained from those for $\rho_0^{aa}(t)$, $\rho_0^{bb}(t)$ by means of the normalization condition $\rho_0^{aa} + \rho_0^{bb} + \rho_0^{cc} = 1$.

The steady-state solution to this equation reads

$$\rho_0^{ee}(t) = \frac{2\lambda}{\gamma}\left[1 - \frac{2\lambda}{\Gamma}\mathrm{Re}\{g(t)e^{-i2\varphi(t)}\}\right], \tag{5}$$

where $\lambda \simeq \Gamma\,(d\mathcal{E}_0/2\hbar\Gamma)^2\,J_1^2(a)$ and $g(t)$ is a solution to

$$\frac{dg}{dt} + 2[i\Delta + \lambda]g = \Gamma e^{i2\varphi(t)}. \tag{6}$$

In equation (6), $\Delta = \delta - \delta_0$, $\delta_0 \simeq 2(a\Omega)^{-1}\,(d\mathcal{E}_0/2\hbar)^2\,J_0(a)J_1(a)$. As $\exp\left(\pm i2\varphi(t)\right) = \sum_{k=-\infty}^{\infty} J_k(2b)\exp(\pm ik\omega_m t)$, the steady state solution to (6), $g(t)$, oscillates at multiples of $\omega_m$, and, consequently, so does $\rho_0^{ee}(t)$, see (5). The oscillations at zero frequency yield the mean absorption,

$$\langle\rho_0^{ee}(t)\rangle = \frac{2\lambda}{\gamma}\left[1 - 4\lambda^2\sum_{k=-\infty}^{\infty}\frac{J_k^2(2b)}{4\lambda^2 + (2\Delta + k\omega_m)^2}\right] \underset{b\ll 1}{\approx} \frac{2\lambda}{\gamma}\frac{\Delta^2}{\lambda^2+\Delta^2}, \tag{7}$$

which gets simplified in the instructive case $b \ll 1$ where it has a Lorentzian profile of width $\lambda$ and centered at $\delta_0$. The non-zero constant $\delta_0$ at which the absorption vanishes is due to the light shift of the atomic levels caused by the entire spectrum of the propagating laser field. For some values of modulation index $a$, the light shift is compensated, $\delta_0 = 0$, see Fig. 3, that may be highly beneficial from the experimental standpoint. We also note that the dependence of $\delta_0$ on $a$ was studied in a number of works, e.g., [2], purely numerically.

Two other spectroscopic signals, the in-phase $M_S$ and quadrature $M_Q$ amplitudes, can be obtained by projecting (5) onto $\cos\omega_m t$ and $\sin\omega_m t$:

$$M_S \propto \sum_{k=-\infty}^{\infty} J_k(2b)\left(\frac{J_{k+1}(2b)}{\lambda'^2 + (\Delta'+k+1)^2} + \frac{J_{k-1}(2b)}{\lambda'^2 + (\Delta'+k-1)^2}\right), \tag{8}$$

$$M_Q \propto \sum_{k=-\infty}^{\infty} J_k(2b)\left(\frac{J_{k-1}(2b)\,(\Delta'+k-1)}{\lambda'^2 + (\Delta'+k-1)^2} - \frac{J_{k+1}(2b)\,(\Delta'+k+1)}{\lambda'^2 + (\Delta'+k+1)^2}\right), \tag{9}$$

where $\Delta' = 2\Delta/\omega_m$, $\lambda' = 2\lambda/\omega_m$ and $M_S$, $M_Q$ are defined up to an irrelevant $\delta$-independent constant. Both the in-phase and quadrature amplitudes turn to zero at $\delta = \delta_0$ whereas $M_Q$ vanishes at two extra points of $\delta$. The latter follows, for example, from expressions (8)-(9) simplified at $b \ll 1$:

$$M_S \propto \frac{\Delta'}{[\lambda'^2 + (\Delta'-1)^2][\lambda'^2 + (\Delta'+1)^2]}, \quad M_Q \propto \frac{\Delta'\left[3\lambda'^2 + 1 - \Delta'^2\right]}{[\lambda'^2 + (\Delta'-1)^2][\lambda'^2 + (\Delta'+1)^2]\,[\lambda'^2 + \Delta'^2]}.$$

## 4. Numerical versus analytical results

In this section we compare the results of the preceding section to those obtained by integrating equations (1) numerically. We employ a Fortran multi-thread ODE solver, yet a typical computational time is as long as 2 to 20 hours depending on the ratio $\omega_m/\Omega$ that crucially affects the time step. We also check the results against a simplified 3-wave model where we keep only three terms in the spectral representation (3): the carrier, $k = 0$, and the first sidebands, $k = \pm 1$. One can easily show that the respective formulae under the 3-wave approximation

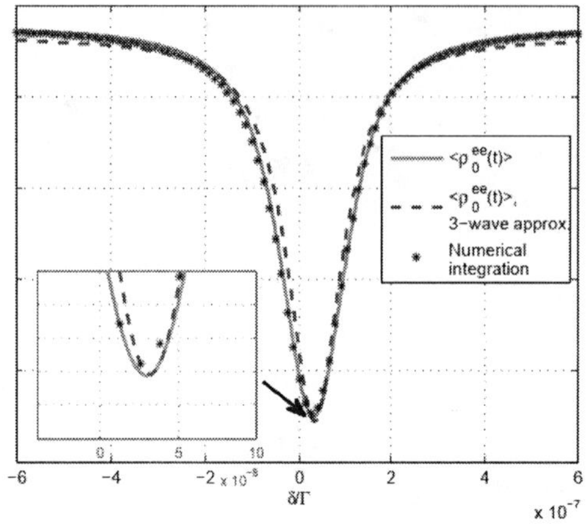


**Figure 2.** The CPT resonance line, arbitrary units along the $y$-axis.

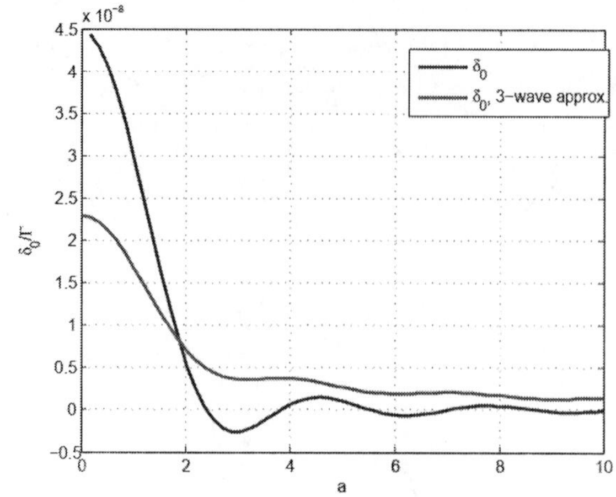


**Figure 3.** Detuning $\delta_0$ as a function of modulation index $a$.

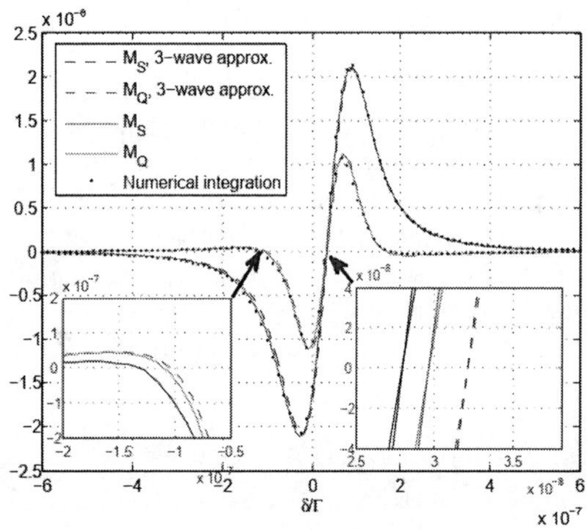


**Figure 4.** The in-phase and quadrature amplitudes (8)-(9), arbitrary units along the $y$-axis. The parameters used throughout Figs. 2-4: $\Gamma/2\pi = 0.265$ GHz, $d\mathcal{E}_0/(2\hbar\Gamma) = 1/1400$, $\omega_m/2\pi = 129$ Hz, $b = 1/2$, $a = 1.5$ (except for Fig. 3)

coincide with those of Section 3 but the displacement of resonance minimum in this case reads: $\delta_0 = \sum_{k=1}^{2} k\Omega V_{k-1}^2/(k^2\Omega^2 + \Gamma^2)$.

In Figs. 2-4 we plot the CPT resonance line, the displacement of resonance, $\delta_0$, versus modulation index $a$ and the profiles of the in-phase and quadrature amplitudes. We see that the full-spectrum model, the 3-wave approximation and the direct numerical integration are in a very good agreement with one another. The 3-wave model, however, overestimates the displacement $\delta_0$ due to lack of the higher sidebands which effectively compress the energy distance between the lower states $|a\rangle$ and $|b\rangle$.

## 5. Conclusion

In summary, we proposed a theoretical framework for the FM spectroscopy of the CPT resonance and attested it using numerical simulations. In particular, we quantified the influence of the light shift of the atomic levels on the location of resonance that may be important for selecting the zero light shift regime making the atomic clocks less sensitive to variations of light intensity.

We would like to add that our results are in good agreement with a recent experiment [5] on the FM spectroscopy of absorption in Rb vapor. However, the experimental in-phase and quadrature amplitudes vanish at two different points of $\delta$ as opposed to formulae (8)-(9) which predict a common zero of $M_S$ and $M_Q$ (we mean the central zero for the latter).

In that regard, we also refer to the technique [4] where a combined use of the in-phase and quadrature signals for stabilizing atomic frequency standards relies on the difference between the zeros of the respective amplitudes. Under that approach, the in-phase response is used to lock the local oscillator to the atomic resonance whereas the difference between the zeros of the in-phase and quadrature signals allows to control the light shift of the atomic levels. The above mentioned splitting of zeros may be attributed to the absorption of laser field in the optically thick atomic cell [4]. Therefore, we feel a need to generalize the proposed model onto a thick optical layer case. We plan to address this important problem in the nearest future.

## References

[1] Shah V and Kitching J 2010 *Adv. At. Mol. Opt. Phys.* **59** 21
[2] Vanier J 2005 *Appl. Phys.* B **81** 421
[3] Zhu M and Cutler L S 2000 *32nd Annual Precise Time and Time Interval (PTTI) Meeting* p 331
[4] McGuyer B H, Jau Y Y and Happer W 2009 *Appl. Phys. Lett* **94** 251110
[5] Vaskovskaya M I, Zibrov S A, Velichansky V L, Tsygankov E A, Petropavlovsky S V and Yakovlev V P 2017 *Proc. 3rd International Conference LaPlas-2017 (Moscow)* p 279

# Dry cryomagnetic system with $MgB_2$ coil

**D A Abin, N A Mineev, M A Osipov, S V Pokrovskii and I A Rudnev**

National Research Nuclear University MEPhI (Moscow Engineering Physics Institute), Kashirskoe sh. 31, 115409 Moscow, Russia

E-mail: Dima_abin@mail.ru

**Abstract**. $MgB_2$ may be the future superconducting wire material for industrial magnets due to it's higher operation temperature and potentially lower cost than low temperature superconductors (LTS) have. We designed a compact cryomagnetic system with the use of $MgB_2$. The possibility of creating a magnet with a central field of 5 T from a commercial $MgB_2$ wire by the "react and wound" method was investigated. The magnetic system is cooled by a cryocooler through a copper bus. The magnet has a warm bore diameter of 4 cm. The design of a magnet consisting of three concentric solenoids is proposed: an internal one of high-temperature superconductor (HTS), an average of $MgB_2$, and an external of NbTi. The operating current of the system is 100 A. Two pairs of current leads are used. A separate pair of current leads for power supplying NbTi coil allows testing of $MgB_2$ and HTS coils in an external field. The load curves for each of the magnets are calculated.

## 1. Introduction

In comparison with the LTS materials, $MgB_2$ is characterized by higher values of the critical temperature, $T_c \approx 39$ K, and the higher critical magnetic field, $H_{c2} \approx 60$ T. This material finds application in energy transport systems, but much more interest is caused by the possibility to use it in magnetic systems. Solenoids made from this superconductor can be used in magnets, superconducting turbine systems, motors, magnetic focusing and separation systems, magnetic energy storage devices. The main development direction for the use of modern magnets based on $MgB_2$ wires are MRI devices [1]. Due to the specific application, the designed coils have a large internal diameter allowing to place inside them a human body or some its parts.

Due to the high critical temperature of $MgB_2$, it becomes possible to design cooling systems using cryocoolers and solid nitrogen [2]. However, there exists a problem of wire performance degradation under bending strain which impedes manufacturing solenoids with small diameters, especially from ex situ $MgB_2$ wires produced by powder-in-tube (PIT) technology. Usually for the production of solenoids with diameters less than 15 cm the "wound and react" technology is used, which avoids the occurrence of bending stress during winding. This technology allows to make a $MgB_2$ coil with a high critical current value and, as a consequence, with a high value of the central magnetic field [3]. However, in this case the obtained parameters are affected by additional stages of the technological process such as a homogeneous annealing. That is acceptable for the wire manufacturer or industrial production, but users, who want to make themselves a $MgB_2$ coil, have some additional difficulties. We have investigated a possibility to create a cryomagnetic system based on modern commercially available $MgB_2$ wire without homogeneous annealing. The following parameters were set for the system design:

- The diameter of the warm bore is not less than 40 mm.

- The shaft height of the magnet is not less than 100 mm.
- The cooling system does not contain liquid refrigerant.
- The maximum magnetic field on the axis of the bore of at least 5 T.
- The maximum operating current 100 A.

Sketch of the cryomagnetic system is shown in Figure 1. The RDK 408D2 cryocooler and the $MgB_2$ wire manufactured by Columbus Superconductors SpA were used as a basis for the system.

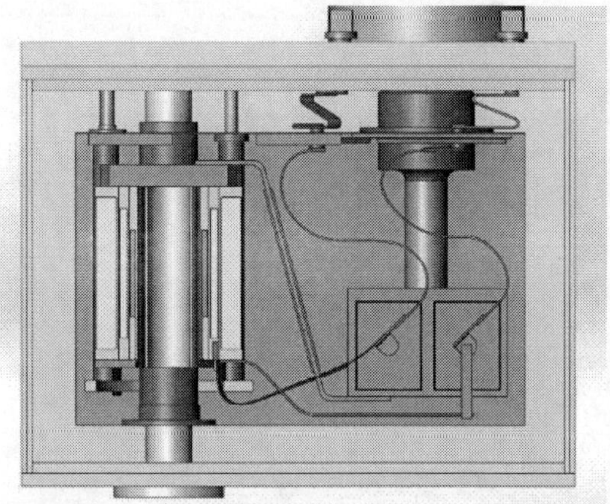

**Figure 1.** The model of the cryomagnetic system

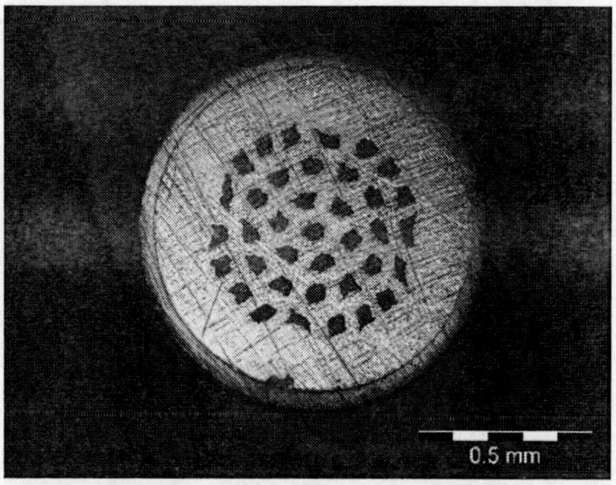


**Figure 2.** Photography of the wire cross section. Diameter 1 mm, 37 $MgB_2$ filaments, filling factor 13,5 %

## 2. $MgB_2$ wire

Photo of a bare wire cross-section is shown in Figure 2. The wire of 1.0 mm diameter contains 37 $MgB_2$ filaments, each of them is enclosed by the Nb barrier and placed in Monel matrix. The $MgB_2$ filaments take up about 13.5 % of total cross-section area of the wire. The wire is stabilized by a copper layer electro-deposited on its surface.

Dependence of critical current on external magnetic field is shown on Figure 3. Measurements were carried out at liquid helium in magnetic field applied perpendicular to the wire axis. The critical current at $H = 5$ T is about 55 A. Despite the fact that the critical temperature of the $MgB_2$ is much higher than the temperature of liquid helium, we have to design a cryomagnetic system with an operating temperature of 4.2 K because of such a low value of the wire critical current. The possibility of using $MgB_2$ wire for coil application at temperatures above 4.2 K was considered in our previous work [4].

As we have intend to wound a coil, we measured the transport properties of the wire under bending deformation with diameter $D = 8$ cm. At $H = 5$ T the critical current decreased about 1.8 times. In our previous work we have shown that critical current is noticeably reduced with reduction of bending diameter [5]. We can expect that for $D = 5$ cm the critical current will be 20% less than for $D = 8$ cm and the current density, which is necessary to obtain the field $H_{coil} = 5$ T at the coil center, is estimated as $J_c = 3 \times 10^3$ A/cm$^2$ (we neglected, that the field on the coil winding is higher than at the center). For the $MgB_2$ coil a current in the inner loop restricts the operating current.

Thus, to make a coil with an internal diameter of 8 cm and a maximum magnetic field in the center of 5 T, we need 34 km of the $MgB_2$ wire (winding density 0.785). The coil should has the height of 30 cm, the outer diameter of 45 cm and the weight of over 300 kg. Obviously, a magnet based on this coil cannot compete with widely used NbTi magnets.

To optimize the space inside the cryostat we proposed magnet design consisting of three concentric solenoids, namely a pancake HTS coil inside, a $MgB_2$ one in the middle and a NbTi solenoid outside. Presence of the inner coil allows us to use a larger diameter for $MgB_2$ solenoid winding and so to deminish decrease of the critical current of the $MgB_2$ wire caused by a bending stress. In addition, this will reduce influence of magnetic field on $I_c$ in the middle coil. As seen in Figure 3, the critical current of the $MgB_2$ wire becomes more than 100 A, when magnetic field is reduced down to 4 T. This value exceeds the operating current of our system.

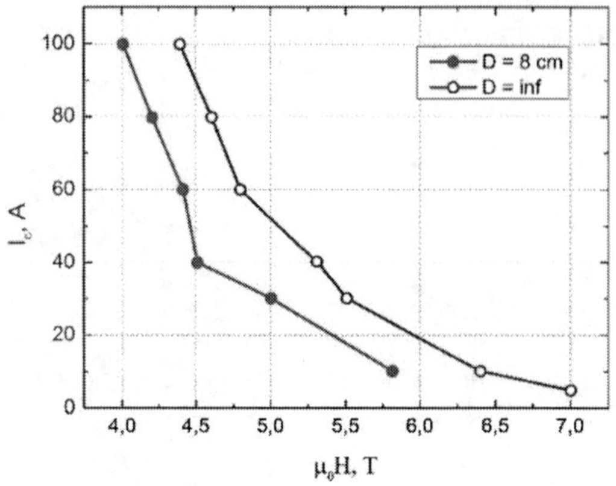


**Figure 3.** Dependence of critical current on external magnetic field for $MgB_2$ wire. The field direction is perpendicular to wire axis. Open circles - in the absence of bending stresses; closed circles - bend on the form with $D = 8$ cm.

Mechanical properties of HTS tapes allow us to make the inner solenoid with a small inner diameter without loss of critical current [6]. Due to high critical currents at 4.2 K in high magnetic fields of up to 5 T, HTS tapes are very suitable to make inserts in magnets [7]. To make the inner solenoid we used the 4 mm HTS tape produced by SuperOx. The critical current of the tape exceeds 100 A in $H$ = 5 T at $T$ = 4.2 K [7]. The coil consists of 10 double pancakes: the two upper and two lower ones have 27 turns, and 4 medium double pancakes have 30 turns. The remaining transition puncakes are asymmetric, a larger layer has 37 turns, a smaller layer 30. Calculation of the field produced by the HTS coil was carried out by the finite element method using the COMSOL multiphysics package.

**Table 1.** Parameters of coils

| | HTS | $MgB_2$ | NbTi |
|---|---|---|---|
| Diameter of wire (mm) | 4 mm width tape | 1 | 0.5 |
| Inner diameter (mm) | 64 | 80 | 100 |
| Winding density | | 0.725 | 0.8 |
| Outer diameter (mm) | 76 | 93 | 138 |
| Height (mm) | 80 | 120 | 140 |
| Total length of wire (m) | | 195 | 3473,7 |
| Total number of turns | 482 | 714 | 19684 |
| Average number of turns in the layer | | 119 | 259 |
| Load factor (mT/A) | 6.9 | 6.12 | 63.6 |

To investigate influence of magnetic field on current in the $MgB_2$ coil we separated current leads of the outer NbTi coil from common leads supplying the inner and outer coils. That allows us to test $MgB_2$ and HTS coils in external magnetic field.

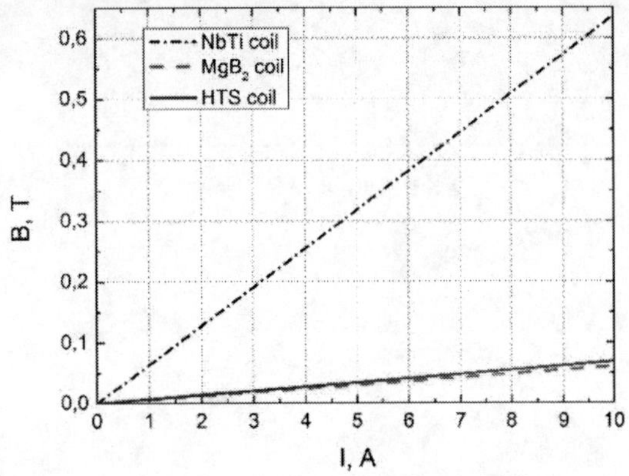


**Figure 4.** The loading curves of the NbTi, $MgB_2$ and HTS coils.

Parameters of the made coils are presented in Table 1. The calculated load curves for each coil are shown in Figure 4. The load factors calculated from the load curves are presented in Table 1.

## 3. Summary

The original design of the cryomagnetic system was propose taking into account the parameters of modern commercially available $MgB_2$ wires. It was shown that for the time being the transport characteristics of commercial wires are insufficient to make (using the "react and wound" technology) a solenoid producing magnetic field of 5 T. We have proposed magnet design consisting of three concentric solenoids. The inner coil is pancake HTS coil, the middle coil is $MgB_2$ soleniod, and the outer coil is NbTi solenoid. The load curves for each of the coils are calculated. Afterwards we plan to manufacture the designed cryomagnetic system and study the characteristics of the $MgB_2$ coil in an external magnetic field.

The main obstacles preventing manufacture of a $MgB_2$ solenoid with the "react and wound" technology are both low values of critical currents of industrial wires in high magnetic fields and a serious degradation of $I_c$ under bends with diameters less than 100 mm. Such a manufacture becomes possible if the critical current in the 1mm wires achieves 240 A (taking into account a degradation under bending). In this case one can make a 5 T solenoid with internal diameter of 8 cm and overall dimensions of about 15 cm.

**Acknowledgments**
The work is supported by the Ministry of Education and Science of the Russian Federation (Agreement №14.575.21.0050 (RFMEFI57514X0050)).

## References

[1] Ling J, Vocco P J, Hahn S, Qu T, Bascuñán J and Iwasa Y 2017 *Supercond. Sci. Tehnol.* **30** 024011
[2] Iwasa Y, Bascuñán J, Hahn S and Park D K 2012 *Phys. Proc.* **36** 1348-53
[3] Sumption M D, Bhatia M, Buta F, Bohnenstiehl S, Tomsic M, Rindfleisch M, Yue J, Phillips J, Kawabata S and Collings E W 2007 *Physica C* **458** 12-20
[4] Abin D A, Mineev N A, Osipov M A, Pokrovskiy S V and Rudnev I A 2016 *J. Phys. Conf. Ser.* **747** 012023
[5] Abin D A, Mineev N A, Osipov M A, Pokrovsky S V and Rudnev I A 2015 *Phys. Proc.* **71** 412-

6
[6] Rudnev I, Mareeva A, Mineev N, Pokrovskiy S and Sotnikova A 2014 *J. Phys. Conf. Ser.* **507** 022029
[7] Chepikov V, Mineev N, Abin D, Petrykin V, Pokrovskii S, Amelichev V, Molodyk A, Lee S, Samoilenkov S, Rudnev I and Kaul A 2017 *IEEE Trans. Appl. Supercond.* **27** 8000905

IOP Conf. Series: Journal of Physics: Conf. Series **941** (2017) 012057 doi:10.1088/1742-6596/941/1/012057

# Simulation of magnetization and levitation characteristics of HTS tape stacks

**I V Anischenko*, S V Pokrovskii, N A Mineev**

National Research Nuclear University MEPhI (Moscow Engineering Physics Institute), Kashirskoe sh. 31, 115409 Moscow, Russia

*E-mail: mephizic@gmail.com

**Abstract**. In this work it is presented a computational model of a magnetic levitation system based on stacks of high-temperature second generation superconducting tapes (HTS) GdBa2Cu3O7-x. Calculated magnetic field and the current distributions in the system for different stacks geometries in the zero-field cooling mode are also presented. The magnetization curves of the stacks in the external field of a permanent NdFeB magnet and the levitation force dependence on the gap between the magnet and the HTS tapes stack were obtained. A model of the magnetic system, oriented to levitation application, is given. Results of modeling were compared with the experimental data.

## 1. Introduction

Currently, high-temperature superconductors are widely applied in manufacturing current-carrying constructing elements. The superconducting ceramics are replaced by the thin second generation high-temperature superconductor tapes having high current-carrying characteristics, strength and simplicity of thermal stabilization. This led to the prospects of their use as the main material for manufacturing of trapped flux magnets, HTS levitation bearings, suspensions and other devices.

As a basic and simple element of a magnetic or levitation system, a stack of HTS tapes can be considered. The magnetic properties and levitation characteristics of the tapes stacks depend not only on critical characteristics of the used materials but its geometry (the number and shape of the tapes), conditions of magnetization and cooling. Therefore to design any devices based on the HTS tapes stacks it is necessary to determine the technical parameters of the elements taking into account the magnetic and transport characteristics of the tapes.

This work aims to develop simulation model that makes it possible to calculate magnetic and levitation properties of a system based on HTS tapes stacks of various configurations.

## 2. Computational model

The simulated magneto-levitation system is a stack of spatially separated planar superconducting layers. The geometry of the system is shown in the Figure 1. The measurements were performed on a stack of superconducting tapes produced by SuperOx (Russia) with the following characteristics. The thickness of the superconducting layer GdBa2Cu3O7-x is 1 micron. The thickness of the silver layer is 1.5 – 2 micron Width / diameter of tapes is 4 - 12 mm. The critical current (at 77 K, own field) - 150 A (criterion 1 μV / cm)

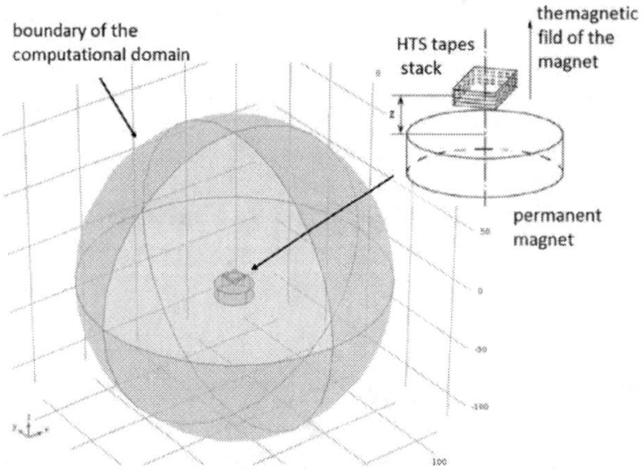


**Figure 1.** The geometry of the magnetic levitation system.

To study the behavior of a HTS tapes stack in an external magnetic field a finite element method was used. The model was developed on the base of the Partial Differential Equations (PDE) form of the Comsol Multiphysics software module.

To describe the superconducting magnetic system we used Faraday and Ampere laws equations. The resistance nonlinear dependence on the current is given by the power law (1) and the current-voltage characteristic can be written by formula (2):

$$\rho = \frac{E_c}{J_c} \cdot \left(\frac{|J|}{J_c}\right)^{n-1} \tag{1}$$

$$\boldsymbol{E} = \rho * \boldsymbol{J} \tag{2}$$

where $E_c$ is the electric field which equals 1 μV/cm, that corresponds to the critical current. The $n$-value is assumed to be 21, which is typical for the samples of HTS tapes. The critical current density $J_c$ is determined on the basis of the transport measurements of the used HTS tapes.

Also, in this model it is taken into account the dependence of the critical current on the magnetic field perpendicular to the wide surface of the tapes. Calculations were performed within the two-exponential model [1]. Using it increases the convergence of the solution and also significantly reduces the computation time in comparison with Kim's model.

In addition, the interaction force of the HTS tapes stack with the permanent magnet (the levitation force) is calculated by formula (3):

$$\boldsymbol{F} = \int_V \boldsymbol{J} \times \boldsymbol{B}\, \mathrm{d}V \tag{3}$$

where $\boldsymbol{B}$ is the magnetic field induction of the permanent magnet. The currents distribution $\boldsymbol{J}$ in the superconducting tapes of stack was calculated in result of the solutions of the PDE equations. Integration is performed by all entire volume of the superconductor.

In Comsol Multiphysics the general PDE standard equation view is (4):

$$e_a \frac{\partial^2 u}{\partial t^2} + d_a \frac{\partial u}{\partial t} + \nabla \cdot \Gamma = F \tag{4}$$

Here the dependent variable, magnetic field $\boldsymbol{u}$=$\boldsymbol{H}(x,y,z)$, is chosen in the entire space. All variables and parameters of equation (4) are given by (5)-(8):

$$\Gamma = \rho \cdot \boldsymbol{J} \cdot \begin{bmatrix} 0 & -1 & 1 \\ 1 & 0 & -1 \\ -1 & 1 & 0 \end{bmatrix} \tag{5}$$

$$d_a = \begin{bmatrix} \mu & 0 & 0 \\ 0 & \mu & 0 \\ 0 & 0 & \mu \end{bmatrix} \tag{6}$$

$$e_a = 0 \tag{7}$$

$$F = 0 \tag{8}$$

The current components are determined from the relations (9):

$$J = \begin{cases} J_x = \frac{\mathrm{d}H_z}{\mathrm{d}y} - \frac{\mathrm{d}H_y}{\mathrm{d}z} \\ J_y = -\frac{\mathrm{d}H_z}{\mathrm{d}x} + \frac{\mathrm{d}H_x}{\mathrm{d}z} \\ J_z = \frac{\mathrm{d}H_y}{\mathrm{d}x} - \frac{\mathrm{d}H_x}{\mathrm{d}y} \end{cases} \tag{9}$$

To solve the problem it is also necessary to determine the boundary conditions. In our case it is convenient to use the Dirichlet conditions on the boundary of the external domain around the stack. The magnetic field is calculated as a superposition of external and internal contributions [2]. The internal contribution $H_s = [H_{sx} \quad H_{sy} \quad H_{sz}]$ is calculated from the solution of the Ampere equation, the external contribution $H_e = [H_{ex} \quad H_{ey} \quad H_{ez}]$ corresponds to the applied magnetic field.

It should be noted that the singularity of the problem requires that the dimensions of the calculated domain outside the superconductor should exceed the sample size by at least 10 times (see Figure 1).

One of the specific questions of developed model is the mesh construction question. In our case the ratio of the lateral dimensions of the stack to the thickness of the superconducting layer is ~ $10^3$. Therefore, using a standard mesh leads to unsatisfactory results. We used multiscale structure: the mesh element's size changes several times as we approach from the external domain boundary to the sample (Figure 2).

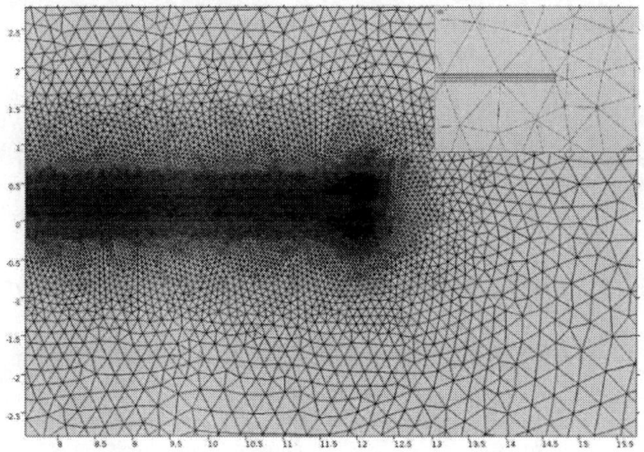

**Figure 2.** Multiscale mesh structure for five HTS tapes stack. Insets: enlarged mesh near the edge of the tape.

In general, the developed model is three-dimensional, however, in order to decrease the calculation time we additionally used the fact that the outer layers of the stack shield the applied magnetic field. Consequently, the inner layers of the HTS stack of tapes can be considered as two-dimensional layers of zero thickness. Our research has shown that with this approach the calculation time is reduced by more than 60 times, the maximum discrepancy between the 2D and the 3D models does not exceed 10% (Figure 3).

## 3. Results and discussion

The magnetic and levitation characteristics of the stacks were calculated for square and round shape of tapes, width/diameter is 12 mm, number of tapes in a stack was varied from 1 to 5.

Gradient magnetic field was produced by the permanent Nd-Fe-B magnet, 30 mm diameter, 10 mm thickness, class n42 (induction on the surface in the range 0.30 - 0.35 T). During the magnetization the HTS stack slowly approaches to the magnet (at a constant speed of 2 mm/s), and then moves away from it. The stack fall into an inhomogeneous magnetic field whose magnitude varies with time. Thus we realized zero field cooling mode magnetization.

Shown in the Figure 3 dependence $B_z(r)$ is for the moment of complete removal of the magnetic field for 2D (the HTS tape thickness is equal zero) and 3D (superconductor thickness is 1 μm) models and experimental data for the one tape in a stack.

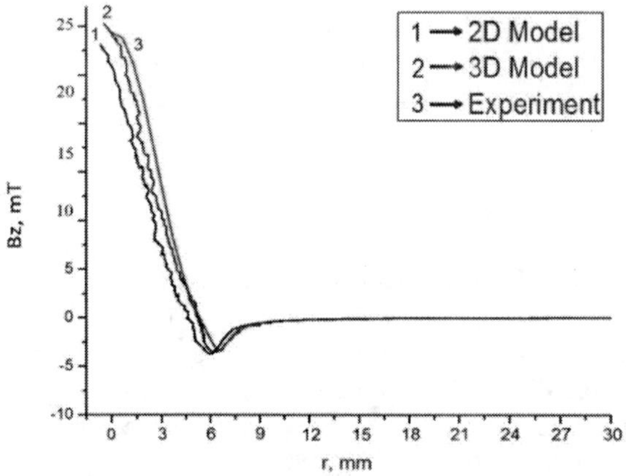


**Figure 3.** Profile of the trapped magnetic flux $B_z(r)$ by the HTS tapes stack. The edge of the sample is at $r = 6$ mm, $r = 0$ mm - center of the sample.

The calculated magnetic field and the current distributions for the tapes stacks of different geometries were obtained. We performed comparison of the calculated distributions of the magnetic flux trapped by the stacks with the experimental dependences of Hall magnetometry measurements. Scanning height was 0.7 mm [3]. The calculated data are in consistence with the experimental data for the samples of circular shape (Figure 4 (a)) and the square shape (Figure 4,(b)).

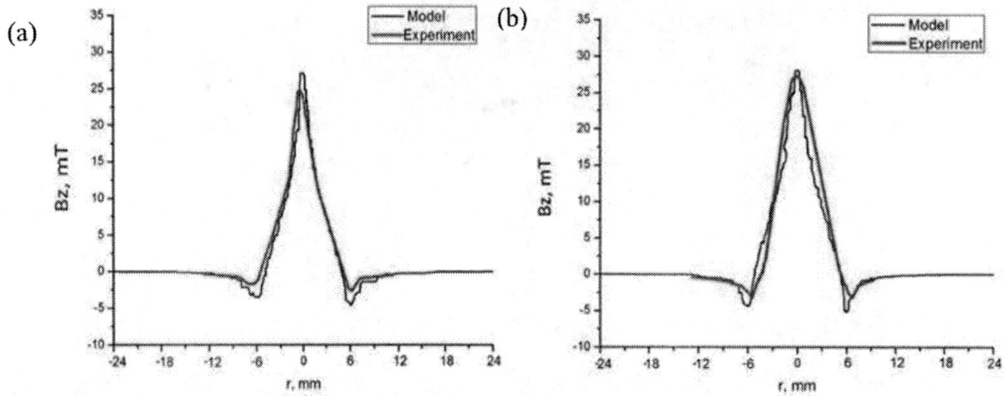


**Figure 4.** Dependence of $B_z$ $(r)$ for the round tape HTS (a) and for the square HTS tape (b) The edges of the sample: $r = 6$ mm and $r = -6$ mm.

The magnetization curves of the tapes stacks were calculated in the external gradient field of the permanent magnet. Comparison of the calculated hysteresis loops for the one and five HTS tapes in the stack with the experimental dependences shows qualitative and quantitative coincidence (Figure 5).

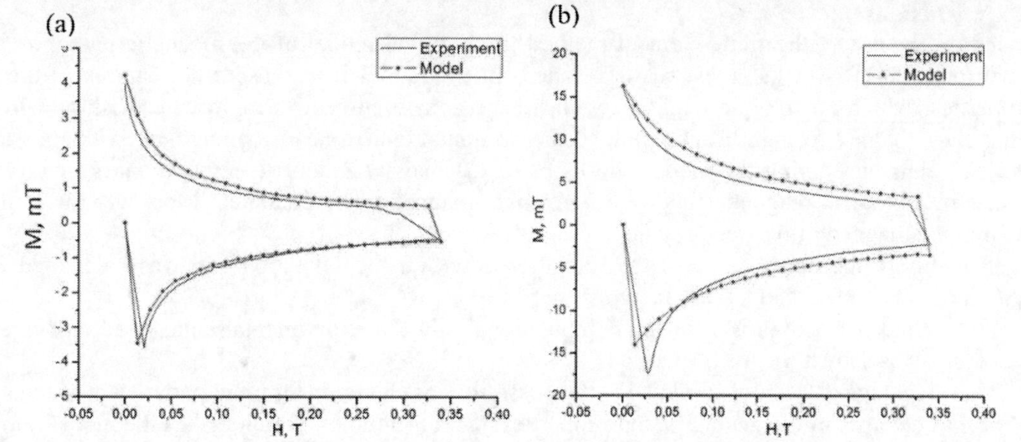


**Figure 5.** Theoretical and experimental magnetization curves for round-shaped tapes stacks in an external field of a permanent magnet for the single tape (a) and for 5 tapes in the stack (b).

The dependence of the levitation force *F* on the levitation gap *z* between the magnet and the tapes stacks was obtained (Figure 6). The maximum calculated levitation force increases in proportion to the number of tapes in the stack from 1 to 5 which corresponds to the experimental data. However, in general, for a large number of tapes (more than 50-60) the dependences of the magnetization and the levitation force on the number of tapes in the stack are not linear [5].

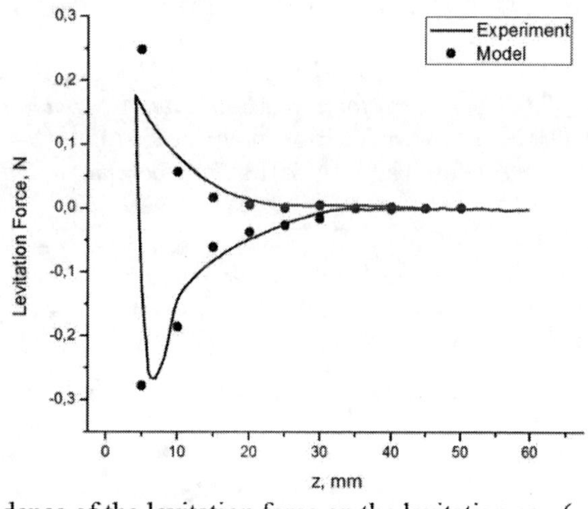


**Figure 6.** Dependence of the levitation force on the levitation gap (a stack of 5 tapes).

The differences in the dependences can be explained by the presence of external factors (in particular, the thermal processes, the angular dependences of the critical current on the magnetic field) that were not taken into account in the modeling, model simplifications and by the features of an experimental method (for more details see [4]). But that discrepancies are minor and including into the model for additional parameters negatively affects on the calculation time. Nevertheless, the analytical model demonstrates a qualitative coincidence with the experimental results.

Obtained results indicate the existence of optimal parameters and system configurations to achieve the maximum performance of devices based on HTS tapes stacks. Thus, the developed model allows to predict the characteristics and optimize the parameters of the levitation and magnetic systems for various applications.

## 4. Conclusions

Based on the use of the finite element method the physical model of the magnetic levitation system based on the HTS stacks tapes was developed. In the model it was used the features of the stacks properties, which on the one hand necessitated the construction of a multiscale mesh and from the other hand helped to significantly reduce the computational time in comparison with the standard mesh of elements. Using the model, the loops of the magnetization of the tapes stacks with various shapes in a gradient magnetic field of a permanent magnet were calculated. Moreover the profiles of the trapped magnetic flux were obtained.

In addition, the dependence of the levitation force on the levitation gap was calculated for the system of HTS tapes stack - permanent magnet.

We carried out comparison of the computational and the experimental data, good coincidence of the results was shown.

The advantage of developed model is the possibility of changing all input parameters of the system, as well as changes in the geometry and properties of HTS samples, which makes the model applicable to the calculation and optimization of magnetic and levitation systems of various configurations (in the fields of building of new transport systems of various scales, electrical rotating machines, energy and wind generators).

## Acknowledgments

This work was supported by the National Research Nuclear University MEPhI Academic Excellence Project (contract 02.A03.21.0005).

## References

[1] Podlivaev A I, Rudnev I A, Shabanova N P2014 *Bulletin of the Lebedev Physics Institute* **41**(12) 351–4

[2] Rodriguez-Zermeno V M , Sorensen M P, Pedersen N F, Mijatovic N, Abrahamsen A B *Fast 2D Simulation of Superconductors: A Multiscale Approach TECHNICAL PAPERS AND PRESENTATIONS* https://www.comsol.com/paper/ 6765

[3] Patel A, Hopkins S C, Glowacki B A 2013 *Superconductor Science and Technology* **26**(3) 032001

[4] Pokrovskii S, Osipov M, Abin D, Rudnev I (2106) *IEEE Transactions on Applied Superconductivity* **26**(3) 7416628

[5] Pokrovskii S, Osipov M, Abin D, Rudnev I, Ermolaev Y S, Mineev N A (2015) *Physics Procedia* **65** 141-4

IOP Conf. Series: Journal of Physics: Conf. Series **941** (2017) 012058 doi:10.1088/1742-6596/941/1/012058

# Low-temperature anomalies of EXAFS at the *K*-edge of As in superconducting $LaFe_{0.89}Co_{0.11}AsO$

**A Bianconi[1,2], A P Menushenkov[1], V G Ivanov[1], A A Ivanov[1] and B Joseph[2,3]**

[1]National Research Nuclear University MEPhI (Moscow Engineering Physics Institute), Kashirskoe sh. 31, 115409 Moscow, Russia

[2] RICMASS, Rome International Center for Materials Science Superstripes, Via dei Sabelli 119A, 00185 Roma, Italy

[3] Elettra-Synchrotron Trieste, 34149 Basovizza, Trieste, Italy

Email: APMenushenkov@mephi.ru

**Abstract**. The temperature dependence of the EXAFS-spectra measured above the *K* absorption edge of As in superconducting ($T_c$ = 13.5 K) single crystals of $LaFe_{0.89}Co_{0.11}AsO$ were investigated. Analysis of the spectra in the harmonic approximation revealed anomalies in the temperature dependence of the Debye-Waller factor for As-Fe interatomic bond which correlated with the temperature dependence of electrical resistivity, the coefficient of thermal expansion and magnetic susceptibility given in literature. Taking into account that similar anomalies were earlier observed in superconducting oxides based on $BaBiO_3$ and in cuprates we conducted the EXAFS spectra analysis in the anharmonic approximation using the potential of arbitrary shape for As-Fe bond vibrations. It was shown that the double-well approximation describes the temperature dependence of the EXAFS spectra better than the harmonic one. The temperature dependence of tunneling frequency and a distance between the wells for double-well potential were obtained. The results indicate that local structural dynamic heterogeneities have a strong impact on macroscopic properties of iron-based superconductors.

## 1. Introduction

A novel unconventional superconductor $LaFe_{1-x}Co_xAsO$, discovered in 2008 [1], belongs to the iron-based family. A comprehensive study of this material is nowadays one of the most attractive, though challenging, problem in physics of high-temperature superconductors (HTSC). Understanding the processes that underlie the formation of the superconducting state in these systems is far from complete and promises to gain insight into the nature of HTSC.

This research concerns the study of the local structure anomalies of superconducting $LaFe_{0.89}Co_{0.11}AsO$ at low temperature.

## 2. Results and discussion

High-quality single crystal of superconducting $LaFe_{0.89}Co_{0.11}AsO$ of rather large size $2\times2\times0.2$ mm$^3$, were grown under ambient pressure in NaAs flux in Ames laboratory, USA [2, 3].

The EXAFS spectra measurements were made in the European Synchrotron Radiation Facility ESRF (Grenoble, France) at BM23 beamline [4]. X-ray absorption spectra were measured in fluorescence mode at the *K*- edge of As in temperature range of 14-300 K with an increment of ~ 10 K. Low-temperature measurements were carried out using a liquid-helium flow cryostat, temperature instability didn't exceed ±1K at 300K and 0.1K at 4K. The double-crystal Si (111) monochromator was detuned to 50% from maximal intensity to eliminate the contributions of higher harmonics.

The polarized As *K*-edge EXAFS measurements were carried out in the geometry when the crystal was oriented so that the As-Fe bond was parallel to the electric field direction of the photon beam. In this case only one of four tetrahedral As-Fe bonds made a main contribution to the spectra.

The fine oscillatory structure of the absorption spectra was converted to the atomic pair distribution function (PDF) in processing the EXAFS spectra by Fourier transform technique. Consideration of the PDF in [3] as a function of temperature revealed anomalies in the behavior of the first peak of Fourier-transform which corresponded to As-Fe bonds. The amplitude of the peak is determined by the Debye-Waller factor value and therefore should increase monotonically with temperature decreasing due to attenuation of thermal vibrations. However, reported results [3] demonstrated a reverse behavior in the vicinity of 60 and 150 K which was interpreted in the framework of harmonic approach to the EXAFS data analysis as abnormal increase of Debye-Waller factor at given temperature points.

It should be noted that the same temperatures correspond to anomalies in the temperature dependencies of electrical resistivity [5], the coefficient of thermal expansion [6] and magnetic susceptibility [7] of these materials. It clearly demonstrates that the local structure influences on the properties of $LaFe_{0.89}Co_{0.11}AsO$ and specially emphasizes the importance of identified anomalies. Therefore, the correct description of these anomalies is a very significant problem.

It's worth mentioning that similar anomalies were observed in cuprates [8-13] and in $BaBiO_3$-based [14-16] high-temperature superconductors. It is believed that vibrations in the double-well potential may be responsible for the formation of the superconducting state [12-16]. Our recent studies of iron chalcogenides $Fe(TeSe)_{1-\delta}$ [17] have been revealed the correlation of local structure peculiarities and superconducting properties of $Fe(TeSe)_{1-\delta}$. Drechsler et al. recently presented a fruitful discussion concerning mechanisms of superconductivity in iron-based superconductors [18]. However, the nature of low-temperature anomalies in these compounds has not been well established so far [20, 21].

In [21] we for the first time successfully applied the concept of a double-well potential to one of the iron-based superconductors – $LaFe_{1-x}Co_xAsO$.

The reason for the observed anomalies is that the standard harmonic approach for EXAFS spectra processing is incorrect if a bond movement is strongly anharmonic, and one should refer to the more general expression:

$$\chi(k) = -\sum_n \frac{1}{k}|f_n(k,\pi)| \int \frac{g_n(r)}{r^2} \sin[2kr + 2\delta_1(k) + \phi_a(k,\pi)]dr \qquad (1)$$

with non-Gaussian PDF $g_n(r) = N_n \dfrac{\sum |\Psi_{n'}(r)|^2 \exp(-E_{n'}/kT)}{\sum \exp(-E_{n'}/kT)}$, where $g_n(r)$ – pair distribution function (PDF); $N_n$ – coordination number; $E_{n'}$ and $\Psi_{n'}$ – $n'$ energy level and wave function in the model potential.

As a result, the process of fitting the experimental spectra by the new model EXAFS function (1) gives us the shape of the unknown potential U and atomic pair distribution function g. The treatment was performed by means of FEFF [23] and VIPER [24] software.

Double-well potential consists of two parabolas $U_1(r) = (a/2)(r - r_1)^2$ and $U_2(r) = (b/2)(r - r_2)^2$. The presence of two equilibrium atomic positions is most likely due to the redistribution of the electron density, which is supposed to occur dynamically, so that there is a shortening or lengthening of the studied bonds. The coefficients of stiffness of parabolas *a* and *b* are equal and fixed in the current consideration, positions of the wells $r_1$ and $r_2$ are varied.

Figure 1(a) demonstrates comparative results of both harmonic and anharmonic approach in EXAFS data processing for superconductor $LaFe_{0.89}Co_{0.11}AsO$. The fits of experimental Back Fourier Transforms (BFT) by harmonic and double-well models for the set of key temperatures are shown. It is seen that double-well model provides a slightly better fit than harmonic one (corresponding experiment-model mismatch factors $R = \left( \frac{\sum_{i=1}^{N} |Data_i - Model_i|^2}{\sum_{i=1}^{N} |Data_i|^2} \right)^{1/2} \cdot 100\%$ are presented in Table 1).

**Table 1.** *R*-factors (quality of the fit) for the treatment of the experimental BFT by conventional harmonic and double-well models for the sets of key temperatures for superconductor $LaFe_{0.89}Co_{0.11}AsO$.

| | *R*-factor, % | |
|---|---|---|
| Temperature, K | Harmonic model | Double-well model |
| 20 | 19.3 | 16.7 |
| 30 | 19.4 | 20.6 |
| 70 | 20.4 | 16.1 |
| 100 | 24.1 | 17.6 |
| 225 | 21.8 | 21.3 |

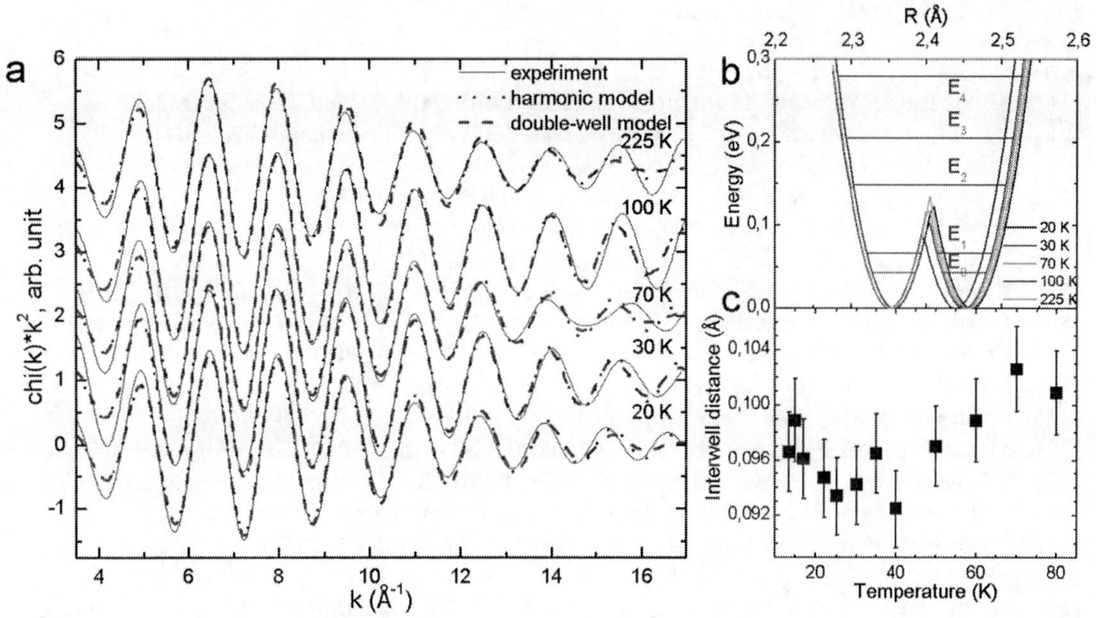


**Figure 1.** Comparative fits of experimental Back Fourier Transforms (BFT) by harmonic and double-well models for the set of key temperatures (a), calculated double-well potentials with energy levels (b) and temperature dependence of the distance between the wells (c) for superconductor $LaFe_{0.89}Co_{0.11}AsO$.

The calculated double-well potentials for $LaFe_{0.89}Co_{0.11}AsO$ are shown in Figure 1(b). One can see the change in the shape of potential when passing through the temperatures where anomalies are observed. Namely, when the temperature goes down, there is the decrease in the height of the potential barrier between wells, which accompanied by the increase in the probability of Fe atoms tunneling through the barrier, reflected in [21] by PDF dependencies.

Numerical characteristic of the tunneling process – its frequency – can be determined from the found energy levels $E_1$ and $E_0$ in a double-well potential according to quantum mechanics as $\omega = E_1 - E_0$. Our recent work [21] reported on the temperature dependence of tunneling frequency, from which it is be clear that in the vicinity of the anomalies under discussion the oscillations of Fe atoms in the double-well potential near As are amplified and become more dynamic. This is due to the decrease of the distance between the wells of the potential (Figure 1(c)), which results in the decrease of the height of the potential barrier, as shown on Figure 1(b).

In addition, it should be noted that tunneling frequencies are in the terahertz range, which is typical for the phonon modes. The study of phonon dispersion may be useful to search for some phonon spectrum features in the temperature range where anomalies are observed. If they are found, this will independently confirm the feasibility of the proposed scenario of dynamical atomic oscillations in the double-well potential.

## 3. Conclusion

Summarizing all the above, we can conclude that the anomalies in the local crystalline structure of iron-based superconductor $LaFe_{0.89}Co_{0.11}AsO$ observed by EXAFS may be due to the excitation of Fe atoms vibrations in double-well potential around As atoms, causing, in turn, changes in the temperature dependencies of electrical resistivity, linear thermal expansion and magnetic susceptibility.

**Acknowledgements**
Authors acknowledge the ESRF Program Committee for providing the opportunity of XAFS measurements at BM23 beamline. APM and VGI acknowledge the support of the Ministry of Education and Science of the Russian Federation within agreements no. 14.587.21.0002, project no. RFMEFI58714X0002.

## References

[1] Kamihara Y, Watanabe T, Hirano M and Hosono H 2008 *J. Am. Chem. Soc.* **130** 3296
[2] Yan J Q, Nandi S, Zarestky J L, Tian W, *et al* 2009 *Appl. Phys. Lett.* **95** 222504
[3] Joseph B, Ricci A, Poccia N, Ivanov V G, *et al* 2016 *J. Supercond. Nov. Magn.* **29** 3041
[4] Mathon O, Beteva A, Borrel J, *et al* 2015 *J. Synchr. Rad.* **22** 1548
[5] McGuire M A, Christianson A D, Sefat A S, Sales B C, *et al.* 2008 *Phys. Rev.B* **78** 094517
[6] Wang L, Köhler U, Leps N, *et al* 2009 *Phys. Rev. B* **80** 094512
[7] Klingeler R, Leps N, Hellmann I, Popa A, Stockert U, *et al* 2010 *Phys. Rev. B* **81** 024506
[8] Mustre de Leon J, Conradson S D, Batistić B and Bishop A R 1990 *Phys. Rev. Lett.* **65** 1675
[9] Conradson S D, Raistrick I D and Bishop A R 1990 *Science* **248** 1394
[10] Mustre de Leon J, Batistić I, Bishop A R, *et al* 1992 *Phys. Rev. Lett.* **68** 3236
[11] Mustre de Leon J, Conradson S D, Batistić I, *et al* 1992 *Phys. Rev. B* **45** 2447
[12] Menushenkov A P, Kuznetsov A V, Chernikov R V, *et al* 2010 *Z. Kristallogr.* **225** 487
[13] Menushenkov A P, Kuznetsov A V, Chernikov R V, Ivanov A A, Sidorov V V and Klementiev K V 2013 *J. Surf. Invest.: X-Ray, Synchrotron Neutron Tech.* **7** 407
[14] Menushenkov A P, Benazeth S, Purans J, *et al* 1997 *Physica* C **277** 257
[15] Menushenkov A P, Klementev K V, Konarev P V and Meshkov A A 1998 *JETP Lett.* **67** 1034
[16] Menushenkov A P, Klementev K V 2000 *J. Phys.: Condens. Matter* **12** 3767
[17] Ivanov V G, Chareev D A, Ivanov A A, *et al* 2016 *Physics of the Solid State* **58** 447
[18] Drechsler S L, Rosner H, Grinenko V and Johnston S 2017 *Phys. Status Solidi B* **254** in press preprint arxiv:1701.00596 URL http://arxiv.org/abs/1701.00596. 1701.00596.
[19] Oyanagi H, Zhang C J, Sun Z H, *et al* 2009 *J. Phys.: Conf. Ser.* **190** 012094
[20] Joseph B, Iadecola A, Malavasi L and Saini N L 2011 *J. Phys.: Condens. Matter* **23** 265701
[21] Ivanov V G, Ivanov A A, Menushenkov A P, Joseph B and Bianconi A 2016 *J. Supercond. Nov. Magn.* **29** 3035
[22] Rehr J J, Mustre de Leon J and Zabinsky S I 1991 *J. Am. Chem. Soc.* **113** 5135
[23] Klementev K V 2001 *J. Phys. D: Appl. Phys.* **34** 209

IOP Conf. Series: Journal of Physics: Conf. Series **941** (2017) 012059 doi:10.1088/1742-6596/941/1/012059

# Temperature dependences of the time of electron-electron interactions in two-dimensional heterojunction

**K V Bukhenskyy[1], A B Dubois[1], S I Kucheryavyy[2], S N Mashnina[1], A S Safoshkin[1], A A Baukov[1] and E Yu Shchigorev[1]**

[1]Ryazan State Radioengineering University, Gagarina Str. 59/1, 390005 Ryazan, Russia

[2]National Research Nuclear University MEPhI (Moscow Engineering Physics Institute), Kashirskoe sh. 31, 115409 Moscow, Russia

E-mail: safoshkin.a.s@rsreu.ru

**Abstract**. The article discusses the joint solution of the Schrödinger and Poisson equations for two-dimensional semiconductor heterojunction. The application of a triangular potential of well approximation for the calculation of the electron-electron interaction is offered in the paper. The influence of the parameters of the selected approximation was analyzed.

## 1. Introduction

A number of experimental studies on the characteristics of the oscillations of the transverse magnetoresistance of the Shubnikov - de Haas effect in a wide range of temperatures and magnetic fields for the two-dimensional volumetric 3D and 2D electrons have detected some anomalies that had a certain affinity in qualitative similarities, but also significant differences [1]. For example, it was found that oscillating Dingle temperature $T_D$ depends on temperature $T$, and therefore the small-angle relaxation time $\tau_q$ also oscillates with $T$. Oscillations $T_D(T)$ and $\tau_q(T)$ were observed for the heavily doped heterostructures ($n_s$>$8.5\cdot 10^{11} cm^{-2}$), which was filled with match with basic and second excited subband [2,3]. The experiment has established a unique correlation of these anomalies with strong (for 3D and 2D degenerate electrons) electron - electron interaction. The attempt to quantitatively describe the experimental curves $\tau_q(T)$ in the existing theoretical ideas was not successful [4,5]. To explain in qualitative terms the observed effects, it would requires an examination of the channels of electron - electronic interactions in a complex system of 2D degenerate electrons. The application of the method of self-consistent solution of the Poisson and Schrödinger equations [4] allowed one to identify the kind of energy structure of the conduction band $E_c(z)$ of the nanostructure and the spatial distribution of the electron density $|\psi(z)|^2$ at the levels (subbands) of size quantization.

## 2. Heterojunction band structure calculation

In [6] we solved the system differential equations of Schrödinger and Poisson:

$$E_{\mathrm{F}}\left(-\frac{1}{k_{\mathrm{F}}^{2}}\frac{\mathrm{d}^{2}}{\mathrm{d}z^{2}}+\frac{V(z)}{E_{\mathrm{F}}}\right)\psi_{j}(z)=\varepsilon_{j}\psi_{j}(z);\quad -\frac{\mathrm{d}^{2}V}{\mathrm{d}z^{2}}=\frac{4\pi\rho(z)}{\chi} \tag{1}$$

where $E_{\mathrm{F}}=\frac{\hbar}{2m^{*}}$ is the Fermi energy, $k_{\mathrm{F}}=\sqrt{2\pi n_{s}}$ is wave vector, $n_{s}$ is electron surface density and

$$\begin{cases}\rho(z)=eN_{0}^{3/2}\left(1-\dfrac{z}{D}\right), & z<D,\\ \rho(z)=-A(z-D)\exp\left(-\dfrac{z-D}{a}\right), & z>D,\end{cases} \tag{2}$$

Here $D\cong\frac{2}{k_{\mathrm{F}}}$, $A=\frac{eN_{0}^{3/2}D}{2a}$ is normalization constant, $N_{0}$ is the doping level, $a$ is the length of Thomas-Fermi screening [2]. The solution of system (1) with (2) is in function and adjustable parameters $\lambda$ and $\beta$:

$$V(z)=c(\beta)\sum_{n=0}^{\infty}(-1)^{n}\beta^{n}\mathrm{e}^{-(n+1)\lambda z};$$

$$\psi_{k}^{+}(z)=\exp(iz)\sum_{j=0}^{\infty}b_{j}(k)\exp(-j\lambda z) \tag{3}$$

On the Figure 1 we use the curves $V(z)$ as a function of the adjustable parameters $\lambda$ and $\beta$. Figure 2 shows a band structure calculation for one of the samples with filling two subzones size quantization and the designation of the parameters that would be used below.

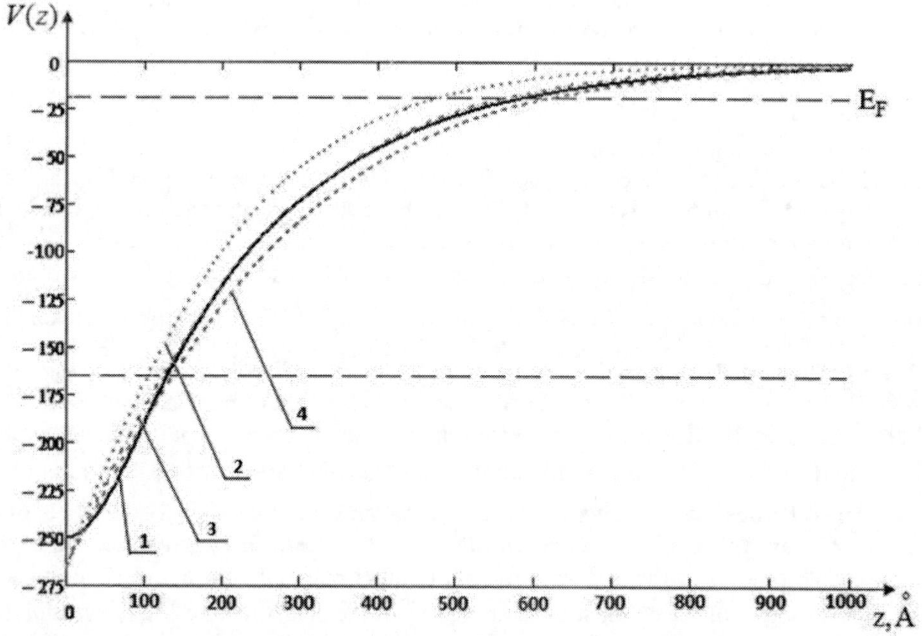


**Figure 1.** The $\lambda$ - profile of the potential well. Curve (1) was calculated using the formula (2), curves (2) - (4) were calculated using the formula (3) with $\lambda=0.65\cdot10^{6}$, $\lambda=0.55\cdot10^{6}$ and $\lambda=0.5\cdot10^{6}$ respectively.

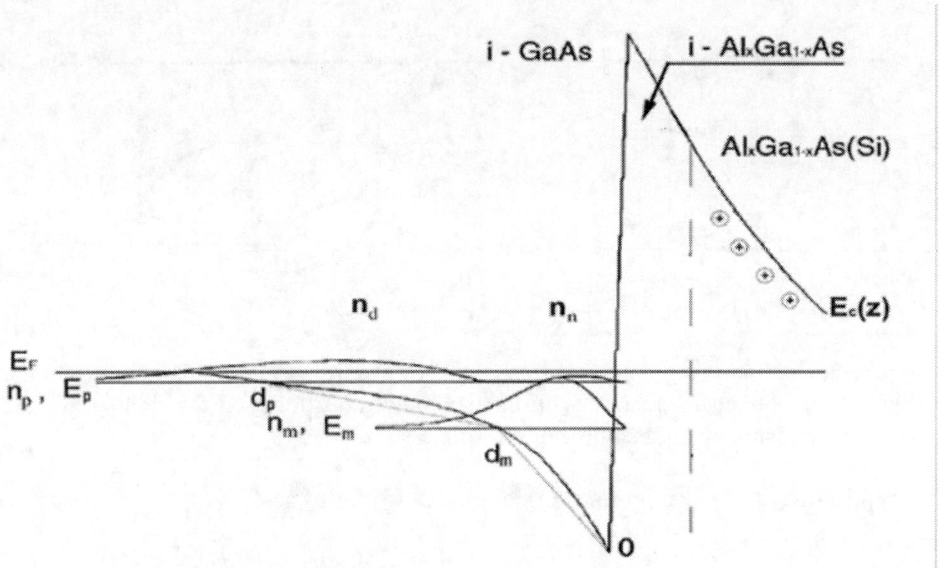


**Figure 2.** Energy diagram of the conduction band $E_c(z)$ for a heterojunction comes with two filled quantized subbands: $E_m$ и $E_p$; $n_m$, $n_n$, $n_d$ —2D electron concentration, $E_{\mathrm{F}}$ — the Fermi energy.

### 3. Electron–electron interactions

We describe the Hamiltonian of the electron-electron interaction in a strongly doped heterojunction:

$$H_{\mathrm{int}}^{ijkl}(t)=\sum_{\mathbf{q}} g(\mathbf{q})V_{\mathrm{tot}}^{ijkl}(\mathbf{q},\omega)a_{\mathbf{p+q}}^{+}(t)a_{\mathbf{p}}(t)+\sum_{\mathbf{p_i},\mathbf{p_j},\mathbf{q_i},\mathbf{q_j}} V_{\mathrm{tot}}^{ijkl}(\mathbf{q},\omega)\delta(\mathbf{p_i}+\mathbf{q_i}-\mathbf{p_j}-\mathbf{q_j})a_{\mathbf{p}}(t)a_{\mathbf{p+q}}^{+}(t) \qquad (4)$$

where $V_{\mathrm{tot}}^{ijkl}(\mathbf{q},\omega)$ is matrix element of Fourier image of the full potential of the Coulomb screening with the field operators of creation and annihilation:

$$a_{\mathrm{p}}(t)=\mathrm{e}^{-\frac{i}{\hbar}E(p)t}a_{\mathrm{p}},\quad a_{\mathrm{p}}^{+}(t)=\mathrm{e}^{-\frac{i}{\hbar}E(p)t}a_{\mathrm{p}}^{+}$$

Then, within the second term of the expansion of the external potential of the perturbation theory expression for the electron-electron interaction can be represented as following (5):

$$\frac{1}{\tau_{ij}^{ee}}=\int_{-\infty}^{\infty}\mathrm{d}\omega\sum_{k,m}\sum_{\mathbf{q}}\left|V_{\mathrm{tot}}^{ijkl}(\mathbf{q},\omega)\right|^2\sum_{\mathbf{k},\mathbf{p}}\delta\left(E_j(\mathbf{k}+\mathbf{q})+E_l(\mathbf{p}-\mathbf{q})-E_i(\mathbf{k})-E_k(\mathbf{p})\right)f_{\mathbf{k}}f_{\mathbf{p}}\left(1-f_{\mathbf{k+q}}\right)\left(1-f_{\mathbf{p-q}}\right) \qquad (5)$$

Thus, the solution to the problem of finding the analytical expressions for the relaxation time is reduced to finding the matrix elements of the screening for these types of interactions:

1) Intrasubband interaction, resulting in intrasubband transitions (matrix elements: $V_{tot}^{mmmm}$, $V_{tot}^{pppp}$, "e-e" times: $\tau_{mm}^{\mathrm{intra}}$, $\tau_{pp}^{\mathrm{intra}}$).
2) Intrasubband interaction, resulting in intersubband transitions (matrix elements $V_{tot}^{mpmp}$, $V_{tot}^{pmpm}$, "e-e" times: $\tau_{mm}^{\mathrm{intra}}$, $\tau_{pp}^{\mathrm{intra}}$).
3) Intersubband interaction, resulting in intrasubband transitions (matrix elements $V_{tot}^{mmpp}$, $V_{tot}^{ppmm}$, "e-e" times: $\tau_{mm}^{\mathrm{inter}}=\tau_{pp}^{\mathrm{inter}}$).
4) Intersubband interaction, resulting in intersubband transitions (matrix elements $V_{tot}^{mppm}=V_{tot}^{pmmp}$, "e-e" times: $\tau_{mp}^{\mathrm{inter}}=\tau_{pm}^{\mathrm{inter}}$).

Figure 3 shows the Feynman diagrams of various types of electron-electron interactions.

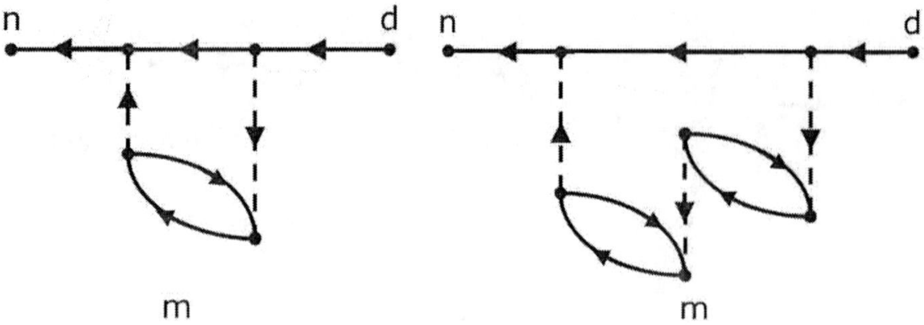


**Figure 3.** Feynman diagrams of various types of electron-electron interaction. On the left side, - intersubband interaction with intersubband transitions, on the right, - the interaction with intrasubband intersubband transitions.

After application of the identity transformations [2] (5) forms as

$$\frac{1}{\tau_{ij}^{ee}}=\int_{-\infty}^{\infty}\frac{d\omega}{\mathrm{ch}^2(\hbar}\sum_{:,l}\sum_{\mathbf{q}}\left|V_{\mathrm{tot}}^{ijkl}(\mathbf{q},\omega)\right|^2\chi_{ik}(\mathbf{q},\omega)\chi_{jl}^*(\mathbf{q},\omega) \tag{6}$$

where $k_B$ - Boltzmann constant, dielectric function in the random phase approximation has this form

$$\chi_{mn}(\mathbf{q},\omega)=\lim_{\delta\to 0}\sum_{\mathbf{k}}\frac{n_{m,\mathbf{k+q}}-n_{n,\mathbf{k}}}{E_{m,\mathbf{k+q}}-E_{n,\mathbf{k}}-\hbar} \tag{7}$$

The matrix elements of the potential total screening within the parameters of the energy diagram for the first and second types of transitions:

$$V_{\mathrm{tot}}^{mmmm}(\mathbf{q},\omega)=\frac{E_m}{2d_m\left(q^3+2\pi e^2q^4\chi_{mm}(\mathbf{q},\omega)\right)} \tag{8}$$

for the third and forth types

$$V_{\mathrm{tot}}^{mmpp}(\mathbf{q},\omega)=\frac{E_p(1-qd_m)-E_m(1+qd_p)}{(d_p-d_m)\left(q^3+2\pi e^2q^4\chi_{mp}(\mathbf{q},\omega)\right)} \tag{9}$$

Relaxation times $\tau_{ee}$ for the first and second types of transitions

$$\frac{1}{\tau^{\mathrm{intra}}}=\frac{E_m^2m^{*2}S}{d_m^2\hbar}P\ (T)\zeta(T,n_m) \tag{10}$$

for the third and forth types

$$\frac{1}{\tau^{\mathrm{inter}}}=\frac{m^{*2}(E_n-E_m)^2}{(d_n-d_p)^2\hbar}\left\{Q_{-n}(T)\zeta(T,n_m)-W_{-n}(T)\zeta(T,n_n)\right\} \tag{11}$$

where $Q_{-n}(T)$, $P_{-n}(T)$ и $W_{-n}(T)$ are polynomials

$$P_{-n}(T)=\frac{B_{p1}}{T}+\frac{B_{p2}}{T^2}+\frac{B_{p3}}{T^3}+\ldots;\ Q_{-n}(T)=\frac{B_{q1}}{T}+\frac{B_{q2}}{T^2}+\frac{B_{q3}}{T^3}+\ldots;\ W_{-n}(T)=\frac{B_{w2}}{T^2}+\frac{B_{w4}}{T^4}+\ldots$$

with the coefficients $B_n$ defined by the Riemann zeta-function $\varsigma(T,n_m,n_p,d_m,d_p)$.

## 4. Conclusions

The calculations for $\tau_{ee}^{th}$ are performed in accordance with the circuit-breaking model of Landau quantization [7,8]. At the same time, considering the "trajectories" of channels (1) - (4), including inter-

and intrasubband transitions, the calculations for $\tau_{ee}^{th}$ are shown in the formulas (6) and (11) are made in accordance with the result of the Matthiessen rule $\tau_{ee}^{-1} = \sum_i \left(\tau_{ee}^{-1}\right)_i$, where the sum of all components circuit the model. Figures 4a and 4b show a comparison of the experimental [3] and calculated curves of the destruction of Landau quantization for multiple heterostructures, in which two lower zones of size quantization are filled. The first thing you should pay attention to is quantitative "hit" of τ in the interval calculation values in the temperature range $2 \le T \le 12K$ and for real densities $n_m \approx 10^{12}\,\text{cm}^{-2}$, $n_d = 0.1 n_m$, $n_n = 0.01 n_m$. In addition, it should be noted that at low temperatures ($T < 5K$) damping Landau quantization is determined by $n_s$-electrons and numerical analysis of the decomposition of the dielectric function (7) shows the appearance of nonmonotonicities of $\tau_{ee}^{th}(T, n_s)$ with $n_m > 8 \cdot 10^{11}\,\text{cm}^{-2}$ and when $T > 5K$. This suggests that the characteristic oscillations occur only after filling the second subband with electrons and the response to thermal effects with $T > 5K$.

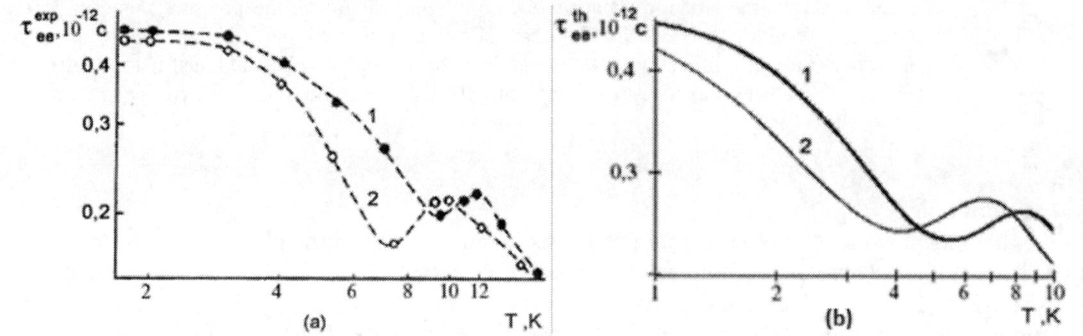


**Figure 4.** Comparison of the experimental $\tau_q^{\exp}(T)$ (a) and calculated $\tau_{ee}^{th}$(T) (b) curves for different values of the m-subband electron density: 1 – 9.1; 2 – 10.0.

**References**

[1] Coleridge P T 1991 *Phys.Rev.* B **44** 3793
[2] Kadushkin V I and Dubois A B 2002 *Phys. Low-Dim. Struct.* **11/12** 27-37
[3] Ambartsumyan V A, Andryushchenko E A, Bukhenskyy K V, Dubois A B, Dvoretskova E A, Gordova T V, Kucheryavyy S I, Mashnina S N and Safoshkin A S 2014 *Nanosystems: phys., chem., math.* **5**(3) 343-53
[4] Slutzky M, Entin-Wohlman O, Berk Y and Palevsky A 1996 *Phys. Rev. B* **53**(7) 4065-72
[5] Goh K E J, Simmons M Y and Hamilton A R 2008 *Phys. Rev. B* 77 235410
[6] Dubois A B, Zilotova M A, Kucheryavyy S I and Safoshkin A S 2013 *Vestnik of RSREU* **3**(45) 88-92
[7] Bukhenskyy K V, Dubois A B, Gordova T V, Kucheryavyy S I, Mashnina S N and Safoshkin A S 2015 *Phys. Procedia* **71** 359-63
[8] Baskakova A V, Bukhenskyy K V, Dubois A B, Kucheryavyy S I, Mashnina S N and Safoshkin A S 2016 *Journal of Physics: Conference Series* **747** 012026

# Metallic atomic hydrogen at a pressure of 300-500 GPa

**N N Degtyarenko and E A Mazur**

National Research Nuclear University MEPhI (Moscow Engineering Physics Institute), Kashirskoe sh. 31, 115409 Moscow, Russia

eugen_mazur@mail.ru

**Abstract**. Atomic metallic hydrogen with a symmetry FDDD of a lattice cell is shown to have a stable crystalline structure under hydrostatic compression in the pressure range of 350-500 GPa. The resulting phase is shown to have a stable structure regarding the collapse of the phonon spectrum. Ab-initio simulation method has been used to calculate the structural, electronic, phonon and other characteristics of the normal FDDD metallic phase of hydrogen at a pressure of 350-500 GPA.

## 1. Introduction

In the theoretical works [1-4] the normal properties of various crystalline phases of metallic hydrogen have been studied. There are definite prospects to use the atomic hydrogen as a substance with a high density of stored energy. These prospects seem to be extremely interesting in the issue of producing atomic hydrogen at high pressures and under normal conditions. In [5] the normal properties of the I41/amd phaseof the atomic metallic hydrogen at a pressure of 500 GPa have been studied. In [6] the superconducting properties of I41/amd stable phase predicted in [5] have been investigated. The superconducting transition temperature $T_C$= 217 K at a pressure of 500 GPa has been obtained in [6] for the I41/amd metallic hydrogen. In [7] the metallic hydrogen at a pressure of 500 GPa has been discovered experimentally.

The phase transformations at the compression of the molecular hydrogen on the basis of theoretical calculations have been predicted. According to calculations by the DFT LDA method [8] the phases P63/m, C2c, Cmca-12 and Cmca should be stable in the pressure ranges $P$ < 70, $P$ = 70-165, $P$ = 165-260, and $P$ >260 GPa, respectively. Later the pressure range limits were defined more precisely as $P$ < 110 GPa for P63/m, $P$ = 110-245 GPa for C2c, $P$ = 245-370 GPa for and Cmca-12 and $P$ > 370 GPa for Cmca.

These results are in excellent agreement with the calculations in [9]. Phase transition under pressure of was of 110 GPa was detected using both Raman scattering spectra and IR spectra [10]. No direct information about the structure of the phases was obtained in these experiments. The ranges of stability of atomic hydrogen phases has been calculated in [11] as 480-1000 GPa for I41/amd structure and 1000-1500 GPa for R-3m one. According to [11], the phase transition of the molecular phase to an atomic phase occurs at the pressure of 500 GPa. As follows from the results [12], metallization of the system occurs at a pressure of about 400 GPa, i.e. there exists a range of pressures exceeding 400 GPa for the metallic molecular phase.

The question arises: what is a lower limit for the pressure at which the dynamic stability of the crystal lattice is still maintained under decompression when the atomic structure is used as an origin.

Such a process corresponds to the movement along the path 2 in the terminology of [7] for the phase diagram in the opposite direction. In our previous work we have shown [5] that pressure less than 450 GPa is critical for stability of the crystal structure of the metallic hydrogen I41/amd phase. In the present paper, we found a new phase with FDDD symmetry which is similar to the I41/amd phase but has a lower critical pressure of about 350 GPa. In addition, the discovered FDDD phase has a significantly larger constant of the electron-phonon interaction. In accordance with calculations by Brovman and Kagan [1-4] the metallic hydrogen atomic phase may be metastable upon decompression down to the atmospheric pressure. A possibility of creating atomic metastable hydrogen phase, which is dynamically stable at the lower pressures, is a matter for the further research.

## 2. Method

Ab-initio calculations of the structural, electronic and phononic characteristics of the metal hydrogen crystal structures at a pressure of 500 GPa has been carried out. The DFT approximation with the plane wave basis and the correlation functional GGA–PBE (Perdew-Burke-Ernzerhof) with a pseudo-potential preserving the norm has been used. All the calculations were performed in the spin-polarized approximation to make possible an adequate comparison of the calculated energy values for different crystalline phases. The method of super cells with the periodic boundary conditions has been used to simulate the crystal. DFPT method with an extended number of unit cells has been used in the calculations of the phonon spectra.

## 3. Results

### *3.1. Structure geometry*

Figure 1 shows a rectangular basic cell structure with the FDDD symmetry as well as the face-centered orthorhombic primitive cell with two atoms, the reciprocal cell and the isosurfaces of the two e-bands crossing the Fermi level. Red lines represent the path connecting points of high symmetry (G-X-Y-Z-G).

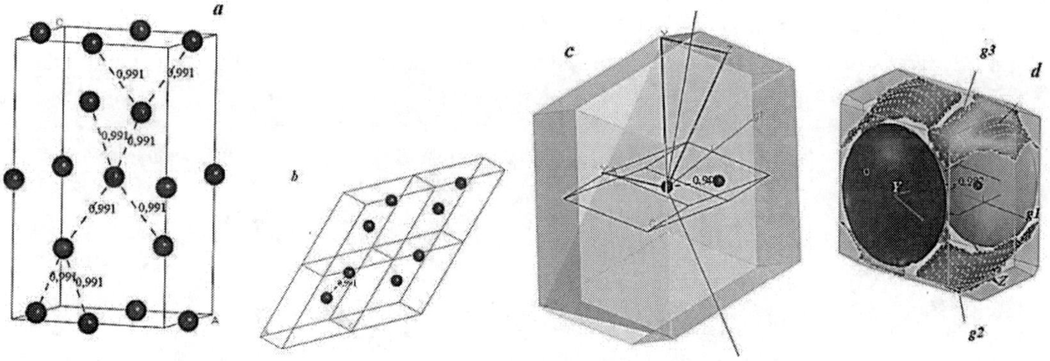


**Figure 1.** (a) Rectangular unit cell of the structure with the FDDD symmetry; (b) four primitive orthorhombic cells; (c) reciprocal cell; (d) isosurfaces for the two bands crossing the Fermi level.

### *3.2. Structure parameters*

The structure with the FDDD symmetry slightly differs from the I41/amd one described by us earlier in [5]. The rectangular cell (Figure 1a) contains 8 atoms and has a rectangular face with the sides difference of about 10% remaining at all pressures. The cell of the I41/amd structure contains only 4 atoms and has a square face. Nevertheless for both structures the primitive cell contains two hydrogen atoms. The common feature of these structures is the configuration of atoms in the first coordination sphere. Both the approximation method and the convergence conditions have been strictly maintained in calculation of these structures performed for the primitive cells. We obtained that the enthalpies of both structures differ only slightly when the zero-point oscillations addition to the phonon contribution is excluded. For example, at $P$ = 500 GPa the enthalpy values are equal to $H$ = - 9,77 eV/atom for the both crystalline phases and a difference occurs only in 3-4 digits after the decimal point. Such an equality holds for all pressures. A relatively larger difference , which is of the order of $10^{-2}$, is observed for the distances between atoms in the second coordination sphere and for relative value of the electron density between the second neighbours. Although the energy characteristics of these two structures are almost degenerate, due to these differences a dynamical stability of the FDDD structure is kept in a larger pressure range in comparison with the I41/amd phase.

*3.3 Electron spectra*

Calculation of the electron spectra were carried out for the FDDD structure with two atoms in the primitive cell. There are two filling bands in the obtained spectra (Figure 2). In the pressure range from 300 to 500 GPa the spectrum of filled states is practically unchanged. In vicinity of $X$ and $Z$ points the states are not filled.

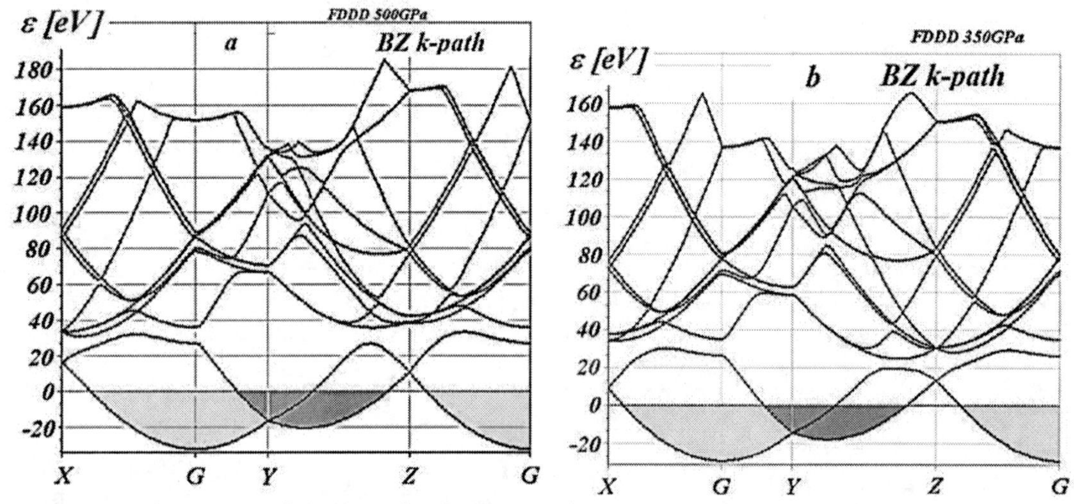


**Figure 2** Band structure of electrons for the FDDD phase at pressures of 500 GPa (a) and 300 GPa (b). The zero energy corresponds to the Fermi level.

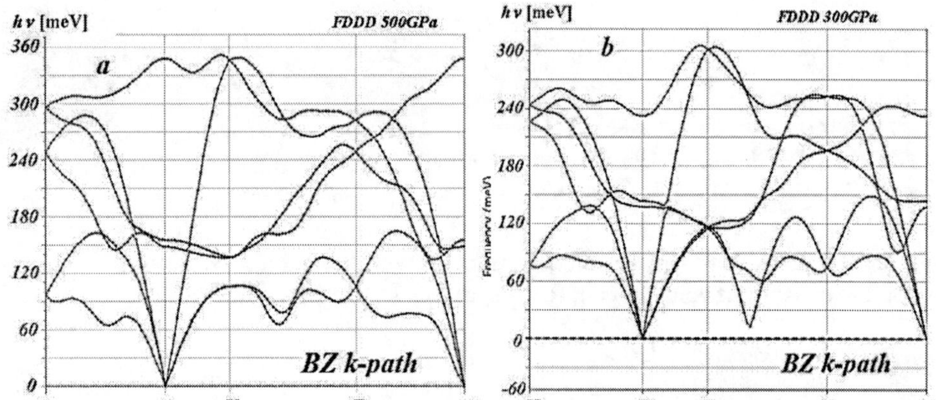


**Figure 3.** Dispersive dependences of the phonons for the FDDD phase at pressures of 500 GPa (a) and 300 GPa (b). Calculations have been carried out for the super-cell FDDD structure with the number of atoms of the order of 128.

### *3.4. Phonon spectra*

As seen from Figure 3, there are no the imaginary frequencies in the phonon spectrum at $P$ = 500 GPa. The imaginary phonon frequencies are also absent along the (X-G-Y-Z-G) path for the pressure of 300 GPa. Influence of pressure on of the phonon dispersion curves is more noticeable than for the electron spectra. One of the phonon branches almost approaches the zero frequency on the X-Z segment. In addition, the estimated spectrum shifts up relative to the zero value with the frequency increase. At $P$ = 300 and 500 GPa the maximum values of the phonon frequency correspond respectively to 300 and 370 meV.

Analysis of the states density curves revealed also that the imaginary phonon frequencies in the FDDD structure do not appear at pressures of 500 and 350 GPa (the curves tend to zero for zero energy). There appears a "tail" in the phonon density of states of the FDDD structure in the range of "negative" frequency at pressure of 300 GPa (the "negative" frequency denotes an imaginary values of the phonon frequency). Though the qualitative behavior of the phonon spectra at the different pressures is quite similar, the spectra become harder with pressure. The maximum value of the phonon energy of 350 meV is achieved at pressure of 500 GPa.

## 4. Conclusions

1. The structures of the metallic atomic hydrogen phase with both FDDD and I41/amd symmetries coexist in the pressure range of 450-500 GPa due to the small difference of their enthalpies even if the zero-point oscillations are taken into account.

2. The phase with the FDDD symmetry is dynamically stable in a wider range of pressures P=350-500 Gpa in comparison with the I41/amd phase.

3. In the range form 350 to 500 GPa the electron spectrum of the phase with the FDDD symmetry changes slightly with the pressure.

4. Value of the maximum phonon energy increases from 300 to 350 meV with pressure increase from 350 to 500 GPa.

5. At pressures below 450 GPa the atomic metallic phase with the FDDD symmetry coexists with a molecular metallic phase of hydrogen. Such a coexistence, in our opinion, represents the physical nature of the PRE phase in the metal hydrogen phase diagrams.

6. Calculation of the critical temperature using the method [13] gives the value of 250 K for the structure of the atomic hydrogen with the FDDD symmetry.

## References

[1] Brovman E, Kagan Yu 1971 *JETP* **61** 2429
[2] Brovman E, Kagan Yu 1972 *JETP* **62** 1492
[3] Brovman E, Kagan Yu, Kholas A, Pushkarev V V 1973 *JETP Letters* **18** 269
[4] Kagan Yu, Pushkarev V V, and Kholas A 1977 *JETP* **73** 967
[5] Degtyarenko N N, Mazur E A 2016 *JETP Letters* **104**(5) 329
[6] Kudryashov N, Kutukov A A, and Mazur E A 2016 *JETP Letters* **104**(7) 488
[7] Dias R P, Silvera I F 2017 *Science* 355(6326) 715-8
[8] Sam Azadi and Foulkes W M C 2013 *Phys. Rev. B* **88**(1) 014115
[9] Pickard C J and R. J. Needs 2007 *Nat. Phys.* **3** 473
[10] Mao H K and Hemley R J 1994 *Rev. Mod. Phys.* **66** 671.
[11] McMahon J M and Ceperley D M 2011 Phys. Rev. Lett. **106** 165302.
[12] Hanyu L, Li Z, Wenwen C, and Yanming M 2012 *J. Chem. Phys.* **137** 074501
[13] Grebenev V N, Mazur E 1987 *Low temperature Physics* **13** 479
[14] Eremets M I, Troyan I A and Drozdov A P 2016 arXiv:1601.04479v1

# Investigation of the sensitivity of MIS-sensor to thermal decomposition products of cables insulation

**D V Filipchuk[1,2], A V Litvinov[1], M O Etrekova[1,2] and D A Nozdrya[2]**

[1]National Research Nuclear University MEPhI (Moscow Engineering Physics Institute), Kashirskoe sh. 31, 115409 Moscow, Russia
[2]OOO "NPF INKRAM", Lyublinskaya ul. 151 109341 Moscow, Russia

Email: AVLitvinov@mephi.ru

**Abstract.** Sensitivity of the MIS-sensor to products of thermal decomposition of insulation and jacket of the most common types of cables is investigated. It is shown that hydrogen is evolved under heating the insulation to temperatures not exceeding 250 °C. Registration of the evolved hydrogen by the MIS-sensor can be used for detection of fires at an early stage.

## 1. Introduction

The problem of fires in Russia and around the world is very acute. More than 35 thousand incidents of ignition have occurred in Russia only during the first quarter of 2014, and more than 3 thousands of people have been perished. According to the official statistics of the Ministry of Emergency Situations, almost 30% of the total number of fires for the above period occurred due to damage of insulation and jacket of cables, which led to a short circuit and a further ignition of the insulation [1].

To ensure the safety of both people's lives and property, we must detect fires at an early stage. Not one of the modern fire detectors can cope with this task, since they reacts to an already existing fire. Therefore development of a device which can detect fires at an early stage is an actual task.

A MIS-sensor can be used as a sensitive element of such a device since it exhibits a very high (at ppm level) sensitivity to hydrogen [2] which, as a rule, is evolved during the thermal decomposition of the cable insulation. Detecting hydrogen appearance the MIS-sensor can warn about an imminent threat of fire. More details on the thermal decomposition of insulation materials used in this work are available in [3,4].

A fire detector based on a MIS-sensor can be used in places of large people accumulation, such as concert halls and shopping centers. Also, this device can be used in works with a high danger.

## 2. MIS-sensor scheme

Sketch of the MIS-sensor is shown in Figure 1. A silicon (4) plate of KEF-15 with thickness of 0.4 mm is used as a basis for the structure. One surface of the plate is coated with $SiO_2$ film (3) obtained by oxidizing in dry oxygen. A $Ta_2O_5$ film (2) of about 100 nm thickness covers the $SiO_2$ layer and a catalytic metal film (palladium) of about 30 nm thickness is deposited from above. The MIS structure is operated at a constant temperature of 100 °C. The temperature is measured by the thermistor (10) and stabilized by the electronic circuit of the gas analyzer.

The MIS structure represents a capacitor whose capacitance depends on the applied voltage (bias voltage). This dependence is called the volt-farad (C-U) characteristic. When hydrogen appears over the palladium film, the C-U characteristic shifts to the left along the voltage axis (Figure 2) and a change of the MIS-sensor capacitance is registered when a bias voltage is fixed.

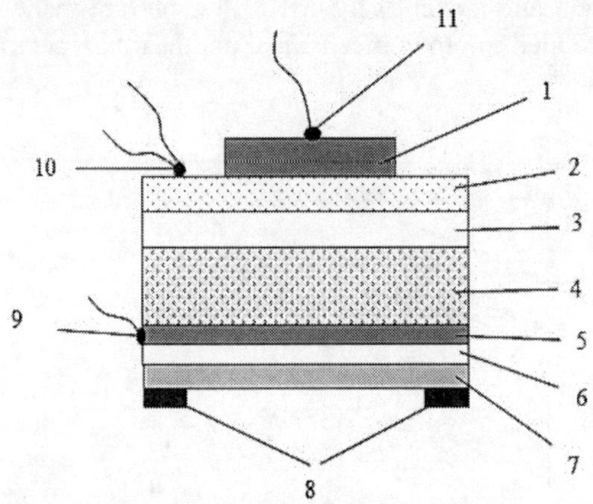


**Figure 1.** MIS-sensor sketch: 1 – Pd film; 2 – $Ta_2O_5$ film; 3 – $SiO_2$ layer; 4 – silicone wafer; 5 – metal electrode; 6 –insulating layer; 7 – heater; 8 – electric contacts of heater; 9,11 – electric contacts of MIS-structure; 10 – thermistor.

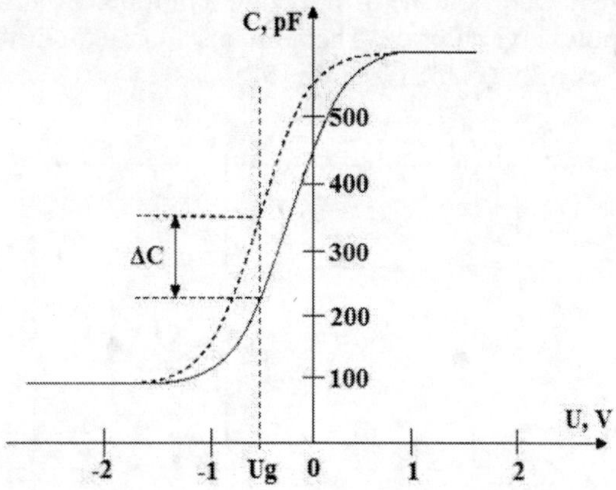


**Figure 2.** C-U characteristic of the MIS-sensor. The dashed line shows the shift of the volt-farad characteristic under hydrogen influence.

## 3. Experimental

The experimental setup shown in Figure 3 was assembled to perform experiments on thermal decomposition of cables.

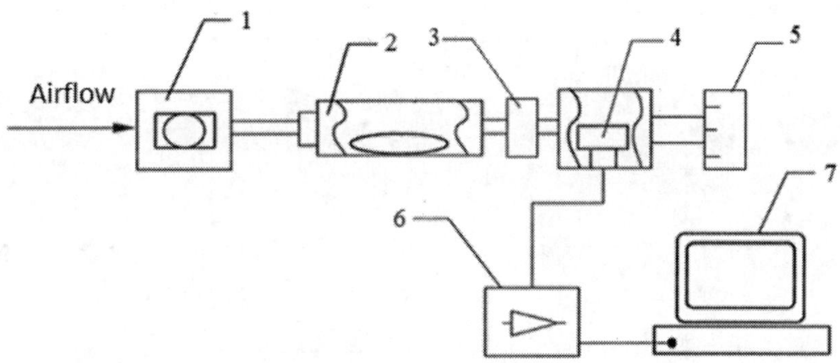


**Figure 3.** Sketch of experimental setup: 1 – air pump, 2 - heat reactor with a sample of cable insulation, 3 – carbon filter, 4 – vessel with MIS-sensor inside, 5 – rotameter, 6 – chip, 7 – PC.

The experimental setup works as follows. The pump (1) supplies air to the reactor (2) made of quartz tube on which a nichrome-wire heater was wound. Thermal insulation of the reactor was provided with glass wool and asbestos cloth. Sample of the cable inside the reactor was heated to the initial temperature of 100 °C and then temperature was raised in steps of 10 °C up to a value at which hydrogen evolution from the cable insulation begins. The air flow supplies all products of the thermal decomposition to the MIS-sensor (4) while only hydrogen detection is the aim of our work. Since the MIS-sensor can be sensitive also to other gases produced under the thermal decomposition they were delayed by two carbon filters situated in front of the sensor housing (3) and directly in the housing. The signal from the MIS-sensor was processed using an electronic recorder and PC (7). The recorder

IOP Conf. Series: Journal of Physics: Conf. Series **941** (2017) 012061 doi:10.1088/1742-6596/941/1/012061

detected capacitance of the MIS-sensor in picofarads. Change of the capacitance with time was measured. Sensitivity of the MIS-sensor to hydrogen (Figure 4) was calibrated before the measurements.

Experiments were carried out for KSPVG and MKESh cables widely used in household and industrial equipment, communication systems. Insulation of the KSPVG cable is made of polyethylene and its jacket is made of polyvinyl chloride. Both insulation and jacket of the MKESh cable are made of polyvinyl chloride. There are also some impurities intended improve mechanical and heat characteristics of the cable isolation [5].

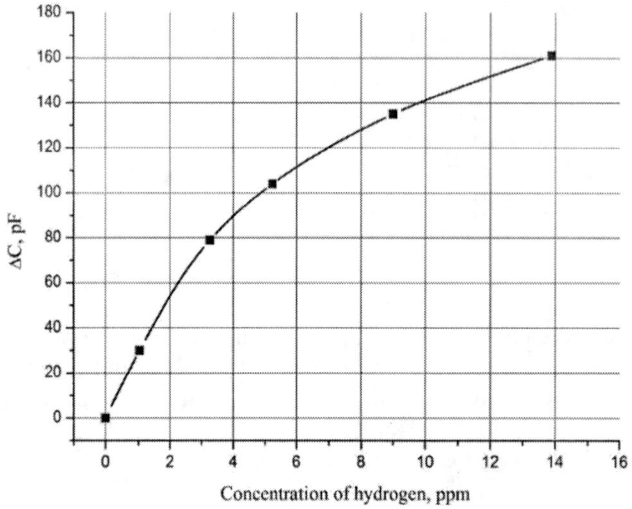


**Figure 4.** Calibration curve of MIS-sensor.

## 4. Results and discussion

Response of the MIS-sensor to thermal treatment of the insulation and jacket of the KSPVG and MKESH cables are presented in Figures 5 and 6.

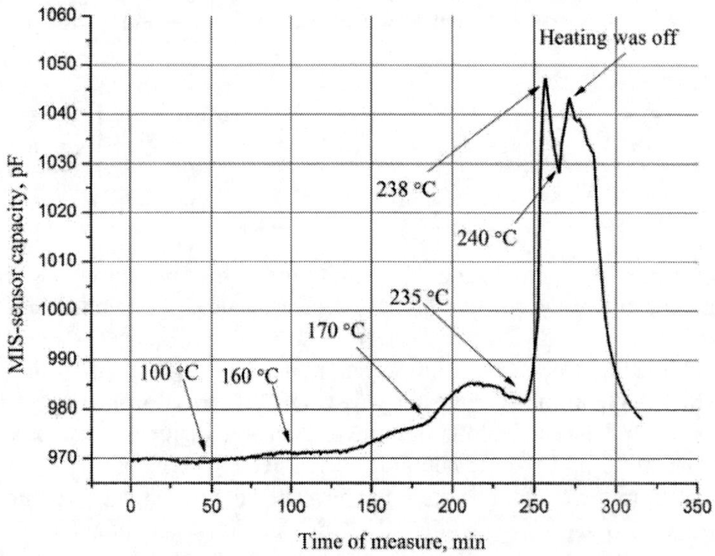


**Figure 5.** MIS-sensor response under heating KSPVG insulation and jacket.

In the experiment with KSPVG cable (Figure 5) of the MIS-sensor response was not found at room temperature. A small increase of the sensor readings was observed in the range from 170 to 235 °C of temperature inside the reactor. At temperature of 235 °C the response began sharply increase and the signal raised during 10 minutes. As known, capacitance of the MIS-sensor increases with increase of hydrogen concentration (see Figure 2). Such a sharp increase of the signal may mean that at 235 °C, an intensive evolution of hydrogen from the insulation and jacket of the cable begins. Appearance of small peaks (small jumps of capacitance) on the response curve can be caused by the evolution of hydrogen from impurities in the cable's insulation and jacket at high temperature.

Increase of temperature to 238 °C leaded to partial relaxation of the MIS-sensor response, however the signal began to grow again when temperature was rised to 240 °C. After 15 minutes the reactor heater was switched off and the capacity of the MIS-sensor began to decrease: the relaxation process began. According to calibration of the sensor (see Figure 2) the overall increase of the capacitance obtained in this experiment, which is equal to 75 pF, correspond to hydrogen concentration of 3 ppm.

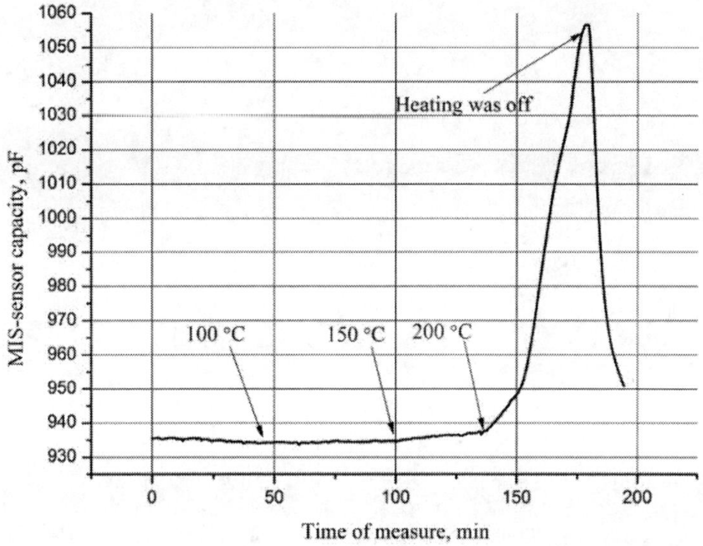


**Figure 6.** MIS-sensor response under heating MKESh insulation and jacket.

In the experiment with the MKES cable (Figure 6) no response of the MIS-sensor was also found at room temperature. A small increase of the the sensor readings was observed in the range from 150 to 200 °C of temperature inside the reactor. Above 200 °C the response sharply increased and the signal raising lasted for half an hour until temperature inside the reactor reached 215 °C and the output signal of the sensor was overloaded. After that the reactor heating was switched off. This experiment has also revealed that the capacity of the MIS-sensor increases with increasing reactor temperature and thermal decomposition of cable's insulation and jacket leads to hydrogen evolution. Increase of the MIS-sensor capacitance by 120 pF observed in this experiment corresponds to hydrogen concentration of 7 ppm.

We should note in addition that in both experiments the sharp increase in the MIS-sensor response began at temperatures not exceeding 250 °C. Such low temperatures are typical for the early stages of a fire.

## 5. Conclusion

Our experiments have revealed that hydrogen evolution caused by thermal decomposition of isolation and jacket of KSPVG and MKES cable begins at temperatures typical for early stages of a fire.

Amount of the evolved hydrogen is enough to safely detect it by the MIS sensor. We concluded that a MIS-sensor can be used as a sensitive element for detection of fires at early stages.

**Acknowledgments**
This work was supported by the National Research Nuclear University MEPhI Academic Excellence Project (contract 02.A03.21.0005).

**References**
[1] Official information of Russian Ministry of Emergency Situations about fires for 2014
[2] Emelin E V, Nikolaev I N and Sokolov A V 2005 *Sensors and systems* **10** pp 37-9
[3] Madorsky S 1967 *Termicheskoye razlozhenie organicheskikh polimerov* ed. Rafikov S R (Moscow: Mir) 111-77
[4] Beyler C L and Hirschler M M 2002 *SFSE Handbook of Fire Protection Engineering 2 ed.* DiNenno P J (Quincy: Springer) sec. 1 chap. 7 p. 125
[5] Askadsky A A, Bulgakov B I, Popova M N and Popov A V 2013 *Izvestiya SFEdU. Engineering sciences* **8** 93-4

IOP Conf. Series: Journal of Physics: Conf. Series **941** (2017) 012062 doi:10.1088/1742-6596/941/1/012062

# Structure, morphology and electrocatalytic properties of $WO_x$ thin films prepared by reactive pulsed laser deposition

**V Y Fominski**[1], **S N Grigoriev**[2], **R I Romanov**[1], **M A Volosova**[2], **D V Fominski**[1] **and A V Irzhak**[3]

[1]National Research Nuclear University MEPhI (Moscow Engineering Physics Institute), Kashirskoe sh. 31, 115409 Moscow, Russia
[2]Moscow State University of Technology "STANKIN", Vadkovskii per. 3a, 127005 Moscow, Russia
[3]Institute of Microelectronics Technology and High-Purity Materials of the Russian Academy of Sciences, 142432 Chernogolovka, Moscow Region, Russia

Email: vyfominskij@mephi.ru

**Abstract.** Nanostructured $WO_{x<3}$ films were prepared by pulsed laser deposition in air from W target. The pressure of air was varied in a range of 10 – 100 Pa and the temperature of a glassy carbon substrate during the film deposition was 22 or 500°C. For the films deposited at 22°C, thermal post-treatment in air of laboratory humidity at 500°C was applied. The increase of air pressure during PLD at 22°C caused both a decrease in packing density of ball-like amorphous nanoclusters of $WO_x$ and decline of their size. Post-treatment in air at 500 °C has initiated crystallization of the nanoclusters but characteristic morphology of the films was preserved in a whole. The PLD of $WO_x$ on a heated substrate resulted in the formation of nanoneedles and nanosheets. Fairly good catalytic properties in hydrogen evolution reaction (HER) and in electrolytic Pt deposition were revealed for the $WO_x$ films that consisted of loosely packed ball-like nanocrystals possessing a size of ~20 nm. The catalytic activity of the films consisted of nanoneedles and nanosheets was not good enough. Adjustable electrolytic deposition of Pt on the nanostructured $WO_x$ support film allowed to prepare effective hybrid HER catalyst containing only ~7 μg/cm$^2$ of Pt.

## 1. Introduction

Nanostructured materials possessing different complex morphology have stimulated great interest due to their importance in basic scientific research and technological applications. Tungsten oxide ($WO_x$), an important n-type semiconductor, has received much attention for its potential application in fabricating the electrochromic and photochromic devices, gas sensors, and solar/hydrogen energy devices. The $WO_x$-containing materials could be successfully used for preparing electrocathodes for hydrogen evolution reaction (HER) in acid solutions as well as for obtaining photoanode for hydrogen generation by photocatalytic splitting of water.

For an area of HER application, the importance demands control over the dimension, size, and crystal structure of $WO_x$ nanoparticles for higher technology applications. Zhuiykov et al. [1] have shown that orthorhombic β-$WO_3$ nanoflakes with thickness of about 7 to 9 nm is a superior material as a hydrogen production electrode of HER compared to hexagonal $WO_3$ nanowires and commercial $WO_3$. Ham et al. [2] prepared monoclinic $WO_3$ (m-$WO_3$) nanoplates and nanorods that showed much higher electrocatalytic activity for HER from water than that of the commercial bulk m-$WO_3$. Rajeswari et al. [3] revealed that $WO_3$ nanorods with monoclinic structure, constitute a better electrocatalytic system for HER in acid medium compared to their bulk counterpart.

Chhina et al. [4], Lori et al. [5] and some other investigators considered $WO_x$ as a candidate material as an alternative to a traditional (e.g., carbon) catalyst support. The interaction between the $WO_x$ support material and the metal catalysts such as Pt could result in an increased HER activity of a hybrid Pt/$WO_x$ film.

Filipescu et al. [6], Bailini et al. [7] and other investigators have shown that pulsed laser deposition (PLD) allowed to prepare thin-film $WO_x$ coatings possessing various structures and morphology. The laser deposited $WO_x$ films are successfully used for preparing the gas sensors (e.g., Zuev et al. [8]). However, catalytic activity of pulsed laser deposited $WO_x$ films for HER hardly investigated. In this work, to obtain the $WO_x$ films with different structures and morphology, pulsed laser deposition was performed under various pressure of reactive gas (air) at low and higher temperatures of glassy carbon (GC) substrates. Electrocatalytic activity of the as-prepared $WO_x$ films and the hybrid Pt/$WO_x$ film for HER was studied. The hybrid Pt/$WO_x$ film were prepared by cathodic deposition of Pt on the $WO_x$ films.

## 2. Experimental details

"Solar" laser with wavelength of 266 nm was used for ablation of W target. A pulse duration was 15 ns, a repetition rate was 25 Hz, and a pulse energy did not exceed 40 mJ. The fluence in the laser spot was ~8 J/cm$^2$. Laser pulses of ~18000 were used to obtain the films. The chamber with targets and substrates was evacuated to residual pressures of 10, 40 and 100 Pa. The laser-initiated plume was directed along the normal toward the surface of the substrate that was located at 4 cm from the target. The substrates were made of polished glassy carbon and cleaned in ultrasonic bath with acetone. The temperature of the substrate during deposition of the films was 22 or 500°C. Some samples prepared at 22°C were annealed at 500°C in 60 min in air.

The obtained films were examined by the methods of scanning electron microscopy (SEM) using secondary electrons (SE) and back scattered electrons (BSE), energy dispersive X-ray spectroscopy (EDS), and micro-Raman spectroscopy (MRS).

The HER activity of the $WO_x$ films was investigated at room temperature using a traditional three-electrode electrochemical cell. An electrochemical cell E-1C has a separated reference electrode container with two electrolytic keys. A Pt wire counter electrode was located in a special chamber that is connected with main working volume through a glass tube, inside which a porous glass plate is installed. It is necessary to divide electrolytic volumes of the working and counter electrodes, while ensuring the electrical contact between them. The aqueous 0.5 M $H_2SO_4$ solution was used for the HER studies. The cathodic polarization curves were measured at a slow change in the voltage (1.0 mV/s). The potential was measured relative to the standard hydrogen electrode.

For electrolytic Pt deposition on the $WO_x$ films, a Pt sheet (counter electrode) was located in front of the working electrode at a distance of 4 mm. A low content of KCl was added in 0.5 M $H_2SO_4$ solution to accelerate dissolution of Pt counter electrode under anodic polarization. The HER activity of the prepared $WO_x$ and Pt/$WO_x$ films was compared with that of the Pt film prepared on GC substrate by PLD.

## 3. Results and discussions

All the obtained $WO_x$ films had blue color that indicates an effective interaction of the laser-deposited W plume with oxygen from the residual gas and, consequently, the formation of substoichiometric film composition ($x<3$). The morphology of the prepared $WO_x$ films is found to be dependent on both the gas pressure and the temperature of substrate at the PLD (Figures 1 and 2). For air pressures of 10 and 40 Pa, the GC substrate was relatively uniformly covered with the $WO_x$ film. Homogeneity and continuity of the films were visibly disturbed when the air pressure was increased up to 100 Pa.

IOP Conf. Series: Journal of Physics: Conf. Series **941** (2017) 012062 doi:10.1088/1742-6596/941/1/012062

The deposition at 22°C caused the formation of ball-like nanoparticles of $WO_x$. The size of the particles and their packing density decrease if the air pressure is increased. The morphology of the film changes from dense/spheroidal to highly porous/fractal. Post-treatment in air at 500 °C resulted in discontinuity of the coating layer (Figure 3). Ball-like particles in the bulk of the film have grown to some extent, however the morphology of the films was not changed significantly (Figure 4).

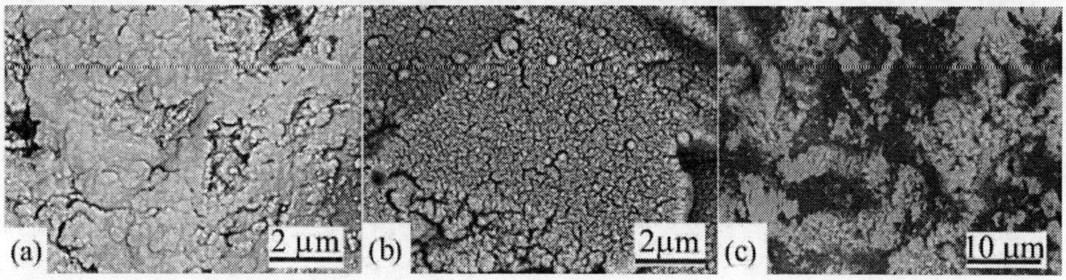


**Figure 1**. BSE SEM images of $WO_x$ films prepared on GC at 22°C by PLD at 10 (a), 40 (b), and 100 Pa (c).

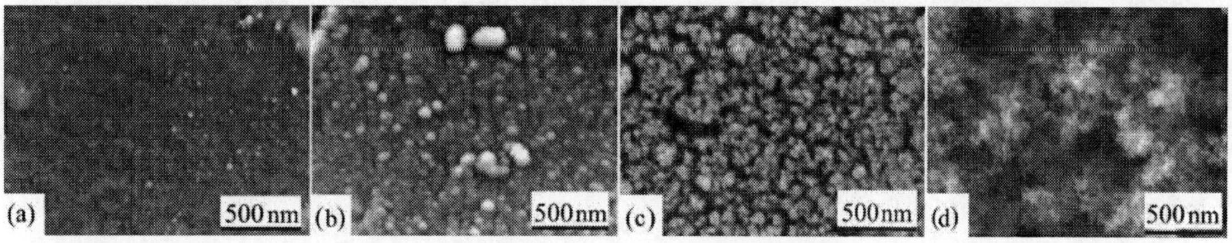


**Figure 2**. SE SEM images of a bare GC (a) and $WO_x$ films prepared on GC 22°C by PLD at 10 (b), 40 (c), and 100 Pa (d).

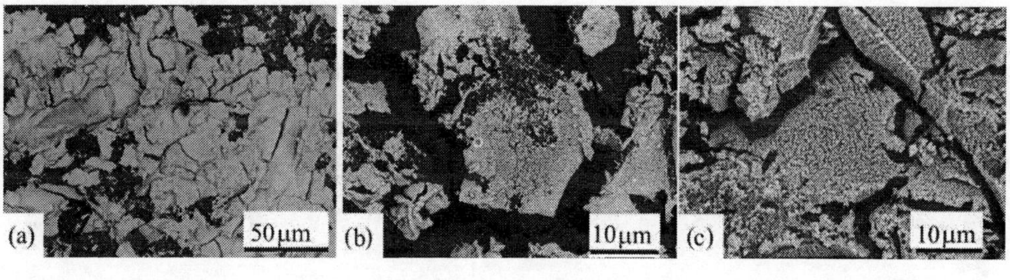


**Figure 3**. BSE SEM images of $WO_x$ films prepared at 22°C on GC at 10 (a), 40 (b), and 100 Pa (c) by PLD with thermal post-treatment.

**Figure 4**. SE SEM images of the $WO_x$ films prepared at 22°C on GC at 10 (a), 40 (b), and 100 Pa (c) by PLD with thermal post-treatment

Figure 5 shows the results of SEM study for the $WO_x$ films prepared by PLD on the hot substrate. Independently on air pressure the growth of nanoneedles and nanosheets was revealed. These nanoelements were predominantly perpendicularly to the substrate surface.

**Figure 5**. SE SEM images of the $WO_x$ films prepared on GC at 500°C by PLD at 10 (a), 40 (b), and 100 Pa (c).

Figure 6 allows to reveal the dependence of $WO_x$ crystallinity on the preparation conditions. Annealing in air caused the crystallization of the $WO_x$ films that were obtained at 22°C. Relatively sharp lines at 270, 327, 700, and 800 $cm^{-1}$ appeared in the spectra of annealed films (Figure 6*b*) instead of broad lines in a range 600 – 900 cm that are characteristic of disordered structure of the $WO_x$ (Figure 6*a*). Two lines in a range 1200 – 1700 cm are due to GC substrate. The PLD on the hot substrates caused the formation of crystalline structure at relatively high pressure of air (40 and 100 Pa). However, for low pressure of 10 Pa, the structure of the nanoneedles was strongly disordered (Figure 6*c*).

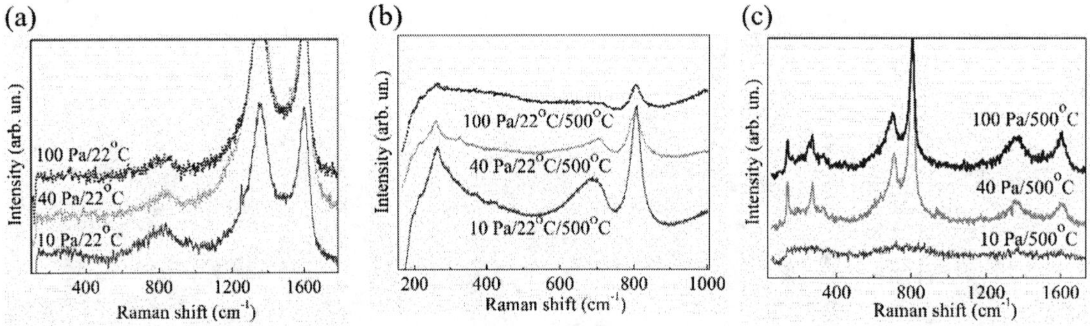


**Figure 6**. MRS spectra of the $WO_x$ films prepared by PLD at various air pressures on GC at 22°C before (a) and after thermal post-treatment (b), and obtained at 500°C (c).

Figure 7*a* shows polarization curves of different $WO_x$ films deposited on GC substrate. Activation of HER was observed only for the $WO_x$ films that were deposited at 22°C in are at pressures of 10 and 40 Pa. The annealing procedure increases the activity of these films. In the case of bare GC, the reduction current density reached -0.03 $mA/cm^2$ at an overpotential around -200 mV. For this overpotential, after the preparation of the $WO_x$(40 Pa/22°C/500°C) film the current density has increased in ~15 times. The deposition of the $WO_x$ film which consisted of nanoneedles and nanosheets adversely affects the HER activity of the samples.

The increased HER catalytic activity of the $WO_x$(40 Pa/22°C/500°C) film facilitates to Pt deposition on the $WO_x$ in Cl-containing 0.5M $H_2SO_4$ solution. Reactive molecules of this solution could cause the decomposition of Pt counter electrode. The Pt-containing ions could migrate in the solution to the $WO_x$ surface and initiate Pt nanoparticles growth due to redox processes. Figure 8 shows SEM images of the $WO_x$(40 Pa/22°C/500°C)/GC sample which was polarized in Cl-containing 0.5M $H_2SO_4$ solution in 6 cycles. The polarization curve of this sample is shown in Figure 7*b*. The density of cathodic current for the Pt/$WO_x$(40 Pa/22°C/500°C)/GC catalyst exceeds that for the Pt/GC catalyst in all range of applied potentials. This effect could be caused by a porous structure of the $WO_x$ supported layer possessing a specific distribution of Pt on the surface of the metal-oxide layer.

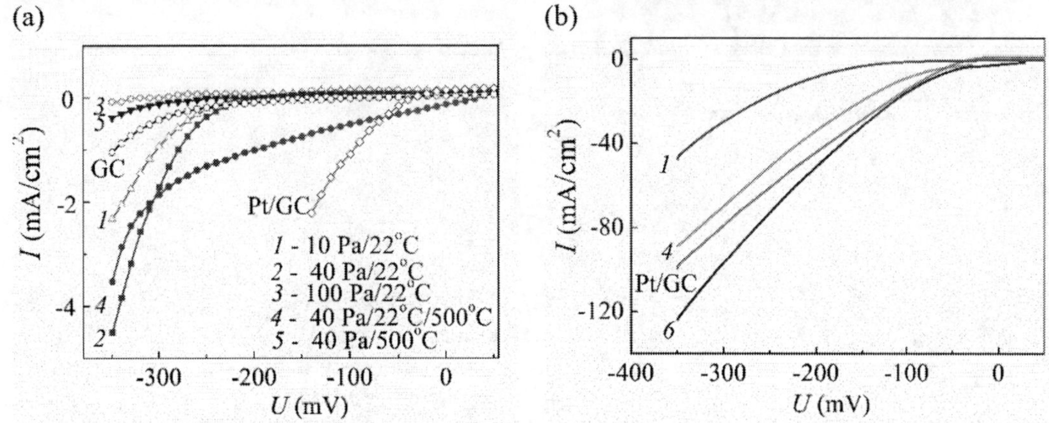


**Figure 7**. (a) polarization curves of the different $WO_x$/GC, bare GC and Pt/GC samples in 0.5M $H_2SO_4$ solution; (b) evolution of the polarization curves for the $WO_x$(40 Pa/22°C/500°C)/GC sample in 0.5M $H_2SO_4$/KCl solution after 1, 4 and 6 cycles of cathodic polarization. The result for Pt/GC sample is presented for comparison.

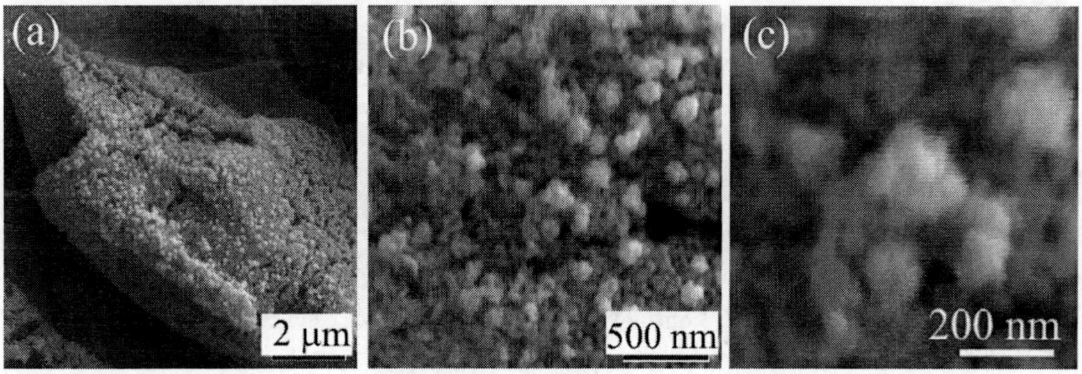


**Figure 8**. (a,b) SEM images of the $WO_x$(40 Pa/22°C/500°C)/GC sample after the electrolytic deposition of Pt; (c) berry-like clusters of $WO_x$ nanoballs (whiter fragments) loaded with Pt.

Figure 8 shows that some ball-like $WO_x$ nanoparticles with a size of ~20 nm could break away the GC surface and migrate on the surface of $WO_x$ film. Berry-like clusters that consisted of weakly bonded nanoballs covered the surface of this catalyst (Figure 8*b*). Platinum was effectively deposited on these clusters that results in a large area of catalytically active sites for HER (Figure 9). Estimated value of the Pt loading was about 7 µg/cm$^2$.

## 4. Conclusions

Pulsed laser deposition allows to prepare thin-film $WO_x$ coatings possessing various structures and morphology. Only particular modifications of the nanostructured $WO_x$ could be used for activation of HER in an acidic solution. These $WO_x$ films consist of weakly bonded ball-like nanoparticles with a defective crystalline structure. Catalytically active $WO_x$ film could be used for preparation of a hybrid Pt/$WO_x$/GC catalyst using the method of electrolytic Pt deposition. The hybrid Pt/$WO_x$/GC catalyst is superior the Pt/GC catalyst in the efficiency of HER activation.

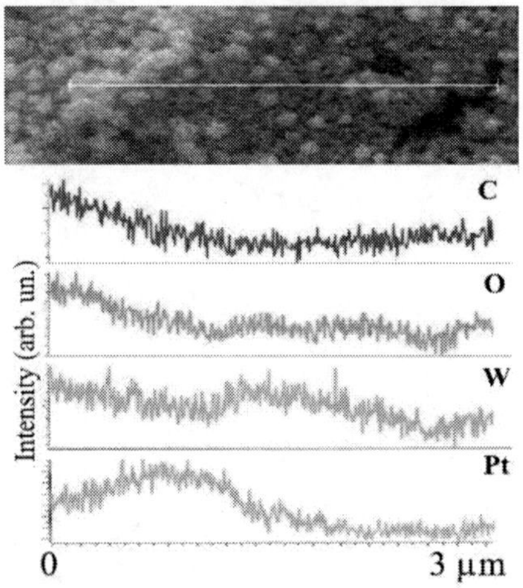


**Figure 9.** EDS distributions of C, O, W, and Pt on the surface of the $WO_x$(40 Pa/22°C/500°C)/GC sample after electrolytic deposition of Pt in 6 cycles of cathodic polarization.

**Acknowledgments**
The present work was supported by the Ministry of Education and Science of the Russian Federation (grant No 3.9750.2017/8.9).

**References**
[1] Zhuiykov S and Kats E 2014 *Nanoscale Res. Lett.* **9** 401
[2] Ham D J, Phuruangrat A, Thongtem S and Lee J S 2010 *Chem. Eng. J.* **165** 365–9
[3] Rajeswari J, Kishore P S, Viswanathan B and Varadarajan T K 2007 *Nanoscale Res. Lett.* **2** 496–503
[4] Chhina H, Campbell S and Kesler O 2007 *J. Electrochem. Soc.* **154** B533-9
[5] Lori O and Elbaz L 2015 *Catalyst* **5** 1445-64
[6] Filipescu M, Ossi P M and Dinescu M 2007 *Appl. Surf. Sci.* **254** 1347–51
[7] Bailini A, Fonzo F D, Fusi M, Casari C S, Bassi A L, Russo V, Baserga A and Bottani C E 2007 *Appl. Surf. Sci.* **253** 8130–5
[8] Zuev V V, Grigoriev S N, Romanov R I, Fominski V Y, Volosova M A and Demin M V 2016 *J. Phys. Conf. Ser.* **747** 012050

IOP Conf. Series: Journal of Physics: Conf. Series **941** (2017) 012063 doi:10.1088/1742-6596/941/1/012063

# Magneto-quantum oscillations in $Bi_2Se_3$ nanowires

**A V Frolov[1,2,*], A P Orlov[1], A A Sinchenko[1], V A Volkov[1,2], Ya A Gerasimenko[3], A Yu Kuntsevich[2,3,4], V M Pudalov[3,4]**

[1]Kotel'nikov Institute of Radio Engineering and Electronics Russian Academy of Sciences, Mokhovaya 11-7, 125009 Moscow, Russia
[2]Moscow Institute of Physics and Technology (State University), Institutskiy per. 9, 141701 Dolgoprudny, Moscow Region, Russia
[3]P. N. Lebedev Physical Institute, Russian Academy of Sciences, Leninskii pr. 53, 119991 Moscow, Russia
[4]National Research University Higher School of Economics, Myasnitskaya 18, 101000 Moscow, Russia

[*]E-mail: fralek@mail.ru

**Abstract**. We report preparation of nanoribbons (crossection ~ 250*25 nm$^2$) by focused ion beam etching of single-crystalline $Bi_2Se_3$ and detailed measurements of their magnetoresistance at temperatures down to 4.2 K, magnetic field up to 9 T. In a magnetic field parallel to the axis of nanowire the magnetoresistance shows up oscillations. Surprisingly, the Fourier analysis shows the presence not only of oscillations with a period corresponding to the flux quantum ($\Phi_0 = hc/e$), but also oscillations with a period of $2\Phi_0$ and $4\Phi_0$. Possible mechanisms of the observed effect are discussed.

## 1. Introduction

The study of topologically non-trivial quantum states is one of the most interesting and developing directions in solid-state physics past decade. The existence of such surface electronic states was theoretically predicted [1] in materials $Bi_2Se_3$, $Bi_2Te_3$ and $Sb_2Te_3$ and then confirmed by several independent experiments: angle resolved photoemission spectroscopy (ARPES) [2-4], scanning tunneling microscopy STM [5], transport measurements [6-8]. However, the nature of surface states is still a subject for investigation. The reason is that the real materials contains impurities, thermal excitations and structural defects. This leads to bulk contribution in conductivity which impedes interpretation of the observed effects. One of the most efficient methods to study the nature of topologically protected surface states is low-temperature magnetotransport measurements. This method is especially efficient in quasi-one-dimensional structures where the contribution of the bulk carriers is reduced due to the large surface-to-volume ratio.

There are several works in which the magnetotransport in grown single-crystalline $Bi_2Se_3$ nanoribbons is studied [9-11]. Most of studied samples are whiskers of topological insulator grown by various methods: vapor–liquid–solid (VLS) growth, chemical vapor deposition (CVD), electrochemical deposition. The object of our study is $Bi_2Se_3$ nanowires prepared from bulk crystals by means of mechanical and focused-ion beam thinning. The group of Bhattacharyya [12,13] which

IOP Conf. Series: Journal of Physics: Conf. Series **941** (2017) 012063 doi:10.1088/1742-6596/941/1/012063

studied similar objects should be mentioned. The comparative analysis of our studies with the studies of other research groups is the subject of this article.

## 2. Methods

### *2.1. Scotch-tape exfoliation*

As was mentioned above, our $Bi_2Se_3$ nanoribbons were fabricated from bulk crystals. We used $Bi_2Se_3$ crystals grown by the standard Bridgman method with a carrier concentration of ~ $10^{19}$ $cm^{-3}$. At first, the initial bulk crystal was thinned down by the modification of classical micromechanical exfoliation technique (scotch tape method) [14,15]: a chosen high quality layered crystal of $Bi_2Se_3$ was attached to the substrate using epoxy glue and then was thinned down to tens nanometers with adhesive tape. The thickness of films at first was estimated optically and then measured more precisely with focused ion beam (FIB): the sample was placed at the certain angle (α) relative to the exposition surface so that the side edge of the film could be seen; then the real thickness was calculated as $D_{real} = D_{measured}/\sin(\alpha)$. Then the area of the future nanostructure was masked and an Au film was sputtered by laser ablation (in order to make the electric contacts for the sample). After that the mask was removed and the sample was ready for FIB milling.

### *2.2. Fib milling*

Gallium focused ion beam (SMI3050, SII NanoTechnology and Crossbeam Neon 40 EsB (Carl Zeiss)) was used to make 1D structure from the $Bi_2Se_3$ 2D flake. Structures have the form of nanoribbon (fig.1) with typical sizes: length of 10 μm, width of 200–500 nm, height of 10–100 nm. We tried to minimize the FIB exposition on the workspace of the sample using scanning electron microscope (SEM) to target on the sample or liquid (collodion) and solid masking as a protection from Ga ions [15-16]. The milling of the structure was made with the low ion beam currents (10 pA). The area at the distance from the structure (the edges of the flake and Au contacts) was milled with higher currents of 100 pA – 10 nA.

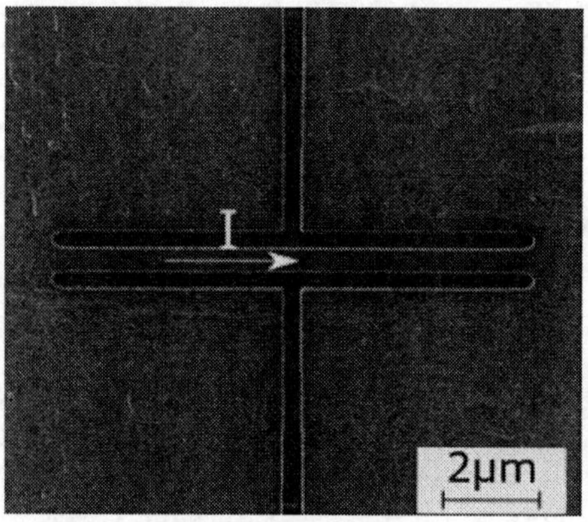


**Figure 1.** FIB-image of typical $Bi_2Se_3$ nanostructure.

### *2.3. Measurement details*

The resistance of the samples was measured using a conventional four-wire scheme, by the Kelvin method at an alternating current of a few tens of Hertz with a low-noise Lock-in Amplifier [17]. Magnetotransport measurements were carried out in a cryostat with a superconducting magnet.

## 3. Results and discussion

The magnetoresistance of $Bi_2Se_3$ nanowires was measured for several samples at temperatures 4.2 – 15 K in magnetic fields up to 9 T parallel to the axis of nanowire (Figure 2). At low magnetic fields 0-1 T for all our samples, a strong magnetoresistance peculiarity was observed, most probably because of weak antilocalization. The observation of this effect points out high quality of our samples.

At higher magnetic fields (0.5–9T), magnetoresistance oscillations are observed. The main period of the oscillations is close to the flux quantum $\Phi_0$ = hc/e through the cross-section of the wire (Figure 2). The existence of such Aharonov-Bohm-type (AB) oscillations is in agreement with the other experiments and is a manifestation of the surface states in $Bi_2Se_3$ insulator. The observation of the AB-oscillations evidences that these states are robust to the Ga ions which are inevitably implanted in the sample during the FIB milling [12].

Besides the oscillations with the period 0.8 T, corresponding to $\Phi_0$ (for the sample #1) we observed unexpectedly other periods: 1.6 T, 3.2 T, corresponding to $2\Phi_0$, and $4\Phi_0$ through the cross-section of the wire. These periods are clearly seen on the fast Fourier transform (FFT) graph (fig2b). The existence of such subharmonic oscillations is rather surprising. Such oscillations were reproduced for our other samples. Their amplitude drops with ~~the~~ temperature increase and almost disappears above 10 K. Signatures of such oscillations may be noticed in the raw data by Peng [9] for a 120 nm-wide sample, but little attention is paid to them. Their existence was accounted for bulk contribution to conductivity.

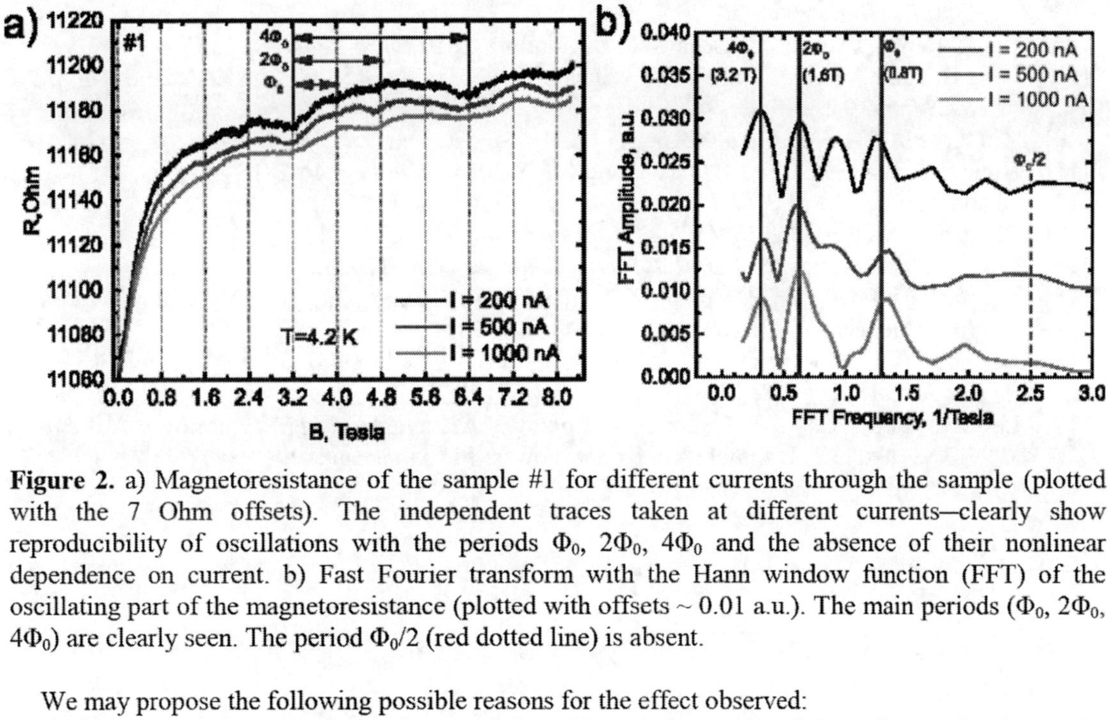


**Figure 2.** a) Magnetoresistance of the sample #1 for different currents through the sample (plotted with the 7 Ohm offsets). The independent traces taken at different currents–clearly show reproducibility of oscillations with the periods $\Phi_0$, $2\Phi_0$, $4\Phi_0$ and the absence of their nonlinear dependence on current. b) Fast Fourier transform with the Hann window function (FFT) of the oscillating part of the magnetoresistance (plotted with offsets ~ 0.01 a.u.). The main periods ($\Phi_0$, $2\Phi_0$, $4\Phi_0$) are clearly seen. The period $\Phi_0/2$ (red dotted line) is absent.

We may propose the following possible reasons for the effect observed:

i) Our sample has a rectangular shape and the subharmonics could originate from reflections at its edges. In this case the trajectories of particle motion become more complex, and this can cause oscillations with long periods.

ii) The influence of implanted Ga ions on the quality of the sample can be stronger than we presume. It is possible that gallium ions damage significant area near the edge of the sample. As a result, some particles change their trajectory due to the reflection. The latter reason however, seems less likely, because random distribution of Ga ions can hardly cause sharp reflection effects.

## 4. Conclusions

We have studied the magnetoresistance of $Bi_2Se_3$ nanowires prepared by focused ion beam and found the presence of the oscillations with abnormally large periods ($2\Phi_0$, $4\Phi_0$) beside the anticipated period $\Phi_0$. We proposed two possible reasons of their existence. In order to clarify the effect, further

IOP Conf. Series: Journal of Physics: Conf. Series **941** (2017) 012063 doi:10.1088/1742-6596/941/1/012063

research is required. The observation of the AB oscillations in our $Bi_2Se_3$ nanoribbons proves that the topologically protected surface states propagate coherently over the nanoribbon perimeter ~1000nm even in the sample whose side walls are prepared by etching with focused beam of Ga-ions.

**Acknowledgments**
The work was supported by the Russian Foundation for Basic Research (no. 16-32-00927 and no. 16-29-03330).

**References**
[1] Zhang H, Liu C-X, Qi X-L, Dai X, Fang Z and Zhang S-C 2009 *Nat. Phys.* **5** 438–42
[2] Xia Y *et al* 2009 *Nat. Phys.* **5** 398–402
[3] Chen C *et al* 2012 *Proc. Natl Acad. Sci. USA* **109** 3694–8
[4] Hsieh D *et al* 2009 *Nature* **460** 1101–5
[5] Hanaguri T, Igarashi K, Kawamura M, Takagi H, Sasagawa T 2010 *Phys. Rev. B* **82** 081305(R)
[6] Analytis J G, McDonald R D, Riggs S C, Chu J-H, Boebinger G S and Fisher I R 2010 *Nature Phys.* **6** 960
[7] Analytis J G, Chu J-H, Chen Y, Corredor F, McDonald R D, Shen Z X and Fisher I R 2010 *Phys. Rev. B* **81** 205407
[8] Eto K, Ren Z, Taskin A A, Segawa K and Ando Y 2010 *Phys. Rev. B* **81** 195309
[9] Peng H, Lai K, Kong D, Meister S, Chen Y, Qi X-L, Zhang S-C, Shen Z-X and Cui Y 2010 *Nat. Mater.* **9** 225–9
[10] Xiu F *et al.* 2011 *Nat. Nano* **6** 216–21
[11] Hong S S, Zhang Y, Cha J J, Qi X-L and Cui Y 2014 *Nano Lett.* **14** 2815–21
[12] Bhattacharyya B *et al* 2017 *J. Phys.: Condens. Matter* **29** 115602]
[13] Bhattacharyya B *et al* 2017 *J. Phys.: Condens. Matter* **29** 07LT01]
[14] Frolov A V, Latyshev Yu I 2012 *J. Phys.: Conf. Ser.* **393** 012025.
[15] Latyshev Yu I, Orlov A P, Volkov V A, Enaldiev V V, Zagorodnev I V, Vyvenko O F, Petrov Yu V, Monceau P 2014 *Scientific Reports* **4** 7578.
[16] Latyshev Yu, Smolovich A, Orlov A, Frolov A, Vlasenko V 2015 *Nanoscience and Nanoengineering* **3**(2) 13-8
[17] Latyshev Yu I, Orlov A P, Frolov A V, Volkov V A, Zagorodnev I V, Skuratov V A, Petrov Yu V, Vyvenko O F, Ivanov D Yu, Konczykowski M, Monceau P 2013 *JETP Lett.* **98**:4, 214–8.

IOP Conf. Series: Journal of Physics: Conf. Series **941** (2017) 012064 doi:10.1088/1742-6596/941/1/012064

# Models of $WO_x$ films growth during pulsed laser deposition at elevated pressures of reactive gas

**A G Gnedovets[1], V Y Fominski[2], V N Nevolin[2], R I Romanov[2], D V Fominski[2] and A A Soloviev[2]**

[1]Baikov Institute of Metallurgy and Materials Science, Russian Academy of Sciences, Leninskii pr. 49, 119991 Moscow, Russia
[2] National Research Nuclear University MEPhI (Moscow Engineering Physics Institute), Kashirskoe sh. 31, 115409 Moscow, Russia

E-mail: vyfominskij@mephi.ru

**Abstract**. The films of tungsten oxides were prepared by pulsed laser ablation of W target in a reactive gas atmosphere (air of laboratory humidity). Optical analysis and ion signal measurements for the laser plume allowed to recognise a threshold gas pressure that suppresses the deposition of non-scattered atomic flux from the plume. When the pressure exceeds about 40 Pa, the films grow due to the deposition of species that could be formed in collisions of W atoms with reactive molecules (e.g., $O_2$). Kinetic Monte Carlo method was used for modelling film growth. Comparison of the model structures with the experimentally prepared films has shown that the growth mechanism of ballistic deposition at a pressure of 40 Pa could be changed on the diffusion limited aggregation at a pressure of ~100 Pa. Thus, a cauliflower structure of the film transformed to a web-like structure. For good correlation of experimental and model structures of $WO_x$, a dimension of structural elements in the model should coincide with W-O cluster size.

## 1. Introduction

Various nanostructures based on semiconductor metal oxides have attracted much attention due to their potential in nanoelectronics and optoelectronics. With salient electrochromic, optochromic, gaschromic, and catalytic properties, tungsten oxides ($WO_x$) are of great interest, being promising in applications such as flat panel displays, 'smart windows', semiconductor gas sensors, and photocatalysts. The functional properties of $WO_x$-based devices strongly depend on the structure/architecture of the metal oxide layer. Fonzo et al. [1], Bailini et al. [2], and Fominski et al. [3] have shown that pulsed laser deposition (PLD) of $WO_x$ thin-film layers could cause the formation of films with different morphology and structures. In a case of the deposition at room temperature, the pressure of reactive/buffer gas atmosphere was observed to be the main factor which influences the mode of the structure growth. An increase of the temperature during PLD as well as thermal post treatment of the deposited films could significantly alter the formed structure and morphology of the $WO_x$ films. Penner et al. [4] revealed that for some regimes of thermal post treatment (e.g., annealing at a temperature lower than 500 °C), the prepared by PLD morphology could be conserved and only crystallization of the film would be realized.

IOP Conf. Series: Journal of Physics: Conf. Series **941** (2017) 012064 doi:10.1088/1742-6596/941/1/012064

Complex morphology of $WO_x$ films was observed to be formed during PLD at a relatively high pressure of reactive or buffer gases. Fonzo et al. [1] and Bailini et al. [2] have shown that a compact structure with a density similar to bulk would be transformed to a film with an open, low density foam-like mesostructure and a high fraction of voids if the pressure of the gas (Ar, $O_2$) exceeds 20 – 40 Pa.

In this study, we tried to identify a boundary pressure of the gas (air) above which characteristics (e.g., reactivity, composition, velocity and angular distributions) of the deposited particles could be changed significantly. The main reason of the changing was associated with the collisions between the deposited particles and molecules of the gas. The laser plume was studied by measuring the ion flux and the visible plume length. We chose two air pressures equal to 40 and 100 Pa. At these pressures, the films grew due to the deposition of scattered species flow, and the gas pressure played essential role in the film morphology. To understand the possible mechanisms of the $WO_x$ films growth, the experimental studies and the modelling of the PLD films growth have been carried out.

## 2. Experimental details

The forth harmonic with a wavelength of 266 nm from an electro-optically Q-switched yttrium-aluminium-garnet laser was used for pulsed laser ablation of W target. A pulse duration was 15 ns and a repetition rate was 25 Hz. The laser energy per pulse was 40 mJ. About 18000 laser pulses were used for the film preparation. The laser energy and focusing conditions were chosen to minimize sparks formation during pulsed laser ablation of the W target.

The deposition chamber was evacuated to pressures that were varied in a range 10 – 100 Pa. The laser plume was directed along the normal toward the substrate surface. Polished glassy carbon substrates were placed at 4 cm from the target. To control the transport of ionized laser plume in various conditions, time-of-flight ion signals were detected in a place of the substrate location. Plasma plume pictures were taken through a viewport (orthogonal to the plume axis) with a digital camera with exposure time of 0.5 s corresponding to an average over 10 plumes. With this choice and assuming different plumes as equivalent, the time integrated shape and dimension of the plume (i.e. its visible part) were captured. Surface morphology of the prepared films was characterized by scanning electron microscopy (SEM).

## 3. Modelling details

Film growth modelling was conducted by the kinetic Monte Carlo (KMC) method using a modified Witten-Sander algorithm [5]. The formation of the model coating using the KMC method is performed by sequential ''stacking'' of the incident particles (atoms, molecules or clusters) of the deposited flux on the substrate. To simplify the calculations, the growth of the film was considered on a cubic spatial lattice with cell size $a$, which is the characteristic dimension of the particle.

The particles perform off-lattice random flights and are tracked in the in the Cartesian coordinates $X$, $Y$, $Z$ (the $X$ and $Y$ axes lie in the substrate plane). Both the film growth and the particle motion are considered in an $L_X \times L_Y \times L_Z$ region. In the model, periodic boundary conditions along the $X$ and $Y$ directions are used. Every particle having left the simulation region, e.g., through its right boundary, is again introduced into this region at the same angle of motion through the left boundary.

Two different deposition regimes were considered: ballistic aggregation (BA) and diffusion-limited aggregation (DLA). In BA regime, the direction of the particle random straight-line trajectory was determined using the standard procedure of sampling from a given angular distribution of the deposited flux. In DLA model, the particles perform off-lattice random flights with limited step length $a$.

In both BA and DLA regimes, the trajectories started at a launching plane $Z = Z_{max} + 20a$ at a point the $X_0$ and $Y_0$ coordinates of which were chosen randomly. Here, $Z_{max}$ is the maximum thickness of the film and the quantity $20a$ warranties that the particle realizes some steps before it reaches the film. If a random particle cross the upper plane $Z = L_Z$ it is discarded.

When a particle reaches a nearest neighbour site of the film, it sticks irreversibly to this site. Finally, the film structure is renewed. The cell occupied by a newly deposited particle is attributed to the set

IOP Conf. Series: Journal of Physics: Conf. Series **941** (2017) 012064 doi:10.1088/1742-6596/941/1/012064

of the elementary volumes composing the film. After that, a new random particle is generated and its motion is tracked as described previously. This algorithm is cyclically repeated until a film of required thickness is grown.

## 4. Results and discussions

It is well known that plume dynamics in vacuum is characterized by a practically free and collisionless propagation regime. The spatial distribution is strongly forward directed and the observed light emission is weak. When a background gas is present in the deposition chamber the light emission increases due to particle collisions producing radiative de-excitation of the ablated species, both in the body of the plume and particularly in the expansion front; the plume edge can be better defined due to the presence of a shock wave front and the plume is slowed down and spatially confined. Bailini et al. [2] proposed that the time integrated visible plume length is related with the maximum position reached by the shock wave front (i.e. the stopping distance). For a substrate, which is located slightly far from this position, the growth mechanism could be significantly changed.

Figure 1 shows time integrated picture of W plume measured during PLD of $WO_x$ films at various pressures of air. For a pressure of 10 Pa, a bright plume surrounds the substrate. The increase of the pressure up to 40 Pa resulted in a visible decrease of the plume volume but the front of shock wave could touch the substrate surface. For a pressure of 100 Pa, the stopping distance of the laser plume should be shorter than the distance from the target to the substrate.

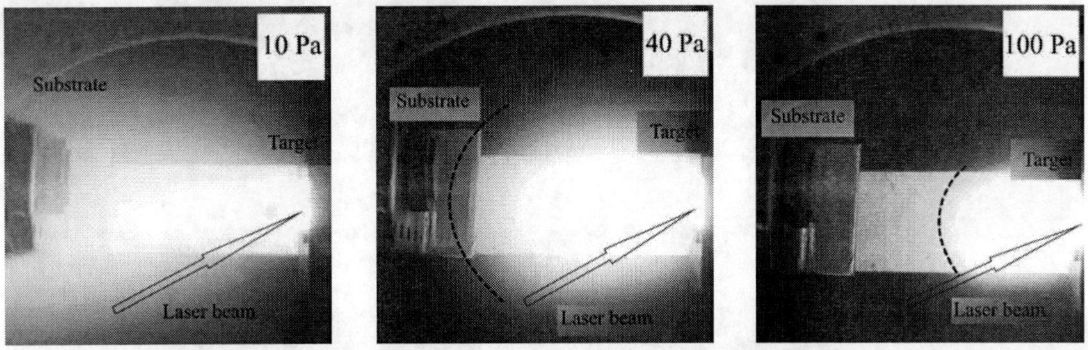


**Figure 1**. Time integrated optical images of laser plume formed during pulsed laser deposition of $WO_x$ films from W target at various pressures of air. Visual front boundary of the plume is shown.

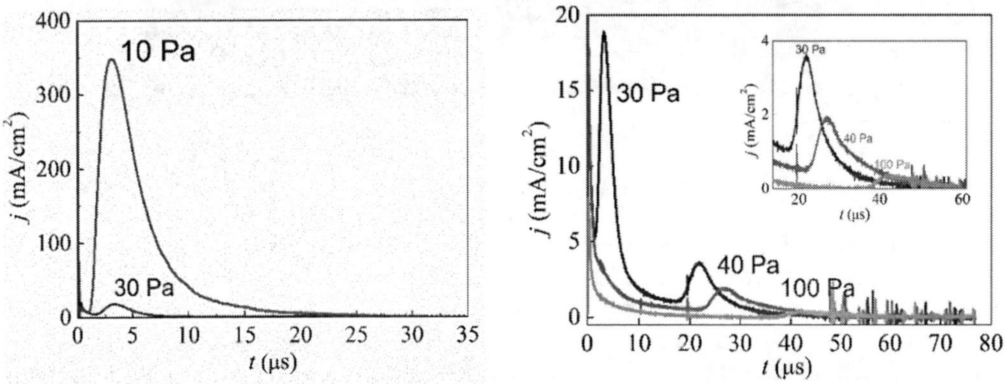


**Figure 2**. Ion flux which bombarded the surface of the substrate during the formation of $WO_x$ films by reactive PLD at various pressures of air. Intensities of ion signals measured at 40 and 100 Pa were very low, so these signals are not presented in the left figure. Insert shows the slow components of the ion fluxes.

The evolution of the measured ion signal, as recorded at a distance of 4 cm for increasing air pressure is shown in Figure 2. Splitting of the plume into a fast and a slow component was observed at a pressure of ~30 Pa. For higher pressure, only a slow component could be detected. Wood et al. [6] proposed that the plume is broken into orders corresponding to the number of collisions made with the background. The first order reaches the detector without any scattering, the second order undergoes one scattering event, and so forth.

For PLD at 40 Pa, analysis of the ion signal and the optical image of the plume allowed to propose that the shock wave front reached the surface of the growing film. The flux of deposited particles included only the scattered atoms. Comparison of the ion signals measured at 30 Pa and 40 Pa shows that a lot of atoms deposited at 40 Pa belong to the second order of scattering event, therefore these atoms could possess relatively high velocities.

For a pressure of 100 Pa, only a weak ion signal from a low-velocity shock wave was detected. This means that prior the deposition on the substrate, the W atoms undergo a lot of scattering events.

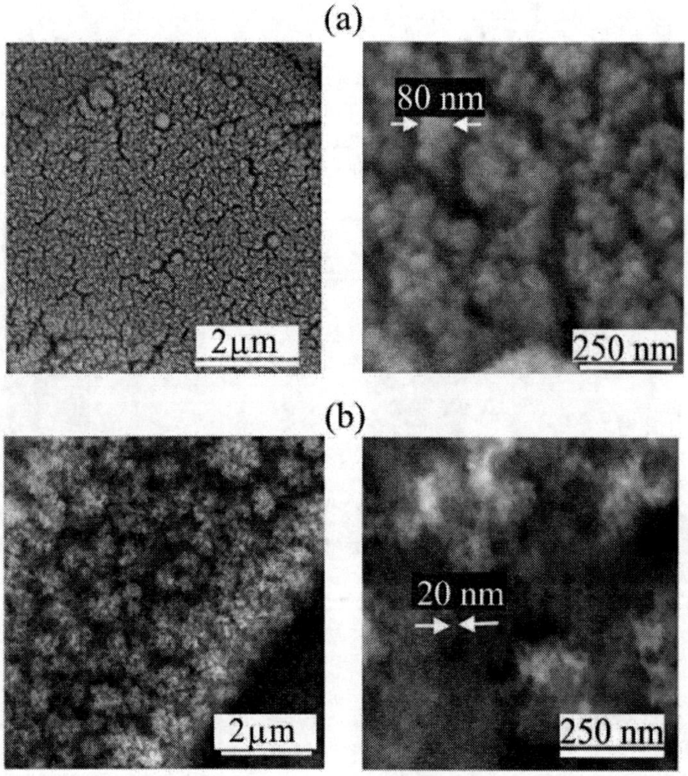


**Figure 3**. SEM images of the $WO_x$ films prepared by PLD in air at a pressure of 40 Pa (a) and 100 Pa (b).

The SEM studies of the $WO_x$ films show that the deposition at a pressure of 40 Pa caused the formation of typical cauliflower morphology (Figure 3*a*). Cross-sectional size of an individual cauliflower was about 50 – 80 nm and its growth in height could exceed 200 nm. Distances between the cauliflowers varied in a range 10 – 50 nm. For films deposited at 100 Pa, the morphology would be characteristic of a web-like structure (Figure 3*b*). The thickness of fibers was about 10 – 20 nm and the size of pores (cells) varied in a range 10 – 100 nm.

The obtained $WO_x$ films had a blue colour that indicates an effective interaction of ablated W atoms with oxygen which is contained in the air. The formation of substoichiometric film composition

($x$<3) could be proposed. Voevodin et al. [7] and Jones et al. [8] revealed the dissociation of $O_2$ in the chamber during the PLD that was accompanied by the formation of metal–oxygen molecules around the substrate.

Figure 4a shows the model structure which could grow due to ballistic deposition of structural blocks. The modelling morphology of the film is greatly similar to cauliflower morphology of the $WO_x$ film which was prepared by PLD at a pressure of 40 Pa. Figure 4b presents the model structure grown due to diffusion motion and deposition of structural blocks. A web-like image of this model structure correlated well with that of the experimental film prepared by PLD at 100 Pa.

A more rigorous analysis of the shape and size of some characteristic fragments for model and experimental structures shows that good similarity could be achieved in a case of block size of about 1 nm. The value of 1 nm correlates with the characteristic distance between the atoms in the clusters formed by chemically bonded W and O atoms. Clusters of W-O could be formed in a gas phase due to reactive collisions of W plasma with oxygen. Kinetic energy and trajectories of these clusters depend on the gas pressure. The studies of plume dynamics allow to propose that W-O clusters could retain relatively high energy of directed/straight transport at a pressure of 40 Pa because the energy of the main shock wave dissipate only in front of the substrate surface. For pressure of 100 Pa, clusters more effectively lose their energy in collisions and move to the substrate due to diffusion in a gas phase.

(a) (b)

**Figure 4**. Cross-sectional (top) and planar (bottom) images of model $WO_x$ films grown due to ballistic (a) and diffusion (b) deposition of structural blocks with size $a$. Surface areas $512a \times 512a$ (a) and $200a \times 200a$ (b) are shown.

## 5. Conclusions

Pulsed laser deposition of $WO_x$ films in a reactive gas atmosphere allows to vary the morphology of the films by regulating the gas pressure if all other parameters of deposition procedure are fixed. The reactive gas changes a transport mechanism of the ablated species and probably to some extent the composition of the deposited flux of species. For the deposition from wave shocks formed at a relatively high gas pressure, there is possibility to grow cauliflower morphology if the mechanism of ballistic deposition dominates. The web-like structure would be grown if the transport mechanism is transformed to diffusion-like under the pressure increase.

**Acknowledgments**
The present work was supported by the Ministry of Education and Science of the Russian Federation (grant No 3.9750.2017/8.9).

**References**

[1] Fonzo F D, Bailini A, Russo V, Baserga A, Cattaneo D, Beghi M G, Ossi P M, Casari C S, Li Bassi A and Bottani C E 2006 *Catalysis Today* **116** 69–73
[2] Bailini A, Fonzo F D, Fusi M, Casari C S, Bassi A L, Russo V, Baserga A and Bottani C E 2007 *Appl. Surf. Sci.* **253** 8130–5
[3] Fominski V Y, Gnedovets A G, Romanov R I and Demin M V *Proc. SPIE* 2011 **7994** 799410
[4] Penner S, Liu X, Klötzer B, Klauser F, Jenewein B and Bertel E 2007 *Appl. Surf. Sci.* **253** 8130–5
[5] Witten T A and Sander L M 1981 *Phys. Rev. Lett.* **47** 1400–3
[6] Wood R F, Leboeuf J N, Chen K R, Geohegan D B and Puretsky A A 1998 *Appl. Surf. Sci.* **127–129** 151–8
[7] Voevodin A A, Jones J G and Zabinski J S 2001 *Appl. Phys. Lett.* **78** 730–9
[8] Jones J G, Voevodin A A and Zabinski J S 2001 *Surf. Coat. Technol.* **146-147** 258–65

IOP Conf. Series: Journal of Physics: Conf. Series **941** (2017) 012065 doi:10.1088/1742-6596/941/1/012065

# Activation of electrocatalytic properties of a-C films by doping with $MoSe_x$ clusters

**S N Grigoriev**[1], **V Y Fominski**[2], **R I Romanov**[2], **M A Volosova**[1] **and D V Fominski**[2]

[1] Moscow State University of Technology "STANKIN", Vadkovskii per. 3a, 127005 Moscow, Russia
[2] National Research Nuclear University MEPhI (Moscow Engineering Physics Institute), Kashirskoe sh. 31, 115409 Moscow, Russia

E-mail: vyfominskij@mephi.ru

**Abstract**. Nanocomposite a-C(Mo/$MoSe_x$) thin films containing amorphous carbon matrix a-C, nano-Mo and $MoSe_{x\geq2}$ clusters were obtained by pulsed laser co-deposition of carbon and $MoSe_2$. The deposition was carried out at room temperature onto a graphite substrate. Atomic content of the $MoSe_{x\geq2}$ phase did not exceed 25%. The use of a buffer gas at a pressure of 10 Pa allowed to obtain the maximum Se/Mo ratio in the films and to increase the concentration of $sp^2$-hybridized C atoms for high conductivity realization. The formation of $MoSe_{x\geq2}$ cluster inclusions was the essential factor for activation of hydrogen evolution reaction (HER) in 0.5 M $H_2SO_4$ aqueous solution. These clusters also promoted cathodic deposition of Pt nanoparticles on the surface of a-C(Mo/$MoSe_x$) in a $H_2SO_4$/KCl solution when a Pt anode was used as a source of Pt. Hybrid Pt/a-C(Mo/$MoSe_x$) thin-film coatings with a low Pt loading (~6 $\mu g/cm^2$) exhibit excellent HER property, which noticeably exceeds that of relatively thick Pt coating prepared on a graphite substrate by pulsed laser deposition.

## 1. Introduction

Theoretical and experimental investigation results have indicated that the edge sites of the sandwich-type structure of transition metal dichalcognides (TMDs), particularly molybdenum/tungsten disulphides/diselenides $(Mo/W)(S/Se)_2$, are catalytically active for hydrogen evolution reaction (HER) (e.g., Jaramillo et al. [1], Lu et al. [2]). The studies have shown that the exchange current density of TMDs nanocrystals is proportional to the length of the edge sites but not to the basal areas in truncated hexagons. However, Morales-Guio et al. [3] and Vrubel et al. [4] revealed that the increase in catalytic activity is possible for amorphous TMDs nanoparticles if they are used in a hybrid catalyst system with conductive carbon particles. Because TMD are semiconductors, the low conductivity of TMDs-based catalysts may limit their catalytic performance. Coupling TMD nanostructures with a highly conductive substrate should alleviate this problem.

The value of chalcogen/metal ratio ($x$) is another essential factor that can influence the HER activity of amorphous TMDs. Klein et al. [5] have shown that the excess of chalcogen atom concentration over the stoichiometric one ($x$ >2) decreases the electron transport resistance in these materials. Merki et al. [6] and Saadi et al. [7] prepared and studied TMDs films enriched with S or Se. The amorphous $MoS_3$ and $MoSe_3$ films was observed to become catalytically active due to cathodic treatment in an acid solution. Merki et al. [6] suspect that prepared films may have specific unsaturated sites thanks to their amorphous nature.

To produce nanoparticles, thin films, and coatings based on TMDs, different methods of chemical synthesis are widely used. Various precursors with a wide spectrum of chemical composition and properties are required for TMDs preparation. As a rule, the used chemical procedures are not environmentally friendly methods. The production of hybrid TMDs-containing films with a specific structure by physical vapor deposition methods including pulsed laser deposition appears to be an intractable problem. This work aimed to investigate the adjustability of structure, chemical composition and catalytic properties of thin films based on molybdenum selenide and carbon, which were obtained using the method of pulsed laser co-deposition from $MoSe_2$ and graphite targets.

## 2. Experimental details

An electro-optically Q-switched Nd:yttrium-aluminium-garnet laser (wavelength 1.06 μm) was used with a pulse duration of 15 ns, a repetition rate of 25 Hz, and a pulse energy of 30 mJ. The fluence in the laser spot was ~8 $J/cm^2$. A laser beam was rapidly transferred from one target to another, so that the amount of deposited material in one cycle of irradiation of the target did not exceed one monolayer. For the graphite target, the total irradiation time was three times more than that for the $MoSe_2$ target. The chamber with targets and substrates was evacuated to a residual pressure below $10^{-4}$ Pa, and then argon (Ar) was introduced into the chamber to a pressure of 10 Pa. The laser-initiated plume was directed along the normal toward the substrate surface.

Deposition of the films was carried out on the substrates at room temperature. Mechanically polished graphite and NaCl substrates were placed at 4 cm from the target. The film thickness was not greater than 70 nm. The obtained thin-film coatings were examined by the methods of scanning electron microscopy (SEM) using secondary electrons (SE) and back scattered electrons (BSE), energy dispersive X-ray spectroscopy (EDS), high resolution transmission electron microscopy and microdiffraction (HRTEM/MD), micro-Raman spectroscopy (MRS), and X-ray photoelectron spectroscopy (XPS).

The catalytic activity of thin-film coating was investigated by the electrochemical method at room temperature with the use of a traditional three-electrode electrochemical cell. Pure 0.5 M $H_2SO_4$ solution and mixture of this solution with low content of KCl were used as aqueous mediums for the electrochemical studies. The cathodic polarization curves were measured at a slow change in the voltage (1.5 mV/s). The potential was measured relative to the standard hydrogen electrode. Platinum foil was used as reference electrode. The catalytic activity of the prepared films was compared to the activity of a Pt coating produced on the graphite by pulsed laser deposition. The thickness of the Pt coating on the graphite electrode was about 80 nm.

## 3. Results and discussions

Figure 1 shows HRTEM and MD images that were obtained for the film prepared by co-deposition of $MoSe_2$ and carbon on NaCl substrate. This film was redeposited on metal grids in water and then placed in the column of the electron microscope.

Diffuse-broadened reflections on the MD and TEM contrast consisting of light and dark dots indicate an amorphous structure of the film matrix. A dark nanoparticle with a round shape corresponds to elemental molybdenum. Monocrystalline structure of this Mo nanoparticle results in a striped HRTEM image of (110) atomic planes. Romanov et al. [8] revealed that Mo nanoparticles could be formed during pulsed laser ablation of the $MoSe_2$ target. Grigoriev et al. [9] have shown that the Mo nanoparticles facilitate the HER activity of $MoSe_x$ films due to structure modification.

XPS spectra that were measured on the surface of the prepared film are shown in Figure 2. Model spectra for various chemical bonds in the film were constructed using characteristic XPS data that were obtained by Saadi et al. [7] for chemically synthesized $MoSe_3$ films.

Three sets of doublet peaks for Mo 3d are observed in Figure 2. The first doublet (binding energy for $Mo3d_{5/2}$ is about 232.5 eV) is characteristic of $MoO_3$. The second doublet (signal for Mo $3d_{5/2}$ at

IOP Conf. Series: Journal of Physics: Conf. Series **941** (2017) 012065 doi:10.1088/1742-6596/941/1/012065

231.6 eV) was located in a region typically associated with $Mo^{6+}$. The third doublet (signal for Mo $3d_{5/2}$ at 228.4 eV) could be due to $Mo^{4+}$ in $MoSe_{x\leq 2}$ compound. The single peak at 230 eV is characteristic of Se 3s. The Se 3d region has two sets of 3d doublets: the higher-energy doublet that can be ascribed to $Se^0$ and the lower-energy doublet that can be ascribed to $Se^{2-}$. It should be noted that for amorphous $MoS_x$ films, S 3d doublet with high binding energy may be attributed to bridging and/or apical Sulphur atoms. Ting et al. [10] revealed that $MoS_x$ films with higher population of these active S sites will ultimately result in a more efficacious HER catalyst.

**Figure 1**. TEM image of the suspended film that was prepared on NaCl by pulsed laser co-deposition from the $MoSe_2$ and graphite targets in the buffer Ar gas at a pressure of 10 Pa. Inserts show MD picture of the film (upper) and HRTEM image (bottom) of Mo nanoparticle.

Deconvolution of C 1s spectrum allowed to reveal some different chemical states of carbon atoms. The main signal at 284.3 eV can be ascribed to C atoms in the film matrix that have predominantly $sp^2$ configuration of C-C bonds. Another intensive signal at 284.8 eV can be arise due to contamination of the film surface with very thin layer containing C-H compounds. The lines at 281.6 and 288.2 corresponded to different C-O bonds on the surface and in the bulk of the film. Any signal which could be attributed to C-Mo bonds (binding energy is about 282.2 eV) was not observed in the C 1s spectrum of the prepared film. Concentration of $MoSe_x$ phase in the film did not exceed 25 at.%.

Thus TEM, MD and XPS studies have shown that pulsed laser co-deposition of $MoSe_2$ and carbon results in the formation of amorphous films that contain the inclusions of nano-Mo and clustered $MoSe_{x\sim 2}$/$MoSe_{x\sim 3}$ phases. MRS studies have shown that the $MoSe_x$ inclusions have not structural ordering and that the amorphous carbon matrix (a-C) is graphitic to a high extent (Figure 3).

Figure 4 shows the results of polarization curves measurements for graphite substrate before and after deposition of a-C(Mo/$MoSe_x$), Pt/a-C($MoSe_x$) and Pt thin-film coatings. The pulsed laser deposition of pure a-C film on the graphite has not changed the electrochemical properties of the graphite cathode. Covering of the graphite substrate with the a-C(Mo/$MoSe_x$) film caused the decrease of HER overpotential from ~330 to ~250 mV (to produce a current density of -10 mA/$cm^2$). The Tafel slope of a-C(Mo/$MoSe_x$) film was determined to be about 86 mV/dec with an exchange current density of ~$10^{-5}$ A/$cm^2$. The a-C(Mo/$MoSe_x$) catalyst exhibits rather good HER property, which is comparable to that of the $MoSe_3$ catalyst which was chemically synthesized by Saadi et al. [7].

IOP Conf. Series: Journal of Physics: Conf. Series **941** (2017) 012065 doi:10.1088/1742-6596/941/1/012065

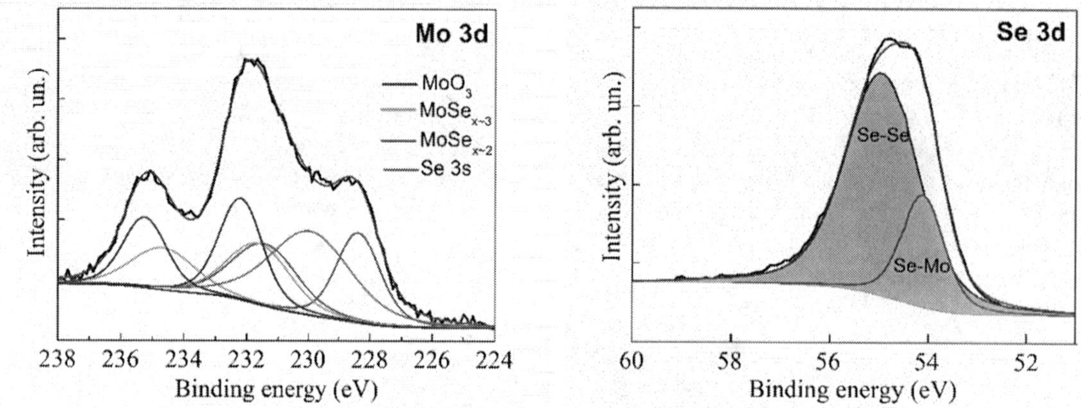


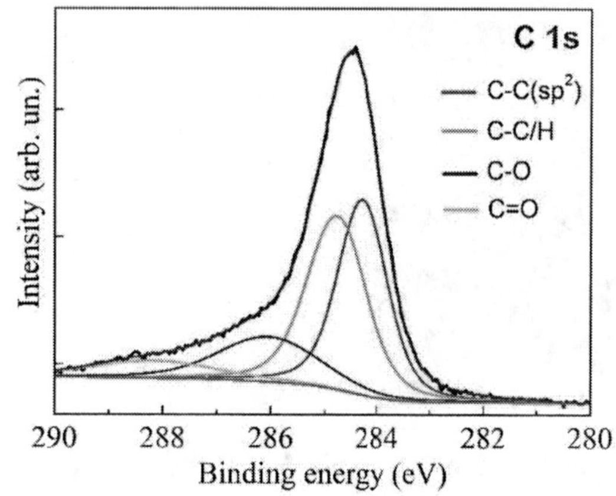


**Figure 2**. XPS spectra of Mo3d, Se3d, and C1s on the surface of the film prepared by pulsed laser co-deposition from the $MoSe_2$ and graphite targets in the buffer Ar gas at a pressure of 10 Pa. Doublets for different chemical states of Se 3d are presented as single colored peaks.

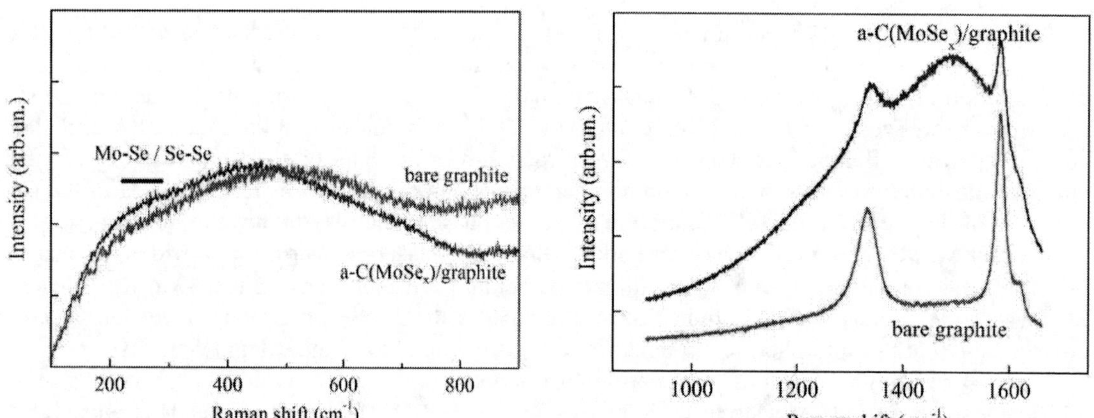


**Figure 3**. MRS spectra of graphite substrate measured before and after deposition of a-C(Mo/$MoSe_x$) film. Only a broad line was observed in a region of 200 – 300 $cm^{-1}$ typically associated with Mo-Se and Se-Se compounds. The spectrum of a-C matrix in a region of 1000 – 1600 $cm^{-1}$ was typical of the amorphous carbon film possessing a high concentration of $sp^2$-hybridizied C atoms.

To form Pt/a-C(Mo/$MoSe_x$) film, the graphite substrate with a-C(Mo/$MoSe_x$) film was undergone the cathodic treatment in Cl-containing sulfuric acid solution at a potential of -350 mV. Ions of Cl interacted with Pt anode and the formed Pt-containing molecules could migrate to cathode for deposition on its surface. The $MoSe_x$ clusters have activated the processes of Pt nanoparticle growth on the surface of a-C(Mo/$MoSe_x$) film (Figure 5). The used time of cathodic treatment allowed to prepare catalytically effective Pt/a-C(Mo/$MoSe_x$) thin-film coating with lower Pt loading.

Figure 5 shows the results of SEM study of the surface of the bare graphite electrode and graphite electrode with deposited thin-film coatings of Pt/a-C(Mo/$MoSe_x$). The graphite consisted of plate microcrystals which formed relatively loose packing of the matrix and provided the large effective surface of the electrode. Pulsed laser ablation of the $MoSe_2$ and carbon in the Ar buffer gas provided conformal deposition of a-C(Mo/$MoSe_x$) film on graphite microcrystals with round-shaped nanoparticles formation. The size of nanoparticles was 50 – 100 nm

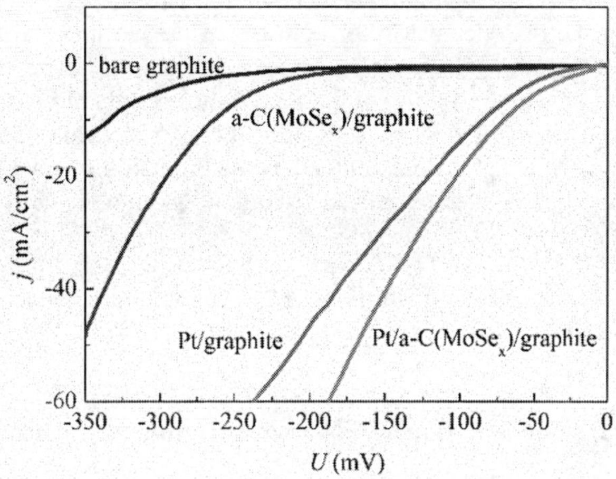


**Figure 4**. Polarization curves for the bare graphite substrate and the graphite substrate covered with a-C(Mo/$MoSe_x$), Pt/a-C(Mo/$MoSe_x$) and Pt thin-film coatings.

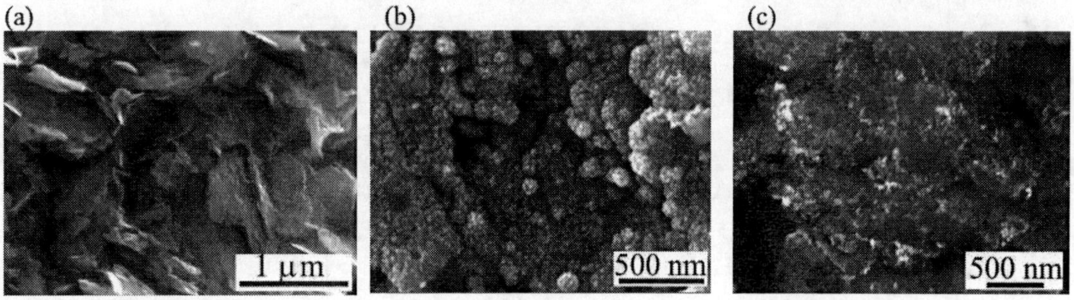


**Figure 5**. SE (a,b) and BSE (c) SEM images of the graphite substrate before (a) and after the Pt/a-C(Mo/$MoSe_x$) deposition (b, c).

BSE SEM (Figure 5*c*) and EDS (Figure 6) studies revealed that Pt nanoparticles have grown on the enriched with Se/Mo areas of a-C(Mo/$MoSe_x$) film. The Pt nanoparticle were quite evenly distributed on the surface without the formation of large conglomerates and the size of Pt nanoparticles was ~5 nm. The HER property of the hybrid Pt/a-C(Mo/$MoSe_x$) catalyst with a Pt loading of ~6 $\mu g/cm^2$ noticeably superior that of Pt/graphite catalyst with a Pt loading of 170 $\mu g/cm^2$ (Figure 4).

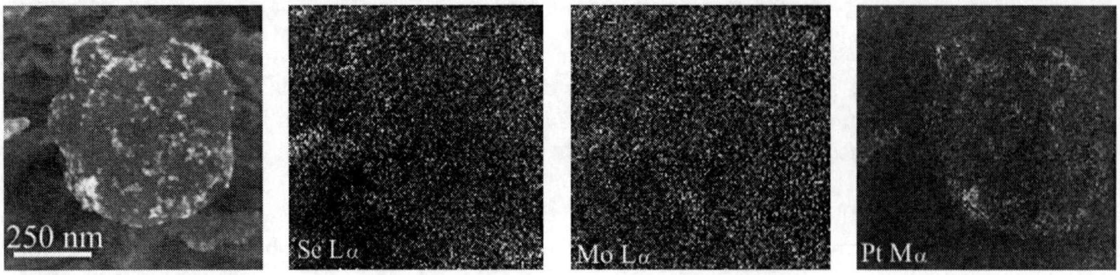


**Figure 6**. SEM image (left) and EDS maps for Mo, Se and Pt distributions on the surface of Pt/a-C(Mo/$MoSe_x$)/graphite sample.

## 4. Conclusions

Pulsed laser co-deposition of carbon and $MoSe_2$ at room temperature resulted in the formation of nanostructured thin-film coating which includes the amorphous a-C matrix with a high $sp^2$ content, the nanocrystalline Mo and the disordered $MoSe_{x\geq 2}$ clusters. The Mo nanoparticle deposition caused an increase of the surface roughness. The $MoSe_{x\geq 2}$ clusters formation is very important for providing a large amount of active sites. The formation of the active sites results in both a significant $H_2$ evolution and an effective cathodic deposition of nano-Pt onto the a-C(Mo/$MoSe_x$) film. Even with a low Pt loading, the hybrid Pt/a-C(Mo/$MoSe_x$)/graphite catalyst exhibits excellent HER properties superior to those of the pure Pt coatings deposited by pulsed laser ablation onto the GC substrate.

## Acknowledgments

The present work was supported by the Ministry of Education and Science of the Russian Federation (grant No 3.9750.2017/8.9).

## References

[1] Jaramillo T F, Jørgensen K P, Bonde J, Nielsen J H, Horch S and Chorkendorff I 2007 *Science* **317** 100-1
[2] Lu Q, Yu Y, Ma Q, Chen B and Zhang H 2016 *Adv. Mater.* **28** 1917–33
[3] Morales-Guio C G, Stern L A and Hu X 2014 *Chem. Soc. Rev.* 43 6555–69
[4] Vrubel H, Moehl T, Gratzel M and Hu X 2013 *Chem. Commun.* **49** 8985–7
[5] Klein A, Dolatzoglou P, Lux-Steiner M and Bucher E 1997 *Sol. Energy Mater. Sol. Cells* **46** 175–86
[6] Merki D, Fierro S, Vrubel H and Hu X 2011 *Chem. Sci.* **2** 1262–7
[7] Saadi F H, Carim A I, Velazquez J M, Baricuatro J H, McCrory C C L, Soriaga M P and Lewis N S 2014 *ACS Catal.* **4** 2866–73
[8] Romanov R I, Fominski V Y, Gnedovets A D, Grigoriev S N and Volosova M A 2016 *J. Phys. Conf. Ser.* **747** 012045
[9] Grigoriev S N, Fominski V Y, Romanov R I, Volosova M A and Shelyakov A V 2015 *Thin Solid Films* **592** 175–81
[10] Ting L R L, Deng Y, Ma L, Zhang Y J, Peterson A A and Yeo B S 2016 *ACS Catal.* **6** 861–7

IOP Conf. Series: Journal of Physics: Conf. Series **941** (2017) 012066 doi:10.1088/1742-6596/941/1/012066

# Momentum distribution in two-orbital Hubbard model: Quantum Monte Carlo approach

**Vladimir Kashurnikov, Andrey Krasavin, Yaroslav Zhumagulov**

[1]National Research Nuclear University MEPhI (Moscow Engineering Physics Institute), Kashirskoe sh. 31, 115409 Moscow, Russia

Email: avkrasavin@mephi.ru

**Abstract.** The two-dimensional two-orbital Hubbard model is studied with the use of world-line quantum Monte Carlo algorithm. Spectral functions and the density of states for various parameters of the model are obtained in both the undoped and low-doped regimes. The invariance of the Fermi surface with respect to the strength of the interaction is testified. Fermi-liquid parameters of the model are derived.

## 1. Introduction

The role of electron correlations in high-temperature superconductors (HTSC) based on iron [1] is decisive in the formation of physical properties of these systems. The effect of strong Coulomb interactions on the nature of superconductivity and the formation of a complex phase diagram including antiferromagnetic, structural and superconducting ordering, is a major focus of interest [2-3]. Like copper HTSC, iron-based superconductors are characterized by a strongly expressed anisotropy, and have a structure of closely spaced atomic planes of Fe and As (for pnictides). The main contribution to the band structure near the Fermi surface is provided by two orbitals, $d_{xz}$ and $d_{yz}$. Therefore, the simplest model Hamiltonian for iron-based HTSC, reflecting their crystal and electronic structure, is a two-dimensional generalized two-orbital Hubbard model [4], which has been intensively studied since the discovery of iron-based HTSC with the use of various approximate analytical and numerical methods. The applicability of finite-size cluster quantum Monte Carlo algorithms that allow, in principle, to obtain exact results, is limited in this case by the sign problem, which exponentially slows calculations at a sufficiently low temperature and/or away from the half-filling. In this work, with choosing specific basis states for the Hamiltonian of the two-orbital model and the use of CTWL-algorithm [5], we were able to overcome partly the sign problem and to obtain data for Matsubara Green's function at sufficiently low temperature in a wide range of model parameters. Using the relationship between Matsubara Green's function and the density of electron states, we have derived the Fermi liquid parameters of the model: quasi-particle weight ($Z$-factor), self-energy $\Sigma$, and scattering rate $\Gamma$, and studied the properties of the model under doping. The asymmetry in the behavior of $Z$-factor was observed at electron and hole doping, which reflects the asymmetry of the Hamiltonian. The evolution of the Fermi surface was studied at changing the model parameters and the level of doping; an increase of the volume of the Fermi surface with the growth of interaction was observed, which shows the deviation from the Luttinger theorem.

IOP Conf. Series: Journal of Physics: Conf. Series **941** (2017) 012066 doi:10.1088/1742-6596/941/1/012066

## 2. Two-orbital 2D Hubbard model

First introduced in [4], the minimum two-orbital model for iron-based HTSC takes into account the real crystal structure of these compounds, as well as the two-dimensional nature of physical properties, and has the following Hamiltonian:

$$H = H_{int} + H_{kin}; \tag{1}$$

$$H_{int} = U\sum_{i\alpha} n_{i\alpha\uparrow}n_{i\alpha\downarrow} + V\sum_{i} n_{ix}n_{iy} - \mu\sum_{i} n_i - J\sum_{i}\left(n_{ix\uparrow}n_{iy\uparrow} + n_{ix\downarrow}n_{iy\downarrow}\right) -$$

$$-J\sum_{i}\left(a^{+}_{ix\downarrow}a_{ix\uparrow}a^{+}_{iy\uparrow}a_{iy\downarrow} + a^{+}_{ix\uparrow}a_{ix\downarrow}a^{+}_{iy\downarrow}a_{iy\uparrow} + a^{+}_{ix\uparrow}a_{iy\downarrow}a^{+}_{ix\downarrow}a_{iy\uparrow} + a^{+}_{iy\uparrow}a_{ix\downarrow}a^{+}_{iy\downarrow}a_{ix\uparrow}\right);$$

$$H_{kin} = -t_1\sum_{i\sigma}\left(a^{+}_{ix\sigma}a_{i+x,x\sigma} + a^{+}_{iy\sigma}a_{i+y,y\sigma}\right) - t_2\sum_{i\sigma}\left(a^{+}_{iy\sigma}a_{i+x,y\sigma} + a^{+}_{ix\sigma}a_{i+y,x\sigma}\right) -$$

$$-t_3\sum_{i\sigma}\left(a^{+}_{ix\sigma}a_{i+x+y,x\sigma} + a^{+}_{ix\sigma}a_{i+x-y,x\sigma} + a^{+}_{iy\sigma}a_{i+x+y,y\sigma} + a^{+}_{iy\sigma}a_{i+x-y,y\sigma}\right) +$$

$$+t_4\sum_{i\sigma}\left(a^{+}_{ix\sigma}a_{i+x-y,y\sigma} + a^{+}_{iy\sigma}a_{i+x-y,x\sigma} - a^{+}_{ix\sigma}a_{i+x+y,y\sigma} - a^{+}_{iy\sigma}a_{i+x+y,x\sigma}\right) + h.c.$$

Here operator $a^{+}_{ix(y)\sigma}(a_{ix(y)\sigma})$ creates (annihilates) an electron with spin projection $\sigma$ on site $\boldsymbol{i}$ and orbital $x(y)$; $t_j, j = 1,\dots,4$ are hopping amplitudes between orbitals $x(y)$; $U$, $V$, and $J$ are Coulomb interaction terms; $\mu$ is the chemical potential. In this work we used the quantum continuous time world-line Monte Carlo algorithm (CTWL-algorithm), adapted for the two-orbital model [5]. The parameters of $H_{kin}$ in (1) were taken the same as in [6], $t_1 = 0.058\ (eV)$; $t_2 = 0.22$; $t_3 = -0.21$; $t_4 = -0.08$; and the relationships between Coulomb terms in $H_{int}$ are $V = 0.5U; J = 0.25U$.

## 3. Momentum distribution

The momentum distribution $n(\boldsymbol{k})$ can be obtained from the relation

$$n(\boldsymbol{k}) = G(\boldsymbol{k}, \tau \to -0); \quad G(\boldsymbol{k},\tau) = \sum_{ij} G_{ij}(\tau)\, e^{i\boldsymbol{k}\boldsymbol{r}_{ij}}. \tag{2}$$

Here $G_{ij}(\tau)$ is Matsubara Green's function, $G_{ij}(\tau) = -\langle T_\tau a_{i\alpha\sigma}(\tau)a^{+}_{j\beta\sigma'}(0)\rangle$.

Direct approximation (2) in a numerical calculation leads to uncontrolled errors; therefore, we have used the following expression:

$$n(\boldsymbol{k}) = \int \frac{A_\sigma(\boldsymbol{k},\omega)}{1 + e^{\beta(\omega-\mu)}}d\omega. \tag{3}$$

To calculate (3), the spectral function $A_\sigma(\boldsymbol{k},\omega)$ and the total density of states $A(\omega) = \sum_{k\sigma} A_\sigma(\boldsymbol{k},\omega)$ were recovered with the use of a stochastic procedure [7] from the integral equation relating the spectral function with Matsubara Green's function,

$$G_\sigma(\boldsymbol{k},\tau) = -\int \frac{A_\sigma(\boldsymbol{k},\omega)e^{-\tau(\omega-\mu)}}{1 + e^{-\beta(\omega-\mu)}}d\omega. \tag{4}$$

## 4. Half-filling

Figure 1 shows the profiles of the momentum distribution depending on the parameter $U$ at half-filling. The data are presented for the inverse temperature $\beta = 1$. This is a reasonably low temperature as $\beta t_{max} = \beta t_2 {\sim} 0.2$; in addition, the sign problem does not allow lowering the temperature significantly. The flattening of the distribution with increase of the interaction is clearly visible, as well as the presence of the hole pockets at the center (Γ) and the periphery (M) of the zone, the filling of which increases with $U$.

The total density of states (DOS) depending on the value of the interaction is shown in Figure 2. The calculations were performed at $\beta = 1$, so a thermal broadening of the zones takes place; however, we believe that it should not prevent to see the evolution of the zones with changing $U$. Such choice of temperature is associated with the convergence of quantum Monte Carlo algorithm, namely, with the sign problem. Only at this temperature we were able to perform the calculations at a series of values of the interaction parameters with sufficient accuracy. The interval between the bands increases with the increase of the interaction, and is close to the value of $U$, but not identical with it, since the interaction part of the Hamiltonian (1) is more complicated than the conventional Hubbard term. With the growth of the interaction the bands shrink turning into narrow peaks; this leads to a reduction of the dispersion and flattening of the momentum distribution.

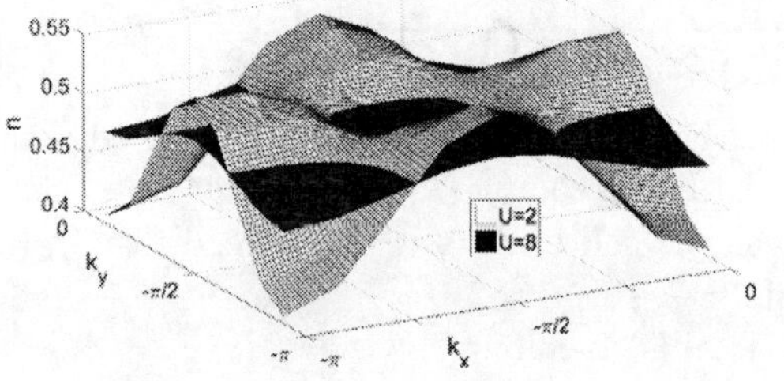


**Figure 1.** Momentum distribution in the first Brillouin zone as a function of $U$.

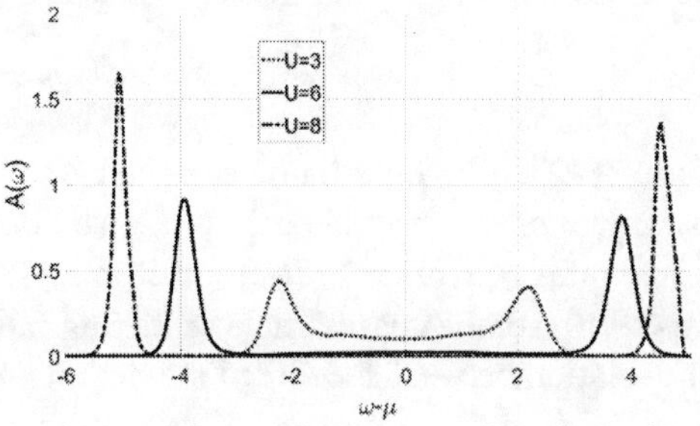


**Figure 2.** The total DOS as a function of $U$. $\beta$=1.

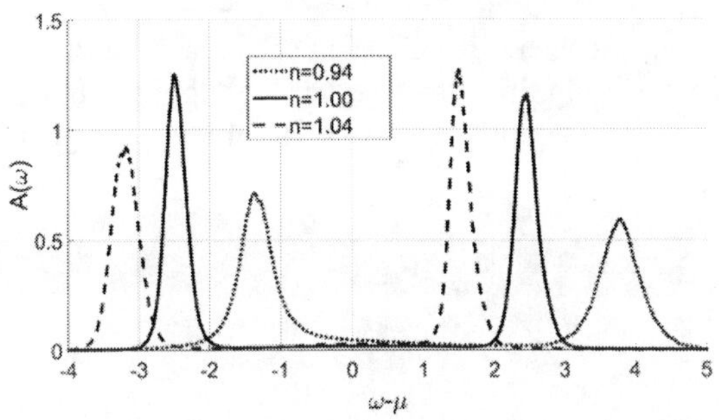


**Figure 3.** The evolution of the density of states with change of doping near the half-filling. $U = 4$; $\beta = 2$.

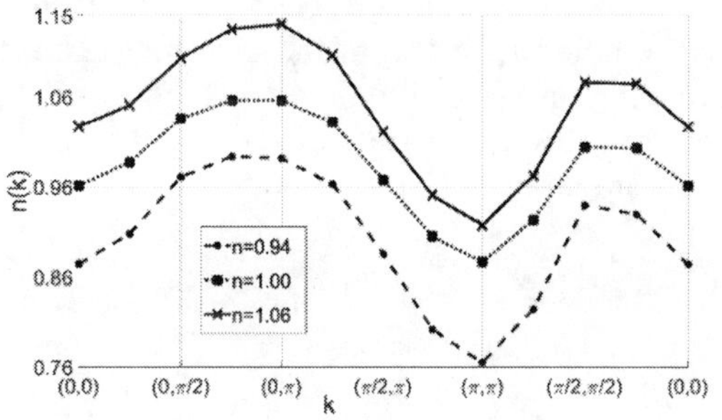


**Figure 4.** Momentum distribution for various doping. $U = 4$; $\beta = 2$.

## 5. The doped case

The quantum Monte Carlo algorithm suffers from the minus-sign problem when calculating a two-dimensional Fermi system. This leads to the fact that the deviation from the half-filling may not afford calculation of thermodynamic average values, including the Green's function, with the reasonable accuracy. We have managed to obtain a good convergence in the doped case only at several parameters of $U$ and at temperatures not lower than $\beta = 2$. Figure 3 shows the total DOS at a low-level of electron and hole doping. The shape of the bands changes as the chemical potential moves from the electron to the hole doping region. The broadening of the bands is observed with increasing the level of doping (both hole and electron). The value of the density of states at the Fermi level increases slightly with increasing the hole doping. Therefore, the change of the number of carriers in the system is not equivalent to a simple filling of the bands (the so-called "hard-band approximation"), i.e., to a simple shift of the chemical potential at a constant density of states, which is the contribution of Coulomb correlations in this typical generalized Hubbard model. Similar correlation effects were also observed in [8] for the Hubbard model in the low-doped regime.

Figure 4 shows the profiles of the momentum distribution along the main crystallographic directions in the first Brillouin zone for the same values of doping, as in Figure 3. The following can be noted. The profiles with different fillings are substantially equidistant, which means that the bands are filled with virtually no distortion as the concentration changes. The momentum distributions vary little with the change of the filling, which is consistent with the statement that the momentum distribution of quasiparticles has the same features (including the Fermi jump) as the total momentum distribution. Interaction blurs distribution gradients as well as at the half-filling case.

## 6. Fermi-liquid parameters

Studies of the Hubbard model [8, 9] and experimental data on FeAs-systems [10] may indicate non-Fermi liquid nature of these strongly correlated systems. To investigate this issue, we extracted the Fermi-liquid parameters of the model (1). In the quasiparticle approximation, the spectral density near the maximum of the peak for electron and hole excitations can be described as follows [11]:

$$A(\boldsymbol{k},\omega) = Z(\boldsymbol{k})\frac{\Gamma(\boldsymbol{k},\omega)}{\pi\left((\omega-\varepsilon(\boldsymbol{k})+\mu)^2+\left(\Gamma(\boldsymbol{k},\omega)\right)^2\right)}. \quad (5)$$

Here $Z(\boldsymbol{k})$ is the quasiparticle weight or $Z$-factor, $\varepsilon(\boldsymbol{k})$ is the excitation energy identified with the maximum of the spectral peak for a given point of the Brillouin zone; $\Gamma(\boldsymbol{k},\omega)$ is the quasiparticle scattering rate, $\Gamma(\boldsymbol{k},\omega) = -Z(\boldsymbol{k})\,\mathrm{Im}\big(\Sigma(\boldsymbol{k},\omega)\big)$; $\Sigma(\boldsymbol{k},\omega)$ is the self-energy; it is assumed that $|(\varepsilon(\boldsymbol{k})-\mu)| \gg |\Gamma|$, so, the quasiparticle scattering rate is small.

The relation between the Green's function and the $Z$-factor is given by

$$Z = \frac{\pi}{\beta}\frac{1}{\mathrm{Im}\left[\left(G(k,i\omega_0)\right)^{-1}\right]}; \quad G(k,i\omega_0) = \int_0^\beta G(k,\tau)e^{i\omega_0\tau}. \quad (6)$$

The renormalization of the $Z$-factor was observed with the increase of the interaction. The effect of doping on Z-factor is shown at Figure 5. Z-factor increases significantly even at low doping; this indicates growth of density of states at the Fermi-level. Note that similar results were obtained in [12] for the single-band Hubbard model.

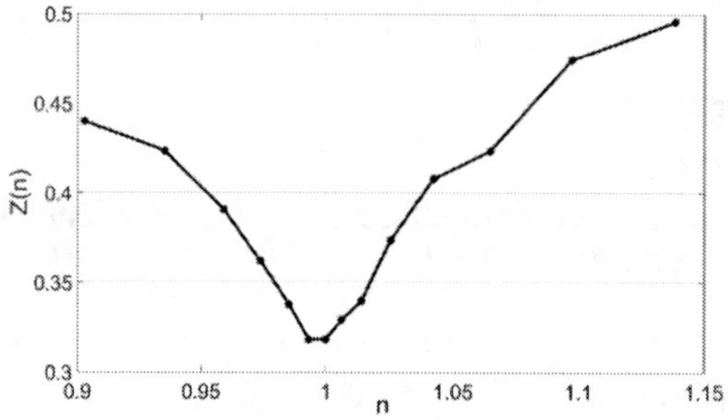


**Figure 5.** $Z$-factor as a function of doping. $U = 4$; $\beta = 2$.

**Acknowledgments**

The work was supported by Russian Found for Basic Research (project #15-02-02764).

## 7. References

[1] Kamihara Y, Watanabe T, Hirano M and Hosono H 2008 *J. Am. Chem. Soc.* **130** 3296
[2] Isyumov Yu, Kurmaev E 2010 *High-$T_c$ Superconductors Based on FeAs Compounds* Springer
[3] Dagotto E 2013 *Rev. Mod. Phys.* **85** 849

[4] Raghu S *et al* 2008 *Phys. Rev. B* **77** 220503(R)
[5] Kashurnikov V A, Krasavin A V 2013 *JETP Lett.* **97** 333
[6] Moreo A, Daghofer M, Riera J A and Dagotto E 2009 *Phys. Rev. B* **79** 134502
[7] Mishenko A S, Prokof'ev N V, Sakamoto A, Svistunov B V 2000 *Phys. Rev. B* **62** 6317
[8] Gröber C, Eder R and Hanke W 2000 *Phys. Rev. B* **62** 4336
[9] Korshunov M M, Ovchinnikov S G 2003 *Phys. Solid State* **45** 1415
[10] Dai Y M *et al* 2015 *Phys. Rev. X* **5** 031035
[11] Abrikosov A A, Gorkov L P, Dzyaloshinski I E 2012 *Methods of Quantum Field Theory in Statistical Physics* Dover Publications
[12] Arsenault L-F, Sémon P and Tremblay A-M S 2012 *Phys. Rev. B* **86** 085133

# Critical current of layered superconductor with defects in tilted magnetic field

**V A Kashurnikov, A N Maksimova, I A Rudnev, D S Odintsov**

[1]National Research Nuclear University MEPhI (Moscow Engineering Physics Institute), Kashirskoe sh. 31, 115409 Moscow, Russia

Email: nastymaksimova@yandex.ru

**Abstract**. The simulation of vortex lattice in layered high-temperature superconductor (HTSC) has been done using Monte Carlo method. The case of highly anisotropic superconductor with crossing lattices of Abrikosov and Josephson vortices under applying of tilted magnetic field has been examined. The critical current as a function of absolute value of dc magnetic field, parallel and perpendicular to the anisotropy axis *c*, has been calculated. The simulations have been done for different anisotropy parameter of HTSC and different concentrations of defects. Configurations of the Josephson vortices in presence of the Abrikosov vortices have been obtained.

## 1. Introduction

High-temperature superconductors (YBCO, BSCCO) are anisotropic materials with layered structure. The vortex line in HTSC can be represented as a stack of interacting pancake vortices. The vortex lattice of layered superconductor in tilted magnetic field had been widely analyzed both theoretically and experimentally. The effect of tilted field on magnetic and transport properties of HTSC is under significant practical interest [1-3].

Theoretical studies [4-10] show that layered superconductors have a rich phase diagram in tilted magnetic field. The structure of vortex lattice is defined by the parameter $\alpha=\lambda/\lambda_J$, which is the ratio of London, $\lambda$, and Josephson, $\lambda_J=\gamma s$, lengths [5]. Here $\gamma$ is an anisotropy parameter and $s$ is an inter-layer spacing. At large $\alpha$ the superconductor represents a tilted vortex lattice and at small $\alpha$ the crossing array of Abrikosov and Josephson vortices is realized. This paper aims to consider the interacting Abrikosov and Josephson vortex lattices in the frame of 3D model of layered HTSC and to obtain numerically the critical current as a function of an angle between the direction of applied magnetic field and an anisotropy axis.

## 2. Expressions for energy in Monte Carlo simulations

According to theoretical studies [4-10], pancake vortices are generated by *c*-axis field and Josephson vortices are generated by in-plane component. Josephson vortices are parallel to superconducting layers (*ab*-plane) and located in the inter-layer spacing.

Using the results [11,12] we can obtain the expressions for energy of Josephson vortices which can be used in Monte Carlo simulations. The self-energy of Josephson vortex (per unit length):

$$\varepsilon_1 = \frac{\Phi_0}{4\pi} H_{c1} = \frac{\Phi_0^2}{16\pi^2 \lambda_b(T)\lambda_c(T)} \left[ \ln \frac{\lambda_b(T)}{s} + 1.12 \right], \ \lambda_b^2 = \frac{s}{d_s}\lambda_s^2, \ \lambda_c^2 = (\gamma s)^2 + \frac{d_s}{s}\lambda_s^2. \quad (1)$$

$d_s = 0.27$ nm is the superconductor layer thickness and the inter-layer spacing $s$=2.7 nm.

The interaction energy of two Josephson vortices is defined as a work of Lorenz force (from the supercurrents forming Josephson vortex) when the vortex moves from infinity to the adjusted point ($y_1$, $sn_1$) in the vicinity of second vortex (at the position ($y_2$, $sn_2$)):

$$U_{1,2}^{jos} = \frac{\Phi_0^2}{8\pi^2 \lambda_b \lambda_c} K_0 \left( \sqrt{\frac{(y_2 - y_1)^2}{\lambda_c^2} + \frac{s^2 (n_2 - n_1)^2}{\lambda_b^2}} \right).$$

$n$ is a number of superconducting layer. Interaction between Abrikosov and Josephson vortex:

$$U_{1,2}^{v} = -\frac{\Phi_0^2}{8\pi^2 \lambda_b \lambda_c} \frac{\frac{z}{\lambda_b}}{\sqrt{\left(\frac{y}{\lambda_c}\right)^2 + \left(\frac{z}{\lambda_b}\right)^2}} \frac{x_0 \pm d/2}{\lambda_b} K_1 \left( \sqrt{\left(\frac{y}{\lambda_c}\right)^2 + \left(\frac{z}{\lambda_b}\right)^2} \right), \quad (2)$$

$d$ is a width of a superconductor, $x_0$ is the $x$-coordinate of Abrikosov vortex and ($x$-axis is perpendicular to the transport current), $y$ and $z$ are the differences between $y$- and $z$-coordinates of Abrikosov and Josephson vortices. The origin is in the centre of the superconductor.

Using (1) we can obtain the first critical field for Josephson vortices:

$$H_{c1} = \frac{\Phi_0}{4\pi \lambda_b(T)\lambda_c(T)} \left[ \ln \frac{\lambda_b(T)}{s} + 1.12 \right].$$

## 3. Simulation results

The Monte Carlo simulation (see [13-17] for details of the method) for a stack of superconducting layers can be done only for few ($N_L \sim 10$) layers in the stack. However, at $\gamma \sim 100$ (typical for BSCCO) the characteristic distances for Josephson vortex interaction $\lambda_b$ и $\lambda_c$ are ~500 nm and the triangular lattice of Josephson vortices can be formed only when $N_L \sim 1000$. For simulations, we can roughly define a stack of 100 pancakes as an elementary object so that vortex line is still consisted of ~10 elementary objects and Josephson vortices can be placed between any of 1000 layers. This approach is approximately correct since the mutual displacements of pancakes are not numerous in the most part of simulations. The simulations have been done for typical parameters of BSCCO. The size of superconducting slab is 5×3 μm.

Figure 1 represents the calculated critical current as a function of the magnitude of magnetic field, parallel and perpendicular to superconducting layers. The obtained dependencies are mostly decreasing which is in agreement with experimental results. Measurements on the YBCO samples with columnar defects show that critical current decreases when magnetic field increases and $j_c$ in **H** || **c** is lower than in **H** || $ab$. However, if we increase the concentration of columnar defects, the critical current in **H** || **c** becomes higher. Both these cases, in fact, can be reproduced in simulations. Figure 2 represents the critical current for $\gamma = 20$ and both concentrations of defects.

Figure 3 represents the configurations of Josephson vortices. Abrikosov vortices are shown schematically in Figure 3b, in the field of transport current $H_I = 100$ Gs Abrikosov vortices do not penetrate into the slab. As seen, the triangular lattice of Josephson vortices breaks down in presence of Abrikosov vortices. Moreover, the highest density of Josephson vortices is observed at the upper boundary of the slab. This fact can be explained with features of interaction between Abrikosov and Josephson vortices (2).

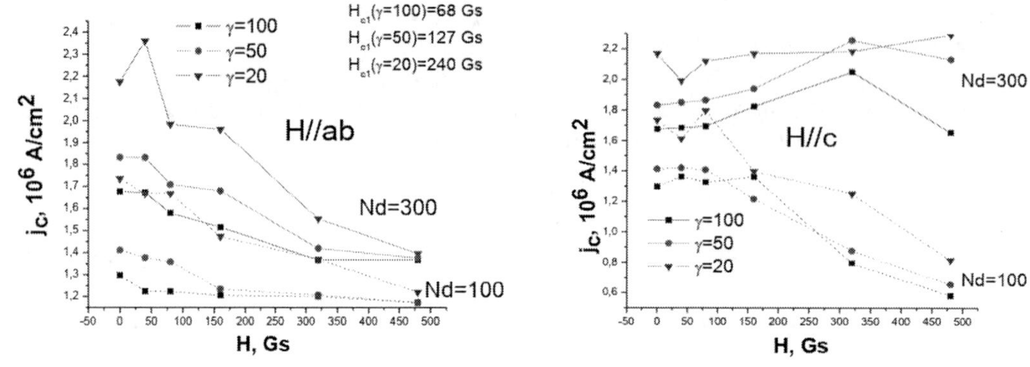


a) б)

**Figure 1.** Critical current as a function of external dc magnetic field. $N_d$ – number of defects in the slab 5×3 μm

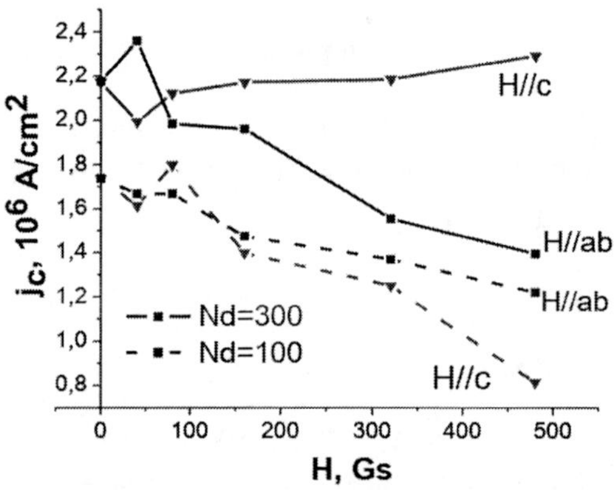


**Figure 2.** Critical current at γ=20

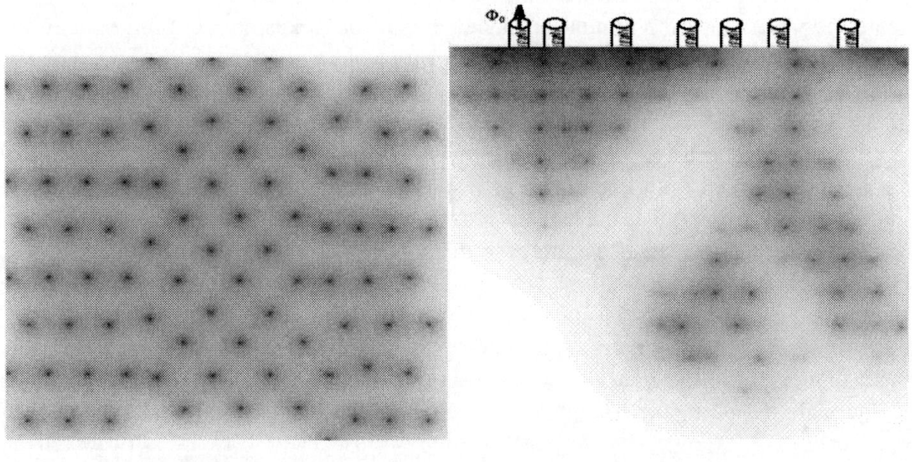


(a) б)

**Figure 3.** Configurations of Josephson vortices. a) Self-field of transport current $H_I$ = 100 Gs, b) $H_I$ = 355 Gs. External dc magnetic field is $H$ = 320 Gs and is tilted to $c$-axis by angle θ=45°.

## 4. Conclusion

The simulation of vortex lattice in HTSC under applying of tilted dc magnetic field has been done by using Monte Carlo method. For this algorithm, the description of interacting Abrikosov and Josephson vortex lattices had been introduced into 3D model of layered HTSC. The critical current as a function of the magnitude of external dc magnetic field, parallel and perpendicular to the superconducting layers has been calculated. The obtained dependences are decreasing and agree qualitatively with experiment. The experimental fact that the critical current of HTSC with columnar defects in field parallel to the anisotropy axis exceeds the critical current in field parallel to the *ab* plane is qualitatively reproduced in our calculations. The configurations of Josephson vortices in HTSC magnetized by self-field of transport current have been obtained. Our method allows the calculation of magnetization curves, current-voltage characteristics and critical current of layered superconductors with high $\gamma$ in arbitrary magnetic fields, with arbitrary configurations of defects and different anisotropy parameters.

**Acknowledgement**

The research was done under financial support of RFBR, grant 15-02-02764

## References

[1] Roas B, Schultz L, Saemann-Ischenko G 1990 *Phys. Rev. Lett.* **64(4)** 479
[2] Konczykowski M, van der Beek C J, Mosser V, Koshelev A E, Li M, Kes P H 2010 *Physica C* **470** S257–S258
[3] Zhang Y, Specht E D, Cantoni C, Christen D K, Thompson J R, Sinclair J W, Goyal A, Zuev Y L, Aytug T, Paranthaman M P, Chen Y, Selvamanickam V 2009 *Physica C* **469** 2044–2051
[4] Burlachkov L I 1989 *Europhys. Lett.* **8(7)** 673-677
[5] Koshelev A E 2005 *Phys. Rev. B* **71** 174507
[6] Koshelev A E 1993 *Phys. Rev. B* **48(2)** 1180
[7] Koshelev A E 1999 *Phys. Rev. Lett.* **83(1)** 187
[8] Koshelev A E 2003 *Phys. Rev. B* **68** 094520
[9] Beleggia M, Pozzi G, Tonomura A, Kasai H, Matsuda T, Harada K, Akashi T, Masui T and Tajima S 2004 *Phys. Rev. B* **70** 184518
[10] Samokhvalov A V, Mel'nikov A S, and Buzdin A I 2012 *Phys. Rev. B* **85** 184509
[11] John R. Clem and Mark W. Coffey 1990 *Phys. Rev. B* **42(10)** 6209
[12] John R. Clem, Mark W. Coffey and Zhidong Hao 1991 *Phys. Rev. B* **44(6)** 2732
[13] Kashurnikov V A, Maksimova A N, Rudnev I A, Odintsov D S 2015 *Physics Procedia* **65** 97-100
[14] Kashurnikov V A, Maksimova A N, Rudnev I A 2014 *Physics of the Solid State* **56 (5)** 894-911
[15] Odintsov D S, Rudnev I A, Kashurnikov V A 2006 *JETP* **103** 66
[16] Rudnev I A, Odintsov D S, Kashurnikov V A 2008 *Phys. Lett. A* **372** 3934-3936
[17] Odintsov D S, Rudnev I A, Kashurnikov V A 2007 *JETP* **105** 253

IOP Conf. Series: Journal of Physics: Conf. Series **941** (2017) 012068 doi:10.1088/1742-6596/941/1/012068

# Pseudospectral method for the quantum state determination of the superconductor in the nonuniform magnetic field

**D S Korchagin, P F Kartsev**

National Research Nuclear University "MEPhI" (Moscow Engineering Physics Institute), Kashirskoe sh., 31, 115409 Moscow, Russia

E-mail: Dmserkor@yandex.ru, PFKartsev@mephi.ru

**Abstract.** In this work, we present the numerical approach to determine the quantum state of a superconductor using the Ginzburg-Landau equations by means of a pseudospectral method. Its convergence is demonstrated by simulation of a test problem. All the analytics is given with details.

## 1. Introduction

A standard approach to find the quantum state of a superconductor placed in a nonuniform magnetic field, is to apply the Ginzburg-Landau (GL) equations which can be solved numerically [1, 2, 3, 4]. In a dimensionless form, GL equations are written as [1]:

$$\left(\frac{i}{\varkappa}\nabla+\mathbf{A}\right)^2\Psi-\Psi+|\Psi|^2\Psi=0, \tag{1}$$

$$\nabla\times(\nabla\times\mathbf{A})+\frac{i}{2\varkappa}(\Psi^*\nabla\Psi-\Psi\nabla\Psi^*)+|\Psi|^2\mathbf{A}=0, \tag{2}$$

where $\Psi$ is the wave function of the Cooper pair condensate, $\mathbf{A}$ is the vector potential of magnetic field inside the superconductor, and $\varkappa$ is the dimensionless Ginzburg-Landau parameter.

The system of equations (1), (2) is complemented by the gauge for vector potential (we choose the Coulomb gauge):

$$\nabla\cdot\mathbf{A}=0, \tag{3}$$

and boundary conditions:

$$\left(\frac{i}{\varkappa}\nabla+\mathbf{A}\right)\Psi\cdot\mathbf{n}=0, \tag{4}$$

$$(\nabla\times\mathbf{A})_\tau=(\mathbf{H})_\tau, \tag{5}$$

where $\mathbf{H}$ is the external magnetic field on the boundary, $\mathbf{n}$ is a normal vector to the surface, and $\tau$ denotes a component tangential to the surface.

The case of nonuniform external field is the most interesting from the practical point of view but makes the computation very labor-intensive. More effective way should involve representing the approximate solution in the form appropriate to simplify the mathematical formulation of the problem.

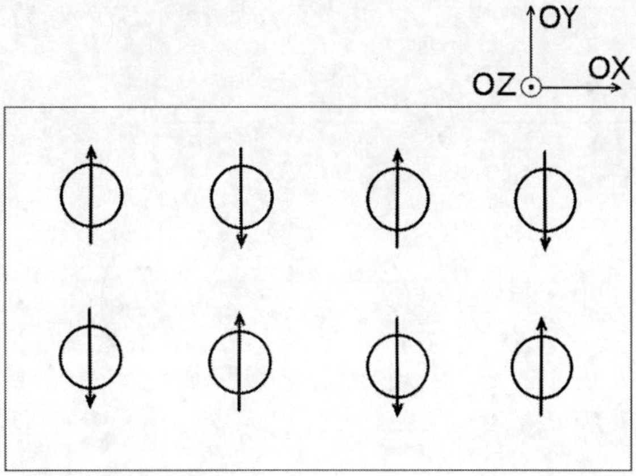


**Figure 1.** Geometry of the problem under consideration

In this report, we present the numerical approach to solve the GL equations using a pseudospectral method.

We analyze the case of highly-nonuniform external field present in the localized area of superconductor, so that far enough from the source, its influence on the superconducting state is negligible. As a model problem, we consider a superconducting film in the field of periodic array of magnetic particles placed in alternating way, as shown in Fig. 1. Alternation is needed to guarantee that no macroscopic current is flowing through the system. As a result, the external magnetic field and the state of the film are made periodic.

With periodic boundary conditions applied, the solution is represented by the series of plane waves:

$$\Psi(x,y,z) = \sum_{k_x,k_y} \psi_{k_x,k_y}(z) e^{i(k_x x + k_y y)} \equiv \sum_{\mathbf{k}} \psi_{\mathbf{k}}(z) e^{i\mathbf{k}\boldsymbol{\rho}}, \tag{6}$$

$$\mathbf{A}(x,y,z) = \sum_{k_x,k_y} \mathbf{a}_{k_x,k_y}(z) e^{i(k_x x + k_y y)} \equiv \sum_{\mathbf{k}} \mathbf{a}_{\mathbf{k}}(z) e^{i\mathbf{k}\boldsymbol{\rho}}, \tag{7}$$

where $\boldsymbol{\rho} = \{x,y\}$, $k_\alpha = \frac{2\pi}{L_\alpha} n_\alpha$, $\alpha = x,y$, and $n_\alpha = N_{\min},\ldots,N_{\max}$. The range of the summation is limited by practical considerations.

Such form of the solution allows us to convert the 3D problem $(x,y,z)$ into 1D $(z)$ and take advantage of Fourier transform.

## 2. Main theory

Substituting equations (6), (7) into initial system (1)...(5) and collecting similar terms, we get for the wave function:

$$-\frac{1}{\varkappa^2}\left(\frac{d^2\psi_{\mathbf{k}}}{dz^2} - \mathbf{k}^2\psi_{\mathbf{k}}\right) - \psi_{\mathbf{k}} +$$
$$+\frac{2i}{\varkappa}\sum_{\mathbf{p}}\left[\left(a_{\mathbf{p}}^{(x)}(k_x - p_x) + a_{\mathbf{p}}^{(y)}(k_y - p_y)\right) i\psi_{\mathbf{k}-\mathbf{p}} + a_{\mathbf{p}}^{(z)}\frac{d\psi_{\mathbf{k}-\mathbf{p}}}{dz}\right] +$$
$$+\sum_{\mathbf{p},\mathbf{q}}\left[(\mathbf{a}_{\mathbf{p}}\cdot\mathbf{a}_{\mathbf{q}})\psi_{\mathbf{k}-\mathbf{p}-\mathbf{q}} + \psi_{\mathbf{p}}\psi_{\mathbf{q}}^*\psi_{\mathbf{k}-\mathbf{p}+\mathbf{q}}\right] = 0, \tag{8}$$

for three components of vector potential:

$$\mathbf{k}^2 a_{\mathbf{k}}^{(x)} - \frac{d^2 a_{\mathbf{k}}^{(x)}}{dz^2} - \frac{1}{2\varkappa}\sum_{\mathbf{p}}(k_x + 2p_x)\psi_{\mathbf{p}}^*\psi_{\mathbf{k}+\mathbf{p}} + \sum_{\mathbf{p},\mathbf{q}} a_{\mathbf{p}}^{(x)}\psi_{\mathbf{q}}\psi_{\mathbf{p}+\mathbf{q}-\mathbf{k}}^* = 0, \tag{9}$$

$$\mathbf{k}^2 a_{\mathbf{k}}^{(y)} - \frac{d^2 a_{\mathbf{k}}^{(y)}}{dz^2} - \frac{1}{2\varkappa}\sum_{\mathbf{p}}(k_y + 2p_y)\,\psi_{\mathbf{p}}^* \psi_{\mathbf{k}+\mathbf{p}} + \sum_{\mathbf{p},\mathbf{q}} a_{\mathbf{p}}^{(y)} \psi_{\mathbf{q}} \psi_{\mathbf{p}+\mathbf{q}-\mathbf{k}}^* = 0, \tag{10}$$

$$\mathbf{k}^2 a_{\mathbf{k}}^{(z)} - \frac{d^2 a_{\mathbf{k}}^{(z)}}{dz^2} + \frac{i}{2\varkappa}\sum_{\mathbf{p}}\left(\psi_{\mathbf{p}}^* \frac{d\psi_{\mathbf{k}+\mathbf{p}}}{dz} - \psi_{\mathbf{k}+\mathbf{p}}\frac{d\psi_{\mathbf{p}}^*}{dz}\right) + \sum_{\mathbf{p},\mathbf{q}} a_{\mathbf{p}}^{(z)} \psi_{\mathbf{q}} \psi_{\mathbf{p}+\mathbf{q}-\mathbf{k}}^* = 0, \tag{11}$$

for boundary conditions:

$$\left(\frac{i}{\varkappa}\left\{ik_x\psi_{\mathbf{k}}, ik_y\psi_{\mathbf{k}}, \frac{d\psi_{\mathbf{k}}}{dz}\right\} + \sum_{\mathbf{p}} \mathbf{a}_{\mathbf{p}}\psi_{\mathbf{k}-\mathbf{p}}\right)\cdot \mathbf{n} = 0, \tag{12}$$

$$\left\{ik_y a_{\mathbf{k}}^{(z)} - \frac{da_{\mathbf{k}}^{(y)}}{dz}, \frac{da_{\mathbf{k}}^{(x)}}{dz} - ik_x a_{\mathbf{k}}^{(z)}, ik_x a_{\mathbf{k}}^{(y)} - ik_y a_{\mathbf{k}}^{(x)}\right\}_{\tau} = (\mathbf{h}_{\mathbf{k}})_{\tau}, \tag{13}$$

and for the gauge:

$$ik_x a_{\mathbf{k}}^{(x)} + ik_y a_{\mathbf{k}}^{(y)} + \frac{da_{\mathbf{k}}^{(z)}}{dz} = 0. \tag{14}$$

Note multiple summations in the nonlinear terms of equations (8)...(11). Direct numerical calculation of these sums is very time-consuming. More effective way is based on the Convolution Theorem [5] allowing to calculate the discrete convolution using two successive Fourier transforms:

$$(f * g)_{\mathbf{k}} = \sum_{\mathbf{p}} f_{\mathbf{p}} g_{\mathbf{k}-\mathbf{p}} = \frac{1}{V}\sum_{\mathbf{r}} e^{-i\mathbf{k}\mathbf{r}} F_{\mathbf{r}} G_{\mathbf{r}}, \tag{15}$$

$$(f * g * h)_{\mathbf{k}} = \frac{1}{V}\sum_{\mathbf{r}} e^{-i\mathbf{k}\mathbf{r}} F_{\mathbf{r}} G_{\mathbf{r}} H_{\mathbf{r}}.$$

Here $F_{\mathbf{r}}$ is the Fourier transform of $f_{\mathbf{k}}$ calculated in the **k**-range extended by factor of 2 (the factor depends on the number of functions in the convolution), with function values $f_{\mathbf{k}}$ outside initial range $K_{\min} \le k_\alpha \le K_{\max}$ manually set to zero (this is the so-called 'zero-padding' known in digital signal processing [5]).

Applying (15) to nonlinear terms in equations (8)...(14), we obtain:

$$\sum_{\mathbf{p}}\left[\left(a_{\mathbf{p}}^{(x)}(k_x - p_x) + a_{\mathbf{p}}^{(y)}(k_y - p_y)\right) i\psi_{\mathbf{k}-\mathbf{p}} + a_{\mathbf{p}}^{(z)}\frac{d\psi_{\mathbf{k}-\mathbf{p}}}{dz}\right] =$$
$$= \frac{1}{V}\sum_{\mathbf{r}}\left(\left[k_x a_{\mathbf{r}}^{(x)} + k_y a_{\mathbf{r}}^{(y)} - \left(p_x a^{(x)}\right)_{\mathbf{r}} - \left(p_y a^{(y)}\right)_{\mathbf{r}}\right] i\psi_{\mathbf{r}} + a_{\mathbf{r}}^{(z)}\left(\frac{d\psi}{dz}\right)_{\mathbf{r}}\right) e^{-i\mathbf{k}\mathbf{r}}, \tag{16}$$

$$\sum_{\mathbf{p},\mathbf{q}}\left[(\mathbf{a}_{\mathbf{p}}\cdot\mathbf{a}_{\mathbf{q}})\psi_{\mathbf{k}-\mathbf{p}-\mathbf{q}} + \psi_{\mathbf{p}}\psi_{\mathbf{q}}^*\psi_{\mathbf{k}-\mathbf{p}+\mathbf{q}}\right] = \frac{1}{V}\sum_{\mathbf{r}}\left(\mathbf{a}_{\mathbf{r}}^2 + |\psi_{\mathbf{r}}|^2\right)\psi_{\mathbf{r}} e^{-i\mathbf{k}\mathbf{r}}, \tag{17}$$

$$\sum_{\mathbf{p}}(k_x + 2p_x)\psi_{\mathbf{p}}^*\psi_{\mathbf{k}+\mathbf{p}} = \frac{1}{V}\sum_{\mathbf{r}}\left[k_x|\psi_{\mathbf{r}}|^2 + 2\,(p_x\psi^*)_{\mathbf{r}}\,\psi_{\mathbf{r}}\right] e^{-i\mathbf{k}\mathbf{r}}, \tag{18}$$

$$\sum_{\mathbf{p}}(k_y + 2p_y)\psi_{\mathbf{p}}^*\psi_{\mathbf{k}+\mathbf{p}} = \frac{1}{V}\sum_{\mathbf{r}}\left[k_y|\psi_{\mathbf{r}}|^2 + 2\,(p_y\psi^*)_{\mathbf{r}}\,\psi_{\mathbf{r}}\right] e^{-i\mathbf{k}\mathbf{r}}, \tag{19}$$

$$\sum_{\mathbf{p},\mathbf{q}} a_{\mathbf{p}}^{(x)}\psi_{\mathbf{q}}\psi_{\mathbf{p}+\mathbf{q}-\mathbf{k}}^* = \frac{1}{V}\sum_{\mathbf{r}} a_{\mathbf{r}}^{(x)}|\psi_{\mathbf{r}}|^2 e^{-i\mathbf{k}\mathbf{r}}, \tag{20}$$

$$\sum_{\mathbf{p},\mathbf{q}} a_{\mathbf{p}}^{(y)} \psi_{\mathbf{q}} \psi_{\mathbf{p}+\mathbf{q}-\mathbf{k}}^{*} = \frac{1}{V} \sum_{\mathbf{r}} a_{\mathbf{r}}^{(y)} |\psi_{\mathbf{r}}|^2 e^{-i\mathbf{k}\mathbf{r}}, \tag{21}$$

$$\sum_{\mathbf{p},\mathbf{q}} a_{\mathbf{p}}^{(z)} \psi_{\mathbf{q}} \psi_{\mathbf{p}+\mathbf{q}-\mathbf{k}}^{*} = \frac{1}{V} \sum_{\mathbf{r}} a_{\mathbf{r}}^{(z)} |\psi_{\mathbf{r}}|^2 e^{-i\mathbf{k}\mathbf{r}}, \tag{22}$$

$$\sum_{\mathbf{p}} \left( \psi_{\mathbf{p}}^{*} \frac{d\psi_{\mathbf{k}+\mathbf{p}}}{dz} - \psi_{\mathbf{k}+\mathbf{p}} \frac{d\psi_{\mathbf{p}}^{*}}{dz} \right) = \frac{1}{V} \sum_{\mathbf{r}} \left[ \psi_{\mathbf{r}}^{*} \left( \frac{d\psi}{dz} \right)_{\mathbf{r}} - \psi_{\mathbf{r}} \left( \frac{d\psi^{*}}{dz} \right)_{\mathbf{r}} \right] e^{-i\mathbf{k}\mathbf{r}}. \tag{23}$$

Finally, we have the system of equations for Fourier-transformed wave function $\psi_{\mathbf{k}}(z)$ and vector potential $\mathbf{a}_{\mathbf{k}}(z)$, consisting of algebraic expressions for these components, and additional Fourier transforms in extended **k**-range with appropriate zero padding.

## 3. Test realization

For the problem shown in Fig. 1, the boundary conditions (12), (13) are written as:

$$\frac{i}{\varkappa} \frac{d\psi_{\mathbf{k}}}{dz} + \sum_{\mathbf{p}} a_{\mathbf{p}}^{(z)} \psi_{\mathbf{k}-\mathbf{p}} = 0, \tag{24}$$

$$\frac{da_{\mathbf{k}}^{(x)}}{dz} = ik_x a_{\mathbf{k}}^{(z)} + h_{\mathbf{k}}^{(y)}, \tag{25}$$

$$\frac{da_{\mathbf{k}}^{(y)}}{dz} = ik_y a_{\mathbf{k}}^{(z)} - h_{\mathbf{k}}^{(x)}. \tag{26}$$

In our realization, the GL equations and boundary conditions are approximated by first-order finite differences. On the first boundary ($z = 0$), the corresponding equations are the following:

$$\frac{i}{\varkappa} \frac{[\psi_{\mathbf{k}}]_1^n - [\psi_{\mathbf{k}}]_0^n}{\Delta z} + \sum_{\mathbf{p}} \left[ a_{\mathbf{p}}^{(z)} \right]_0^n [\psi_{\mathbf{k}-\mathbf{p}}]_0^n = 0, \tag{27}$$

$$\frac{\left[ a_{\mathbf{k}}^{(x)} \right]_1^n - \left[ a_{\mathbf{k}}^{(x)} \right]_0^n}{\Delta z} = ik_x \left[ a_{\mathbf{k}}^{(z)} \right]_0^n + \left[ h_{\mathbf{k}}^{(y)} \right]_0^n, \tag{28}$$

$$\frac{\left[ a_{\mathbf{k}}^{(y)} \right]_1^n - \left[ a_{\mathbf{k}}^{(y)} \right]_0^n}{\Delta z} = ik_y \left[ a_{\mathbf{k}}^{(z)} \right]_0^n - \left[ h_{\mathbf{k}}^{(x)} \right]_0^n, \tag{29}$$

$$ik_x \left[ a_{\mathbf{k}}^{(x)} \right]_0^n + ik_y \left[ a_{\mathbf{k}}^{(y)} \right]_0^n + \frac{\left[ a_{\mathbf{k}}^{(z)} \right]_1^n - \left[ a_{\mathbf{k}}^{(z)} \right]_0^n}{\Delta z} = 0, \tag{30}$$

while on the second boundary ($z = z_d$):

$$\frac{i}{\varkappa} \frac{[\psi_{\mathbf{k}}]_d^n - [\psi_{\mathbf{k}}]_{d-1}^n}{\Delta z} + \sum_{\mathbf{p}} \left[ a_{\mathbf{p}}^{(z)} \right]_d^n [\psi_{\mathbf{k}-\mathbf{p}}]_d^n = 0, \tag{31}$$

$$\frac{\left[ a_{\mathbf{k}}^{(x)} \right]_d^n - \left[ a_{\mathbf{k}}^{(x)} \right]_{d-1}^n}{\Delta z} = ik_x \left[ a_{\mathbf{k}}^{(z)} \right]_d^n + \left[ h_{\mathbf{k}}^{(y)} \right]_d^n, \tag{32}$$

$$\frac{\left[ a_{\mathbf{k}}^{(y)} \right]_d^n - \left[ a_{\mathbf{k}}^{(y)} \right]_{d-1}^n}{\Delta z} = ik_y \left[ a_{\mathbf{k}}^{(z)} \right]_d^n - \left[ h_{\mathbf{k}}^{(x)} \right]_d^n, \tag{33}$$

$$ik_x\left[a_{\mathbf{k}}^{(x)}\right]_d^n + ik_y\left[a_{\mathbf{k}}^{(y)}\right]_d^n + \frac{\left[a_{\mathbf{k}}^{(z)}\right]_d^n - \left[a_{\mathbf{k}}^{(z)}\right]_{d-1}^n}{\Delta z} = 0. \tag{34}$$

We see that the equations for the vector potential components are independent of the wave function and linear, thus they can be solved analytically. The values of the wave function, in their turn, can be calculated using the procedure similar to calculation of nonlinear terms of GL equations. It should be noted that the order of approximation can be increased using the method of fictitious areas.

Our realization is based on the pseudoviscosity method [1]. Time steps are made with explicit Euler method, the space is discretized with uniform mesh with step $\Delta z$. To test the stability and convergence of the suggested approach, we simulated the system with the external field $\mathbf{H} = 0$ starting from the arbitrary chosen state $[\psi_{\mathbf{k}}]_m^0 = 0.25$, $[\mathbf{a}_{\mathbf{k}}]_m^0 = 0$. The exact solution of GL equations in this case is $\Psi_{\text{exact}}(x,y,z) = 1$, $\mathbf{A}_{\text{exact}}(x,y,z) = 0$. Other parameters of the problem are: $\varkappa = 0.2$ (type-I superconductor), $L_x = L_y = L_z = 1.6$.

The convergence to the exact solution is demonstrated in Fig. 2: during the simulation, the free energy approaches the exact value $F_{\text{exact}} = -L_xL_yL_z/2$.

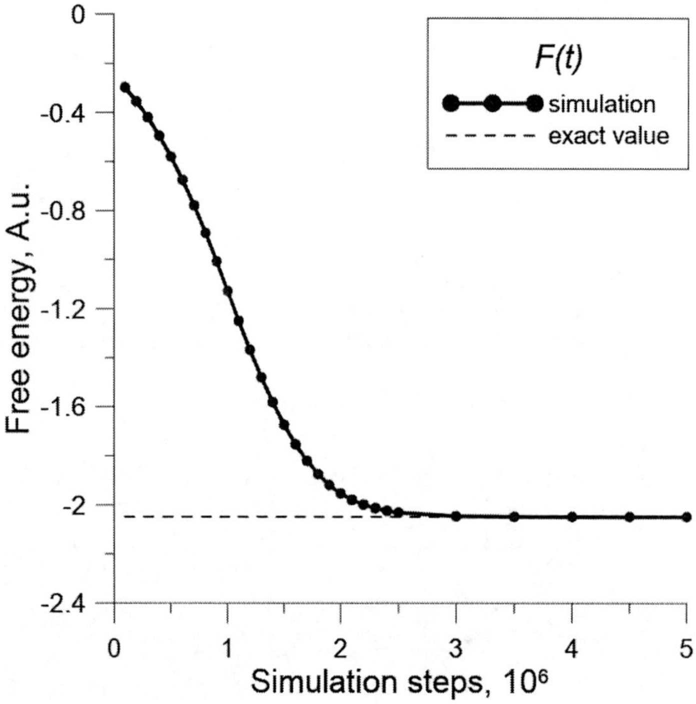


**Figure 2.** Test of stability and convergence of the suggested approach using the pseudoviscosity method. Parameters of the calculation are $\Delta t = 10^{-6}$, $\Delta z = 0.1$, $N_{\text{min}} = 0$, $N_{\text{max}} = 1$. It can be seen that the free energy of the film approaches exact value $F_{\text{exact}} = -2.048$ during the calculation.

## 4. Conclusion

In this article, we presented the approach for numerical determination of quantum state of superconductor using the pseudospectral method to solve the Ginzburg-Landau equations. Due to reduction of the problem dimensionality and application of efficient numerical algorithms, this method can sufficiently reduce the amount of calculations needed, and can speed-up the process of solution. All the relations needed for realization are shown. The suggested approach was applied to the problem with known exact solution, to demonstrate the stability and convergence of the algorithm.

IOP Conf. Series: Journal of Physics: Conf. Series **941** (2017) 012068 doi:10.1088/1742-6596/941/1/012068

## Acknowledgments

P. F. Kartsev is grateful for support to Russian Foundation for Basic Research, Grant 15-02-02764.

## References

[1] Richardson W B, Pardhanani A L, Carey G F and Ardelea A 2004 *Int. Journal for Numerical Methods Engineering* **59** 1251-1272
[2] Sadovskyy A, Koshelev A E, Phillips C L, Karpeyev D A and Glatz A 2015 *J. Comput. Phys* **294** 639-654
[3] Du Q 2005 *J. Math. Phys.* **46** 095109
[4] Du Q 1994 *Comput Math Appl* **27** (12) 119-133
[5] Smith S W 1999 *The Scientist and Engineers Guide to Digital Signal Processing, Second Edition* (California Technical Publishing) p 118

IOP Conf. Series: Journal of Physics: Conf. Series **941** (2017) 012069 doi:10.1088/1742-6596/941/1/012069

# Evaluation of solution stability for two-component polydisperse systems by small-angle scattering

**A E Kryukova[1,2], P V Konarev[1,2] and V V Volkov[1]**

[1]Laboratory of Reflectometry and Small-angle Scattering, A.V. Shubnikov Institute of Crystallography of FRC "Crystallography and photonics" RAS, Leninsky pr. 59, 119333 Moscow, Russia
[2] NRC "Kurchatov Institute", pl. Kurchatova 1, 123098 Moscow, Russia

E-mail: KrukovaAE@yandex.ru

**Abstract.** The article is devoted to the modelling of small-angle scattering data using the program MIXTURE designed for the study of polydisperse multicomponent mixtures. In this work we present the results of solution stability studies for theoretical small-angle scattering data sets from two-component models. It was demonstrated that the addition of the noise to the data influences the stability range of the restored structural parameters. The recommendations for the optimal minimization schemes that permit to restore the volume size distributions for polydisperse systems are suggested.

## 1. Introduction

A number of nanoparticle systems that form polydisperse mixtures in solution are of great interest to industrial, medical and biological applications [1-5]. However, the analysis of these systems is not always straightforward. Small-angle scattering (SAS) is the effective structural method to quantitatively investigate the behavior of complex multicomponent systems in solution. Earlier the program MIXTURE [4] that belongs to the package ATSAS [2, 3] was developed. This program allows one to analyze multicomponent systems containing various types of polydisperse particles and take into account the interparticle interactions.

Normally the analysis of experimental SAS curves is carried out using the following scheme: an appropriate theoretical model that depends on a finite number of structural parameters is selected and an iterative minimization process is used to optimize a given set of parameters that minimizes the difference between the experimental data and the calculated data from the theoretical model.

However, the non-linear dependence of the parameters complicates the search of the global minimum of the target minimization function. In a standard approach the range of possible values is defined for each parameter and the solutions are searched within this range. The solutions may often differ significantly from each other. It depends on the complexity of the model, the initial search values of the parameters and the noise of the experimental data. Therefore, the analysis of the solution stability plays an important role for the reliable restoration of the theoretical model.

In this study we performed the simulation of the theoretical data sets using the program MIXTURE and evaluated the stability of the solutions from the systems consisting of two types of spherical particles. The structural parameters describing the system were randomly changed and the correctness of the restored solution was checked. In addition, the noise with different relative levels was added to the theoretical data sets and its influence on the solution stability was studied.

## 2. Small-angle scattering intensity from multicomponent polydisperse system

The scattering intensity of the multicomponent system (without the interparticle interference) can be represented as a linear combination of the individual components:

$$I(s) = \sum_k v_k I_k(s), \tag{1}$$

where the sum goes over different components, $v_k$ and $I_k(s)$ are the volume fraction and the scattering intensity from the $k$–th component, respectively, $s$ is the momentum transfer $s = (4\pi/\lambda)\sin\theta$, $2\theta$ is the scattering angle, $\lambda$ is the wavelength.

For a polydisperse system of interacting particles, the intensity of each component can be calculated as:

$$I_k(s) = S_k(s)\int D_k(R)V_k(R)[\Delta\rho_k(R)]^2 i_{0k}(s,R)dR, \tag{2}$$

where $R$ is the size of nanoparticles, $D_k(R)$ is the normalized volume size distribution of the particles for the k-th component, $V_k(R)$ is the effective volume of k-th component, $\Delta\rho_k(R)$ is the scattering contrast of k-th component, $i_{0k}(s,R)$ is the normalized form factor of the k-th component, $S_k(s)$ is the structural factor of the k-th component, responsible for the interparticle interactions.

In this study we modelled the diluted solutions of the particles, therefore the effect of interparticle interference was neglected.

The program MIXTURE uses the minimization procedure due to Broyden-Fletcher-Goldfarb-Shanno (B-F-G-S) with simple bounds on the parameters. It allows one to model mixtures that contain up to ten components of different body types (spherical core-shells, cylinders, ellipsoids, dumbbells etc.). The input command file for the program MIXTURE should contain the following structural parameters, describing system: the number of components, the type of each component, the relative volume fraction of the component ($v_k$), the average size of the component ($R_k$), its polydispersity ($dR_k$) and the type of the size distribution function.

## 3. The ranges of stable solutions for theoretical model systems of nanoparticles

The systems under investigation contain two types of spherical nanoparticles characterized by their radii ($R_1$ and $R_2$), polydispersities ($dR_1$ and $dR_2$) and volume fractions ($v_1$ and $v_2$). Three typical cases were selected (see Table 1) for the theoretical data sets without noise (Figure 1, curve 1) and with added Gaussian and Poisson noise (Figure 1, curves 2 and 3, respectively). The relative level of the noise was equal to 25%. The corresponding volume size distributions are shown in Figure 1 (right panel). As one can see from Figure 1 the volume size distribution of the first model has two narrow separated peaks, whereas $D_V(R)$ functions of other models contain one wide and one narrow separated peaks.

**Table 1.** The "true" parameters of three theoretical model systems.

| №, model number | $v_1$, relative | $v_2$, relative | $R_1$, nm | $R_2$, nm | $dR_1$, nm | $dR_2$, nm |
|---|---|---|---|---|---|---|
| 1 | 0.25 | 0.75 | 5.0 | 15.0 | 0.1 | 0.1 |
| 2 | 0.25 | 0.75 | 5.0 | 15.0 | 2.0 | 0.1 |
| 3 | 0.25 | 0.75 | 5.0 | 15.0 | 0.1 | 4.0 |

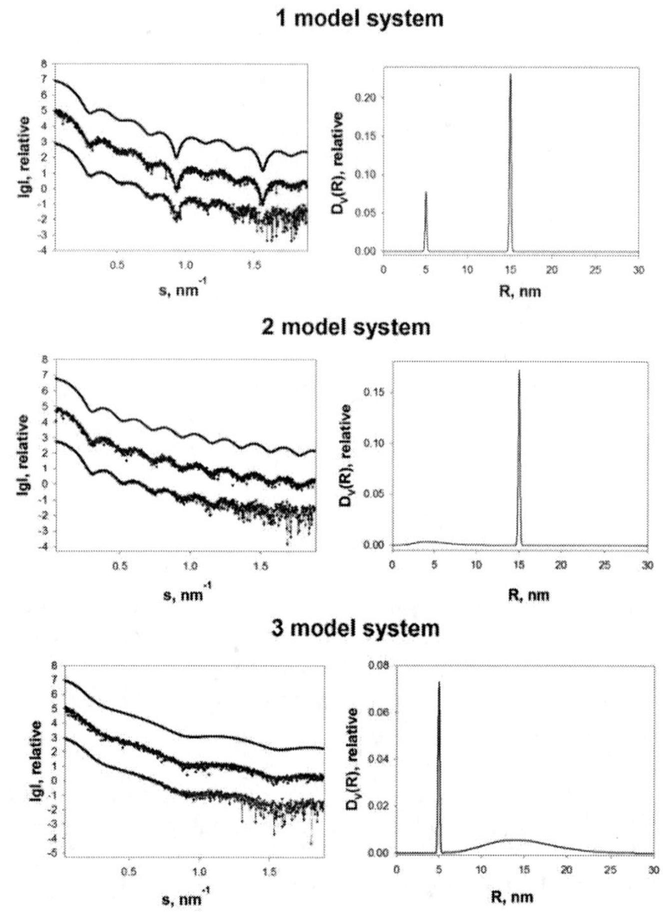


**Figure 1.** The left panel contains the simulated theoretical SAS curves with the parameters from Table 1 (curve 1 - with relative noise level 0%, curve 2 - Gaussian noise 25%, curve 3 - Poisson noise 25%). The curves are vertically displaced for better visualization. The right panel displays the volume size distribution functions $D_V(R)$.

The ranges of the solution stability for each structural parameter were obtained when only one parameter was varied while the others were fixed (see Table 2). It appears that the range of stable solutions always expands in the case of Poisson noise data, whereas it becomes normally shorter in the case of Gaussian noise data (with the exception for the second model). Thus, the data with added moderate level of Poisson noise (that become closer to real experimental data) can have a wider range of convergence to the correct solution in comparison with pure (free of noise) theoretical data.

Further, for each model system the values of two parameters have been changed at the same time while the rest of the parameters were fixed. The results are shown in Figures 2-5. They display the contour plots of solution stability ranges for the following pairs of parameters: ($R_1$, $R_2$) – Figure 2, ($R_1$, $dR_2$) – Figure 3, ($dR_1$, $R_2$) – Figure 4, ($dR_1$, $dR_2$) – Figure 5. The interception of the dotted lines corresponds to the theoretical ('true') solution. If the restored parameters are located within 5% in the vicinity of the "true" solution, the solution is treated as successful (red areas), otherwise the solution is treated as false (blue areas).

As one can see from Figures 2-5 the dependence of the solution stability ranges is similar to the cases when only one parameter is been varied. At the same time the areas of successful solutions (red areas) depend not only on the selected pair of parameters but also on the presence of the noise in the data. They form fractal-like boundaries with separated areas of instabilities. It demonstrates a high sensitivity of the minimization method to different model parameters.

**Table 2.** The ranges of solution stability (only one parameter has been varied, others are fixed).

| №, model number | 1 | 2 | 3 |
|---|---|---|---|
| Relative noise level is 0% | | | |
| $dR_1$, nm | 0.01÷4.0 | 0.1÷5.0 | 0.01÷5.0 |
| $dR_2$, nm | 0.01÷15.0 | 0.01÷15.0 | 0.2÷15.0 |
| $R_1$, nm | 3.0÷6.0 | 2.1÷10.0 | 2.9÷6.1 |
| $R_2$, nm | 13.8÷16.2 | 14.0÷16.0 | 10.0÷30.0 |
| Relative level of Gaussian noise is 25% | | | |
| $dR_1$, nm | 0.01÷4.3 | 0.1÷5.0 | 0.01÷4.6 |
| $dR_2$, nm | 0.2÷15.0 | 0.01÷18.8 | 0.2÷15.0 |
| $R_1$, nm | 3.9÷6.0 | 1.0÷10.0 | 3.3÷6.1 |
| $R_2$, nm | 14.0÷16.0 | 14.0÷16.2 | 10.0÷30.0 |
| Relative level of Poisson noise is 25% | | | |
| $dR_1$, nm | 0.01÷5.0 | 0.1÷6.0 | 0.01÷5.0 |
| $dR_2$, nm | 0.01÷20.0 | 0.01÷20.0 | 0.2÷20.0 |
| $R_1$, nm | 2.4÷8.0 | 1.0÷10.0 | 1.5÷8.0 |
| $R_2$, nm | 10.0÷30.0 | 10.0÷30.0 | 10.0÷30.0 |

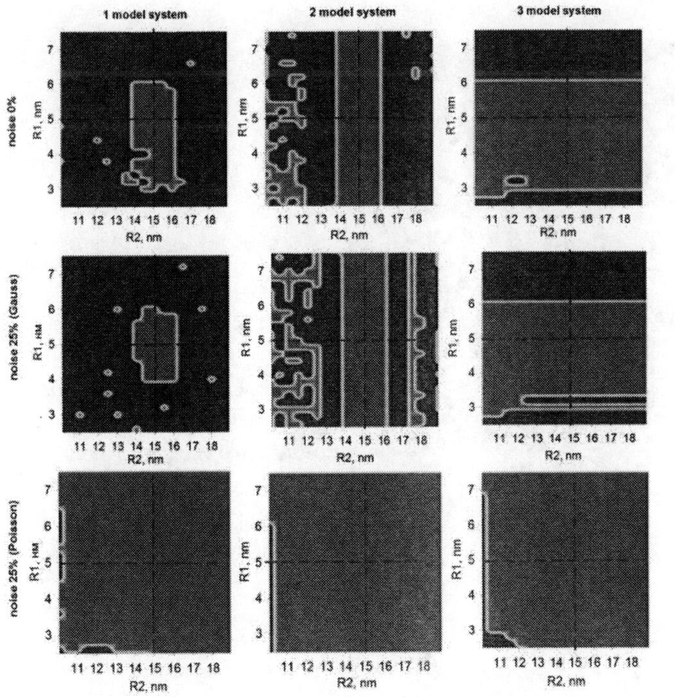


**Figure 2.** The contour plots of solution stability ranges for three model systems depending on the pair of parameters ($R_1$, $R_2$). The left column corresponds to the model 1, the central column – the model 2, the right one – the model 3. The plots for the curves without noise (relative noise level 0%) are located in the first row, with Gaussian noise (relative noise level 25%) - in the second row, with Poisson noise (relative noise level 25%) - in the third row. The interception of the dotted lines corresponds to the theoretical ('true') solution. If the restored parameters are located within 5% in the vicinity of 'true' solution, then the solution is treated as successful (red areas), otherwise the solution is treated as false (blue areas).

IOP Conf. Series: Journal of Physics: Conf. Series **941** (2017) 012069 doi:10.1088/1742-6596/941/1/012069

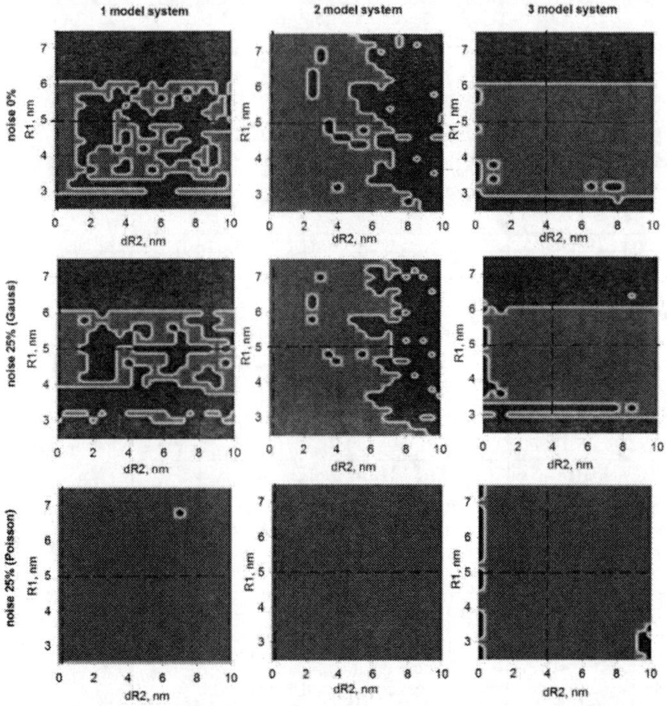


**Figure 3.** The contour plots of solution stability ranges for three model systems depending on the pair of parameters ($R_1$, $dR_2$). The notations and colours are the same as in Figure 2.

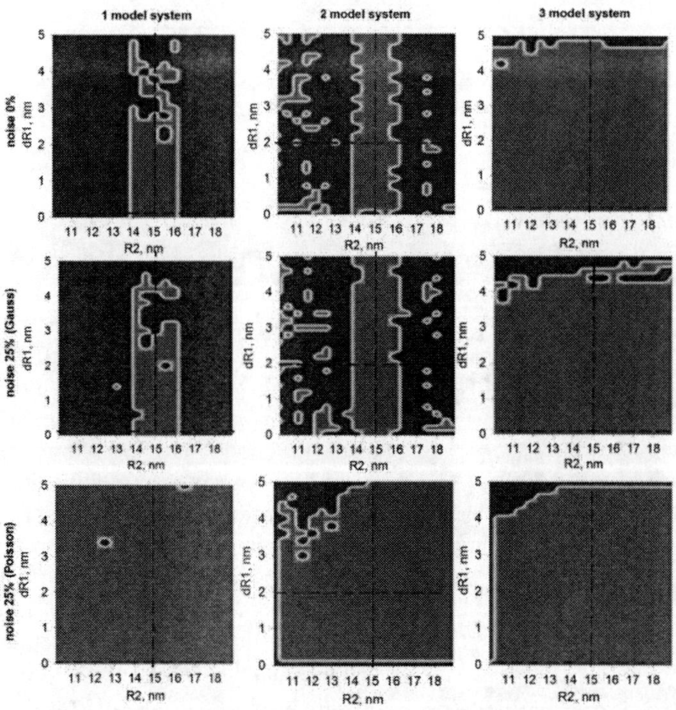


**Figure 4.** The contour plots of solution stability ranges for three model systems depending on the pair of parameters ($dR_1$, $R_2$). The notations and colours are the same as in Figure 2.

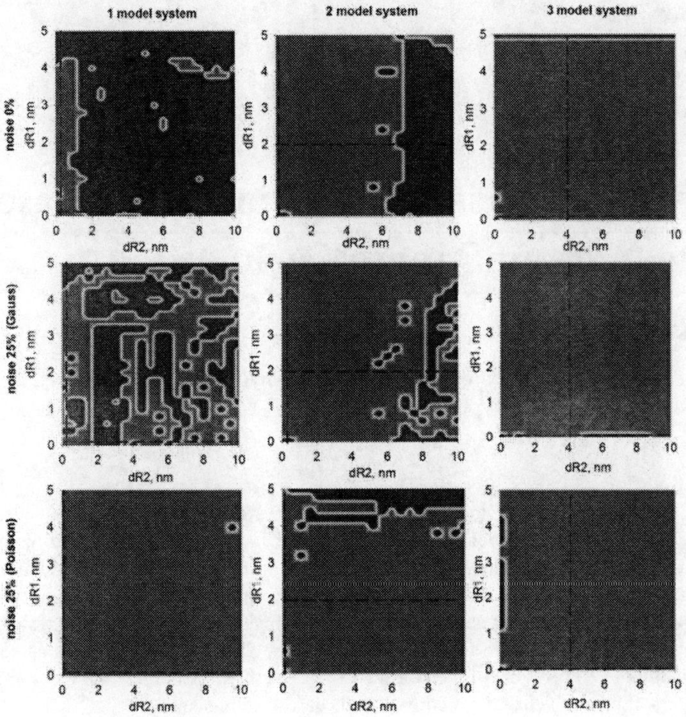


**Figure 5.** The contour plots of solution stability ranges for three model systems depending on the pair of parameters ($dR_1$, $dR_2$). The notations and colours are the same as in Figure 2.

## 4. Conclusion

1) The obtained results demonstrate the possibilities for the correct model restoration and depend on the choice of the initial parameter values as well as the type and the level of the data noise.

2) It was found that the presence of Poisson noise in small-angle scattering data extends the ranges of solution stability. Thus, the addition of the moderate Poisson noise (up to 25%) to real experimental data with a high signal-to-noise ratio can increase the probability of the 'true' solution.

3) In the present study the two-component system of spherical particles was considered. In future we plan to analyze the multicomponent systems of cylindrical particles as well as the mixtures of spherical and cylindrical particles.

**Acknowledgments**
This work was performed using the equipment of the Shared Research Center FSRC "Crystallography and Photonics" RAS and was supported by the Russian Ministry of Education and Science.

**References**

[1] Svergun D I, Konarev P V, Volkov V V, Koch M H J, Sager W F C, Smeets J and Blokhuis E M 2000 *J. Chem. Phys.* **113** 1651-65

[2] Konarev P V, Petoukhov M V, Volkov V V and Svergun D I 2006 *J. Appl. Cryst.* **39** 277-86

[3] Petoukhov M V, Franke D, Shkumatov A V, Tria G, Kikhney A G, Gajda M, Gorba C, Mertens H D T, Konarev P V and Svergun D I 2012 *J. Appl. Cryst.* **45** 342-50

[4] Konarev P V, Volkov V V, Sokolova A V, Koch M H J and Svergun D I 2003 *J. Appl. Cryst.* **36** 1277-82

[5] Feigin L A and Svergun D I *Structure Analysis by Small-Angle X-Ray and Neutron Scattering* 1987 (New York: Plenum Press)

# Simulation of equilibrium particle distribution of the Bose gas of polaritons using quantum Monte Carlo

**I O Kuznetsov and P F Kartsev**

National Research Nuclear University "MEPhI" (Moscow Engineering Physics Institute) , Kashirskoe sh. 31, 115409 Moscow, Russia

E-mail: PFKartsev@mephi.ru

**Abstract.** We present a numerical approach to study the system of exciton polaritons in an optical resonator under pumping. First task in this problem, is to determine the stationary particle distribution. In the case of low enough temperature, to describe the system correctly one cannot neglect the correlations between particles (including the Bose-Einstein condensate) and should take into account all interaction types present in the model. One of the most promising approaches to describe many-particle interacting systems, is the Quantum Monte Carlo (QMC) method. In this article, we describe a modification of the trajectory algorithm based on the Continuous Time World-Line (CTWL) QMC method to simulate the polaritonic system for various parameters of the problem. The modification is described in details.

## 1. Introduction

The weakly-interacting Bose gas has generated new wave of interest last years, due to new experimental problem of the excitonic polaritons in semiconductor under laser pumping in optical cavity [1, 2]. The interest is explained by the possibility of reaching the state of Bose-Einstein condensation (BEC) at high enough temperature.

The state of BEC is the result of the thermal equilibrium distribution of particles. To reach the equilibrium state, the interparticle interaction should be strong enough. The rate of relaxation and BEC formation must be faster than harmful processes such as exit of photons and decay of excitons [3].

The polaritonic system is a subject for large count of theoretical [2, 4] and experimental [5] studies. Among others, appearance of superfluidity [6] and quantized vortices [7] can be named. The spontaneous transitions in the polaritonic system, including laser applications, are studied in [7, 8]. The so-called "polaritonic laser" is discussed when the coherent generation due to spontaneous recombination of polaritons leads to the condensate deterioration [9, 10]. A significant number of theoretical [11, 12] and experimental works is dedicated to the BEC dynamics in this system [13].

Considering the interest to the system, reliable theoretical model should be useful to determine the optimal parameters of the experiment. Complex model of interactions leads to the need of taking into account the perturbation terms of high orders. The method of Quantum Monte Carlo [11, 12, 14] is one of the promising numerical approaches to interacting quantum systems.

In this report, we describe the modification of the Worm algorithm [15] based on the diagrammatic QMC method [16] designed for the polaritonic system.

## 2. Model

The system under study consists of two subsystems — excitons and photons — connected with Rabi coupling $\Omega$. Photonic subsystem is coupled to the external electromagnetic field $G_k$ reflecting optical pumping and finite lifetime of photons in the resonator. Excitonic subsystem includes interparticle interaction $U_0$ and amplitude of particle decay $R_k$. The Hamiltonian of the composite system is written as:

$$H_{pol} = \sum_k \left(E_k^{(cav)} - \mu\right) \hat{a}_k^+ \hat{a}_k + \sum_k \left(E_k^{(exc)} - \mu\right) \hat{b}_k^+ \hat{b}_k + \hbar\Omega \sum_k (\hat{a}_k^+ \hat{b}_k + \hat{a}_k \hat{b}_k^+) + \\ + U_0 \sum_{klp} \hat{b}_k^+ \hat{b}_l^+ \hat{b}_p \hat{b}_{k+l-p} + \sum_k (G_k \hat{a}_k^+ + G_k^* \hat{a}_k) + \sum_k (R_k \hat{b}_k^+ + R_k^* \hat{b}_k). \quad (1)$$

where the operators of photonic and excitonic parts are denoted by $\hat{a}^+$,$\hat{a}$ and $\hat{b}^+$,$\hat{b}$, respectively, and $E_k^{(cav)}$, $E_k^{(exc)}$ are the dispersion laws for the cavity photon and exciton, respectively. The chemical potential $\mu$ is introduced to control the total number of particles.

## 3. Quantum Monte Carlo

In diagrammatic QMC, each perturbation diagram is represented by the configuration of particle trajectories (worldlines) in the imaginary time $\tau = 0 \ldots \beta$, where $\beta \equiv 1/k_B T$. Each term of the Hamiltonian gives either the corresponding worldline change (the so-called *kink*), for example hopping between sites, or the propagation of the worldline along the imaginary time axis. During the simulation, the Monte Carlo configurations are transformed into each other using stochastic changes of the worldlines (*updates*), with their correct statistical weight guaranteed by the Metropolis algorithm.

The numerical simulation of such system with fixed temperature will give the established distribution of Bose particles under condition of constant external pumping and outflow from the system. Here we neglect the interaction between excitons and lattice phonons which can result in cooling of the polaritons, but with characteristic time much longer compared to the times of other processes in the system. This term, nevertheless, can be introduced into model as the third subsystem – phonons. New terms in the Hamiltonian (1) would then enter the QMC simulation with the appropriate types of kinks and updates.

In order to simplify the simulation process, we transform the Hamiltonian (1) to real space:

$$H_{pol} = -\mu \sum_i (\hat{a}_i^+ \hat{a}_i + \hat{b}_i^+ \hat{b}_i) - t^{(cav)} \sum_{<ij>} \hat{a}_i^+ \hat{a}_j - t^{(exc)} \sum_{<ij>} \hat{b}_i^+ \hat{b}_j + \hbar\Omega \sum_i (\hat{a}_i^+ \hat{b}_i + \hat{a}_i \hat{b}_i^+) + \\ + \tilde{U}_0 \sum_i \hat{n}_i(\hat{n}_i - 1) + \sum_i (G_i \hat{a}_i^+ + G_i^* \hat{a}_i) + \sum_i (R_i \hat{b}_i^+ + R_i^* \hat{b}_i). \quad (2)$$

The coefficients in (2) are the Fourier transforms of the corresponding coefficients in (1). For simplicity, we take the nearest-neighbour hopping:

$$E^{(cav)}(k) = -2t^{(cav)} \sum_r \cos(kr),$$

$$E^{(exc)}(k) = E_0 - 2t^{(exc)} \sum_r \cos(kr),$$

where $E_0$ is the offset between photonic and excitonic dispersion laws, and the summation is made on the nearest neighbours of zero point in the $r$-lattice.

## 4. Features of simulation

We note that the Rabi frequency in Eq. (2) plays the role of hopping between subsystems. The interparticle interaction term is diagonal allowing to use the Worm algorithm.

IOP Conf. Series: Journal of Physics: Conf. Series **941** (2017) 012070 doi:10.1088/1742-6596/941/1/012070

One can discuss the alternative approach including the diagonalization of the first three terms of the Hamiltonian (all the single-particle parts) and working in the basis of hybridized particles [10, 17]. This approach, however, has a certain disadvantage due to more complex form of the interaction term in the Hamiltonian.

The last remark applies to the terms with single operators (photon pumping and exciton decay). These terms generate kinks which add or remove particle from the worldline configuration, i.e. create worm 'heads' and 'tails'. In the original Worm algorithm, these coefficients are added artificially and the count of worms is limited by algorithm. In our case, however, the worldline configuration may contain many worms simultaneously. The count of worms is determined by the actual model parameters (pumping intensity and lifetime of excitons) and temperature.

The example of the worldline configuration specific for the model (2) is shown in Figure 1.

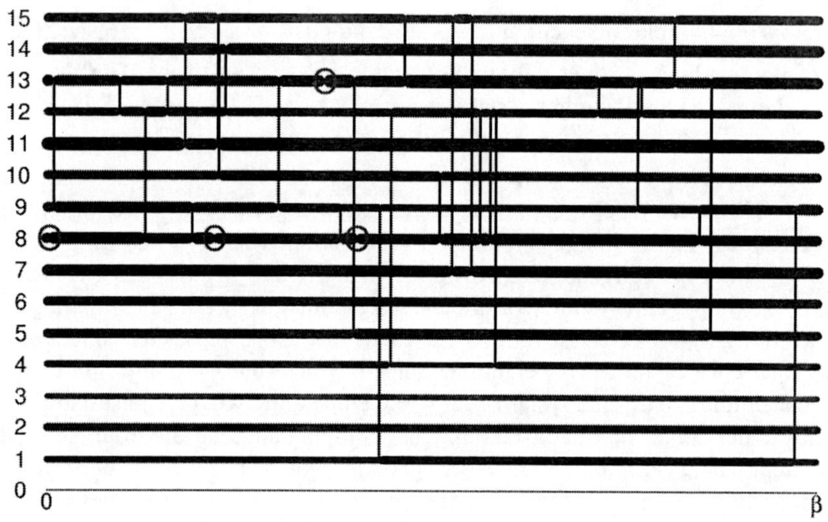


**Figure 1.** The typical configuration of worldlines in simulation of polaritonic system (2) with diagrammatic QMC. The sites (placed vertically) correspond to two subsystems: excitons at the bottom (sites 0–7), photons at the top (sites 8–15). Particle worldlines ('worms') consist of horizontal lines (particle propagation in time), vertical changes (hopping), and worm 'heads' and 'tails' (circles). Width of the horizontal line shows the site occupation at the corresponding time range.

## 5. Calculation of average values

During the CTWL simulation, the average values are calculated. Mean energy $\langle E \rangle$ and average occupation numbers on sites $\langle n_i \rangle$ are obtained using standard averaging procedure [16]. To calculate occupation in momentum space $\langle n_k \rangle$, one needs to obtain also the Green function $\langle a_i^+ a_j \rangle$, and apply the Fourier transform $r \to k$. Calculation of more complex correlation function is straightforward.

## 6. Conclusion

To simulate the system of exciton polaritons described by the Hamiltonian (1), we developed a numerical calculation algorithm based on the modified Worm algorithm [15] of diagrammatic QMC [16]. The model parameters include Rabi splitting, dispersion laws of photons and excitons, and coupling to external field (optical pumping and exciton decay).

The algorithm allows to obtain mean energy, distribution of particles, and various correlation functions needed to study the state of the system. Numerical simulation of polaritonic system with various values of parameters would allow to find optimal parameters of the experiment, including the Bose-Einstein condensation of polaritons. The modifications of Worm algorithm are described in details.

**Acknowledgments**
P. F. Kartsev is grateful for support to Russian Foundation for Basic Research, Grant No. 15-02-02764.

## References

[1] Lambert K. van Vugt *et al.* 2011 *PNAS* **108**(2) 10050
[2] Rahimi-iman A *et al.* 2012 *Phys. Rev.* B **86** 153308
[3] Sun Y and Wen P *et al.* 2017 *Phys. Rev. Lett* **118** 016602
[4] Voronova N S and Lozovik Yu E 2012 *Phys. Rev.* B **86** 193305
[5] Das A and Bhattacharya P *et al.* 2012 *Phys. Rev.* B **85** 195321
[6] Amo A and Lefrere J *et al.* 2009 *Nat. Phys.* **5** 805
[7] Bhattacharya A and Baten M Z *et al.* 2016 *Phys. Rev.* B **94** 035203
[8] Lagoudakis K G and Ostatnicky T *et al.* 2009 *Science* **326** 974
[9] Colombelli R and Manceau J-M 2015 *Phys. Rev.* X **5** 011031
[10] Takemura N and Anderson M D *et al.* 2016 *Phys. Rev.* B **94** 195301
[11] Wouters V 2012 *Phys. Rev.* B **85** 165303
[12] Verger A and Carusotto I and Ciuti C 2007 *Phys. Rev.* B **76** 115324
[13] Wertz E and Amo A and Solnyshkov D D *et al.* 2012 *Phys. Rev. Lett.* **109** 216404
[14] Yang Kai-Yu and Kozik E and Wang Xin and Troyer M 2011 *Phys. Rev.* B **83** 214516
[15] Prokof'ev N V and Svistunov B V and Tupitsyn I S 1998 *Phys. Lett.* A **238** (4-5) 253
[16] Prokof'ev N V and Svistunov B V and Tupitsyn I S 1998 *J. Exp. Theor. Phys.* **87** 310
[17] Berman O L and Lozovik Yu E and Snoke D W 2008 *Phys. Rev.* B **77** 155317

# Temperature dependence of critical current in $YBa_2Cu_3O_{7-\delta}$ films

**A V Kuznetsov, I I Sannikov, A A Ivanov**

National Research Nuclear University "MEPhI" (Moscow Engineering Physics Institute) , Kashirskoe sh. 31, 115409 Moscow, Russia

E-mail: AVKuznetsov@mephi.ru

**Abstract.** Analysis of temperature behavior of the critical current in $YBa_2Cu_3O_{7-\delta}$ films is presented. We demonstrate that the overall current can be separated into two components caused by weak pinning on oxygen vacancies in $CuO_2$ planes and strong pinning on defects in volume of the superconductor. Temperature dependences of the components are obtained and discussed.

## 1. Introduction

A large variety of defects act as pinning centers in $YBa_2Cu_3O_{7-\delta}$ films, such as vacancies, twin boundaries, dislocations, segmentation faults, non-superconducting inclusions and so on [1]. Vacancies produce a weak pinning [2, 3] while other defects produce a strong pinning [2–4]. Parameters of pinning centers strongly affect value of the critical current density $J_c$ in superconducting state and its dependence on temperature and applied magnetic field. Any $YBa_2Cu_3O_{7-\delta}$ film contains oxygen vacancies but presence and concentration of other defects strongly depend on both technology and conditions of the film synthesis. Therefore different samples contain different sets of defects and as a sequence their critical current demonstrates different temperature and field behavior.

Roughly exponential decrease of the critical current at low and elevated temperatures has been observed for $YBa_2Cu_3O_{7-\delta}$ films in numerous experiments [5–12]. The dependence [5–8, 12]

$$J_c = J_c(0)\exp(-T/T_0) \tag{1}$$

was attributed to presence of pinning centers with small energies $T_0 = 17 - 32$ K. Such small pinning energy causes thermally activated depinning of vortices and rapid decay of the critical current with temperature [13]. This dependence produces rather well approximation for $J_c(T)$ in the range from about 10 to 50–60 K. The dependence

$$J_c = J_w\exp(-T/T_w) + J_s\exp[-3(T/T_s)^2] \tag{2}$$

was used to take into account the presence of both weak and strong pinning centers and the temperatures $T_w = 8$–13 K and $T_s = 78$–93 K were found [9–11]. Here $w$ and $s$ mark the current components produced by weak and strong pinning. For some samples this expression extends the higher range of the exponential approximation for $J_c(T)$ up to $T \lesssim 75$ K.

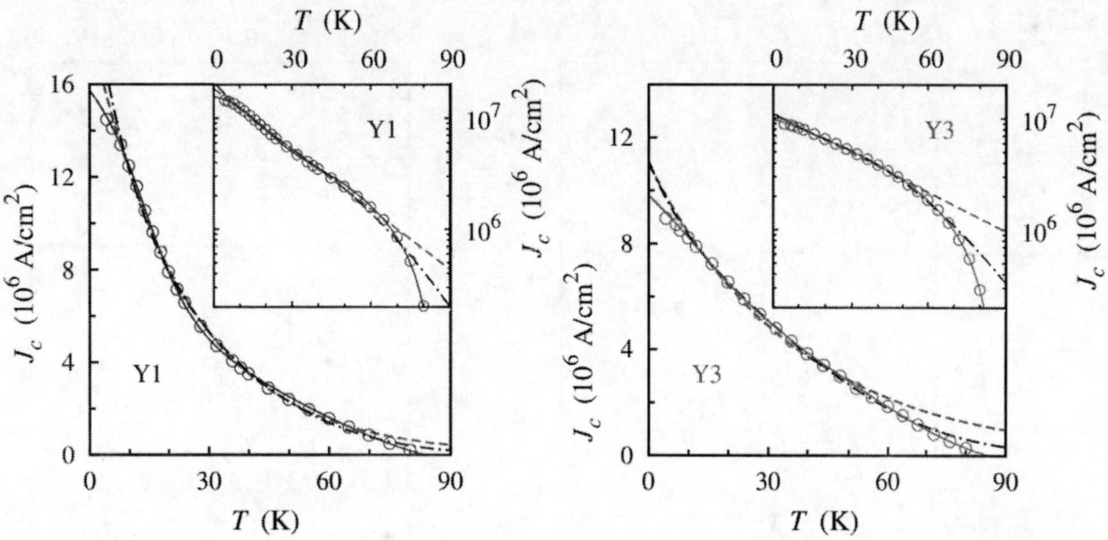


**Figure 1.** Color online. Temperature dependences of the critical current. The curves are shown in standard and semilogarithmic scales in order to bring out both low and high temperature behavior of $J_c$. Symbols are experimental data. Lines are fits by dependences (1) — dashed , (2) — dash-dotted, as well as sum of dependences (4) and (5) — solid.

Exponential approximation cannot describe $J_c(T)$ behavior at high temperatures since neither eq. (1) nor eq. (2) become zero in vicinity of the critical temperature $T_c$. The power laws

$$J_c = J_c(0)(1 - T/T_c)^{\alpha}, \tag{3a}$$
$$J_c = J_c(0)[1 - (T/T_c)^2]^{\beta}, \tag{3b}$$

were found in this temperature range [14–19]. Values $\alpha \simeq 0.9$–$2$ [14–16, 18] and $\beta \simeq 1.2$–$1.5$ [17, 19] were obtained. The observed power law was explained by temperature dependence of the coherence length $\xi$ and the penetration depth $\lambda$.

As follows from this brief analysis an approach to describe $J_c(T)$ behavior in the whole temperature range is still absent. In this paper we show that the critical current of $YBa_2Cu_3O_{7-\delta}$ films can be separated into two components caused by weak and strong pinning, obtain their temperature dependences and precisely describe $J_c(T)$ in the range from 4.21 K to $T_c$.

## 2. Results and discussion

Thin epitaxial $YBa_2Cu_3O_{7-\delta}$ films were prepared on (100) $SrTiO3_3$ substrates by pulsed laser deposition technique [20] using KrF laser. In resistivity measurements the films demonstrated sharp superconducting transitions at $T_c$ = 90–91 K with transition width about 1 K. Temperature dependence of saturated remanent magnetic moment $M$ of films was measured by means of original SQUID-magnetometry method described in [21, 22]. The critical current density was calculated from the moment as $J = 24M/(\pi D^3 d)$ (in SI units), where $D$ and $d$ are the film's diameter (~ 2 mm) and thickness (300–500 nm). Typical temperature dependences of the critical current obtained for two samples are shown in Figure 1. As seen, experimental data are rather well approximated by the dependence (1) in the range from 10 to 50-60 K though curvatures of experimental and approximating curves slightly differ in the range from 20 to 40 K. The dependence (2) provides a better approximation and extends the higher limit of the range up to 75 or 68 K for samples Y1 and Y3. The obtained characteristic temperatures $T_0$ = 24 and 37 K, $T_w$ = 12 K, $T_s$ = 84 and 90 K well agree with published data.

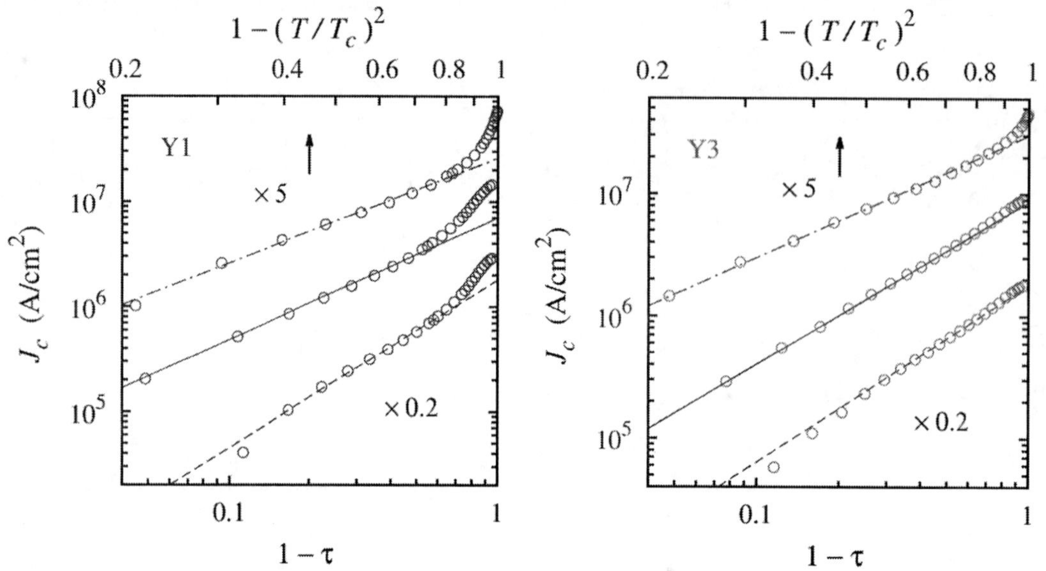


**Figure 2.** Color online. Temperature dependences of the critical current. $\tau = T/T_{dp}$ and $\tau = T/T_c$ respectively for middle and bottom curves. Top and bottom curves are shifted (multiplied by shown factors) to avoid overlapping. Symbols are experimental data. Lines are fits by dependences (3a) — dashed , (3b) — dash-dotted and (4) — solid.

High temperature behavior is shown in Figure 2 where $J_c$ is plotted in logarithmic scales versus $1 - T/T_c$ and $1 - (T/T_c)^2$. Note that $T_c = 90$ and 90.5 K for samples Y1 and Y3. In a wide temperature range the power law (3a) well approximates experimental data for sample Y3. More rapid decrease of the measured current is observed only at $T \gtrsim 72$ K ($1 - T/T_c \lesssim 0.2$). The dependence (3b) agree with $J_c(T)$ behavior at $T \gtrsim 35$ K [$1-(T/T_c)^2 \lesssim 0.85$]. At lower temperatures the measured current rises more rapidly and at 4.21 K becomes twice more than the approximating $J_c$. More stronger deviation from the power laws (3) is observed for sample Y1 at both high and low temperatures.

Thermal fluctuations strongly affects superconductivity in $YBa_2Cu_3O_{7-\delta}$ at high temperatures. They destroy the critical state at the depinning temperature $T_{dp}$ which is lower than $T_c$ [2, 23–25]. Thus the critical current disappears at $T_{dp} < T_c$ and the power laws (3) cannot correctly describe $J_c$ suppression in vicinity of $T_{dp}$. We suppose that at high temperatures only strong pinning centers are effective and

$$J_c = J_2 = J_2(0)(1 - T/T_{dp})^{\alpha}, \tag{4}$$

where $J_2$ is the critical current produced by strong pinning. We fit the measured current by this dependence and obtained $J_2(0) = 7.8$ and 8.8 MA/cm², $T_{dp} = 84$ and 86.7 K, $\alpha = 1.16$ and 1.33 for samples Y1 and Y3. As seen in Figure 2, for sample Y3 the measured current is well fitted by (4), however at low temperatures a small "excessive" increase of $J_c$ is observed. For sample Y1 such an increase at $T < 37$ K ($1 - T/T_{dp} \lesssim 0.55$) is distinctly seen.

Calculating the critical current of a layered superconductor, Ovchinnikov and Ivlev have shown that the overall current $J_c = J_1 + J_2$ consists of two components produced by pinning in superconducting planes, $J_1$, and pinning in superconductor volume, $J_2$ [4]. Following to Ovchinnikov and Ivlev we subtracted from the measured current the component $J_2$ described by (4) to separate the component $J_1$ produced by pinning on oxygen vacancies in superconducting $CuO_2$ planes. Obtained $J_1(T)$ dependence is presented in Figure 3. The component $J_1$ exists only at temperatures below 30–40 K. We found that it

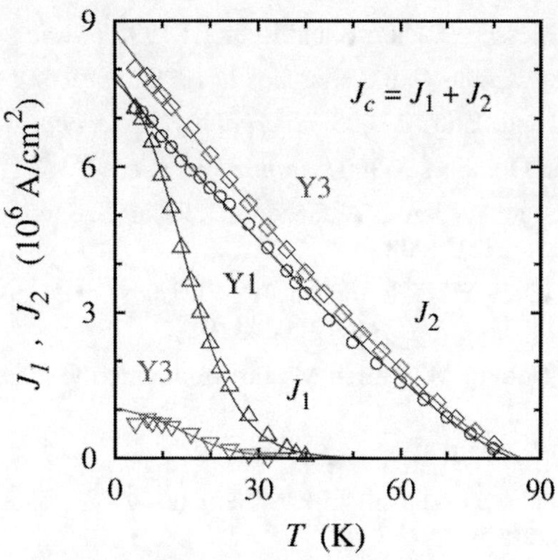


**Figure 3.** Color online. Temperature dependences of the critical current components. Symbols are experimental data: triangles — $J_1$, circles and diamonds — $J_2$. Lines are fits by dependences (4) and (5).

is well described by the dependence

$$J_1 = \frac{J_1^*}{1 + \exp(T/T_1)/2T_1}, \tag{5}$$

where $J_1^*$ is a characteristic current and the parameter $T_1$ is in Kelvins in the exponent power and dimensionless in its divisor. The values $T_1 = 5.8$ K, $J_1^* = 8.6$ and 1.1 MA/cm$^2$ were obtained for samples Y1 and Y3. We attribute the quasi exponential dependence (5) to influence of flux creep which is strong because of low energy of pinning on oxygen vacancies. Fast relaxation of $J_c$ was observed at low temperatures [26].

Dependence $J_2(T)$ obtained in the whole temperature range via subtraction of $J_1(T)$ from the measured current is also presented in Figure 3. As seen, amplitudes of $J_1$ and $J_2$ components at low temperatures are equal each other for sample Y1, while for sample Y3 $J_2$ is eight times more than $J_1$.

Approximation of measured $J_c(T)$ by sum of dependences (4) and (5) is shown in Figure 1. Analysis of the separated current components $J_1$ and $J_2$ allows us to describe precisely $J_c(T)$ behavior in the whole temperature range.

Summing up we conclude that different ratio of $J_1$ and $J_2$ components in combination with different values of both the depinning temperature and the power $\alpha$ in samples with different set of defects provide a large variety of $J_c(T)$ dependences in $YBa_2Cu_3O_{7-\delta}$ films.

**Acknowledgments**
Authors thank Prof. A.P. Menushenkov and DLTCT GPI RAS for support of this work.

**References**

[1] Foltyn S R, Civale L, MacManus-Driscoll J L, Jia Q X, Maiorov B, Wang H and Maley M 2007 *Nat. Mater.* **6** 631–42

[2] Blatter G, Feigel'man M V, Geshkenbein V B, Larkin A I and Vinokur V M 1994 *Rev. Mod. Phys.* **66**(4) 1125–388

[3] Blatter G, Geshkenbein V B and Koopmann J A G 2004 *Phys. Rev. Lett.* **92**(6) 067009

[4] Ovchinnikov Y N and Ivlev B I 1991 *Phys. Rev. B* **43**(10) 8024–9

[5] Lairson B M, Sun J Z, Bravman J C and Geballe T H 1990 *Phys. Rev. B* **42**(1) 1008–11

[6] Wen H, Schnack H, Griessen R, Dam B and Rector J 1995 *Physica C* **241** 353–74

[7] Moraitakis E, Pissas M, Kallias G and Niarchos D 1999 *Supercond. Sci. Technol.* **12** 682–9

[8] Peurla M, Huhtinen H and Paturi P 2005 *Supercond. Sci. Technol.* **18** 628–33

[9] Gutiérrez J, Puig T and Obradors X 2007 *Appl. Phys. Lett.* **90** 162514

[10] Puig T, Gutiérrez J, Pomar A, Llordés A, Gázquez J, Ricart S, Sandiumenge F and Obradors X 2008 *Supercond. Sci. Technol.* **21** 034008

[11] Polat O, Sinclair J W, Zuev Y L, Thompson J R, Christen D K, Cook S W, Kumar D, Chen Y and Selvamanickam V 2011 *Phys. Rev. B* **84**(2) 024519

[12] Senatore C, Barth C, Bonura M, Kulich M and Mondonico G 2016 *Supercond. Sci. Technol.* **29** 014002

[13] Christen D K and Thomson J R 1993 *Nature* **366** 98

[14] Fedotov Y V, Ryabchenko S M, PashitskiĭE A, Semenov A V, Vakaryuk V I, Pan V M and Flis V S 2002 *Low Temp. Phys.* **28**(3) 172–83

[15] Pashitskiĭ É A, Vakaryuk V I, Ryabchenko S M and Fedotov Y V 2001 *Low Temp. Phys.* **27**(2) 96–102

[16] Djupmyr M, Cristiani G, Habermeier H U and Albrecht J 2005 *Phys. Rev. B* **72** 220507(R)

[17] Ijaduola A O, Thompson J R, Feenstra R, Christen D K, Gapud A A and Song X 2006 *Phys. Rev. B* **73**(13) 134502

[18] Albrecht J, Djupmyr M and Brück S 2007 *J. Phys.: Condens. Matter* **19** 216211(7)

[19] Miura M, Maiorov B, Baily S A, Haberkorn N, Willis J O, Marken K, Izumi T, Shiohara Y and Civale L 2011 *Phys. Rev. B* **83**(18) 184519

[20] Ivanov A A, Galkin S G, Kuznetsov A V and Menushenkov A P 1991 *Physica C* **180** 69–72

[21] Kuznetsov A V, Ivanov A A, Eremenko D V and Trofimov V N 1995 *Phys. Rev. B* **52**(13) 9637–46

[22] Sannikov I I, Ivanov A A, Kuznetsov A V, Menushenkov A P and Churkin O A 2014 *Bull. of the Lebedev Phys. Inst.* **41**(8) 215–7

[23] Feigel'man M V and Vinokur V M 1990 *Phys. Rev. B* **41**(13) 8986–90

[24] Nelson D R and Vinokur V M 1993 *Phys. Rev. B* **48**(17) 13060–97

[25] Brandt E H 1995 *Rep. Prog. Phys.* **58** 1465–594

[26] Kuznetsov A V, Sannikov I I and Ivanov A A 2016 Two components of critical current in $YBa_2Cu_3O_{7-\delta}$ films *arXiv:*1612.05454v1 cond-mat.supr-con

IOP Conf. Series: Journal of Physics: Conf. Series **941** (2017) 012072 doi:10.1088/1742-6596/941/1/012072

# XMCD and TEM studies of as-cast and rapidly quenched $Fe_{50}Nd_{50}$ alloys

**V P Menushenkov[1,*], A P Menushenkov[2], I V Shchetinin[1], F. Wilhelm[3], A A Ivanov[2], I A Rudnev[2], V G Ivanov[2], A Rogalev[3], A G Savchenko[1], D G Zhukov[1], A V Rafalskiy[1], S V Ketov[4]**

[1]National University of Science and Technology "MISiS" 119049 Moscow, Russia
[2]National Research Nuclear University MEPhI (Moscow Engineering Physics Institute), Kashirskoe sh. 31, 115409 Moscow, Russia
[3]European Synchrotron Radiation Facility (ESRF), CS40220, F-38043 Grenoble Cedex 9, France
[4] WPI-Advanced Institute for Materials Research, Tohoku University, 2-1-1 Katahira, Aoba-ku, Sendai, 980-8577 Japan

[*]E-mail: menushenkov@gmail.com

**Abstract**. We present the XMCD analysis of as-cast and melt spun $Fe_{50}Nd_{50}$ samples performed at $L_{2,3}$-Nd and $K$-Fe absorption edges at 5 and 50 K in comparison with macroscopic data of XRD, TEM and magnetic properties measurements. In addition, we have measured the magnetic field dependence of XMCD signal for both types of the samples in magnetic fields up/down to 17 T. The obtained results pointed to the strong difference between structure and magnetic properties of the as-cast and melt spun $Fe_{50}Nd_{50}$ alloys for both macroscopic and local measurements. The element selective XMCD loops for melt spun alloy show almost identical value of the coercive force $H_{ci}$ for $L_2$-Nd and $K$-Fe edges and practically do not depend on temperature. XMCD loop at $K$-Fe edge is a sum of contributions of the Fe-based phases. The main Fe-rich phase has high $H_{ci} \approx 2,4$ T as a highly anisotropic phase. The absence of the $K$-Fe XMCD loop saturation in the field up to 17 T points to presence of the second Nd-rich Nd-Fe phase which is ferromagnetic at temperature lower than 50 K. In accordance to the TEM results these both phases may coexist as the mixture of nanocrystals which was formed as a result of decomposition of the amorphous-like matrix phase. The XMCD loop at $L_2$-Nd edge with $H_{ci} \approx$ 1,9 T is the sum of contributions from two Nd-based phases: hard Fe-rich phase ($H_{ci} \approx 2,4$ T) and Nd-Fe matrix phase of medium hardness with $H_{ci} \approx 1,3$ T. The macroscopic loop showed the higher $H_{ci}$ compared to XMCD loops. Such discrepancy may be caused by the fact that XMCD signal is collected from a 5-10 mcm thick surface layer, which contains many defects that reduce anisotropy and coercivity.

## 1. Introduction

The interest in the study of the Fe-Nd alloys was renewed during last decade owing to the important role of the Nd-rich intergranular phase in the formation of high coercivity in the sintered Nd-Fe-B-based magnets. The effect of cooling rate and composition on the magnetic properties and microstructure of the melt-spun Nd-Fe alloys was thoroughly investigated. High coercivity values in as-cast Nd-rich Fe-Nd alloys were found to correlate to a metastable eutectic microstructure which includes hard magnetic phase ($A_x$) with unidentified structure and composition [1-5]. The formation mechanism of hard magnetic phase during cooling of the Fe-Nd alloys and the structure-property relationship remain unknown. The study of magnetization of the Fe-Nd alloys is complicated due to the presence of several magnetic phases observed by SEM and TEM: primary Nd or $Nd_2Fe_{17}$ phase, Nd and hard magnetic phase as part of eutectic structure. X-ray magnetic circular dichroism (XMCD) technique gives possibility to separate the contribution of Nd and Fe-Nd phases in magnetic behavior of the Fe-Nd alloys. Therefore, the aim of presented study was to characterize the magnetic phases in as-cast and rapidly quenched $Fe_{50}Nd_{50}$ alloys by XMCD and to distinguish their contribution in magnetization process of these alloys.

## 2. Experimental

The ingot of $Fe_{50}Nd_{50}$ alloy was prepared be melting pure metals in an argon-arc furnace. The rapidly quenched ribbons were made by melt-spinning on a copper-wheel speeded with $v \sim 40$ m·s$^{-1}$ velocity. X-ray diffraction (XRD) studies were carried out using the diffractometer Rigaku Ultima-IV with a graphite monochromator focusing $Co_{K\alpha}$-radiation beam in the Bragg-Brentano geometry. The grain size in the micro- and nanostructure phases of the as-cast and melt spun $Fe_{50}Nd_{50}$ alloys was determined by the JEOL JSM-6610LV scanning electron microscope (SEM) and the JEM-1400 transmission electron microscope (TEM) operated a 120 kV. Measurements of magnetic properties were performed using the physical property measurement system PPMS (EverCoolII, Quantum Design) in fields up to 9 T.

The XMCD studies were carried out at the beamline ID 12 [6] of the European Synchrotron Radiation Facility (ESRF), Grenoble, France. The high-field XMCD end-station was equipped with superconducting solenoid NbTi/$Nb_3Sn$ of Cryogenic Ltd producing a magnetic field up to 17 T in wide temperature range (5 - 300 K). The solenoid axis is set parallel to the incoming X-ray beam. The disks of as-cast or melt spun $Fe_{50}Nd_{50}$ alloy were inserted in a bore of superconducting solenoid so that the disk plane was perpendicular to the solenoid axis. X-ray absorption (XANES) spectra were recorded at the *K*-Fe (7113 eV) and the $L_2$-Nd (6725 eV) edges in the total fluorescence yield detection mode using Si photodiode mounted on the nitrogen screen of the liquid helium cryostat with solenoid. The APPLE-II type undulator ($h\nu \geq$ 5keV) was a source of circularly polarized X-rays and allowed to easily change the polarization of the beam. The XMCD spectra were obtained as a direct difference between consecutive XANES recorded with opposite helicities of the incoming X-ray beam. To ensure that the XMCD spectra are free from any experimental artifacts the data were collected for both directions of the applied magnetic field of 17 T (parallel and antiparallel to the X-ray beam).

In addition, we have measured magnetic field dependence of XMCD signal of as-cast and melt spun $Fe_{50}Nd_{50}$ samples in magnetic fields up/down to 17 T. For that purpose the energy of X-ray monochromator was tuned to the positions of the main maxima of previously recorded XMCD spectra at either Fe *K*- or Nd- absorption edges as in [7,8]. These so-called element selective magnetization curves were compared with macroscopic ones that have measured in the temperature range from 5 to 650 K using a PPMS EverCool-II magnetometer in magnetizing field up to 9 T.

## 3. Results and discussion

Figure 1 demonstrates the X-ray diffraction patterns for the as-cast and melt spun $Fe_{50}Nd_{50}$ alloy. The XRD pattern of as-cast sample can be indexed by 40% Nd (P63/mmc) and 60% $Nd_2Fe_{17}$ (R-3m).

Figure 1 b shows that the diffraction peaks in the spectra of the ribbon sample are broadened and amorphous background takes place, which should be the result of structural refinement and partial amorphization during melt spinning. The SEM images of the as-cast and melt spun $Fe_{50}Nd_{50}$ alloy are presented in Figure 2. The microstructure of the as-cast sample consists of the primary $Nd_2Fe_{17}$ particles surrounded by the Nd-rich cover with eutectic region between them (left panel). At the same time, the SEM image of the ribbon sample shows a single-phase homogeneous microstructure (right panel).

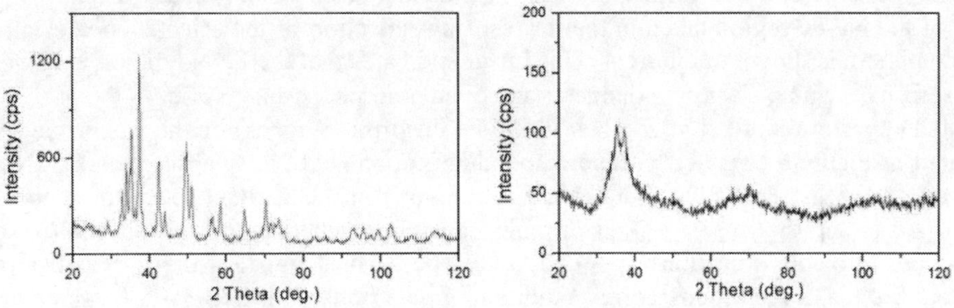


**Figure 1**. XRD patterns for the as-cast and melt spun $Fe_{50}Nd_{50}$ alloys

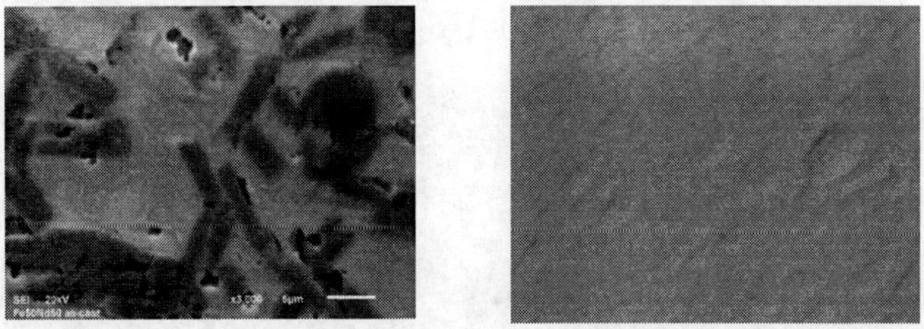

**Figure 2**. SEM micrographs of the as-cast and melt spun $Fe_{50}Nd_{50}$ samples

The hysteresis loops for the as-cast and melt spun $Fe_{50}Nd_{50}$ alloys are shown in Figure 3. The very low coercive force of the as-cast sample results from the high-volume fraction of the soft $Nd_2Fe_{17}$ phase at all measured temperatures. However, the visible width of the loops at high magnetic fields at temperature lower than 50 K points to the presence of the hard magnetic phase in the microstructure of as-cast alloy, and also to the presence of the Nd-based phase which is not saturated up to 17 T. Measurements of the melt spun $Fe_{50}Nd_{50}$ ribbon showed the kink in the low demagnetization field. This fact is evidence that some part of the sample volume is magnetically soft, which can arise from the presence of a soft magnetic phase $Nd_2Fe_{17}$ in the melt spun ribbon.

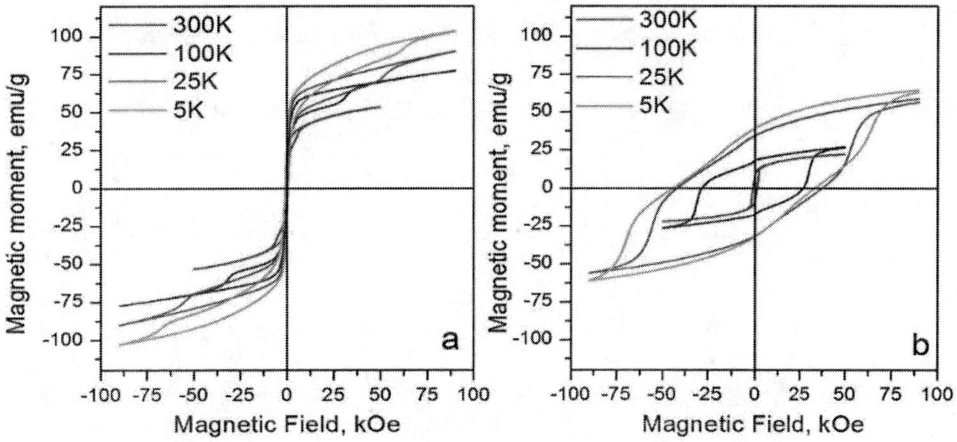


**Figure 3**. Hysteresis loops for the as-cast and melt spun $Fe_{50}Nd_{50}$ samples at 5 - 300 K

The dark field images of the ribbon $Fe_{50}Nd_{50}$ are presented in Figure 4(a). The microstructure of the ribbon consists of amorphous-like regions with nanocrystals of the unknown phases. The darkfield image of the Nd-Fe region taken in the nearest ring reflection of the selected area electron diffraction (SAED) pattern is shown in Figure 4. This image indicates that the Fe-Nd region is composed of a fine dispersed phase with 5–10 nm in diameter and of an amorphous-like phase.

XMCD spectra recorded at $L_{2,3}$-Nd and *K*-Fe absorption edges for the ribbon $Fe_{50}Nd_{50}$ alloy are presented in Figure 5 together with corresponding normalized X-ray absorption (XANES) spectra. It has been shown [4,5] that the first feature (minimum) in the XMCD spectrum at the $L_3$ absorption edge of Nd is due to $2p_{3/2} \rightarrow 4f_{5/2}$ quadrupole transitions, whereas the second one (major maximum) corresponds to dipole transitions $2p_{3/2} \rightarrow 5d_{3/2,5/2}$. The XMCD signal of the $L_2$-Nd-absorption edge has a pronounced major maximum, corresponding to dipole transitions $2p_{1/2} \rightarrow 5d_{3/2}$ and a very weak negative feature at lower photon energies that could be attributed to quadrupole transitions. The observed spectra at the *K*-Fe -edge are due to dipole transitions from 1*s* core level of Fe to 4*p* states in the conduction band that are hybridized with the 3*d* states of iron. The *K*-Fe edge XMCD signal is much smaller in comparison with $L_{2,3}$-Nd edge XMCD signal because of low magnetization of the 4*p* states.

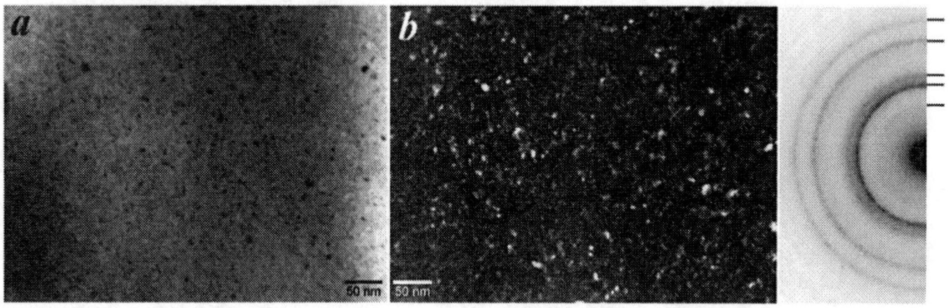


**Figure 4.** Bright-field (*a*) and dark-field (*b*) images of the ribbon x = 50 sample taken in the second ring reflection of the SAED pattern

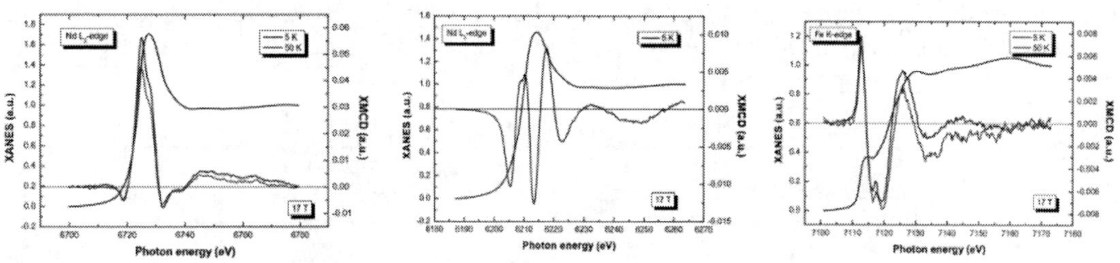


a) b) c)

**Figure 5.** Comparison of XANES and XMCD spectra measured at $L_{2,3}$-Nd (a,b) and Fe-*K* (c) edges of melt spun $Fe_{50}Nd_{50}$ alloy in the magnetic field H = 17 T at 5 and 50 K

In addition to XMCD spectra at the $L_2$-Nd and at the *K*-Fe edges, we have measured so-called XMCD magnetization curves that are shown in Figure 6. These element-selective magnetization curves have been recorded at temperatures 5 and 50 K in fields up/down to 17 T by monitoring the magnitude of XMCD signal at the photon energy corresponding to a major maximum as a function of an applied magnetic field. For the case of $L_2$-Nd absorption edges the measurements were performed at the positions of first maximum 6.725 keV, while for *K*-Fe -edge the major maximum of XMCD spectra was observed at 7.113 keV. These measurements are fundamentally different from the conventional macroscopic magnetization curves measurements, which give information about the integrated characteristics of the magnetization reversal.

Figure 6 shows the XMCD magnetic loops measured at $L_2$-Nd and *K*-Fe edges at 5 and 50 K, respectively. One can see that XMCD loops for $L_2$-Nd and *K*-Fe edges show almost identical value of the coercive force and are practically independent of temperature. The element selective XMCD loops were compared with macroscopic ones measured using a magnetometer at 5 K (Figure 7).

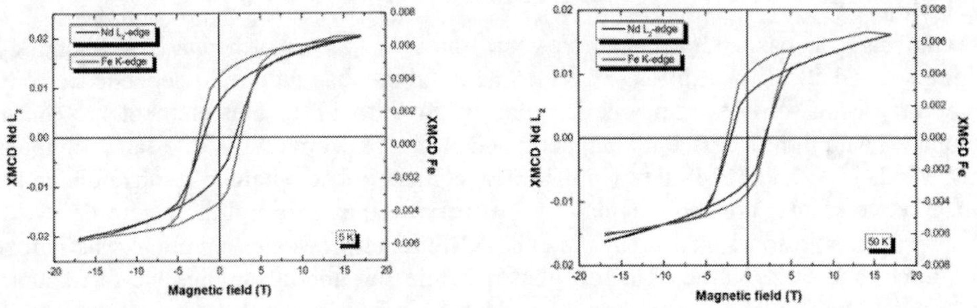


**Figure 6.** XMCD hysteresis loops measured at $L_2$-Nd and *K*-Fe edges for ribbon $Fe_{50}Nd_{50}$ at 5 K (left panel) and 50 K (right panel)

The XMCD loop at *K*-Fe edge is a sum of contributions of the Fe-based phases. The main Fe-rich phase has high $H_{ci} \approx 2{,}4$ T (Figure 7) hence it is a highly anisotropic phase. Since the XMCD loop at *K*-Fe edge is not saturated in the field up to 17 T there exists the second Nd-rich Nd-Fe phase which is ferromagnetic at temperature lower than 50 K.

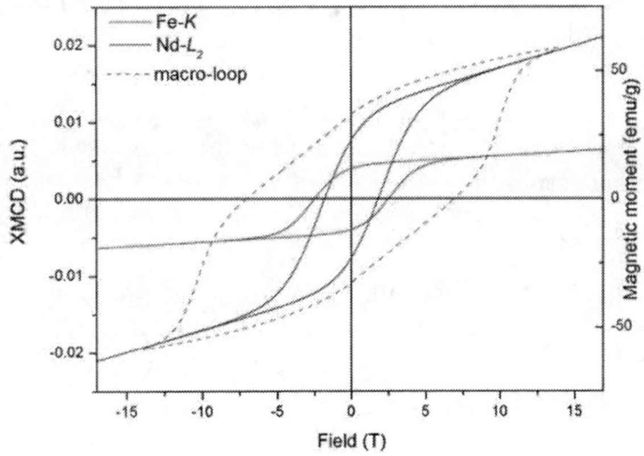


**Figure 7.** Macroscopic hysteresis loop for as-cast $Fe_{50}Nd_{50}$ sample in comparison with HMCD hysteresis loops, measured on $L_2$-Nd and *K*-Fe edges at 5 K

According to the TEM results (Figure 4) both phases ($A_x$ and Nd-rich) may coexist as the mixture of nanocrystals which was formed because of decomposition of the amorphous-like matrix phase. The XMCD loop at $L_2$-Nd edge with $H_{ci} \approx 1{,}9$ T is resulted from contributions of the Nd-based phases: these include hard $A_x$ phase ($H_{ci} \approx 2{,}4$ T) and medium hard Nd-Fe phase with $H_{ci} \approx 1{,}3$ T. Presumably, in the ribbon $Fe_{50}Nd_{50}$ the last phase can be a matrix.

It worth noticing that the macroscopic loop in the Figure 7 normalized to the saturation of the XMCD $L_2$-Nd loop showed a higher coercive force ($H_{ci} = 4{,}7$ T) compared to XMCD loop ($H_{ci} \approx 1{,}9$ T). At the same time, as can be seen in Figures 3 and 7, the coupling of both macroscopic and

XMCD $L_2$-Nd hysteresis loops occurs in a field of about 7,5 T which is higher than coercive force. Therefore, this value of field should be used to compare the difficulty of the magnetization reversal process. The reason of the difference between values of $H_{ci}$ may be that macroscopic loop results from remagnetization of the bulk of alloy while the XMCD loop is obtained from a 5-10 mcm thick surface layer. The surface contains many defects that may reduce anisotropy and coercivity. Indeed, both the coercivity of XMCD loop and penetration depth of X-ray at *K*-Fe edge (7113 eV) are bigger than corresponding quantities at $L_2$-Nd edge (6725 eV).

## 4. Conclusion

XMCD analysis of as-cast and melt spun $Fe_{50}Nd_{50}$ samples was performed at $L_{2,3}$-Nd and *K*-Fe absorption edges at 5 and 50 K. In addition, we have measured the magnetic field dependence of XMCD signal for both types of samples in magnetic fields up/down to 17 T. This element-specific information was compared with the macroscopic magnetic and structure properties of the same samples measured by XRD, SEM, TEM and PPMS EverCool-II magnetometer. The obtained results pointed to the strong difference between structure and magnetic properties of the as-cast and melt spun $Fe_{50}Nd_{50}$ alloys for both macroscopic and local measurements. The XRD pattern of as-cast sample was indexed as 40% Nd (P63/mmc) and 60% $Nd_2Fe_{17}$ (R-3m) phases, while for ribbon samples the diffraction peaks are broadened and amorphous background take place as a result of structural refinement and partlial amorphization during melt spinning. The TEM images show that the microstructure of the as-cast sample consists of the primary $Nd_2Fe_{17}$ particles surrounded by the Nd-rich cover with eutectic region between them, while the ribbon sample has a single-phase homogeneous microstructure. The element selective XMCD loops for melt spun alloy show the almost identical value of the coercive force for $L_2$-Nd and *K*-Fe edges and practically don't depend from temperature. XMCD loop at *K*-Fe edge is a sum of contributions of the Fe-based phases. The main Fe-rich phase has high $H_{ci} \approx 2{,}4$ T as a highly anisotropic phase. The absence of the *K*-Fe XMCD loop saturation in the field up to 17 T points to presence of the second Nd-rich Nd-Fe phase which is ferromagnetic at lower than 50 K temperature. According to the SEM results both phases may coexist as the mixture of nanocrystals which was formed as a result of decomposition of the amorphous-like matrix phase. The XMCD loop at $L_2$-Nd edge with $H_{ci} \approx 1{,}9$ T is the sum of contributions from two Nd-based phases: hard Fe-rich phase ($H_{ci} \approx 2{,}4$ T) and Nd-Fe matrix phase of medium hardness with $H_{ci} \approx 1{,}3$ T. The macroscopic loop normalized to the saturation point of the XMCD $L_2$-Nd loop showed a higher $H_{ci}$ compared to XMCD loops, since the XMCD loop is obtained from a thin surface layer of 5-10 mcm thickness, which may contain many defects that reduce anisotropy and coercivity.

## Acknowledgements

The authors acknowledge the ESRF Program Committee for providing the opportunity of XMCD measurements at ID12 beamline in according with MA-3314 project. This work was supported by the Ministry of Education and Science of the Russian Federation (agreement no. 14.587.21.0028, unique number of project RFMEFI58716X0028).

## References

[1] Delamare J, Lemarch D, P. Vigier P 1994 *J. Alloys Compd.* **216** 273
[2] Menushenkov V P, Anderson S J, and Hoier R 1998 Proc. of the 20-th Int. Symp. on Magnetic Anisotropy and Coercivity in Rare-Earth Transition Metal Alloys, ed. By L. Schultz and K.-H. Müller, Werkstoff-Infromationsgesellschaft, Frankfurt, p. 97
[3] Menushenkov V P, *et al* 1999 *J. Magn. Magn. Mater.* **203** 149
[4] Kumar G, J. Eckert J, Roth S, Löser W, Schultz L and Ram S2003 *Acta Mater.* **51** 229
[5] Kumar G, Filip O, Löser W, Schultz L, Eckert J 2006 *Intermetallisc* **14** 47
[6] Sarafidis C, Wilhelm F, Rogalev A, *et al* 2009 *J. Phys.: Condens. Matter***21** 236001
[7] Menushenkov A P, Ivanov V G, Shchetinin I V, *et al* 2017 *JETP Letters* **105** 38
[8] Menushenkov A P, Ivanov V G, Shchetinin I V, *et al* 2016 *J. Phys.: Conf. Ser* **747** 012039

# The influence of the shock treatment under heating on the structure and properties of HTS tapes

**B P Mikhailov[1,*], A B Mikhailova[1], I V Borovitskaya[1], V Ya Nikulin[2,4], P V Silin[2], E N Peregudova[2], S N Polukhin[2], S V Shavkin[3], N A Mineev[4], V F Shamray[1], V N Kolokoltsev [1], N A Krutskih[5] and S Y Alibekov[5]**

[1] A.A. Baikov Institute of Metallurgy and Materials Sciences, Russian Academy of Sciences, Leninskiy pr. 49, 119991 Moscow, Russia
[2] P.N. Lebedev Physical Institute, Russian Academy of Sciences, Leninskii pr. 53, 119991 Moscow, Russia
[3]NRC «Kurchatov Institute», pl. Kurchatova 1, 123098 Moscow, Russia
[4] National Research Nuclear University MEPhI (Moscow Engineering Physics Institute), Kashirskoe sh. 31, 115409 Moscow, Russia
[5] Volga State University of Technology, Lenin Sq. b. 3, 424000 Yoshkar-Ola, Russia

[*]E-mail: borismix@yandex.ru

**Abstract**. The influence of shocks of different intensity on the structure and properties of multifilamentary superconducting tapes of $Bi_2Sr_2Ca_2Cu_3O_{10+x}$ (Bi-2223) and $MgB_2$ compounds was studied. The Plasma Focus setup was used to produce the plasma shock waves, and a specially designed setup was utilized for the mechanical shock treatment. The experiments have shown a possibility to increase the critical current of $MgB_2$ tapes by more than 60% in magnetic fields of 1.5-2.0 T due to the treatment. The critical current increase is caused by homogeneity improvement, densification of superconducting filaments and the pinning enhancement.

## 1. Introduction

An interest to study an influence of plasma shock-waves and mechanical shocks on the structure and superconducting parameters of high-temperature superconductors (HTS) permanently grows. This interest is caused by a possibility of directional formation of structure of the HTS ceramic filaments enclosed in the metal shell due to increase of density and texture of the filaments, as well as creation of additional effective pinning centers. An increase of both the critical temperature $T_c$ and the critical current $I_c$ in the external magnetic fields was demonstrated in such a study. An increase of $T_c$ by 15 K was observed after ion irradiation of films, consisting of mixture of Bi-2212 and Bi-2223 phases, at the Plasma focus setup [1]. In addition, $I_c$ increase by 15% was observed for Y-123 ceramic based tapes produced by Superpower (USA) [2]. Similar increase by more than 60% in the magnetic field up to 8T was also found for the HTS tapes in [3,4].

The shock-wave impact leads to densification of superconducting filaments due to short-time heating and high pressure in the impact zone, formation of additional nano-sized defects (dislocations) acting as pinning centers, crushing and grinding grains and to a number of other factors. The improvement of pinning characteristics under the shock treatment is explained in [5-9] by formation of vacancies, point defects and interstitial atoms (collective Frenkel pairs) at the shock-wave front. Concentration of vacancies produced by the plasma jet shock is much more than that in virgin material.

This work aims to find new ways and conditions for purposeful formation of microstructure of HTS ceramic filaments, to enhance the critical characteristics and other functional properties of these materials by means of various shock effects.

## 2. Experiment

The multifilamentary HTS tapes on the base of Bi-2223 or $MgB_2$ ceramics were investigated. Dimensions of the samples were as follows: the cross-section of 0.2×4 mm$^2$ and length of 35-40 mm for Bi-2223 tapes and 0.65×3.65 mm$^2$ $MgB_2$ ones.

Original setups shown in Figure 1 were used for the mechanical shock treatment made at temperatures up to 500°C. Specific energy of mechanical shocks was varied in a wide range from 0.5 to 100 J/cm$^2$ via change of the hummer weight (0.5 and 0.8 kg), height of the hammer's fall (10-40 mm), number of shocks, width of the impact zone, and etc. The hammer width was larger than the tape width. The impact-zone areas were of 1.6 mm$^2$ for Bi-2223 tapes and 1.46 mm$^2$ for $MgB_2$ tapes. The neighboring impact-zones overlapped by about 0.05 -0.1 mm. Single shock with the specific energy from 0.5 to 5.0 J/cm$^2$ was used in experiments on Bi-2223-tapes heated up to 500 °C in an electric furnace. A thermocouple was used to control temperature. After a series of mechanical shocks, the samples were annealed in air at temperature of 830-835°C for 5-30 hours.

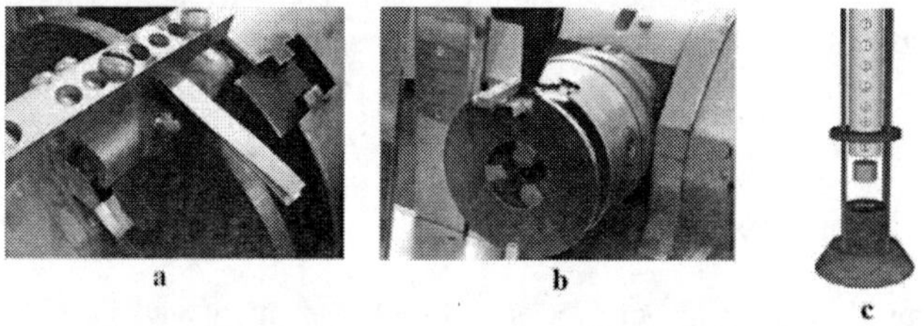

**Figure 1.** Setups for mechanical shock treatment: (a) - the disc with fixed HTS tape; (b) – device producing the tape shift after applying mechanical shocks, (c) - an installation for shock-treatment of heated samples.

To treat tapes with the shock-waves produced by cumulative high-speed plasma streams (CHSPS), the plasma focus setup PF-4 (P.N. Lebedev Physical Institute, RAS) was used [10,11]. The samples were protected from the direct exposition of the plasma, by 0.2 mm thick molybdenum plate used as a target in which shock-waves were created. Sketch of the PF-4 setup and photograph of the plasma jet in the discharge space are shown in Figure 2. The PF-4 parameters are as follows: the energy in the plasma stream hitting the target is about 100 J, the duration of the impact is about $10^{-7}$ s, the energy flux density is up to $10^9$ W/cm$^2$. The distance from the PF-4 anode to the tape surface was 25, 30, 35, and 40 mm. Either 3 or 5 shocks were applied to one side of the tape with time interval of 1.5 min.

The surface microstructure of the tapes was examined using the scanning electron microscope EVA-40 (Zeiss). The elemental analysis of Bi-2223 and $MgB_2$ filaments was performed on a scanning electron microscope JSM-35 with a Link attachment. Scanning Hall magnetometry was used to study a magnetic field distribution frozen in the Bi-2223 tapes at 77K. The current-voltage characteristics of the $MgB_2$ tapes before and after treatment were measured at 4.2 K in magnetic fields from 0 to 4 T in NRC “Kurchatov Institute”. The 1 μV/cm criterion was used to obtain the critical current.

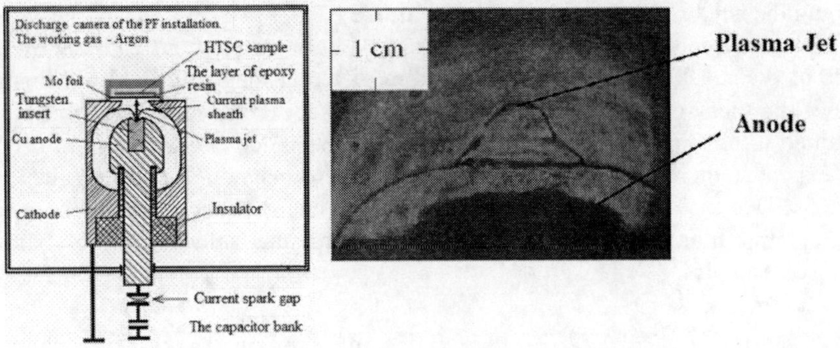


**Figure 2.** The PF-4 sketch and photograph of the plasma jet in the discharge space.

**3. Results and discussion.**

Photographs of the surface of the $MgB_2$ tape before (a) and after mechanical shocks with specific energy of 0.11 J/cm$^2$ (b, c) are shown in Figure 3. The shocks were applied at room temperature and the tape was shifted by 0.4 mm after each shock. The relief of the tape surface was investigated after the shock treatment. Comparing the panels (b) and (c) in Figure 3 we clearly see transverse stripes produced by the shocks on the treated side of the tape (b) The opposite side (c) demonstrates traces of the initial rolling while the stripes are less noticeable.

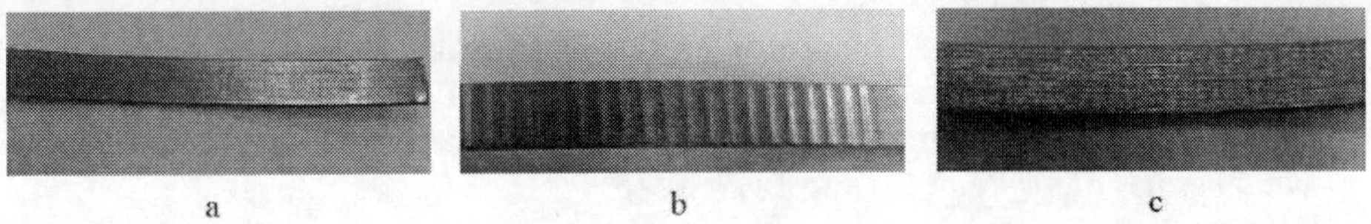


**Figure 3.** Surface of $MgB_2$ tape before (a) and after mechanical shocks with the specific energy of 0.11 J/cm$^2$. (b) - treated side, (c) - the opposite side. After each shock the tape was shifted by 0.4 mm.

Increasing the shift length to 0.8 or 1 mm leads to change of the relief (Figure 4). On the treated side the transverse stripes become more pronounced and deeper, while the longitudinal stripes appear on the opposite side. With all this going on, such transformations are more pronounced for the longer shift (Figure 4 c, d).

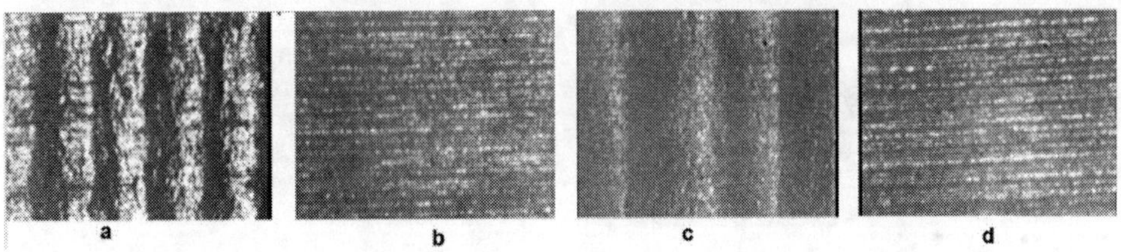


**Figure 4.** Surface of $MgB_2$ sample after mechanical shocks with energy of 0.11 J/cm$^2$. The tapes were shifted by 0.8 mm (a,b) or 1 mm (c,d) after each shock. Surface of both treated (a,c) and opposite (b,d) sides is shown.

Similar transformations for the similar conditions (specific energy of 0.1 J/cm$^2$) were found for Bi-2223 tapes. Clearly defined strips along the tape appeared when the tape was shifted by 0.4 mm, while the transverse stripes were hard to notice. When the shift was increased to 0.6 mm, the longitudinal strips still remained, but the transverse ones became more noticeable. With increase of the shift up to 0.8 and 1.0 mm the relief completely changed and only transverse stripes were observed.

IOP Conf. Series: Journal of Physics: Conf. Series **941** (2017) 012073 doi:10.1088/1742-6596/941/1/012073

The obtained results indicate that the specific energy, the shift length, the impact-zone area and overlapping the neighboring zones must be optimized.

The critical current of $MgB_2$ tapes was measured before and after treatment by mechanical shocks. Influence of the shock energy on $I_c$ measured in different magnetic fields is illustrated in Figure 5a. Increase of the shock energy from 14 J/cm$^2$ up to 21 J/cm$^2$ leads to degradation of the critical current. The obtained results have shown the need to reduce the shock energy down to 0.1 - 0.5 J/cm$^2$. In this case we expect an increase of $I_c$ up to 250-300 A in zero magnetic field. $I_c(H)$ dependences are shown in Figure 5b. Decrease of the critical current under mechanical shock treatment is caused by fragmentation of ceramic filaments. $I_c$ has been restored to previous values and even has increased after annealing of the samples.

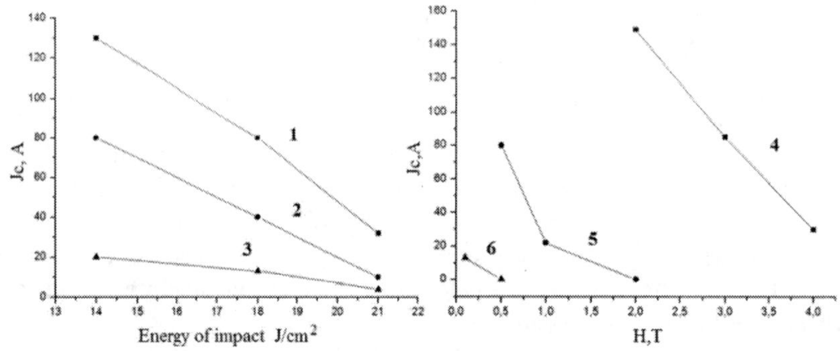


**Figure 5.** (a) Influence of energy of mechanical shocks on the critical current of $MgB_2$ tape in different magnetic fields: 1 - 0, 2- 0.5 T, 3 - 1 T. (b) Field dependence of the critical current before (4) and after treatment with mechanical shocks with energies of 14 J/cm$^2$ (5) and 18 J/cm$^2$ (6).

Figures 6(a,b) show the surface of Bi-2223 tapes after single mechanical shocks to the sample heated to 500 ° C. The drop height was 3 cm (a) and 9 cm (b) in these experiments. With an increase of shock energy up to 0.91 J/cm$^2$, the impact zone became more noticeable, and the tape thickness decreased for several microns.

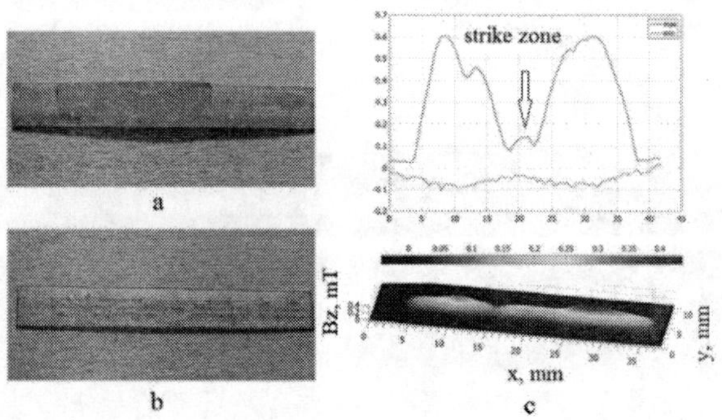


**Figure 6.** Surface of Bi-2223 tapes after mechanical shocks with energies of 0.91 J/cm$^2$ (a) and 0.32 J/cm$^2$ (b). Distribution of the frozen magnetic field measured at 77 K for the tape treated at 500°C with strokes with energy of 0.32 J/cm$^2$ (c).

Analyzing magnetic field frozen in the tape at 77 K we found that the field decreases in the treated zone (Figure 6c). However, this effect is significantly reduced in comparison with the case when the tape was treated without heating. Annealing at 835 °C for 5 hours led to the complete recovery of $I_c$.

Unlike mechanical shocks, treatment with the plasma shock-waves (PSW) leads to significant increase of the critical current of $MgB_2$ tapes in the magnetic fields of 1.5-3.0 T. As seen in Figure 7, the critical current in magnetic field of 2T, which was 150 A in virgin tape, has increased up to 200 or 225 A after 5 or 3 impacts respectively, produced by CHSPS generated at the distance of 40 or 45 mm from the plasma anode. Noticeable difference in $I_c(H)$ behavior is caused by both different number of shocks and different distances between the sample and the plasma anode.

Analysis of the cross section of virgin and PSW treated tapes revealed noticeable difference in microstructure of the samples (Figure 8). Larger grains and cracks were observed in $MgB_2$ interlayers in virgin tape. After the PSW treatment the grains became smaller while the cracks disappeared. These structural transformations under PSW treatment cause the critical current increase.

Results of microanalysis of the $MgB_2$ interlayers in the tapes are presented in Table 1. Comparing data for virgin and PSW treated tapes one can see that stoichiometry is changed by the treatment, namely magnesium and boron content increases while carbon and oxygen content decreases. The highest boron content was found in precipitates with dimensions about 5 μm (spectrum 1) contained about 33 wt% of magnesium and 46.39 wt% of boron. The homogeneity has been improved under PSW treatment.

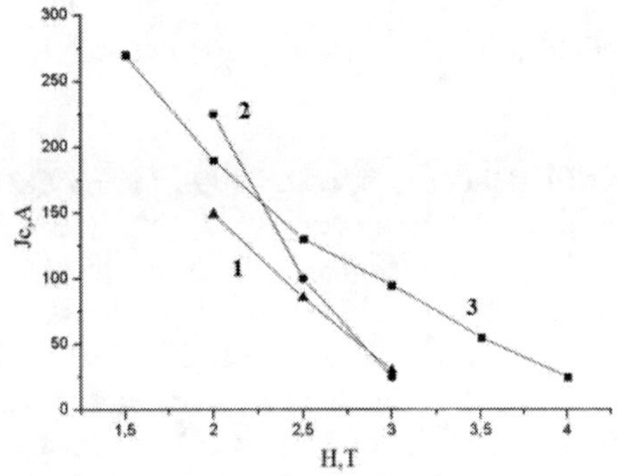


**Figure 7.** $I_c(H)$ dependences for $MgB_2$ tapes. (1) – virgin tape, (2) and (3) – the tapes treated with PSW. Number of impacts and distances between the sample and the plasma anode were 3 and 45 mm for (2), 5 and 40 mm for (3).

**Table 1.** Chemical composition of $MgB_2$ interlayers in virgin and PSW treated tapes.

| After PSW impact | B | C | O | Mg | Virgin state | B | C | O | Mg |
|---|---|---|---|---|---|---|---|---|---|
| Spectrum 1 | 46.39 | 18.44 | 2.21 | 32.96 | Spectrum 1 | 5.38 | 42.76 | 36.22 | 15.64 |
| Spectrum 2 | 40.02 | 20.04 | 8.22 | 31.72 | Spectrum 2 | 10.70 | 35.92 | 23.21 | 30.17 |
| Spectrum 3 | 40.97 | 26.04 | 10.32 | 22.67 | Spectrum 3 | 8.80 | 42.83 | 31.98 | 16.39 |
| Spectrum 4 | 27.69 | 25.88 | 12.96 | 33.47 | Spectrum 4 | 22.00 | 29.21 | 16.60 | 32.19 |
| Spectrum 5 | 31.54 | 40.93 | 11.74 | 15.79 | Spectrum 5 | 18.43 | 30.67 | 21.38 | 29.51 |

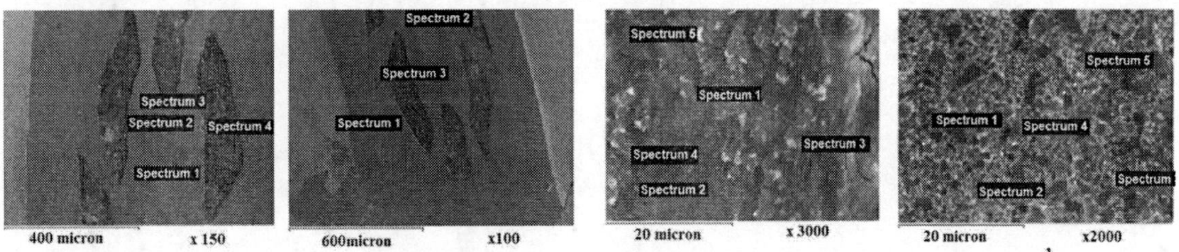


**Figure 8**. Macro- and microstructure of the cross-section of the $MgB_2$ interlayer in virgin state (a, b) and after the PSW impact (c, d).

## 4. Conclusions

1. Critical current increase from 150 to 225 A in the magnetic field of 2.0 T was obtained for $MgB_2$ tapes treated by plasma shock-waves at the distances of 40-45 mm from the surface of the tape to the plasma anode.

2. The study of influence of mechanical shocks on critical current of $MgB_2$ tapes has revealed decrease of $I_c$ with increase of impact energy from 14 to 21 J/cm$^2$. We have observed crushing the grains in the $MgB_2$ interlayers and violation of the electrical conductivity of grain boundaries.

3. It was shown that increase of the shock step from 0.4 to 1.2 mm leads to significant change of the surface relief from both treated and opposite sides of $MgB_2$ and Bi-2223 tapes. Clearly defined transverse traces of impacts were observed on the surface of the tapes when shock step exceeded 0.4 mm.

4. Microanalysis of superconducting layers performed before and after plasma impacts has revealed increase of homogeneity of the superconducting core microstructure.

5. Optimization of the impact treatment allows to improve the functional characteristics of superconducting tapes on the base of $MgB_2$ and Bi-2223.

## Acknowledgments
The work was supported by RFBR grants No. 15-08-04045a and No. 15-02-05995a

## References

[1] Aragwala P, Srivastava M P, Dheer P N, *et a.* 1999 *Physica C* **313** 87-92
[2] Mikhailova G N, Antonova L, Troitsky A V, *et al* 2008 Proc. of the 3rd Int. Conf. "*Fundamental Problems of High-Temperature Superconductivity*" Moscow FIAN 270-3
[3] Antonova L Kh, Borovitskaya I V, *et al* 2009 *Doklady Physics* **54**(10) 463-5
[4] Mikhailov B P, Shamray V F, Nikulin V Ya *et al* 2009 *Perspectivnie Materiali* **6** 57-60
[5] Antonova L Kh, Borovitskaya I V, *et al* 2011 *Physica metallov I metallurgia* **111**(2) 162-8
[6] Mikhailov B P, Mikhailova A B, Mineev N A *et al* 2016 *Perspectivnie Materiali* **11** 61-9
[7] Mezoh Z I, Yanushkevich VA, Ivanov L I *et al* 1971 *FHOM* **4** 163-5.
[8] Grindnev V N, Dekhtyar I Ya, Ivanov L I, *et al* 1973 *Pisma v ZHETF* **18**(4) 258-60.
[9] Dekhtyar I Ya, Ivanov L I, *et al* 1973 *Kvantovaya Electronika* **3**(4) 844-947.
[10] Mikhailov B P, Ivanov L, Borovitskaya I V, *et al* 2012 *Reports of the Academy of Sciences, Technical Physics* **442**(5) 614-6
[11] Polukhin S N, Dzhamankulov A M, Gurei A E, Nikulin V Ya *et al* 2016 *Plasma Physics Reports* **42**(12) 1127-1133
[12] Pokrovskiy S V, Rudnev I A, Podlivaev A I 2009 *J. Phys: Conf. Ser.* **150**(5) 052211-4

IOP Conf. Series: Journal of Physics: Conf. Series **941** (2017) 012074 doi:10.1088/1742-6596/941/1/012074

# Influence of strain on transport characteristics of HTSC

**A N Moroz, A N Maksimova, V A Kashurnikov and I A Rudnev**

National Research Nuclear University MEPhI (Moscow Engineering Physics Institute), Kashirskoe sh. 31, 115409 Moscow, Russia

Email: moroz.anna@hotmail.com

**Abstract**. By using the Monte-Carlo method, the dependencies of transport characteristics of a type-II superconductor on the degree of bending strain applied to the sample have been numerically studied. Multiple series of voltage-current characteristics have been acquired for a sample of a high-temperature superconductor for different numbers of strain-induced cracks, as well as various point defects concentrations and different values of external magnetic field. Some peculiarities of the voltage-current characteristics have been detected. Average vortex configurations have been acquired showing that in the points of peculiarities the vortex lattice "freezes" in spaces between cracks. The results for the increasing number of cracks have shown a significant degradation of critical current density.

## 1. Introduction

Today, type-II superconductors are becoming widely spread. They have successfully found application in various areas, mostly in the form of tapes. However, during the fabrication process the tapes experience different strains, most importantly, bending strains that occur when a tape is wound into coils [1]. Because of this, it is important to study the influence of bending strain on transport characteristics of superconducting materials

Various studies have been conducted on how bending, compressing, and tensile strains affect the value of critical current density in YBCO films [2-5], and the results indicate that the occurrence of cracks inside the material due to strain leads to the decrease in critical current density.

In this paper we present the results of numerical simulation of influence of cracks and point defects on critical current density of a high-temperature superconductor (HTSC).

## 2. Model for calculations

All computations have been carried out within the limits of the model of layered HTSC by using the method developed by the authors of [6-7]. The Monte-Carlo method, being based on the theory of probability and mathematical statistics, provides more accurate results for a system with alternating number of particles that take part in different kinds of interactions, and allows to take all the necessary properties of a vortex system into account. The system's geometry within the limits of this model is shown in Figure 1.

The Gibbs potential for a two-dimensional system of interacting vortices can be described by (1). It is composed of a sum of the vortices' own energies (2), as well as the energies of vortex-vortex (3), vortex-surface (4), vortex-transport and Meissner current (5), and vortex-defect (6) interactions.

$$G = N\varepsilon + \frac{1}{2}\sum_{i\neq j} U(r_{ij}) + \sum_{i,j} U_{\text{pn}}(r_{ij}) + \sum_{i,j} U_{\text{surf}}(|r_i - r_j|) + \sum_i U_{\text{m}}(x_i) \qquad (1)$$

IOP Conf. Series: Journal of Physics: Conf. Series **941** (2017) 012074 doi:10.1088/1742-6596/941/1/012074

$$\varepsilon = \delta \frac{\Phi_0^{\,2}}{(4\pi\lambda)^2}\left(\ln\left(\frac{\lambda_0}{\xi_0}\right) + 0.52\right) \quad (2)$$

$$U(r_{ij}) = \delta \frac{\Phi_0^{\,2}}{8\pi^2\lambda^2} \mathrm{K}_0\left(\frac{r_{ij}}{\lambda}\right) \quad (3)$$

$$U_{\mathrm{pn}}(r_{ij}) = -\alpha\left(1 + \frac{r_{ij}}{\xi}\right)^{-1} \exp\left(-\frac{r_{ij}}{2\xi}\right) \quad (4)$$

$$U_{\mathrm{surf}}(|r_i - r_j|) = -\frac{1}{2}\delta \frac{\Phi_0^{\,2}}{8\pi^2\lambda^2} \mathrm{K}_0\left(\frac{|r_i - r_j|}{\lambda}\right) \quad (5)$$

$$U_{\mathrm{m}}(x_i) = \frac{\Phi_0}{4\pi}\left\{H_0\left[\cosh\left(\frac{x_i}{\lambda}\right)\left(\cosh\left(\frac{dL}{2\lambda}\right)\right)^{-1} - 1\right] + H_{\mathrm{I}}\left[-\sinh\left(\frac{x_i}{\lambda}\right)\left(\sinh\left(\frac{L}{2\lambda}\right)\right)^{-1} \pm 1\right]\right\} \quad (6)$$

A 5×5 μm sample of BSCCO superconductor ($\lambda$=180 nm, $\xi$=2 nm, $T_c$=84 K) was chosen for calculations. The degree of strain was simulated by varying the number and length of cracks made in the form of chains of numerous closely spaced point defects with the width of potential well ~$\xi$, as shown in Figure 2.

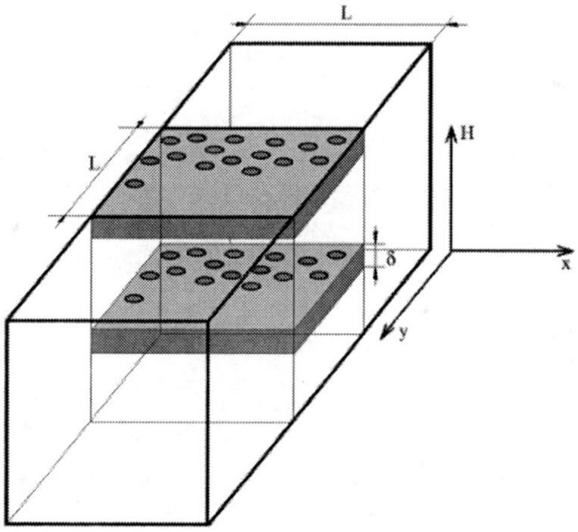


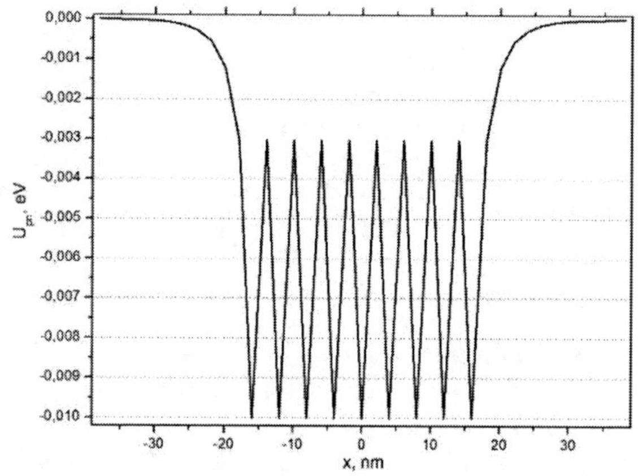


**Figure 1.** Geometry of computations. L=5μm, δ=0.27 nm.

**Figure 2.** Energy potential of a crack. The depth of potential well α=0.01 eV.

Calculations have been carried out for different values of external magnetic field $H$ (0 and 600 Oe). The number of point defects randomly distributed within the sample varied from 100 to 1000. The length of cracks varied from 0.2 to 0.6 of the sample length, and their width equaled that of a single point defect (~$\xi$). The number of cracks varied from 0 to 1500.

## 3. Results

Figure 3 shows a series of voltage-current characteristics for various numbers of cracks from 5 to 20. This sample contained 200 randomly distributed point defects and the external magnetic field value was equaled 600 Oe. It can be seen that the curve for 15 cracks (the curve with upright triangles as markers) has a peculiarity: at some point, while the current density increases, the change in the voltage is very slight.

This happened again for a sample with 500 point defects at the same conditions as above. Only this time, for the case of 12 cracks (the curve with filled circles as markers) the voltage showed a slight decrease.

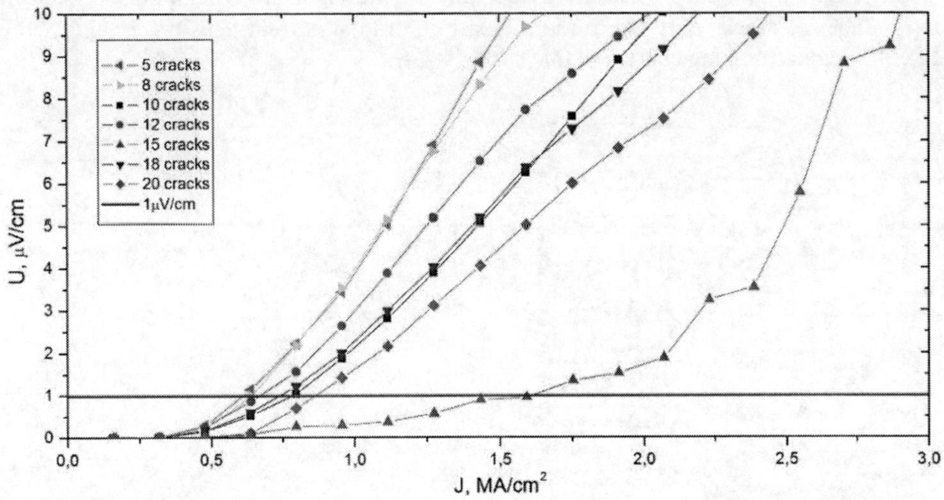


**Figure 3.** A series of voltage-current characteristics for a sample with 200 point defects and various numbers of cracks in external magnetic field $H = 600$ Oe.

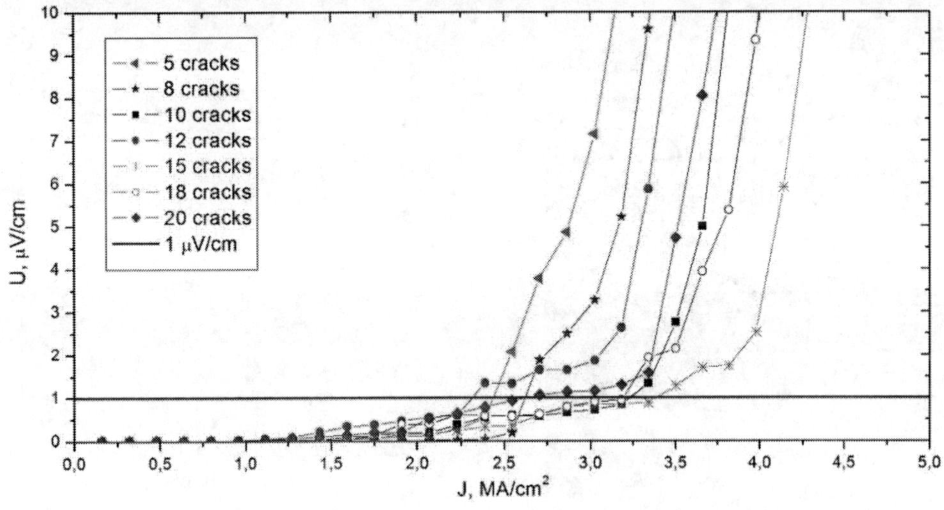


**Figure 4.** A series of voltage-current characteristics for the same sample as above, only for 500 point defects.

In order to understand the reasons behind such peculiarities, we conducted a detailed study of the behavior of the vortex system, and found out that before (Figure 5 (a)) and after (Figure 5 (c)) the point of voltage step the vortices flow along the cracks. However, in the point of peculiarity (Figure 5 (b)) the vortex flux stops, and the lattice "freezes" in the parts between the cracks. The corresponding voltage-current characteristic is shown in Figure 5 (d). In this case, the sample contained 100 point defects and 21 cracks, and the magnetic field was 600 Oe.

We presume that a particular arrangement of cracks, point defects, and pinned vortices creates an equilibrium-like state that prevents the vortices from moving towards the center of the sample where they can annihilate vortices of the opposite sign (flux-flow).

Figure 6 shows a series of voltage-current field characteristics for a wide range of numbers of cracks from 10 to 700. This time the sample had a fixed number of point defects (500), and the external magnetic field was zero. It can be seen, that the peculiarities occur at much higher numbers of cracks in this case compared to when the magnetic field was nonzero. It can also be said that the increasing number of cracks only led to the increase in critical current density, in other words, the cracks acted as regular pinning centers in this case.

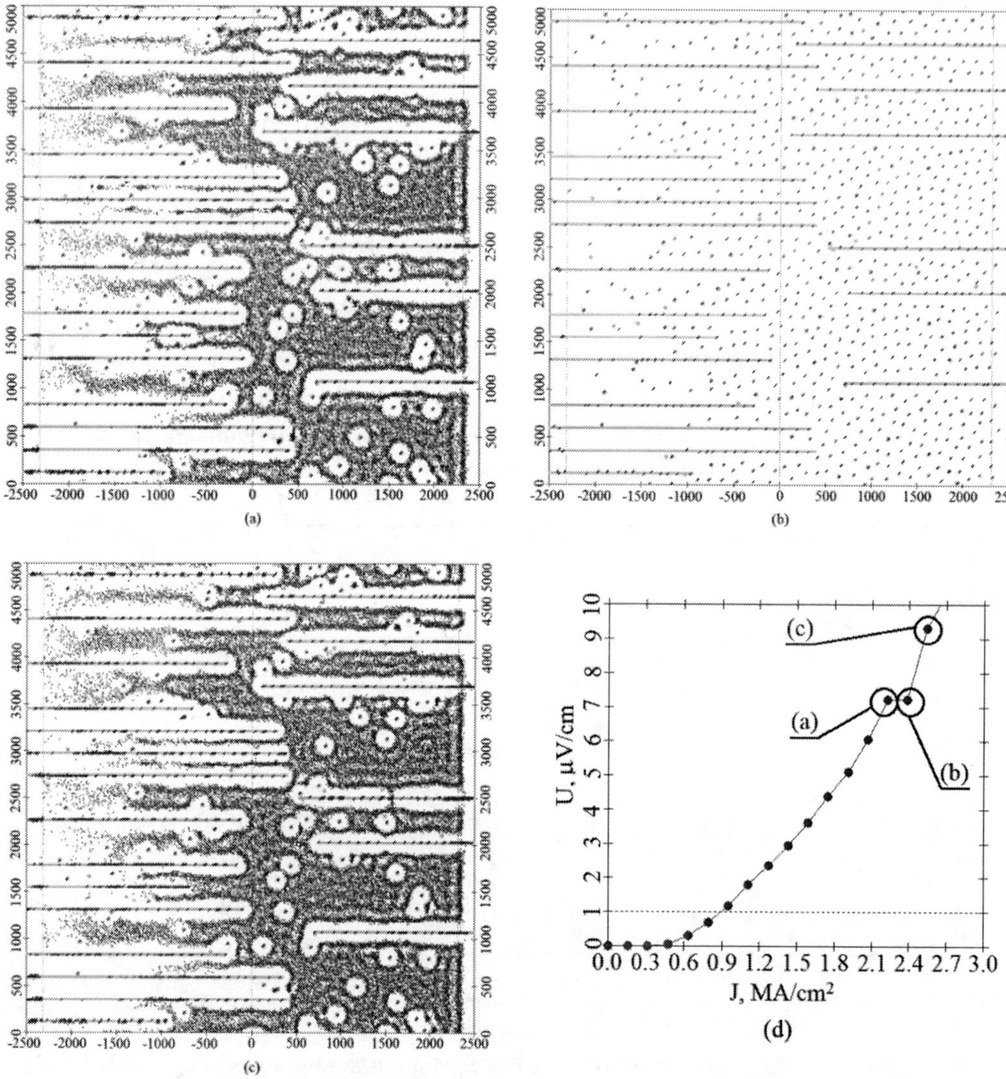


**Figure 5.** Average vortex lattice configurations for a sample containing 100 point defects and 21 cracks in an external magnetic field $H = 600$ Oe before (a), in (b) and after (c) the point of peculiarities. The corresponding voltage-current characteristic is shown in (d).

However, the increase in the critical current density stopped after a certain point, and $J_c$ started to decrease instead. Figure 7 shows the dependency of critical current density (shown as the field induced by the transport current) on the number of cracks extracted from the previous figure. There is a clear significant decrease in the value of $J_c$ at 830 cracks. We have estimated this number to correspond to a strain with a 15 mm bending radius.

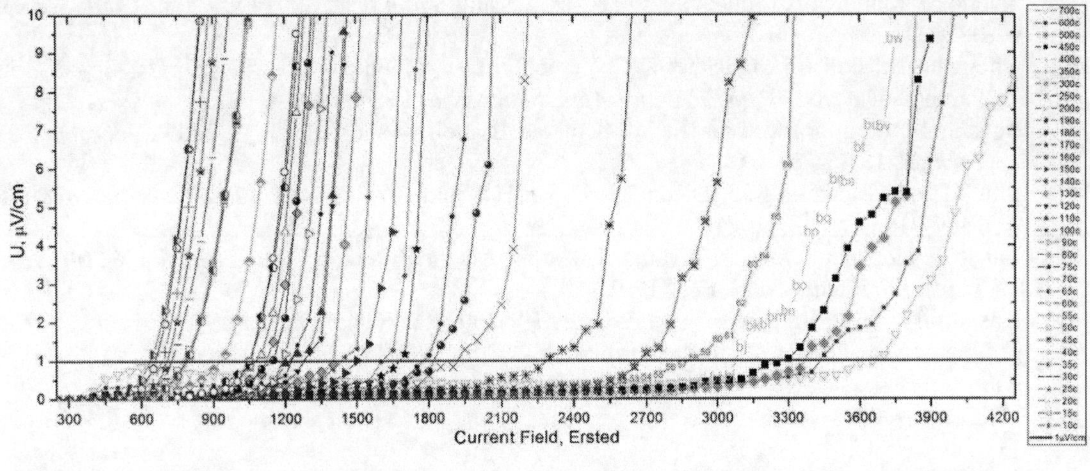


**Figure 6.** A series of voltage-current field characteristics for a sample containing 500 point defects and various numbers of cracks from 10 to 700, while the external magnetic field equals zero.

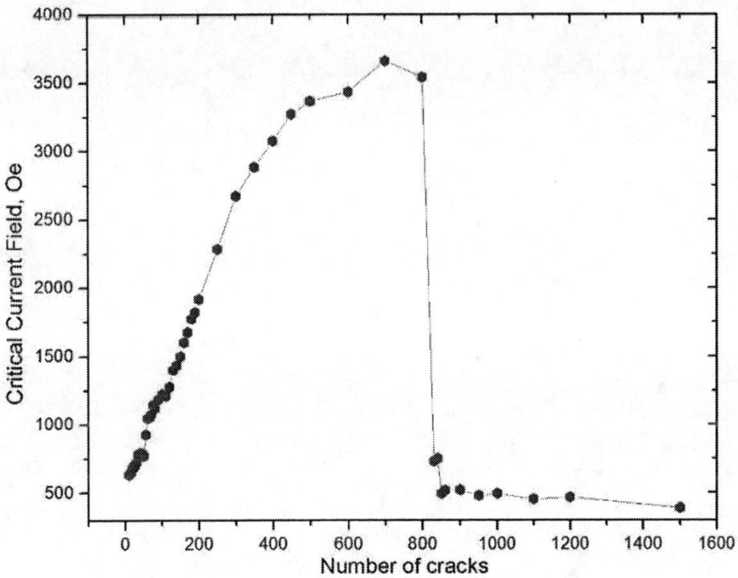


**Figure 7.** The dependency of critical current on bending strain.

## 4. Conclusion

In this paper, a 5×5 μm sample of high-temperature superconductor has been numerically studied. Voltage-current characteristics have been calculated for different degrees of strain simulated by the number of cracks, as well as for different point defect concentrations and external magnetic field values. Some peculiarities of the voltage-current characteristics (voltage steps) have been detected, and it has been shown that they correspond to a "freezing" of the vortex lattice for small amounts of cracks. A further increase of degree of strain led to a significant decrease in critical current density.

**Acknowledgements**
The research was done under financial support of RFBR, grant 15-02-02764.

IOP Conf. Series: Journal of Physics: Conf. Series **941** (2017) 012074 doi:10.1088/1742-6596/941/1/012074

**References**

[1] Rudnev I, Mareeva A, Mineev N, Pokrovskiy S and Sotnikova A 2014 *J. Phys.: Conf. Ser.* **507** 022029

[2] Cheggour N, Ekin J W, Clickner C C, Verebelyi D T, Thieme C L H, Feenstra R, Goyal A and Paranthaman M 2003 *IEEE Trans. Appl. Supercond.* **13** 3530-33

[3] Sugano M, Osamura K, Prusseit W, Semerad R, Itoh K and Kiyoshi T 2005 *Supercond. Sci. Technol.* **18** 364-8

[4] Shin H S, Dedicatoria M J, Dizon J R C, Ha H S and Oh S S 2009 *Phys. C: Supercond.* **469** 1467-71

[5] Takao T, Koizuka S, Oi K, Ishiyama A, Kato J, Machi T, Nakao K and Shiohara Y 2007 *IEEE Trans. Appl. Supercond.* **17** 3517-9

[6] Kashurnikov V A, Rudnev I A and Zyubin M V 2002 *JETP* **94** 377

[7] Zyubin M V, Rudnev I A and Kashurnikov V A 2003 *JETP* **96** 1065

# Structure and electroconductivity of nanostructured ceramics $BiFeO_3$

**Zh Kh Murlieva[1,2,*], D K Palchaev[1], N M-R Alikhanov[1], M Kh Rabadanov[1], S A Sadykov[1]**

[1]Dagestan State University, Gadzhieva 43-A, 367000 Makhachkala, Russia
[2]Dagestan State University of National Economy, Ataeva 5, 367008 Makhachkala, Russia

[*]E-mail: zariyat@mail.ru

**Abstract.** We present results of study of structure and electrical conductivity of the nanostructured ceramics $BiFeO_3$ measured for direct and alternating currents (from 1 kHz to 10 MHz) in the temperature range 25-500 °C. The power-law nature of the frequency dependence of the conductivity is discussed in the framework of the model of the correlated barrier jumps of charge carriers, presumably between two- and trivalent iron ions $Fe^{2+} \rightarrow Fe^{3+}$.

## 1. Introduction

Ceramics based on $BiFeO_3$ (BFO) were obtained by the method of spark plasma sintering (SPS) of a nanopowder synthesized by burning a precursor on the basis of aqueous solutions of $Bi(NO_3)_3$ and $Fe(NO_3)_3$ with the addition of glycine. The SPS method was chosen since BFO nanoparticles do not achive to 62 nm (size of the antiferromagnetic cycloid) under crystallization in this process because of its short short duration and relatively low temperatures. This size of the antiferromagnetic cycloid is essential for the practical application of BFO multiferroic [1-3]. Relatively high values of electrical conductivity, which reduce the contribution of the ferroelectric phase to the linear magnetoelectric effect [1.2], are disadvantages of this ceramics. Some researchers relate the high conductivity of BFO to presence of extrinsic phases [4,5], other ones - to transport of the oxygen vacancies [6,7] or transformation of the iron ions $Fe^{2+} \rightarrow Fe^{3+}$ [8].

The structure and morphology of the nanostructured BFO ceramics were studied in a transmission electron microscope Titan 80-300 and the PANalytical Empyrean series 2 X-ray diffractometer. Active conductivity studies (*G*) were carried out using the LCR-78110G meter from Good Will Instrument Co. A silver paste was used to make electrodes for disk-shaped samples with a thickness of ~1 mm and a diameter of ~ 4 mm.

## 2. Results and discussion

The nanopowder used for sintering of the ceramics was synthesized by a new technology [9], which differs from the method given in [10], by the amount of glycine and nitric acid. The nanopowder consisted of BFO nanocrystals and amorphous phase. The amount of the former was decreased down to 6% or 4% after heat treatment at 600 °C for 0.5 hours. The X-ray diffraction pattern of the obtained ceramics (Figure 1) and particle sizes (50–60 nm) estimated using the Wolf-Scherer formula did not differ significantly from the data for the heat-treated powder. The density of the BFO ceramic obtained by the SPS- method (~ 7.6 g/cm$^3$) is close to the theoretical value.

Figure 2 illustrates the crystal structure and the morphology of the cleavage surface of BFO ceramics, which indicates a fairly high packing density of the grains. Note that extrusion of voids between particles onto the grain surface is observed under pressing. Defects in the form of pores are observed at the boundaries between the grains (see Figure 2a). Presence of nanoscale defects in the crystal structure of the BFO grains themselves was established by TEM method. These defects may be formed as a result of the transition from the agglomerate of nanoparticles (in the original powder) to the crystalline grain after pressing and sintering the ceramic. As seen in Figure 2(b), these defects represent the crystal lattice distortions extended from about atomic level to 50 nm and more.

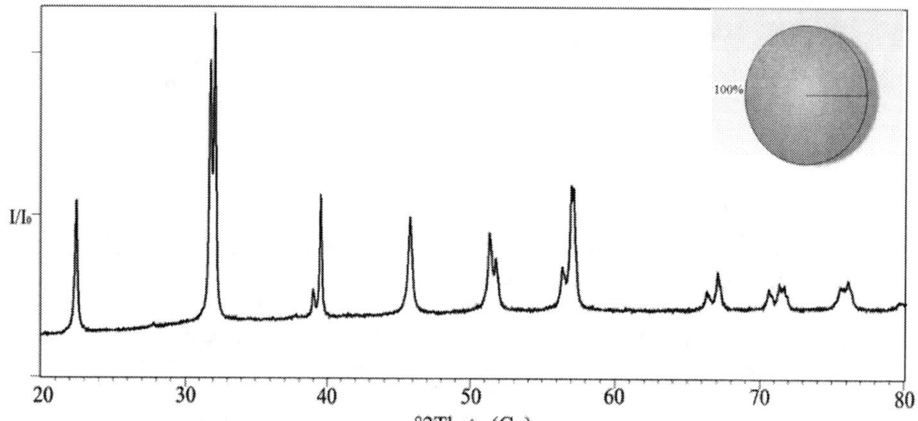


**Figure 1.** Diffractogram of nanostructured ceramics BFO.

**Figure 2.** The morphology of the chip surface (a) and the crystal structure of grain (b) for BFO ceramics.

Temperature dependences of conductivity of the ceramic (Figure 3) in the temperature range 25-500 °C were investigated for direct $\sigma_{dc}$ and alternating $\sigma_{ac}$ currents in the frequency range 1 kHz–10 MHz. These dependences demonstrate two ranges separated by a temperature of about 350 °C, with different temperature coefficients of the conductivity.

Below 350 °C the $\sigma_{dc}$ and $\sigma_{ac}$ values practically coincide (~ $10^{-8}\ \Omega^{-1}cm^{-1}$) for frequencies <100 kHz. At higher frequencies, the conductivity is described by a power function. On the conductivity isotherms obtained in the range 1 kHz – 10 MHz (Figure 4) a frequency dispersion is observed above ~ 350 °C, which is amplified with frequency increase. Beginning from 1 MHz (Figures 3 and 4) the anomaly is noticeably manifested, i.e. maximum of σ appears at some temperature T* exceeding 480 °C. It was assumed [11, 12] that the temperature dependence of the BFO conductivity is affected by an oxygen non-stoichiometry of the bismuth ferrite which The oxidation-reduction reaction leads to increase of number of both the $Fe^{2+}$ ions and lattice defects with temperature increase [12]:

$$O \leftrightarrow \frac{1}{2}O_2\uparrow + V_O^{\cdot\cdot} + 2e^-; \qquad Fe^{3+} + e^- \leftrightarrow Fe^{2+} \quad (1)$$

Formation of the oxygen vacancies causes appearance of free electrons, so the BFO ceramics may be considered as *n*-type semiconductor. Electron hopping occurs between $Fe^{2+} \rightarrow Fe^{3+}$ ions located in the nodes of the octahedra of the BFO crystal lattice.

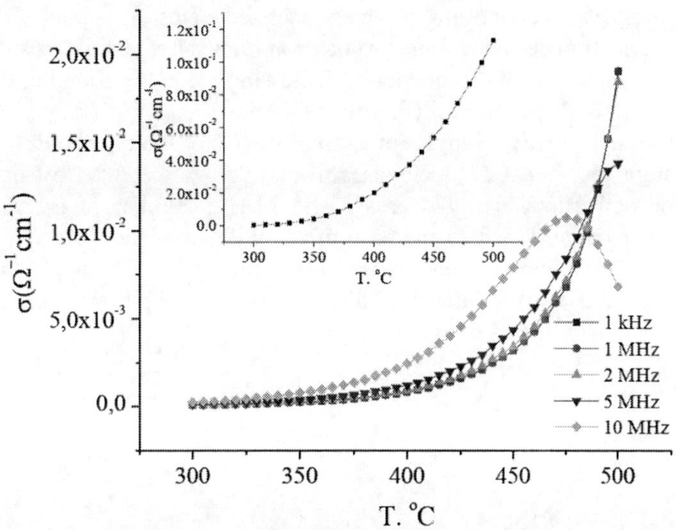


**Figure 3.** Temperature dependences of conductivity $\sigma_{dc}$ (inset) and $\sigma_{ac}$.

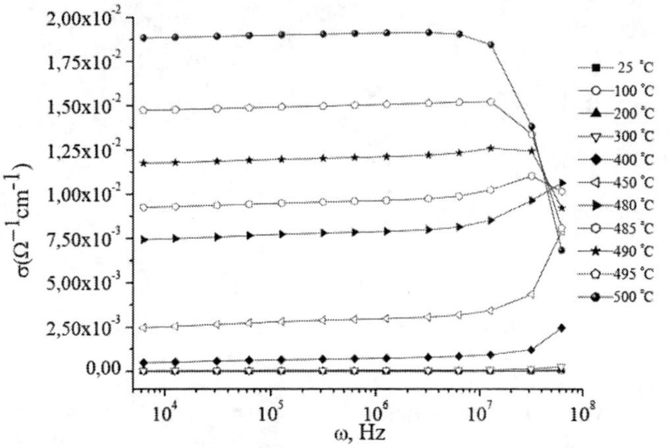


**Figure 4.** Frequency dependence of the conductivity $\sigma_{ac}$ of the BFO ceramic.

Both concentration and mobility of the oxygen vacancies increase when temperature increases above 350 °C. Closeness of this temperature to the Neel temperature (370 °C,) can be explained by increase of the oxygen vacancies concentration due to increase of volume under transition from an antiferromagnetic (ordered) to a paramagnetic (disordered) state. Frequency dependence of the conductivity of our ceramics, as well as in [13], is described by the double power law:

$$\sigma(\omega) = \sigma(0) + A\omega^{s1} + B\omega^{s2}, \qquad (2)$$

where σ(0) is the frequency-independent conductivity which can be taken as $\sigma_{dc}$ , $\omega = 2\pi\nu$ is the circular frequency, *A, B, s*1 and *s*2 are the coefficients depending on temperature and frequency, and the $B\omega^{s2}$ component appears at frequencies above 1 MHz. The power-law dependence σ(ω) indicates the hopping nature of the conductivity in BFO [14]. According to our data, frequency indices *s*1 and *s*2 decrease with increasing temperature. Moreover, the *s*2 sign undergoes an inversion above 480 °C. Such a behavior of the *s*1 and *s*2 indices corresponds to the correlated barrier hopping in the model for *ac* conductivity [14]. Estimate of the conductivity activation energy $E_A$ revealed that in the temperature range below 150 °C $E_A$ decreases from 0.46 eV to 0.042 eV with frequency increase from 1 kHz to 10 MHz (Figure 5). Probably, raising the applied field frequency promotes increase of attempt frequency of charge carriers hopping between localized states.

This means that a charge transfer over short distances at higher frequencies costs a less energy than the transfer over long distances at low frequencies. Following [15], the complex dependence $\ln\sigma = f(1/T)$ can be separated into three ranges (Figure 5). Above 350 °C (range III), the frequency dependence is weak, and $E_A$ is relatively high.For example, $E_A$ is of the order of 0.8 eV for frequencies above 1 MHz in the range 350-480 °C. $E_A$ decreases from ~ 0.7 eV to ~ 0.11 eV in the interval below 350 °C (range II). The activation energy decrease should rise number of carriers overcoming the barrier and, as a sequence, should lead to increase of $\sigma_{ac}$ with frequency. At a fixed frequency, the activation energy is higher for a higher temperature. Calculations according to [16] showed that the the minimum hop length increases with frequency at low temperatures while with increasing temperature it tends to zero. Such a behavior agrees with data in [16].

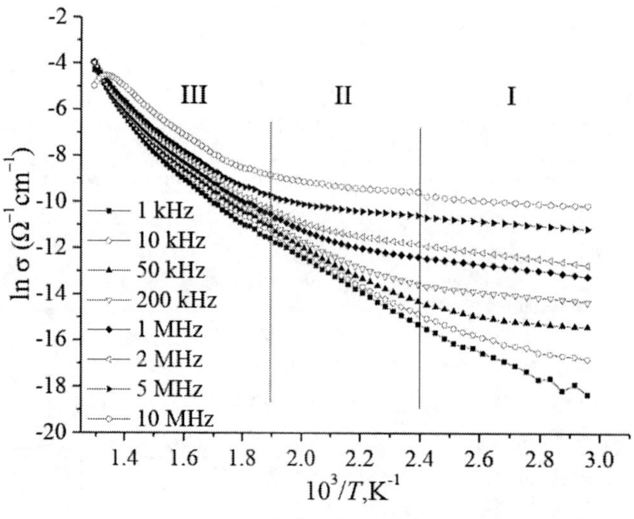


**Figure 5.** Temperature dependence of the conductivity in the coordinates $\ln\sigma = f(10^3/T)$

We conclude that several competing mechanisms, depending on applied field frequency, contribute into conductivity of the BFO nanoceramics when temperature raises. In the range of medium and high frequencies (> 100 kHz), the hopping mechanism with a short jump length is dominant. Both successful (forward) and unsuccessful (backward) jumps affect hopping conductivity with short-range hopping in the range of intermediate frequencies from 100 kHz to 1 MHz (the $A\omega^{s1}$ component). Increase of the successful jumps number strengthens the conductivity dispersion in this frequency range. Difference in conductivity of grains and boundaries should be taken into account in this case. With frequency increase, the contribution from in-grain conductivity becomes dominant due to the charge transfer between $Fe^{2+}$ and $Fe^{3+}$ ions [17]. The contribution from conductivity along the grain boundaries, where oxygen vacancies are located, is essential at relatively low frequencies.

The anomalous behavior of the high-frequency conductivity in the temperature range above $T^*$ can be explained if we take into account the relation between the hopping frequency $\omega_h$ and the applied alternating field frequency $\omega$. The $\sigma_{ac}$ value (Figure 4) increases with both temperature and $\omega$ passing through a maximum at frequencies higher than 1-2 MHz. As follows from Figure 4, $\omega_h$ is shifted towards low temperatures with $\omega$ increase, while this behavior contradicts to published data [18]. At high frequencies $\omega > \omega_h$, delay of the charge carriers hoppings relatively to the electric field change leads to the appearance of dielectric losses tan δ, which occurs at the onset of the conductivity decrease. As seen in inset of Figure 6, the $\sigma_{ac}$ relaxation is observed at $\omega >$ 100 kHz and at frequencies higher than 1 MHz it is sharply enhanced. In the frequency range 1–10 MHz, the temperature corresponding to $\sigma_{ac}$ maximum is shifted on about 25 °C. This effect is more pronounced for the *ac* losses, shift of the tgδ maximum with frequency reaches 45 °C.

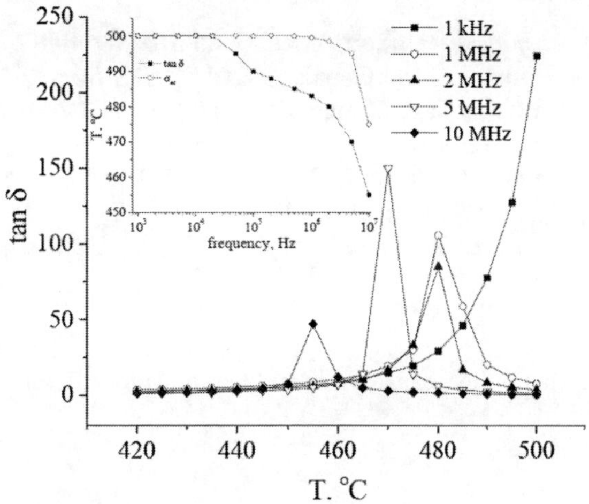


**Figure 6.** Temperature dependence of the dielectric loss tgδ at different frequencies. Inset: frequency dependence of the temperature at which the maxima of $\sigma_{ac}$ (open symbols) and tan δ (closed symbols) are observed.

The above results indicate that the short relaxation time of charge carriers is the main factor which determines frequency dependence of the conductivity ($B\omega^{s2}$) The gradual “shutdown” of the hopping mechanism of the charge redistribution between $Fe^{2+}$ ions to $Fe^{3+}$, along with the transport of oxygen vacancies, lead to a decrease in the electrical conductivity of BFO ceramics with frequency observed in the studied frequency range.

## 3. Conclusion

At low frequencies and temperatures, the conductivity of the $BiFeO_3$ ceramics is almost frequency independent. Above ~350°C it rises with frequency following to a power function, and the conductivity frequency dispersion strengthens with $\omega$. For $\omega >$ 1 MHz, $\sigma_{ac}(\omega)$ passes a maximum at a some temperature $T^*$. At a fixed temperature the conductivity activation energy decreases with frequency. Such a behavior of σ results from competition of various conduction mechanisms. Above

100 kHz, the conductivity behavior corresponds to the model of correlated barrier jumps of charge carriers between $Fe^{2+}$ and $Fe^{3+}$ ions.

**Acknowledgments**
The work was supported by the State tasks №№ 3.5982.2017/8.9 and 16.1103.2014 / K.

**References**

[1] Catalan G and Scott J 2009 *Adv. Mater.* **21** 2463
[2] Bernardo M 2014 *Bol. Soc. Esp. Ceram. Vidr.* **53**(1) 1-14
[3] Pyatakov A and Zvezdin A 2012 *J. Progress in Physical Sciences* **182** 593-620
[4] Bea H *et al.* 2005 *Appl. Phys. Lett.* **87**(7) 072508
[5] Lahmar, Zhao K, Habouti S, Dietze M and Solterbeck C 2011 *Solid State Ionics* **202**(1) 1-5
[6] Rao T and Asthana S 2014 *J. Appl. Phys.* **116** 164102
[7] Palaimiene E, Macutkevic J, Karpinsky D, Kholkin A., Banys 2015 *J Appl. Phys. Lett.* **106** 012906
[8] Ke H, Wang W, Wang Y, Zhang H, Jia D, Zhou Y, Lu X, Withers P 2012 *J. Alloys Compd.* **541** 94-8
[9] Alikhanov N, Palchaev D, Rabadanov M, MurlevaZh, Sadykov S and Emirov R. 2016 *Patent* 2016117571
[10] Rabadanov M, Palchaev D, Akhmedov Sh, Faradzheva M, MurlevaZh, Kallaev S and Sadykov S 2015 *Patent* 2556181
[11] Sitchai H, Prasit T, Teerapon Y, Rattikorn Y and Santi M 2009 *Appl.Phys.Lett.* **94**(6) 062904
[12] Kolte J, Salame P, Daryapurkar A and Gopalan P 2015 *AIP Advances* **5** 097164
[13] Ortega N, Kumar A, Bhattacharya P, Majumder S and Katiyar R 2008 *Phys. Rev. B* **77** 014111
[14] Ghosh A 1990 *Phys. Rev. B* **42** 1388
[15] Kumari S, Ortega N, Kumar A, Pavunny S, Hubbard J, Rinaldi C, Srinivasan G, Scott J and Katiyar Ram S 2015 *J. Appl. Phys.* **117** 114102
[16] Prasad K, Bhagat S, Amarnath K, Choudhary S and Yada K *Materials Science-Poland* **28**(1) 317
[17] Markiewicz E, Hilczer B, Blaszyk M, Pietraszko A and Talik E 2011 *J. of Electroceramics.* **27** 154-161
[18] Taher Y, Oueslati A, Maaloul N, Khirouni K and Gargouri M 2015 *Appl. Phys. A* **120** 1537

# Structure and conductivity of nanostructured YBCO ceramics

**D K Palchayev[1,*], S Kh Gadzhimagomedov[1], Zh Kh Murlieva[1,2], M Kh Rabadanov[1] and R M Emirov[1]**

[1]Dagestan State University, Gadzhieva 43-A, 367000 Makhachkala, Russia
[2]Dagestan State University of National Economy, Ataeva 5, 367008 Makhachkala, Russia

[*]E-mail: dairpalchaev@mail.ru

**Abstract.** Superconducting nanostructured ceramics based on $YBa_2Cu_3O_{7-\delta}$ were made of nanopowder obtained by burning nitrate-organic precursors. The structure, morphology, electrical resistivity, and density of ceramics were studied. Various porosity values of the ceramics were achieved by preliminary heat treatment of the nanopowder. The features of conductivity and the reason for increase of the of the superconducting transition temperature in these materials are discussed.

## 1. Introduction

The paper is devoted to the problem of establishing a connection between the structure of nanostructured ceramics based on $YBa_2Cu_3O_{7-\delta}$ (YBCO) compounds and the effect of increasing the superconducting transition temperature. The structure and morphology of the obtained nanopowder and nanostructured ceramics of various densities were studied on X-ray diffractometer "PANalytical Empyrean series 2", TEM "Titan 80-300". Measurements of the electrical resistivity of high-temperature superconductors (HTSC) were carried out by the four-probe method.

## 2. Results and discussion

### *2.1. Structure and morphology of nanostructured ceramics $YBa_2Cu_3O_{7-\delta}$.*

Nanostructured superconducting ceramic materials with different densities based on $YBa_2Cu_3O_{7-\delta}$ were obtained from nanopowders. Nanopowders were produced [1] by burning nitrate-organic precursors obtained after evaporation of an aqueous solution of yttrium, barium and copper nitrates with the addition of various amounts of glycerin. Variation of the solution viscosities and the precursor burning speed allows one to obtain predetermined particle sizes as well as size distribution in the powders which contain also some amount of glycerin. Ceramics with different porosity were provided from these powders via appropriate preliminary heat treatment of the nanopowders. An optimal oxygen content in the ceramics was reach due to the high specific surface of nanopowders used for their sintering.

Figure 1 shows the results of diffraction analysis for four samples of HTSC ceramics containing ~ 95% phase of $YBa_2Cu_3O_{6.9}$. These samples were obtained from two batches of powders. The first batch differs from the second one by more glycerin amount used for the powder producing. The batches were thermally treated at 350 °C for 1 hour (Figures 1a and 1c) and at 910 °C for 20 hours (Figures 1b and 1d), respectively. Four briquettes were sintered at 920 °C for 1 hour after pressing the powders from both batches at a pressure of 100 MPa. Density of the ceramics obtained from the powders ther-

IOP Conf. Series: Journal of Physics: Conf. Series **941** (2017) 012076 doi:10.1088/1742-6596/941/1/012076

mally treated at 350 °C (Figures 1a and 1c) was equal to 2.5 and 3.0 g/cm$^3$. Whereas the ceramics sintered from powders thermally treated at 910 °C (Figures 1b and 1d) have higher density of 6.0 and 5.7 g/cm$^3$. Figure 1e also shows the reference diffraction pattern of $YBa_2Cu_3O_{7-\delta}$ from the PAN-ICSD database. As 5% of the copper oxide crystalline phase is observed in the samples (Figures 1a and 1b) so they should also contain Y2BaCuO$_5$ phase in the X-ray amorphous state. Indeed, this phase was observed (211) after its crystallization under prolonged heat treatment. The $Y_2BaCuO_5$ nanoparticles uniformly distributed in the $YBa_2Cu_3O_{7-\delta}$ matrix produce pinning centers for the magnetic flux [2].

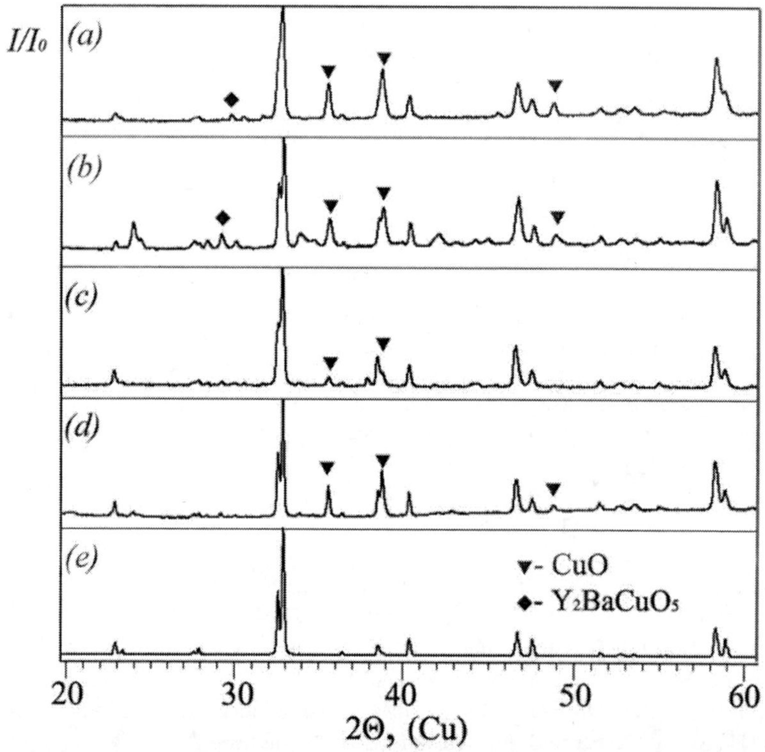


**Figure 1.** Diffraction patterns for ceramics made of two batches of thermally treated nanopowders. The first batch treated at 350 °C (a) and 910 °C (b). The second batch treated at 350 °C (c) and 910 °C (d). Reference $YBa_2Cu_3O_{7-\delta}$ structure from the database PAN-ICSD (e).

As shown in [3], both highly porous (2.5 g/cm$^3$) and dense (6.0 g/cm$^3$) nanostructured ceramics made from the first batch of nanopowder consist of micron grains formed by the nanoparticles. According to the diffraction data, an average nanoparticle size was about 46 and 53 nm respectively. Slightly larger particle size of about 53 nm and 65 nm, respectively, was obtained for nanostructured ceramics made from the second batch of nanopowder which had the density of 3.0 g/cm$^3$ and 5.7 g/cm$^3$ [1].

The morphology inside the grains of nanostructured ceramics of 3.0 and 5.7 g/cm$^3$ density are respectively shown in Figures 2a and 2b. The TEM have revealed presence of nanoscale defects in the nanocrystalline structure of YBCO grains, which are formed as a result of a "topological" transition from the nanoparticle agglomerate in the original powder to the crystalline grains consisting of nano-sized domains and containing defects. Nano-domains in micro-sized grains are oriented chaotically. Defects are distortions of the crystal lattice at the atomic level. As seen in Figures 2a and 2b, the structure defectiveness of the grains at the atomic level is higher for ceramics with lower density. Similar results were obtained for the ceramics made from both batches of the nanopowder.

Reducing the size of particles and increasing the role of surface tension forces obviously should lead not only to the structure distortion at the atomic level, but also affect other properties of the ceramics.

**Figure 2.** Morphology inside grains of the nanostructured $YBa_2Cu_3O_{7-\delta}$ ceramics: (a) porous, (b) dense. The insets show electron diffraction patterns for corresponding grains

*2.2. Electrical resistivity of nanostructured $YBa_2Cu_3O_{7-\delta}$ ceramics having various porosity.*

According to data given in [1,3], transition to the superconducting state begins at temperature $T_c$ of about 96 K in samples from batches of dense and porous ceramics. Values of electrical resistivity ρ of ceramics with greater porosity are an order of magnitude higher. The temperature coefficient of resistivity $\partial\rho/\partial T$ is lower and the transition is stretched for these ceramics. That apparently is associated with a smaller contact area between the grains. As shown in [1,3] samples from both batches of dense and porous ceramics have the same size of the grains. Higher values of electrical resistivity for porous ceramics are caused not only by a smaller contact area between the grains, but also by a higher defectiveness at the atomic level of the grains themselves (Figure 2a). Size of agglomerates and nanoparticles in the nanopowders from both batches, as well as size of grains in porous and dense ceramics are approximately the same. Therefore we can expect also the same values for transition temperatures $T_c$. In this case, the $T_c$ increase up to ~ 96 K in the nanostructured ceramics appears to be caused by decrease of the crystalline-domain size by about twenty times in comparison with that for ordinary ceramics having micro-sized grains and $T_c$ = 92 K. The decrease of the nanoparticles size leads to a corresponding increase in the interatomic interaction forces. As known [4], YBCO is characterized by loose packing and negative thermal expansion. Therefore, an increase of the interatomic interaction forces averaged over the lattice can compensate growth of the interatomic distances due to the thermal expansion and lead to the ordering of the oxygen ions in the *b* direction of the unit cell in a wider (up to ~ 96 K) temperature range.

Figures 3 and 4 show the experimental data on temperature dependences of the electrical resistivity and their temperature coefficient for both batches of dense and porous ceramics. As seen, the onset of the superconducting transition occurs at a temperature of about 96 K.

According to the location of the most intense peak, the average transition temperature $T_c$ for the batch with a larger particle size (from 65 to 53 nm) is about 95.5 K (see Figure 4), while for the batch (Fig. 3) with a smaller particle size (from 53 nm to 46 nm) it varies from 92.6 to 94.3 K. At the same time, as seen in Figures 3 and 4, there is a tendency to reduction of average $T_c$ values when size of the nanoparticles decreases from 65 to 46 nm.

Several peaks in the $\partial\rho/\partial T$ dependencies indicate presence of phases with different values of oxygen index. As known [5], presence of superconducting phases with different $T_c$ is associated with precipitates of extrinsic nanostructured phases of different stoichiometry [6-8] and a violation of the cationic composition of YBCO.

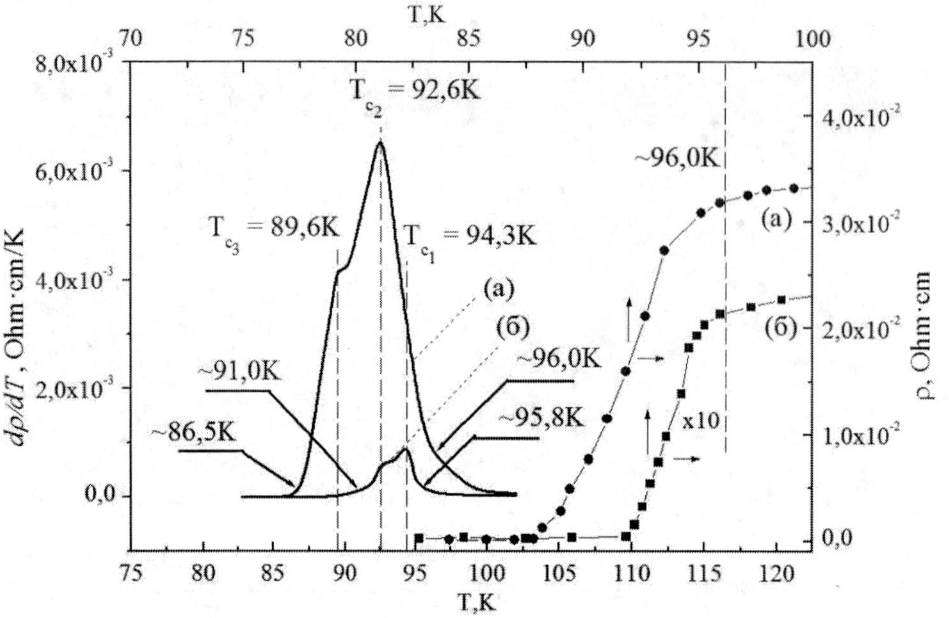


**Figure 3.** Temperature dependences of the electrical resistivity and the temperature coefficient of electrical resistivity for porous (a) and dense (b) ceramics made of nanopowder from the first batch. Resistivity of dense sample (b) is multiplied by ten.

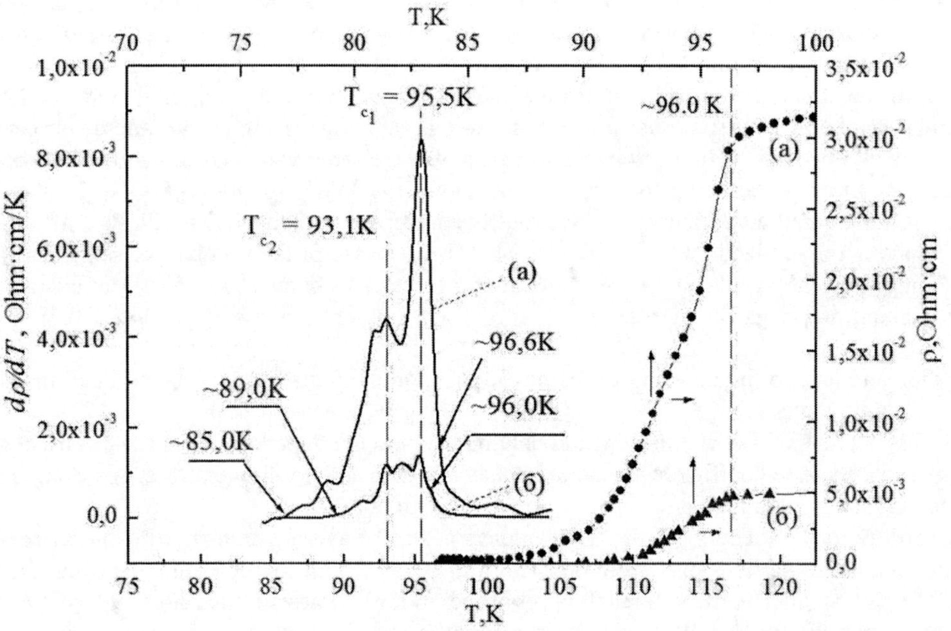


**Figure 4**. Temperature dependences of the electrical resistivity and the temperature coefficient of electrical resistivity for porous (a) and dense (b) ceramics made of nanopowder from the second batch.

## 3. Conclusion

Developed chemical technology of production and heat treatment of nanopowders allow us to produce nanostructured $YBa_2Cu_3O_{7-\delta}$ ceramics with predetermined density and optimal oxygen content in one step process. Increase of the transition temperature up to ~ 96 K in these ceramics is apparently caused by decrease of particle size and a corresponding increase of the interatomic interaction forces. At the same time, when the dimensions of the nanoparticles decrease below ~ 50 nm, there is a tendency to reduction of average $T_c$ values.

## Acknowledgments

The work was supported by the State tasks № 2560, 3.5982.2017/8.9 and 16.1103.2014/K

## References

[1] Gadzhimagomedov S, Palchaev D, Rabadanov M, Murleva Zh, Shabanov N, Palchaev N, Murlev E and Emirov R 2016 *J. Technical Physics Letters* **42**(1) 4–7

[2] Pavan Kumar Naik S, Missak Swarup Raju P, Rajasekharan T and Seshubai V 2014 *J. Supercond. Nov. Magn.* **27** 1211

[3] Palchaev D, Murlieva Zh, Rabadanov M, Gadzhimagomedov S, Alikhanov N, Emirov R and Asvarov A 2016 *Journal of Physics: Conference Series* **747** 012040

[4] Titova S, Lukoyanov A, Pryanichnikov S, Cherepanova L, Titov A 2016 *J. of Alloys and Compounds* **658** 891–7

[5] Klinkova A and Nikolaichik V 2014 *Bulletin of the Russian Academy of Sciences. Physics* **78**(8) 751–3

[6] Duarte E A, Rudawski N G , Quintero P A, Meisel M W and Nino J C 2015 *Supercond. Sci. Technol.* **28** 015006

[7] Rejith P P, Vidya S and Thomas J K 2015 *IOP Conf. Series: Materials Science and Engineering* **73** 012146

[8] Hannachi E, Slimani Y, Ben Salem M K, Hamritaa A, Mania D K, Ben Salema M and Ben Azzouza F 2015 *Materials Chemistry and Physics* **159** 185-93

IOP Conf. Series: Journal of Physics: Conf. Series **941** (2017) 012077 doi:10.1088/1742-6596/941/1/012077

# Investigation of semiconductor heterostructures HgTe/HgCdTe under magnetic fields up to 50 T

**V V Platonov, Yu B Kudasov, I V Makarov, D A Maslov and O M Surdin***

Sarov State Physics and Technical Institute, National Research Nuclear University MEPhI (Moscow Engineering Physics Institute), 607186 Sarov, Russia
National Research Nuclear University MEPhI (Moscow Engineering Physics Institute), Kashirskoe sh. 31, 115409 Moscow, Russia
Russian Federal Nuclear Center – VNIIEF, pr. Mira 37, 607188 Sarov, Russia

*E-mail: mossom1@rambler.ru

**Abstract**. Our paper is devoted to a compact economic magnetic facility, which is equipped by a 300 kJ capacitor bank. The facility is designed for the exploration of magnetic and optical substance properties under magnetic fields up to 50 T and under cryogenic temperatures. The cyclotron resonance and quantum Hall effect measurement results are presented for semiconductor heterostructures HgTe/HgCdTe with quantum wells.

## 1. Introduction

High magnetic fields are an important instrument for a new material research and searching new effects. A majority of recent discoveries are connected with high magnetic fields. The generation of such fields is a difficult technical problem [1,2]. The pressure on a solenoid shell reaches the ultimate breaking strength of basic constructional materials at the magnetic induction of 50 T. This limit can be overcomed by a pulse time reduction or by a load distribution in a compound sectional solenoid [1,3]. The increase of sections quantity leads to a solenoid size increase and magnetic energy growth. High-energy sources (50-100 MJ) are necessary, they are available for small amount of large magnetic laboratories [4]. Record magnetic fields up to 2800 T [5] can be generated in several laboratories with explosive techniques.

The majority of high magnetic fields experiments refer to solid state physics: investigation of upper critical magnetic field in high-temperature superconductors, magnetic phase transitions, magnetic resonance phenomena and etc. One of the most efficient methods of semiconductor zone structure research is a cyclotron resonance (CR). Its observation is possible when a cyclotron frequency exceeds a carrier scattering frequency. This demands high quality samples or high magnetic fields. A sample quality in new technologies is usually not high. Sample parameters measuring are important for the determination of the technology development prospect. That is why, higher magnetic fields allow samples of lower quality.

## 2. Experimental facility

A wire-wound coil with the internal diameter of 20 mm and external diameter of 80 mm is developed for the generation of magnetic fields up to 50 T. A small coil size leads to a high magnetic energy density and appearance of giant pressures on coil shell up to 1-1.2 GPa. Usual copper or steel wires fail under such load. A microcomposite Cu-Nb wire with a cross-section of 2×3 $mm^2$ developed by VNIINM Bochvar was used for coil winding. The wire is characterized by a high elastic modulus and giant breaking strength more than 1.3 GPa.

Special technological equipment [6] is developed for a coil winding by a rigid wire. It allows winding with a gradual increase of the wire tension according to the layer radius increase. This leads to the production of the preliminary mechanical tension in a coil [7]. The wire is covered by a glasscloth isolation before winding. The wire coil is reinforced by a bandage of 20-30 mm from high-tensile fiber ARMOS. The coil is impregnated by an epoxy compound with a powdered quartz filler, and is dried layer-by-layer.

The electrical circuit of the facility is thoroughly described earlier [8]. One of the main component of the facility is a commutator based on the reversely connected dynistors. The advantage of semiconductor commutators is the absence of electric noise at the beginning of a pulse. The dynistors can endure high current overloads. At the nominal conditions the current is equal to 30 kA with 10 ms duration and can reach 90 kA with 2 ms duration of a pulse. Such overload ability is useful in a breakdown when the facility approaches a short circuit regime.

**Figure 1.** Coil immersed into the plastic cryostat filled by the liquid nitrogen.

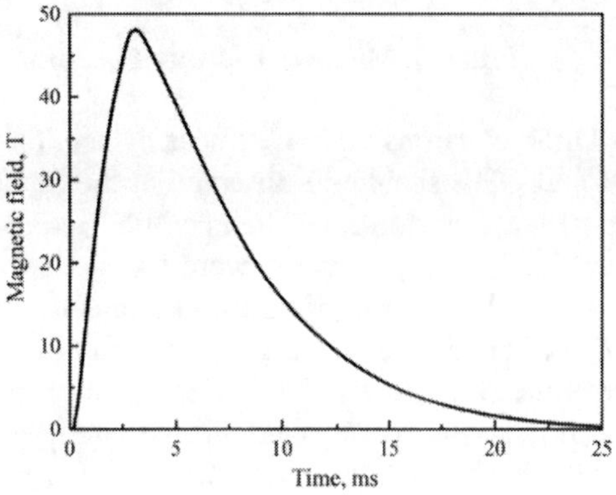


**Figure 2.** Time dependence of the magnetic flux density for this coil.

The coil is preliminary cooled by the liquid nitrogen in the plastic cryostat (Figures 1, 2). The vacuum optical channel is placed along the coil axis and extends over the cryostat. The sample is attached to the heat-conducting ceramic holder in the coil center. On the holder a thermocouple and induction sensor are also mounted. The thermoregulator is used to set up a sample temperature in the range of 77-250 K. Optical windows are fixed at the ends of a vacuum channel, and they are used for optical measurements. They are carried out from a cooling region and provide a free transmission of the visible and infrared probing radiation. A different radiation sources and sensors can be used during the magnetic absorption investigation.

The sample cooling up to a liquid helium temperature is performed in a flow cryostat. The liquid helium is delivered under the pressure from the transport Dewar vessel. The sample is cooled by the helium vapor or by immersing into the liquid helium. The cryostat is used for measurements of the magnetoresistance and quantum Hall effect (QHE).

## 3. Experiment results and discussion

The results of the cyclotron resonance and QHE measurements are discussed for the semiconductor heterostructures $HgTe/Cd_xHg_{1-x}Te$. These structures have remarkable properties caused by the invert band and spin-orbital coupling [9]. The band-gap energy tends to zero for the heterostructures $HgTe/Cd_xHg_{1-x}Te$ with a critical width of quantum well (QW) ($d_c$ = 6.3 nm, $x$ = 0.7) and the energy dependence of the charge carrier on the wave vector becomes linear for the valence and conduction bands [9,10], similarly to the graphene spectrum.

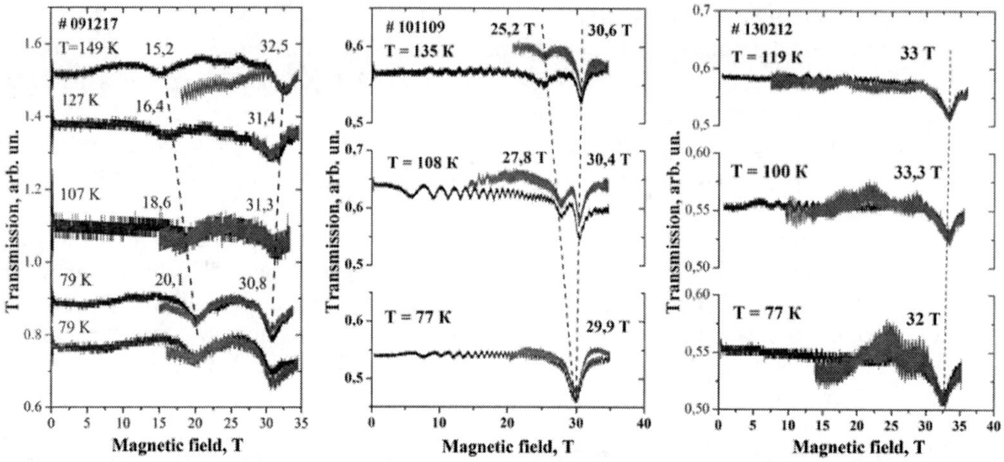


**Figure 3.** Magneoabsorption oscillograms for the samples under different temperatures.

Dirac electrons with the linear dispersion law should be light with a high mobility. Consequently, CR and QHE should be observed at the higher temperatures than those for the common semiconductors. First liquid nitrogen temperature experiments have been carried out in Refs. [11,12]; and two lines of magnetoabsorption were observed with a wavelength of 10.6 μm. These lines survive till 200 K. The magnetoabsorption oscillograms for the samples with a different width of QW are presented in Fig. 3. Our samples were grown by a molecular beam epitaxy (MBE) on semi-insulating GaAs (013) substrates [13]. A CdTe buffer, ~ 40 nm lower $Cd_xHg_{1-x}Te$ barrier, HgTe QW, and ~ 40 nm $Cd_xHg_{1-x}Te$ top barrier were grown one by one. A 40 nm CdTe cap layer was also grown above the structure. The samples parameters are presented in the Table 1.

**Table 1.** Samples parameters.

| № | Quantum well-width, nm | Concentration of Cd | Carrier concentration, $cm^2$ | Carrier mobility, $cm^2/V*s$ |
|---|---|---|---|---|
| 091217 | 7 | 0.72 | 2.2 | 50000 |
| 091223-1 | 8 | 0.62 | 1.5-1.6 | 20000 |
| 101109 | 8 | 0.77 | 4.5 | 85000 |
| 130212 | 8.3 | 0.78 | 10.2 | 69000 |

The location of one of the lines is temperature independent. The location of another line has a strong temperature dependence. To interpret the experimental results, the temperature-dependent band structure and Landau level calculations based on the eight-band *kp* Hamiltonian for (013)-oriented heterostructures HgCdTe were performed in Ref. [14,15].

It can be assumed that one of the lines is associated with an interband transition, and its location is determined by a band-gap energy, which depends on the temperature. The second line is associated with an intraband transition, that is why, its temperature dependence is weak. The band structure of the heterostructure with QW ~8 nm changes from the invert to normal structure near T = 50 K, when the temperature rises. The detailed analysis of the temperature dependence of the magnetoabsorption lines is presented in Ref. [16].

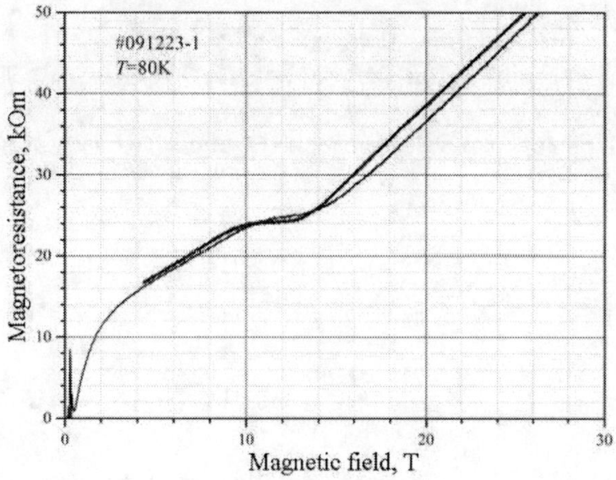


**Figure 4.** Longitudinal resistance measured at the liquid nitrogen temperature.

## 4. Conclusions

The results of the magnetoabsorption investigation and CR of heterostructures HgTe/HgCdTe with QW in pulsed magnetic fields up to 40 T are presented for a wide temperature region. The intraband transition (cyclotron resonance) and interband transition (from the valence band Landau level to the conduction band) are observed. The quantum limit of the longitudinal resistance QHE is registrated (Figure 4). The transition from the invert band structure to the normal one is observed for the structures with QW ~8 nm.

## Acknowledgments

This work was partially supported by Russian Foundation of Basic Research (Project No. 16-08-00337).

## References

[1] Knoepfel H 1970 *Pulsed High Magnetic Fields* (Amsterdam: North–Holland)
[2] Lagutin A S, Ozhogin V I 1988 *Sil'nyeimpul'snyemagnitnyepolya v fizicheskomeksperimente* (*Strong Pulsed Magnetic Fields in Physical Experiment*) (Moscow: Energoatomizdat)
[3] Date M 1975 *J. of Phys. Soc. of Jap.* **39** 89
[4] The National High Magnetic Field Laboratory www.magnet.fsu.edu
[5] Boriskov G V, Bykov A I, Dolotenko M I, Egorov N I, Kudasov Yu B, Platonov V V, Selemir V D, Tatsenko O M 2011 *Physics-Uspekhi* **54** 421
[6] Kudasov Yu B, Popov E Ya, Popova T I, Makarov I V, Sirotkina A G, Evlanov K I 2013 *Byull. Izobret.* **30** RF Patent 2497218
[7] Koblova O N, Kudasov Yu B 2009 *Tech. Phys.* **35** 34
[8] Kudasov Yu B, Makarov I V, Maslov D A, Platonov V V, Popov E Ya, Surdin O M, Voronov S L, MalyshevA Yu, KorotkovS V,. VodovozovV M 2015 *Instruments and Experimental Techniques* **58** 781
[9] Konig M, Wiedmann S, Brune C, Roth A, Buhmann H, Molenkamp L W, Qi X-L, Zhang S-C 2007 *Science* **318** 766
[10] Buttner B, Liu CX, Tkachov G, Novik E G, Brune C, Buhmann H, Hankiewicz E M, Recher P, Trauzettel B, Zhang S C, Molenkamp LW 2012 *Nature Physics* **8** 491
[11] Platonov V V, Kudasov Yu B, Filippov A V, Makarov I V, Maslov D A, Surdin O M 2015 *IEEE Trans. Plasma Sci.* **43** 365
[12] Platonov V V, Kudasov Yu B, Makarov I V, Maslov D A, Surdin O M, Zholudev M S, Ikonnikov A V, Gavrilenko V I, Mikhailov N N, Dvoretskii S A 2015 *Semiconductors* **49** 1611

IOP Conf. Series: Journal of Physics: Conf. Series **941** (2017) 012077 doi:10.1088/1742-6596/941/1/012077


[13] Dvoretsky S, Mikhailov N, Sidorov Y, Shvets V, Danilov S, Wittman B, Ganichev S, 2010 *J. Electron. Mater.* **39** 918

[14] Zholudev M *et al.* 2012 *Phys. Rev. B* **86**, 205420

[15] Krishtopenko S S, Yahniuk I, But D B, Gavrilenko V I, Knap W, Teppe F, *arXiv*:1607.03083 [condmat.mtrl-sci]

[16] Ikonnikov A V, Krishtopenko S S, Drachenko O, *et al.* 2016 *Phys.Rev.B* **94** 155421

IOP Conf. Series: Journal of Physics: Conf. Series **941** (2017) 012078 doi:10.1088/1742-6596/941/1/012078

# Effect of picosecond laser exposure on the transport characteristics of second-generation HTS tapes

**S V Pokrovskii, O B Mavritskii, A N Egorov, N A Mineev, A A Timofeev, I A Rudnev**

National Research Nuclear University MEPhI (Moscow Engineering Physics Institute), Kashirskoe sh. 31, 115409 Moscow, Russia

E-mail: sergeypokrovskii@gmail.com

**Abstract.** The main purpose of the work is to improve the magnetic and transport characteristics of 2G HTS tapes by creating an ordered array of artificial pinning centers. Using a picosecond laser exposure, a local modification of the HTS film were performed. The pulse energy varied from 300 to 4000 nJ, the pulse duration was 25 ps. A triangular array of micron size defects with a period of 50 μm was created. The dependences of the critical current and *n*-value change on the pulse energy were obtained. The parameters of the exposure resulting in an increase of the HTS tape critical current and tune of the *I-V* characteristic slope were found.

## 1. Introduction

The work is devoted to the investigation of the modification of high-temperature superconductor (HTS) $GdBa_2Cu_3O_{7-\delta}$ thin film using picosecond laser exposure. The main goal is to increase the magnetic and transport characteristics of the second-generation (2G) HTS tapes by creating an artificial pinning centers array. There are many papers describing researches and methods of artificial pinning centers creation in superconducting materials in literature. Most of them is devoted to the study of bulk superconductors or thin films on the single-crystalline substrates. In that cases an array of pinning centers is obtained by drilling the superconducting material, by irradiation or by addition of nonsuperconducting nanopowders [1-3]. The typical size of one element is of the order of several nanometers. These techniques have a number of disadvantages such as difficulty to use for industrial tapes on metal substrates, high cost and inability to scale on long-length tapes. There are also works on the use of laser action to create pinning centers to fix the vortex structure in the superconductor [4,5]. However, in most studies low temperature superconductors (for example Nb) have been used as samples materials. The developed method allows us to perform a controlled change in the parameters of industrial HTS tapes and locally improve the current-carrying characteristics of the material.

## 2. Experimental method

While selecting the laser equipment for the experiment, the following considerations were taken into account. It is well-known, that the precise lasers micromachining is closely related to the use of laser radiation pulses with durations shorter than the time of electron-phonon relaxation. This time is material-dependent and has the order of 0.1-20 ps [6]. So, pico- or even femtosecond lasers are required for the processing with the minimal thermal load to the material.

The diameter of holes in the HTS film should be as small as possible (ideally tens of nanometers). For the TEM00 lasers working in the visible or near-infrared wavelength range, the respective focused beam diameter can be an order of wavelength, i.e. about 1 μm. To tightly focus the laser beam into the smallest spot, the high numeric aperture (NA) objective with minimum aberrations for the working wavelength must be used. The laser beam quality $M^2$ is also very important parameter, guaranteeing the beam distortions are small enough to get the best results.

The micromachining application requires the precise adjustment of the laser pulse energy, allowing to control the regime of material processing. For example, by slightly varying the energy near the threshold of the material damage, it is possible to get not a through hole, but a crater, providing specific defects that may exhibit some interesting properties. Also, the laser energy stability from pulse to pulse should be good enough to reduce the energy adjustment error.

It is desirable to position the focused laser beam on the HTS film of various sizes (up to 12 mm width) with highest possible accuracy. Typical accuracy values for commercially available translation stages are about 1 μm or better with travel ranges of 25…100 mm.

According to the mentioned above requirements, the experiments were made using FEMTO-T laser facility designed in NRNU MEPhI [7]. Figure 1 presents the general configuration of this facility, the main part of which is the AVESTA's MPAP 2500 laser source. The output pulse frequency can be set from single shot with external firing (pulse-on-demand) up to 100 Hz.

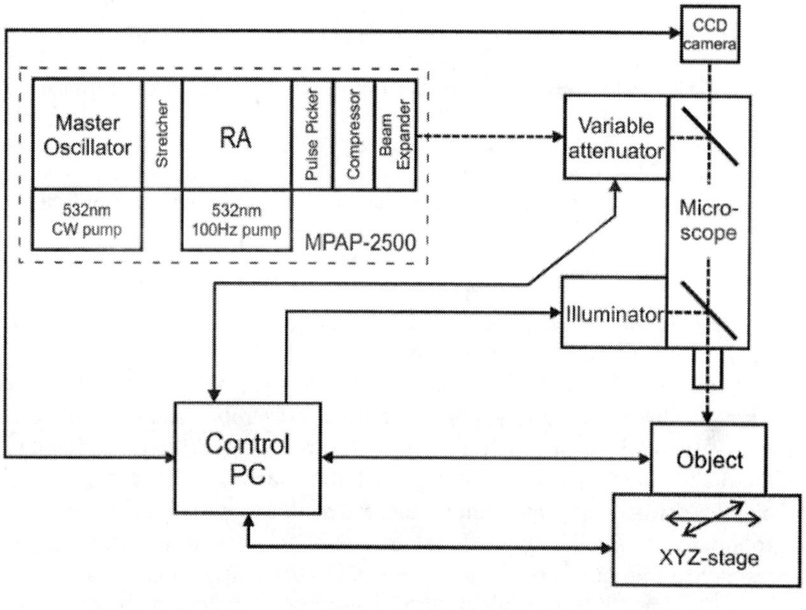


**Figure 1.** Schematic diagram of FEMTO-T laser facility.

Femtosecond pulses of low energy are generated by TiF-50 Ti:Sapphire master oscillator, pumped with second harmonic of built-in CW diode-pumped Nd3+:YAG laser. From the output of master oscillator, the pulses pass to the stretcher, which increases their duration up to 200 ps for further amplification. The amplification subsystem consists of regenerative amplifier (RA), combined with pulse picker, producing the output pulses with up to several hundreds of μJ with controlled frequency or in single shot mode. To pump the RA the additional frequency-doubled Q-switched diode-pumped Nd3+:YAG laser is used. After amplification, the output pulses' duration is reduced in tunable compressor to the value necessary for specific application. Beam expander, located at the output of laser source, matches the laser beam diameter to the input pupil of the focusing unit.

Laser pulses from the laser source output travel through the variable attenuator to focusing unit. The maximum attenuation coefficient is as much as $10^5$, allowing to precisely adjust laser pulse energy on the object from tens of pJ to several μJ.

The focusing unit is a high-resolution optical microscope equipped with an input laser port, LED illuminator and a set of large working distance infinity corrected NIR objectives from Mitutoyo. The CCD camera, attached to the microscope, produces the images of the object surface and helps to target the focused laser beam. The minimum laser spot diameter on object surface is about 1.2 μm, that is very close to the optical diffraction limit for the working wavelength.

The three-dimensional translation stage is used to precisely move the object relative to the focused laser beam. Its maximum travel range in both horizontal (X and Y) directions is 100 mm, while in vertical (Z) direction travel range is 25 mm, with minimum step of 0.15 μm. Motion control is performed by step-motor controller connected to PC via USB interface. The same PC is also used to control laser source, variable attenuator, microscope illuminator and CCD camera.

**3. Results and discussion**

In this paper we used the second generation industrial HTS tapes of 12 mm width manufactured by SuperOx, Russia. Samples were pre-selected and prepared for study. First of all, silver layer has been partially removed from the samples, which allowed to perform laser exposure directly on the superconducting film. In addition, since the critical current of the tapes exceeded 300 A at 77 K in self field (SF), it was dangerous to perform transport current measurements and don't burn the samples without silver layer. We divided the tapes into 2 pieces and the samples critical current in the pieces waas dropped to a value little less than 150 A. The critical current measurements were carried out before and after laser exposure in the same measurement conditions. To test the absence of degradation one sample was not subjected to modification and passed the whole set of measurements.

Using a picosecond laser exposure, a local triangular array of defects of ~ 1 μm size with a period of 50 μm was created in the samples. We used single pulse regime. Figure 2 shows a SEM photography of the modified area. When the radiation energy is increased from 300 to 4000 nJ, the defect size also increases from 3 to 16 μm. In addition, the elemental composition in the melt region was investigated. There were no significant changes in stoichiometry in comparison with the initial material (Figure 3).

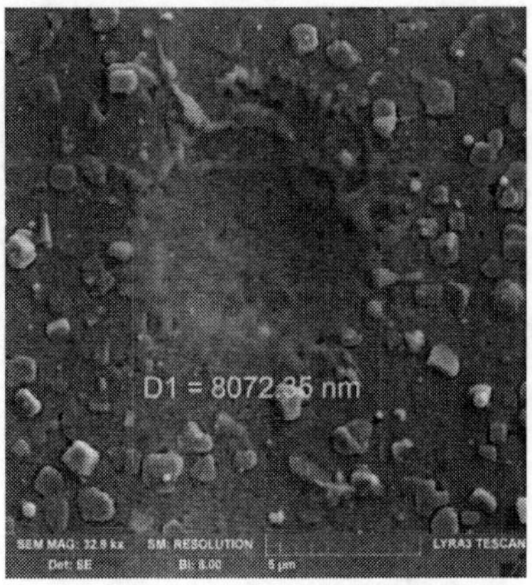


**Figure 2.** SEM photography of the modified area, E=500 nJ.

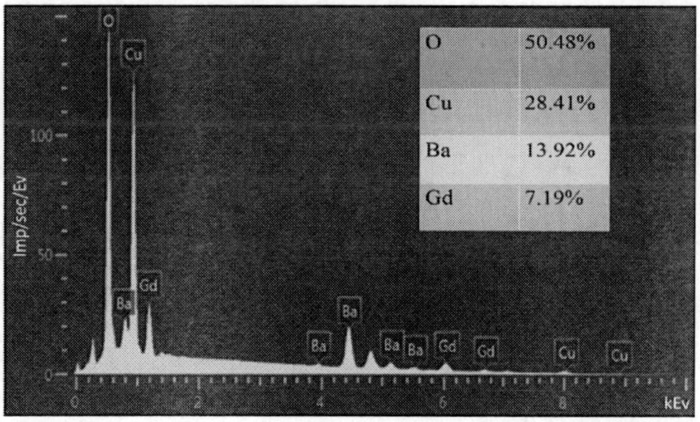


**Figure 3.** Elemental composition in the center of the irradiated region.

We measured the current-voltage characteristics of the samples before and after the laser exposure (Figure 4) and used a typical approximation of the $I$-$V$ characteristic using the power law function. Dependences of the critical current and the $n$-value of the $I$-$V$ characteristics on the energy of the laser pulse were obtained (Figure 5).

The obtained dependences have a domed shape that indicates the existence of optimal pulse parameters for the maximum increase the critical current. In our work we have found two special points: 500

nJ (sample 5, single defect size - 8.072 μm) and 2000 nJ (sample 7, single defect size - 14.024 μm). At that parameters we achieved almost the same enhancement of the critical current, but sharpness of the *I-V* curve changed in different way.

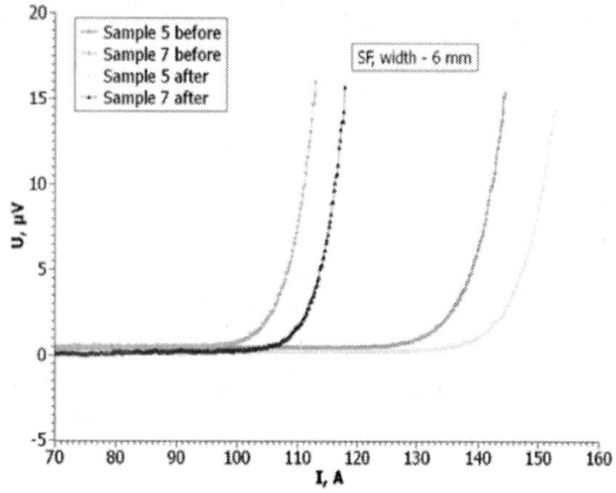


**Figure 4.** I-V characteristics of HTSC tapes before and after laser exposure.

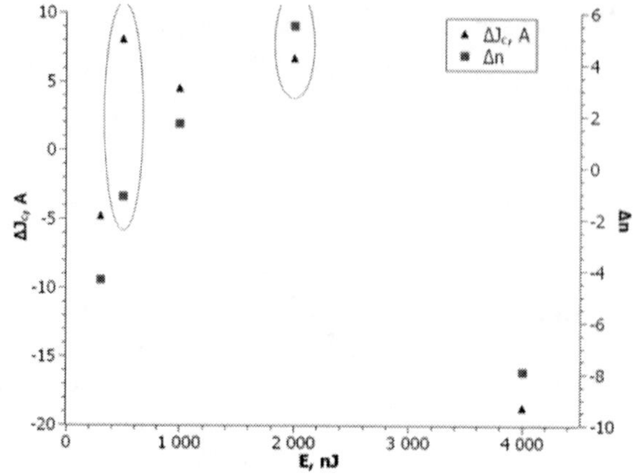


**Figure 5.** Dependences of the critical current (criterion 1 μV/cm) and the *n*-value of the *I-V* characteristics on the laser pulse energy.

This behavior can be explained by a change in the pinning force of the material in the region of the laser exposure. We created defects of large size that can't play role of pinning centers but at the edges could be generated cracks or point defects that pin vortex system. On the other hand the edge of a big defects itself is a good pinning center. This leads to an increase the critical current and a change in the *n*-value. But since we damage the film material the effective width of the superconducting tape reduces and the critical current also decreases. Thus, it is possible additionally to tune slope of the *I-V* curve. In more detail these questions will be considered in further works.

**Acknowledgments**

Research was done with the financial support of RSF under grant № 17-19-01527

**References**

[1] Noudem J G, Meslin S, Horvath D *et al* 2007 *Physica C* **463–465** 301–7
[2] Swiecicki I, Ulysse C, Wolf T *et al* 2012 *Phys. Rev. B* **85** 224502
[3] Palau A, Monton C, Rouco V *et al* 2012 *Phys. Rev. B* **85** 012502
[4] Wang Y L, Latimer M L *et al* 2012 *Phys. Rev. B* **87** 22050
[5] Welling Marco S, Wijngaarden Rinke J *et al* 2004 *Physica C* **404** 410–4
[6] Breitling D, Ruf A, Dausinger F 2004 *Proc. SPIE* **5339** 49–63
[7] Gordienko A V, Mavritskii O B, Egorov A N, Pechenkin A A, Savchenkov 2014 *Quantum Electronics* **44**(12) 1173–8

IOP Conf. Series: Journal of Physics: Conf. Series **941** (2017) 012079 doi:10.1088/1742-6596/941/1/012079

# Features of formation and evolution of crystal and local structures in nanocrystalline $Ln_2Zr_2O_7$ (Ln = La - Tb)

**V V Popov[1,2], A P Menushenkov[2], B R Gaynanov[2], Ya V Zubavichus[1], R D Svetogorov[1], A A Yastrebtsev[2], A A Pisarev[2], L A Arzhatkina[3], K V Ponkratov[4]**

[1]NRC “Kurchatov Institute”, pl. Akademika Kurchatova 1, 123182 Moscow, Russia
[2]National Research Nuclear University MEPhI (Moscow Engineering Physics Institute), Kashirskoe sh. 31, 115409 Moscow, Russia
[3]JSC “Scientific Research Institute of Chemical Technology”, Kashirskoe sh. 33, 115309 Moscow, Russia
[4]Renishaw plc, Kantemirovskaya st.58, 115477 Moscow, Russia

E-mail: BRGaynanov@gmail.com

**Abstract.** The formation of nanocrystal powders of $Ln_2Zr_2O_7$ (Ln=La - Tb) compounds upon calcinations up to 1500°C has been investigated by using of X-ray absorption fine structure spectroscopy combined with X-ray powder diffraction and Raman spectroscopy. It was found that the fluorite-pyrochlore phase transition occurs at calcination temperature in the range of 900-1200 °C (except Tb, T~1500 °C). The appearance of the ordered pyrochlore structure at calcination temperatures above 900 °C was fixed as an observation of superlattice peaks in diffraction and splitting of the first oxygen coordination shell in the local lanthanides environment in EXAFS and Raman spectra. In addition the scheme of fluorite-pyrochlore transition temperature dependence on radii of lanthanide cations was constructed.

## 1. Introduction

Rare-earth zirconates $Ln_2Zr_2O_7$ (Ln=La-Tb) have received considerable attention in recent years [1-5]. These systems contain several phases: pyrochlore (*Fd-3m*), fluorite (*Fm-3m*) types of crystalline structure and a phase with rhombohedral structure (*R-3*), formulated as $Ln_4M_3O_{12}$ [6]. The thermodynamic stability of the pyrochlore type compounds can be determined using empirical rule, which identifies the effective radii ratio $\gamma = r_{Ln3+}/r_{M4+}$ of REE and metal ions ratio as $1.46 < \gamma < 1.78$ [1]. Earlier we investigated and refined the temperature dependence of the fluorite-pyrochlore phase transition in a “boundary” zirconates $Dy_2Zr_2O_7$ ($\gamma = 1.43$) [7-8], $Gd_2Zr_2O_7$ ($\gamma = 1.46$) [9] and $Pr_2Zr_2O_7$ ($\gamma = 1.56$) [10].

The aim of the present work is to study the crystallization processes of an amorphous precursor upon isothermal calcination and to study the specific features of fluorite - pyrochlore phase transition in $Ln_2Zr_2O_7$ (Ln=La-Tb) ($\gamma$=1.44-1.61), using combined X-ray diffraction (XRD), X-ray absorption fine structure spectroscopy (EXAFS) at synchrotron radiation and Raman spectroscopy.

## 2. Experimental section

The starting materials were $Ln(NO_3)_3{\cdot}6H_2O$ (more than 99,8% pure), $ZrOCl_2{\cdot}8H_2O$ (99% pure). Precursor (mixed Ln-Zr hydroxide) was prepared by co-precipitation of metal salts solution with ammonia $NH_3{\cdot}H_2O$ (analytical grade). The $Ln_2Zr_2O_7$ powders were prepared by isothermal calcination of amorphous precursors at 600-1500°C for 3 h in air. A more detailed experimental procedure was described in Refs. [11-14].

X-ray diffraction investigations were carried out at the "Structural Materials Science" beamline of the Kurchatov synchrotron radiation source. The measurements were implemented in the Debye-Scherrer geometry (transmission mode) using an environment optimized for high quality experimental data: the distance the sample - detector - 200 mm; beam size 200×200 mm; the exposure time of 15-30 minutes, the wavelength $\lambda$ = 0.68886 Å. Six series of $Ln_2Zr_2O_7$ compounds with Ln=La-Tb were examined. Each series included 8-10 samples obtained by calcination of initial precursor at different temperatures in a temperature range 600 – 1500°C. The Rietveld full profile analysis of X-ray diffraction patterns was performed with the Jana2006 software [15].

EXAFS spectra of $Ln_2Zr_2O_7$ (Ln=La-Tb) above the $L_3$-Ln (for La 5483 eV) and $K$-Zr (17998 eV) absorption edges, were measured at i811 beamline of MAXLab (Lund, Sweden) and P65 beamline of DESY (Gamburg, Germany) in a transmission mode at room temperature. Processing and modeling of XAFS-spectra were carried out using the VIPER [16] and Demeter [17] program packages. The amplitudes and phases of photoelectron back scattering in the local environment of zirconium, REE and hafnium atoms were calculated by program FEFF-8.20 [18] using the crystal structure parameters of pyrochlore and fluorite, taken from the diffraction data for $Ln_2Zr_2O_7$ (where Ln= La-Tb).

The Raman spectra were collected on a inVia Reflex confocal Raman microscope (Renishaw) (laser $\lambda_1$=532 nm and $\lambda_2$=785 nm) at room temperature [19].

## 3. Results and discussion

XRD study showed that precursors $La_2O_3{\cdot}2ZrO_2{\cdot}12{,}6H_2O$ calcinated at temperatures below 800°C have the X-ray amorphous structure. An increase in the atomic number along the La-Tb series leads to a decrease in the crystallization temperature. However, structure can't be considered as completely disordered, since from one to several broad peaks were observed in the region 13-21° of double Bragg angle. The increasing of the calcination temperature above 800 °C leads to appearance of about ten X-ray reflections on XRD patterns for all samples, which corresponds to the fluorite structure formation. Besides at the annealing temperature above 1000°C we observed the gradual narrowing of the diffraction reflexes and the appearance of superlattice peaks (Figure 1), which point to the increase of the structure ordering. This is also correlated with increase of crystallite size and decrease of microstrain values obtained from the analysis of the diffraction patterns (see Figure 2).

These superstructure reflections are substantially wider than the main peaks of the fluorite-type structure. It suggests that cationic ordering with formation of the pyrochlore phase upon calcination occurs inside the separate regions (nanodomains) distributed over well-crystallized fluorite matrix.

Using EXAFS spectroscopy we have studied the evolution of the $Ln_2Zr_2O_7$ (Ln=La-Tb) local atomic structure upon the gradual heat treatment from initially amorphous precursors to well-crystallized powders.

Figure 3 shows the Fourier transform moduli of $L_3$-Nd EXAFS-functions $\chi(k)k^2$ of $Nd_2Zr_2O_7$ powders calcinated at different temperatures. The first maximum at $R$~1.9 Å corresponds to the nearest oxygen coordination shell around Nd. For the sample calcination at 600 °C, which is still amorphous and for the samples calcination at temperatures <1000 °C with the distinct fluorite structure, the shoulder at ~2.2 Å on the right slope of the main peak is quite small. But with an increase of calcination temperature the oxygen coordination shell starts to split into two components.

The observed splitting peak corresponds to two shorter Nd-O(2) and six longer Nd-O(1) distances. The second large features at 3.1 Å and 3.5 Å include contributions from Nd-Zr and Nd-Nd interatomic bonds. Amplitude of this features increases significantly with calcination temperature and the position drifts to larger radius (Figure 3). This indicates gradual ordering of the overall crystal structure. At calcination temperature exceeding 1300°C the peak shape and amplitude substantially stabilize, which

IOP Conf. Series: Journal of Physics: Conf. Series **941** (2017) 012079 doi:10.1088/1742-6596/941/1/012079

corresponds to the end of the pyrochlore structure formation. Analysis of EXAFS data showed that spectra for samples prepared at calcination temperature <800°C are well reproduced by the model of fluorite structure (Figure 4). However, at higher calcination temperatures the fluorite structure model can no longer provide an adequate simulation of EXAFS-function and the analysis was conducted using the pyrochlore structure model. Thus above 1000 °C there is clearly observed splitting of the Nd-O bond length (Figure 4). This directly points to the restructuring of the local oxygen environment of rare-earth ions as a result of the observed fluorite-pyrochlore phase transition.

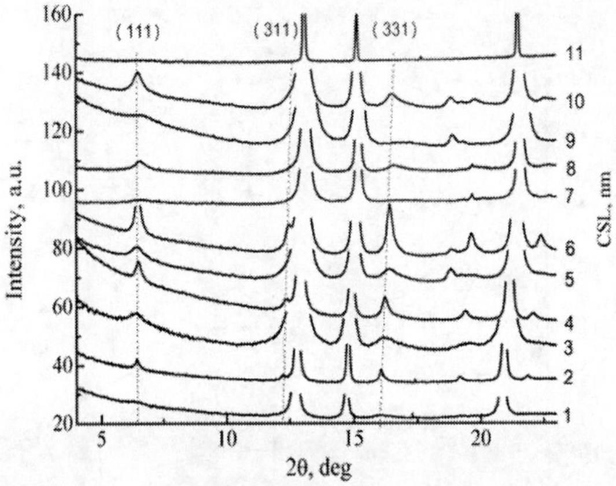


**Figure 1.** XRD patterns of $Ln_2Zr_2O_7$ powders obtained by precursor annealing in vicinity of the phase transition. Diffraction reflexes (111), (311) и (331) indicate presence of the pyrochlore phase. $La_2Zr_2O_7$ (1- 1000°C, 2- 1100°C), $Pr_2Zr_2O_7$ (3- 1000°C, 4- 1100°C), $Nd_2Zr_2O_7$ (5- 1100°C, 6- 1200°C), $Sm_2Zr_2O_7$ (7- 1100°C, 8- 1200°C), $Eu_2Zr_2O_7$ (9- 1100°C, 10- 1200°C), $Tb_2Zr_2O_7$ (11- 1500°C).

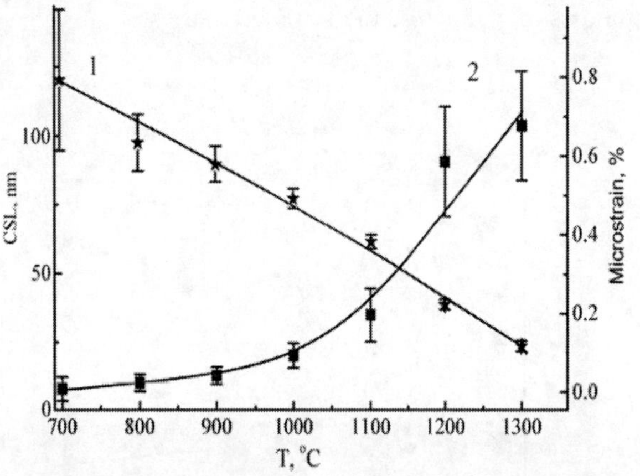


**Figure 2.** Microstrains (1) and coherent scattering lengths (2) of the series $Eu_2Zr_2O_7$ established by Rietveld method. Similar results were obtained for other Ln compounds.

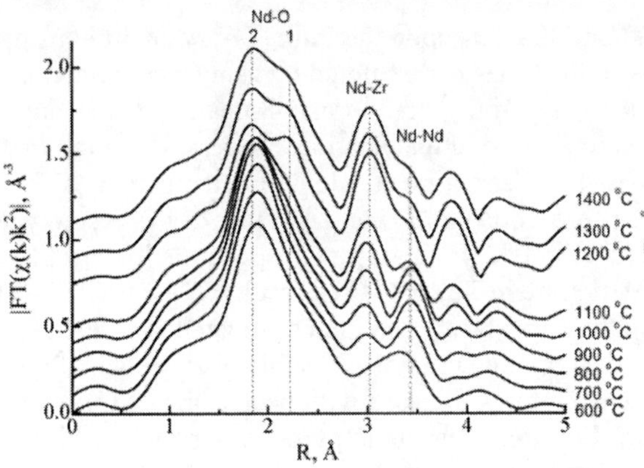


**Figure 3.** Fourier transform moduli of EXAFS experimental spectra measured above the $L_3$-Nd edge for $Nd_2Zr_2O_7$ calcinatated at different temperatures.

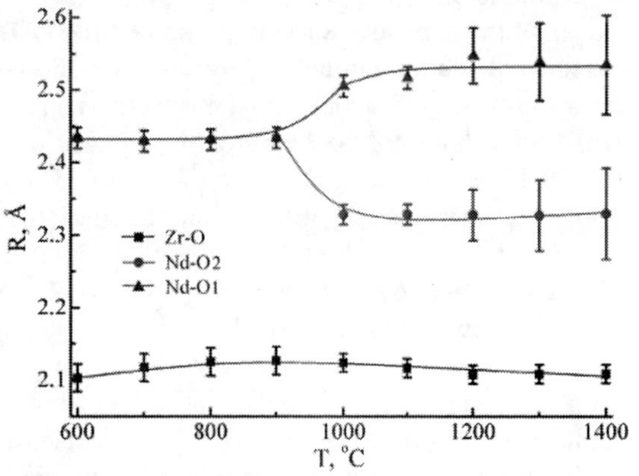


**Figure 4. Figure 4.** The interatomic bond lengths Nd-O and Zr-O in $Nd_2Zr_2O_7$ extracted from the EXAFS date.

Also, we have conducted a Raman spectroscopy study for each series of samples $Ln_2Zr_2O_7$. The effect of the type of rare-earth ions on the Raman spectra is represented in Figure 5. As seen in Figure 5,

IOP Conf. Series: Journal of Physics: Conf. Series **941** (2017) 012079 doi:10.1088/1742-6596/941/1/012079

Raman mode peaks for compounds from La to Tb become wider and shift to higher wavenumbers. Thus the most intense $F_{2g}$ mode (~305 cm$^{-1}$) in the Raman spectra of $Tb_2Zr_2O_7$ is approximately one order of magnitude broader than that of $La_2Zr_2O_7$.

The results of our complex study of the phase transition in the $Ln_2Zr_2O_7$ series using EXAFS, XRD and Raman spectroscopy methods are summarized in Figure 6. EXAFS curve shows the beginning of the fluorite-pyrochlore transition, while Raman data point to the end. The initial crystallization of amorphous precursors leads to the formation of nanocrystalline oxides with the fluorite structure, regardless the $Ln^{3+}$ cation type. The crystallization onset temperature is lower for $Ln_2Zr_2O_7$ with a smaller $Ln^{3+}$ cation radius (or γ ratio). In contrast, the fluorite-pyrochlore phase transition temperature becomes higher across the $Ln_2Zr_2O_7$ series.

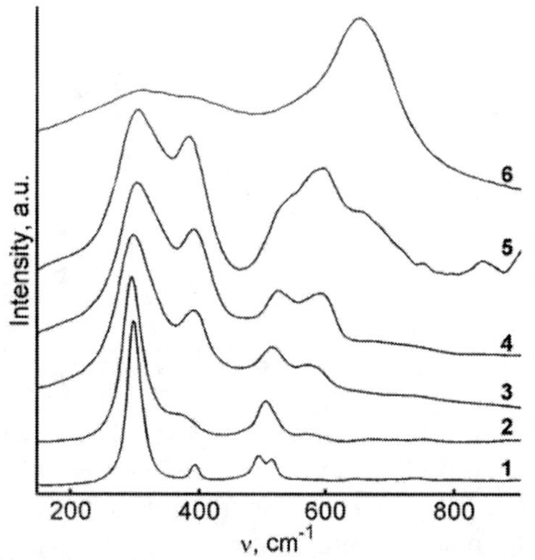


**Figure 5.** Raman spectra of $Ln_2Zr_2O_7$ (Ln=1-La, 2-Pr, 3-Nd, 4-Sm, 5-Eu, 6-Tb) powders with calcination temperature 1400 $^0$C measured using 532 nm excitation.

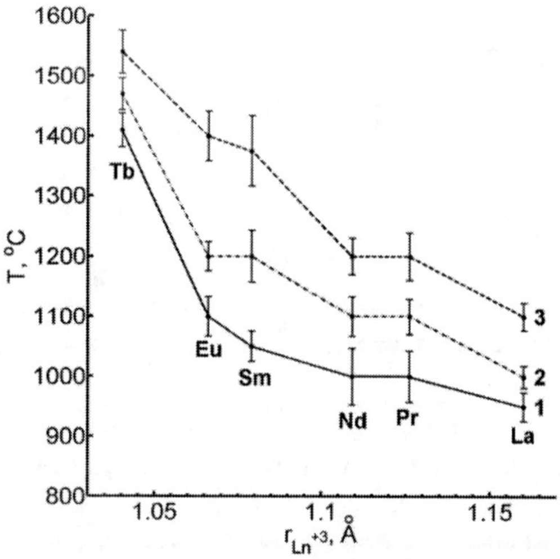


**Figure 6.** Scheme of the phase transition temperature for $Ln_2Zr_2O_7$ (Ln=La-Tb) compounds obtained from EXAFS-data analysis (1), XRD (2) and Raman spectra (3).

## 4. Conclusion

Local and crystal structure evolution of $Ln_2Zr_2O_7$ (Ln = La-Tb) upon the initial X-ray amorphous precursors calcination in the temperature range 600 – 1600 °C has been studied by using a combination of X-ray powder diffraction, XAFS-spectroscopy at synchrotron radiation and Raman spectroscopy. It was found that the fluorite-pyrochlore phase transition occurs at calcination temperature in the range of 900-1200 °C. The appearance of the ordered pyrochlore structure at calcination temperatures above 900 °C was fixed as an observation of superlattice peaks in diffraction and splitting of the first oxygen coordination shell in the local lanthanides environment in EXAFS and Raman spectra. Although X-ray diffraction data does not show crisp signs of superstructure in $Tb_2Zr_2O_7$, the results of the EXAFS data analysis show traces of pyrochlore ordering also in this boundary compound calcinated at 1500 °C.

Characteristic temperatures for the ordering in the cationic and anionic sublattices are slightly different (lower for the anionic sublattice) but both temperatures increase with decreasing of $Ln^{3+}$ cation radius (or γ ratio). It was shown that the phase transition from fluorite to pyrochlore structure for Ln = La-Tb proceeds via the intermediate emergence and gradual growth of pyrochlore-type nanodomains in a fluorite matrix. In addition the scheme of fluorite-pyrochlore transition temperature dependence on radii of lanthanide cations was constructed.

**Acknowledgements**
The present work was supported by the Ministry of Education and Science of the Russian Federation (grant No 3.9750.2017/8.9).

**References**

[1] Subramanian M, Aravamudan G, Subba Rao G 1983 *Prog. Solid State Chem.* **15** 55
[2] Shamblin J, Feygenson M, Neuefeind J, Tracy C, Zhang F, Finkeldei S, Bosbach D, Zhou H, Ewing R, Lang M 2016 *Nat. Mater.* **15** 507
[3] Blanchard P, Clements R, Kennedy B, Ling C, Reynolds E, Avdeev M, Stampfl V, Zhang Z, Jang L-Y 2012 *Inorg. Chem.* **51** 13237-44.
[4] Karthik C, Anderson T, Gout D, Ubic R 2012 *J. Solid State Chem.* **194** 168
[5] Blanchard P, Liu S, Kennedy B, Ling C, Avdeev M, Aitken J, Cowie B, Tadich A 2013 *J. Phys. Chem. C* **117** 2266
[6] Andrievskaya E R 2008 *J. Eur. Ceram. Soc.* **28** 2363
[7] Popov V V, Menushenkov A P, Yastrebtsev A A, Zubavichus Ya V, Svetogorov R D, Kolyshkin N A 2016 *J. Phys.: Conf. Ser.* **747** 012042(1)
[8] Popov V V, Menushenkov A P, Zubavichus Ya V, *et al.* 2015 *Russ. J. Inorg. Chem.* **60** 16-22
[9] Popov V V, Zubavichus Ya V, Menushenkov A P, Yaroslavtsev A A, Kulik E S, Petrunin V F, Korovin S A and Trofimova N N 2014 *Russ. J. Inorg. Chem.* **59** 279
[10] Menushenkov A P, Popov V V, Gaynanov B R, Zubavichus Y V 2016 *J. Phys.: Conf. Ser.* **747** 012038(1)
[11] Popov V V 2015 *Russ. J. Inorg. Chem.* **60** 420
[12] Popov V V, Zubavichus Ya V, Petrunin V F, Menushenkov A P, Kashurnikova O V, Korovin S A, Chernikov R V and Yaroslavtsev A A 2011 *Glass Phys. Chem.* **37** 512
[13] Popov V V, Petrunin V F, Korovin S A, *et al.* 2011 *Russ. J. Inorg. Chem.* **56** 1538(10)
[14] Popov V V, Menushenkov A P, Zubavichus Ya V, *et al.* 2013 *Russ. J. Inorg. Chem.* **58** 1400(12).
[15] Petricek V, Dusek M, Palatinus L 2014 *Z. Kristallogr.* **229** 345
[16] Klementiev K V 2001 *J. Phys. D: Appl. Phys.* **34** 209
[17] Ravel B, Newville M 2005 *J. Synchr. Rad.* **12** 537
[18] Newville M 2001 *J. Synchr. Rad.* **8** 322
[19] Popov V V, Menushenkov A P, Yaroslavtsev A A, Zubavichus Ya V, Gaynanov B R, Yastrebtsev A A, Leshchev D S and Chernikov R V 2016 *J. Alloys Compd.* **689** 669

# Influence of synthesis conditions on the crystal structure of the powder formed in the "$HfO_2$ - $CeO_2/Ce_2O_3$" system

**V V Popov[1,2,*], A P Menushenkov[2], R M Khubbutdinov[2], A A Yastrebtsev[2], R D Svetogorov[1], Ya V Zubavichus[1], A L Trigub[1], A S Sharapov[2], A A Pisarev[2], V V Kurilkin[3], N A Tsarenko[4], L A Arzhatkina[4]**

[1]NRC "Kurchatov Institute", pl. Akademika Kurchatova 1, 123182 Moscow, Russia
[2]National Research Nuclear University MEPhI (Moscow Engineering Physics Institute), Kashirskoe sh. 31, 115409 Moscow, Russia
[3] Peoples' Friendship University of Russia, Miklukho-Maklaya st. 6, 117198 Moscow, Russia
[4]JSC "Scientific Research Institute of Chemical Technology", Kashirskoe sh. 33, 115309 Moscow, Russia

[*]E-mail: victorvpopov@mail.ru

**Abstract**. Influence of synthesis conditions (type of atmosphere: reduction or oxidation, annealing temperature) on the chemical composition and structure of the compounds formed in the "$HfO_2$ - $CeO_2/Ce_2O_3$" system has been investigated by X-ray absorption fine structure spectroscopy combined with Raman spectroscopy, X-ray diffraction and thermogravimetric analysis. It was revealed that isothermal annealing of precursor at temperatures less than 1000°C in air leads to formation of $Ce_{0.5}Hf_{0.5}O_2$ powders with cubic fluorite-type structure (space group *Fm*-3*m*). Further increase of annealing temperatures above 1000°C causes decomposition of formed crystal structure into two phases: cubic and monoclinic. Annealing in reduction hydrogen atmosphere causes formation of $Ce^{4+}{}_{2x}Ce^{3+}{}_{2-2x}Hf_2O_{7+x}$ compounds with intermediate oxidation state of cerium, where value of $x$ depends on the reducing conditions and treatment parameters. Annealing in vacuum at 1400°C strongly reduces the content of $Ce^{4+}$ in a powder samples and leads to formation of pyrochlore structure (space group *Fd*-3*m*) with predominant +3 oxidation state of cerium.

## 1. Introduction

Cerium based mixed oxides are of great interest from both theoretical and practical points of view due to their physical-chemical properties (firstly, the ability of cerium cation to change an oxidation state depending on the synthesis conditions) for applications in three-way catalysts and as material for anodes in Solid Oxide Fuel Cells (SOFC) [1]. Despite on the large number of investigations of Zr-based $Ce_2Zr_2O_8/Ce_2Zr_2O_7$ compounds [1-3] the information concerning hafnium analogue remains insufficient. The phase relation and the thermal expansion studies of Hf-based materials in the "$HfO_2$ - $CeO_2$" system are of considerable interest from the point of view of the use in nuclear reactors, oxygen storage capacitors and thermal barrier coatings [4]. Moreover $CeO_2$ is commonly used as a surrogate material for $PuO_2$, which is a nuclear fuel material. So that the phase relation studies in the "$HfO_2$ - $CeO_2$" system could be used to simulate the phase relation in the "$HfO_2$ - $PuO_2$" system [4]. Accordingly to [4, 5] the phase diagram of "$HfO_2$ - $CeO_2$" system involves not only the stable phase boundaries (between monoclinic, tetragonal and cubic phases) but also some metastable boundaries which appear under prolonged heating at higher temperatures and quenched conditions. Unfortunately information about properties of $Ce_2Hf_2O_{7+x}$ compounds is very limited [6].

The aim of the present work is to study the influence of the synthesis conditions and annealing in reduced atmosphere on chemical composition, oxidation state of cerium and structure of the powders formed in "$HfO_2$ - $CeO_2/Ce_2O_3$" system by means of X-ray diffraction, XAFS, Raman spectroscopy and thermogravimetric analysis.

## 2. Experimental

The starting materials were $Ce(NO_3)_3{\cdot}6H_2O$ (99.99%), $HfOCl_2{\cdot}8H_2O$ (99.7%). Precursor was prepared by co-precipitation of metal salts solution with ammonia $NH_3{\cdot}H_2O$ (analytical grade) [3, 7]. Powders of $Ce_{0.5}Hf_{0.5}O_2$ were prepared by calcination of precursor at 600-1000°C for 3 h in air. Two different reducing procedures were carried out: isothermal annealing in vacuum at 1400°C for 3h and isothermal annealing in hydrogen at 1000°C for 4h (for more detailed experimental procedure see [3]).

Thermogravimetric analysis (TGA) of powders was measured at temperature range 30 – 1400°C using thermogravimetric analyzer SDT Q600 (TA Instruments) with a heating rate of 10 K/min in air flow. The composition and the oxidation degree of synthesized powders was determined by TGA measurements of sample mass during heating to 1000°C and further isothermal annealing for 4 hours in air flow [3].

Crystal structures of synthesized powders were studied by X-ray powder diffraction at "Structural Materials Science" beamline of the Kurchatov synchrotron radiation source. Measurements were carried out in the transmission mode at $\lambda$ = 0.68886 Å, using XY detector Imaging Plate Fuji Film BAS-5000 [8, 9]. The Rietveld full-profile analysis of X-ray diffraction patterns was performed with the Jana2006 software [10].

The Raman spectra were collected on a Nicolet iS50 FT-IR spectrometer (Thermo Fisher Scientific Inc.) with Thermo Scientific iS50 Raman module ($\lambda$=1064 nm) at room temperature [11].

X-ray absorption fine structure (EXAFS) spectra were measured above $L_3$-Ce and $L_3$-Hf edges at i811 beamline of MAX-lab (Lund, Sweden) and at "Structural Materials Science" beamline of the Kurchatov synchrotron radiation source in transmission mode at room temperature. The processing and simulation of the EXAFS spectra were performed using IFEFFIT [12] program packages. The back scattering amplitudes and phases were calculated with FEFF-8.20 [13] based on the known fluorite and pyrochlore crystal structure parameters and the diffraction data. The XANDA program [14] was used to fit the near-edge region of X-ray absorption spectra.

## 3. Results and discussion

It is revealed that isothermal annealing of precursor at 600-1000°C temperatures in air leads to formation of $Ce_{0.5}Hf_{0.5}O_2$ powders with FCC fluorite-type structure (space group *Fm-3m*) (Figure 1) with the lattice parameter $a \sim 5.26$ Å (low temperature cubic phase). Increasing synthesis temperature leads to the narrowing the main peaks, due to increase in the crystal size (or coherent scattering length - CSL) and to decrease of microstrain ($\varepsilon$) values.

In samples annealed at 1000°C we observed three phases: low temperature cubic phase (55 %), monoclinic (7 %) and high temperature cubic phase with $a \sim 5.36$ Å (38 %). Further increase of annealing temperature ≥ 1000°C led to formation of biphasic system: high temperature cubic $Ce_{0.6}Hf_{0.4}O_2$ and monoclinic $Ce_{0.12}Hf_{0.88}O_2$ (with content 40 - 43 %, at annealing temperature 1200 - 1400°C). The formation of biphasic region for Ce-Hf-O system is in good agreement with reported in [4]. The XRD study showed that increase of annealing temperature in the range 1000-1400°C causes increase of CSL and decrease of microstrains for both types of the formed phases (Figure 2).

The structural information was confirmed by Raman-spectroscopy data. It was observed that Raman spectra of powders with fluorite structure (prepared at < 1000°C) have only one Raman mode ($F_{2g}$) around 470 $cm^{-1}$ originating from the symmetric vibrations of the oxygen anions around the $Ce^{4+}$ cation in $Ce$-$O_8$ cubic cell. Increasing the annealing temperature to > 1000°C leads to the appearance of the some additional modes due to the formation of additional monoclinic phase in prepared samples (Figure 3). The concentration of monoclinic phase increases with the annealing temperature increase (Figure 4). It is worth to note that Raman-spectra analysis have shown that the samples prepared at low temperature (< 1000°C) may contain an admixture of the tetragonal phase (Figure 4). Thus these samples with low temperature FCC fluorite-type structure can be characterized by not clear cubic structure but the *t''*-structure. Analogous *t''*-structure (according to Raman data) was identified in cubic (according to XRD) $(1\text{-}x)HfO_2{\cdot}xY_2O_3$ [15].

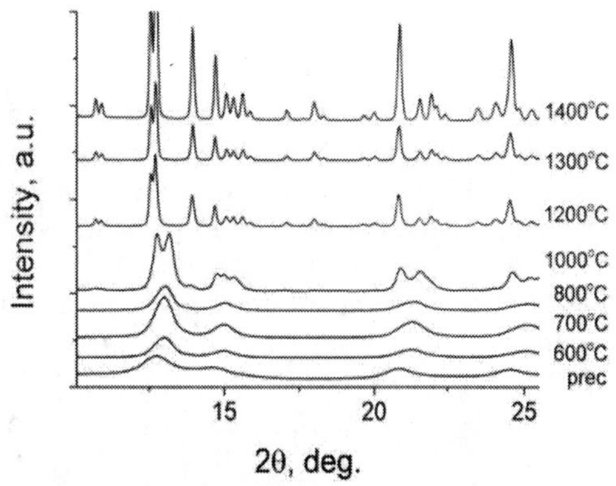


**Figure 1.** XRD patterns of $Ce_{0.5}Hf_{0.5}O_2$ powders obtained by calcination of the precursor at different temperatures.

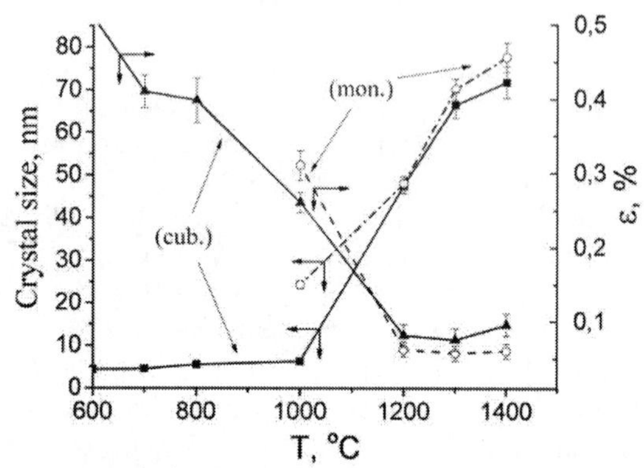


**Figure 2.** Crystal size and microstrains of $Ce_{0.5}Hf_{0.5}O_2$ powders.

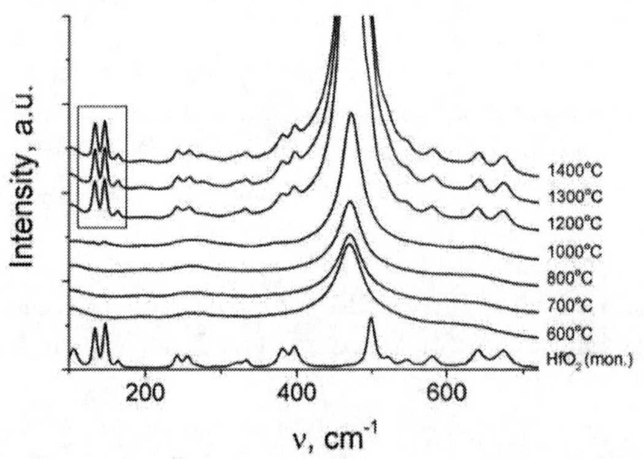


**Figure 3.** Raman spectra of $Ce_{0.5}Hf_{0.5}O_2$ powders prepared at different temperature using 1064 nm excitation. The most intensive monoclinic phase modes indicated by rectangle.

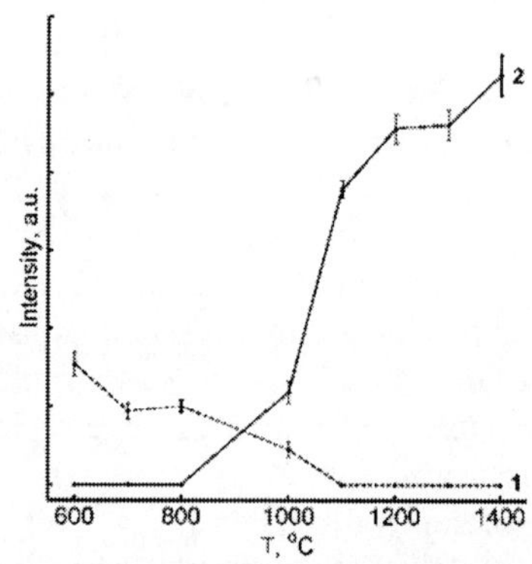


**Figure 4.** The temperature dependence of ratio $I_{(112\ cm\text{-}1)}/I_{(470\ cm\text{-}1)}$ (for tetragonal phase) (1) and $I_{(146\ cm\text{-}1)}/I_{(470\ cm\text{-}1)}$ (for monoclinic phase) (2).

Using XAFS spectroscopy we have traced the $Ce_{0.5}Hf_{0.5}O_2$ local atomic structure evolution upon the progressive heat treatment from initially amorphous precursors to well-crystallized powders. XANES-analysis of *L₃*-Ce absorption edge of precursors and samples synthesised by annealing in air have shown that all these compounds contain only $Ce^{4+}$ cations (similar to $CeO_2$ [16]). It was observed that the width and magnitude of the splitting of the XANES peak increases with the annealing temperature increase (Figure 5) probably due to the formation of an additional monoclinic phase.

The analysis of EXAFS-data showed that spectra of samples calcinated at temperature < 1000°C are well fitted by using the fluorite structure model. It was found that the difference between the Ce-O and Hf-O bond lengths appears to be about 0.15 Å in full range of annealing temperature, which significantly exceeds the standard error of the interatomic distances determination by EXAFS (0.01-0.02

Å) (Figure 6). It means that the local environment of the $Ce^{4+}$ and $Hf^{4+}$ cations in fluorite powders (< 1000°C) cannot be described by the ideal fluorite model, and is rather characterized by the defect-fluorite structure with nonequivalent positions of cations. Further increasing of the annealing temperature > 1000°C leads to a remarkable splitting of the both Ce-O and Hf-O coordination shell into two components due to the formation of the additional monoclinic phase.

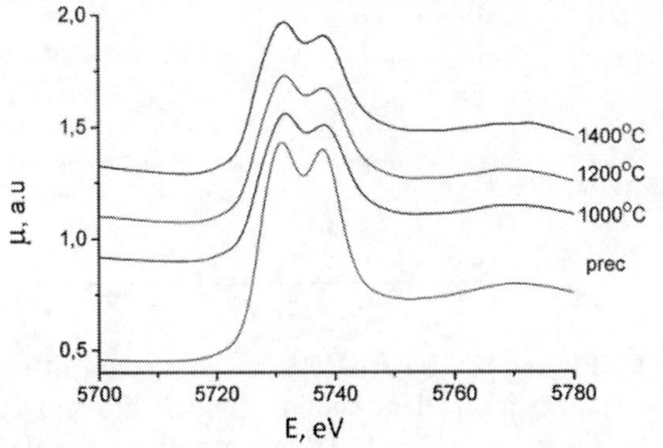


**Figure 5.** The $L_3$-Ce XANES spectra of $Ce_{0.5}Hf_{0.5}O_2$ powders annealed at different temperature on air.

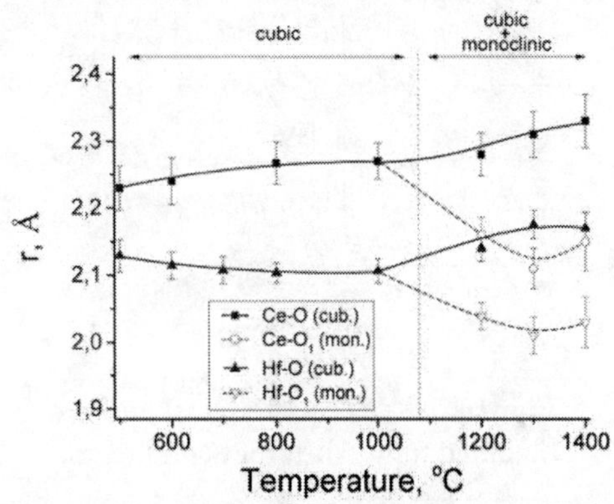


**Figure 6.** The interatomic bond lengths Ce-O and Hf-O in $Ce_{0.5}Hf_{0.5}O_2$ extracted from the EXAFS date.

Annealing in reduction hydrogen atmosphere causes formation of $Ce^{4+}{}_{2x}Ce^{3+}{}_{2-2x}Hf_2O_{7+x}$ compounds with intermediate oxidation state of cerium, where value of $x$ depends on the reducing conditions and treatment parameters of precursor (Table 1 and Figure 7). Annealing in vacuum at 1400°C strongly reduces the content of $Ce^{4+}$ in a powder samples and leads to formation of pyrochlore structure (space group *Fd-3m*) (Figure 7) and with predominant +3 oxidation state of cerium (Table 1).

**Table 1.** Structural parameters of $Ce^{4+}{}_{2x}Ce^{3+}{}_{2-2x}Hf_2O_{7+x}$ reduced samples.

| Sample | Structure | CSL (nm) | ε (%) | Composition (TGA) | Composition (XANES) |
|---|---|---|---|---|---|
| Precursor / $H_2$ 1000°C | *P2₁/c*<br>*Fm-3m* | 23<br>7 | 0.199<br>0.094 | $Ce^{4+}{}_{0.876}Ce^{3+}{}_{1.124}Hf_2O_{7.438}$ | $Ce^{4+}{}_{0.68}Ce^{3+}{}_{1.32}Hf_2O_{7.34}$ |
| Air 1100°C / $H_2$ 1000°C | *P2₁/c*<br>*Fm-3m* | 27<br>19 | 0.13<br>0.036 | $Ce^{4+}{}_{1.250}Ce^{3+}{}_{0.750}Hf_2O_{7.625}$ | $Ce^{4+}{}_{1.36}Ce^{3+}{}_{0.64}Hf_2O_{7.68}$ |
| Air 1400°C / $H_2$ 1000°C | *P2₁/c*<br>*Fm-3m* | 57<br>41 | 0.05<br>0.075 | $Ce^{4+}{}_{1.538}Ce^{3+}{}_{0.462}Hf_2O_{7.769}$ | $Ce^{4+}{}_{1.50}Ce^{3+}{}_{0.50}Hf_2O_{7.75}$ |
| Air 1100°C / Vac 1400°C | *Fd-3m* | 220 | 0.035 | $Ce^{4+}{}_{0.162}Ce^{3+}{}_{1.838}Hf_2O_{7.081}$ | $Ce^{4+}{}_{0.18}Ce^{3+}{}_{1.82}Hf_2O_{7.09}$ |

Oxidation state of cerium in synthesized $Ce^{4+}{}_{2x}Ce^{3+}{}_{2-2x}Hf_2O_{7+x}$ powders excluded from TGA and XANES data analysis is presented in Table 1. It worth to notice the good agreement between the data obtained by these two methods.

The presented results (Table 1) point to the reducing ability of $Ce_2Hf_2O_8$ powders in hydrogen decreasing with increasing of the annealing temperature of the preliminary treatment. Moreover the reducing ability of $Ce_2Hf_2O_8$ powders (after thermal treatment) are less than that of Zr-based compounds $Ce_2Zr_2O_8$ [3]. These results are in good agreement with reported in [6].

As seen in Figure 8, the powders obtained by heat treatment in air at 1100 and 1400°C/3h has the the $L_3$-Ce XANES spectrum splitted “white line” corresponding to compounds containing $Ce^{4+}$ cations, particularly to $CeO_2$ [16]. Herewith the decrease in the degree of oxidation of cerium cations causes shifting the position of the absorption edge to lower energies, corresponding to +3 oxidation state of cerium (Figure 8). Similar results were obtained by us for cerium aluminate [17] and cerium zirconate [3].

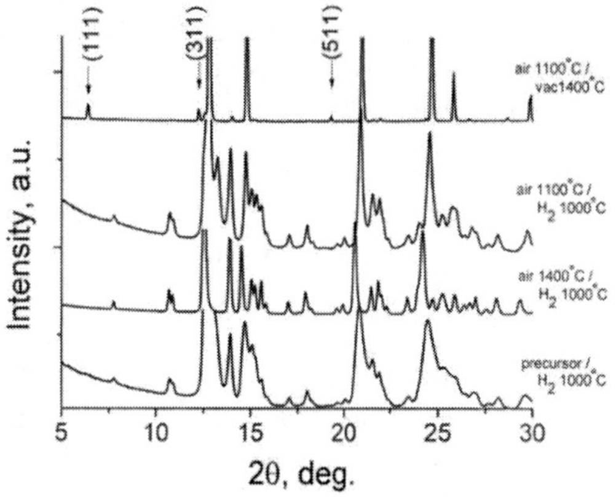


**Figure 7.** XRD patterns of reduced samples obtained under different conditions.

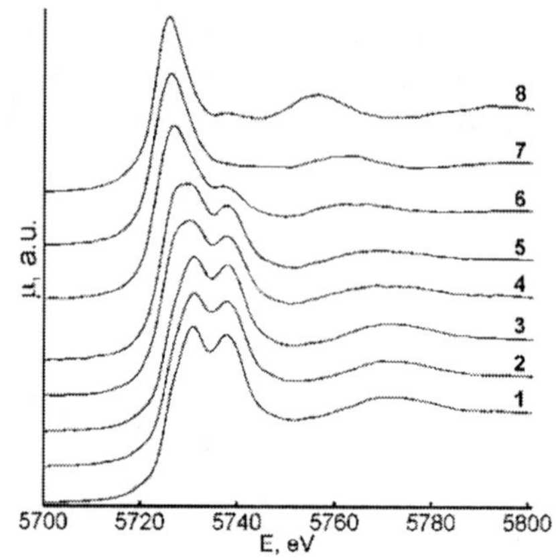


**Figure 8.** The $L_3$-Ce XANES-spectra of cerium hafnate powders obtained under different conditions: 1 – $CeO_2$ (standard); 2 – $Ce_{0.5}Hf_{0.5}O_2$ (1100°C); 3 – $Ce_{0.5}Hf_{0.5}O_2$ (1400°C); 4 – Air 1400°C / $H_2$ 1000°C; 5 - Air 1100°C / $H_2$ 1000°C; 6 - Precursor / $H_2$ 1000°C; 7 - Air 1100°C / Vac 1400°C; 8 – $CeAlO_3$ (standard).

## Conclusion

Isothermal annealing of mixed hydroxides of cerium and hafnium at 600-1000°C in air leads to the formation of $Ce_{0.5}Hf_{0.5}O_2$ powders with FCC fluorite-type structure (space group *Fm*-3*m*). Increase of annealing temperatures above 1000°C causes decomposition of formed crystal structure into the two phases: cubic $Ce_{0.6}Hf_{0.4}O_2$ and monoclinic $Ce_{0.12}Hf_{0.88}O_2$. It is shown that type and parameters of crystal structure of obtained powders strongly depends on the reducing conditions and treatment parameters of precursor. It was found that the reduction in vacuum or in hydrogen atmosphere causes formation of $Ce^{4+}{}_{2x}Ce^{3+}{}_{2-2x}Hf_2O_{7+x}$ compounds with intermediate oxidation state of cerium. In addition it was observed that the reduction in vacuum leads to formation of a compound with predominant +3 oxidation state of cerium which has the pyrochlore-type structure (space group *Fd*-3*m*).

## Acknowledgments

The present work was supported by the Ministry of Education and Science of the Russian Federation (grant No 3.9750.2017/8.9).

## References

[1] Kašpar J, Graziani M and Fornasiero P 2000 *Handbook on the physics and chemistry of rare earths: The role of rare earths in catalysis* vol **29**, ed K A Gschneidner Jr and L Erwin (Amsterdam: Elsevier Science) 159
[2] Omata T, Kishimoto H, Otsuka-Yao-Matsuo S, Ohtori N and Umesaki N 1999 *J. Solid State Chem.* **147** 573
[3] Popov V V, Menushenkov A P, Khubbutdinov R M, Svetogorov R D, Zubavichus Ya V, Sharapov A S and Kurilkin V V 2016 *J. Phys.: Conf. Ser.* **747** 012041
[4] Chavan S V and Tyagi A K 2006 *Mater. Sci. Eng. A* **433** 203
[5] Fujimori H, Yashima M, Sasaki S and Kakihana M 2001 *Phys. Rev. B* **64** 134104
[6] Baidya T, Hegde M S and Gopalakrishnan J 2007 *J. Phys. Chem. B* **111** 5149
[7] Popov V V 2015 *Russ. J. Inorg. Chem.* **60** 420

IOP Conf. Series: Journal of Physics: Conf. Series **941** (2017) 012080 doi:10.1088/1742-6596/941/1/012080

[8] Popov V V, Zubavichus Ya V, Petrunin V F, Menushenkov A P, Kashurnikova O V, Korovin S A, Chernikov R V and Yaroslavtsev A A 2011 *Glass Phys. Chem.* **37** 512
[9] Popov V V, Menushenkov A P, Zubavichus Ya V, Veligzhanin A A, Yaroslavtsev A A, Chernikov R V, Leshchev D S, Petrunin V F, Korovin S A and Bednarcik J 2013 *Russ. J. Inorg. Chem.* **58** 331
[10] Petricek V, Dusek M and Palatinus L 2014 *Z. Kristallogr.* **229** 345
[11] Popov V V, Menushenkov A P, Yaroslavtsev A A Zubavichus Ya V, Gaynanov B R, Yastrebtsev A A, Leshchev D S and Chernikov R V 2016 *J. Alloys Compd.* **689** 669
[12] Newville M 2001 *J. Synchrotron Rad.* **8** 322
[13] Ankudinov A, Bouldin C, Rehr J and Conradson S 1998 *Phys. Rev. B* **58** 7565
[14] Klementiev K V 2001 *J. Phys. D: Appl. Phys.* **34** 209
[15] Yashima M, Takahashi H, Ohtake K, Hirose T, Kakihana M, Arashi H, Ikuma Y, Suzuki Y and Yoshimura M 1996 *J. Phys. Chem. Solids* **57** 289
[16] Safonova O V, Guda A A, Paun C, Smolentsev N, Abdala P M, Smolentsev G, Nachtegaal M, Szlachetko J, Soldatov M A, Soldatov A V and van Bokhoven J A 2014 *J. Phys. Chem. C.* **118** 1974
[17] Popov V V, Menushenkov A P, Zubavichus Ya V, Sharapov A S, Kabanova V A, Yastrebtsev A A, Arzhatkina L A, Tsarenko N A, Strel'nikova A M and Kurilkin V V 2016 *Russ. J. Inorg. Chem.* **61** 225

IOP Conf. Series: Journal of Physics: Conf. Series **941** (2017) 012081 doi:10.1088/1742-6596/941/1/012081

# Influence of the concentration and lanthanoide type on the phase transitions in $Ln_2O_3$-$MO_2$ system (Ln = La, Gd, Y; M = Zr, Hf)

**V V Popov[1,2], A P Menushenkov[2], A A Yastrebtsev[2,*], A A Pisarev[2], I V Shchetinin[3], M V Zheleznyi[3], Ya V Zubavichus[1], L A Arzhatkina[4], N A Tsarenko[4]**

[1] NRC "Kurchatov Institute", pl. Akademika Kurchatova 1, 123182 Moscow, Russia
[2] National Research Nuclear University MEPhI (Moscow Engineering Physics Institute), Kashirskoe sh. 31, 115409 Moscow, Russia
[3] National University of Science and Technology MISiS (Moscow Institute of Steel and Alloys), Leninskiy pr. 4, 119049 Moscow, Russia
[4] JSC "Scientific Research Institute of Chemical Technology", Kashirskoe sh.33, 115309 Moscow, Russia

[*]E-mail: alexyastrebtsev@mail.ru

**Abstract.** Influence of the concentration and type of rare-earth oxide on the chemical composition and structure of the compounds formed in the "$Ln_2O_3$-$MO_2$" system has been investigated by X-ray powder diffraction and Raman spectroscopy. It was found that crystallization temperature of synthesized powders rises with an increase in the concentration of $Ln_2O_3$. Besides, it was revealed that the efficiency of stabilization of high-temperature phases (tetragonal and cubic) in $Ln_2O_3$-$ZrO_2$ systems increases with the introduction of RE cations in the $La^{3+} \rightarrow Gd^{3+} \rightarrow Y^{3+}$ series in accordance with the increase in the solubility of $Ln_2O_3$ in $ZrO_2$. A similar behavior was also observed for the $Ln_2O_3$-$HfO_2$ systems. Finally, we established that the compounds formed in $Ln_2O_3$-$HfO_2$ systems have a greater ability to crystallize in the monoclinic phase.

## 1. Introduction

Complex oxides forming in the $Ln_2O_3$-$MO_2$ system (Ln is a lanthanoide and M is a titanium subgroup metal) represent a large group of inorganic materials, that demonstrate a variety of outstanding physical effects, such as order-disorder phase transitions [1-4], geometrically frustrated magnetism [5] and ionic conductivity [6]. These materials have wide technological applications, which include solid electrolytes in high temperature solid oxide fuel cells, neutron absorbing [7], thermal barrier coatings [8], etc.

Despite a large number of investigations concerned with $Ln_2O_3$-$MO_2$ compounds, the vast majority of publications are dedicated to the properties of well-crystallized samples [1-3] prepared by the solid-state synthesis method at elevated temperature. In this paper, we use another synthesis methodic, based on the annealing of precursors with highly homogeneous cation distribution, prepared by reverse co-precipitation [9]. This method is more suitable for studying the evolution of crystalline phases and polymorphic transformations during the process of $Ln_2O_3$-$MO_2$ calcination.

The aim of the present work is to study the influence of annealing temperature and molar ratio of initial oxides in precursors on structure and phase composition of the powders formed in "$Ln_2O_3$-$MO_2$" system by means of X-ray powder diffraction (XRD) and Raman spectroscopy.

## 2. Experimental

The lanthanide hafnate and zirconate series with composition $Ln_2O_3$-$MO_2$ (Ln = La, Gd, Y; M = Zr, Hf) have been synthesized from lanthanide nitrates $Ln(NO_3)_3$•$6H_2O$ and zirconium (hafnium) oxychloride octahydrate $MOCl_2$•$8H_2O$. Initial metal salts were taken in stoichiometric amounts affording samples $x$$Ln_2O_3$•(1-$x$)$MO_2$ with the molar content $x$ = 3, 8, 18, 33 mol. %. Precursor (mixed Ln-M hydroxide) was prepared by co-precipitation of metal salts solution with ammonia $NH_3{\cdot}H_2O$ (analytical grade) [9-10]. The final powders of $Ln_2O_3$-$MO_2$ were prepared by calcination of precursor in a muffle furnace annealed isothermally in air for 3 h. From each synthesized precursor with a certain content of initial oxides 5 powder samples were annealed at 600, 800, 1000, 1200 and 1400°C.

Crystal structures of synthesized powders were studied by X-ray powder diffraction at "Center for X-ray structural research and diagnostics of materials" MISiS using a Rigaku MiniFlex 600 diffractometer with Cu $K_\alpha$ radiation (λ = 1.54178 Å). Also, XRD experiments for $33Ln_2O_3$•$67MO_2$ series were carried out at "Structural Materials Science" beamline of the Kurchatov synchrotron radiation source. Measurements were performed in the transmission mode at λ = 0.68886 Å, using XY detector Imaging Plate Fuji Film BAS-5000 [10, 11]. The Rietveld full-profile analysis of X-ray diffraction patterns was performed with the Jana2006 software [12].

The Raman spectra were collected on a Nicolet iS50 FT-IR spectrometer (Thermo Fisher Scientific Inc.) with Thermo Scientific iS50 Raman module (λ = 1064 nm) at room temperature [4].

## 3. Results and discussion

X-ray powder diffraction analysis of precursor demonstrated that, at any molar ratio of the components, all synthesized samples were X-ray amorphous [4, 11]. At calcination temperature 600°C, order of crystallization in sample with certain molar content x in $xLn_2O_3$ • (1-x)$MO_2$ compound was slightly different for each powder sample, and depended both on the type and concentration of the rare-earth cation.

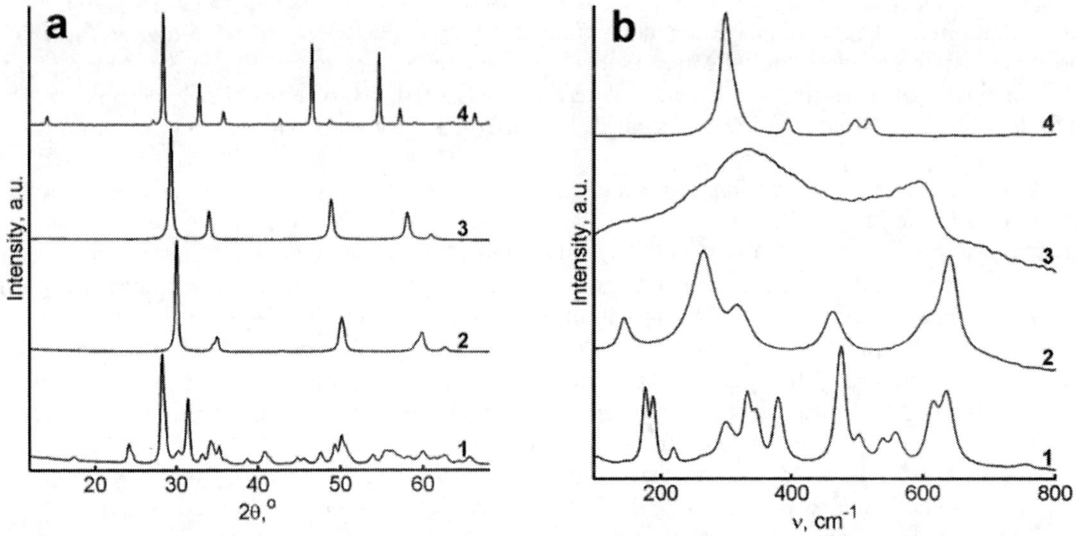


**Figure 1.** XRD patterns (a) and Raman spectra (b) of $x$$La_2O_3$•(1-$x$)$ZrO_2$ powders obtained by calcination of the precursor at 800 and 1200°C: 1 - 3LaSZ (1200°C), 2 - 3LaSZ (800°C), 3 - 18LaSZ (800°C), 4 - 33LaSZ (1200°C).

For example, at molar concentration of $Ln_2O_3$ equals 3 and 8 mol. %, we observed weakly crystallized (CSL ~ 20-30nm) tetragonal structure (space group *P4₂/nmc* (137)) and *t''*-structure (according to Raman) for each concentration, respectively. Figure 1 (curve No. 2) shows approximate XRD patterns and Raman spectra for structures with these types of ordering. The feature of powders with *t''*-ordering mainly consists in the absence of a clear cubic structure in these compounds with presence an admixture of the tetragonal phase [13]. So, the X-ray diffractograms of such compounds represent analogous to the XRD patterns of the structure with fluorite ordering (space group *Fm-3m* (225), shown in Figure 1(a) (curve No. 3). The sample labels are given in the generally accepted abbreviated form: LnSZ = lanthanoide stabilized zirconia. Similarly, an abbreviated form (LnSH) is also accepted for powders based on $HfO_2$.

For higher concentration of $Ln_2O_3$, the presence of a crystal structure in a sample strongly depends on the type of the rare-earth cation. At a concentration of the dopant impurity $x = 18$ mol. %, the lanthanum compounds remain X-ray amorphous at 600°C, while the gadolinium and yttrium compounds have a fluorite-type structure with similar values of CSL ~ 20 nm. When the molar ratio of the initial oxides becomes $Ln_2O_3 : MO_2 = 1 : 2$, only yttrium compounds have a weakly crystallized crystalline structure with very broad peaks (CSL ~ 4-5 nm). Other compounds (with concentration of lanthanum and gadolinium oxides x = 33 mol. %) remain X-ray amorphous at an annealing temperature of 600°C. Moreover, in the $La_2O_3$-$MO_2$ system at $x = 33$ mol. %, the crystallization begins only at temperatures above 800°C.

Thus, the behavior of the crystallization temperature of studied systems can be described by two common tendencies:

1) An increase in the rare-earth cation radius, used as a dopant in the $Ln_2O_3$-$MO_2$ compounds, increases the crystallization temperature.

2) An increase in the content of rare-earth oxide in the $Ln_2O_3$-$MO_2$ compound raises the crystallization temperature of the sample as well.

These results well agree with the results of analysis of the DSC data for $xY_2O_3{\bullet}(1\text{-}x)ZrO_2$ ($x = 0.5$ - 18 mol. %), reported in [14].

At annealing temperatures from 800 to 1000°C, in the $Ln_2O_3$-$MO_2$ systems a gradual transition between low-temperature phases was observed. It was found that an increase in the fraction of $Ln_2O_3$ in the precursor from 3 to 18 mol. % causes a smooth transition from a tetragonal to a fluorite ordering with an intermediate *t''*-structure. In ref. [14] it was also reported that the compound $xY_2O_3{\bullet}(1\text{-}x)ZrO_2$ crystallizes in the tetragonal phase at a molar content $x = 2$ mol. %, while at $x < 1.5$ mol. % these compounds crystallizes mainly into the pure monoclinic structure (space group *P2₁/c* (14). Suming up we can conclude that at relatively low annealing temperatures < 1000°C a series of polymorphic transformations $P2_1/c \rightarrow P4_2/nmc \rightarrow Fm\text{-}3m$ occur in the $Ln_2O_3$-$MO_2$ systems with an increase in the $Ln_2O_3$ concentration.

A further increase of annealing temperatures ≥ 1000°C causes more complex polymorphic transformations in the $Ln_2O_3$-$MO_2$. An example of a such polymorphic behavior in the studied systems is shown in Figure 2 for the $xLa_2O_3{\bullet}(1\text{-}x)HfO_2$ compounds. As can be seen from Figure 2, the specific phase composition of a single sample strongly depends both on the annealing temperature and on the molar concentration of the $La_2O_3$. Along with the tetragonal and fluorite phases, the presence of both monoclinic and pyrochlore (space group *Fd-3m* (227) phases was observed in the synthesized at high temperatures samples. XRD patterns and Raman spectra for all observed phases located in the $xLa_2O_3{\bullet}(1\text{-}x)ZrO_2$ samples have shown in Figure 1. All the experimental curves are given in order of increasing crystal structure symmetry from monoclinic *P2₁/c* (curve No. 1) to pyrochlore *Fd-3m* structure (curve No. 4). It should be noted that maps of the phase composition for $La_2O_3$-$ZrO_2$ compounds are quite similar to those shown in Figure 2. The difference between zirconium and hafnium based $La_2O_3$-$MO_2$ complex oxides concludes in the ratio of the tetragonal and monoclinic phases at higher concentrations of rare-earth oxide (at $x > 8$ mol. %).

In contrast to the compounds doped with lanthanum oxide, the systems with $Gd_2O_3$ and $Y_2O_3$ doping do not have such a large variety of polymorphic phases. This is in particular due to the restriction on the formation of the pyrochlore structure in gadolinium hafnates [15] and zirconates [16], and absence of this possibility in $Y_2O_3$-$MO_2$ system. Thus, for systems based on $ZrO_2$, only the tetragonal and fluorite phases were revealed throughout the annealing temperature range from 600 to 1400°C. It should be noted, that in $8Gd_2O_3{\bullet}92ZrO_2$ samples annealed at 1000 and 1200°C, a higher content of the tetragonal phase (20% and 40%, respectively) was observed in comparison with other $ZrO_2$-based samples.

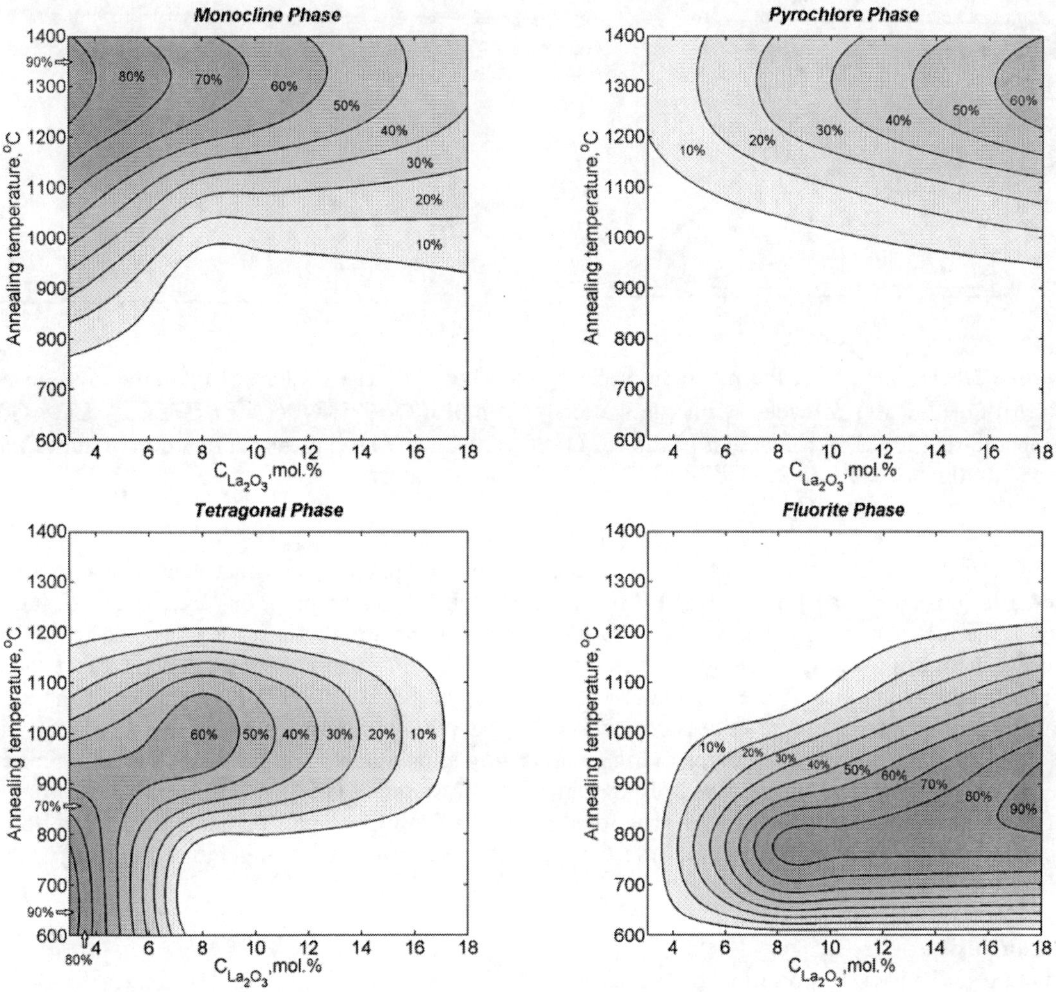


**Figure 2.** Maps of the phase composition for multiphase compounds $xLa_2O_3{\bullet}(1\text{-}x)HfO_2$ ($x$ = 3 - 18 mol. %).

On the other hand, in $HfO_2$ based compounds doped by $Gd_2O_3$ and $Y_2O_3$ the phases with monoclinic and pyrochlore (only in $Gd_2O_3$ containing samples) ordering were also observed in addition to the above-mentioned structures (fluorite and tetragonal). Moreover, the pyrochlore *Fd-3m* structure was detected only in the $xGd_2O_3{\bullet}(1\text{-}x)HfO_2$ samples calcinated at 1200°C, which once again indicates the peculiarity of this temperature for $Gd_2O_3$-$MO_2$ systems.

So, using the obtained results, we can single out several general rules concerning the behavior of the $Ln_2O_3$-$MO_2$ systems. First of all, it should be noted that compounds based on hafnium oxide have a greater ability to form a monoclinic phase at high annealing temperatures in comparison with compounds based on zirconium oxide. Differences in the behavior of systems based on $HfO_2$ and $ZrO_2$ can be seen from Figure 3. It is obvious that the curves for the compounds $xLn_2O_3{\bullet}(1\text{-}x)HfO_2$ are located above the corresponding curves for $x\mathrm{Ln_2O_3}\cdot(1\text{-}x)\mathrm{ZrO_2}$ with the same doping of the $Ln^{3+}$ ions. In addition, Figure 3 shows another common trend for all $Ln_2O_3$-$MO_2$ compounds, which consists in a decrease in the monoclinic phase content with an increase in the concentration of the doped ion.

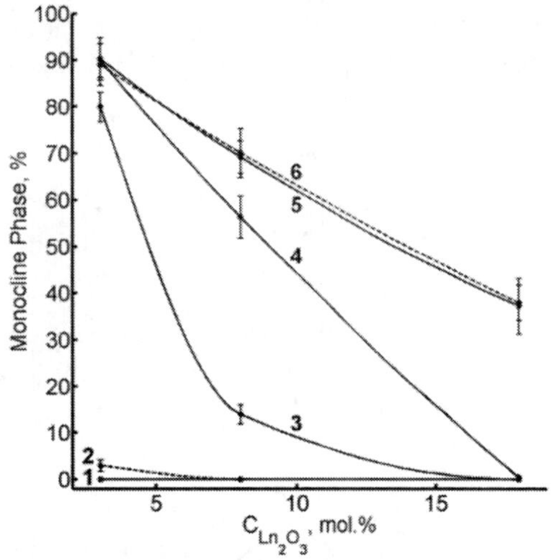


**Figure 3.** The content of the monocline phase in $x\mathrm{Ln_2O_3}\cdot(1\text{-}x)\mathrm{MO_2}$ powders with calcination temperature 1200°C as a function of the $Ln_2O_3$ concentration (1 - YSZ, 2 - GdSZ, 3 - YSH, 4 - LaSZ, 5 - LaSH, 6 - GdSH).

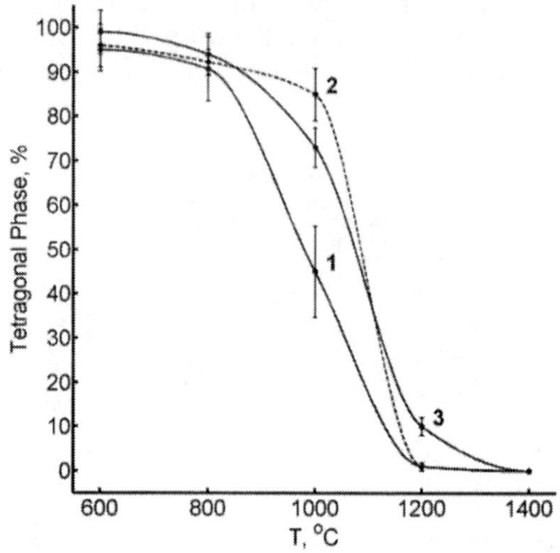


**Figure 4.** The content of the tetragonal phase in $3\mathrm{Ln_2O_3}\cdot97\mathrm{HfO_2}$ (1 - Ln = La, 2 - Ln = Gd, 3 - Ln = Y) samples as a function of annealing temperature.

Finally, it is necessary to point out on another remarkable phenomenon that occurs when various rare-earth ions are replaced in the $Ln_2O_3$-$MO_2$ system, while all other conditions are preserved. It consists in the fact that the efficiency of stabilization of high-temperature phases, namely tetragonal and fluorite, increases in $Ln_2O_3$-$MO_2$ systems as the radius of the $Ln^{3+}$ cation decreases. This effect can be demonstrated by comparing the tetragonal phase content in the $3\mathrm{Ln_2O_3}\cdot97\mathrm{HfO_2}$ samples at the same annealing temperature (Figure 4). It is clearly seen that doping with rare-earth ions in accordance with $La^{3+} \rightarrow Gd^{3+} \rightarrow Y^{3+}$ shifts the curves to higher annealing temperatures and increases the region of existence of the tetragonal phase. The only exception is the sample $3\mathrm{Gd_2O_3}\cdot97\mathrm{HfO_2}$, for which the tetragonal phase content is higher than that for the $3\mathrm{Y_2O_3}\cdot97\mathrm{HfO_2}$ sample. Perhaps, this is due to the peculiarity of the unusual behavior of the $Gd_2O_3 \cdot HfO_2$ system in the temperature range of 1000 - 1200°C, which was noted above.

## 4. Conclusion

Structure and phase composition of the zirconium and hafnium based complex oxides formed in "$Ln_2O_3$-$MO_2$" system has been studied using a combination of X-ray powder diffraction and Raman spectroscopy. As a result, it was found that the crystallization temperature of the synthesized powders increases both with an increase in the $Ln_2O_3$ concentration and an increase in the radius of the $Ln^{3+}$ cation. Also, it was revealed that the formation of polymorphic forms in $x\mathrm{Ln_2O_3}\cdot(1\text{-}x)\mathrm{MO_2}$ compounds depends strongly on both the ratio of initial oxides $Ln_2O_3$ and $MO_2$, and the annealing temperature of precursor. The observed behavior of the $Ln_2O_3$-$MO_2$ systems can be reduced into two general rules:

1) Efficiency of stabilization of high-temperature phases (tetragonal and cubic) in $Ln_2O_3$-$ZrO_2$ systems with the introduction of $Ln^{3+}$ cations increases in the $La^{3+} \rightarrow Gd^{3+} \rightarrow Y^{3+}$ series in accordance with the increase in the solubility of $Ln_2O_3$ in $ZrO_2$. A similar effect is also valid for the $Ln_2O_3$-$HfO_2$ systems.

2) Compounds based on $HfO_2$ have a greater ability to form a monoclinic phase at high annealing temperatures in comparison with compounds based on $ZrO_2$.

**Acknowledgments**
The present work was supported by the Ministry of Education and Science of the Russian Federation (grant No 3.9750.2017/8.9).

**References**

[1] Subramanian M, Aravamudan G, Subba Rao G 1983 *Prog. Solid State Chem.* **15** 55
[2] Blanchard P, Clements R, Kennedy B, Ling C, Reynolds E, Avdeev M, Stampfl V, Zhang Z, Jang L-Y 2012 *Inorg. Chem.* **51** 13237
[3] Blanchard P, Liu S, Kennedy B, Ling C, Avdeev M, Aitken J, Cowie B, Tadich A 2013 *J. Phys. Chem. C* **117** 2266
[4] Popov V V, Menushenkov A P, Yaroslavtsev A A, Zubavichus Ya V, Gaynanov B R, Yastrebtsev A A, Leshchev D S and Chernikov R V 2016 *J. Alloys Compd.* **689** 669
[5] Gardner J, Gingras M, Greedan J 2010 *Rev. Mod. Phys.* **82** 53
[6] Jiang S P and Chan S H 2004 *J. Mater. Sci.* **39** 4405
[7] Risovany V D, Zakharov A V, Muraleva E M, Kosenkov V M, Latypov R N. 2006 *J. Nucl. Mater.* **355** 163
[8] Wan C, Qu Z, Du A, Pan W 2011 Order–Disorder *J. Am. Ceram. Soc.* **94** 592
[9] Popov V V 2015 *Russ. J. Inorg. Chem.* **60** 420
[10] Popov V V, Petrunin V F, Korovin S A, Menushenkov A P, Kashurnikova O V, Chernikov R V, Yaroslavtsev A A and Zubavichus Y V 2011 *Russ. J. Inorg. Chem.* **56** 1538
[11] Popov V V, Menushenkov A P, Zubavichus Y V, Yaroslavtsev A A, Leshchev D S, Kulik E S, Bednarcik J, Petrunin V F, Korovin S A and Chernikov R V 2013 *Russ. J. Inorg. Chem.* **58** 1400
[12] Petricek V, Dusek M and Palatinus L 2014 *Z. Kristallogr.* **229** 345
[13] Yashima M, Takahashi H, Ohtake K, Hirose T, Kakihana M, Arashi H, Ikuma Y, Suzuki Y and Yoshimura M 1996 *J. Phys. Chem. Solids* **57** 289
[14] Popov V V, Yastrebtsev A A, Korovin S A, Tsarenko N A and Arzhatkina L A 2016 *Russ. J. Inorg. Chem.* **61** 1378-1386
[15] Popov V V, Zubavichus Y V, Menushenkov A P, Yaroslavtsev A A, Kulik E S, Pisarev A A and Kolyshkin N A 2015 *Russ. J. Inorg. Chem.* **60** 16
[16] Popov V V, Zubavichus Y V, Menushenkov A P, Yaroslavtsev A A, Kulik E S, Petrunin V F Korovin S A, Trofimova N N 2014 *Russ. J. Inorg. Chem.* **59** 279

IOP Conf. Series: Journal of Physics: Conf. Series **941** (2017) 012082 doi:10.1088/1742-6596/941/1/012082

# High-pressure cells for study of condensed matter by diffraction and inelastic neutron scattering at low temperatures and in strong magnetic fields.

**R A Sadykov[1,2], Th Strassle[3], A Podlesnyak[3,4], L Keller[3], B Fak[5,6], J Mesot[3]**

[1] Institute for Nuclear Research, Russian Academy of Sciences, Prospekt 60 Letiya Oktiabria 7a, 117312 Moscow, Russia
[2] Vereschagin Institute for High Pressure Physics, Russian Academy of Sciences, 142190 Troitsk, Moscow, Russia
[3]Laboratory for Neutron Scattering, ETH&PSI, 5232 Villigen PSI, Switzerland.
[4] Oak Ridge National Laboratory, Oak Ridge, Tennessee 37831, USA
[5] Commissariat à l'EnergieAtomique, Département de Recherche Fondamentale sur la Matière Condensée, SPSMS, 38054 Grenoble, France
[6] Institut Laue-Langevin 71 avenue des Martyrs 38000 Grenoble, France

E-mail: rsadykov@inr.ru

**Abstract.** We have developed and implemented series of new original clamp high-pressure cells for neutron diffraction and inelastic neutron scattering at low temperatures. The cells design allows one to place them in the standard cryostats or cryomagnets used on neutron sources. Some results obtained for ZnCr2Se4 are demonstrated as an example.

## 1. Introduction
The increased interest in quantum critical phenomena, including under pressure caused the need for direct neutron scattering experiments performed at low temperatures in strong magnetic fields and at high pressure. We have made various cylinder-piston high pressure clamp cells (HPCC) for diffraction and inelastic neutron scattering experiments under pressure up to 1.5 GPa at neutron diffractometers DMC (SINQ, PSI, Switzerland) and ILL.

## 2. Two clamp high pressure cells from zero TiZr alloy
The cells made of nonmagnetic alloys TiZr (zero matrix), as in paper [1], and hard of Al-alloys (V95T, V96T) were successfully used in studies of spiral magnetic structures $ZnCr_2Se_4$ as well as magnetic structures in $NdAl_3$ and $Er_{0.57}Y_{0.43}Co_2$ compounds under pressure up to 15 kbar at temperature $T = 4.4$ K [2,3,4].We present some results obtained in neutron powder diffraction study of the magnetic spiral $ZnCr_2Se_4$ structure under high pressure. $ZnCr_2Se_4$ undergoes a transition from the paramagnetic state to a simple spiral structure at the Neel temperature $T_N$ of 21K [5,6]. The cube spinel crystal structure has a small tetragonal distortion (aprox.$10^{-4}$) at this temperature [7]. This distortion increases with decreasing temperature and the period of the spiral decreases correspondingly [6]. To investigate the pressure dependence of the spiral magnetic structure we have performed a neutron powder diffraction study using two high pressure clamp cells (Figure 1 a,b).

IOP Conf. Series: Journal of Physics: Conf. Series **941** (2017) 012082 doi:10.1088/1742-6596/941/1/012082

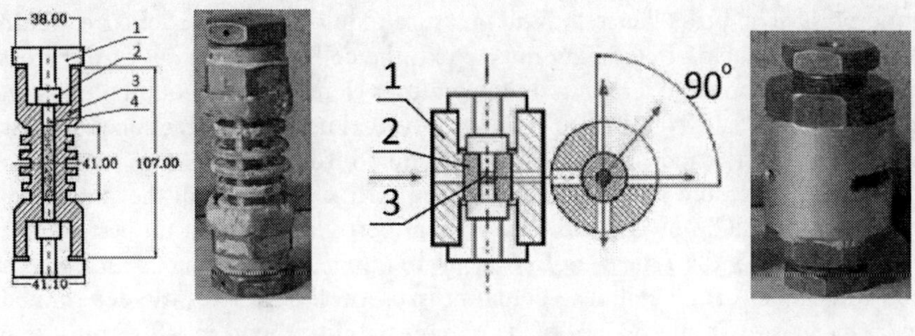


(a) (b)

**Figure 1.** Sketches and photos of the two zero-matrix (TiZr alloy) high pressure clamp cells used in PSI-LNS. (a) - small clamp HPCC for use of up to 12kbar: 1- nut-steel, 2- and 4-support and piston-hard steel, 3- TiZr, zero matrix alloy. (b) - big clamp HPCCfor use of up to 15 kbar: 1- support from hard steel, 2- supported inner part -zero alloy TiZr, 3- sample. Diameters of the inner hole are 6mm for (a)-cell and 1 cm for (b)-cell. Heights of the samples are 25mm for (a)-cell and 6mm for (b)-cell.

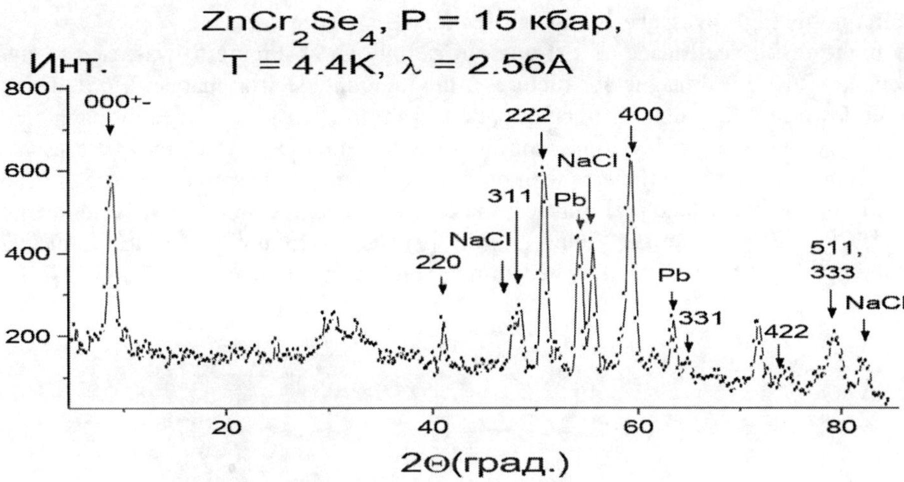


**Figure 2.** Neutron diffraction pattern of $ZnCr_2Se_4$ at $P$ = 15 kbar and $T$ = 4.4 K).

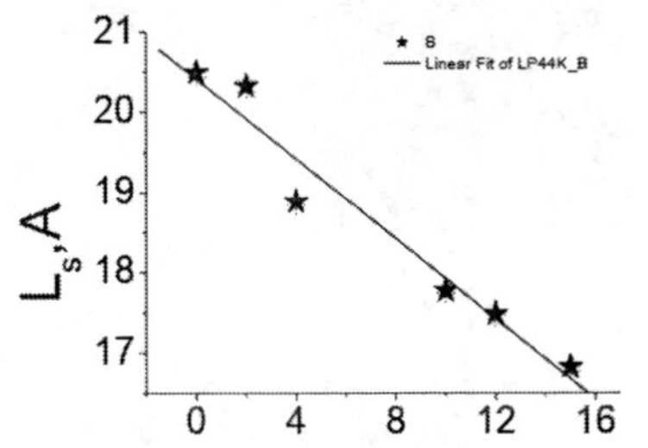


**Figure 3**. The length of the magnetic spiral *vs* pressure at $T$ = 4.4 Kin $ZnCr_2Se_4$

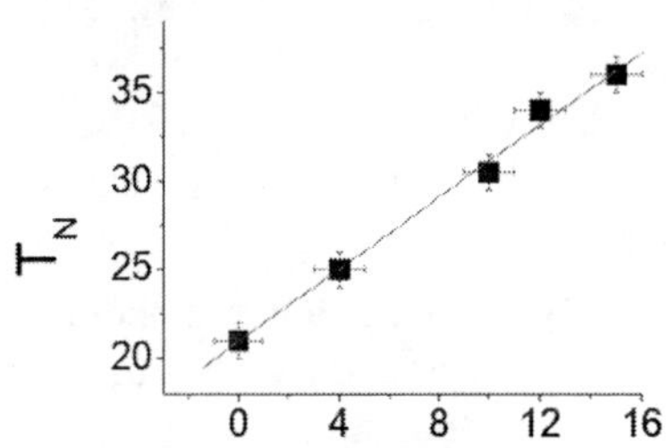


**Figure 4.** The Neel temperature *vs* pressure in $ZnCr_2Se_4$

A powder sample of semi-conducting materials was placed into the cell, then the cell was mounted in a hydraulic press. The press generated the pressure required for a particular experiment. This pressure was fixed with the help of clamping nuts. Next, the cell was taken out of the press and put into a cryostat, where it was cooled to a required temperature. Neutron diffraction patterns were obtained on the diffractometer DMC at SINQ. Figure 2 shows a typical neutron diffraction pattern at $T = 4.4$ K and $P = 15$ kbar. The length (period ) of the spiral determined from shift 000(+-) satellite is plotted in Figure 3 as a function of pressure at 4.4 K. The neutron diffraction study of the simple spiral structure of $ZnCr_2Se_4$ mixed with NaCl powder and liquid Fluorinert was made in the temperature and pressure ranges 4.4-50 K and 0-15 kbar.The sample was put in thin lead foil cylinder for decreasing the friction of the piston in channel of the cell. The actual pressure in the sample was determined by comparing the NaCl peak position with the values of its compressibility found in [8]. From the obtained results we can conclude that the period of the spiral structure decreases (Figure 3) and consequently the angle of the spiral magnetic structure increases under pressure up to 15 kbar at $T = 4.4$ K. The Neel temperature increases under pressure up to 36 K at $P = 15$ kbar (Figure 4).

The lattice parameter of $ZnCr_2Se_4$ at $T = 4.4$ K changes with increase of the pressure from 10.4712(0.0013) Å at $P = 0$ up to 10.3991(0.0011) Å at $P = 15$k bar. Accordingly, the increase of both $T_N$ and the spirals angle may be understood if the antiferromagnetic superexchange interactions between the $Cr^{3+}$ ions increase more stronger than the ferromagnetic ones with decrease of the sample volume under pressure.

## 3. Nonmagnetic high-pressure clamp cell from NiCrAl alloy

A new high-pressure cell made of a strong NiCrAl alloy [9] (Figure 5) was used to study a pressure dependence of the spiral magnetic structure of the itinerant electron magnet MnSi. This cell was nonmagnetic down to 2 K and no diffraction pattern from the cell was noticed in magnetic fields up to 6 Tesla. A high sensitivity to deviations from hydrostatic conditions which may be present in the clamping cells is one of the main problems with measuring the magnetic spiral structure of MnSi under pressure. To improve hydrostatic performance, the crystal was placed inside the Teflon cell directly in the 50-50% FC87 / FC84 Fluorinert liquid mixture [9] (the maximum hydrostatic limit is 23 kbar). The cell of the nonmagnetic NiCrAl alloy had an inner and outer diameters of 4.7 and 12.7 mm.

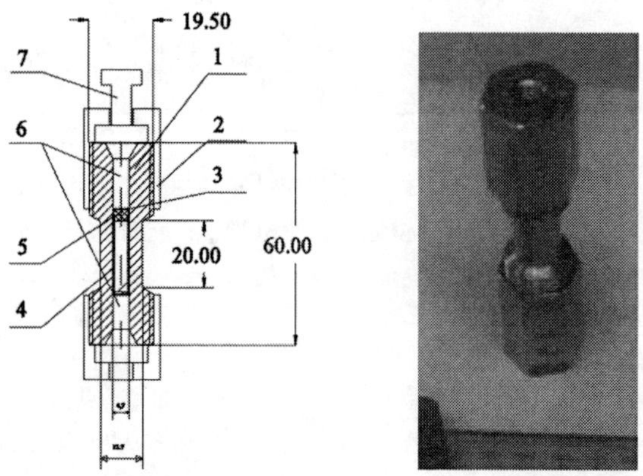


**Figure 5.** Sketch and photo of the clamp pressure cell (HPCC20-NiCrAl) for neutron diffraction studies in magnetic fields and under pressure up to 20 kbar. 1- cell body (nonmagnetic alloy 40HNU (NiCrAl)), 2- nut (nonmagnetic Ti alloy), 3- extrusion ring (CuBe), 4- sample can (Pb), 5- can caps (Pb), 6- pistons (nonmagnetic alloy 40HNU), 7- pistons.

Pistons of non-magnetic NiCrAl were used excepting for the highest pressure for which weakly magnetic pistons of tungsten carbide were used. A magnetic field in the horizontal scattering plane

IOP Conf. Series: Journal of Physics: Conf. Series **941** (2017) 012082 doi:10.1088/1742-6596/941/1/012082

was used to induce a controlled magnetic region in the sample, which is especially important at high pressures. The crystallographic axis $[11\bar{1}]$ was parallel to the magnetic field with precision better than $0.2^{\circ}$. The small size of the pressure cell allowed us to thermalize the sample for 15 min in the actual temperature range 1.5-35 K. All $M(T)$ measurements were carried out with increasing temperature and thermal hysteresis was not controlled.

## 4. High-pressure clamp cell for inelastic scattering on the TOF spectrometer FOCUS

A clamping cell made of a strong Al alloy was developed and tested to perform measurements by inelastic scattering (INS) of neutrons under pressure. The sketch of the cell is shown in Figure 6. The cell is a clamp-type pressure cell, however some features differ from elements with zero matrix pressure. The INS signal is significantly smaller in comparison with an elastic signal, for example, in neutron diffraction measurements. The absorption associated with the cell must be minimized for successful INS experiments under pressure while a sample volumes should be as maximum as possible. The geometry of the TOF spectrometer FOCUS (PSI, SINQ) requires also optimization of the cell geometry in order to minimize data corrections due to anisotropic cell absorption and the presence of a parasitic inelastic intensity arising from elastic neutrons multiply scattered by the sample and the cell. The developed cell meets these requirements. The axial symmetric cell with an internal diameter of 70 mm and an inner diameter of 7 mm was made of a reinforced aluminum type. It provides a sample volume of 2500 mm$^3$ at the maximum pressure of 10 kbar. Aluminum absorbs some few neutrons, but it is not so much strong material. To strengthen the cell an inner tube with a core of 1.5 mm thick, made of strong hardened steel (or an alloy with a zero matrix) was used, and the nut was screwed into a hardened threaded steel insert allowing to support the steel insert (tube) and inner Al-cylinder. Losses due to absorption and elastic scattering on the cell were estimated as about 6 times for the incoming neutron wavelength $\lambda = 2.09$ Å. No inelastic features originating from the cell were observed in the spectra measured for $\lambda = 2.09$, 4.3 and 5.0 Å above and below the Bragg limit. The cell was filled in the same way as for cells with zero matrix. Measuring the INS shift of the pressure calibrator lattice parameter were supplemented by measurements on a powder diffractometer before or after the INS experiment. Experience with clamping cells has shown that the pressure in the cell is maintained for several days and does not depend on repeated thermal cycling.

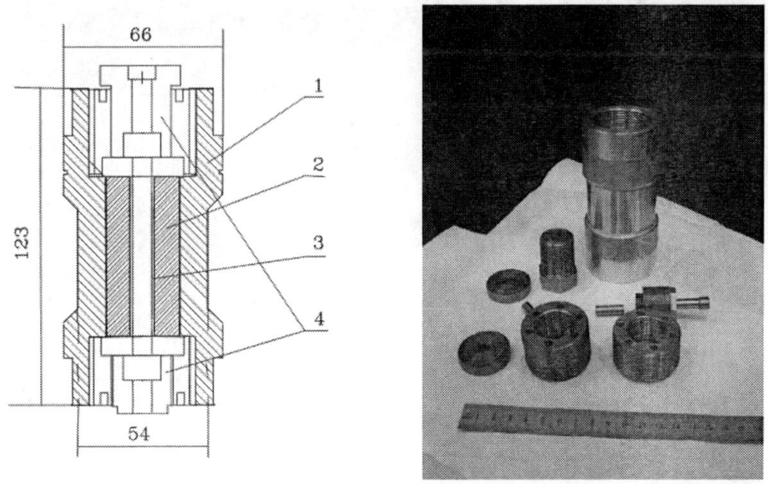


**Figure 6**. Sketch and photo of the clamp pressure cell (HPCC10-Al) for inelastic neutron scattering studies on the TOF spectrometer FOCUS(SINQ) up to 10 kbar ($T$ = 2-300K, $V$ = 1.6 cm$^3$). 1 - Al alloy B95T, 2 - insert part Al alloy B96T, 3 - inner part steel 45XMHΦA, 4 - nuts from steel 45XMHΦA.

IOP Conf. Series: Journal of Physics: Conf. Series **941** (2017) 012082 doi:10.1088/1742-6596/941/1/012082

**References**

[1] Sadykov R A, Gruzin P L, Suhoparov V L 1995 *High Pressure Research* **14** 199-202
[2] Podlesnyak A, Strassle Th, Mirmelstein A, Pirogov A and Sadykov R 2002 *Eur. Phys. J. B* **29** 547-52
[3] Strassle T, Divis M, Rusz J, Janssen S, Juranyi F, Sadykov R and Furrer A 2003 *J. Phys.: Condens. Matter* **15** 3257-66
[4] Fak B, Sadykov R, Flouquet J, Lapertot G 2005 *J. Phys.: Condens.Matter* **17** 1635-44
[5] Plumier R J 1966 *J. Appl. Phys.* **37**(3) 964
[6] Jun Akimitsu *at all.* 1978 *J . Phys. Soc. Japan* **44**(1) 172
[7] Kleinberger R and R. de Kouchkovsky 1966 *CR Acad .Sc.Paris* **262**(Ser.B) 628-30
[8] Decker D L 1971 *J.Appl.Phys.* **42** 3239
[9] Sidorov V A and Sadykov R A 2005 *J. Phys.: Condens. Matter* **17** S3005–8

# Statistical precision in charged particle EDM search in storage rings

**A E Aksentev**[1,2] **and Y V Senichev**[2]

[1] IKP, Forschungszentrum Jülich, Jülich, Germany

[2] National Research Nuclear University "MEPhI," Moscow, Russia

E-mail: `a.aksentev@fz-juelich.de`

**Abstract.** Currently, the "Jülich Electric Dipole moment Investigations" (JEDI) collaboration, together with present EDM experiments at the COSY ring, is developing the conceptual design of a ring specifically for the search for the deuteron electric dipole moment (dEDM). One of the main problems in the EDM study is the spin precession in the vertical plane caused by the non-ideal positioning of accelerator elements through the magnetic dipole moment (MDM). The idea of how to separate the EDM from MDM is based on measuring the spin tune in different processes and comparing the results. The high precision of the spin tune measurement is achieved by collecting huge amounts of data. The JEDI collaboration aims at detecting the EDM at a level better than $10^{-29}$ $e \cdot cm$, for which one requires a precision in the frequency estimate $\sim 10^{-9}$ $rad/sec$. An estimate's statistical precision is conditional on the following factors: the total measurement time, determining the independent variable spread; measurement error; temporal modulation and spacing of sample points. In this paper we analyze the interplay between these factors, and estimate the best achievable precision under given conditions.

## 1. Detector counting rate model

We assume the following model for the detector counting rate:

$$N(t) = N_0(t) \cdot \left(1 + P \cdot e^{-t/\tau_d} \cdot \sin(\omega \cdot t + \phi)\right), \tag{1}$$

where $\tau_d$ is the decoherence lifetime, and $N_0(t)$ is the counting rate from the unpolarized cross-section.

Since the beam current can be expressed as a function of time as

$$I(t) \equiv N^b(t)\nu = I_0 \cdot e^{\lambda_b t},$$

$\lambda_b$ the beam lifetime, the expected number of particles scattered in the direction of the detector during measurement time $\Delta t_c$ is

$$\begin{aligned} N_0(t) &= p \cdot \int_{-\Delta t_c/2}^{+\Delta t_c/2} I(t+\tau)\mathrm{d}\tau \\ &= p \cdot \frac{\nu N_0^b}{\lambda_b} e^{\lambda_b t} \cdot \left(e^{\lambda_b \Delta t_c/2} - e^{-\lambda_b \Delta t_c/2}\right) \\ &\approx \underbrace{p \cdot \nu N_0^b e^{\lambda_b t}}_{\text{rate } r(t)} \cdot \Delta t_c, \end{aligned} \tag{2}$$

where $p$ is the probability of "useful" scattering (approximately 1%, according to Y.V. Senichev, Dr. Sci. (written communication, December 2016)).

The actual number of detected particles will be distributed as a Poisson distribution

$$P_{N_0(t)}(\tilde{N}_0) = \frac{(r(t)\Delta t_c)^{\tilde{N}_0}}{\tilde{N}_0!} \cdot e^{-r(t)\Delta t_c},$$

hence $\sigma^2_{\tilde{N}_0}(t) = N_0(t)$.

We are interested in the expectation value $N_0(t) = \mathrm{E}\left[\tilde{N}_0(t)\right]$, and its variance $\sigma_{N_0}(t)$. Those are estimated as summary statistics:

$$\langle \tilde{N}_0(t) \rangle_{\Delta t_\epsilon} = \sum_{i=1}^{n_{c/\epsilon}} \tilde{N}_0(t_i), \; n_{c/\epsilon} = \Delta t_\epsilon / \Delta t_c,$$

and

$$\sigma_{\tilde{N}_0(t)|\Delta t_\epsilon} = \sum_{i=1}^{n_{c/\epsilon}} \left(\tilde{N}_0(t_i) - \langle \tilde{N}_0(t_i) \rangle_{\Delta t_\epsilon}\right)^2.$$

($\Delta t_\epsilon$ is the event measurement time, $\Delta t_c$ is the polarimetry measurement time.) Being a sum of random variables, $N_0(t)$ is normally distributed.

The standard error of the mean then is

$$\sigma_{N_0}(t) = \sigma_{\tilde{N}_0}(t)/\sqrt{n_{c/\epsilon}} = \sqrt{N_0(t)\frac{\Delta t_c}{\Delta t_\epsilon}}$$
$$\approx \sqrt{\frac{p \cdot \nu N_0^b}{\Delta t_\epsilon} \cdot \Delta t_c \cdot \exp\left(\frac{\lambda_b}{2} \cdot t\right)}.$$

Relative error grows:

$$\frac{\sigma_{N_0}(t)}{N_0(t)} \approx \frac{A}{\sqrt{\Delta t_\epsilon}} \cdot \exp\left(-\frac{\lambda_b}{2} t\right) = \frac{A}{\sqrt{\Delta t_\epsilon}} \cdot \exp\left(\frac{t}{2\tau_b}\right), \; A = \frac{1}{\sqrt{p \cdot \nu N_0^b}}. \tag{3}$$

## 2. Figure of merit

A measure of the beam's polarization is the relative asymmetry of detector counting rates: [1, p. 17]

$$\mathcal{A} = \frac{N(\frac{\pi}{2}) - N(-\frac{\pi}{2})}{N(\frac{\pi}{2}) + N(-\frac{\pi}{2})}. \tag{4}$$

In the simulation to follow, the function fitted to the asymmetry data is:

$$\mathcal{A}(t) = \mathcal{A}(0) \cdot e^{\lambda_d \cdot t} \cdot \sin(\omega \cdot t + \phi), \tag{5}$$

with three nuisance parameters $\mathcal{A}(0)$, $\lambda_d$, and $\phi$.

Due to the decreasing beam size, the measurement of the figure of merit is heteroscedastic. From [1, p. 18], the heteroscedasticity model assumed is

$$\sigma^2_{\mathcal{A}}(t) \approx \frac{1}{2N_0(t)}. \tag{6}$$

## 3. Conditions for maximum precision

Assuming a Gaussian error distribution with mean zero and variance $\sigma_\epsilon^2$, the maximum likelihood estimator for the variance of the frequency estimate of the cross-section asymmetry $\mathcal{A}$ can be expressed as

$$\mathrm{var}\left[\hat{\omega}\right] = \frac{\sigma_\epsilon^2}{X_{tot} \cdot \mathrm{var}_w\left[t\right]},$$

with

$$X_{tot} = \sum_{j=1}^{n_\epsilon} x_j = \sum_{s=1}^{n_{zc}} \sum_{j=1}^{n_{\epsilon/zc}} x_{js},$$

$$\mathrm{var}_w\left[t\right] = \sum_i w_i \left(t_i - \langle t\rangle_w\right)^2, \quad \langle t\rangle_w = \sum_i w_i t_i,$$

$$w_i = \frac{x_i}{\sum_j x_j}, \quad x_i = (\mathcal{A}(0)\exp(\lambda_d t_i))^2 \cos^2(\omega t_i + \phi) = \left(\mu'_\phi(t_i)\right)^2.$$

In the expression above, $X_{tot}$ is the total Fisher information of the sample, and $\mathrm{var}_w\left[t\right]$ is a measure of its time-spread. It can be observed that by picking appropriate sampling times, one can raise the $X_{tot}$ term, since it is proportional to a sum of the signal's time derivatives. If the oscillation frequency and phase are already known to a reasonable precision, further improvement can be achieved by the application of a sampling scheme in which measurements are taken only during rapid change in the signal (sampling modulation). Improvement here is limited by the polarimetry sampling rate.

Both the $\mathrm{var}_w\left[t\right]$ and $X_{tot}$ terms are bounded as a result of spin tune decoherence. We can express $\sum_{j=1}^{n_{\epsilon/zc}} x_{js} = n_{\epsilon/zc} \cdot x_{0s}$, for some mean value $x_{0s}$ at a given node $s$. $n_{\epsilon/zc}$ is the number of asymmetry measurements per node. The period of time during which measuring takes place, $\Delta t_{zc}$, is termed *compaction time*. The value of the sum $\sum_{j=1}^{n_{\epsilon/zc}} x_{js}$ falls exponentially due to decoherence, hence $x_{0s} = x_{01}\exp\left(\lambda_d \cdot \frac{(s-1)\cdot\pi}{\omega}\right)$. Therefore,

$$X_{tot} = n_{\epsilon/zc} \cdot x_{01} \cdot \frac{\exp\left(\frac{\lambda_d \pi}{\omega} n_{zc}\right) - 1}{\exp\left(\frac{\lambda_d \pi}{\omega}\right) - 1} \equiv n_{\epsilon/zc} \cdot x_{01} \cdot g(n_{zc}); \tag{7}$$

$$x_{01} = \frac{1}{\Delta t_{zc}} \int_{-\Delta t_{zc}/2}^{+\Delta t_{zc}/2} \cos^2(\omega \cdot t) \mathrm{d}t = \frac{1}{2} \cdot \left(1 + \frac{\sin \omega \Delta t_{zc}}{\omega \Delta t_{zc}}\right), \tag{8}$$

$$n_{\epsilon/zc} = \frac{\Delta t_{zc}}{\Delta t_\epsilon}. \tag{9}$$

Eq. (7) provides us with a means to estimating the limits on the duration of the experiment. In Table 1, the percentage of the total Fisher information limit, the time in decoherence lifetimes by which it is reached, and the signal-to-noise ratio by that time, are summarized. The signal-to-noise ratios are computed according to:

$$\mathrm{SNR} \triangleq \frac{\mathcal{A}(0) \cdot e^{-t/\tau_d}}{\sigma_{\mathcal{A}}(t)} \approx \sqrt{2 \cdot p \cdot \nu N_0^b \cdot \Delta t_c} \cdot \mathcal{A}(0) \cdot \exp\left[-\frac{t}{\tau_d} \cdot \left(1 + \frac{1}{2}\frac{\tau_d}{\tau_b}\right)\right], \tag{10}$$

in which, from $\sigma_{\mathcal{A}(0)}/\mathcal{A}(0) \approx 3\%$, the factor before the exponent is 33.

IOP Conf. Series: Journal of Physics: Conf. Series **941** (2017) 012083 doi:10.1088/1742-6596/941/1/012083

Table 1: Amount of Fisher information (in percents of the available limit) contained in a sample collected for the duration specified in decoherence lifetimes, and the corresponding signal-to-noise ratio.

| FI limit (%) | Duration ($\times\tau_d$) | SNR |
|---|---|---|
| 95 | 3.0 | 0.4 |
| 90 | 2.3 | 1.1 |
| 70 | 1.2 | 5.5 |
| 50 | 0.7 | 11.7 |

## 4. Simulation

We simulated data from two detectors with parameters gathered in Table 2 for $T_{tot} = 1000$ seconds, sampled uniformly at the rate $f_s = 375$ Hz. These figures are chosen for the following reason: the beam size in a fill is on the order of $10^{11}$ particles; if we want to keep the beam lifetime equal to the decoherence lifetime, we cannot exhaust more than 75% of it; only 1% of all scatterings are of the sort we need for polarimetry, so we're left with $7.5 \cdot 10^8$ useful scatterings. A measurement of the counting rate $N_0(t)$ with a precision of approximately 3% requires somewhere in the neighborhood of 2000 detector counts, which further reduces the number of events to $3.75 \cdot 10^5 = f_s \cdot T_{tot}$. One thousand seconds is the expected duration of a fill, hence $f_s = 375$ Hz.

Relative measurement error for the detector counting rates is depicted in Figure 1; the cross-section asymmetry, computed according to Eq. (4), is shown in Figure 2. To these data we fit via Maximum Likelihood a non-linear heteroscedastic model given by Eq. (5), with the variance function for the weights given by Eq. (6). The fit results are summarized in Table 3.

Table 2: Detector counting rates' model parameters.

| | Left | Right | |
|---|---|---|---|
| $\phi$ | $-\pi/2$ | $+\pi/2$ | rad |
| $\omega$ | 3 | | rad/sec |
| $P$ | 0.4 | | |
| $\tau_d$ | 721 | | sec |
| $\tau_b$ | 721 | | sec |
| $N_0(0)$ | 6730 | | |

Table 3: Asymmetry fit results.

| Parameter | Estimate | SE | Unit |
|---|---|---|---|
| $\mathcal{A}(0)$ | 0.400 | $9.03 \cdot 10^{-5}$ | |
| $\lambda_d$ | −0.001 | $7.86 \cdot 10^{-7}$ | 1/sec |
| $\omega$ | 3.000 | $7.55 \cdot 10^{-7}$ | rad/sec |
| $\phi$ | −1.571 | $2.25 \cdot 10^{-2}$ | rad |

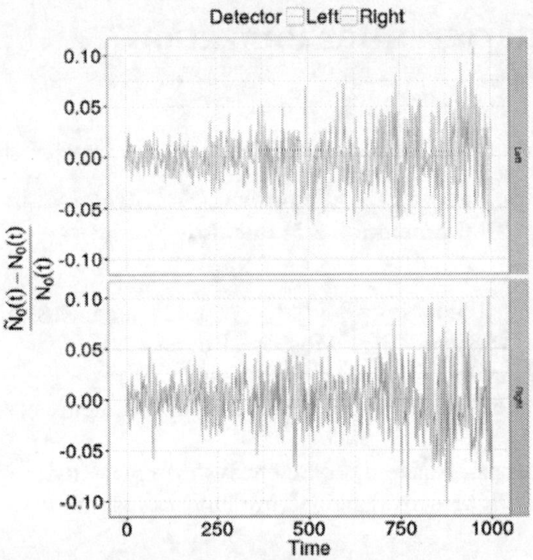


Figure 1: Simulated relative counting rate measurement error for the left and right detectors as a function of time.

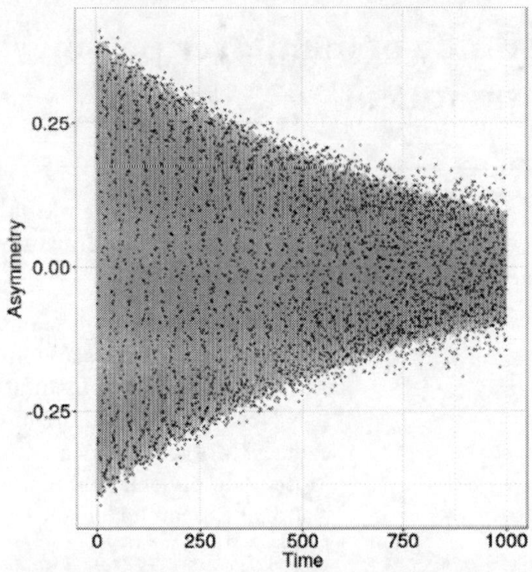


Figure 2: Expectation value (red line) and sample measurements (black dots) of the cross-section asymmetry in a simulation.

### *4.1. Modulation gains*

If the initial frequency estimate obtained from a time-uniform sample has a standard error on the order of $1 \cdot 10^{-6}$ rad/sec, simulation shows the standard error of the estimate can be improved to $\approx 5.8 \cdot 10^{-7}$ rad/sec.

## References

[1] Eversmann D. Analysis of the Spin Coherence Time at the Cooler Synchrotron COSY [master's thesis on the Internet]. [Aachen (Germany)]: Rheinisch-Westfälische Technische Hochschule Aachen (RWTH); 2013 [cited 2017 Feb 28]. Available from: `http://wwwo.physik.rwth-aachen.de/fileadmin/user_upload/www_physik/Institute/Inst_3B/Mitarbeiter/Joerg_Pretz/DEMasterarbeit.pdf`

# Study of multilayer polymer materials after ionization treatment

**V T Tarasyuk[1], A A Semkina[1], V I Solovyeva[1], D D Fedotova[1], N E Strokova[2], D M Malenko[3], O V Baranov[4], A V Bakumenko[5], S N Puchkov[5] and A V Prokopenko[6]**

[1] Russian Research Institute of Canning Technology, Vidnoye, Moscow area, Russia
[2] M.V. Lomonosov Moscow State University (MSU), Moscow, Russia
[3] All-Russian Research Institute of Dairy Industry, Moscow, Russia
[4] A.N.Nesmeyanov Institute of Organoelement Compounds of Russian Academy of Sciences, Moscow, Russia
[5] Moscow Radiotechnical Institute Russian Academy of Sciences, Moscow, Russia
[6] National Research Nuclear University - Moscow Engineering Physics Institute, Moscow, Russia

E-mail: tara@vniitek.ru; pav14@mail.ru

**Abstract**. Electron-beam technologies of food products processing involves the use of modern packaging materials in form of polymer films of different composition. The objective of the research is to study the impact of accelerated electrons on the structure of the polymeric packaging materials used for storage of agricultural products. It was investigated radiation exposure on film material PE/PA (80/20) with a thickness of 80 mkm. This film used for storage of vegetables and fruits and has the necessary indicators for gas and vapor permeability. Electron beam treatment of the films was performed on a compact radiation sterilization installation with local bio-protection with electron energy of 5 MeV. A polymer films were irradiated with doses from 1 to 10 kGy. Changing the structure of the film composition was monitored by IR spectrometry. As a result of irradiation by accelerated electrons with doses up to 18 kGy is established that the polymer film is modification of the polymeric material in the form of a partial degradation with subsequent intra-molecular cross-linking. This improves the physico-mechanical properties in the transverse direction, and such film can be used for food packaging before electron-beam treatment.

## 1. Introduction

Radiation technologies of food processing and food packing is accurate radiation dosing of radiation, the possibility of exposure of packaged products, a high degree of efficiency and performance, and low operating costs [1]. High-energy electrons beam treatment reduces level of pathogenic bacteria and viruses. The applied dose is relatively different from 0.05 – 0.15 kGy to inhibit sprouting of potatoes and onions to 2.0 – 10 kGy to improve technological properties of food products and to 30 - 50 kGy for sterilizing meat products.

In the industry to maintain sterilization effects and increase the shelf life of foods often radiation sterilization is carried out in the package. Basically multilayer film materials of different composition are used. By combining multiple layers of different polymers, the manufacturer is able to create film materials with desired properties for each type of agricultural products, taking into account the respiratory processes.

In radiation sterilization process it is possible to observe in polymers the processes of crosslinking and degradation. The impacts of electron beams in the presence of oxygen could be form of oxygen-containing groups: -C=O, -OH, -O-O-H, -C-O-O-C-. Groups C=O are formed of different types: acid, ketone, aldehyde, ester, percolate, perepisnye. These groups have characteristic absorption bands in the IR spectrum [2]. When exposed to radiation on the macromolecules of the polymers, they can break down into free radicals with rupture of the C-H, C-C etc [3].

There are works on studying of influence of irradiation on the structure of single-component film materials and their physical and mechanical properties [4-6]. At this work the radiation impact on the suitability of multi-component film materials for packaging agricultural products before irradiation in order to increase its shelf life are studied.

## 2. Research methods

Researched polymer packing film polyamide/polyethylene (PA/PE) at a ratio of 20:80, thickness 80 $\mu$m manufacture by Dmitrov factory of flexible packaging was studied.

Installation of X-ray sterilization plant with local bio-protection is located in the radiation technologies center of Moscow Radiotechnical Institute Russian Academy of Sciences. X-ray sterilization plant has the following characteristics: energy of accelerated electrons - 5 MeV, power of the electron beam - 1.5 kW radiation performance - 3600 Mrad· kg/hour, maximum dimensions of sterilized products - 57 cm × 38 cm × 27 cm.

IR spectra of polymer packing film are recorded on device FMS-1201 (Russia) at M.V. Lomonosov Moscow State University. The device has the following characteristics: spectral range of 400-4000 $cm^{-1}$; spectral resolution – 1 $cm^{-1}$; source of infrared flux – nichrome; IR-receiver flow – pyroelectric. The film samples are placed in the experimental cell perpendicular to the light flux. IR absorption spectrum of polymer packing film is recorded before and after irradiation.

Water vapor permeability of packing film before and after irradiation is determined according to GOST 21472-81 "Sheet materials. Gravimetric method for determining water vapor permeability" at room conditions of temperature (25.0 ± 0.5) °C and humidity (75 ± 2) %.

## 3. Research results

When storing fresh agricultural produce best results showed the multilayer film materials composition of polyamide/polyethylene in a percentage ratio of 20:80 (PA/PE) [7-8].

Samples of the polymer film PA/PE has been exposed to radiation at a dose of from 3 to 18 kGy in 10 replicates. Actual doses were determined in comparison with film detectors which were irradiated together with the samples. Film detectors were manufactured by "Scientific Production Enterprise "Doza" and are accredited in Rostest Moscow. The absorbed dose is determined by comparing the optical density of the film detector relative to the control sample on spectrophotometer Varian Cary 100 Scan. Spectrophotometer is in the Russian Research Institute of Canning Technology. The results are shown in table 1.

**Table 1.** Absorbed doses of fast electrons.

| Established dose (kGy) | Optical density of film | Absorbed dose, (kGy) |
|---|---|---|
| 3 | 0.061 | 3.3 |
| 6 | 0.117 | 5.3 |
| 9 | 0.177 | 8.3 |
| 12 | 0.261 | 12.8 |
| 15 | 0.302 | 14.9 |

| | | |
|---|---|---|
| 18 | 0.375 | 18.9 |

The study of the structure of samples before and after irradiation was conducted. It was removed IR spectra in the range of 400-5000 $cm^{-1}$. The infrared absorption spectrum of the polymer film PA/PE after electrons irradiation with 18 kGy is shown in figure 1.

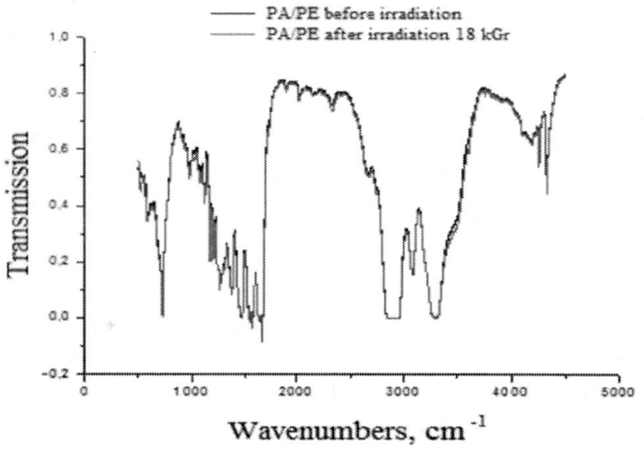


**Figure 1.** The infrared absorption spectrum of the polymer film, PA/PE in the range from 400 – 4500 $cm^{-1}$ before and after electrons irradiation with 18 kGy.

The infrared spectrum from 400 to 4500 $cm^{-1}$ are characteristic basic intense absorption bands (see figure 1.) to the PE-layer related to the stretching (2820-2980 $cm^{-1}$) and the deformation (1480 $cm^{-1}$, 725-740 $cm^{-1}$) fluctuations -$CH_2$- groups. Absorption bands related to deformation vibrations of -$CH_3$ groups are observed in the region 1380-1370 $cm^{-1}$. Characteristic absorption band for the PA-layer: the deformation vibration of the N-H in the region of 3040 $cm^{-1}$, 1550 - 1570 $cm^{-1}$ and the carbonyl group bending vibrations in the 1620-1680 $cm^{-1}$ [2]. The most characteristic absorption of functional groups in the test sample is presented at table 2.

**Table 2.** Characteristic absorption in IR spectra of the sample PA/PE.

| Functional group | Characteristic absorption frequency, $cm^{-1}$ | | |
|---|---|---|---|
| | PE-layers | PA-layers | PA/PE |
| | Stretching vibrations | | |
| -C-C- | | | 580 (sr.) |
| | | | 928 (sr.) |
| -$CH_2$- | 2820-2980 (s.) | | |
| -C=O | | 1623-1680 (s.) | |
| | | 2020 (sl.) | |
| -C-O-C- | | 1120 (s.) | |
| | | 1257-1275 (s.) | |
| -COO- | | | 2340 (sl.) |
| -CH- | | | 3268-3338 (s.) |
| | Deformation vibrations | | |
| -$CH_2$- | 1480 (s.) | | 718-733 (s.) |
| | 725-740 (s.) | | |
| -$CH_3$ | 1380-1371 (s.) | | 1170 (s.) |

After irradiation, the characteristic absorption bands of PE- and PA- layers do not undergo radical changes in the structure of the packaging material. Explore the intensity of absorption bands of functional groups present the change of their intensity. The intensities of the absorption bands of functional groups at absorbed doses 0 kGy and 18 kGy are in table 3. It is possible to speak about degradation or crosslinking in samples of films. We have found that almost no change occurs in the structure of multilayered sample PA/PE.

**Table 3.** Intensities of absorption bands in films PA/PE.

| Absorption band, $cm^{-1}$ | Functional group | Intensities of the absorption band |
|---|---|---|
| | Dose 0 kGy | |
| 584 | -C-C- | 0.36 |
| 1371-1366 | $-CH_3$ | 0.09 |
| 2340 | -COO- | 0.75 |
| 3085 | -NH- | 0.16 |
| | Dose 18 kGy | |
| 584 | -C-C- | 0.36 |
| 1371-1366 | $-CH_3$ | 0.10 |
| 2340 | -COO- | 0.77 |
| 3085 | -NH- | 0.18 |

On the basis of the obtained results have been constructed the intensity of absorption bands in the sample PA/PE on electron irradiation at different dose. These dependences are shown in figure 2. and figure 3.

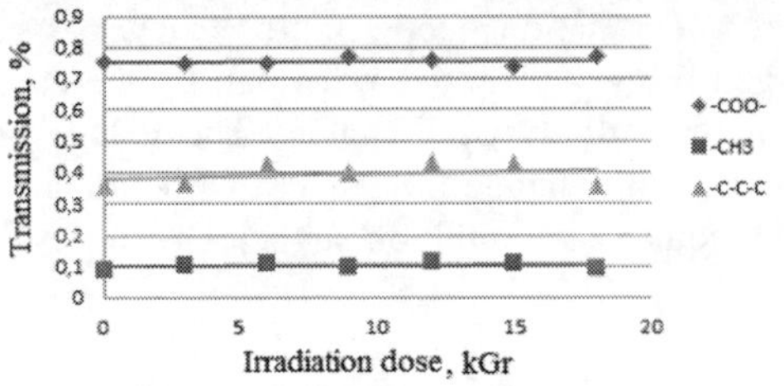


**Figure 2.** The dependence of the bands intensity of functional groups (-COO-; -$CH_3$; -C-C-C-) of different doses of fast electrons irradiation.

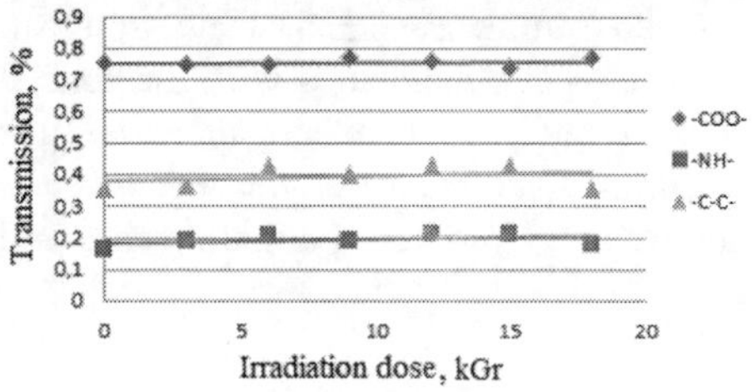


**Figure 3.** The dependence of the bands intensity of functional groups (-COO-; -NH; -C-C-) of different doses of fast electrons irradiation.

According to the curve of functional groups -NH and -C-C- the number of groups are reduction with growth of irradiation dose, but groups -COO- are increase. The functional groups -NH and -COO- are decreased at samples irradiation dose 18 kGy (see figure 3).

Thus, during irradiation of the sample of PA/PE film with 3 kGy in the PA-layer is degradation, and in PE-layer is cross-linking between molecules with the formation of oxygen bridges. With increasing doses of irradiation up to 20 kGy is noted the degradation in the PA-layer and the reduction of the ultimate methyl groups in PE-layer. Analysis of IR spectra show, that there are a slight change in the structure of the sample, which can result in changing its barrier properties.

We obtained the dependence of the parameter wetting contact angle from the doses of irradiation with fast electrons from the inner (figure 4) and external (figure 5) layers. From figure 4 it is seen that the wetting angle of the irradiated sample is not in PE-layer (inner) is 80° and begins to decrease starting from 15 kGy 18 kGy and when the wetting angle is 70°. Therefore, the hydrophilicity of the PE-layer increases due to the increase in oxygen in the PE-layer.

In the PA-layer before irradiation is 73° and after radiation dose of 18 kGy – 76°, hence the parameter wetting contact angle practically does not change [9].

Thus, minor changes in the structure with increasing radiation doses cause changes in the properties of the polymer material, and therefore significantly affect the shelf life of agricultural products.

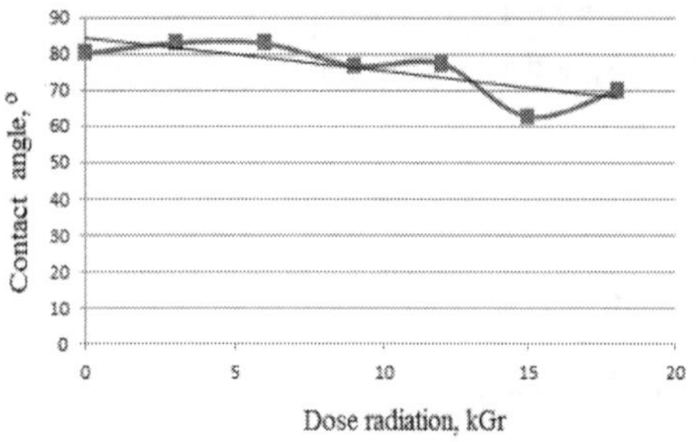


**Figure 4.** Dependence of the contact angle of PA/PE sample on the irradiation dose by fast electrons of the PE side.

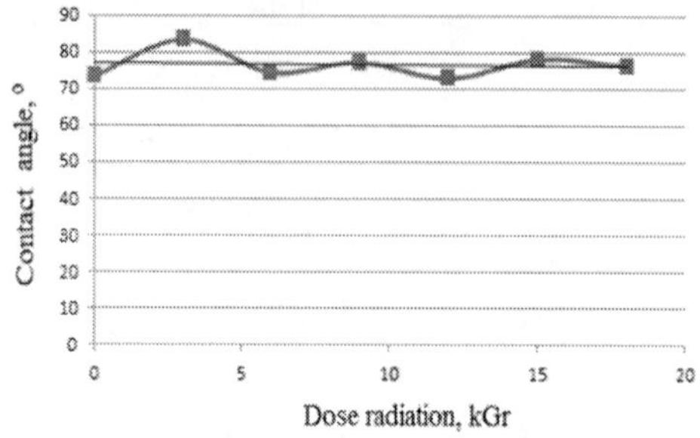


**Figure 5.** Dependence of the contact angle of the PA/PE sample on the irradiation dose by fast electrons of the PA side.

## 4. Conclusion

In studying the structure of the polymer packaging PA/PE as a percentage of 20:80 film before and after irradiation by 5 MeV electrons doses of 3 to 20 kGy significant changes in the samples does not occur (there is little change in the structure of the material). Infrared spectroscopy showed that the irradiated sample PA/PE fast electrons in the PA-layer degradation occurs, and PE-layer is cross-linking between molecules with the formation of oxygen bridges. This is confirmed by the study of wetting contact angle where there is an increase in hydrophilicity of PE-layer. The values of the wetting contact angle in the PA-layer have practically not changed. Hence the degradation is insignificant. However, the findings are preliminary and require accumulation statistics.

## References

[1] Gelfand S Yu, Zavialov M A, Petrov A N, Prokopenko A V, and Filipovich V P 2013 *J. Storage and Processing of agricultural products* (Moscow: Publishing house Food industry) **2** pp 25-31
[2] Bellamy L J 2012 *The Infrared Spectra of Complex Molecules – Vol. Two Advances in Infrared Group Frequencies* (New York: Springer Science & Business Media)
[3] McKeen L 2012 *The effect of sterilization on plastics and elastomers* (Atlanta: Elsevier Inc.) p 363
[4] Bovey F 1959 *Effect of ionizing radiation on natural and synthetic polymers* (Moscow: Foreign literature) p 295
[5] Neiman M B 1964 *Aging and stabilization of polymers* (Moscow: Nauka) p 332
[6] Vereschinsky I V, Pikaev A K 1963 *Introduction to radiation chemistry* (Moscow: Academy of Sciences USSR) p 409
[7] Gracheva A Yu, Zav'yalov M A, Ilyukhina N V, Kukhto V A, Tarasyuk V T, Filippovich V P, Egorkin A V, Chasovskikh A V, Pavlov Yu S, Prokopenko A V, Strokova N E, Artem'ev S A and Polyakov S P 2016 *Physics of Atomic Nuclei* vol **79** N 14 pp. 1682–87.
[8] Karastoyanova O, Klyueva O, Korovkina N, Megerdichev E and Shishkina N 2014 *Packaging* (Moscow: Tarkom) **1** pp 20-22
[9] Subedi D P 2011 *The Himalayan Physics* (Pokhara: NepJOL) vol **2** pp 1-4

IOP Conf. Series: Journal of Physics: Conf. Series **941** (2017) 012085 doi:10.1088/1742-6596/941/1/012085

# Design of robust microlinacs for wide replacement of radioisotope sources

**A V Smirnov[1], R A Agustsson[2], S Boucher[1], M Harrison[1], K Junge[2], E Savin[3] and A Yu Smirnov[2]**

[1]RadiaBeam Systems, LLC, Santa Monica, CA 90404, USA;
[2]RadiaBeam Technologies, LLC, Santa Monica, CA 90404, USA
[3]National Research Nuclear University "MEPhI", Moscow, 115409, Russian Federation

asmirnov@radiabeam.com

**Abstract.** To improve public security and prevent the diversion of radioactive material for Radiation Dispersion Devices, development of an inexpensive, portable, easy-to-manufacture linac system is very important. The bremsstrahlung X-rays produced by relativistic electron beam on a high-Z converter can mimic X-rays radiated from various radioactive sources. Here we consider development of two designs: one matching a Ir-192 source used in radiography with ~1-1.3 MeV electrons, and another one Cs137 source using 3.5-4 MeV electrons that can be considered for borehole logging. Both designs use standing wave, high group velocity, cm-wave, accelerating structure. The logging tool conceptual design is based on KlyLac concept combining a klystron and linac operating in self-oscillating mode and sharing the same vacuum envelop, and electron beam.

## 1. Introduction

Linear accelerators of electrons are employed in a wide variety of applications, including radiography and sterilization, radiolysis and imaging, modification of materials and borehole logging as well as testing of semiconductor electronic components for space applications. Among emerging applications is driving table-top sub-mm wave sources [1] and replacement of radionuclide sources [2] to improve public security and prevent the diversion of radioactive material for radiation dispersion devices. A challenging specific of such a replacement is inexpensive, portable, easy-to-manufacture linac system delivering relativistic electrons. The bremsstrahlung X-rays produced by the electron beam on a high-Z converter at the end of the linac may match the penetration and dose rate of a radionuclide source to be substituted using beam energies, e.g., ~(1-1.3) MeV for Ir192 and ~(3-3.5)MeV for Cs137.

As a rule classical linac system is rather expensive and not portable. MicroLinac technology originally proposed at SLAC [3] employs a compact X-band linear accelerator powered by an inexpensive, low power, magnetron [4,5]. However, to make the MicroLinac concept suitable for a wider scope of applications, the compact linac technology needs to be advanced to reduce further cost, weight, and dimensions. One such modification may employ of a light-weight, compact modulator, using, e.g., a transformer-free Marx scheme with array of isolated gate bipolar transistors (IGBTs).

However, fabrication of a multi-cell, tapered MicroLinac structure (usually side coupled periodic or circular disk loaded) remains rather expensive and time consuming. It includes high precision machining, cold-testing and re-machining of the cells, testing of the clamped assembly, multi-step brazing, and individual tuning of the cells and of the brazed assembly.

A relatively unexplored candidate for a slow wave structure of a robust MicroLinac is a so-called cross-bar ("jungle-gym" or "loaded easitron") structure. Such an accelerating structure have been first considered among variants for the SLAC two-mile accelerator [6] and analysed later numerically [7].

Among attractive features of this structure for MicroLinac is a high group velocity (>0.2c), possibility for simplified fabrication (including low-beta cells), and compactness (transverse dimensions are about half a wavelength).

In this paper we focus solely on resonant, standing wave (SW) mode of operation as being more effective for implementation in miniature RF linacs, though the cross-bar structure can be employed in a traveling wave mode as well.

A single section is a preferable option for a MicroLinac to avoid problems of inter-section detuning and RF network complexity. On the other hand, requirements imposed by cost, weight, or compactness may seriously limit the RF power available. That dictates a significant number of cells of linac section. For a SW single-section linac the solution is high enough group velocity which also determines the structure robustness in terms of sensitivity to tolerances related to production cost.

For borehole logging the conventional linacs with their automatic frequency control (AFC) system, ferrite-based circulator, and magnetron are not suitable because of very limited borehole diameter (3.5" is of practical interest), temperatures as high as 150°C at the depths ~kilometre, and vibrations with accelerations ~2G. KlyLac concept [8,9,10,11] offers an alternative, rugged approach to this problem. Both KlyLac and KlyNac [12,13] concepts are based on a combination of klystron and linac sharing the same vacuum envelop and electron beam.

Below we discuss some of the developments related to ~1 MeV MicroLinac and 3.8 MeV KlyLac with emphasis made to performance of the long "jungle-gym" accelerating structure.

## 2. Linac section of 1 MeV for Ir192 replacement

A 36-cell, cross-pin, X-band structure shown in Figure 1 has been designed using CST Suite™ [14] frequency domain solver and SolidWorks. The structural design takes into account specifics related to brazing of the ∅1.75 mm pins (which are available from third parties).

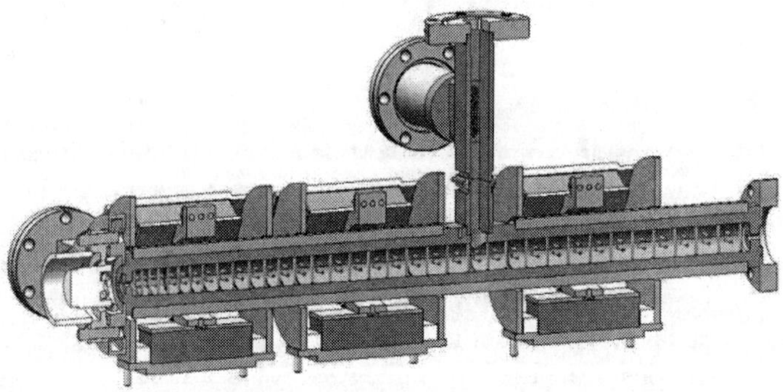

**Figure 1.** Cut-view of linac design with elliptical RF coupler, two vacuum ports, electron gun, and three-section alternating solenoid focusing.

S11 curves and the operating mode profile along the section are shown in Figure 2 and Figure 3 respectively. The results are obtained with a very high degree of confidence due to fast and stable convergence relatively insensitive to meshing (vs., e.g., DaR [15] or circular disk loaded structures).

Note as large as ~(70-100) MHz separation between adjacent modes exceeds by about two orders that for conventional circular disk loaded cavity having a comparable number of cells and resonant frequency.

Beam dynamics have been simulated for ~14 keV injection energy and ~0.15 T magnitude of on-axis longitudinal magnetic field. Complex RF fields (with magnitude shown in Figure 3) have been exported from CST Suite™ results into ASTRA code [16]. Simulation results are shown in Figure 4.

One can enhance capture if the pins are made from copper-plated mild steel. One configuration with the "passive" magnetic pins included into the full magnetic model with the focusing PM-based solenoid is shown in Figure 5. As it can be seen from Figure 5 presence of the iron pins results in two effects: a) magnetic field magnitude increase by ~16%; and b) strong derivatives of the local fields (i.e. local enhancement of transverse fields).

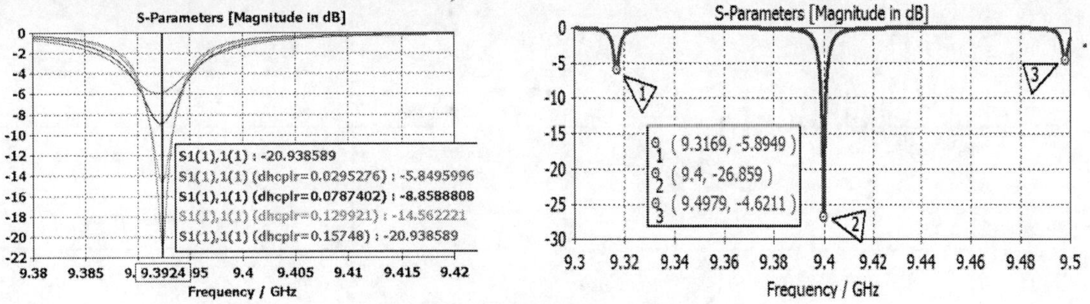


**Figure 2.** S11 curve simulated for the finalized RF model of Figure 1.

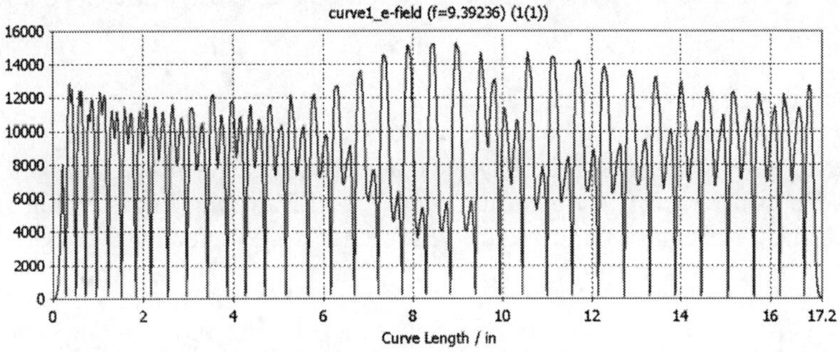


**Figure 3.** Magnitude for longitudinal RF electric field profile along the axis of RF model of Figure 1.

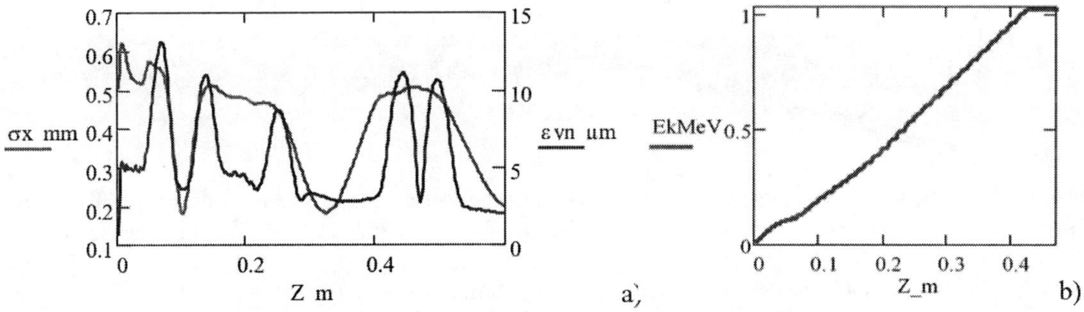


**Figure 4.** RMS Emittance (blue, right ordinate) and beam rms dimensions (red, left ordinate) (a) simulated with ASTRA code for the field profile of Figure 3. Beam energy gain plotted along the structure for ~100 kW RF power (b). Beam capture is ~17 %.

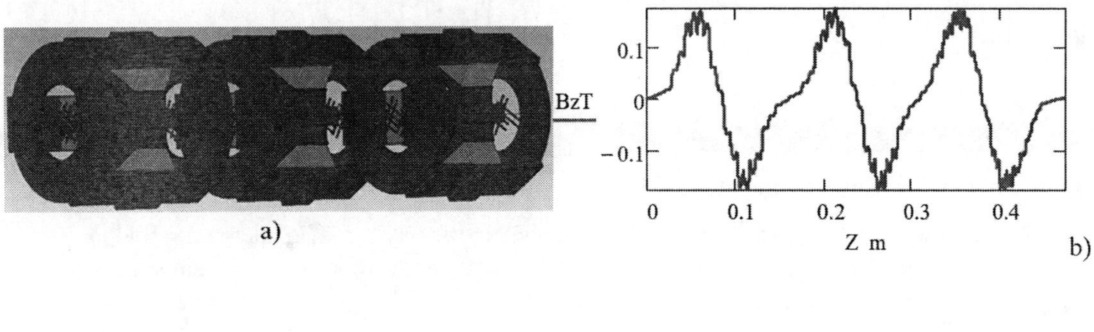

**Figure 5.** Magnetic model with RF structure pins made from mild steel and copper plated inside the PM focusing system (a) and magnetic field profile along the axis perturbed by the pins (b).

Beam dynamics simulations results are given in Figure 6. The ASTRA results indicate significant increase of beam capture by 32% (at the expense of lower energy).

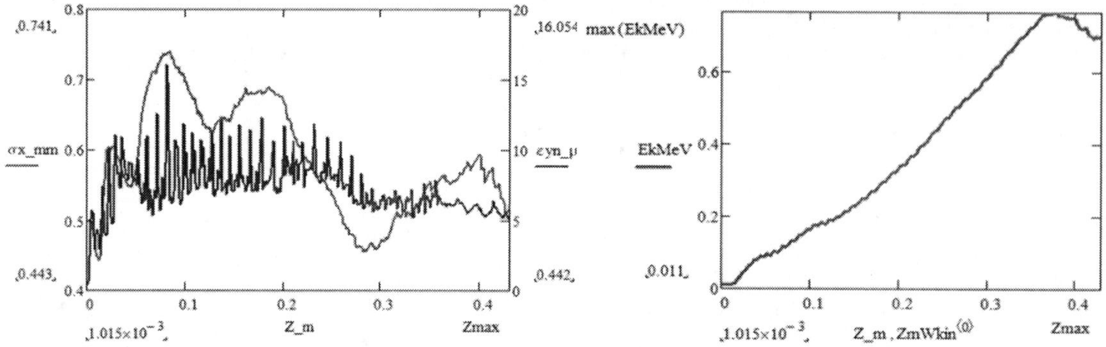


**Figure 6.** LEFT: RMS Emittance (blue, right ordinate) and beam rms dimensions (red, left ordinate) simulated with ASTRA code. RIGHT: Energy gain along the MicroLinac. Focusing field profile is given in Figure 5.

## 3. A conceptual design of a few MeV KlyLac

The results above are encouraging to apply the cross-bar type structure for a much longer section required for replacement of Cs137. The KlyLac concept using a sheet beam klystron (SBK) combined with the cross-bar structure is illustrated in Figure 7. A 1.3 m long linac section with about hundred cells is characterized in Figure 8 and Figure 9. It demonstrates excellent RF performance and rather uniform field distribution along the section at as large as 31 MHz separation between the adjacent modes.

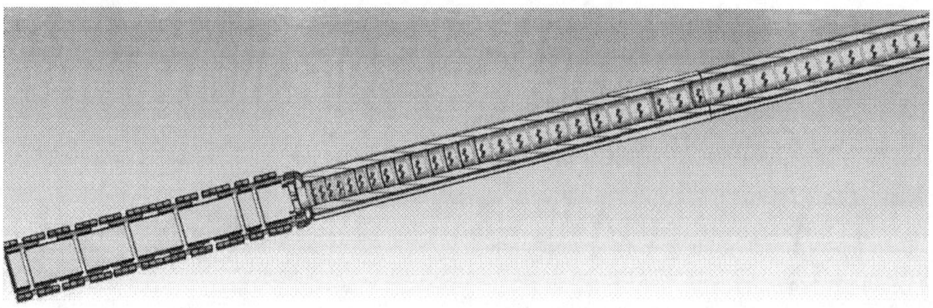

**Figure 7.** A cut-view fragment of a KlyLac generating-accelerating conceptual design. SBK cavities are given in gray. The maximum diagonal dimension of the structure is 1.5”. Sheet beam electron gun is not shown.

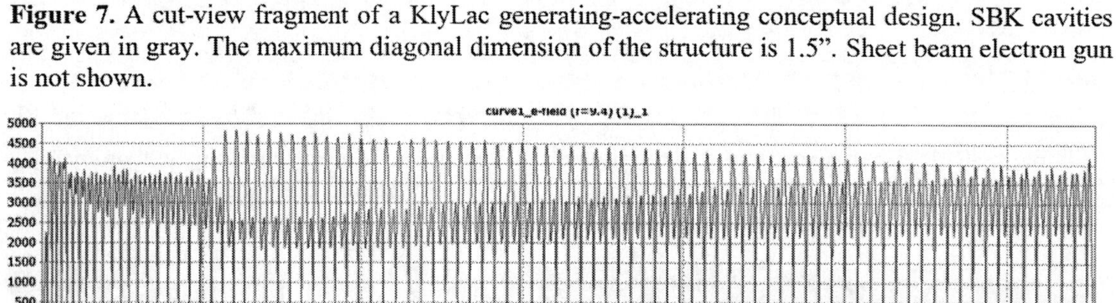


**Figure 8.** Field magnitude simulated along the 1.3 m long cross-pin linac section.

One variant of the linac focusing system we applied to the KlyLac is an alternating solenoid focusing system formed simply by three stacked focusing systems of Figure 1. To simulate and analyse beam dynamics we applied the RF fields simulated above (see Figure 8). As it can be seen from Figure 10a,b the ASTRA simulations resulted in 3.72 MeV output energy and ~15% capture coefficient.

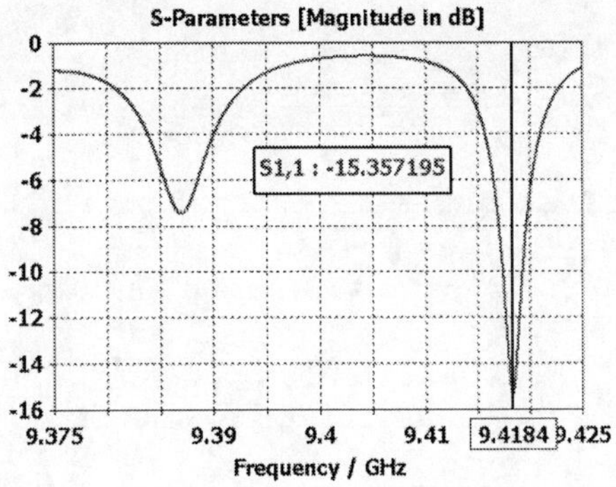


**Figure 9.** S11-curve for two-port, 1.3 m long section.

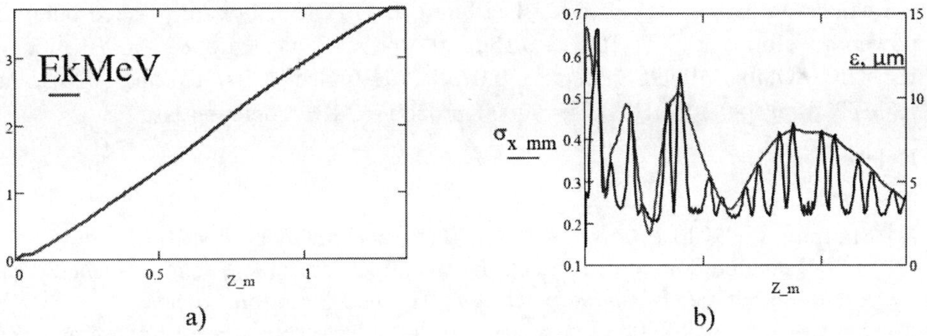


**Figure 10.** Kinetic energy [MeV] (a), the beam rms size and transverse emittances (b) plotted along the linac section [m]. The pulsed microwave power is 575 kW at ~9.4 GHz frequency.

### *3.1 Modelling of feedback operation*

A number of CW and pulsed klystrons operate in a self-oscillating mode (some of them being employed even without circulator). However, to understand and design such a self-adjusted system a corresponding model is vital, because in general a positive feedback is vulnerable to instabilities. Such a transient model with delayed feedback loop starting from noise has been built. The model is based on classical Slater [17] and Vainshtein [18] theories of cavity excitation and includes a) feedback with adjustable gain; b) variable group delay Δt delay of the feedback; c) realistic modulator envelope with adjustable duration tHVpulse, raise Δtraise, and fall times; d) realistic non-linear gain as a function of input power in a wide range imported from sheet-beam klystron simulations; e) variable attenuation Kprobe in the feedback loop; f) frequency detuning and loop phase shift; and g) a conservatively estimated noise level and noise figure as a seed triggering the self-oscillation. The model has been extensively benchmarked using known analytical solutions and experimental data (including RF compression setups such as SLED).

As it is shown in Figure 11, left improper choice of the feedback parameters may result in instabilities and bifurcations. One of the main limitations imposed on VSWR in ordinary high power RF sources is RF window. Since the KlyLac system does not employ circulator and RF windows we conservatively determined that the safe margin for the peak reflected power would be ~33% level of full power of (which corresponds to peak VSWR=3.7). We found several stable solutions with satisfactory moderate reflections (given in Figure 11 on the middle and right).

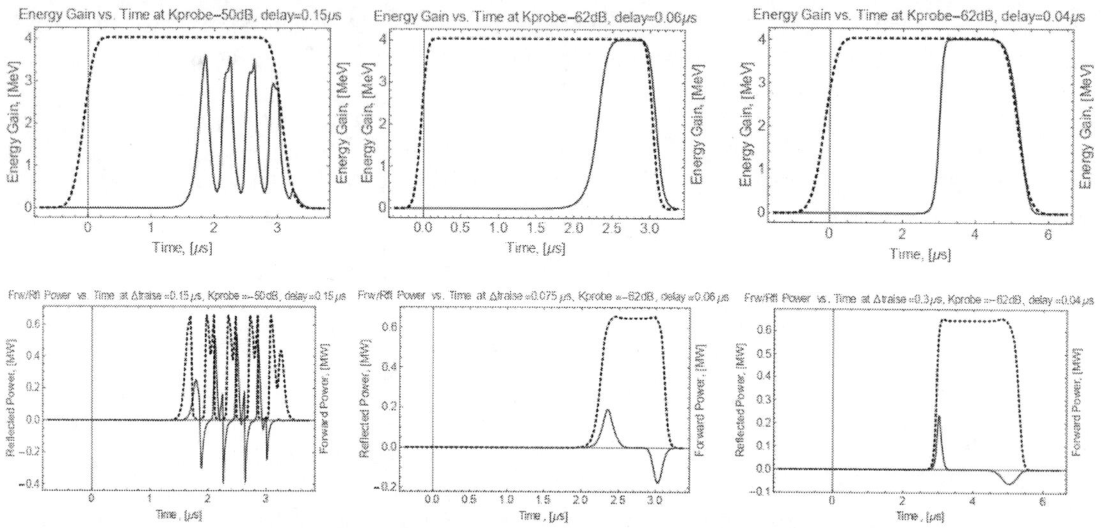


**Figure 11.** TOP: Waveforms for accelerating voltage at feedback loop (red, top) and HV envelope normalized to max energy gain (dashed blue). BOTTOM: Forward (dashed blue) and reflected (red) power waveforms. LEFT: Δtraise=0.15μs, tHVpulse=3 μs, Kprobe= -50dB, Δtdelay=150ns. Cavity Rsh=25 MOhm. MIDDLE: Δtraise=0.075μs, tHVpulse=3 μs, Kprobe=-62dB, Δtdelay=(30-60)ns. RIGHT: Δtraise=0.3μs, tHVpulse=5 μs, Kprobe= -62dB, Δtdelay=40ns.

## References

---

[1] Smirnov A V 2013 *Proc. of NA-PAC'13* (Pasadena, CA USA) p 1421
[2] 2008 Radiation Source Use and Replacement: Abbreviated Version. *Sciences Committee on Radiation Source Use and Replacement, National Research Council*
[3] Ackerman N et al. 2006 MicroLinac – A Portable Accelerator for Radiography *The 12th Advanced Accelerator Concepts Workshop*
[4] Yamamoto T et al. 2008 Development of portable X-band linac X-ray source for non-destructive testing, *Proceedings of the Joint International Workshop: Nuclear Technology and Society–Needs for Next Generation* (Berkeley, California).
[5] Boucher C, Ding X, Murokh A 2010 *Proceedings of Intern. Particle Accelerator Conf.* (IPAC'10 Kyoto, Japan) p 178
[6] Borghi R P et al. 1965 Design and Fabrication of the Accelerating Structure for the Stanford Two-Mile Accelerator. *Preprint SLAC-PUB-71*
[7] Loo J et al. 1988 *Particle Accelerators* v 23 pp 279-287
[8] Nyguard J 1954 *Compact linear accelerator* U.S. Patent #2,922,921
[9[ Post R 1955 *Linear accelerator* U.S. Patent #2,940,001
[10] Schriber S O 1978 *Klystron-accelerator system* Canadian patent #1040309
[11] Bogdanovich B Yu et al. 1997 *Proc. of the 1997 Part. Acc. Conf.* p 3869
[12] Potter J M et al. 2013 *AIP Conference Proceedings* v 1525 p 178
[13] Malyzhenkov A Carlsten B Nichols K 2016 *KLYNAC (Combined klystron and linac). Proc. of NAPAC'16* (Chicago)
[14] https://www.cst.com/
[15] Smirnov A V Savin E 2016 *Nuclear Inst. and Meth.* NIM A v 820 N1 p 48
[16] Floettmann K 2011 *Astra - A space charge tracking algorithm. Version 3.0* (DESY, Hamburg)
[17] Slater J C 1950 *Microwave Electronics* (Van Nostrand*)*
[18] Vainshtein L A 1988 *Electromagnetic waves* (Moscow, Izdatel'stvo Radio i Sviaz', In Russian)

IOP Conf. Series: Journal of Physics: Conf. Series **941** (2017) 012086 doi:10.1088/1742-6596/941/1/012086

# Thermometry system development for thermoradiotherapy of deep-seated tumours

**A M Fadeev[1], S M Ivanov[1,2], E A Perelstein[3], and S M Polozov[1]**

[1] National Research Nuclear University - Moscow Engineering Physics Institute, Moscow 115409, Russia
[1] N.N. Blokhin Russian Cancer Research Center, Moscow 114478, Russia
[3] Joint Institute for Nuclear Research, Dubna 141980, Russia
SMPolozov@mephi.ru

**Abstract**. Therapeutic hyperthermia (including RF hyperthermia) in combination with radiotherapy (called thermoradiotherapy) is one of widely used contemporary cancer treatment methods. The independent electron linac and RF system or their combinations are necessary for effective therapy. Whole-body hyperthermia is used for treatment of metastatic cancer that was spread throughout the body, regional one is used for treatment of part of the body (for instance leg or abdominal cavity). Local hyperthermia with characteristic size of heating volume of 20-100 mm permits to heat tumour without overheating of healthy tissues. The thermometry of deep suited tissues during the hyperthermia process is an important and complex task. Invasive methods as thermistors, optical sensors or thermo-couples can not be widely used because all of them are able to transport tumor cells to the healthy region of the patient body. Distant methods of the temperature measurement such, as radiothermometry and acoustic thermometry can not be used for tissues seated deeper than 5-7 cm. One of possible ways to solve the problem of temperature measurement of the deep suited tissues is discussed in this article: it was proposed to use the same electrodes for RF hyperthermia and thermometry. As known electrodynamics characteristics of tissues are sufficiently depends on temperature. It was proposed to use this effect for active radiothermometry in local hyperthermia. Two opposite RF dipoles can be used as generator and receiver of pick-up signal.

## 1. Introduction

Hyperthermia is an adjuvant methods of cancer treatment in which tumor temperature is increased to high values (40-44 °C). Many researches have shown that high temperature can damage and kill tumor cells, thus reduces tumor size. The thermal therapy combined with the radiation (thermoradiotherapy, TRT) has been applied in N.N. Blokhin Russian Cancer Research Center (RCRC) since 1980th [1]. More than 1000 patients have been treated to date. Such program allows to sufficient and authoritative reduce of the regional cancer recrudescence and metastases comparatively to the surgery or independently radiotherapy (RT). For instance, the rectum cancer recrudescence was observed for 0.9±0.6 % (2 form 220 patients) and metastases for 5% (11 from 220 patients) comparatively with 16.2±1.9% (64 from 395) for surgery and 9.6±2,1% (26 from 272) for surgery and before radiotherapy [1]. Frequency of full regression after TRT also increased: for prostate cancer form 69±8.2 % for RT to 94±2.3 % for TRT, for soft tissue sarcoma form 14±5.1 to 45±5.2 %, for regional metastasis of neck epidermoid cancer from 12±7.8 to 57±6 % etc. [2].

RCRC clinical studies demonstrate improving results of treatment by combined using of hyperthermia and radiation for the several tumour localizations. But only applicators for superficial hyperthermia were used in RCRC. Common RCRC-MEPhI-JINR project is pointed to expand the range of utilizing devices, i.e. using of devices for the regional hyperthermia gives more advantages for an oncological diseases treatment [2-7]. The most evident approach is using an annular array of applicator situated around the patient body [8]. Arrays of applicators with variations in frequency, phase, amplitude and orientation in space give more possibilities to control heating pattern during hyperthermia treatment [3]. Top view of this structure is shown in the figure 1. Thus the phased array provides deeper tissue penetration of electromagnetic waves in comparison with single applicator, reduce undesirable heating of healthy tissues situated between applicator and tumor and improve local control for heating area. Also using array of applicators gives ability to control and to plan heating process without changing of patient position. Suggested phased array consists of eight copper dipoles, attached on the inner side of the dielectric cylinder, and surrounds a patient body. The variation of phases $\Phi_i$ and amplitudes $A_i$ of each dipole provides to control the heating volume size and localization (see figure 1). The operating frequency choice is also important because the higher frequency allows to decrease the heating volume. As an example modeling of head and neck hyperthermia with operating frequency of 433 MHz produces heated area size of 15-20 mm comparatively with 30-40 mm with 150 MHz antennae array.

The RF power feeding scheme of the designed local hyperthermia facility is presented in figure 2 and discussed in [4]. The operating principle of such layout is the following. The RF signal at an operating frequency from signal source splits into eight channels. Then by means of controlled eight phase shifters and eight solid state amplifiers we can adjust phase and amplitude of every signal. Due to these adjustments electric field is focused indesirable region. Directional couplers prevent reaching reflected wave to generator. Reflected wave will be absorbed by 50 Ohm load.

The first experimental prototype was constructed at MEPhI and numbers of heating localization experiments were carried out [7, 9]. Comparison of simulation and experiments shows that the phased array solve the local heating problem and deep suited tumorous can be locally heated. Experimental heating results are compared with CST Studio Suite simulations and have very good accuracy.

## 2. Problem of the thermometry for deep seated tissues

It is clear, RF hyperthermia is a way to increase the efficiency of the combined cancer therapy. But the problem of the temperature control inside of the patient body especially for heating tissues should be solved before routine clinical usage of the local RF hyperthermia. Thermometry is more difficult for deep-seated tissues. It is possible to use invasive or contact thermometry systems as thermistors and thermo-couples for deep-seated tissues and tumours. Optical fiber sensors based on Bregg's grids became more popular last years. The measured parameter as temperature or mechanical displacement is converted to the length of the light shift. But all of such sensors provide only invasive temperature measurement. Russian hyperthermia protocol permits installation of thermocouples inside the patient body. But this causes pain and temperature probe can transport tumor cells to the healthy region of the patient body. European hyperthermia protocol directly forbids the installation of any temperature probes inside the patient body. They prefer to simulate any radiation process with phantoms. Moreover noninvasive thermometry of human tissues is important for other cases. It is suggested to determine tissues temperature by means of measurements effects that could be observed during heating. Noninvasive control is possible by using of the magnetic resonance imaging (MRI) as it is proposed and realized by BSD Medical Systems (now Pyrexar Medical). But such way has serious difficulties and the price of MRI system is higher than the same of the hyperthermia system. Acoustic thermometry is one of new technologies. 2D and 3D temperature distributions were successfully imaged by means of acoustic thermometry experiments. But such technique can be used only for tissues located not deeper than 3-5 cm (for mammography as an example). Well-known radiothermometry (RTM) also can be used for deep-seated tissues. As known, the specific heat of tumour is directly proportional to its growth velocity. The faster growing tumours will be "hotter" and will be brighter on the thermograph. The possibility to find the fast growing tumours is an unique advantage of the RTM. Main sufficient disadvantage is inherent as for acoustic thermometry: RTM can not be used for the deep suited tissues. The depth of temperature anomaly localization will not be greater than 3-7 cm depending of the tissues humidity.

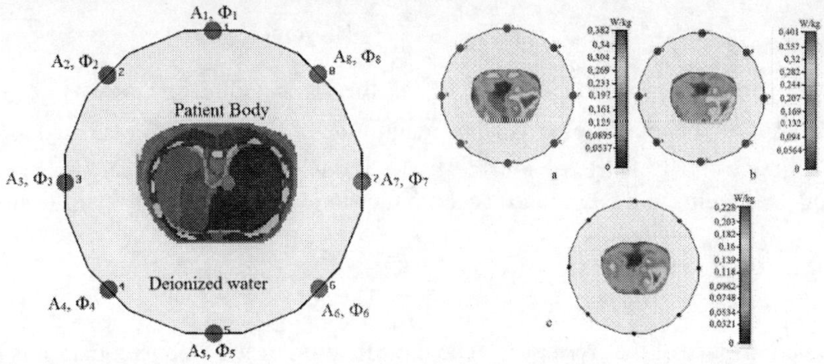


**Figure 1.** Top view of phased array surrounding patient body (left) and cross-section patterns of the SAR distribution with different operation frequencies (a) 150 MHz; (b) 100 MHz; (c) 80 MHz and with input phases of 50°, 50°, 50°, 50°, 0°, -30°, -40°, -10° applied to channels 1, 2, ..., 8 respectively.

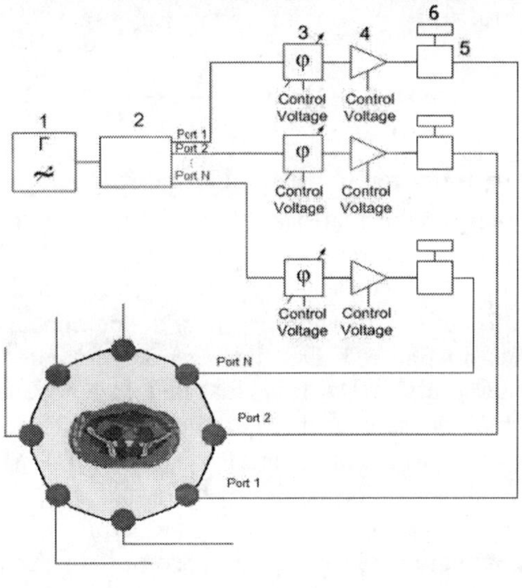


**Figure 2.** RF power system of the hyperthermia facility schematic layout, here 1 – master generator with input signal's frequency range 100 – 400 MHz and input impedance 50 Ohm; 2 – 8-out power splitter; 3 –voltage-controlled phase with phase range 180°; 4 - solid state amplifier with output power up to 100 W, 5 – 10 dB unidirectional coupler 6 – load absorbed, 50 Ohms, 7 – phased array with patient body inside.

## 3. Active radiothermometry with RF hyperthermia dipoles

As known dielectric properties of the tissues (complex dielectric permittivity and tangent of the dielectric loss) depend of temperature. It is suggested to determine tissues temperature by means of measurements effects that could be observed during heating process. Thus dielectric properties of the human tissues sufficiently varies with temperature increasing (fat tissue and skin are not heated during hyperthermia with the phased array). This thermometry system could be used as addition to the other regional or local hyperthermia facilities.

The detail study of dielectric properties of the human body tissues was done in [10-12]. Let us discuss the main equations which can be used to determine the temperature and the frequency dependences of the complex dielectric permittivity of tissues. It can be described using the Debye expression

$$\hat{\varepsilon} = \varepsilon_\infty + \frac{\varepsilon_s - \varepsilon_\infty}{1 + j\omega\tau} \tag{1}$$

where $\varepsilon_\infty$ is the permittivity at $\omega\tau >> 1$ and $\varepsilon_s$ is the same value for $\omega\tau << 1$, $\omega$ is the operating angular frequency, $\tau$ is the relaxation time. The magnitude of the dispersion is equal $\Delta\varepsilon = \varepsilon_s - \varepsilon_\infty$. The other way was proposed by Hurt for muscle tissue model [13] were dispersion present as a set of approximating coefficients and taking into account the static ionic conductivity of tissues $\sigma_i$:

$$\hat{\varepsilon}(\omega) = \varepsilon_\infty + \sum_{k=1}^{5} \frac{\Delta\varepsilon_k}{1 + j\omega\tau_k} + \frac{\sigma_i}{j\omega\varepsilon_0}. \tag{2}$$

Here $\varepsilon_0$ is the permittivity of the free space. The modification of the Debye equation is known as Cole-Cole model:

$$\hat{\varepsilon}(\omega) = \varepsilon_\infty + \frac{\Delta\varepsilon}{1 + (j\omega\tau)^{(1-\alpha)}}. \tag{3}$$

where the distribution parameter $\alpha$ is a measure of the dispersion broad. The Hurt and the Cole-Cole models can be generalized due to:

$$\hat{\varepsilon}(\omega) = \varepsilon_\infty + \sum_k \frac{\Delta\varepsilon_k}{1 + (j\omega\tau_k)^{(1-\alpha_k)}} + \frac{\sigma_i}{j\omega\varepsilon_0}. \tag{4}$$

The values of $\Delta\varepsilon_k$, $\sigma_i$, $\tau_k$ and $\alpha_k$ for different tissues and for the frequency band above 400 MHz can be founded in [12]. As an example, the real part of dielectric permittivity of the muscle tissue or tumour tissue growths by 0.2-0.5 per 1 degree (absolute value is 70-100). It leads to the special absorption rate variation and to RF power scattering modification in tissues. Such variation can be registered. A number of examples of real $\varepsilon'$ and imaginary $\varepsilon''$ components of the complex dialectical permittivity $\varepsilon=\varepsilon'-i\varepsilon''$ are shown in figure 3. Dependences are modelled for tissues using data from [10-12]. The operating frequency is 150 MHz. It is clear that such dependences are close to linear for temperature range 36-43°C which is used for thermoradiotherapy. The analyses of experimental and simulated data form [10-12] shows that for operating frequency of 150 MHz absolute values of $\varepsilon'$ and $\varepsilon''$ are close for all tissues. Only dielectric properties of fat tissues differ very significantly.

Each dipole antennae feeds independently in the hyperthermia facility proposed in [2-7]. Each feeding channel includes RF circulator (see figure 4) to prevent the back wave penetration to the other feeding channels. Such system with minor modifications can be also used for thermometry. In the temperature measurement regime two feeding channels (RF input #1 and #5 in figure 4) will be switched to low RF power regime and the feeding of all other dipoles will be turned off. The output of the circulator directs the transmitted back wave to the band-pass filter and further to the measuring PIN-diode. PIN-diode measures the time dependence of the RF power flux for further A/D conversion. Digital signal processes by the especially developed Fourier analyses code and time dependences of Fourier coefficients define. Current spectral series compares with "non-perturbed" signal which was previously defined for not heated patient body. Proposed radiothermometry technique differs very sufficiently from the conventional one: the thermometry system is combined with hyperthermia one and dipole antennae are used both for heating and thermometry; the thermometry can be realized for deep situated tumours and tissues; the noninvasive thermometry can be realized and the active tissues scanning can be used for the 2D temperature imaging.

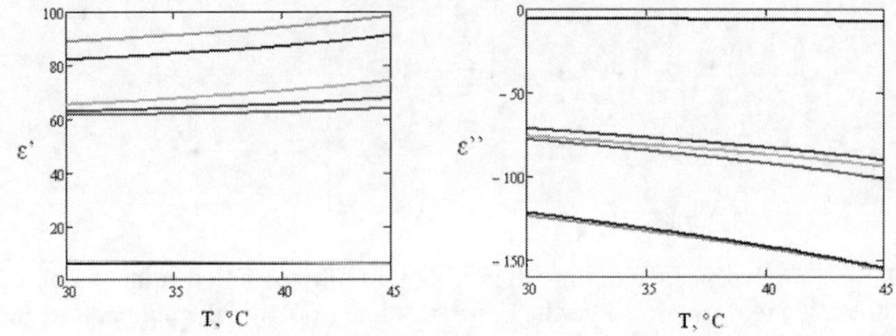


**Figure 3.** Dependences of real ε′ and imaginary ε′′ components of the complex dialectical permittivity for muscle tissue (red), liver (blue), kidney (green), spleen (brown), not infiltrated fat (black) and dry skin (cyan). The operating frequency is 150 MHz.

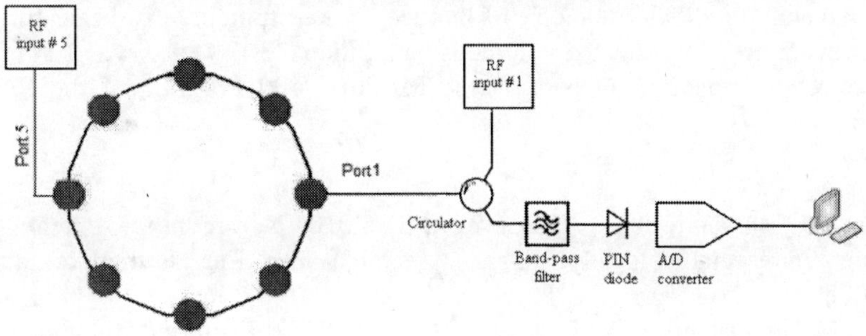


**Figure 4**. The scheme for measurement and analysys of the transmitted signal.

The CST simulation of the heating process with temperature control was done to verify the proposed thermometry technique. The fast Fourier analysis by the especially designed code was done to define the spectral distribution of the transmitted signal as the function of temperature. The main signal harmonic amplitude was controlled and the temperature distribution and its time dependence were calculated. The TRT facility was developed to heat only limited volume inside the patient's body. Only characteristics of such local area will influence to the transmitted signal. Its harmonic distribution and temperature control will be also provided for local volume where heating is taking place. Numerical simulations show that the amplitude of the main Fourier harmonic of the transmitted signal depends of the temperature linearly. The simulation model is shown in [6], it includes voxel model of the human body with a number of tissues and organs. The dielectric characteristics of tissues were varied during simulation with the temperature increases. The detected transmitted signal is shown in figure 5. Such detected signal expands to the Fourier series and its coefficients are compared with the "base" distribution which was done and safe for non-perturbed (not heated, T=37°C) model. It is clear from Figure 5 that the main signal harmonic amplitude depends vs. temperature linearly. Such result simplifies the back problem of the temperature reconstruction for heated volume. About 90 % of RF power absorbs by patient body during local or regional hyperthermia as it was shown by numerical simulation. The RF power necessary for effective treatment is up to 100 W/dipole for facility proposed in [2-7]. The simulation of the heating process with temperature control shows that for the first Fourier harmonic of the transmitted signal the variation is about $\sim 10^{-3}$ form coefficient value for the local hyperthermia and $\sim 10^{-2}$ for the regional one. Temperature increasing range is about 37-43°C. It is clear that the transmitted power measurement accuracy should be better than 100 μW (for measurement signal power is ~1 W). The accuracy of the temperature measurement for the local heated volume will be better than 0.3°C due to such RF power accuracy measurement. Such result is better than for conventional radiothermometry and close to tolerance for invasive thermo-couples or thermistors. The process of local thermoratiotherapy takes 15-30 minutes usually and it is quite enough to measure the temperature 1-3 times/minute.

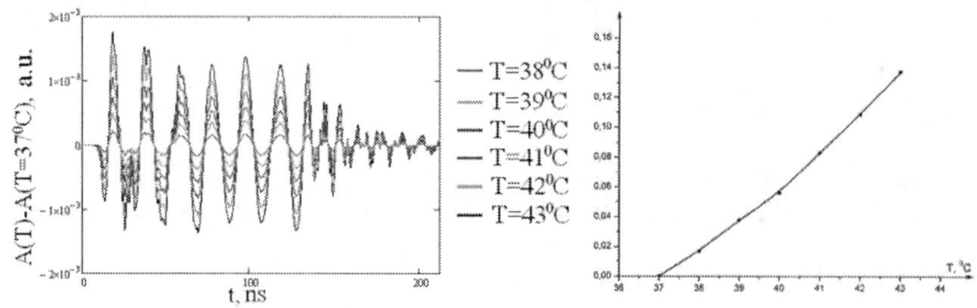


**Figure 5.** Temperature dependence of the difference (a.u.) of transmitted signal amplitudes for current and "base" (T=37°C) temperatures into the heated volume (top) and the temperature dependence of the first Fourier harmonic amplitude.

## 4. Conclusions

It was shown that dipole antennae of the TRT facility can be used both for heating and for temperature measurement and active radiothermometry of the deep-seated tumours can be realized. The accuracy of the thermometry for the local heated volume is better than 0.3°C with low (~1 W) measuring signal value. This project is supported in part by the MEPhI 5/100 Program of the Russian Academic Excellence Project.

## References

[1] Tkachev S.I., Barsukov Y.A., Gordeyev S.S., 2010. New combined treatment regimen for primary unresectable locally advanced rectal cancer. Eur. Journal of Surgical Oncology 36(9), 877.

[2] Ivanov S.M., Perelshtein E.A., Fadeev A.M. et al., 2012. Facility development for a combined cancer therapy. Problems of Atomic Science and Technology. Series: Nuclear Physics Investigations, 3 (79), 191-194.

[3] Fadeev A.M., Polozov S.M., Belyaev V.N. et al., 2012. Cylindrical Phased Dipoles Array for Hyperthermia of Deep-situated Tumors. Proc. of RUPAC'2012, 521-523.

[4] Polozov S.M., Perelshtein E.A., Fadeev A.M., et al., 2012. RF Power and Control Systems for Phased Dipoles Array System for Hyperthermia. Proc. of RUPAC'2012, 524-525.

[5] Ivanov S.M., Perelshtein E.A., Polozov S.M., Fadeev A.M., 2013. Facility for the electromagnetic hyperthermia based on the phased array. Problems of Atomic Science and Technology. Series: Nuclear Physics Investigations 6 (88), 220-224.

[6] Fadeev A.M., Ivanov S.M., Polozov S.M., Blinnikov A.A., 2014. A prototype of a Phased Array for Deep Thermoradiotherapy Proc. of RuPAC'14, 240-242.

[7] Fadeev A.M., Ivanov S.M., Perelshtein E.A., Polozov S.M., 2016. Facility for thermoradiotherapy of deep-seated tumors Journal of Physics: Conference Series, 747, 012076.

[8] Wust P., Seebass M., Nadobny J. et al., 1996. Simulation studies promote technological development of radiofrequency phased array hyperthermia. Int. J. Hyperthermia 12 (4), 477-494.

[9] Fadeev A.M., Ivanov S.M., Perelshtein E.A., Polozov S.M., 2016. Laboratory Model of Thermoradiotherapy Facility: Experimental Results Proc. of RUPAC'2016, 477-479.

[10] Gabriel C., Gabriel S., Corthout E., 1996. The dielectric properties of biological tissues: I. Literature survey. Phys. Med. Biol., 41, 2231–2249.

[11] Gabriel S., Lau R.W., Gabriel C., 1996. The dielectric properties of biological tissues: II. Measurements in the frequency range 10 Hz to 20 GHz. Phys. Med. Biol., 41, 2251–2269.

[12] Gabriel S., Lau R.W., Gabriel C., 1996. The dielectric properties of biological tissues: III. Parametric models for the dielectric spectrum of tissues. Phys. Med. Biol., 41, 2271–2293.

[13] Hurt W.D., 1985. Multiterm Debye dispersion relations for permittivity of muscle IEEE Trans. Biomed. Eng. 32, 60–4.

# Lagrange Equations of Envelopes

**V K Baev**

National Research Nuclear University MEPhI, Russia

baev.valeriy2010@yandex.ru

**Abstract.** The dynamics of charged particles in a traveling wave field with the uniform - magnetic focusing has been studied. A Lagrange function for this motion is constructed and Lagrange equations describing geometrical and kinematic envelopes of the bunch are derived.

## 1. Introduction

This work is devoted to description of the dynamics of charged particles in a traveling wave field with the uniform - magnetic focusing. Its principal purpose is a transition from the dynamics of separate charges to a bunch of charges as a single whole. For this purpose, a Lagrange functions for the longitudinal and transverse motions of a bunch of charged is constructed under following assumptions that (i) charges are uniformly distributed over longitudinal emittance of the bunch and (ii) the bunch is approximated by an ellipsoid of rotation in the configuration space, semi-axes of which are used as the generalized coordinates *q*, *r*.

Let us consider bunch of charged particles and approximate its emittace by an ellipsoid with vertical and horizontal semi-axes $N$ and $M$ respectively. For the bunch consisting of $N_b$ particles uniformly "smeared" over the emittance, the phase density $f$ is constant within the bunch, which can be written in a normalized form as $f = N_b/\pi MN$. In configuration space, we obtain a distribution function $\rho(x)$ defined within the bunch, which can be written in a normalized form as follows:

$$\rho(x) = \frac{8N_b}{3\pi^2 r^2 q}\sqrt{1-\left(\frac{x}{q}\right)^2}, \tag{1}$$

where $x$ is the coordinate measure from the bunch center along the axis of motion and $r$ is transverse semi-axis of the ellipsoid of rotation that approximates the bunch. Both $x$ and $r$, as well as all other distances and sizes used below, are normalized to the wavelength $\lambda$ of a generator that pumps the accelerating system.

Kinetic energy of the bunch $T_b$ we obtain from the following expression for the kinetic energy of the charges particles

$$T = \pi\gamma_s^3\dot{x}^2 + \frac{1}{2}\gamma_s\dot{y}^2, \tag{2}$$

where the first summand taken from [1]; $\dot{x}$, $\dot{y}$ is the longitudinal and transverse velocity of the charges particles correspondingly, expressed in units of the speed of light; $\gamma_s$ is the total energy of a synchronous particle in units of the energy of resting charges bunch.

The kinetic energy of an elementary fragment of the bunch, representing the thin ring, is given by the following formula:

$$dT_b = 2\pi y dx dy \rho(x)\left(\pi\gamma_s^3\dot{x}^2 + \frac{1}{2}\gamma_s\dot{y}^2\right). \tag{3}$$

Let us assume the dependence of the longitudinal and transverse velocity of the ring from the coordinates can be written by the following formula: for the longitudinal velocity $\dot{x}(x) = \dot{q}\chi^{\alpha/2}$; for the transverse velocity $\dot{y}(y) = \dot{r}\eta^{æ/2}$, where $\chi = x/q$; $\eta = y/r$; $\alpha$, æ – the parameters of the dependences.

Integration of expression (3) over the bunch volume yields the kinetic energy of the whole bunch:

$$T_b = \frac{8N_b}{9\pi}\gamma_s\left(3\pi\frac{3-\alpha}{(1+\alpha)(3+\alpha)}\gamma_s^2\dot{q}^2 + \frac{3-æ}{2+æ}\dot{r}^2\right). \quad (4)$$

The potential energy of the longitudinal oscillations of the bunch in the accelerating wave defined by the expression [2]

$$U_{b1} = \frac{\beta_s A\sin\varphi_s}{24\pi}N_b(\psi^2 - \frac{1}{32}\psi^4 + \frac{1}{1920}\psi^6), \quad (5)$$

where $\psi = 2\pi q/\beta_s$; $\beta_s$ – the velocity of the synchronous particle, expressed in units of the speed of light; $A=Z_e\mathcal{E}\lambda/E_0$; $\mathcal{E}$ – the amplitude of the accelerating wave; $E_0$ – the rest energy; $\varphi_s$– is the phase of the synchronous particle.

Integration of the potential energy of the transverse motion of the charge in the accelerating waves field [1] and the potential energy of the thin ring in the uniform - magnetic focusing field

$$dU_{br} = 2\pi y^3 dydx\frac{8N_b}{3\pi^2 r^2 q}(1-\chi^2)^{1/2}\frac{B_f^2}{8\gamma}, \quad (6)$$

which followed from the equations of the motion, yields the potential energy of the transverse motion of the whole bunch:

$$U_{b2} = \frac{5}{24}N_b\frac{\pi A\sin\varphi_s}{\beta_s^2}(1-\beta_s^2)r^2 + \frac{5}{96}N_b\frac{B_f^2}{\gamma}r^2. \quad (7)$$

Here $B_f$ is the non-dimensional magnetic focusing field, connected with the dimensional field $B_f^*$ by the expression $B_f = Z_e B_f^*/m_0 c$.

The Lagrange equations may also naturally include the intrinsic Coulomb field of the bunch. This can be achieved using the Vlasov ellipsoid model [3], which leads to the appearance of the following term on the right-hand side of the Lagrange equations:

$$\mathcal{E}_b = \frac{90Z_e}{r^2\beta_s^2 E_0}(1-\beta_s^2)^{3/2}MI. \quad (8)$$

Here $I$ is the pulsed current of a train of bunches (expressed in amperes) and $M$ is the ellipsoid form factor defined by the expressions [2].

Once the kinetic and potential energies of the bunch are known, the corresponding Lagrange function can be constructed and the Lagrange equations can be written as follows: the equation of the longitudinal motion

$$\frac{d^2q}{dz^2} = \frac{A\sin\varphi_s}{32\beta_s^2\gamma_s^3}\frac{(1+\alpha)(3+\alpha)}{3-\alpha}P(\psi) - \frac{3\gamma_s^2-2}{\gamma_s(\gamma_s^2-1)}\frac{d\gamma_s}{dz}\frac{dq}{dz} + \frac{D}{\beta_s^2}(1-\beta_s^2)^{3/2}, \quad (9)$$

where $D = 90ZMI/r^2E_0$; $Z$ is the order of the charge of the bunch,

$$P(\psi) = \psi - \frac{1}{16}\psi^3 + \frac{1}{640}\psi^5; \quad (10)$$

the equation of the transverse motion

$$\frac{d^2r}{dz^2} = -\frac{15\pi}{256}\left(\frac{2+æ}{3-æ}\right)\left(\frac{4\pi A\sin\varphi_s}{\beta_s^4\gamma_s}(1-\beta_s^2) + \frac{B_f^2}{\gamma_s^2-1}\right)r - \frac{\gamma_s}{\gamma_s^2-1}\frac{d\gamma_s}{dz}\frac{dr}{dz} + $$
$$+D_r(1-\beta_s^2), \quad (11)$$

where $D_r = 90ZM_rI/rqE_0$; $M_r = 0{,}5(1-M)$.

The solutions of those equations are the generalized coordinates and velocities. The generalized coordinates define the sizes of the bunch and the generalized velocities are the boundaries of its spectrums. According to the terminology of the mechanics the generalized coordinates are the geometrical envelopes and the generalized velocities are the kinematic envelopes.

In this way the Lagrange equation contains two envelopes, a namely geometrical and kinematic.

The obtained equations are describes longitudinal and transverse geometrical and kinematic envelopes of the bunch. Those equations are advantageous to the well-known equations [4], primarily,

IOP Conf. Series: Journal of Physics: Conf. Series **941** (2017) 012087 doi:10.1088/1742-6596/941/1/012087

in being simple. Additional advantages are as follows. First, the Lagrange equations determine not only the sizes of a bunch (its geometrical envelopes) , but also the boundaries of its energy spectrums (kinematic envelopes). Second, the Lagrange equations have been derived without any restrictions imposed on the amplitude of the oscillations, so this mathematical model is nonlinear and, hence, more general. Third, the equations take into account the acceleration of a bunch. Finally, the Lagrange equations take into account the force of the Coulomb field of a bunch in terms of the naturally included ellipsoid model, so that those envelope equations are self-consistent.

Figures 1 – 3 illustrates the results of calculations the envelopes as determined from obtained Lagrange equations and calculated by the traditional method of individual particles.

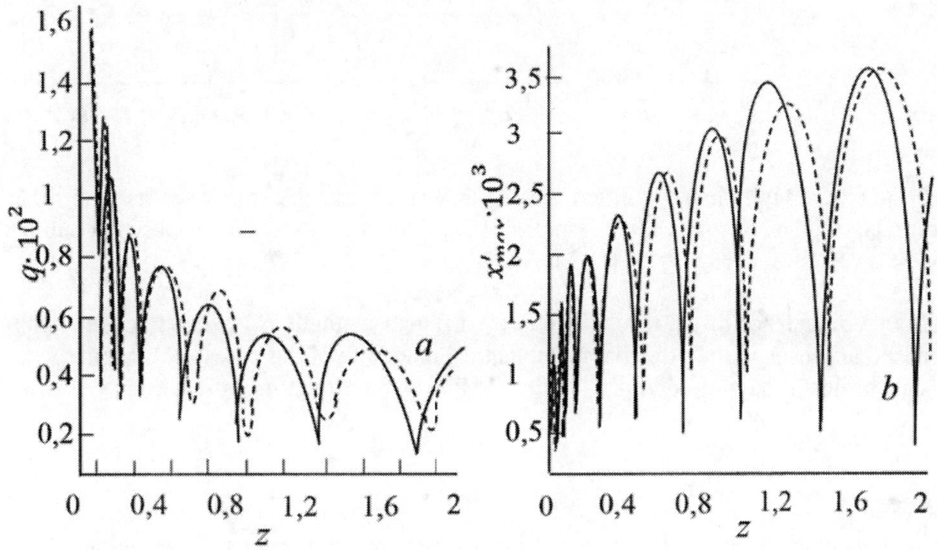


**Figure 1.** Longitudinal geometrical (a) and kinematic (b) envelopes as determined using the traditional method of individual particles (dash – dot curve) and the proposed equations of envelopes (solid curve).

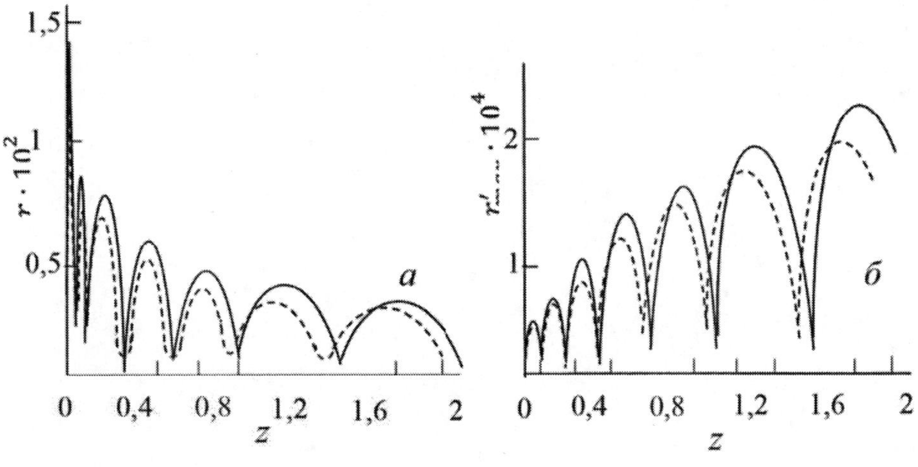

**Figure 2.** Transverse geometrical (a) and kinematic (b) envelopes as determined using the traditional method of individual particles (dash – dot curve) and the proposed equations of envelopes (solid curve).

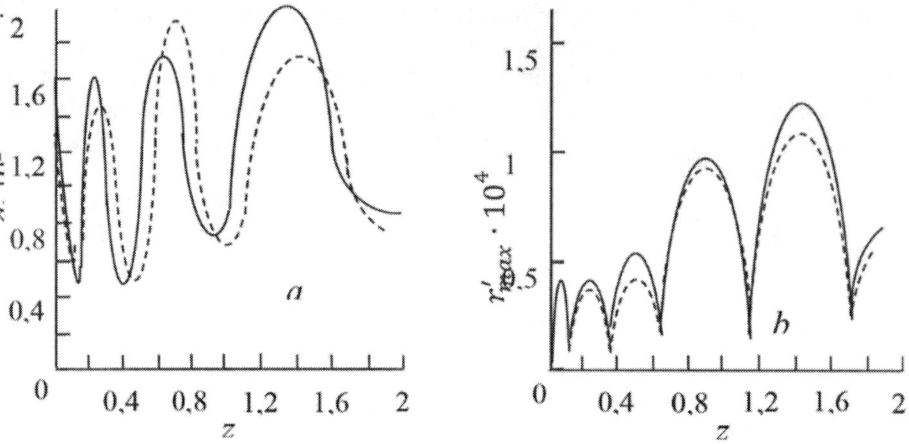


**Figure 3.** Longitudinal geometrical (a) and kinematic (b) envelopes as determined using the traditional method of individual particles (dash – dot curve) and the proposed equations of envelopes (solid curve); the pulsed current is 32 mA.

Good coincidence of the results obtain by different methods is evidence for the applicability of the proposed approach to the effective mathematical modeling of the bunches dynamics.

This project is supported in part by the MEPhI 5/100 Program of the Russian Academic Excellence Project.

## References

[1] Kapchinski I.M. 1982 Theory of Linear Resonant Accelerators (Energoatomizdat, Moscow).
[2] Baev V.K. 2013 Technical Physics Letters, Vol.39, No. 2, pp.193-195.
[3] Vlasov A.D. 1965 Theory of Linear Accelerators (Atomizdat, Moscow).
[4] Chishachev A.S. 2011 Kinetic Theory of the Quasi-Stationary State of the Charge Particles Bunches (Fizmatgiz, Moscow).

# RF design of 324 MHz superconducting (SC) CH cavity for 0.21 beta.

**K Taletskiy[1], D Surkov[1] and M Gusarova[2]**

[1] Student, NRNU "MEPhI", Moscow, Russia
[2] Docent, PhD, NRNU "MEPhI", Moscow, Russia

E-mail: kirill.taletski@gmail.com

**Abstract.** The results of RF optimizations for 324 MHz SC cross-bar H-mode (CH) cavity for 0.21 beta are presented. Maximum surface electric field of 36 MV/m and a corresponding effective accelerating gradient of 7 MV/m have been achieved.

## 1. Introduction

Currently planned project of the new linac injector for Nuclotron-NICA (Figure 1) demand high-gradient SC cavities for medium energy range [1]. The most types of superconducting cavities developed for $\beta \leq 0.5$ have less than 4 cells. But in many cases the RF linac efficiency can be increased significantly by the use of multicell cavities. The possibility of using the five-gap accelerating CH resonator is considered in this paper.

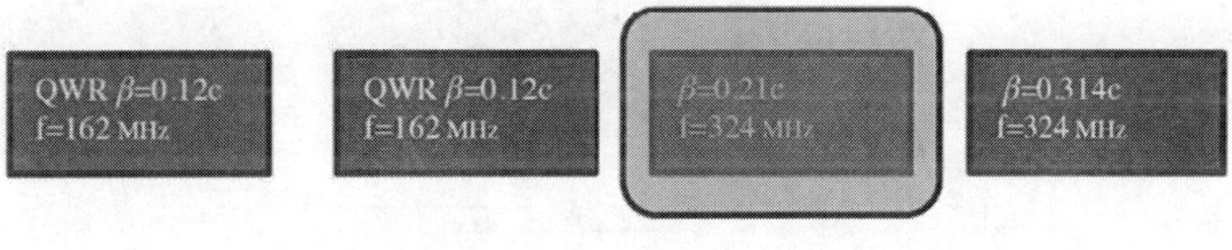


**Figure 1.** Schematic layout of the new linac injector for the Nuclotron-NICA.

## 2. Time transition factor optimization.

Optimization of the TTF increases efficiency of acceleration in given structure. The TTF analysis of two-gap cavity proposed in [2] shows that only one value of period length $d$ (for defined accelerating gap width $g$) corresponds to maximum TTF value. Similar analytical approach was applied to find $d(g)$ dependence that leads to maximum TTF value. The obtained dependence can be sufficiently described by second-order polynomial. It is presented on Figure 2.

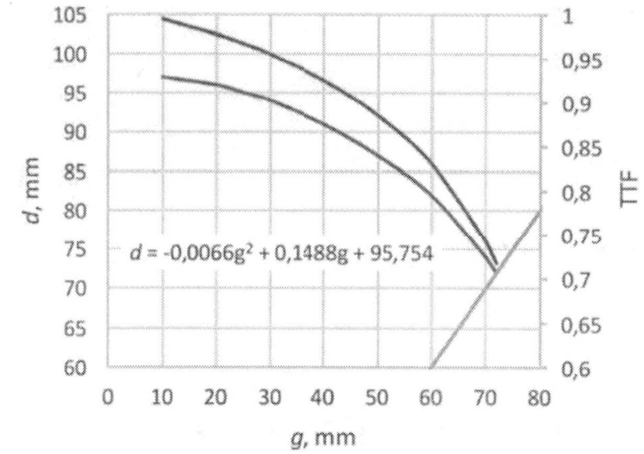


**Figure 2.** Optimal period length $d$ (blue) and corresponding TTF (red) vs. gap length $g$. Green line limits $g < d$ set of values, which has no physical meaning.

### 3. Single period model optimization.

Once TTF was optimized the next step was to develop single period geometry to minimize peak surface electric and magnetic fields ratios ($E_P/E_{ACC}$ and $B_P/E_{ACC}$). Firstly, drift tubes geometry was optimized as the maximums of electric field were concentrated on it. Figure 3 and Figure 4 show the main geometrical parameters that were used for optimization.

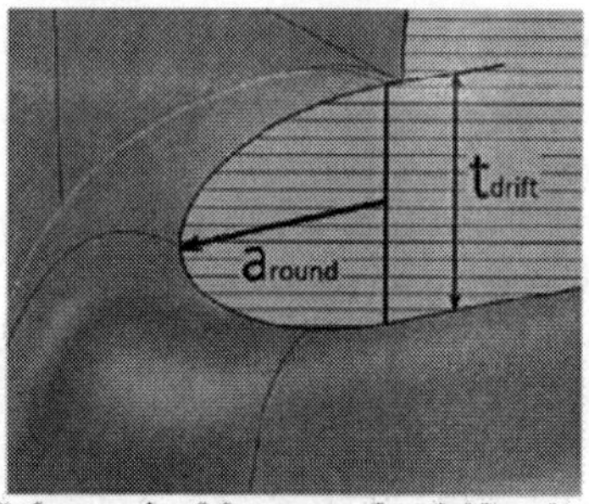


**Figure 3.** Schematical layout of a drift tube geometry.

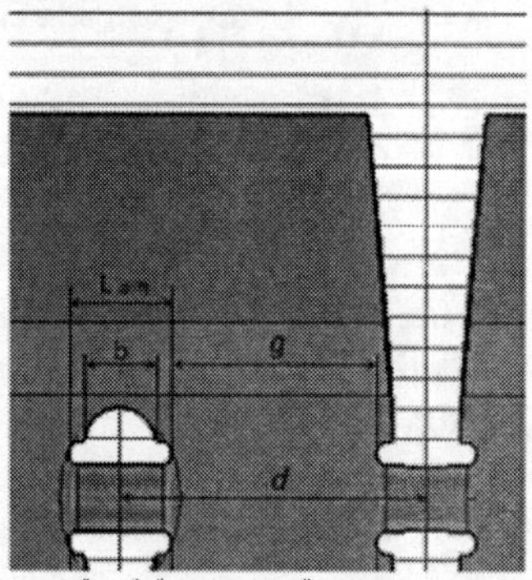


**Figure 4.** Schematical layout of a two-gap periodic model.

Three major types of drift tube geometries were investigated (Figure 5). As it may be seen from Figure 6 decreasing $l_{DRIFT}/b$ ratio allows to lower relative peak electric field $E_P/E_{ACC}$ down to 7.

Geometry that corresponds to $l_{DRIFT}/b = 1$ (i. e. with no "door knops") was considered to be the most effective.

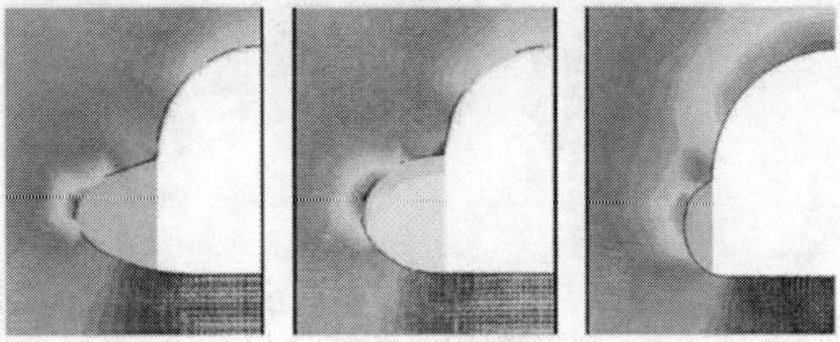

**Figure 5.** Three major types of drift tube geometries ($k = a_{round}/t_{drift}$): $k > 1$ (left), $k = 1$ (middle), $k < 1$ (rigth).

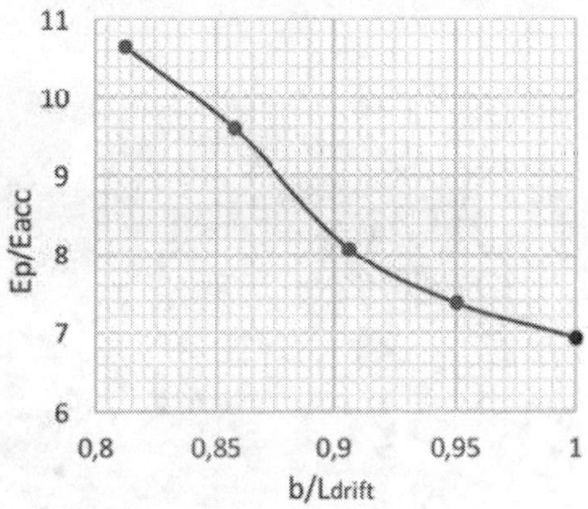


**Figure 6.** Relative surface peak field vs. $b/L_{DRIFT}$ ratio.

## 4. Full cavity model tuning

The main challenge of finite structure optimization is the end-cells geometry design. As was proposed in [3] inclined geometry of the end bars helps to tune end-cells to work frequency. Figure 7 shows end cells design concept proposed in [3] with its main parameters used for tuning. Surface electric and magnetic field distributions of the CH cavity are presented on Figure 8. The parameters of tuned cavity are shown in Table 1.

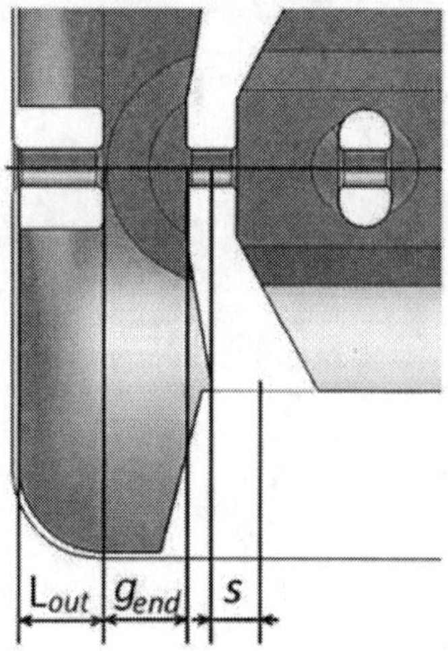


**Figure 7.** Schematic layout of the end-cell geometry.

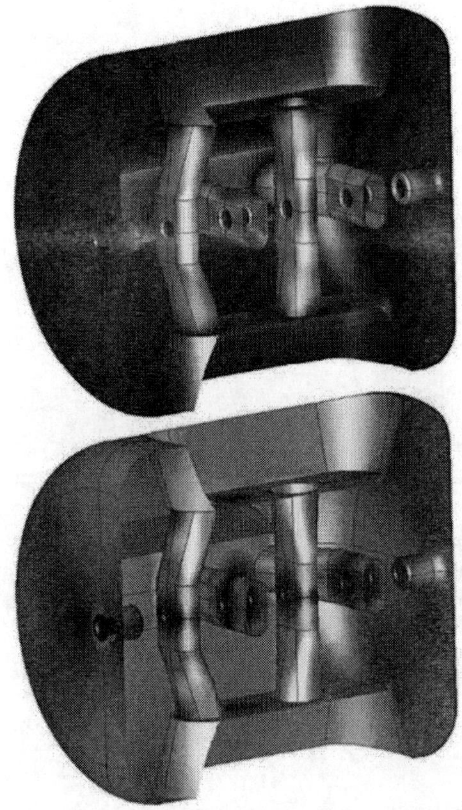

**Figure 8.** Surface electric (top) and magnetic (bottom) fields in CH resonator.

**Table 1.** The geometry and RF parameters of tuned CH cavity.

| | |
|---|---|
| **β** | 0.21 |
| **Frequency [MHz]** | 324 |
| **Length [mm]** | 500 |
| **Radius [mm]** | 204 |
| $E_p/E_a$ | 5.1 |
| $B_p/E_a$ **[mT/(MV/m)]** | 7.5 |
| **G [Ω]** | 84 |
| $R_a/Q_0$ | 615 |

## 5. Conclusion

We presented the RF design of the superconducting CH cavity intended for the proposed proton and deuteron injector for the NICA project. The electromagnetic optimization was performed to show that desired effective accelerating gradient of 7 MV/m can been achieved with this type of cavity.

This project is supported in part by the MEPhI 5/100 Program of the Russian Academic Excellence Project.

## References

[1] M.A. Batouritski, A.V. Butenko, M.A. Gusarova et al., The Perspective of JINR LU-20 Replacement by a Superconducting Linac. Proc. of RuPAC'16.

[2] J.R. Delayen. Design of low velocity superconducting accelerating structures using quarter-wavelength resonant lines. NIM A259 (1987) 341-357.

[3] F. Dziuba, M. Busch, M. Amberg et. al., Development of superconducting crossbar-H-mode cavities for proton and ion accelerators. Phys. Rev. **13**, 041302 (2010).

IOP Conf. Series: Journal of Physics: Conf. Series **941** (2017) 012089 doi:10.1088/1742-6596/941/1/012089

# Electrodynamic characterisitcs measurements of higher order modes in S-band cavity

**R Donetsky , M Lalayan, N P Sobenin, A Orlov and A Bulygin**

National Research Nuclear University, "MEPhI", Moscow, Russia

E-mail: yvshashkov@mephi.ru

**Abstract:** The 800 MHz superconducting cavities with grooved beam pipes were suggested as one of the harmonic cavities design options for High Luminosity LHC project. Cavity simulations were carried out and scaled aluminium prototype having operational mode frequency of 2400 MHz was manufactured for testing the results of simulations. The experimental measurements of transverse shunt impedance with error estimation for higher order modes $TM_{110}$ and $TE_{111}$ for S-band elliptical cavity were done. The experiments using dielectric and metallic spherical beads and with ring probe were carried out. The Q-factor measurements for two-cell structure and array of two cells were carried out.

## 1. The influence of various factors on frequency and Q-factor measurement in the 2400 MHz prototype of superconducting harmonic cavity

Despite the high accuracy of modern electromagnetic field dispersion in cavities simulation software it is still necessary and important to manufacture a prototype of the structure under development in order to measure its electrodynamics characteristics (EDCs) of both operational and higher order modes [1-4]. In order to provide precise frequency $f_0$, Q-factor and shunt impedance measured values various factors should be taken into account, namely the specific pressure at the junction points of the sections, the surface roughness, the temperature and the immersion depth of the exciting and receiving RF pickups. This paper consists of three parts: the first one is devoted to the experiments analysis of the pressure, position and penetration of probes into the cavity influence. All measurements were made on the scaled 2400 MHz harmonic cavity model built for testing the damping efficiency of the method proposed for the High luminosity LHC project [5-8]. The second part presents the experimental and calculated field dispersion analysis results obtained by bead pull measurements. Results of the Q-factor measurements for the two cell cavity structure are presented in third section.

### *1.1. Pressure influence*

Measurements were done with Vector network analyzer Agilent 8753ET. The modes of interest are $TM_{010}$, $TE_{111}$ and $TM_{110}$ (figure 1). Pickups used for modes excitation are shown on figure 2. The Q-values were measured by common 3 dB bandwidth method (1), where $f_0$ is the frequency of the mode and $\Delta f$ is the bandwidth corresponding to -3 dB drop in transmission:

$$Q_L = f_0/\Delta f \tag{1}$$

The accuracy of the frequency measurements with analyser is less than 1 kHz resulting in precision of$10^{-4}$ %.

IOP Conf. Series: Journal of Physics: Conf. Series **941** (2017) 012089 doi:10.1088/1742-6596/941/1/012089

In order to ensure the good contact between different parts of the model it was tested in the test bench equipped with hydraulic press. On figure 3 operational mode $TM_{010}$ specific pressure on Q-factor dependency is presented, on figure 4 frequency dependency is presented. Saturation region can be found on the graph between the values of $\sigma = 160 \div 200$ kg/cm$^2$. All measurements were done at the pressure close to 180 kg/cm$^2$.

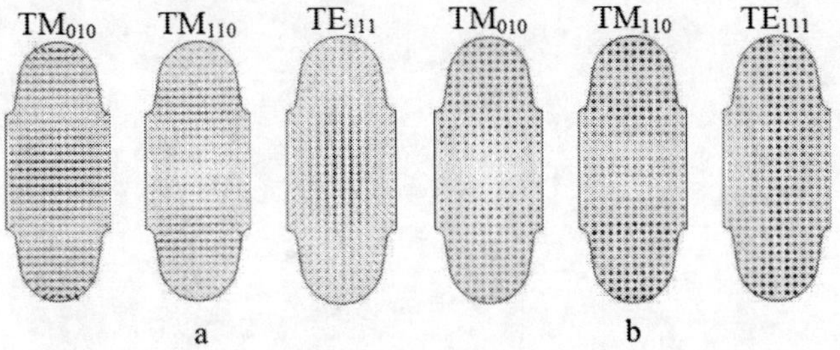


**Figure. 1.** Field dispersion of $TM_{010}$, $TE_{111}$ and $TM_{110}$ modes: (a) Electric field; (b) Magnetic field.

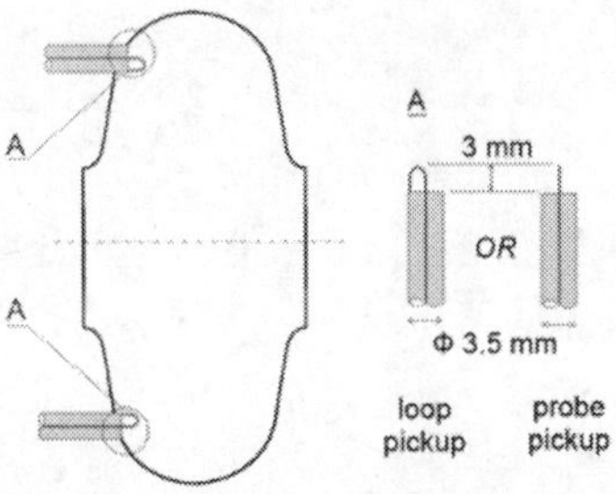


**Figure. 2.** Cavity with the location of the pickups.

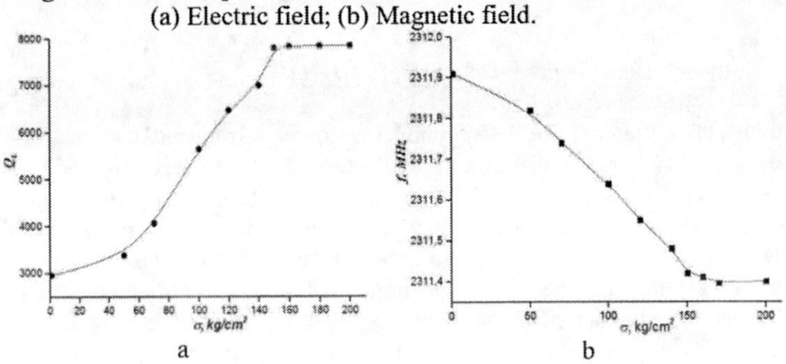


**Figure. 3.** $TM_{010}$ mode Q-factor (a) and Frequency (b)vs. specific pressure.

*1.2. Electric (probe) and magnetic (loop) pickups penetration depth influence*

The influence of the probe penetration depth on $f$ and $Q_L$ was determined, as well as the dependence of the loaded Q-factor on the loop rotation angle with respect to its axis. Measurements were made for a loop and a probe. $TM_{110}$ and $TE_{111}$ modes are weakly excited by the probe. In the place where the probe is located it is not possible to create the necessary configuration of the electric field lines (figure 2). Therefore the electric probe was used for fundamental mode study whereas the loop pickup was used for dipole waves EDCs measurements. The loaded Q-factor measured using a loop features non-linear variation. So the influence of the loop rotation angle relative to the axis, on the loaded Q-factor was measured (figure 4).

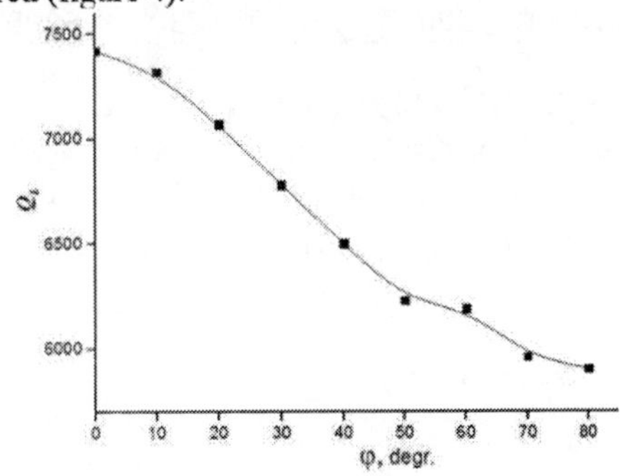


**Figure. 4.** Dependence of the loaded Q-factor on the loop rotation angle for $TM_{010}$ (0° - the plane of the loop is perpendicular to the lines of the magnetic field).

IOP Conf. Series: Journal of Physics: Conf. Series **941** (2017) 012089 doi:10.1088/1742-6596/941/1/012089

The dependence shown on figure 5 is close to linear one with the slope coefficient of the straight line $\frac{Q_L}{\varphi} = 30 \frac{1}{degree.}$. Figure 5 (a) illustrates the frequency $f_0$ dependence on the probe penetration depth $l$ for $TM_{010}$, $TM_{110}$ and $TE_{111}$ modes at different locations of the probe (0 and 2 mm penetration depth). At Fig. 5b penetration depth dependence on $Q_l$ is presented. Rotation angle precision is $\Delta\varphi = \pm 5^{\circ}$ resulting in $\Delta Q_L = \pm 150$ and the length depth $\Delta l = \pm 0{,}05$ mm. At figure 5 $\Delta f(l)$ and $Q_L(l)$ graphs taking into the account precision of the probe position are shown.

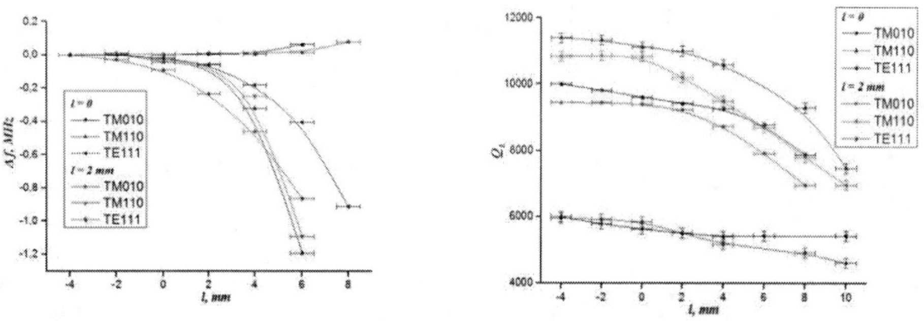


a b

**Figure. 5.** Dependences: $(a)\Delta f(l)$, (b) $Q_L(l)$.

Small perturbation theorem states that the perfectly conducting body introduced into the cavity could affect the frequency in two ways: first by disturbing the electric field and decreasing the cavity frequency, and second by disturbing the magnetic field leading to frequency increase. As it could be seen on Figure 5 the operational mode frequency increases with the introduction of a probe into the cavity, and the dipole waves frequency decreases. As a consequence of the theorem on small perturbations and from the dispersion of the electromagnetic field for the operational mode consideration, the probe interacts with magnetic field and with the electric field for the dipole waves. With decreasing of the coupling pickup penetration the external quality factor increases; in accordance with the known expression $\frac{1}{Q_L} = \frac{1}{Q_0} + \frac{1}{Q_{ext}}$ the measured loaded Q-factor will be closer to $Q_0$, which we compare with the simulation results. Table. 1 shows the values of the loaded Q-factors $Q_L$ and the frequency of the modes under study with the minimal influence of the coupling elements.

**Table 1.** Results of the experiment and simulation

| Mode | $TM_{010}$ | $TM_{110}$ | $TE_{111}$ |
|---|---|---|---|
| Frequency, MHz (calculated) | 2311,4 | 3457,0 | 4000,9 |
| Frequency, MHz (measured) | 2315,9 | 3455,7 | 4006,7 |
| $Q_0$ (calculated) | 13900 | 16500 | 16700 |
| $Q_l$ (measured) | $10000 \pm 200$ | $11400 \pm 230$ | $6000 \pm 120$ |

## 2. Shunt impedance measurements in single cell 2400 MHz cavity

The transverse impedance $r_{\perp}/Q_0$ measurements could be done with two methods: direct integration of transverse magnetic $H_{tr}$ and electric $E_{tr}$ field components (2) and Panofsky-Wenzel theorem (3):

$$\frac{r_{sh\perp}}{Q_0} = \frac{\left| i \cdot c \cdot \mu_0 \int_0^l H_{tr}(z) \exp(ik_z z)\, dz - \int_0^l E_{tr}(z) \exp(ik_z z)\, dz \right|^2}{(\omega \cdot W) \cdot l} \quad (2)$$

$$\frac{r_{sh\perp}}{Q_0} = \frac{\left| \int_0^l \frac{\partial E_z(z)}{\partial z} \exp(ik_z z)\, dz \right|^2}{(\omega \cdot W) \cdot l} \quad (3)$$

where c is the speed of light, $\mu_0$ is the magnetic constant, $k_z = 2\pi/\lambda$ is the longitudinal wave factor, $\omega = 2 \cdot \pi \cdot f$ is the wave frequency, $W$ is the stored energy, $l$ is the length of the structure.

The field distribution can be found using the perturbation theorem [9]:

$$\frac{\Delta f_{cs}(z)}{f_0} = -k_{c\perp}^{E}\frac{E_{tr}^2(z)}{W} - k_{c\|}^{E}\frac{E_{tr}^2(z)}{W} \quad (4)$$

$$\frac{\Delta f_{ms}(z)}{f_0} = -k_{m\perp}^{E}\frac{E_{tr}^2(z)}{W} + k_{m\perp}^{H}\frac{H_{tr}^2(z)}{W} - k_{m\|}^{E}\frac{E_z^2(z)}{W} + k_{m\|}^{H}\frac{H_z^2(z)}{W} \quad (5)$$

$\Delta f_{cs}$and $\Delta f_{ms}$ – are the frequency perturbation by ceramic and metal beads, $k$ – transverse or longitudinal form factor for electric or magnetic field.

The electric field component can be easily measured by the ceramic bead because they do not perturb magnetic field. The magnetic field can be obtained as a difference between measurement results with metal and ceramic beads. The simulations were done for several perturbing beads in order to define the optimal combination of bodies to establish the most precise values of $r_{sh\perp}/Q_0$. Form factor values for different beads are presented in Table 2.

**Table 2.** Formfactors of the perturbing beads

| Formfactor | ceramic sphere | metal sphere | metal ring |
|---|---|---|---|
| $k_{\perp}^{E} \cdot 10^{-20}$, $m^2$s/Ohm | 1,97±0,02 | 2,32±0,03 | 1.92 |
| $k_{\|}^{E} \cdot 10^{-20}$, $m^2$s/Ohm | 1,97±0,02 | 2,32±0,03 | 0.09 |
| $k_{\perp}^{H} \cdot 10^{-15}$, $m^2$ ·s·Ohm | NA | 1,73±0,03 | 1.36 |
| $k_{\|}^{H} \cdot 10^{-15}$, $m^2$ ·s·Ohm | NA | 1,73±0,03 | 0.01 |

While evaluating the measurement results the frequency, Q-factor and phase measurement error, bead form factor uncertainty and bead off-axis misalignment were taken into account. The total measurement uncertainty lies well below 8% of those bead off-axis misalignment introduces the most contribution. The simulation and measurement results of $\Delta f/f_0$ are presented in figure 6. The comparison of measured and simulated field distribution of field components presented in figure 7.

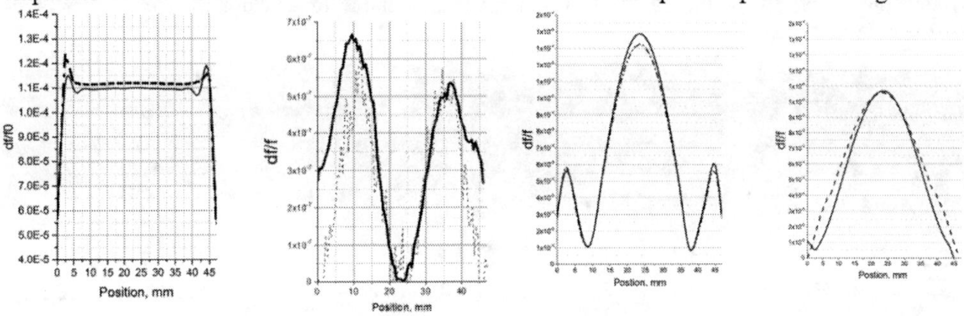


a b c d

**Figure 6.** $\Delta f/f_0$for metal sphere and ceramic sphere for $TM_{110}$ (a,b) and $TE_{111}$ (c,d). Straight line – measurements, dotted – simulations.

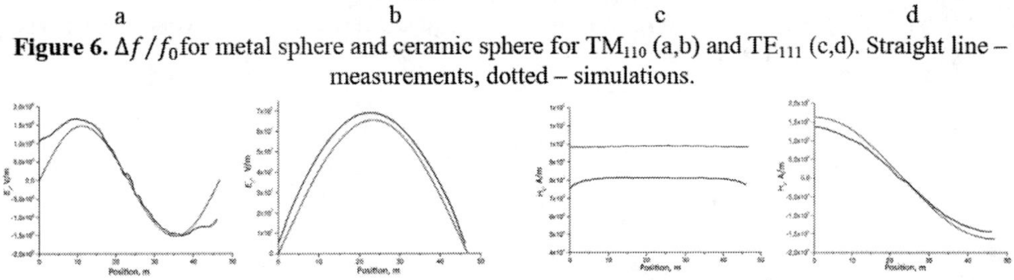

a b

**Figure 7.** (a) Magnetic and (b) Electric field distributions for ring bead for $TE_{111}$ and $TM_{110}$. Red – simulation, black – measurements.

Shunt impedance values obtained experimentally using two methods mentioned above (namely Panofsky-Wenzel referred as P.-W. hereafter and direct integration one) are summarized in Table 3. As one could see all these results agree well.

**Table 3.** $r_{sh\perp}/Q$ simulation and measurements results on $TM_{110}$ and $TE_{111}$ modes

| Mode | $TM_{110}$ | $TE_{111}$ | |
|---|---|---|---|
| | | MS+CS | MR+CS |
| Simulation (direct integration) | 985 | 43,4 | |
| Simulation (P.-W.) | 985 | 42,9 | |
| Measurements (direct integration) | 788 | 35,4 | 48,2 |
| Measurements (P.-W.) | 717 | 32,9 | NA |

## 3. Q-factor measurements for two-cell and array of cells structure

The 800 MHz superconducting cavities with grooved beam pipes were suggested as one of the harmonic cavities design options [10] for HL-LHC project. Simulations showed that it is possible to achieve Q-factor values in a structure with grooved beam pipe lower than 100 for the most dangerous dipole modes and 1000 for modes with higher frequency. The results obtained in Multi-P program [11] showed that this structure is free of multipacting discharge [12].

Q-factor measurements were made for a structure consisting of two cells connected by a narrow drift tube and two corrugated tubes at the sides (figure 8). Q-factor measurements were also made for two-cell cavity with corrugated drift tube (figure 9) which might be an interesting option for Future circular collider . To reduce the Q-value of dangerous dipole HOMs, damping ferrite rings were added at the ends of the drift tubes. Figure 10 presents the results of an experiment on loaded Q measurement in a structure with ferrite rings and without rings.

The damped Q-factor for most of HOMs is less than 1000, except for a few modes indicated by the numbers on the chart. However, those HOMs have low $R/Q$ values and therefore could be treated as inessential with respect to their influence on beam dynamics. All other HOMs were damped well below the sensitivity of the analyzer. The fundamental mode loaded quality factor after damping changed insignificantly.

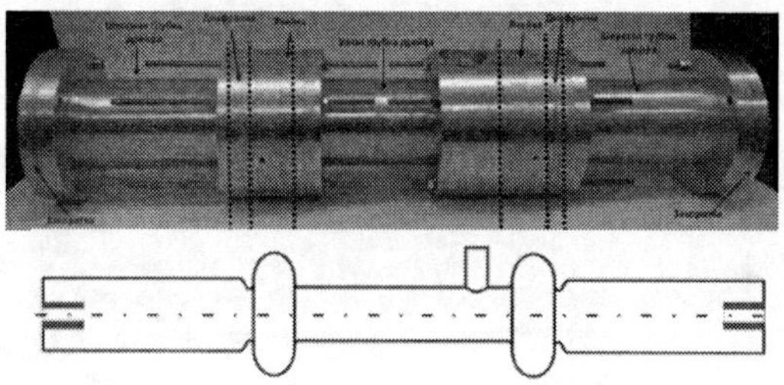

**Figure 8.** Two separated cells structure with corrugated beam pipes.

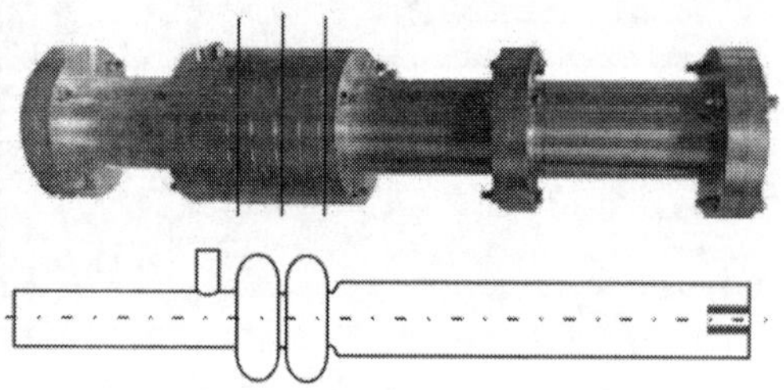

**Figure 9.** Two-cell cavity with corrugated beam pipe.

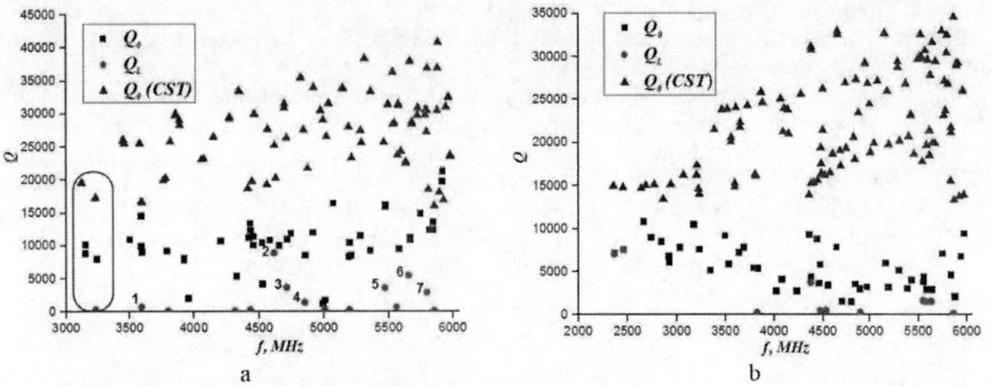


**Figure 10.** Q values for (a) two cells separated by beam pipe; (b) two-cell cavity with drift tubes connected. Blue – Q-factor without damping ferrite rings in the drift tube, red - Q-factor with ferrite rings. 1 – dipole mode trapped in beam pipe between cells; 2,3,4 – quadrupole HOMs; 5,6,7 – sextupole HOMs.

## Conclusion

The field dispersions obtained with bead pull measurements as well as measurements of transverse shunt impedance for higher order modes $TM_{110}$ and $TE_{111}$ for S-band elliptical cavity are in a good convergence with simulation results. For the manufactured two-cell and array of two cells aluminium prototypes results of loaded Q-factor measurements of higher order modes are also agree with a simulation results.

This project is supported in part by the MEPhI 5/100 Program of the Russian Academic Excellence Project.

## References

[1] Navarro-Tapia M and Calaga R 2015 Bead-pull measurements of the main deflecting mode of the double quarter-wave cavity for the HL-LHC *Proceedings of 17th International Conference on RF Superconductivity (SRF2015): Whistler, Canada, September 13-18, 2015,* p. 1105-1109

[2] Hahn H and Calaga R, Jain P, Johnson E C, and Xu W 2014 HOM identification by bead pulling in the Brookhaven ERL cavity, Nucl. Instrum. Methods Phys. Res., Sect. A 734, 72 (2014)

[3] Zavadtsev A A , Zavadtsev D A , Nikolskiy K I , Savin E A and Sobenin N P 2013 Design and tuning of a 40-MeV electron linear accelerator Instrum. Exp. Tech. 2013; 56(5), pp. 506-515

[4] Savin E A, Matsievskiy S, Shashkov Ya V, Sobenin N P and Zavadtsev A A 2014 Multi-bunch generator cavity, Proceedings of the 5th International Particle Accelerator Conference, pp. 2267-2269

[5] Shashkov Ya V, Sobenin N P, Gusarova M A, Lalayan M V, Bazyl D S, Donetskiy R V, Orlov A I, Zobov M M and Zavadtsev A A 2016 Measurement of electrodynamics characteristics of higher order modes for harmonic cavity at 2400 MHz, Journal of Physics: Conference Series, Volume 747, Number 1

[6] Shashkov Ya V, Sobenin N P, Petrushina I I and Zobov M M 2014 Comparison of higher order modes damping techniques for 800 MHz single cell superconducting cavities, Nuclear Instruments and Methods in Physics Research A, v. Volume 767, pp. 271–280

[7] Shashkov Ya V, Sobenin N P and Zobov M M 2014 Comparison of High Order Modes damping techniques for 800 mhz single cell superconducting cavities Proc. IPAC'14,

Dresden, Germany (2014) pp 2258-2261

[8] Ya. V. Shashkov, A.A. Mitrofanov, N.P. Sobenin, V.L. Zvyagintsev, Analysis of higher order modes damping techniques in 9 cell cavity with modified drift tubes, Journal of Physics: Conference Series, Volume 74, p. 124-131

[9] B.V. Zverev B V and Sobenin N P 1993 Electrodynamic Characteristics of Accelerating Cavities Energoatomizdat, Moscow

[10] Shashkov YaV, Sobenin N P, Bazyl D S, Kaminskiy V I, Mitrofanov A A, Zobov M, 2016 Suppression of higher order modes in an array of cavities using waveguides, Journal of Physics: Conference Series, Volume 74, p. 116-123

[11] Gusarova M, Khudyakov S, Petrushina I I and Shashkov Ya V 2014 New Possibilities of MultP-M Code Proc. IPAC'14, Dresden, Germany (2014), pp 433-435

[12] Petrushina I I and Gusarova M A 2014 Multipactor in elliptical cavities 800 MHz Proceedings of 24th Russian Particle Accelerator Conference 2014, pp. 365-367

IOP Conf. Series: Journal of Physics: Conf. Series **941** (2017) 012090 doi:10.1088/1742-6596/941/1/012090

# Numerical simulation of electrons dynamics in a microtron on 6 – 10 MeV

**Y A Bashmakov[1,2], V S Dyubkov[1] and Y Y Lozeev[1]**

[1] National Research Nuclear University MEPhI, Kashirskoe shosse 31, Moscow 115409, Russian Federation

[2] P.N. Lebedev Physical Institute of the Russian Academy of Sciences, Leninskiy prospekt 53, Moscow 119991, Russian Federation

**Abstract.** Electron dynamics in 6.5 MeV classic microtron of the Lebedev Physics Institute (LPI) is investigated by means of numerical methods. Particular emphasis is placed on the formation mechanism of electron bunches at the first circular orbits. An effect of microtron main parameters such as accelerating RF field amplitude, DC magnetic field, as well as a geometry and a position of a thermal emitter on characteristics of electron beam extracted from the microtron are studied. In the space of mentioned parameters a region corresponding an optimal microtron operation mode is found. It is noted that the unique geometric and energy characteristics of accelerated beam makes use of microtron attractive not only as injector into a synchrotron, but also as a driver in experiments on generation of coherent terahertz electromagnetic radiation.

## 1. Introduction

V.I. Veksler proposed two methods of relativistic particles acceleration in 1944 [1]. One of those was implemented in synchrotrons and the other one in the microtrons. In LPI microtron electrons are accelerated in RF cavity located inside the magnet that forms a uniform time-constant field. Electrons, captured in synchronous acceleration mode, move in circles with stepwise increasing radii. Note all these circles theoretically have a common point located inside the cavity [2]. The main advantages of the microtron are extremely narrow energy spectrum and small sizes of accelerated beams [2,3]. Additionally, the microtron has relatively simple design and low cost of operation. Classical microtrons are used in variety application of science (e.g. nuclear physics research), industry (e.g. industrial inspection, sterilization) and medicine (e.g. radioisotopes production, radiotherapy). Thus, microtron is used as injector for electron synchrotrons with average energy is equal to (0.250 – 1.5) GeV [4]. It can be an effective source of high energy photon radiation in photon and neutron activation analysis. Microtrons can be used for photonuclear reaction production [5], as well as for neutron production for pulsed fast neutron reactor [6]. The 6.5 MeV LPI microtron is used to study bremsstrahlung characteristics of relativistic electrons in complex structures. Also it is an injector of 1.3 GeV "Pakhra" LPI synchrotron [7].

## 2. LPI microtron cavity design

Acceleration of the electrons occurs in the cylindrical cavity with operating frequency is equal to 2.856 GHz. Electron source is a thermionic cathode $LaB_6$ placed inside the accelerating cavity. The cavity cross section in microtron median plane is shown in figure 1. There are basic geometric dimensions of the cavity in figure: radius ($R$), cavity length ($d$), distance between center of thermal emitter and the resonator axis ($r_{emit}$). We consider that cavity length $d$ was equal to 19 mm, $g$ and $R$ were equal to 7 mm and 40.3 mm correspondingly. $TM_{010}$ mode is the main one because ratio $d/R < 2.03$. The microtron is operated with the Russian first type of acceleration. It means that the first orbit of particles places inside cavity. Furthermore, cavity geometry has a deviation from axially symmetric because of electron dynamics feature at the first turns. On the one hand magnetic component of RF cavity field as well as output cavity window cause defocusing the particles. On the other hand particle losses can be reduced by proper choosing the input cavity window geometry.

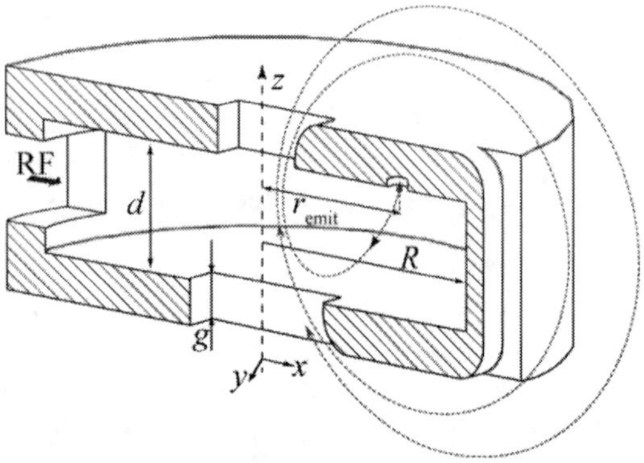


**Figure 1.** Cross-sectional view of RF cavity.

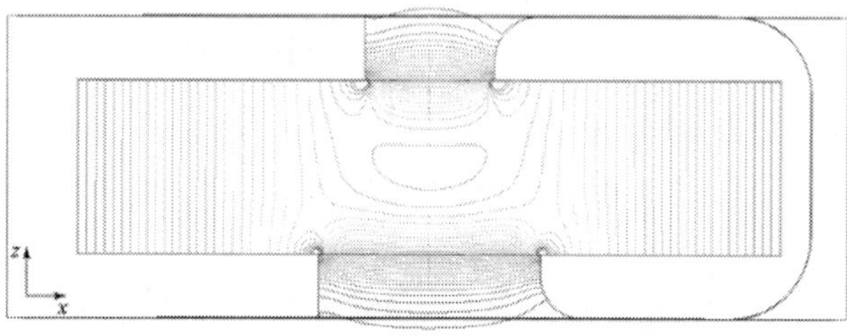


**Figure 2.** Typical spatial distribution of absolute electric field value in median plane.

3D distribution of RF electromagnetic field (electric and magnetic components) in microtron cavity was found numerically with the use of specialized code [8]. Calculated field distribution in described cavity is shown in figure 2. One can see that obtained field topology differs from that one in usual cylindrical cavity. A penetration of the field beyond cavity boundaries through its windows shown in the figure 3 for $x = 0$. There are accelerating component of the cavity field as a function of longitudinal and transversal coordinates in figure 4 and figure 5 correspondingly. There are two curves in the last figure: one of that is numerically obtained curve (solid) and another one is plotted in accordance with formulae for usual cylindrical cavity (dashed).

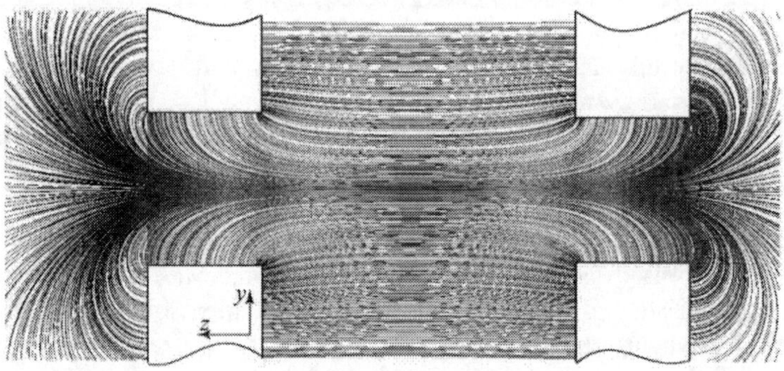


**Figure 3.** A penetration of the RF field beyond cavity boundaries through its windows.

IOP Conf. Series: Journal of Physics: Conf. Series **941** (2017) 012090 doi:10.1088/1742-6596/941/1/012090

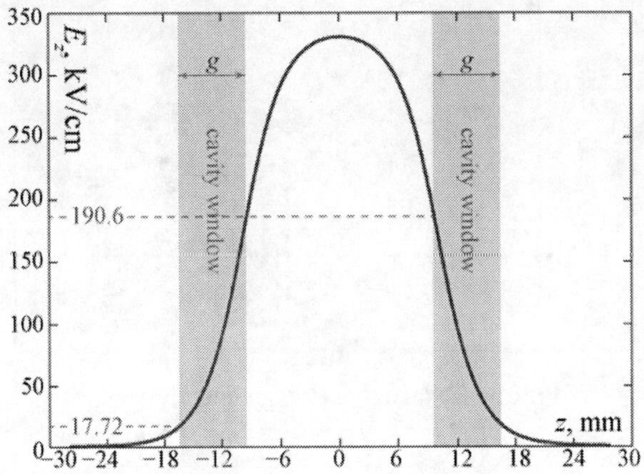


**Figure 4.** $E_z$ vs $z$ for $y = x = 0$.

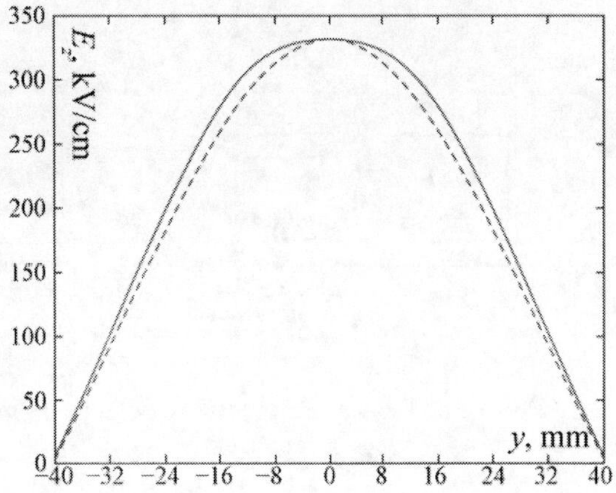


**Figure 5.** $E_z$ vs $y$ for $z = x = 0$.

## 3. Simulation of electron dynamics

Dependence of output beam parameters on microtron operation mode, namely on amplitude of accelerating voltage, was investigated. In the first step a special MICRO code was written [9] that allowed us to carry out a two-dimensional one particle electron dynamics simulation in the field, calculated above, in microtron median plane. The energy of the emitted electrons was equal to 5 eV and $r_{emit}$ was equal to 27 mm (shifted-emitter microtron). On simulating the electron dynamics we found that uniform magnetic field for circular motion is 0.12 T under magnet pole diameter is equal to 75 cm. Energy gain for one period is defined as follows

$$\Delta W_s = eU_0 T \cos\varphi_s,$$

where $U_0$ is an accelerating voltage amplitude:

$$U_0 = \int_{-\infty}^{\infty} E_z(0,0,z)dz$$

and transit-time factor is

$$T = 2\sin(\theta/2)/\theta, \quad \theta = \pi d/\lambda.$$

Thus, it was obtained that an amplitude of accelerating voltage should be equal to 678 kV under equilibrium particle phase $\varphi_s$ is equal to 15° in order to satisfy resonant acceleration condition.
There are several electron trajectories in figure 6. Each of them corresponds to different moments of electron emission from the thermionic cathode in the cavity field. One can see that a significant part of the electrons fall on the inner cavity wall and some part are lost on the outer cavity side. It was found that the maximal energy value is typical for particles that were emitted in phases of the RF field in the range approximately from −90° to −70° (for cosine voltage variation). This is because the time of field-particle interaction inside cavity for RF field phase range at first orbit before RF field changes direction is enough to ensure that electrons obtain the required energy to skirt cavity wall.

Further 3D electron dynamics simulation in LPI microtron assuming continuous emission from thermionic cathode was carried out. Emitter spot diameter was presumed to be equal to 4 mm. It should be noted that the emitter was assumed to be a point in [9].

Calculated longitudinal beam size is equal to 15 mm, vertical and radial beam sizes are equal to 2.2 mm and 0.6 mm correspondingly for the 9th orbit. The beam particles energy spectrum on the 9th orbit is shown in figure 7. There is a typical view of forming the electron bunches in figure 8.

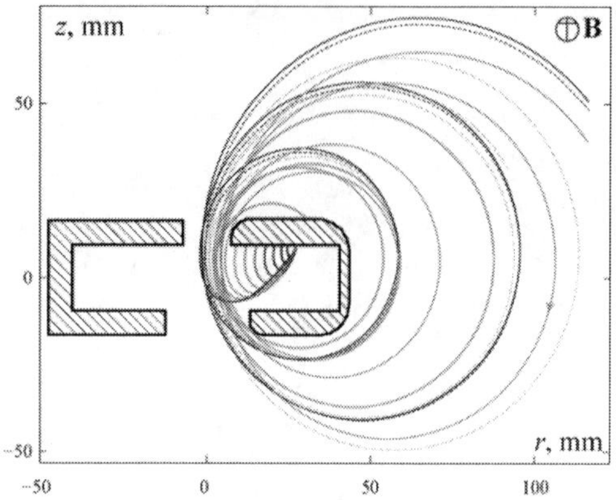


**Figure 6.** Typical trajectories of electrons in microtron.

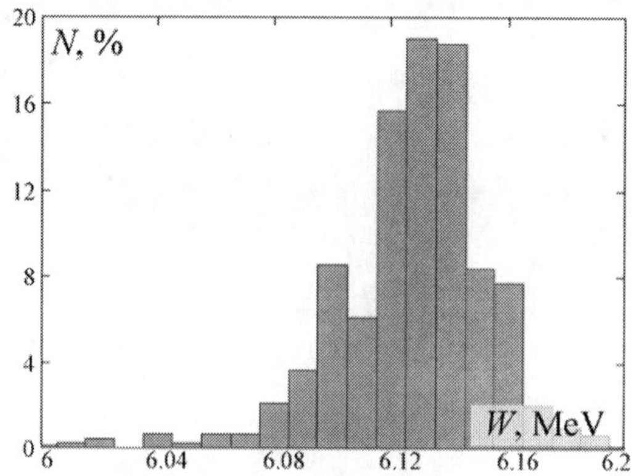


**Figure 7.** Particles energy spectrum on the 9th orbit.

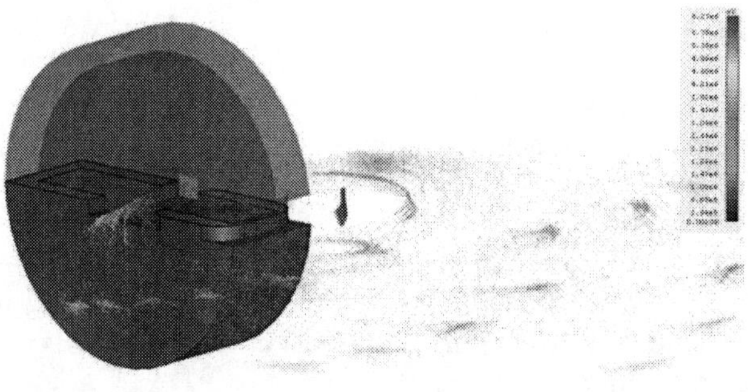

**Figure 8.** Typical view of forming the electron bunches in 3D space.

Note also that approximately 65% of emitted particles hits the internal cavity wall during one period of RF field and only 20% are accelerated up to final energy.

## 4. Conclusion

3D cavity model was developed and electromagnetic field distribution was simulated. Dynamics of electrons in LPI microtron was studied. Dependence of output beam parameters on microtron operation mode was investigated. Optimal values of microtron parameters were found.

**Acknowledgments**
This work was supported by Competitiveness Program of National Research Nuclear University MEPhI.

## References

[1] V.I. Veksler, Dokl. Ak. Nauk., J. Phys. USSR 9 (1945) 153
[2] Kolomensky A A and Lebedev A N 1966 *Theory of cyclic accelerators* (Amsterdam: North-Holland Publication Co.)
[3] Kapitza S P and Melekhin V N 1978 *The Microtron* (London: Harwood Academic Publishers)
[4] Tsipenyuk Y M and Kapitza S P 2002 *The microtron: Development and applications* (London: Taylor & Francis)
[5] Bizzarri U et al. 1969 Lettere al Nuovo Cimento **1**(16) 820
[6] Kralik M, et al. 2012 RSI **83** 083502
[7] Bashmakov Y A, Karpov V A and Yarov A S 1984 Technical Physics **29** 539
[8] www.cst.com
[9] Dyubkov V S, Bashmakov Y A and Lozeev Y Y 2017 Proc. 25th Russian Particle Accelerator Conf. (St. Petersburg) p 428

IOP Conf. Series: Journal of Physics: Conf. Series **941** (2017) 012091 doi:10.1088/1742-6596/941/1/012091

# Investigation of accelerating ion triode with magnetic insulation for neutron generation

**A E Shikanov, K I Kozlovskij, E D Vovchenko, V I Rashchikov, V L Shatokhin and A A Isaev**

National Research Nuclear University - Moscow Engineering Physics Institute, Moscow 115409, Russia

VIRashchikov@mephi.ru

**Abstract**. Vacuum accelerating tube (AT) for neutron generation with the secondary electron emission suppressed by helical line pulse magnetic field which allocated inside accelerating gap in front of hollow conical cathodeis discussed. The central anode was covered by the hollow cathode. This technical solution of AT is an ion triode in which helical line serve as a grid. Computer simulation results of longitudinal magnetic field distributional along the axis are presented.

## 1. Introduction

Laser-plasma deuteron source was used in ion triode to increaseAT resource (up to$10^8$ pulses), which is determined by target mass and target mass spraying per laser pulse(~$5.10^{-12}$ kg).The magnetic field in triodeis considered as superposition of helical line field and field of oppositely directed azimuthal induction current in the cathode electrode (the "magnetic condenser" effect). Cathode has a form of truncated cone with height $L$ and pitch angle $\alpha$ of cone generator to axis. The magnetic field of helical current with $M$ coils simulated by a number of thin currents rings inscribed in the cone perpendicular to it axis with equal stepsand calculated according to the method, described in [2].

## 2. Model of the vacuum accelerating tube with magnetic insulation

To increase efficiency of neutron production at AT along with accelerating voltage gain a number of various methods to suppress secondary electron emission from cathode are used. Among them two scheme with magnetic isolation may be used: permanent magnetic field with azimuthal symmetry and pulse magnetic fields of helical inductive coil. The value of critical magnetic field induction at accelerating gap near the cathode must be $0.2 \div 0.6$ T for the cathode diameter $0.05 \div 0.1$ m and accelerating voltage $100 \div 500$ kV.

Experimental investigation of deuteron acceleration and neutron generation with secondary electron emission from the cathode suppressed by magnetic field was fulfilled on working sectional vacuum AT models, described in [3,4]. On the base of experimental data analyzes two advanced AT models were educed, which corresponds to technical solution [5,6]. The construction of those AT are presented at Fig. 1 and Fig. 2. Accelerating ion triode made in coaxial electrode geometry with interior anode on which laser plasma is produced and external hollow cathode. Cathode has the cylindrical or cone surface form and cover laser-plasma anode. In the case of pulsemagnetic isolation cone helical inductive coilis installed near the inner cathode surface to produce pulse magnetic field for magnetic isolation.The main peculiarity of technical solution discussed is using of laser-plasma source of deuterons. Together with laser beam scanner which scan target surface it allows to improve AT lifetime. For adequate AT parameters: target size ~$10^{-7}$ $m^3$, laser target density ~$5 \cdot 10^3$ $kg/m^3$ (TiD), the number of ion source pulses is ~ $10^8$. That is why AT lifetime will be determined only by process

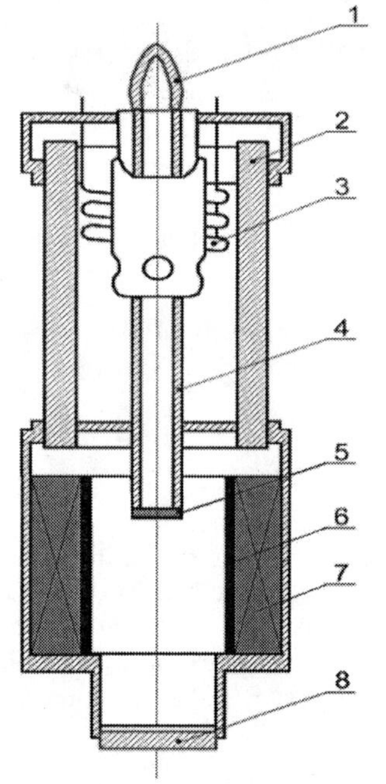


**Figure 1.** Diagram schematic view on the AT setup with laser induced plasma and electron conductivity suppressed by magnetic field of permanent magnet:
1- tubulation for tube sealing-off; 2- insulator; 3- getter pump; 4- anode; 5- plasma-forming target, 6- neutron-forming target; 7- permanent NdFeB magnet, 8- optical window.

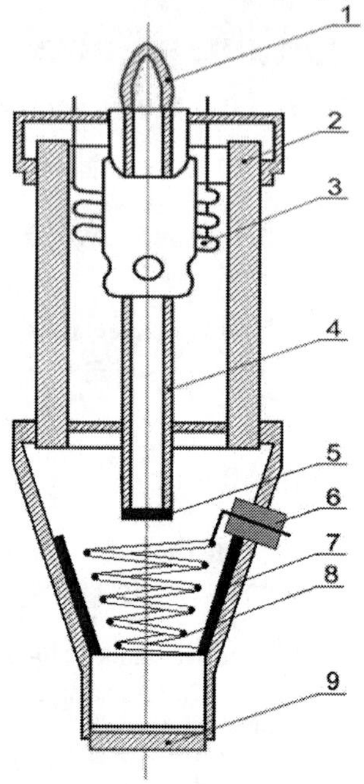


**Figure 2.** Diagram schematic view on the AT setup with laser induced plasma and electron conductivity suppressed by magnetic field of helical line:
1- tubulation for tube sealing-off; 2- insulator; 3- getter pump; 4- anode; 5- plasma-forming target, 6- power input, 7- neutron-forming target (cathode); 8- helical conductor;; 9- optical window.

of tritium replacement with deuterium on neutron-forming target. To obtain laser plasma Nd:YAG laser (wave length λ = 1060 nm) which produce pulses with energy 0.85 J and pulse duration ~ 10 ns. AT on Fig.1 and Fig. 2 may be used with a vacuumarc deuteronsource.The number of this source operations is about ~ $10^6$, and it determine this type AT lifetime. The conducted analysis shown that the main limitations deals withirregularity of source electrode materials burningwhich saturated with deuterium, unpredictable electrode spot moving, strong outgassing and electrode distraction.

## 3. Isolation magnetic field configuration

The efficiency of magnetic isolation was investigated first on the pulse ion diode with permanent magnet, made of a material based on NdFeB compounds, comprised a hollowcylinder with outer diameter 8 cm, inner diameter 4.5 cm, and height 4 cm. Experimental results with magnetic field induction value $B$=0.4 T and laser energy irradiation $W$=0.1 J show possibility of electron current partial suppression. However, this magnet type has typical magnetic field configuration whenpower lines closed on poles which leads to break down for accelerating voltages acceding 300 kV. For magnetic isolation with the help of helical line pulse magnetic field longitudinal component destroyed not so critical. That is why the upper limit of accelerating voltage value 450 kV was achieved.

Let us discussin detail AT with pulse magnetic isolation shown on figure 2. The magnetic field will be considered as superposition of *M* thin rings with cylindrical center coordinates

$$\{0, t(m) = h_1 + (m-1)(\mathrm{L} - h_1 - h_2)/(M-1)\}$$

and radius

$$a(m,\alpha) = r_S - (m-1)\mathrm{tg}\alpha(\mathrm{L} - h_1 - h_2)/(M-1)$$

(m-ring number), with current value *I*, and current *M·I* in opposite direction on the cathode surface, which form a truncated conewith height *L* and pitch angle α of cone generator to axis.

The parameters $h_{1,2}$ determinering longitudinal location with respect to cathode. This rings system simulate helical with *M* coilsallocated inside AT. Magnetic field modeling was fulfilled with the method, described in [2].

Components of alone ring magnetic field induction may be described as

$$b_r(r,z,\alpha,m) = \lambda G[a(m,\alpha),r,z-t(m)];\ b_Z(r,z,\alpha,m) = \lambda H[a(m,\alpha),r,z-t(m)],$$

where

$$G(a,r,z) = \{E[f(a,r,z)]\frac{a^2+r^2+z^2}{(a-r)^2+z^2} \quad K[f(a,r,z)]\}\frac{z}{\sqrt{r^2+\delta^2}\sqrt{(a+r)^2+z^2}};$$

$$H(a,r,z) = \{E[f(a,r,z)]\frac{a^2-r^2-z^2}{(a-r)^2+z^2} + K[f(a,r,z)]\}\frac{1}{\sqrt{(a+r)^2+z^2}};\ \lambda = \frac{\mu_0 I}{2\pi};$$

$$f(a,r,z) = \frac{4ar}{(a+r)^2+z^2}.$$

Functions *K*(*x*) and *E*(*x*) – total elliptic integralsfirst and second kind accordingly. For the cathode field components we have the following approximate expression:

$$B_{rK}(r,z,\alpha) \approx \lambda\frac{M}{N}\sum_{n=1}^{N} G(b(n,\alpha),r,z-t_K(n))\ B_{zK}(r,z,\alpha) \approx \lambda\frac{M}{N}\sum_{n=1}^{N} H(b(n,\alpha),r,z-t_K(n)),$$

where

$$b(n,\alpha) = r_k - (2n-1)\mathrm{tg}\alpha L/2N\ ,\quad t_k(n) = (2n-1)L/2N$$

Those expressions were obtained by numerical calculation cone surface integral of Biot-Savarts formula, *N* – number of quadrature formula nodes. Using superposition principle and suggesting that power input 6 (Fig. 2) is under positive potential, we obtain the final expression for the magnetic field induction components in AT with magnetic insulation:

$$B_r(r,z,\alpha) \approx -\sum_{m=1}^{M} b_r(r,z,\alpha,m) \quad B_{rK}(r,z,\alpha)\ B_z(r,z,\alpha) \approx -\sum_{m=1}^{M} b_z(r,z,\alpha,m) \quad B_{zK}(r,z,\alpha)$$

## 4. Computer simulation

Computer code on the base of the above algorithm has been written. Our experience of constructing and working with pulse neutron generators with ATlet us conclude that all this devises consist of diode with azimuthal symmetry and longitudinal or transverse acceleration of heavy hydrogen nuclides [7].There sizes are in the range of ~($10^{-2}$-$10^{-1}$) m. Therefore, those values were used for computer simulation.

Fig. 3 shows typical longitudinal magnetic field distributions on the axis $B_z(0,z,\alpha)$.

Fig. 4 shows distributions of longitudinal magnetic field projection on cathode surface between the cathode and helical electrode using next relation:

$$B_t(z,\alpha) = \cos\alpha B_z(\frac{r_k + r_S}{2} - z\mathrm{tg}\alpha, z, \alpha) - \sin\alpha B_r(\frac{r_k + r_S}{2} - z\mathrm{tg}\alpha, z, \alpha)$$

Computer simulation shows that best results may be achieved for AT with $\mathrm{tg}\alpha \le r_S/L$. This condition is in a good agreement with experimental results [4].

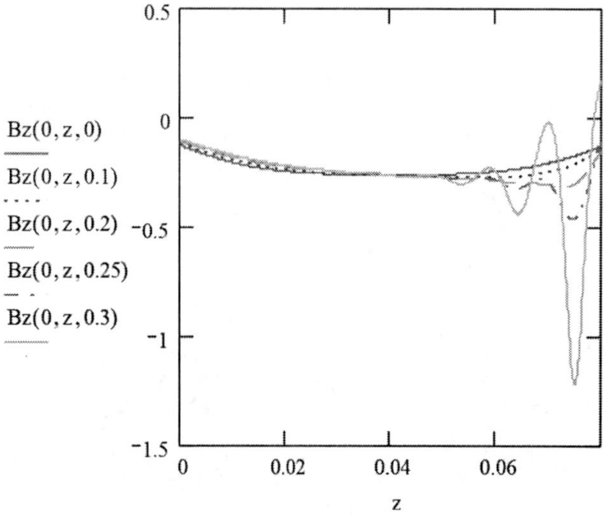


**Figure 3.** Longitudinal magnetic field distributions on the axis for the system sizes: $r_S$=0.025 m, $r_K$=0.03 m, $L$=0.08 m, $h_1$=0.005 m, $h_2$=0.005 m,$M$=8, $N$=30, $I$=$10^4$ A.
Parameter α= 0, 0.1, 0.2, 0.3.

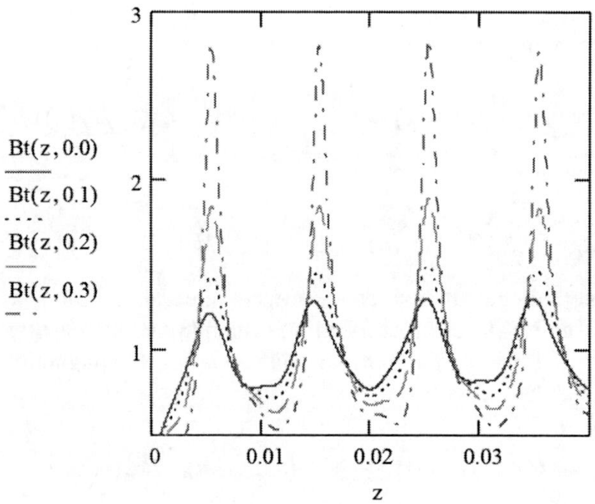


**Figure 4.** Longitudinal magnetic field projection on cathode surface distributions for the system sizes: $r_S$=0.025 m, $r_K$=0.03 m, $L$=0.08 m, $h_1$=0.005 m, ,$h_2$=0.005 m,$M$=8, $N$=30, $I$=$10^4$ A,
Parameter $\alpha$= 0, 0.1, 0.2, 0.3.

According [8] total suppression of electron current obtained if magnetic field induction accede the critical value from thefollowing expression:

$$B_{cr} \cong \frac{2r_p}{r_k^2 - r_p^2}\sqrt{\frac{m}{e}U(2+\frac{eU}{mc^2})} = 10^{-7}\frac{2r_p I_0}{r_k^2 - r_p^2}\sqrt{\frac{eU}{mc^2}(2+\frac{eU}{mc^2})}, \mathrm{T},$$

where

$$I_0 = \frac{4\pi\varepsilon_0 mc^3}{e} \cong 17,\ \text{kA}$$

-Alphen current, $m$ is the electron mass, $U$ is themaximum voltage across the diode gap, $r_k$ is the radius of the cathode (target) of the accelerating tube, $r_p$ is the maximum radius of the plasma anode, and $\varepsilon_0$ is theelectric constant. Further computer simulation shows possibility to use following approximate expression for the critical current value:

$$\frac{I_0 \mathrm{L} r_p (r_S - \mathrm{Ltg}\alpha)^2}{2\pi \mathrm{M}(r_k + r_S - \mathrm{Ltg}\alpha)(r_k - r_S)(r_S^2 - r_p^2)} \sqrt{\frac{eU}{mc^2}(2 + \frac{eU}{mc^2})} \ .$$

## 5. Conclusion

On the base of experimental data simulation of magnetic field configuration from helical inductor in accelerating triode with laser-plasma source was fulfilled. Analyse of longitudinal magnetic field distributions on the axis $B_z(0,z,\alpha)$ allows to find the condition $\mathrm{tg}\alpha \le r_S/\mathrm{L}$ for optimal AT mode. Then for AT with laser power 0.5 – 1 J the transverse and longitudinal sizes of accelerating ion triode may be reduced up to 0.08 m.

This project is supported in part by the MEPhI 5/100 Program of the Russian Academic Excellence Project.

## References

[1] Ananyin O.B., Afanasiev Yu.V., Bykovsky Yu.A, Krohin O.N., 2003. Laser Plasma. Physics and Application.Moscow, MEPHI, 400.
[2] Dobrov R.V, RyzhkovV.I., Ponomarev D.D. et al., 2011. Voprosi Atomnoy Nauki iTekhniki., Series: Yadernoe Priborostroenie, 1(28), 26-31.
[3] Shikanov A E, Vovchenko E D, Kozlovskii K I, 2015. Neutron Generation in a Plasma Diode with Electrons Insulated by a Constant Magnetic Field. Atomic Energy, 119( 4), 258-264.
[4] Kozlovskii K.I., Ryzhkov V.I., TsubinA.S. et al., 1990. Atomic Energy, 112 (3), 182-184.
[5] Gul'ko V.M., Kozlovskii K.I., Kolomiets N.F. et al., 1987. Pulse Neutron Tube. USSR Invention Certificate No. 1468275.
[6] Vovchenko E.D., Isaev A.A., Kozlovskii K.I. et al. ,2016. Ion Triod for Neutron Generation.Patent RF on an useful model № 149963.
[7] Proceedings of the International Scientific–Technical Conference on Portable Neutron Generators and RelatedTechnologies, 2005. Moscow, (VNIIA).
[8] Didenko A.N., Shikanov A.E., Kozlovskii K.I. et al.,2016. Small-size Magnetically Insulated Plasma Diodes for Neutron Generation., Plasma Physics, 40 (11), 910-918.

IOP Conf. Series: Journal of Physics: Conf. Series **941** (2017) 012092 doi:10.1088/1742-6596/941/1/012092

# Intense quasiperiodic beam dynamics in accelerating system: mathematical model and optimization method

**I D Rubtsova[1] and D A Ovsyannikov[1]**

[1]St. Petersburg State University, 7/9 Universitetskaya nab., St. Petersburg, 199034, Russia

E-mail: i.ribtsova@spbu.ru, d.a.ovsyannikov@spbu.ru

**Abstract**. The paper is devoted to quasiperiodic intense beam dynamics is accelerating system. Particle interaction accounting is conducted with the use of well-known cloud-in-cell method. This numerical method is formalized, and mathematical model of space charge field intensity is suggested. Coulomb field intensity is presented in the form of integral over the domain of particle phase states. This permits beam evolution to be described by integro-differential equations. Controlled process quality is characterized by integral functional values. Beam dynamics optimization problem is formulated as trajectory ensemble control problem for dynamic system. Analytical expression of quality functional variation is obtained. It makes possible directed methods using for beam dynamics optimization.

## 1. Introduction

This paper is devoted to intense quasiperiodic beam dynamics modeling and optimization. Beam evolution mathematical model contains integral representation of beam space charge field.

It should be mentioned that the problems of modeling of self-consistent charged particle distributions are urgent and widely discussed by the researchers [1-11]. This paper deals with analytical representation of space charge field. Such an approach is applied by many authors e.g. [1,4-13]; it expands the potentialities of beam dynamics analysis and optimization. Axially symmetric beam Coulomb field is often expressed in the form of Fourier-Bessel series [6-8,11].

In this paper the version of clouds-in-cells method is used for space charge field determination [7]. This convenient for computation and time-saving method is usually considered to be exclusively numerical. In the article the formalization of this method is performed and the analytical expression of Coulomb field intensity is obtained in the form of integral over particle phase states domain.

The integral space charge field representation makes it possible to introduce integro-differential beam dynamics model [1,4,6,9,10,13] and to propose mathematical optimization method. This method is based on the approach formulated in [1,12,14,15] and successfully applied by a number of researchers for investigation of different problems of beam dynamics optimization [10,13,16-19].

In the article presented we consider beam evolution to be a complex of bunch center motion and the motions of particles of bunch. Beam quasiperiodicity is taken into account. Controlled dynamic process is described by a system of integro-differential equations. Beam quality criterion is presented in the form of integral functional defined at beam trajectories. Beam dynamics optimization problem is formulated as trajectory ensemble control problem. An analytical representation of quality functional variation is obtained; it provides the possibility of directed optimization methods applying.

## 2. Beam dynamics equations

Consider a model of intense beam longitudinal dynamics in some accelerating system. We will investigate the processes at beam quasiperiod (i.e. one bunch evolution) and determine space charge field under the assumption of beam spatial periodicity.

Particle beam is supposed to be axially symmetric. A bunch is presented by $N$ model particles moving in conducting tube of radius $a$. Model particles are supposed to be "thick" disks (disks-clouds) of radius $R$ and thickness $2\Delta$ (for resting particle).

Let us introduce independent variable $\tau = ct$ (where $c$ is the speed of light, $t$ is the time) and cylindrical coordinates $r, \theta, z$; may $Oz$ axis be aligned with channel axis. Let particle phase state be presented by vector $(z, p)$, where $z$ is longitudinal coordinate, $p$ is reduced impulse.

Controlled beam evolution is considered to be a complex of bunch center motion and model particle motions. Bunch center phase state is characterized by vector $(z_c, p_c)$. Hence longitudinal beam dynamics is described by the equations:

$$\frac{dz_i}{d\tau} = \frac{p_i}{\sqrt{1+p_i^2}}, \quad \frac{dp_i}{d\tau} = \frac{\zeta}{c^2} E^{(RF)}(\tau, z_i, \boldsymbol{u}) + \frac{|\zeta|}{c^2} E^{(int)}(\tau, z_i, p_i, p_c), \tag{1}$$

$$\frac{dz_c}{d\tau} = \frac{p_c}{\sqrt{1+p_c^2}}, \quad \frac{dp_c}{d\tau} = \frac{\zeta}{Nc^2} \sum_{i=1}^{N} E^{(RF)}(\tau, z_i, \boldsymbol{u}) \tag{2}$$

with initial conditions

$$z_i(0) = z_{0i}, \ p_i(0) = p_{0i}, \ i = \overline{1,N}, \ z_c(0) = \frac{1}{N}\sum_{i=1}^{N} z_{0i}, \ p_c(0) = \frac{1}{N}\sum_{i=1}^{N} p_{0i}. \tag{3}$$

Here $i$ is model particle number, $(z_i, p_i)$ is $i$-th particle phase vector; $\zeta$ is particle specific charge; $E^{(RF)}$, $E^{(int)}$ are the longitudinal intensity components characterizing respectively the effect of RF field and space charge field on the disk-cloud as a whole; $\boldsymbol{u}$ is program control vector function.

Note that model particles are considered to be positively charged in space charge field accounting; $E^{(int)}$ is Coulomb field intensity created by positively charged particles.

## 3. Particle interaction account

Let some control function $\boldsymbol{u}$ and $\tau$ value of independent variable be fixed. Let us describe $i$-th model particle form-factor in laboratory reference frame by piecewise continuous even function $\chi(z - z_i, p_i)$ taking nonzero values only at interval $(-d_i, d_i)$; $2d_i = 2\Delta/\sqrt{1 + p_i^2}$ is cloud size.

Spatial quasiperiod length in laboratory reference frame is $2H = p_c\lambda/\sqrt{1 + p_c^2}$, where $\lambda$ is RF field first harmonic wavelength.

### *3.1. Phase domain partitioning into elements*

Let us enclose the set of model particles phase states in some rectangle $G$ and partition the rectangle into the elements $G_s, s = \overline{1,S}$ by any way. Suppose the elements to be connected compact subsets of positive measure. Assume $q_{is}$ to be the charge value contributed by $i$-th model particle to $s$-th element. Charge values of elements are $q_s = \sum_{i=1}^{N} q_{is}$, $s = \overline{1,S}$. Let us choose the points $(\xi_s, \eta_s) \in G_s$, $s = \overline{1,S}$ arbitrarily.

Let us approximate the action of the element $G_s$ (with all its periodic images) on $i$-th model particle by the force $F_{is}$ presenting the action of uniformly charged "thick" disk $D_s$ with radius $R$, thickness $2h$, charge $q_s$,center position $\xi_s$ at structure axis and reduced impulse $\eta_s$ (and all periodic images of this disk) on $i$-th model particle. Note that $2h = 2H/J$, $J$ is natural number, $10 \le J \le N$.

We will describe the action of periodic beam on $i$-th model particle by the intensity $E_i = \sum_{s=1}^{S} F_{is}/q$, where $q$ is model particle charge assumed to be positive.

### 3.2. Space charge field intensity

The intensity expression is obtained on the basis of the formulae presented in [7]. The dependence $2H = 2H(p_c)$ is taken into account. The result is:

$$E_i = \sum_{s=1}^{S} \frac{q_s}{Q} V(z_i, p_i, p_c, \xi_s, \eta_s), \tag{4}$$

$$V(z,p,p_c,\xi,\eta) = \sum_{k=1}^{\infty} \frac{I_0 \lambda a J_1^2(\mu_k R/a)}{c\varepsilon_0 \gamma h \mu_k^3 J_1^2(\mu_k) sh(\mu_k \gamma H/a)} \int_{-\infty}^{\infty} C_k(y,p_c,\xi,\eta)\, \chi(y-z,p) dy, \tag{5}$$

$$C_k(y,p_c,\xi,\eta) = cosh\left(\frac{\mu_k \gamma}{a}(H - |y-\xi-h|)\right) - cosh\left(\frac{\mu_k \gamma}{a}(H - |y-\xi+h|)\right) \tag{6}$$

Here $Q = I_0\lambda/c$ is bunch charge value; $I_0$ is average beam current; $\varepsilon_0$ is electric constant; $\gamma = \sqrt{1+\eta^2}$, $J_0(x)$ and $J_1(x)$ are Bessel functions of the first kind of order 0 and 1 correspondingly, $\mu_k, k = 1,2,\ldots$ are the zeros of Bessel function $J_0(x)$. Notice that the formulae (4)-(6) are valid under the assumption $|z_i - \xi_s| \le H,\ s = \overline{1,S}$; in the case $|z_i - \xi_s| > H$ the coordinate $\xi_s$ should be replaced by its periodic image nearest to $z_i$.

### 3.3. Particle phase density and integral Coulomb field representation

Now let us suppose particle distribution to be continuous; particle phase states compose the domain $M_{\tau,u}$. Consider a smooth function $\rho(\tau,z,p)$ satisfying the condition:

$$\int_{G_s} \rho(\tau,\xi,\eta) d\xi d\eta = q_s/Q,\ s = \overline{1,S} \tag{7}$$

for any mode of rectangle $G$ partitioning. The relation (4) right side can be viewed as integral sum and

$$\lim_{d_{max}\to 0} \sum_{s=1}^{S} \frac{q_s}{Q} V(z_i,p_i,p_c,\xi_s,\eta_s) = \int_{M_{\tau,u}} V(z_i,p_i,p_c,\xi,\eta)\rho(\tau,\xi,\eta)\, d\xi\, d\eta,$$

where $d_{max}$ is the maximum of element diameters. So mathematical model of Coulomb field may be suggested in the form:

$$E^{(int)}(\tau,z,p,p_c) = \int_{M_{\tau,u}} V(z,p,p_c,\xi,\eta)\rho(\tau,\xi,\eta)\, d\xi\, d\eta. \tag{8}$$

### 3.4. Particular case

Let us partition the rectangle $G$ into identical cells by straight lines parallel to coordinate axes ($J$ cells along $Oz$ axis and $L$ cells along $Op$ axis, $S = JL$). Let $q_{j,l},\ j = \overline{1,J},\ l = \overline{1,L}$ be the cell charges and $(\xi_j, \eta_l),\ j = \overline{1,J},\ l = \overline{1,L}$ be the points chosen in the cells arbitrarily. Besides, let us neglect the difference between cell impulses and assume $\eta_l = p_c,\ l = \overline{1,L}$. In addition, suppose model particles to have the equal thickness $d = \Delta/\sqrt{1+p_c^2}$ (in laboratory reference frame). In this case integral expression of space charge field intensity takes the form:

$$E^{(int)}(\tau,z,p_c) = \int_{M_{\tau,u}} \hat{V}(z,p_c,\xi)\rho(\tau,\xi,\eta)\, d\xi\, d\eta, \tag{9}$$

$$\hat{V}(z,p_c,\xi) = \sum_{k=1}^{\infty} \frac{I_0\lambda a J_1^2(\mu_k R/a)}{c\varepsilon_0 \gamma_c h \mu_k^3 J_1^2(\mu_k) sh(\mu_k \gamma_c H/a)} \int_{-\infty}^{\infty} \hat{C}_k(y-\xi,p_c)\, \chi(y-z,p_c) dy, \tag{10}$$

$$\hat{C}_k(v,p_c) = cosh((\mu_k\gamma_c/a)(H - |v-h|)) - cosh((\mu_k\gamma_c/a)(H - |v+h|)) \tag{11}$$

Here $2h = 2H/J$ is the cell size along $Oz$ axis; $\gamma_c = \sqrt{1+p_c^2}$; the condition $|z-\xi| \le H$ is imposed in view of the assumption of beam periodicity (similarly to the formulae (4)-(6)).

Computational formula (4) is now represented as follows:

$$E_i = \sum_{j=1}^{J} Q_j/Q\, \hat{V}(z_i,p_c,\xi_j), \tag{12}$$

where $Q_j = \left(\sum_{l=1}^{L} q_{j,l}\right)$, $j = \overline{1,J}$. Now it is enough to introduce the grid on variable $z$ only, and $Q_j$, $j = \overline{1,J}$ are the charges of grid cells. The formulae (12),(10)-(11) correspond to Coulomb field calculation with the use of clouds-in-cells method based on spatial quasiperiod partitioning into the cells [7,8].

## 4. Integro-differential beam dynamics model

Let us generalize beam dynamics model (1)-(3). Primarily, let us take into account integral representation (8) of space charge field.

Besides, we have an approximate equality

$$\int_{M_{\tau,u}} \frac{\zeta}{c^2} E^{(RF)}(\tau,\xi,\boldsymbol{u})\rho(\tau,\xi,\eta)d\xi d\eta \approx \frac{\zeta}{Nc^2}\sum_{i=1}^{N} E^{(RF)}(\tau,z_i,\boldsymbol{u}) \qquad (13)$$

presenting Monte Carlo method of calculation of the integral in left-hand side. The approximation accuracy increases with increasing model particles number. Consequently, the integral in relation (13) can be viewed as mathematical model of averaged RF force in right-hand side.

Consider dynamic controlled process (beam evolution) described by the equations:

$$\frac{d\boldsymbol{x}}{d\tau} = \boldsymbol{f}(\tau,\boldsymbol{x},\boldsymbol{x}_c,\boldsymbol{u}) = \boldsymbol{f}_1(\tau,\boldsymbol{x},\boldsymbol{u}) + \int_{M_{\tau,u}} \boldsymbol{f}_2(\tau,\boldsymbol{x},\boldsymbol{x}_c,\boldsymbol{y}_\tau)\rho(\tau,\boldsymbol{y}_\tau)\mathrm{d}\boldsymbol{y}_\tau, \qquad (14)$$

$$\frac{d\boldsymbol{x}_c}{d\tau} = \boldsymbol{f}_c(\tau,\boldsymbol{x}_c,\boldsymbol{u}) = \boldsymbol{f}_{c1}(\tau,\boldsymbol{x}_c) + \int_{M_{\tau,u}} \boldsymbol{f}_{c2}(\tau,\boldsymbol{y}_\tau,\boldsymbol{u})\rho(\tau,\boldsymbol{y}_\tau)\mathrm{d}\boldsymbol{y}_\tau, \qquad (15)$$

$$\frac{\partial\rho(\tau,\boldsymbol{x})}{\partial\tau} + \frac{\partial\rho(\tau,\boldsymbol{x})}{\partial\boldsymbol{x}}\boldsymbol{f}(\tau,\boldsymbol{x},\boldsymbol{x}_c,\boldsymbol{u}) + \rho(\tau,\boldsymbol{x})div_{\boldsymbol{x}}\boldsymbol{f}(\tau,\boldsymbol{x},\boldsymbol{x}_c,\boldsymbol{u}) = 0 \qquad (16)$$

with initial conditions

$$\boldsymbol{x}(0) = \boldsymbol{x}_0 \in M_0, \ \ \boldsymbol{x}_c(0) = \int_{M_0} \boldsymbol{x}_0\,\rho_0(\boldsymbol{x}_0)\mathrm{d}\boldsymbol{x}_0\,, \qquad (17)$$

$$\rho\big(0,\boldsymbol{x}(0)\big) = \rho_0(\boldsymbol{x}_0), \ \ \boldsymbol{x}_0 \in M_0. \qquad (18)$$

Here $\tau \in [0,T]$ is independent variable; $T$ is a constant; $\boldsymbol{x},\boldsymbol{x}_c$ are phase vectors of systems (14) and (15) respectively; $\boldsymbol{u}(\tau)$ is a control; vector functions $\boldsymbol{f}_1,\boldsymbol{f}_{c1},\boldsymbol{f}_{c2}$ are determined by the method of external fields modeling; vector function $\boldsymbol{f}_2$ is determined by the method of particle interaction account; $\rho(\tau,\boldsymbol{x})$ is phase density defined on system (14) trajectories; $M_0$ is initial domain in phase space of the system (14); $\rho_0(\boldsymbol{x})$ is initial phase density; $M_{\tau,u} = \{\boldsymbol{x}_\tau = \boldsymbol{x}(\tau,\boldsymbol{x}_0,\boldsymbol{u}) : \boldsymbol{x}_0 \in M_0\}$. All the functions in the equations (14)-(16) are supposed to be rather smooth to use mathematical optimization methods [1]. Control $\boldsymbol{u}(\tau)$ is supposed to be piecewise-continuous vector-function taking values in a compact.

Assuming $\boldsymbol{x} = (x_1,x_2)^* = (z,p)^*$, $\boldsymbol{x}_c = (x_{c1},x_{c2})^* = (z_c,p_c)^*$ (where $^*$ denotes transposition),

$$\boldsymbol{f}_1(\tau,\boldsymbol{x},\boldsymbol{u}) = \begin{pmatrix} x_2/\sqrt{1+x_2^2} \\ (\zeta/c^2)E^{(RF)}(\tau,x_1,\boldsymbol{u}) \end{pmatrix}, \ \boldsymbol{f}_2(\tau,\boldsymbol{x},\boldsymbol{x}_c,\boldsymbol{y}) = \begin{pmatrix} 0 \\ (|\zeta|/c^2)\hat{V}(x_1,x_{c2},y_1) \end{pmatrix},$$

$$\boldsymbol{f}_{c1}(\tau,\boldsymbol{x}_c) = \begin{pmatrix} x_{c2}/\sqrt{1+x_{c2}^2} \\ 0 \end{pmatrix}, \ \boldsymbol{f}_{c2}(\tau,\boldsymbol{y},\boldsymbol{u}) = \begin{pmatrix} 0 \\ (\zeta/c^2)E^{(RF)}(\tau,y_1,\boldsymbol{u}) \end{pmatrix},$$

one can obtain the system (1)-(2), where $E^{(int)}$ is presented in terms of integral (9) and averaged value of external force is replaced by integral in accordance with (13). In addition, initial bunch center phase state is presented in integral form (see (3), (17)).

Following the approach [1,4], it is not hard to prove that smooth function $\rho(\tau,z,p)$ satisfying the condition (7) for any mode of rectangle $G$ partitioning at any $\tau \in [0,T]$, satisfies the equation (16) with initial condition (18).

## 5. Trajectory ensemble control problem

Quality criterion of dynamic control process (14)-(16) is presented by integral functional

$$I(\boldsymbol{u}) = \int_0^T \int_{M_{\tau,u}} \Phi\left(\tau, \boldsymbol{x}_\tau, \boldsymbol{x}_c(\tau)\right)\rho(\tau, \boldsymbol{x}_\tau)\mathrm{d}\boldsymbol{x}_\tau \mathrm{d}\tau$$

$$+ \int_0^T \Phi_c\left(\tau, \boldsymbol{x}_c(\tau)\right)\mathrm{d}\tau + \int_{M_{T,u}} g(T, \boldsymbol{x}_T, \boldsymbol{x}_c(T))\rho(T, \boldsymbol{x}_T)\mathrm{d}\boldsymbol{x}_T + g_c(T, \boldsymbol{x}_c(T)). \quad (19)$$

All integrands and free term in right-hand side of formula (19) are supposed to be smooth functions.

Choosing the proper functions in quality criterion (19), one can formulate the problem of beam dynamics optimization as the problem of trajectory ensemble control of dynamic process (14)-(16).

Let us consider the problem of quality criterion (19) minimization with respect to control $\boldsymbol{u}$.

## 6. Quality functional variation

Quality functional (18) variation is obtained on the basis of the approach [1]:

$$\delta I(\boldsymbol{u}, \Delta\boldsymbol{u}) = -\int_0^T \int_{M_{\tau,u}} (\boldsymbol{\Psi}^*(\tau, \boldsymbol{x}_\tau, \boldsymbol{x}_c)\Delta_u f_1(\tau, \boldsymbol{x}_\tau, \boldsymbol{u}) +$$

$$+\boldsymbol{\Lambda}^*(\tau, \boldsymbol{x}_c)\Delta_u f_{c2}(\tau, \boldsymbol{x}_\tau, \boldsymbol{u}))\rho(\tau, \boldsymbol{x}_\tau)\mathrm{d}\boldsymbol{x}_\tau \mathrm{d}\tau. \quad (20)$$

Here $\Delta\boldsymbol{u}$ is control $\boldsymbol{u}$ variation; $\Delta_u$ designates the increment of any function with respect to argument $\boldsymbol{u}$ only; vector functions $\boldsymbol{\Psi}(\tau, \boldsymbol{x}, \boldsymbol{x}_c)$ and $\boldsymbol{\Lambda}(\tau, \boldsymbol{x}_c)$ satisfy on the trajectories of dynamic process (14)-(16) the auxiliary system of integro-differential equations

$$\frac{\mathrm{d}\boldsymbol{\Psi}(\tau,\boldsymbol{x},\boldsymbol{x}_c)}{\mathrm{d}\tau} = \left(\frac{\partial\Phi(\tau,\boldsymbol{x},\boldsymbol{x}_c)}{\partial\boldsymbol{x}}\right)^* - \left(\frac{\partial f(\tau,\boldsymbol{x},\boldsymbol{x}_c,\boldsymbol{u})}{\partial\boldsymbol{x}}\right)^* \boldsymbol{\Psi}(\tau, \boldsymbol{x}, \boldsymbol{x}_c) -$$

$$-\int_{M_{\tau,u}} \left(\frac{\partial f_2(\tau,\boldsymbol{y}_\tau,\boldsymbol{x}_c,\boldsymbol{x})}{\partial\boldsymbol{x}}\right)^* \boldsymbol{\Psi}(\tau, \boldsymbol{y}_\tau, \boldsymbol{x}_c)\rho(\tau, \boldsymbol{y}_\tau)\mathrm{d}\boldsymbol{y}_\tau -$$

$$-\left(\int_{M_{\tau,u}} \left(\frac{\partial f_{c2}(\tau,\boldsymbol{x}_\tau,\boldsymbol{u})}{\partial\boldsymbol{x}}\right)^* \rho(\tau, \boldsymbol{x}_\tau)\mathrm{d}\boldsymbol{x}_\tau\right) \boldsymbol{\Lambda}(\tau, \boldsymbol{x}_c),$$

$$\frac{\mathrm{d}\boldsymbol{\Lambda}(\tau,\boldsymbol{x}_c)}{\mathrm{d}\tau} = \left(\frac{\partial\Phi(\tau,\boldsymbol{x},\boldsymbol{x}_c)}{\partial\boldsymbol{x}_c}\right)^* + \left(\frac{\partial\Phi_c(\tau,\boldsymbol{x}_c)}{\partial\boldsymbol{x}_c}\right)^* - \left(\frac{\partial f_{c1}(\tau,\boldsymbol{x}_c)}{\partial\boldsymbol{x}_c}\right)^* \boldsymbol{\Lambda}(\tau, \boldsymbol{x}_c) -$$

$$-\left(\int_{M_{\tau,u}} \left(\frac{\partial f_2(\tau,\boldsymbol{x},\boldsymbol{x}_c,\boldsymbol{y}_\tau)}{\partial\boldsymbol{x}_c}\right)^* \rho(\tau, \boldsymbol{y}_\tau)\mathrm{d}\boldsymbol{y}_\tau\right) \boldsymbol{\Psi}(\tau, \boldsymbol{x}, \boldsymbol{x}_c).$$

with the following conditions at $\tau = T$:

$$\boldsymbol{\Psi}(T, \boldsymbol{x}(T), \boldsymbol{x}_c(T)) = -\left(\frac{\partial g(T,\boldsymbol{x}(T),\boldsymbol{x}_c(T))}{\partial\boldsymbol{x}}\right)^*,$$

$$\boldsymbol{\Lambda}(T, \boldsymbol{x}_c(T)) = -\left(\frac{\partial g(T,\boldsymbol{x}(T),\boldsymbol{x}_c(T))}{\partial\boldsymbol{x}_c}\right)^* - \left(\frac{\partial g_c(T,\boldsymbol{x}_c(T))}{\partial\boldsymbol{x}_c}\right)^*.$$

The analytical representation (20) of quality criterion variation provides the possibility of directed optimization methods application in beam dynamics optimization problems. It may be beneficial to combine the gradient optimization with random search [20].

## References

[1] Ovsyannikov D A 1990 *Modeling and Optimization of Charged Particle Beam Dynamics* (Leningrad: Leningrad State University Press) p 312

[2] Drivotin O I and Ovsyannikov D A 2009 Self-consistent distributions for charged particle beam in magnetic field *Int. J. Mod. Phys.* A **24** (5) pp 816-42

[3] Bondarenko T V Polozov S M and Sumbaev A P 2016 Numerical simulation of the beam loading effect at the LUE-200 accelerator *Physics of Particles and Nuclei Letters* **13** (7), pp 919-22

[4] Ovsyannikov D A and Edamenko N S 2013 Modeling of charged particle beam dynamics *Vestnik St. Petersburg University, Ser. 10: Applied Mathematics, Informatics, Control Processes* **2** 61–66

[5] Kozynchenko V A 2014 The modeling of charged particle interactions in the elliptic beam *Proc. 20th Int. Workshop: Beam Dynamics & Optimization (BDO) (30 June-4 July 2014 St. Petersburg)* (St. Petersburg: IEEE) pp 94-95 DOI: 10.1109/BDO.2014.6890040

[6] Rubtsova I D 2014 On quasiperiodic beam of interacting particles dynamics modeling *Vestnik St. Petersburg University, Ser. 10: Applied Mathematics, Informatics, Control Processes* **1** 104–19

[7] Ovsyannikov D A, Rubtsova I D and Kozynchenko V A 2013 *Some Problems of Intense Charged Particle Beams Modeling in Linear Accelerators* (St. Petersburg: VVM publishing office) p 144

[8] Rubtsova I D and Suddenko E N 2012 Investigation of program and perturbed motions of particles in linear accelerator *Proc. XXIII Russian Particle Accelerator Conf. RuPAC-2012 (24-28 September 2012 St. Petersburg)*(Geneva: JACoW http://www.JACoW.org) pp 367-69

[9] Rubtsova I D 2014 Integral-differential model of quasi-periodic beam longitudinal dynamics *Proc. 20th Int. Workshop: Beam Dynamics & Optimization (BDO) (30 June-4 July 2014 St. Petersburg)* (St. Petersburg: IEEE) p 144

[10] Rubtsova I D 2016 On modeling and optimization of intense quasiperiodic beam dynamics *Proc. XXV Russian Particle Accelerator Conf. RuPAC-2016 (21-25 November 2016 St. Petersburg)*(Geneva: JACoW http://www.JACoW.org) pp 363-66

[11] Rubtsova I D 2016 Analytical Approach to Quasiperiodic Beam Coulomb Field Modeling, *II Conference on Plasma&Laser Research and Technologies (2016), Journal of Physics: Conference Series* **747** No 1 012074 http://iopscience.iop.org/1742-6596/747/1/012074

[12] Ovsyannikov D A 2012 Mathematical modeling and optimization of beam dynamics in accelerators *Proc. XXIII Russian Particle Accelerator Conf. RuPAC-2012 (24-28 September 2012 St. Petersburg)*(Geneva: JACoW http://www.JACoW.org) pp 68-72

[13] Rubtsova I D 2014 Mathematical optimization model of longitudinal beam dynamics in klystron-type buncher *Proc. XXIV Russian Particle Accelerator Conf. RuPAC-2014 (6-10 October 2014 Obninsk)*(Geneva: JACoW http://www.JACoW.org) pp 66-68

[14] Ovsyannikov D A 1997 Modeling and Optimization Problems of Charged Particle Beam Dynamics *Proc. European Control Conf. ECC'97 (1-4 July 1997 Brussels Belgium)* 4 pp 1463-67

[15] Ovsyannikov D A, Ovsyannikov A D, Vorogushin M F, Svistunov Yu A and Durkin A P 2006 Beam dynamics optimization: models, methods and applications *Nuclear Instruments and Methods in Physics Research* A **558** (1) 11-19

[16] Rubtsova I D 2015 Optimization of iterative beam dynamic process *Proc. III Int. Conf. "Stability and Control Processes" in Memory of V.I.Zubov (SCP) (5-9 October 2015 St. Petersburg*) ed L A Petrosyan and A P Zhabko (St. Petersburg: IEEE) pp 198-200 DOI: 10.1109/SCP.2015.7342092

[17] Zavadsky S V, Ovsyannikov D A and Chung S-L 2009 Parametric Optimization Methods for the Tokamak Plasma Control Problem *Int. J. Mod. Phys.* A **24** No 5 pp.1040–47

[18] Ovsyannikov A D, Ovsyannikov D A, Altsybeyev V V, Durkin A P and Papkovich V G 2014 Application of Optimization Techniques for RFQ Design *Problems of Nuclear Science and Engineering* **91** No 3 pp 116-19

[19] Ovsyannikov D A and Altsybeyev V V 2013 Mathematical Optimization Model for Alternating-Phase Focusing (APF) Linac *Problems of Nuclear Science and Engineering* No 4 p 93

[20] Vladimirova L V 2014 Global Extremum Search on the Basis of Density and Its Mode Estimation *Proc. 20th Int. Workshop: Beam Dynamics & Optimization (BDO) (30 June-4 July 2014 St. Petersburg)* (St. Petersburg: IEEE) p 186

IOP Conf. Series: Journal of Physics: Conf. Series **941** (2017) 012093 doi:10.1088/1742-6596/941/1/012093

# On finding the analytic dependencies of the external field potential on the control function when optimizing the beam dynamics

**A D Ovsyannikov, S A Kozynchenko and V A Kozynchenko**

Saint-Petersburg State University, Universitetskaya nab., 7–9, Saint-Petersburg, 199034, Russia

**Abstract.** When developing a particle accelerator for generating the high-precision beams, the injection system design is of importance, because it largely determines the output characteristics of the beam. At the present paper we consider the injection systems consisting of electrodes with given potentials. The design of such systems requires carrying out simulation of beam dynamics in the electrostatic fields. For external field simulation we use the new approach, proposed by A.D. Ovsyannikov, which is based on analytical approximations, or finite difference method, taking into account the real geometry of the injection system. The software designed for solving the problems of beam dynamics simulation and optimization in the injection system for non-relativistic beams has been developed. Both beam dynamics and electric field simulations in the injection system which use analytical approach and finite difference method have been made and the results presented in this paper.

## 1. Introduction

The paper mostly focuses on exploring the ways of algorithmic and software realization of the optimal design methodology in the beam dynamics area, which is proposed in [1-3] and intended to be applied in the injection systems producing the high-precision beams. The development of the optimal design methodology for beam dynamics is considered to be rather complex and laborious problem. It has given rise to a large and growing body of research [4-14].

The main issue in the optimal design techniques that we try to address in the paper consists in finding an analytical expression of the control potential function $\phi(\eta)$ defined over the domain contour, together with developing an algorithm that could compute the electrostatic field $U_{\xi}$ inside the working domain using this control function. The control function is to be used in further optimization procedures. The paper starts with some algorithmic study relating to the numerical solution of the integral of Cauchy type that is applied in the given problem to calculate the electrostatic field. Then we consider a case study model of the axial symmetrical field in an injection system. Some numerical data obtained in the C++ computer simulation are presented.

## 2. A possible basic algorithm of finding the integral of Cauchy type

Let us consider a three-dimensional simply connected bounded domain having the axial symmetry and let $G$ be its diametric cross section. This two-dimensional domain $G$ is bounded by a contour $L$ that is supposed to be a smooth closed curve. Hereinafter the real plane $R^2$ containing the domain $G$ will be

IOP Conf. Series: Journal of Physics: Conf. Series **941** (2017) 012093 doi:10.1088/1742-6596/941/1/012093

identified with the complex plane *C*. Let a continuous complex function $\phi(\eta)$ be defined over the contour *L*. The complex potential of the external electrostatic field inside the domain *G* can be evaluated with the use of an integral of Cauchy type

$$F(\xi,\phi)=\frac{1}{2\pi i}\int_L \frac{\phi(\eta)}{\eta-\xi}d\eta, \quad \xi\notin L, \tag{1}$$

where $\xi = z_G + i\cdot r_G \in G,\ \eta = x_L + i\cdot y_L \in L$. Then the complex potential of the three-dimensional external field can be represented as follows (see [3], p. 97):

$$H(z,r,\phi)=\frac{1}{2\pi}\int_0^{2\pi} F(z+i\cdot r\cos\alpha,\phi)d\alpha \ .$$

The real part of *H*, i.e. the function $U=\operatorname{Re}H$, will be considered as a function determining the electrostatic field in the three-dimensional domain. The complex contour integral (1) can be written as

$$F(\xi,\phi)\equiv\frac{1}{2\pi i}\int_L f(z)dz=\frac{1}{2\pi i}\left[\int_L u(x,y)dx-v(x,y)dy+i\int_L v(x,y)dx+u(x,y)dy\right], \tag{2}$$

where $f(z)=u(x,y)+i\cdot v(x,y)$ is a function of a complex variable $z=x+i\cdot y$. Therefore, the potential function *U* can be defined as a real line integral

$$U(x,y)=\frac{1}{2\pi}\int_L v(x,y)dx+u(x,y)dy$$

taken over the same contour *L* as the complex integral. Let us consider the simplest case when the function $\phi\equiv\phi_0$ is a constant. Then the following expressions can be written

$$f(z)=\frac{\phi_0}{z},\ u(x,y)=\frac{x}{x^2+y^2},\ v(x,y)=-\frac{y}{x^2+y^2},$$

and the real part of the integral (1) is evaluated as follows

$$U\equiv\operatorname{Re}F(\xi,\phi_0)=\frac{\phi_0}{2\pi}\left[\int_L -\frac{ydx}{x^2+y^2}+\int_L\frac{xdy}{x^2+y^2}\right], \tag{3}$$

where $x=x_L-z_G,\ y=y_L-r_G$.

Let us take the domain *G* as a rectangle *ABCD* shown in Fig.1.

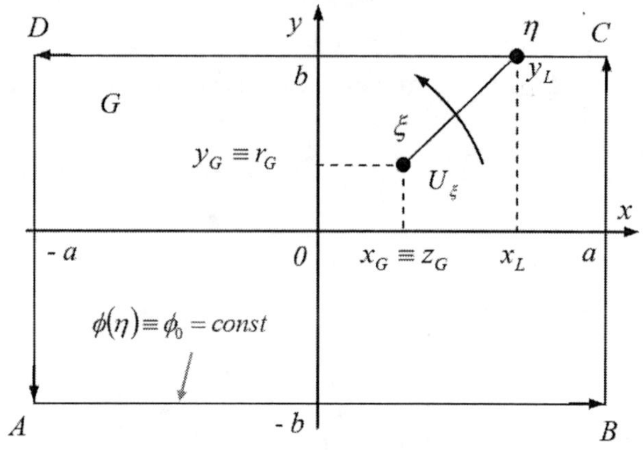


Fig. 1. Finding the potential $U_\xi$ inside the conducting rectangle with the use of integral of Cauchy type

The integral (3) can be represented as a sum of four integrals to be evaluated counter-clockwise over the horizontal and vertical segments *AB*, *BC*, *CD*, *DA* of the closed contour *L*. Actually, these segments can be treated as vectors. Thus, on the vector $\overrightarrow{AB}$ it turns out that $dy = 0$ and the integral (3) takes the form

$$U_{AB} = -\frac{\phi_0}{2\pi}\int_{-a}^{a}\frac{ydx}{x^2+y^2}.$$

Given an arbitrary position of the point $\xi$ in the domain *G*, the variables $x$ and $y$ are defined as $x = x_L - x_G$, $y = y_L - y_G$.

So, we have

$$U_{AB} = -\frac{\phi_0}{2\pi}\int_{-a}^{a}\frac{ydx}{x^2+y^2} = -\frac{\phi_0}{2\pi}\int_{-a}^{a}\frac{d(x/y)}{(x/y)^2+1} = -\frac{\phi_0}{2\pi}\arctan\left(\frac{x}{y}\right)\Bigg|_{-a}^{a} =$$
$$= -\frac{\phi_0}{2\pi}\left[\arctan\left(\frac{a-x_G}{-b-y_G}\right) - \arctan\left(\frac{-a-x_G}{-b-y_G}\right)\right]. \tag{4}$$

In a similar manner, we obtain other constituents of potential $U_{BC}, U_{CD}, U_{DA}$:

$$U_{BC} = \frac{\phi_0}{2\pi}\int_{-b}^{b}\frac{(x_L-x_G)\,d(y_L-y_G)}{(x_L-x_G)^2+(y_L-y_G)^2} = \frac{\phi_0}{2\pi}\left[\arctan\left(\frac{b-y_G}{a-x_G}\right) - \arctan\left(\frac{-b-y_G}{a-x_G}\right)\right], \tag{5}$$

$$U_{CD} = -\frac{\phi_0}{2\pi}\int_{a}^{-a}\frac{(y_L-y_G)\,d(x_L-x_G)}{(x_L-x_G)^2+(y_L-y_G)^2} = -\frac{\phi_o}{2\pi}\left[\arctan\left(\frac{-a-x_G}{b-y_G}\right) - \arctan\left(\frac{a-x_G}{b-y_G}\right)\right], \tag{6}$$

$$U_{DA} = \frac{\phi_0}{2\pi}\int_{b}^{-b}\frac{(x_L-x_G)\,d(y_L-y_G)}{(x_L-x_G)^2+(y_L-y_G)^2} = \frac{\phi_0}{2\pi}\left[\arctan\left(\frac{-b-y_G}{-a-x_G}\right) - \arctan\left(\frac{b-y_G}{-a-x_G}\right)\right]. \tag{7}$$

So, at the point $\xi$ the potential $U_\xi$ equals to

$$U_{\xi} = U_{AB} + U_{BC} + U_{CD} + U_{DA} \tag{8}$$

This expression evidently contains the pairs of terms in which the arguments of the function *arctan* are inverse ones. Taking into account the following trigonometric identity

$$\arctan x + \arctan\frac{1}{x} = \begin{cases} \pi/2, if\ x > 0 \\ -\pi/2, if\ x < 0 \end{cases},$$

we can represent (8) as a sum of binomials forming such a pair of terms. For instance, taking the first term from (5) and the second from (6), we get

$$\frac{\phi_0}{2\pi}\arctan\left(\frac{b-y_G}{a-x_G}\right) + \frac{\phi_0}{2\pi}\arctan\left(\frac{a-x_G}{b-y_G}\right) = \frac{\phi_0}{2\pi}\frac{\pi}{2} = \frac{\phi_0}{4},\ \text{as } b-y_G > 0 \text{ and } a-x_G > 0.$$

Similarly, summing the first term from (4) and the second from (5) gives

$$-\frac{\phi_0}{2\pi}\arctan\left(\frac{a-x_G}{-b-y_G}\right) + \left[-\frac{\phi_0}{2\pi}\arctan\left(\frac{-b-y_G}{a-x_G}\right)\right] = -\frac{\phi_0}{2\pi}\left(-\frac{\pi}{2}\right) = \frac{\phi_0}{4},\quad \text{as } \frac{a-x_G}{-b-y_G} < 0.$$

Finally, the summation of all four pairs in (8) gives us the potential value in an arbitrary point $\xi$ of the domain *G*:

$$U_{\xi} = 4 \cdot \phi_0 / 4 = \phi_0 .$$

This result undoubtedly matches the electrostatics fundamentals: if a plane closed contour (e. g., made from a conducting wire) has been charged to the potential $\phi_0$, then any point inside the contour has the same potential.

It seems to be worth attempting to apply the above scheme in the case of non-constant potential function $\phi(\eta)$ over the closed contour *L*. The possible solution could consist in representing the smooth contour by a sequence of horizontal and vertical directed segments - vectors, together with making the approximation of the function $\phi(\eta)$ by a step function $\phi_s(\eta)$ that is a constant at each segment. A contribution of each segment into the whole potential at a point $\xi$ in *G* can be computed using one of the formulas (4), (5), (6), or (7) depending on the orientation of the segment.

Needless to say, considering more exact, linear, approximation of the boundary potential seems to be appropriate as well. Let us assume that the boundary potential on some horizontal line segment *ab*, $x_{ab} \in [x_{Li}, x_{L(i+1)}]$ is defined by a linear function $\phi = k_0 + k_1 x_L$, $y_{Li} = y_{L(i+1)} \equiv y_L$. The integrand $f(z)$ takes the following form

$$f(z) = \frac{\phi}{z} = \frac{k_0 + k_1 x_L}{x + i \cdot y} \equiv u(x, y) + i \cdot v(x, y),$$

where $u(x, y) = \dfrac{(k_0 + k_1 x_L)x}{x^2 + y^2}$, $v(x, y) = -\dfrac{(k_0 + k_1 x_L)y}{x^2 + y^2}$.

By analogy with the derivation of the formula (4), and taking into account that $x = x_L - x_G$, $y = y_L - y_G$, we can get the following expression chain

$$U_{ab}(x_G,y_G)\equiv \mathrm{Re}\left[\frac{1}{2\pi i}\int_{x_i}^{x_{i+1}} f(z)dz\right]=\frac{1}{2\pi}\int_{x_i}^{x_{i+1}} v(x,y)dx=-\frac{k_0}{2\pi}\int_{x_i}^{x_{i+1}}\frac{ydx}{x^2+y^2}-\frac{k_1}{2\pi}\int_{x_i}^{x_{i+1}}\frac{x_L ydx}{x^2+y^2}=$$
$$-\frac{k_0}{2\pi}\int_{x_i}^{x_{i+1}}\frac{ydx}{x^2+y^2}-\frac{k_1}{2\pi}\int_{x_i}^{x_{i+1}}\frac{xydx}{x^2+y^2}-\frac{k_1 x_G}{2\pi}\int_{x_i}^{x_{i+1}}\frac{ydx}{x^2+y^2}=$$
$$-\frac{(k_0+k_1x_G)}{2\pi}\left[\arctan\left(\frac{x_{L(i+1)}-x_G}{y_L-y_G}\right)-\arctan\left(\frac{x_{Li}-x_G}{y_L-y_G}\right)\right]-$$
$$-\frac{k_1(y_L-y_G)}{4\pi}\ln\left[\frac{(x_{L(i+1)}-x_G)^2+(y_L-y_G)^2}{(x_{Li}-x_G)^2+(y_L-y_G)^2}\right] \quad . \quad (9)$$

Note that the formula (4) can be thought as a special case of (9) if the boundary potential $\phi$ is constant, i.e. when $k_1=0$.

## 3. Algorithm description and a case study simulation

An algorithm considered in the paper includes the following stages:

- Computing the electrostatic field in the working domain of an injection system under the given initial configuration and potentials of electrodes. The computation is carried out using, e.g., the iterative Liebmann's procedure on a square grid, [15], as is done in the paper.
- On moving along the piece-wise contour bounding the working domain, the reference potential profile $\varphi(\eta_i)$ over the boundary of the working domain is formed. This set of exact values is used for finding the approximate boundary control function $\phi(\eta)$.
- A suitable type of an approximating function is chosen and its parameters are computed by one of the methods of solving the systems of linear equations.
- In order to find a potential at any point of the domain, the integrals of Cauchy type are computed at each grid cell side (horizontal or vertical) around the domain contour using the formulas (4-7), where $\phi_0=0.5\cdot[\varphi(\eta_i)+\varphi(\eta_{i+1})]$.

Let us consider these stages in greater detail. As an example in the case study we take an axially-symmetric working domain for an injection system with three electrodes, which cross-section is shown in Fig. 2. The rectangle $AFF_1A_1$ of length $AF=400\ mm$ and height $AA_1=80\ mm$ is covered with a square grid with a cell side $h=0.5\ mm$. Potentials of electrodes are $U_1=100\ kV,\ U_2=45\ kV,\ U_3=87\ kV.$

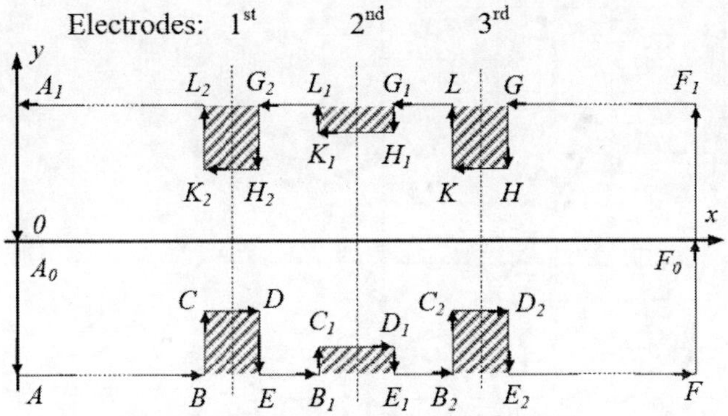


Fig. 2. Cross-section of the axially-symmetric working domain of the 3-electrode injection system

The simulation model has been developed in C++, where a class *Field* defines the model parameters and contains all necessary member functions. One of them computed the electrostatic field in a cylindrical coordinate system by the iterative Liebmann's method with the maximal error $\varepsilon = 0.001\%$. It takes ca. 88360 iterations. Another function scans the grid nodes that bound the top half of the contour, namely $A_0A_1L_2K_2H_2G_2L_1K_1\,H_1G_1LKHGF_1F_0$, and collects the potentials into an array of size $n_L = 1173$. The reference profile $\varphi(\eta_i)$ is shown in Fig. 3, curve 1.

To approximate the profile of the contour potential, we have chosen a rational-fraction approximation:

$$\phi(\eta) = \left(a_0 + \sum_{i=1}^{i=4} a_i \eta^i\right) \Big/ \left(1 + \sum_{j=1}^{j=5} b_j \eta^j\right) .$$

For finding the unknown parameters $a_i, b_j$, the least squares method that minimizes a weighted mean-square approximation error has been used. It handles the over-determined matrix with $M = 15$ rows (i.e., equations) and $N = 10$ columns (unknowns). The approximating function $\phi(\eta)$ is shown in Fig. 3, curve 2.

Finally, the potential function $U_{axis}(x)$ over the domain axis has been calculated using the reference profile $\varphi(\eta_i)$ (see Fig. 4, curve 1), the approximate profile $\phi(\eta)$ (curve 2) and based on the Liebmann's method (curve 3).

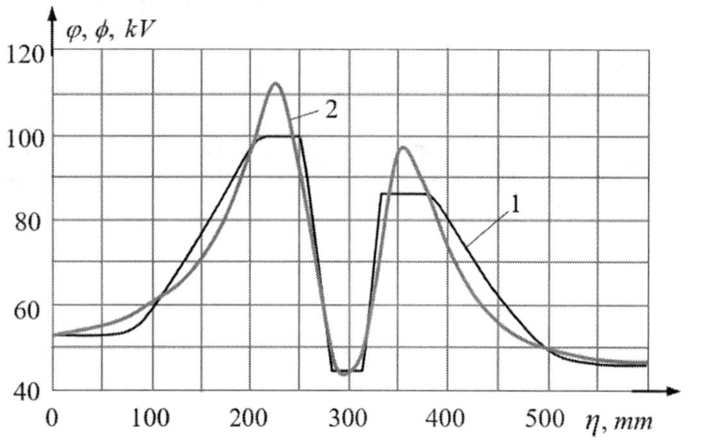


Fig. 3. The reference and approximate potential profiles over the top boundary of the domain

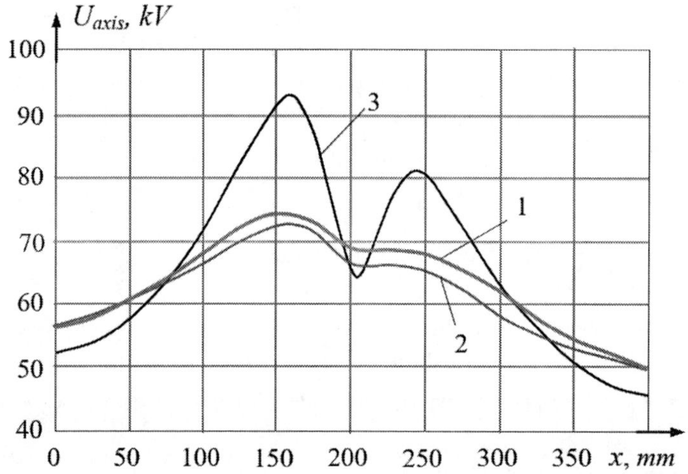


Fig. 4. Different potential functions computed on the domain axis $A_0F_0$

Calculations were performed using both stepwise and piecewise linear approximations of the boundary potential. The corresponding values of the potential function on the axis, calculated on the basis of these approximations, differ very slightly (less than 0.1%), that is, apparently, due to using the fine mesh.

## 4. Conclusions

The numerical technique using the integral of Cauchy type and intended for calculating the electrostatic field inside the axially symmetric domain has been considered. The simulation results have shown that the given approach seems to be practicable but requires some theoretical and algorithmic studies to achieve better accuracy. The algorithm proposed in the paper can be used in computing and optimization of beam dynamics in the injection systems (see, e.g., [16]).

## 5. References

[1] Ovsyannikov A D 2013 *Vestnik St. Petersburg State University Ser* 10 On optimization of

charged particle dynamics in electrostatic field Issue 2 54-9
[2] Ovsyannikov A D 2013 *Problems of Atomic Sci. and Techn.* Beam dynamics optimization in electrostatic field № 4(86) 90-2
[3] Ovsyannikov A D 2014 *Mathematical Models of the Beam Dynamics Optimization* St. Petersburg State University BBM p 181
[4] Masunov E S and Polozov S M 2008 *Phys. Rev. ST Accel Beams* High intensity ion beams in rf undulator LINAC **11** 074201
[5] Ovsyannikov A D, Ovsyannikov D A, Durkin A P and Chang S-L 2009 *Techn. Phys.* Optimization of matching section of an accelerator with a spatially uniform quadrupole focusing **54** 1663–66
[6] Ovsyannikov A D, Ovsyannikov D A, Balabanov M Yu and Chung S-L 2009 *Int. J. Modern Phys.* A On the beam dynamics optimization problem **24** 941–51
[7] Ovsyannikov A D, Ovsyannikov D A and Chung S-L 2009 *Int. J. Modern Phys.* A Optimization of radial matching section **24** 952-58
[8] Svistunov Y A and Ovsyannikov A D 2010 *Problems of Atomic Sci. and Techn.* Designing of compact accelerating structures for applied complexes with accelerators № 2 48-51
[9] Ovsyannikov D A and Ovsyannikov A D 2010 New approach to optimization of RFQ radial matching section *Proc. 1st Int. Particle Accelerator Conf. (IPAC'2010)* May 23-28 Kyoto 1351-53
[10] Ovsyannikov A D 2012 *Problems of Atomic Sci. and Techn.* Transverse motion parameters optimization in accelerators № 4 74-6
[11] Ovsyannikov A D 2013 *Doklady Mathematics* On a class of optimization problems in electric field **88** 751-53
[12] Ovsyannikov A D, Ovsyannikov D A, Altsybeyev V V, Durkin A P and Papkovich V G 2014 *Problems of Atomic Sci. and Techn.* Application of optimization techniques for RFQ design **91** 116-9
[13] Kozynchenko S A 2015 Development of the Win 32 API-based software in C++ for beam dynamics simulation and optimization in the injection systems *Proc. Int. Conf. on "Stability and Control Processes" in Memory of V I Zubov (SCP-BDO'2015)* Oct 5-9 St. Petersburg ed L A Petrosyan and A P Zhabko (IEEE) 172-4
[14] Ovsyannikov D A, Ovsyannikov A D, Antropov I V and Kozynchenko V A 2016 *J. Phys. Conf.: Series* BDO-RFQ program complex of modelling and optimization of charged particle dynamics **747** №1
[15] Sadiku M N O 2001 *Numerical Techniques in Electromagnetics* 2nd ed CRC Press p 760
[16] Kozynchenko V A and Kozynchenko S A 2014 Parallel beam dynamics simulation in injection systems taking into account particle interactions *20th Int. Workshop on Beam Dynamics and Optimization (BDO)* Jun 30- Jul 4 St. Petersburg

# The method of accelerating structure tuning and manufacturing quality control

**I V Rybakov and V V Paramonov**

Institute for Nuclear Research of the RAS, 117312, 60-th October Anniversary Prospect, 7a, Moscow, Russia

E-mail: irybakov@inr.ru

**Abstract**. For the multipactor discharge (MP) damping in the new CDS structure a method of forced excitation of high voltage oscillations in the coupling cell is proposed. It can be realized by application of an alternating frequency shift in the neighboring accelerating cells. For assured MP damping a control of both operating frequencies and the alternating frequency shift obtained is required. A method for determination of shifted frequencies in cells by the measurements in full setup (FS) and minimal setup (MS) of the structure was proposed. The substantiation of the technique arising from the equivalent circuits method and numerical simulations of the operating frequencies in FS and MS is presented.

## 1. Introduction

The CDS structure was proposed to replace the first cavity of the INR RAS linear proton accelerator. It has four sections, consisting of 18 -21 periods of structure [1].

After manufacturing the structure cells some frequency dispersions of the accelerating and coupling modes can be expected. Their values depend on the chosen manufacturing tolerances. If the value of manufacturing tolerances is 30μm the values of frequency dispersion are expected to be $\delta_{fa}=\pm3\sigma_{fa}=\pm0.675$ MHz and $\delta_{fc}=\pm3\sigma_{fc}=\pm5.414$ MHz, respectively [2].

The acceptable value of the stop bandwidth along the cavity is assumed to be 200 kHz. During the measurements in the cells it is necessary to control the mean stop bandwidth value for the full cavity section.

For the multipactor discharge (MP) damping an alternating frequency shift in the accelerating cells is applied. The value of the required frequency shift is estimated using the equivalent circuits method and amounts to ±800 kHz. The numerically calculated value of the required alternating frequency shift for neighboring cells appeared to be ±1.2 MHz. For cavity sections 1 and 4 which are distant from the feeding waveguide a control of alternating frequency shift for MP damping is also required [3].

A technique is proposed for determining the eigenfrequencies of the half cells that make up the cavity section. The application of the technique is illustrated on the example of the first cavity section, consisting of 18 periods of the CDS structure, which will be obtained from 36 half cells during production.

## 2. The half cells tuning devices

For tuning the frequency of the half cells accelerating mode it is proposed to introduce into the structure an additional ring of the trapezoidal cross-section. It is shown in Figure 1.

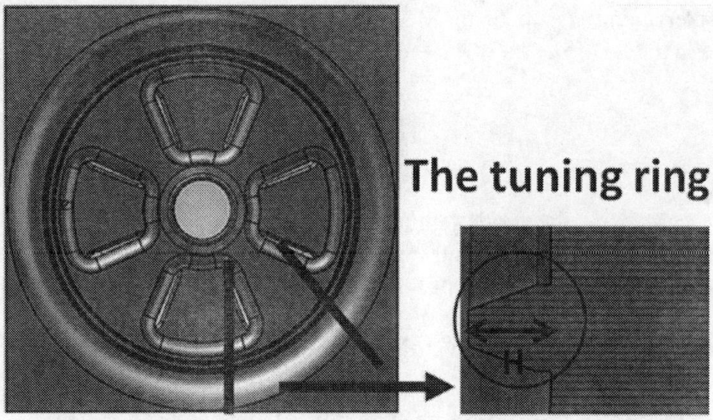


**Figure 1**. The ring for accelerating mode frequency tuning

The coefficient of influence of the tuning ring length on the accelerating mode frequency is $k_{fa}$ = 1.160 MHz/mm. For the assurance of accelerating mode shift for MP damping in neighboring cells and with regard to frequencies dispersion the initial values of geometrical parameters are set to reach the accelerating mode frequency of 989.000 MHz. In this case the height of the tuning ring is selected to set the half-cell operating mode frequency to 993.000 MHz.

In order to provide the required alternating frequency shift the tuning ring is ground off until the operating frequency is 989.800 MHz for the odd-numbered cells and 992.200 MHz for the even-numbered cells. Thus a pair of half cells provides the required frequency shift of 1.200 MHz.

The tuning of the coupling mode frequency in the half cells is possible only by boring the coupling cell radius. For the cell radius the calculated value of the coefficient of influence on frequency is $k_{fc}$ = 14.270 MHz / mm. With regard to the coupling mode dispersion the initial radius of the coupling cell was set in a such way as to reach the coupling mode frequency of 997.000 MHz

Due to nonlinear dependence of the coefficient of influence of the tuning ring length on the accelerating mode frequency, the adjustment of the half cells to the operating frequency value should be performed in several stages. At each stage, the frequencies of the section and half cells are measured to clarify the actual value of the tuning ring influence coefficient and its possible approximation.

## 3. Measurements in full section (Full Setup)

At the first stage of measurements it is supposed to find the values of the accelerating mode frequency, the stop bandwidth and the coupling mode frequency in full setup of the cavity section consisting of 36 half cells. The mean value of the accelerating mode frequency fa in full setup is measured directly [4].

For experimental definition of the stop bandwidth we can use a linear expression stipulating equal deviations of the modes from the operating frequency in the upper and lower branches of the Brillouine diagram (BD). For this purpose let us use the mode closest to the operating one:

$$\delta f = \Delta F_n = f_n^t + f_n^b - 2f_a \tag{1}$$

where $f^{t,b}_n$ are the modes frequencies in the upper and lower branches of the BD, fa is the operating mode frequency. Thus in this case the calculated stop bandwidth amounts to δf = 4.578 MHz with a relative error of the stop bandwidth $(\delta f\text{-}\delta f_{calc})/\delta f$ = 13.5%. To increase the accuracy of determining δf let's proceed to expansion into a series in terms of $\xi=\theta_0\text{-}\theta$ for the linear zone of the BD:

$$f_0^{t,b}(\xi) = f_a + \frac{\delta f}{2} \pm \frac{\beta g \xi}{\pi\beta} + \frac{\partial^2 f^t}{\partial \xi^2}\frac{\xi^2}{2} \pm \frac{\partial^3 f^t}{\partial \xi^3}\frac{\xi^3}{6} + \frac{\partial^4 f^t}{\partial \xi^4}\frac{\xi^4}{24} \pm ... + ... = \frac{\delta f}{2} + f_0^{t,b}(\xi) \tag{2}$$

where $\beta_g$ is a group velocity and $\beta$ is a phase velocity.

If we consider members up to the 3rd order the δf is determined by two modes closest to the operating mode in the BD [5]:

$$\delta f = \frac{m^2 \Delta F_n - n^2 \Delta F_m}{m^2 - n^2} \tag{3}$$

where n, m = 1, 2 are the modes numbers closest to the operating mode, $\Delta F_{n,m}$ are the modes frequencies dispersions in the upper and lower BD branches calculated from (1). Thus in this case the calculated stop bandwidth amounts to δf = 4.162 MHz with a relative error of the stop bandwidth ($\delta f - \delta f_{calc})/\delta f$ = 3.2%. To increase the accuracy in δf determination the members up to the 5th order can be considered in (2). Thus δf is determined using three modes closest to the operating mode in the BD [6]:

$$\delta f = \frac{\Delta F_n m^2 j^2 (j^2 - m^2) + \Delta F_m n^2 j^2 (n^2 - j^2) + \Delta F_j n^2 m^2 (m^2 - n^2)}{n^2 m^2 (m^2 - n^2) + m^2 j^2 (j^2 - m^2) + n^2 j^2 (n^2 - j^2)} \tag{4}$$

where n, m, j = 1, 2, 3 are the modes numbers closest to the operating mode, $\Delta F_{n,m,j}$ are the modes frequencies dispersions in the upper and lower BD branches calculated from (1). Thus in this case the calculated stop bandwidth amounts to δf = 4.041 MHz with a relative error of the stop bandwidth ($\delta f - \delta f_{calc})/\delta f$ = 0.2%.

Thus the mean value of coupling mode frequency along the section could be determined as:

$$\overline{f_c} = \overline{f_a} + \delta f \tag{5}$$

The mean value of coupling mode frequency calculated using (4) and (5) amounted to $f_c$ = 997.092 MHz. The section coupling coefficient amounted to $k_c$ = 15.576 %.

In the first section of the cavity an alternating accelerating cavity frequency shift of ±1.200 MHz is applied for MP damping. Such a shift can cause perturbations of the field amplitude in the cavity section. In this case a field amplitude distribution measurement along the section is required.

## 4. Measurements in minimal setup

To control the obtained values of accelerating and coupling modes frequencies for individual half cells a set of eigenfrequencies measurements in the minimal setup (MS), consisting of two half cells, is proposed. The MS is shown in figure 2.

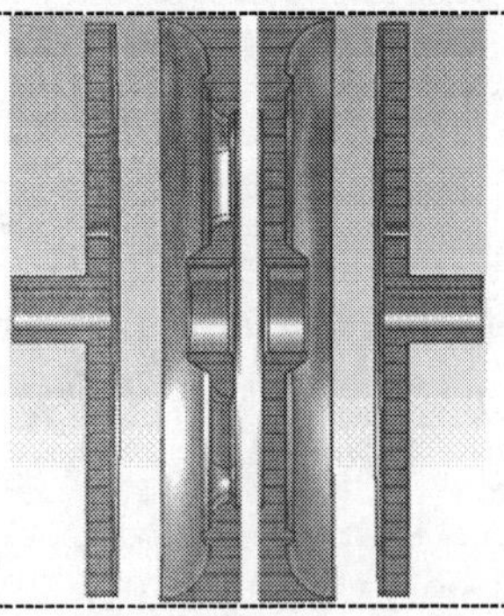

**Figure 2.** Minimal setup of CDS structure

In such a setup, if some electrical shorting plates are implemented, three mode frequencies can be measured directly: the operating π-mode frequency and two 0-mode frequencies.

The coupling mode frequency is calculated from the equations introduced under the condition of frequency shift conservation with a constant coupling coefficient:

$$\delta f_c^{\pi} + \delta f_a^{\pi} = \delta f_0^{(1)} + \delta f_0^{(2)} \tag{6}$$

$$\delta f_c^{\pi} = \delta f_0^{(1)} + \delta f_0^{(2)} - \delta f_a^{\pi} \tag{7}$$

$$f_c^{\pi} = \overline{f_c^{\pi}} + \delta f_c^{\pi} \tag{8}$$

Thus, the coupling mode frequency in the MS is determined by calculating its deviation from the average value over the section which is determined in Paragraph 1.

To determine the accelerating mode frequencies in each half cell, a series of measurements using three half cells is necessary. The frequency of the accelerating mode in the MS is the average value of the frequencies of its constituent half cells. Therefore, if three MCs are composed of three half cells, we can construct a system of equations:

$$\begin{cases} \dfrac{f_a^{(1)} + f_a^{(2)}}{2} = f_a^{(1,2)} \\ \dfrac{f_a^{(1)} + f_a^{(3)}}{2} = f_a^{(1,3)} \\ \dfrac{f_a^{(2)} + f_a^{(3)}}{2} = f_a^{(2,3)} \end{cases} \Rightarrow \begin{matrix} f_a^{(1)} = f_a^{(1,2)} - f_a^{(2,3)} + f_a^{(1,3)} \\ f_a^{(2)} = f_a^{(1,2)} - f_a^{(1,3)} + f_a^{(2,3)} \\ f_a^{(3)} = f_a^{(1,3)} - f_a^{(1,2)} + f_a^{(2,3)} \end{matrix} \tag{9}$$

where $f_a^{(n)}$ is the frequency of the accelerating mode in the corresponding half-cell, $f_a^{(n,\ m)}$ is the measured average frequency of the accelerating mode in the corresponding MS. Thus, solving system (9) we get the values of the frequencies in each of the half cells.

Thereafter, the found frequency value of the accelerating mode for the first half cell fa1 can be used as a reference for determining the frequencies of the rest half cells of the section, forming a MS with them. Similarly, the values of the coupling mode frequency in the half cells are determined.

For the method checkout a numerical experiment was performed with a conducting ball implemented in the volume of one of the MS half cells. Its radius was chosen so that the MS frequency corresponded to a half cell frequency shift of $\delta f = 3\sigma_{fa} = 0.675$ MHz and amounted to 6.20 mm. The MS with a shifted half-cell is shown in figure 3.

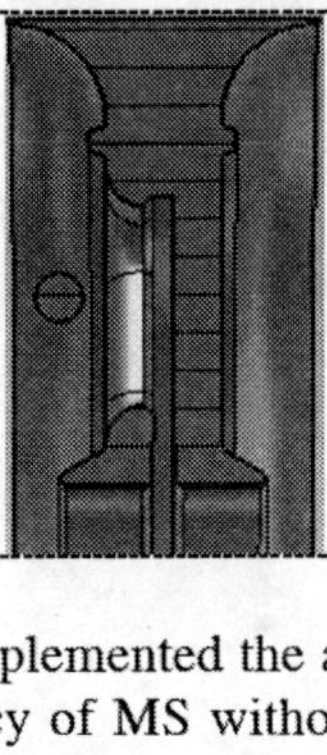

**Figure 3.** MS with a shifted half cell

With the ball implemented the accelerating mode frequency of such a MS amounts to $f_a = 992.670$ MHz. The frequency of MS without shift is $f_a = 993.007$ MHz. The calculated with (9) value of the shifted half-cell accelerating mode frequency is $f_{a\_sh} = 992.333$ MHz.

If the ball is placed near the coupling window as is shown in figure 3 the coupling mode is also shifted. To determine the coupling mode frequency in the shifted MS a FS, including this MS, was composed. The mean value of the coupling mode frequency in this FS, according to (4) and (5), amounted to $f_c$ = 997.046 MHz. The MS coupling mode frequency shift calculated with (7) and (8) was $\delta f_c$ = 0.229 MHz, accordingly the coupling mode frequency itself was $f_c$ = 997.275 MHz. The numerically determined coupling mode frequency amounted to $f_c$ = 997.257 MHz.

So that is the way how the accelerating and coupling mode frequencies in the half-cells of the cavity section are found.

With the known influence coefficients for the tuning ring and coupling cell radius the half cells are tuned to the coupling mode frequency of 991.000 MHz, the odd-numbered half cells are tuned to the accelerating mode frequency of 989.800 MHz and the even-numbered half cells – to 992.200 MHz. In this case the mean value of the accelerating mode frequency along the section is 991.000 MHz and the stop bandwidth doesn't exceed 200 kHz.

## 5. Conclusion

A method has been proposed that allows tuning the accelerating structure to the required values of operating frequencies while maintaining an acceptable level of field dispersion. The application of the method is considered for the new CDS structure of the first cavity of the main part of the INR linac. In the full section setup the mean value of the accelerating mode frequency is measured directly. The mean coupling mode frequency over the section is determined indirectly using the measured modes frequencies nearest to the operating frequency in the BD. The mean coupling mode frequency in the full setup of the cavity section is determined with a high precision using the expressions including members up to the 3$^{rd}$ or 5$^{th}$ order. To calculate the eigenfrequencies of the half cells that compose the cavity section, a series of measurements in the MS is performed. The mean value of the operating mode frequency in the MS is measured directly. The average coupling mode frequency is determined indirectly by its deviation from the mean value over the section.

A specific feature of the described method is that it allows one to adjust both the frequencies of the half cells and the section as a whole, and the alternating frequency shift of the neighboring accelerating cells for multipactor discharge damping.

## 6. References

[1] Rybakov I V, Feschenko A V, Kalinin Y Z, Kravchuk L V, Leontiev V N, Naboka A N, Paramonov V V and Serov V L 2016 *Proc. 25th Russian Particle Accelerator Conf.* pp 216-218

[2] Rybakov I V, Paramonov V V and Skasyrskaya A K 2016 *Proc. 25th Russian Particle Accelerator Conf.* pp 291-293

[3] Rybakov I V and Isaev I I 2016 *Proc. 25th Russian Particle Accelerator Conf.* pp 291-293

[4] Naboka A N, Paramonov V V, Floettmann K 2008 *Problems of Atomic Science and Technology. Series Nuclear Physics Investigations* **№5** pp 35-39

[5] Paramonov V V, Kravchuk L V, Puntus V A 1998 *Proc Linac Conf.* ANL-**98/28** p 579

[6] Paramonov V V 2001 *The Annular Coupled Structure optimization for JAERY/KEK Joint Project. KEK report* Japan **14**

IOP Conf. Series: Journal of Physics: Conf. Series **941** (2017) 012095 doi:10.1088/1742-6596/941/1/012095

# Modular compact solid-state modulators for particle accelerators

**A A Zavadtsev, D A Zavadtsev and D V Churanov**

Nano Invest, LLC, Moscow, 11394, Russian Federation

azavadtsev@yandex.ru

**Abstract.** The building of the radio frequency (RF) particle accelerator needs high-voltage pulsed modulator as a power supply for klystron or magnetron to feed the RF accelerating system. The development of a number of solid-state modulators for use in linear accelerators has allowed to develop a series of modular IGBT based compact solid-state modulators with different parameters. This series covers a wide range of needs in accelerator technology to feed a wide range of loads from the low power magnetrons to powerful klystrons. Each modulator of the series is built on base of a number of unified solid-state modules connected to the pulse transformer, and covers a wide range of modulators: voltage up to 250 kV, a peak current up to 250 A, average power up to 100 kW and the pulse duration up to 20 μsec. The parameters of the block with an overall dimensions 880×540×250 mm are: voltage 12 kV, peak current 1600 A, pulse duration 20 μsec, average power 10 kW with air-cooling and 40 kW with liquid-cooling. These parameters do not represent a physical limit, and modulators to parameters outside these ranges can be created on request.

## 1. Introduction

The solid-state modulators are increasingly being used instead of tube modulators to feed the klystrons and the magnetrons. The main advantages of the solid-state modulator are low voltage of each switch connected in series, long lifetime, and easy control.

## 2. Different solid-state modulator schematics

Extensive development of the solid-sate modulators leads to different schematics, main of which are [1]

- modulator with adding magnetic flow (AMF),
- direct switch modulator (DC),
- modulator with the pulse transformer (PT).

### *2.1. Modulator with adding magnetic flow*

The first type of the modulator includes the complex pulse transformer with several primary windings and one secondary one. The high-voltage solid-state modules are connected to the primary windings as a loads. The current of each module creates the magnetic flow in the transformer core. The magnetic flows of all primary windings are added in the pulse transformer core. So the secondary voltage is a sum of primary voltages times transformer ratio. The modulator with adding magnetic flow in the pulse transformer has been built for HV feeding of S-band magnetron in 5 MeV electron linac [2, 3].

The modulator includes eleven 1 kV modules connected to eleven primary windings of the pulse transformer. The whole modulator is located in the oil-tank as this is shown in Figure 1.

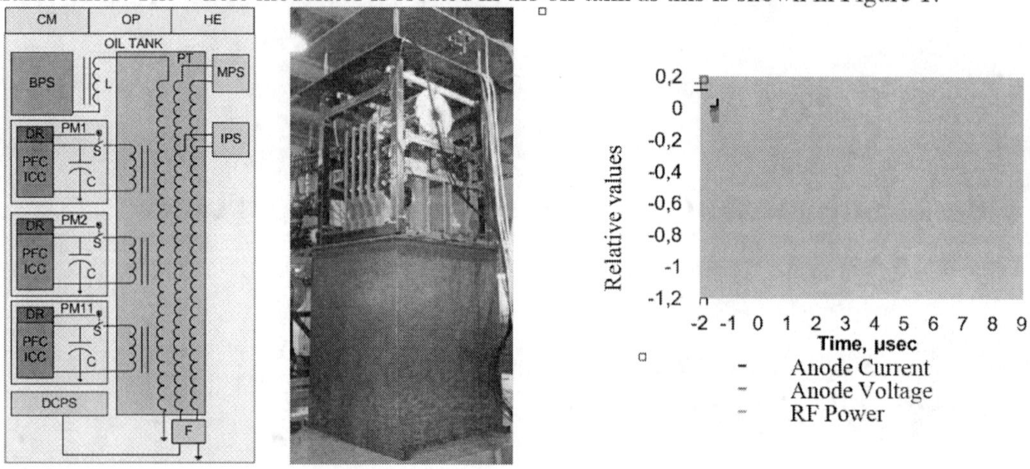


**Figure 1.** Schematics, view of the modulator with adding magnetic flow and the pulse shape.

## *2.2. Direct switch modulator*

The second modulator type includes several high-voltage modules built on the Arkadiev schematics with solid-state switches. Such modulator has been developed for 6 MW multi-beam klystron. Two modulators for 6 MW multi-beam klystron have been built for the 40 MeV electron linac [4]. Each modulator includes six 10 kV modules. . The modulator allows switching off the load voltage within the pulse with limitation of the current in case of break-down. The modulator and the klystron are located in the cabinet.

The modulator is shown in Figure 2.

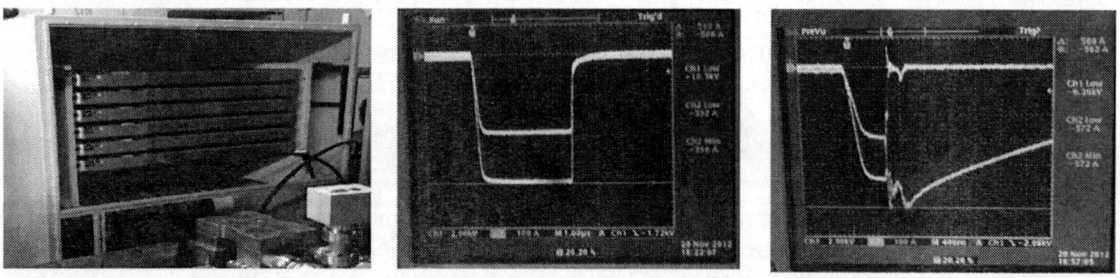

**Figure 2.** 60 kV direct switch modulator and the pulse shape in normal operation (top) and in case of break-down (bottom).

110 kV direct switch modulator with eleven 10 kV modules has been built for the S-band klystron [5]. The modulator is shown in Figure 3.

## *2.3. Modulator with pulse transformer*

130 kV modulator with the pulse transformer has been built to feed the klystron [1]. The modulator includes six 10 kV modules in the cabinet and produces dual output voltage ±24 kV, which is transmitted to the transformer primary (48 kV). Figure 4 shows (from left to right) the pulse transformer in the oil-tank (with klystron and local shielding on the top), the modulator cabinet and the control cabinet.

**Figure 3.** 110 kV direct switch modulator and the pulse shape in normal operation (top) and in case of break-down (bottom).

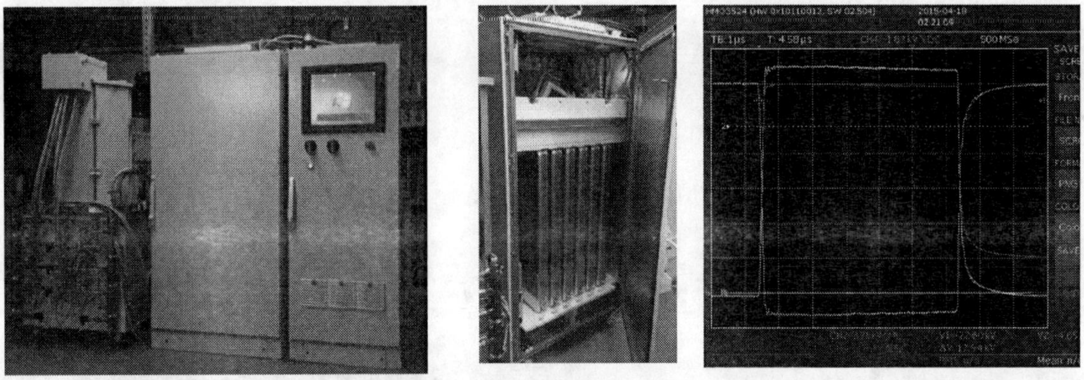

**Figure 4.** 130 kV modulator with pulse transformer and the pulse shape (BLUE is positive voltage, GREEN is negative voltage, RED is differential voltage).

250 kV modulator with the pulse transformer has been built to feed the klystron [1]. The modulator includes two dual 24 kV modules in the cabinet and the pulse transformer in the oil-tank with the klystron on the top. 3D design of the modulator, 3D design of 12 kV block and assembled 12 kV block are shown in Figure 5. The modulator is under construction.
The parameters of the described modulators are summarized in Table 1.

### 3. 24 kV unified module

The new high-voltage module has been designed during development of the 250 kV modulator with the pulse transformer [1]. The module includes two 12 kV blocks with eleven Arkadiev levels. The number of parallel IGBT in each level can be one or two. The number of parallel capacitors in each level can be from one to five. The number of IGBT and the capacitors in each level can be chosen depending on the required peak current, pulse length and flat-top slope. 3D design and assembled block are shown in Figure 5.

The assembled block has been tested at full designed voltage and current. The test results are shown in Figure 6.

The module includes two blocks, connected in series, and one common control unit. The module parameters are summarized in Table 2.

**Figure 5.** 250 kV modulator with pulse transformer.

**Table 1.** Parameters of the modulators.

| Parameter | Unit | AMF modulator | DC modulator | DC modulator | PT modulator | PT modulator |
|---|---|---|---|---|---|---|
| Voltage | kV | 55 | 60 | 110 | 130 | 250 |
| Current | A | 100 | 300 | 80 | 100 | 250 |
| Pulse length | µsec | 6 | 6 | 6 | 6 | 6 |
| Repetition rate | Hz | 300 | 160 | 10 | 10 | 10 |
| Number of modules | - | 11 | 6 | 11 | 6 | 2 |
| Transformer coefficient | - | 6 | - | - | 2.3 | 5.7 |
| Average power | kW | 4.5 | 18 | 6 | 6 | 6 |
| Load | - | Magnetron | Klystron | Klystron | Klystron | Klystron |
| Peak RF power | MW | 3 | 6 | 3 | 3-5 | 24 |

| Parameter | Unit | Value |
|---|---|---|
| Level voltage | kV | 1.1 |
| Block voltage | kV | 12 |
| Block current | kA | 1.5 |
| Pulse length | µsec | 6 |
| Rise time | µsec | 0.6 |
| Flat-top slope | % | <0.5 |

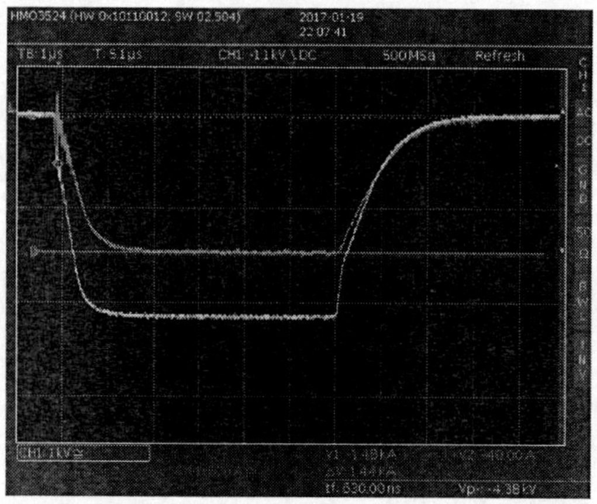

**Figure 6.** Test result of the 12 kV block.

## 4. Development of the modulator series

Developed 24 kV module allows us to develop the series of the solid-state modulators with the module as a common base. Parameters of the series modulators are shown in Table 3.

The block-diagrams of the modulators are shown in Figure 7.

The block-diagram of the Pulse transformer oil tank and the photo of the pulse transformer are shown in Figure 8.

IOP Conf. Series: Journal of Physics: Conf. Series **941** (2017) 012095 doi:10.1088/1742-6596/941/1/012095

**Table 2.** Parameters of 24 kV unified module.

| Parameter | Unit | Value |
|---|---|---|
| Max operating voltage | kV | 24 |
| Max operating voltage in accordance with IGBT specs | kV | 37 |
| Max operating current in accordance with IGBT specs | A | 2400 |
| Max pulse length | μsec | 20 |
| Flat-top slope (22kV@1467A@6 μsec) | % | 0,84 |
| Max average power with air-cooling | kW | 10 |
| Max average power with liquid cooling | kW | 40 |
| Number of levels | - | 2×11=22 |
| Number of IGBT in the level | - | 2 |
| Total number of IGBT | - | 44 |
| Capacitance of each capacitor | μF | 420 |
| Total number of capacitors in the level | - | 5 |
| Total number of capacitors | - | 110 |

**Table 3.** Parameters of the module-type modulators.

| Parameter | Unit | Value | | | | |
|---|---|---|---|---|---|---|
| Modulator Ind. | - | M1 | M2 | M3 | M4 | M5 |
| Voltage | kV | 250 | 175 | 135 | 60 | 60 |
| Current | A | 250 | 125 | 84 | 250 | 100 |
| Module impedance | Ohm | 15,5 | 50 | 50 | 50 | 50 |
| Number of modules | - | 2 | 2 | 2 | 2 | 1 |
| Trans impedance | Ohm | 31 | 100 | 100 | 100 | 50 |
| Tran coefficient | - | 5,7 | 3,7 | 4,0 | 1,5 | 3,5 |
| Trans input voltage | kV | 44 | 47 | 34 | 39 | 17 |
| Module voltage | kV | 22 | 23 | 17 | 19 | 17 |
| Module current | A | 1420 | 468 | 337 | 387 | 346 |
| Load | - | Klystron | Klystron | Klystron | Magnetron /Klystron | Magnetron |
| Peak RF power | MW | 24 | 10 | 5,5 | 7.5/5,5 | 5,5 |

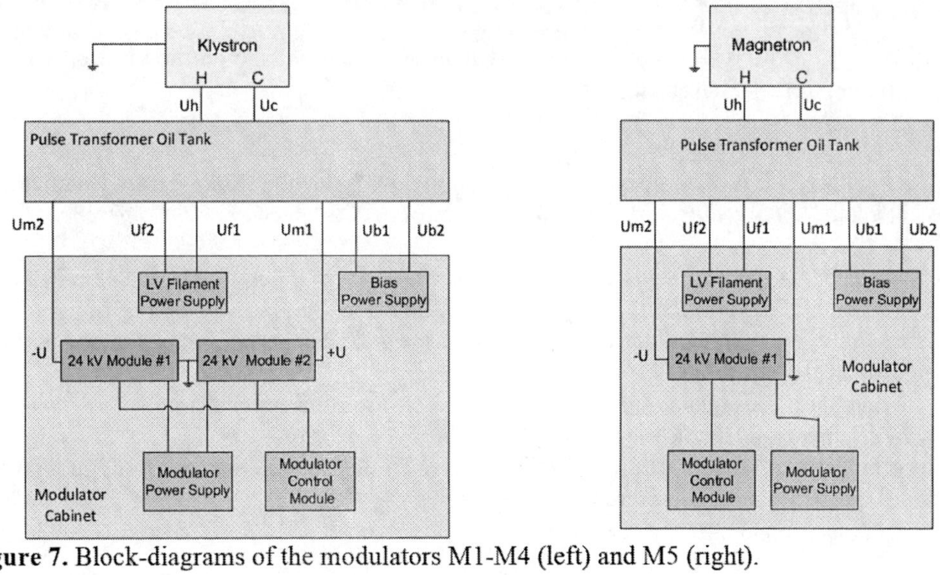


**Figure 7.** Block-diagrams of the modulators M1-M4 (left) and M5 (right).

IOP Conf. Series: Journal of Physics: Conf. Series **941** (2017) 012095 doi:10.1088/1742-6596/941/1/012095

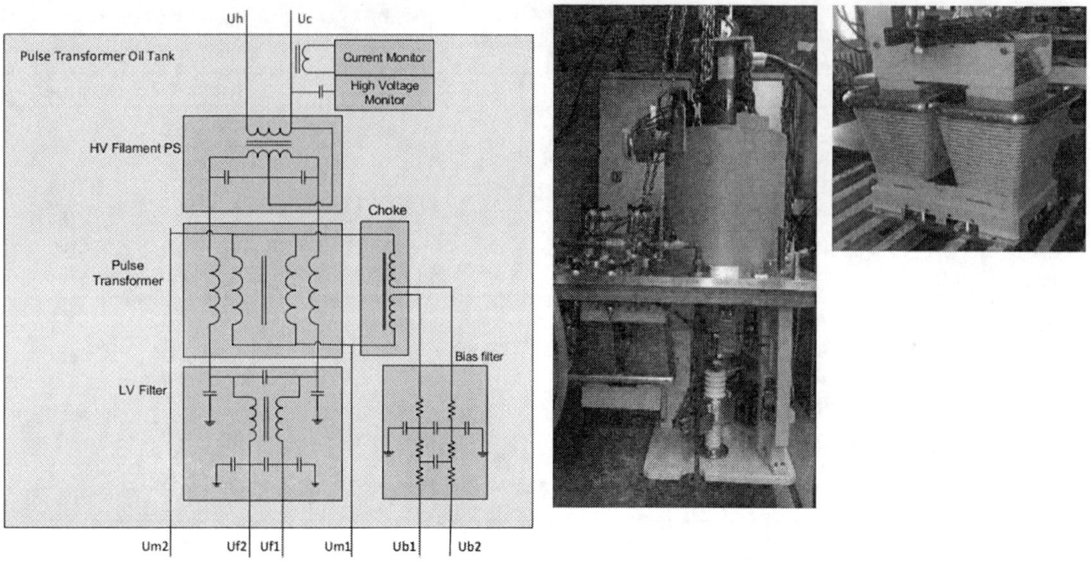


**Figure 8.** Block-diagram of pulse transformer.

## 5. Conclusion

The solid-state modulator series for klystrons and magnetrons has been developed on base of built 24 kV unified module.

The modulators cover a wide range of the parameters: voltage up to 250 kV, current up to 250 A, pulse length up to 20 μsec.

The work has been done on base the experience of the building of the solid-state modulators with different schematics: with adding the magnetic flow, with direct switch, with or without pulse transformer.

The solid-state modulators have following advantages over the tube-based modulators:

- module type of the building;
- easy parameter scaling on base of module type;
- low voltage of the single switch;
- long lifetime of the semiconductor switch;
- easy control of the voltage, pulse length, rise time;
- availability of the fast break-down interlock, switching off the voltage on the klystron within the pulse with limitation of the klystron current.

## References

[1] A. A. Zavadtsev, D. A. Zavadtsev, D. V. Churanov, D. A. Zybin. Solid-State Modulators for Particle Accelerators. *Proceedings of the XXV Russian Particle Accelerator Conference (RuPAC 2016)*

[2] D.A.Zavadtsev et al., "Compact Electron Linear Accelerator RELUS-5 for Radiation Technology Application". *Proceedings of EPAC 2006*, Edinburgh, Scotland, p.2385-2387.

[3] D. Churanov et al., "Solid State Modulator for Linear Accelerators". *Proceedings of RuPAC 2008*, Zvenigorod, Russia, p.175-177.

[4] E.A. Savin et al., "Design and tuning of a 40-MeV electron linear accelerator". *Instruments and Experimental Techniques*, 2013, 56 (5), p. 506-515.

[5] L. Kravchuk et al., "Layout of the PITZ Transverse Deflect ing System for Longitudinal Phase Space and Slice Emittance Measurement". *Proceedings of Linear Accelerator Conference LINAC2010*, Tsukuba, Japan, 2010, p. 416-418.

# Compact Storage Ring for an X-Ray Source

L Ovchinnikova[1,2], V Shvedunov[1,2] and K Ivanov[3]

[1] M.V.Lomonosov Moscow State University, Skobeltsyn Institute of Nuclear Physics (SINP MSU), 1(2), Leninskie gory, GSP-1, Moscow 119991, Russian Federation

[2] Laboratory of Electron Accelerators MSU Ltd, Leninskie Gory 1(77), 119992, Moscow, Russia

[3] Faculty of Physics, M.V.Lomonosov Moscow State University, Leninskie Gory, Moscow 119991, Russia

E-mail: lub.ovch@yandex.ru

**Abstract.** We propose a new design of a compact storage ring for a source of X-ray radiation on the basis of reverse Thomson scattering of laser radiation by electrons with the energy of 35–50 MeV, which has small number of optical elements and a significant clear space for the placement of a beam injection-extraction system and a RF cavity. The original laser cavity layout has been considered. The ring dynamic aperture after correction of chromaticity and a second-order dispersion function is sufficient for the injection and stable circulation of an electron bunch in the ring.

## 1. Introduction

This work continues a cycle of the previously conducted studies of an X-ray radiation source — the laser-electron generator (LEG) based on Thomson scattering of laser radiation by electrons [1–5]. The new design of the storage ring, which is a part of LEG, considered in the present work is intended to retain electron bunches with the energy adjustable within the range of 35 to 50 MeV with the maximum charge of 1 nC and a root-mean-square normalized emittance of about 4 mm mrad at the moment of injection. Intrabeam scattering of particles and their interaction with laser radiation at such a low electron's energy and a large charge of the bunches leads to the growth of transverse emittance and a momentum spread, which requires replacement of the bunches by means of a fast beam injection-extraction system with the period of 20 ms, which is much smaller than the time constant of radiation damping due to synchrotron radiation.

Our previously considered [5] version of the ring "C" was taken as the basis. We have conducted an optimization of the ring lattice, minimization of chromatic aberrations, performed calculations of dynamic aperture and the growth of emittance due to intrabeam scattering. The calculations were performed with the help of the MAD-X [6] and OPA [7] codes. Consideration of the optimal RF cavity operating frequency has been conducted. Since the laser resonator is an integral part of the LEG ring, we have considered an option for its placement, which does not lead to an increase of the ring size and allows to obtain a small value of the $\beta$-function at the interaction point, with its values in other parts of the ring being moderate.

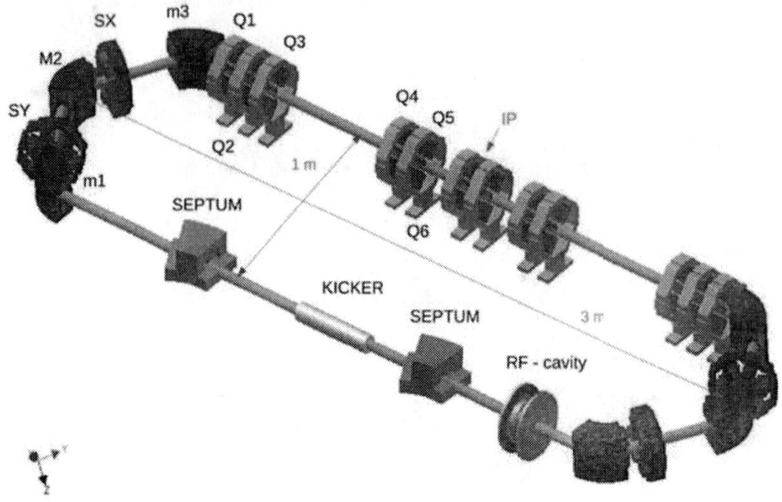


**Figure 1.** LEG storage ring lattice. $Q_1$, $Q_2$, $Q_3$, $Q_4$, $Q_5$, $Q_6$ — quadrupoles, $S_x$, $S_y$ — sextupoles, $M_1$, $M_2$, $M_3$ — dipoles, IP — interaction point.

## 2. The ring lattice

The ring lattice (Fig. 1) is formed by two achromatic 180° turns composed of three 60° dipole magnets: $M_1$, $M_2$, $M_3$, providing the beam focusing in horizontal and vertical planes. The beam focusing at the interaction point is achieved with the help of two pairs of six quadrupoles $Q_1$ – $Q_6$. The ring lattice has a mirror symmetry with regard to the plane passing through the interaction point, IP.

**Table 1.** LEG storage ring parameters

| Parameter | Value |
|---|---|
| RF cavity operating frequency, $f_{RF}$, MHz | 2856 |
| Bunch revolution frequency, $f_{rev}$, MHz | 30 |
| Orbit length, $L$, m | 9.762 |
| Momentum compaction factor, $\alpha$ | 0.136 |
| Voltage across the RF cavity gap, $V_{RF}$, kV | 300 |
| Momentum of the synchronous particle, $p_s$, MeV/c | 50.5 |
| Number of bunches at orbit, $M$ | 1 |
| Average beam current, $I_b$, mA | 35.7 |
| Synchrotron tune, $\nu_z$ | 0.05 |
| Betatron tunes, $\nu_x/\nu_y$ | 2.7/1.8 |
| Betatron functions at the interaction point, $\beta_x/\beta_y$, m | 0.05/0.05 |
| Relative RMS momentum spread at the injection, $\Delta p_{rms}/p_s$ | 0.001 |

The ring lattice was optimized taking into account the following correction of chromatic aberrations and the second-order dispersion function, achieving large dynamic aperture, and a feasibility of the laser cavity placement. The betatron functions and the dispersion function behavior are shown in Fig. 2, and the ring parameters are given in Tab. 1. An important feature of the ring is the absence of quadrupoles in most part of orbit except for the section containing

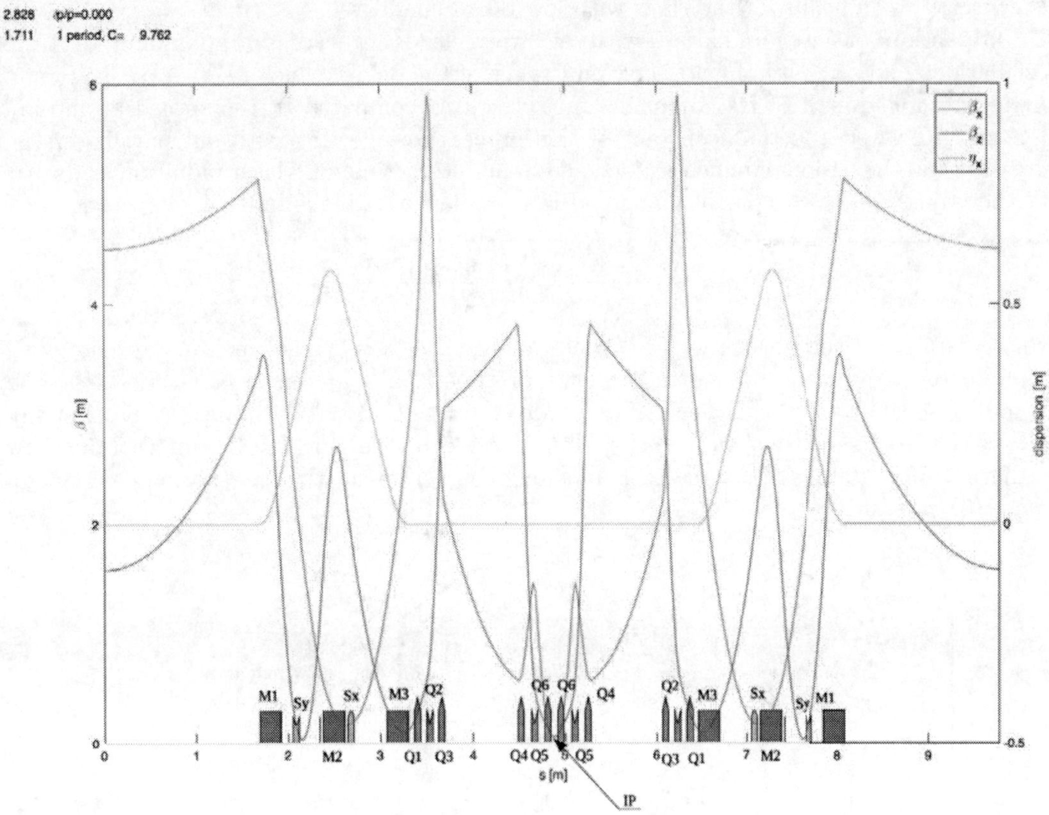


**Figure 2.** Betatron and dispersion functions of the ring. $Q_1$ – $Q_6$ — quadrupoles, $S_x$, $S_y$ — sextupoles, $M_1$, $M_2$, $M_3$ — dipoles, IP — interaction point.

the IP. At that, the maximum value of the betatron functions does not exceed 4 m, which corresponds to the maximum RMS beam radius equals to about 0.4 mm. At the IP, the value of the betatron function equals 5 cm, and respectively, the RMS beam radius equals 45 $\mu$m. The other feature of the ring is a large value of the momentum compaction factor, which requires a high value of the operating RF frequency; the discussion of this point is given below.

## 3. Intrabeam scattering and interaction with laser radiation

In the course of the bunch revolution in the ring, the growth of both the RMS transverse beam emittance in the bending plane and the relative RMS momentum spread occurs as a result of intrabeam scattering. The reason for the emittance growth in the bending plane is a change of longitudinal momentum within a dispersion path due to particles scattering within a bunch, which leads to violation of the achromatic conditions for these particles. The growth of relative momentum spread is related to the transverse momentum transfer into the longitudinal plane; at that in the laboratory reference frame the change of longitudinal momentum grows proportionally to the relative energy of particles [8–11]. Time constant for the emittance and the momentum spread growth depends on the ring lattice functions and is inversely proportional to the charge density in the phase volume occupied by the beam. In this connection, as the emittance and the momentum spread grow, the growth rate is slowing down. It should also be noted that the time constant for the emittance growth contains an addend, which is inversely proportional to the fourth power of the relative beam energy. In this regard, the given problem is especially significant for the rings with low beam energies.

Interaction of laser radiation with electrons leads to radiation damping of transverse oscillations and decrease of transverse emittance of the beam, which in our case is accompanied by significant growth of RMS momentum spread, as compared to the spread at the injection [12, 13]. However, the time constant of this process for the ring with our parameters is much greater than the period of bunch replacement, and interaction with laser radiation leads primarily to the growth of RMS momentum spread in accordance with the following expression [14]:

$$\frac{\delta p_{rms}(t)}{p_s} = \sqrt{\left\{\frac{\langle E_\gamma \rangle}{2E_s}\right\}^2 \frac{N_\gamma}{N_e} t}, \tag{1}$$

where $\langle E_\gamma \rangle$ — average energy of X-ray radiation, $E_s$ — ring synchronous energy, $N_\gamma$ — number of X-ray photons emitted in unit of time, $N_e$ — number of electrons in a bunch. In our case $\langle E_\gamma \rangle \approx 20$ keV, for $E_s = 50$ MeV, $N_e \approx 6.25 \times 10^9$, immediately after injection $N_\gamma \approx 5 \times 10^{13}\ \mathrm{s}^{-1}$, thus, over the operating cycle of the ring due to interaction with laser radiation $\delta p_{rms}(20\ \mathrm{ms})/p_s \approx 0.25\%$. It should be reminded that at the moment of injection $\delta p_{rms}(0)/p_s \approx 0.12\%$.

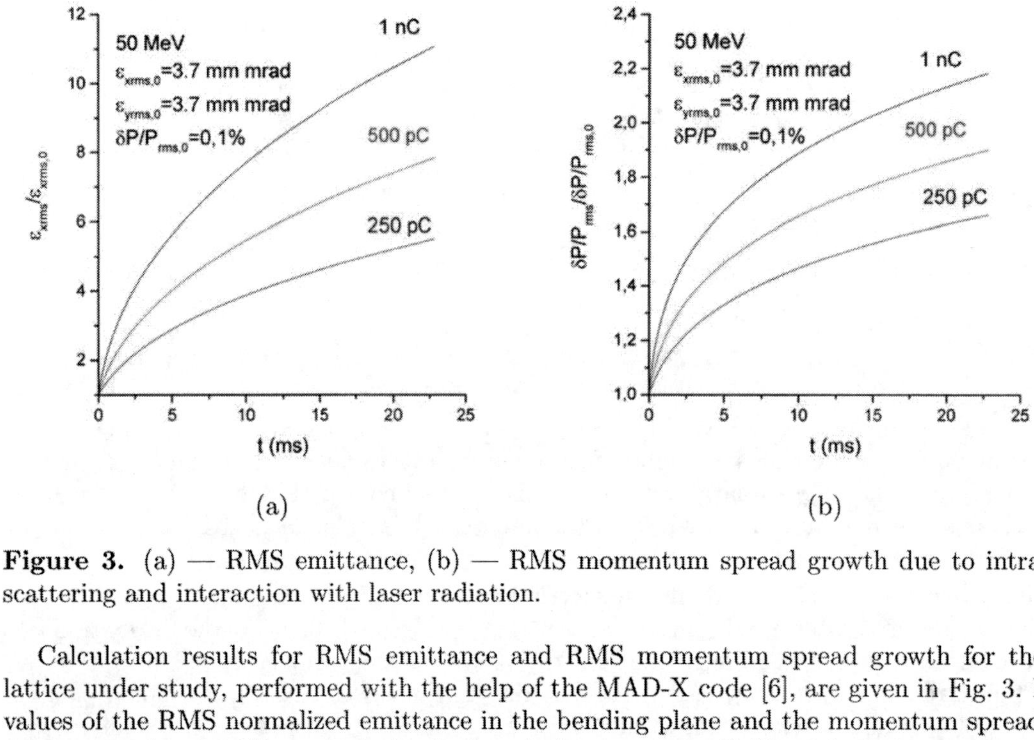


(a) (b)

**Figure 3.** (a) — RMS emittance, (b) — RMS momentum spread growth due to intrabeam scattering and interaction with laser radiation.

Calculation results for RMS emittance and RMS momentum spread growth for the ring lattice under study, performed with the help of the MAD-X code [6], are given in Fig. 3. Initial values of the RMS normalized emittance in the bending plane and the momentum spread were 3.8 mm mrad and 0.1% respectively, the bunch charge was 1 nC. The calculations accounted for the growth of momentum spread described by expression (1).

As one can see from the given graphs, over the time of 20 ms the beam emittance grew 10.5 times (from 3.7 to 14.2 mm mrad), whereas the momentum spread grew 2.14 times (from 0.1 to 0.257%). It needs to be taken into account that at the moment of injection from a linear accelerator, the phase extent of the bunch is much less than the equilibrium value in the ring for the given momentum spread, and during about few thousand revolutions it approaches the equilibrium one.

IOP Conf. Series: Journal of Physics: Conf. Series **941** (2017) 012096 doi:10.1088/1742-6596/941/1/012096

## 4. Correction of chromaticities and second-order dispersion function

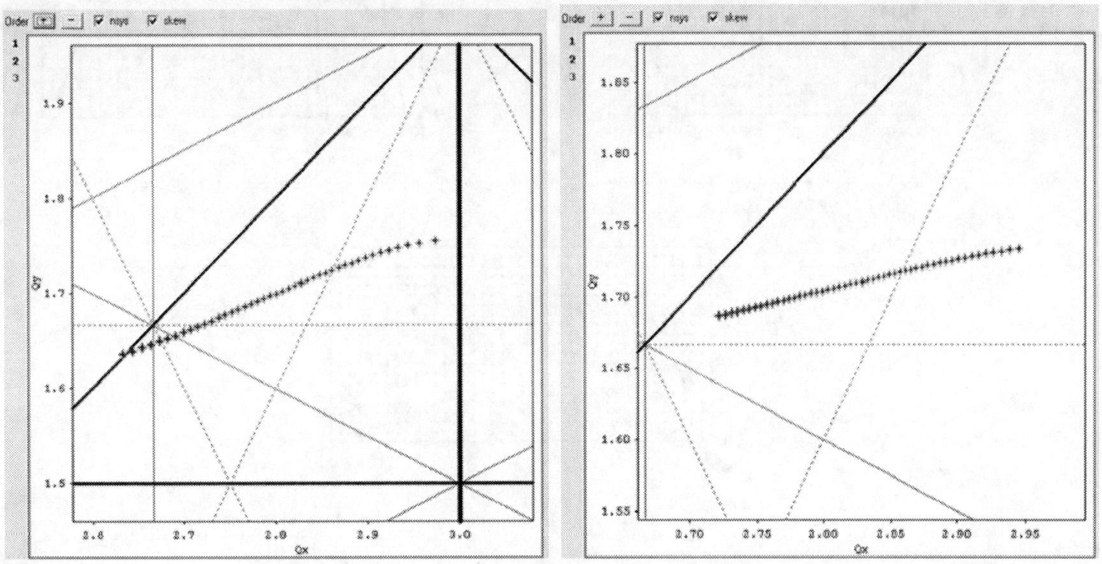

(a) (b)

**Figure 4.** Position of operating point at the tunes diagram (momentun spread ±1.5%). (a) — without correction, (b) — with chromaticity correction. Point color on the diagram means the momentum deviation value, yellow color $\Delta p/p = 0$, blue $\Delta p/p = -1.5\%$, pink $\Delta p/p = +1.5\%$.

Values of natural chromaticities of the considered lattice are $\xi_x = -7$, $\xi_y = -4.5$. Fig. 4 shows position of the operating point at the ring tunes diagram, which was optimized using the OPA code [7]. Momentum spread, including that caused by intrabeam scattering and interaction with laser radiation, leads to an intersection with the strong first and third order resonances (Fig. 4 (a)), which requires the chromaticity correction. Correction of natural chromaticities was carried out with the help of two families of sextupoles ($S_x$, $S_y$, Fig. 1), located at dispersion paths. Chromaticity values after the correction are $\xi_{x,\,\mathrm{corr}} = -3$, $\xi_{y,\,\mathrm{corr}} = -3$. These chromaticity values, driven also by the need to correct the second-order dispersion function (see below), allow one to avoid intersections with strong resonances (Fig. 4 (b)).

The achromat structure we have selected allows one to carry out chromaticity correction using weak sextupole fields not resulting in strong nonlinear effects, which could limit the dynamic aperture value. Calculation results for the dynamic aperture at the injection point performed for 1000 revs with the help of the MAD-X code [6] for various relative pulse spreads are shown in Fig. 5. Note that further increase in the number of revs will insignificantly change the dynamic aperture dimensions. As one can see, the dynamic aperture almost completely covers geometric dimensions of the vacuum chamber shown with a solid line.

In Fig. 6 one can see the behavior of the first and second order dispersion functions. The second order dispersion function at the IP was -1.3 m prior to correction (Fig. 6 (a)). As shown above, relative RMS momentum spread reaches 0.257% due to intrabeam scattering and interaction with laser radiation. In our case this results in an increase of RMS beam size at the IP due to second order dispersion by 81 $\mu$m, which is significantly more that the design value of 45 $\mu$m. We have corrected the second order dispersion with the help of sextupole lenses placed at the nonzero dispersion paths. Fig. 6 (b) shows dispersion functions after the correction: the second order dispersion function value was decreased to 0.08 m, which resulted in an increase of the RMS beam size at the IP by only 8 $\mu$m.

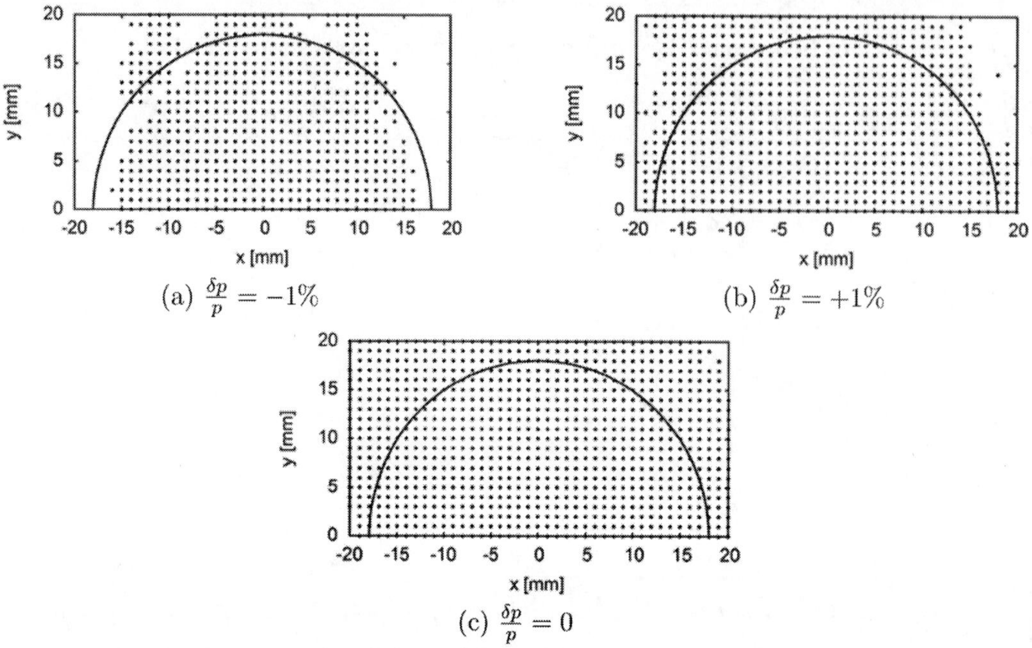


**Figure 5.** Dynamic aperture of the LEG ring. Calculation for 1000 revs for three relative momentum spread values.

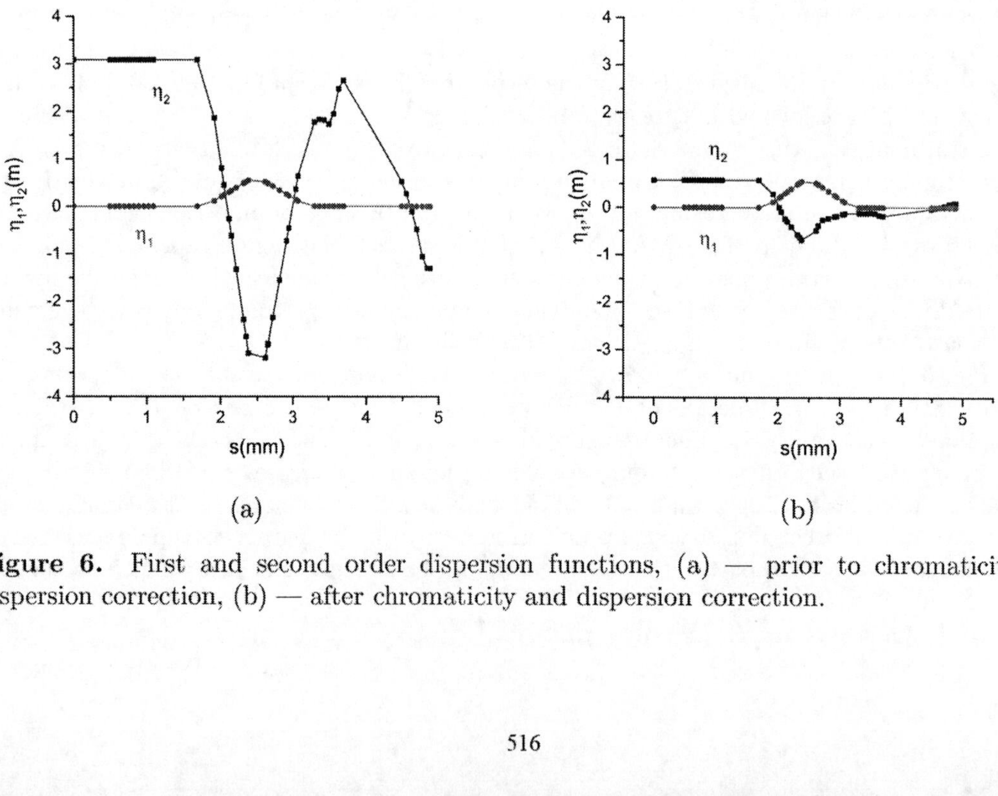


**Figure 6.** First and second order dispersion functions, (a) — prior to chromaticity and dispersion correction, (b) — after chromaticity and dispersion correction.

## 5. Optical cavity placement

An optical cavity in which laser radiation pulse with the energy of about $W_{ph} = 20$ mJ circulates with a bunch revolution frequency $f_{rep} = 30$ MHz, which corresponds to the average accumulated radiation power of over 600 kW, must be integrated into the ring design.

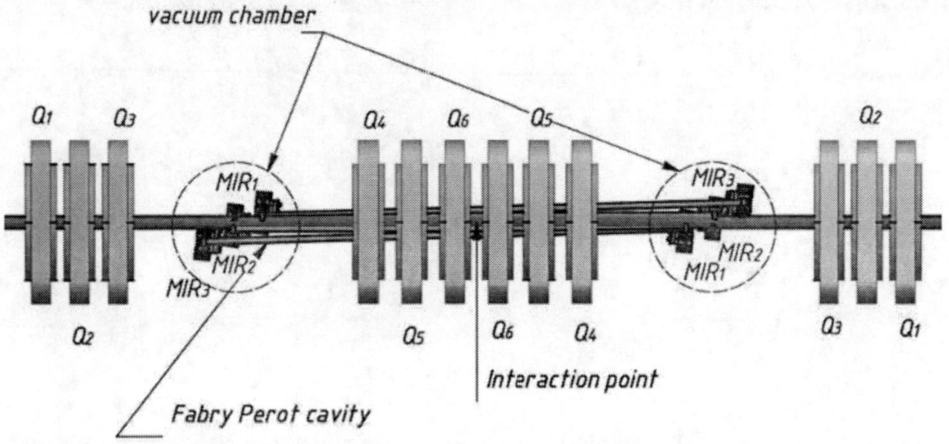


**Figure 7.** Positional relationship of the six-mirror optical cavity laser beams layout and quadrupoles before the interaction point, $Q_1$ – $Q_6$ — quadrupoles, $MIR_1$ – $MIR_3$ — mirrors.

Known optical cavity placement methods are not suitable for the ring lattice we selected. Therefore, we have considered an option of six-mirror cavity with laser beams passing within the aperture of central quadrupoles ($Q_4$, $Q_5$, $Q_6$). Fig. 7 shows a scheme of the optical cavity ensuring an intersection of electron bunches and a laser pulse at the angle of 2°. Placement of the optical cavity is feasible at the lens aperture of not less than 70 mm and the shape of the pole pieces allowing for placement of the cavity's vacuum tubes. The problem is complicated by the fact that in order to achieve a small value of the beta-function at the interaction point, the quadrupoles gradient value must be not less than 22 T/m, and the multipole component contribution must be minimal.

The quadrupoles geometry found as a result of optimization is shown in Fig. 8 (a), and the distribution of magnetic field for the design value of the gradient is shown in Fig. 8 (b). The main problem for the achieving of high gradient is presented by saturation of the pole pieces material. In this connection we have considered two options for the material: super-permendur and vanadium-permendur. Fig. 8 (c) shows gradient as a function of current in the coils. The pole saturation effect begins to appear at the gradient's value above 22–24 T/m. Note that the lens aperture value, which is greater than the beam size by at least two orders of magnitude, guarantees small influence of multipole components on the beam dynamics in the ring.

## 6. RF cavity frequency selection. Estimate of X-ray photon yield

As was noted above, the feature of the considered ring is a large value of the momentum compaction factor $\alpha = 0.136$, which determines, along with other parameters, the RMS bunch length (two standard deviations) in equilibrium conditions:

$$\Delta l_{rms} = 2\frac{\Delta p_{rms}}{p_s}\sqrt{\frac{cL\alpha E_s}{2\pi f_{RF} V_{RF}\cos\varphi_s}}, \tag{2}$$

where $c$ — speed of light, $\varphi_s$ — equilibrium phase, other notations are explained in Tab. 1.

IOP Conf. Series: Journal of Physics: Conf. Series **941** (2017) 012096 doi:10.1088/1742-6596/941/1/012096

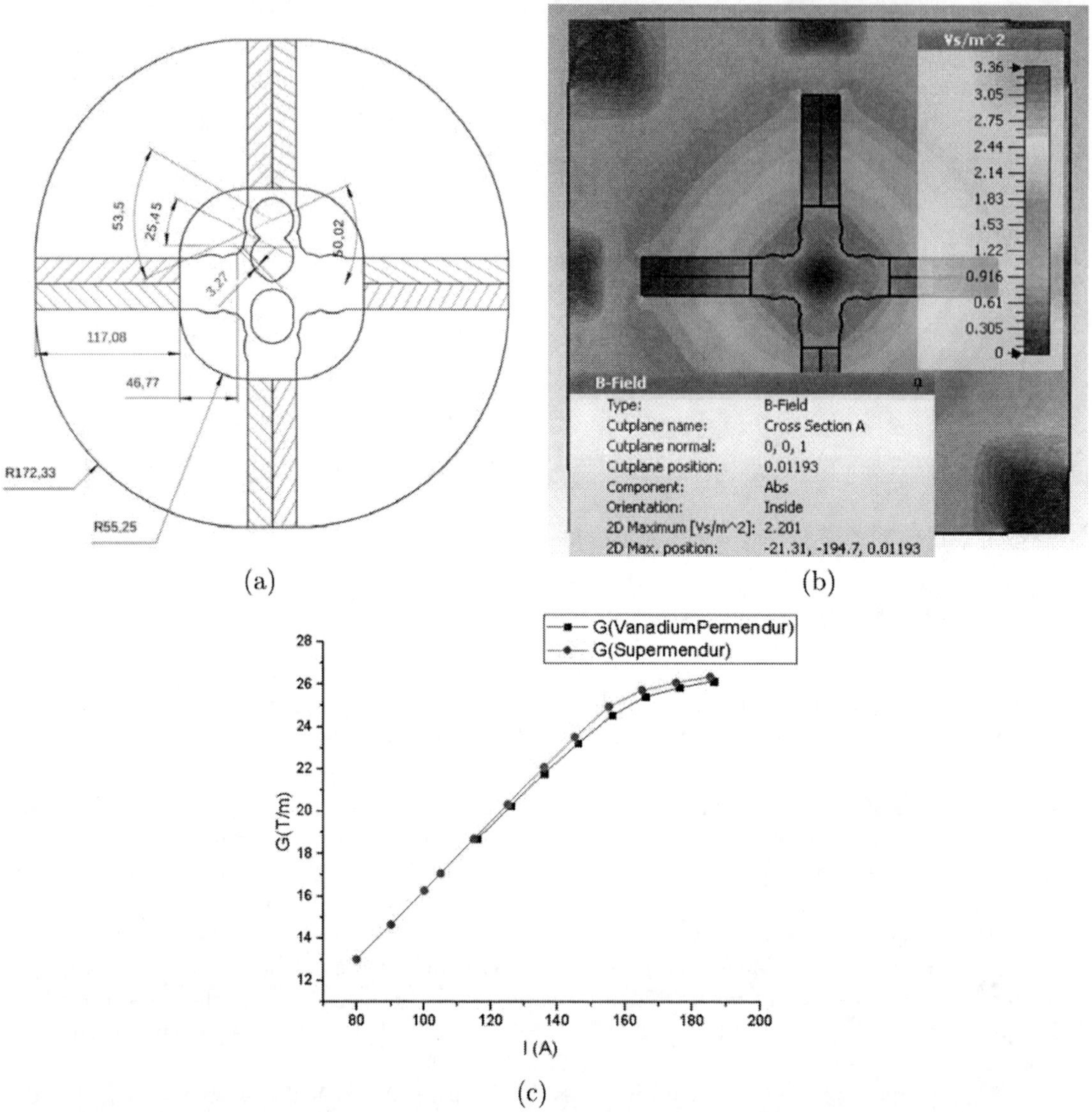


**Figure 8.** (a) — quadrupoles geometry. (b) — magnetic field distribution, (c) — dependence of quadrupole gradient on current in the coils: red plot — super permendur magnetic material, black plot — vanadium permendur.

When an electron bunch collides with a laser pulse at a nonzero angle, as it is in our case, the yield of X-ray photons drops significantly with the increase of the bunch length, which is true when the orbit expansion factor increases. To reduce bunch length, along with decreasing the orbit length, we have selected a rather high operating frequency of RF cavity of the ring, $f_{RF} = 2856$ MHz.

Thus, for RMS relative momentum spread at the injection of $\Delta p_{rms}/p_s = 0.001$, and for the ring parameters given in Tab. 1, the RMS length of the bunch will be $\Delta l_{rms} \approx 3.86$ mm (the RMS duration and phase length are about 13 ps and 13°, respectively). Based on the above data, data from Tab. 1 and optical cavity parameters, one can estimate that at the laser photon energy of $E_{ph} = 1.03$ eV, and electron energy of $E_e = 50$ MeV, total yield per unit of time of scattered photons with threshold energy of $E_\gamma \approx 40$ keV just after injection will amount to $dN_\gamma/dt \approx 5 \times 10^{13}$ phot/s.

## 7. Beam injection-extraction system and RF cavity placement

The beam injection-extraction system, which includes two septum magnets and a kicker (Fig. 1) is placed at a free path with zero dispersion. These elements are described in a separate paper, as well as the RF cavity, which is also placed within that path.

## 8. Conclusion

In this work we offer an option of a compact storage ring for an X-ray radiation generator, which has a large dynamic aperture allowing one to place an optical and a RF cavities, as well as a beam injection-extraction system.

## Acknowledgments

The authors express their gratitude to professors A.B. Saveliev-Trofimov for the discussion of the laser cavity placement, E.G. Bessonov and A.A. Mikhailichenko for the consultations with regard to the storage ring calculations.

## References

[1] Artiukov I, Bessonov E, Vinogradov A, Gorbunkov M *et al.* Laser-electronic generator of x-ray radiation URL `http://www.sinp.msu.ru/en/system/files/preprints/pp-806.pdf`

[2] Bessonov E G, Gorbunkov M V, Maslova Y Y, Kostryukov P V, Tunkin V G, Ishkhanov B S, Shvedunov V I and Vinogradov A V 2007 *Proceedings of SPIE - The International Society for Optical Engineering* vol 6702 pp 6702–1–6702–9 URL `http://proceedings.spiedigitallibrary.org/proceeding.aspx?articleid=818886`

[3] Bessonov E G, Gorbunkov M V, Ishkhanov B S, Kostryukov P V, Maslova Y Y, Shvedunov V I, Tunkin V G and Vinogradov A V 2008 *Laser and Particle Beams* **26** 489–495 ISSN 0263-0346; 1469-803X URL `https://www.cambridge.org/core/journals/laser-and-particle-beams/article/laserelectron-generator-for-xray-applications-in-science-and-technology/55CA149E32B35E2B52C38BD6A877E22C`

[4] Bessonov E G, Gorbunkov M V, Kostryukov P V, Maslova Y Y, Tunkin V G, Postnov A A, Mikhailichenko A A, Shvedunov V I, Ishkhanov B S and Vinogradov A V 2009 *Journal of Instrumentation* **4** P07017 ISSN 1748-0221 URL `https://arxiv.org/ftp/arxiv/papers/1604/1604.08839.pdf`

[5] Ovchinnikova L, Bessonov E, Gorbunkov M, Mikhailichenko A and Shvedunov V 2016 *Proc. of International Particle Accelerator Conference (IPAC'16), Busan, Korea, May 8-13, 2016* (*International Particle Accelerator Conference* no 7) (Geneva, Switzerland: JACoW) pp 3308–3310 ISBN 978-3-95450-147-2 doi:10.18429/JACoW-IPAC2016-THPMB035 URL `http://jacow.org/ipac2016/papers/thpmb035.pdf`

[6] MAD-X URL `https://madx.web.cern.ch/madx`

[7] OPA URL `https://ados.web.psi.ch/opa`

[8] Piwinski A 1974 *Proceedings, 9th International Conference on High-Energy Accelerators, Stanford, California, 2-7 May 1974* pp 405–409 URL `http://inspirehep.net/record/94147/files/HEACC74_426-430.pdf`

[9] Bjorken J D and Mtingwa S K 1983 *Part. Accel.* **13** 115–143 URL `http://lss.fnal.gov/archive/1982/pub/Pub-82-047-T.pdf`

[10] Kubo K, Mtingwa S K and Wolski A 2005 *Phys. Rev. ST Accel. Beams* **8** 081001 URL `http://inspirehep.net/record/694236/files/PhysRevSTAB.8.pdf`

[11] Venturini M 2001 *Conf. Proc.* **C0106181** 2961–2963 [,2961(2002)] URL `http://accelconf.web.cern.ch/AccelConf/p01/PAPERS/RPAH048.PDF`

[12] Huang Z and Ruth R D 1998 *Phys. Rev. Lett.* **80** 976–979 URL `http://slac.stanford.edu/pubs/slacpubs/7500/slac-pub-7556.pdf`

[13] Chaikovska I and Variola A 2014 *Phys. Rev. ST Accel. Beams* **17** 044004 URL `https://journals.aps.org/prab/pdf/10.1103/PhysRevSTAB.17.044004`

[14] Bruni C, Haissinski J, Loulergue A and Nagaoka R 2011 *Conf. Proc.* **C110904** 715–717 URL `http://accelconf.web.cern.ch/AccelConf/IPAC2011/papers/mops050.pdf`

# Development of PET projection data correction algorithm

**P V Bazhanov and E D Kotina**

Saint Petersburg State University, 7/9 Universitetskaya nab., Saint Petersburg, 199034, Russia

E-mail: e.kotina@spbu.ru

**Abstract.** Positron emission tomography is modern nuclear medicine method used in metabolism and internals functions examinations. This method allows to diagnosticate treatments on their early stages. Mathematical algorithms are widely used not only for images reconstruction but also for PET data correction. In this paper random coincidences and scatter correction algorithms implementation are considered, as well as algorithm of PET projection data acquisition modeling for corrections verification.

## 1. Introduction

PET method is based on 511 keV photon pairs registration: radionuclide injected to patient before the procedure generates positrons annihilating with electrons in internal tissue after a short free passage. The main forms of interaction between photons with energy value 511 keV and tissues involve absorption in the result of photoelectric effect and Compton scattering. Both effects affect on images obtained during the procedure. Random coincidences factor, radiation attenuation during decay process, geometric factors connected with detectors and patient motion also influence on images quality. Random coincidences arise when pair of photons from different annihilation acts are registered in the same time window.

Various mathematical algorithms are using in PET data processing [1]. Radiation attenuation during decay process and geometric factors are usually corrected under reconstruction, for example in iteration maximum likelihood estimation method [2]. For random coincidences correction delayed time window method is generally used [3]. Patient motion can be corrected by using an algorithm based on cross-correlation function method [4]; the approaches based on the determination of the velocity field can be helpful for the motion correction as well [5]. For scatter correction the following approaches are used: multiple energy-window (spectral-analytic) approaches and methods based on scatter coincidences distribution estimation from linear attenuation coefficients and activity distributions obtained during emission and transmission tomography data reconstruction [6].

## 2. PET projection data acquisition modeling algorithm

PET projection data acquisition modeling algorithm is based on Monte-Carlo methods and can be presented in the following diagram.

IOP Conf. Series: Journal of Physics: Conf. Series **941** (2017) 012097 doi:10.1088/1742-6596/941/1/012097

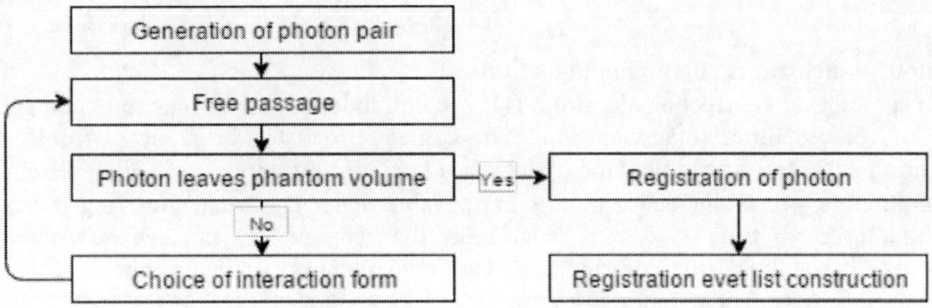


**Figure 1.** PET projection data acquisition modeling algorithm diagram.

A free passage length $\varepsilon$ is taken in the following way [7]:

$$\varepsilon = -\frac{1}{\mu} \cdot ln\gamma, \gamma \in [0,1].$$

An equally probable direction of photon flight $\vec{\omega}$ is chosen as

$$\vec{\omega} = \vec{i} \cdot \omega_i + \vec{j} \cdot \omega_j + \vec{k} \cdot \omega_k,$$

$$\omega_i = \sqrt{1 - cos^2\theta} \cdot cos\,\varphi,$$

$$\omega_j = \sqrt{1 - cos^2\theta} \cdot sin\,\varphi,$$

$$\omega_k = cos\,\theta,$$

$$cos\,\theta = 2 \cdot \gamma_1 - 1, \gamma_1 \in [0,1],$$

$$\varphi = 2\pi\gamma_2, \gamma_2 \in [0,1].$$

The form of interaction between particle and tissue is defined as following: $\alpha = \max_{x \in V}(\mu(x))$, random value $\gamma \in [0,1]$ is chosen. Now if

- $\gamma \leq \frac{\mu_e}{\alpha}$, then photon is absorbed;
- $\frac{\mu_s}{\alpha} < \gamma \leq \frac{\mu_s + \mu_e}{\alpha}$, then photon is scattered;
- $\gamma > \frac{\mu_s + \mu_e}{\alpha}$, Then photon keeps moving in the same direction.

The considered algorithm is implemented by C# program. The input data for the algorithm are activity and linear attenuation distributions given as three-dimensional arrays, data acquisition time, radionuclide decay half-life and detectors configuration. PET projection data acquisition modeling results for different phantoms are given below.

**Figure 2.** Sinogram for acquisition without attenuation account, uniform cylinder as a phantom is used.

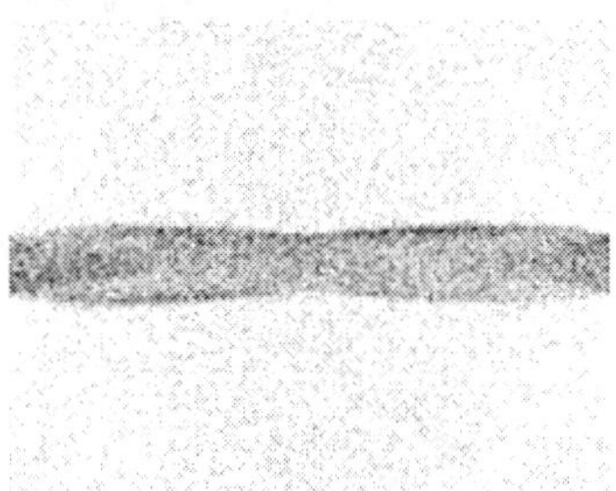

**Figure 3.** Sinogram for acquisition with attenuation account, uniform cylinder as a phantom is used.

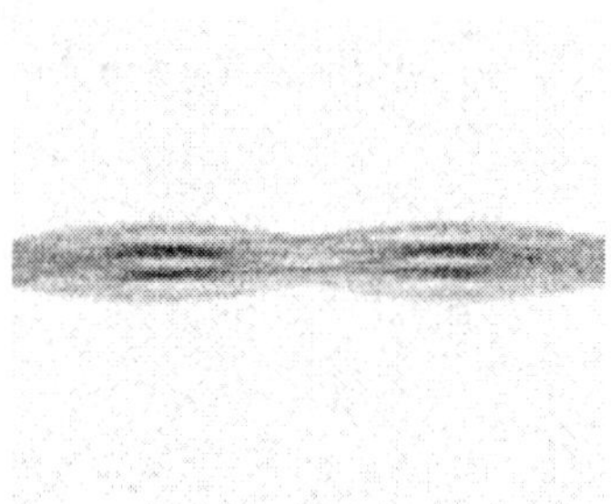

**Figure 4**. Sinogram for acquisition without attenuation account, set of five uniform cylinders as a phantom is used.

## 3. Random coincidences correction algorithm

Random coincidences correction algorithm is based on random coincidences sinogram generation with the use of delayed time window method. The coincidence list is constructed from a single event registration list and a given time window which is maximal difference among single events from the same annihilation act. Coincidence time window value depends on detectors ring radius and usually takes about 10 ns. If time window is much more than coincidence time window then we obtain a coincidences list without true coincidences. That coincidences list is an approximation for random coincidences sinogram construction [3].

Random coincidences correction algorithm consists of three steps:

1. Coincidences list obtaining with the use of delayed time window;
2. Random coincidences sinogram contruction;
3. Subtraction of obtained sinogram from uncorrected sinogram.

The considered algorithm is implemented by C# program, the results are given below.

**Figure 5.** Uncorrected sinogram.

**Figure 6.** Random coincidence sinogram.

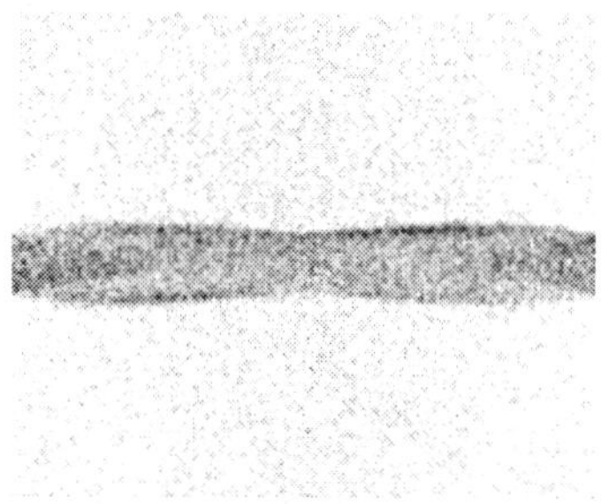

**Figure 7.** Corrected sinogram.

## 4. Single scatter correction algorithm

Single scatter correction algorithm for sinogram consists of the following steps [8]:

- Scattering point distribution in reconstructed volume;
- Single scatter quantity estimation for each sinogram line of response;
- Scattered sinogram construction;
- Subtraction of obtained sinogram from initial sinogram.

Single scatter quantity estimation for a given line of response is obtained as follows: $N = N_{ij}+N_{ji}$, where $i$ and $j$ are detectors indexes identifying line of response,

$$N_{ij} = P_{D_iM} \cdot \int_{D_i}^{M} \lambda(\tau)d\tau \cdot P_s \cdot P_{D_jM},$$

$$P_{D_iM} = exp\left(\int_{D_i}^{M} \mu_e(\tau)d\tau\right)$$

$$P_j = \frac{\mu_e(M)}{\sigma} \cdot \frac{d\sigma}{d\Omega} \cdot \Omega_j,$$

where $\lambda(r)$ is a uncorrected activity distribution, $D_i, D_j$ are detectors centers, M is one of scattering points, $\mu_e(r), \mu_s(r)$ are linear attenuation coefficients of attenuation and scattering, $\sigma$, $\frac{d\sigma}{d\Omega}$

are total and differential scattering cross sections, $\Omega_j$ is a solid angle for detector $j$ and with vertex in point M.

The considered algorithm is implemented by C# program, the results are given below.

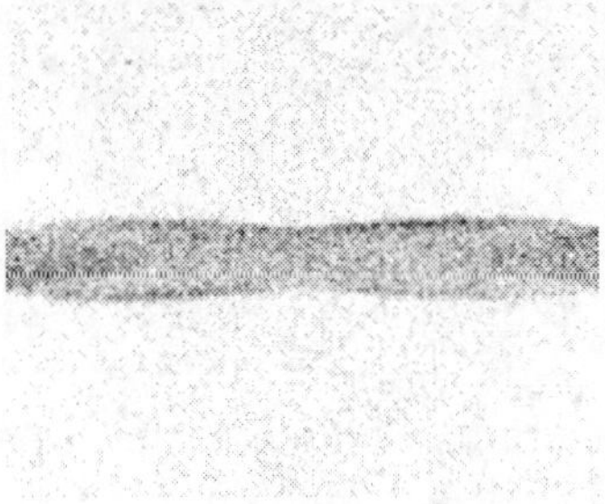

**Figure 8.** Initial sinogram.

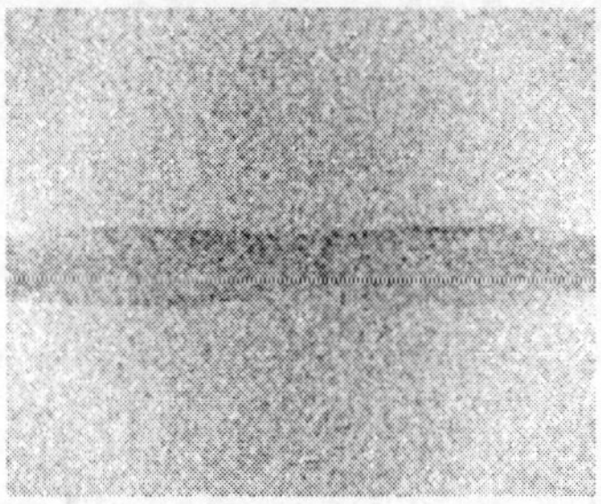

**Figure 9.** Scattered sinogram.

**Figure 10.** Corrected sinogram.

Hence the sinograms have less noise level with the use of this algorithm.

## 5. Conclusion

In the course of this work PET projection data acquisiton modeling algorithm which considers radiation attenuation and scattering is implemented. The obtained results may be used for PET scanners constructions and verification of corrrection algorithms. Also single scattering and random coincidences correction algorithms are implemented and verified by using model PET projection data.

## References

[1] Kotina E D, Ovsyannikov D A, Ploskikh V A, Latypov V N, Babin A V and Shirokolobov A Y 2014 Data Processing in Nuclear Medicine *20th International Workshop on Beam Dynamics and Optimization* vol 3 no 2 pp 55–61

[2] Kotina E D, Latypov V N and Ploskikh V A 2013 Universal system for tomographic reconstruction on GPUs *Problems of Atomic Science and Technology* 88(6) pp 175–178

Kotina E D, Babin A V, Bazhanov P V, Ovsyannikov D A, Ploskikh V A and Shirokolobov A Y 2016 Mathematical and Computer Methods of Data Processing in Nuclear Medicine Studies *Proc. of RuPAC2016* (Saint Petersburg) pp 480–482

[3] Badawi R D, Miller M P, Bailey D L and Marsden P K Randoms variance reduction in 3D PET 1999 *Phys. Med. Biol.* 44(4) pp 941–954

[4] Ovsyannikov D A, Kotina E D and Shirokolobov A Y Mathematical methods of motion correction in radionuclide studies 2013 *Problems of Atomic Science and Technology* 88(6) pp 137–140

[5] Ovsyannikov D A and Kotina E D Determination of velocity field by given density distribution of charged particles 2012 *Problems of Atomic Science and Technology* 79(3) pp 122–125

[6] Habib Zaidi and Marie-Louise Montandon 2007 Scatter compensation techniques in PET *PET Clin.* vol 2(2), pp 219–234

[7] Bazhanov P V 2015 PET projection data modeling and scatter correction *Proc. of SCP 2015* Saint-Petersburg pp 516–517

[8] Accorsi R, Adam L E, Werner M E and Karp J S 2004 Optimization of a fully 3D single scatter simulation algorithm for 3D PET *Phys. Med. Biol.* no 49 pp 2577–98

Watson C C 2000 New, faster, image-based scatter correction for 3D PET *IEEE Trans. Nucl. Sci.* no 47 pp 1587–1594

Ollinger J M 1996 Model-based scatter correction for fully 3D PET *Phys. Med. Biol.* no 41 pp 153–176

IOP Conf. Series: Journal of Physics: Conf. Series **941** (2017) 012098 doi:10.1088/1742-6596/941/1/012098

# Electron beam agrobionanotechnologies for agriculture and food industry enabled by electron accelerators

**Y S Pavlov[1], A A Revina[2], O V Souvorova[1], N L Voropaeva[3], D V Chekmar[1], E V Abkhalimov[1], M A Zavyalov[4,5] and V P Filippovich[4]**

[1] Laboratory of Radiation Technologies, A.N. Frumkin Institute of Physical Chemistry and Electrochemistry Russian Academy of Sciences (IPCE RAS), Moscow, 119071, Russian Federation
[2] Laboratory of Electronic and Photonic Processes in Polymer Nanomaterials, A.N. Frumkin Institute of Physical Chemistry and Electrochemistry Russian Academy of Sciences (IPCE RAS), Moscow, 119071, Russian Federation
[3] FSBSI All-Russian Rapeseed Research Institute, Federal Agency for Scientific Organizations (FSBSI ARRI of Rapeseed, FASO Russia), Lipetsk, 398037, Russia, Lipetsk, Russian Federation
[4] FSBSI Russian Research Institute of Canning Technology, Federal Agency for Scientific Organizations (FASO Russia), Vidnoe, 142703, Russian Federation
[5] FSUE Russian Electrotechnical Institute, SC Rosatom, Moscow, 111250, Russian Federation

E-mail: rad05@bk.ru

**Abstract**. Electron beam (EB) radiation technologies have been employed to increase efficiency of biologically active nanochips developed for agricultural plants seed pre-treatment with purpose of enhancing crop yield and productivity. Iron-containing nanoparticles (NPs), synthesized in reverse micelles following known radiation-chemical technique, have served as a multifunctional biologically active and phytosanitary substance of the chips. Porous chip carriers activation has been performed by EB ionization (doze 20kGy) of the active carbons (AC) prepared from agricultural waste and by-products: Jerusalem artichoke (*Helianthus tuberosus*) straw, rape (*Brassica napus L. ssp. oleifera Metzg)* straw, camelina *(Camelina sativa (L.) Crantz*) straw, wheat (*Triticum aestivum*) straw. Three methods, UV-VIS spectrophotometry, Electron Paramagnetic Resonance (EPR) spectroscopy, cyclic voltammetry (CV) have been used for process control and characterization of radiation-activated and NPs-modified ACs. The results show a notable effect of ACs activation by electron beam radiation, evidenced by FeNPs-adsorption capacity increase. Studies of the impact of Fe NPs-containing nanochip technology on enhancement of seeds germination rate and seedlings vigour suggest that reported electron beam radiation treatment techniques of the ACs from selected agricultural residues may be advantageous for industrial application.

## 1. Introduction

Development of enhanced agricultural bionanotechnologies is becoming vital to sustainable crop production [1, 2]. Focused integration of research approaches is unavoidable for successful introduction of new pre-sowing seed treatment procedures involving multifunctional and multicomponent (nano)systems, deriving, preferably, from renewable resources, including

IOP Conf. Series: Journal of Physics: Conf. Series **941** (2017) 012098 doi:10.1088/1742-6596/941/1/012098

agricultural residuals[3-5]. Biologically active nanochips (BANchips), when used to treat seeds of agricultural plants, allow application of desired ultra small amounts of active substances for the purpose of concurrent achievement of phytosanitary plant protection, nutrition or microelements feed-in, seed germination enhancement, and crop yield increase under adverse conditions.

The BANchip usually consists of porous carrier matter and nanoparticles of biologically active and phytosanitary substances, and nutrients. Multiple nano pores of the carrier accommodate and hold nanoparticles of the impregnated substances selected for further seeds treatment. The carrier is usually mixed with mineral, polymer, clay, etc in a variety of proportions. BANs are then applied to the seed covers.

Electron beam (EB) radiation technologies[6-10] can offer a variety of solutions capable of increasing BANchip efficiency. They provide unique tools and methods for fundamental research of both homogeneous and heterogeneous systems, precise modification techniques, wide range of possibilities for varying parameters of radiation-chemical production of nanostructured composites with predetermined properties.

**Figure 1.** IPCE RAS Radiation Center. UELV-10-10-C-70 EB-accelerator with energies up to 10 MeV.

**Figure 2.** UELV-10-10-C-70 Operation Control Panel.

In earlier papers [12-14] suggested methods for activation of carbonaceous sorbent from natural materials and pre-sowing seed treatment using developed BANchips with AgNPs content, and related crop yield increase issues were discussed [15-18]. Current studies are focused on MeNPs introduction to BANchips, metal precursors are being chosen from the microelements of vital importance for plant growth, preferably, not included within noble metals range. Iron-containing NPs, synthesized in "microreactors" of reverse micelles pools following known radiation-chemical technique [7, 19-21] could be offered as a multifunctional biologically active and phytosanitary substance for the chips [22-25]. Porous chip carrier's treatment is being presently investigated and tried using EB ionization (doze 20kGy) in order to further activate the carbons (ACs), derived from agricultural waste and byproducts.

## 2. Radiation-chemical metal NPs synthesis (RadChem)

Radiation-chemical reduction of metal ions and stable NP formation in reverse micelles systems (RMS) [7, 19-21,34] – such as $Fe^{2+}/H_2O/0.15M$ AOT/isooctane – can be induced and implemented by EB radiation straight after solubilization of the components and solution's saturation with inert gas. Metal ions reduction then occurs due interaction with short-lived reductant products of water radiolysis: solvated electrons $e^-_{solv}$, hydrogen H-atom and the radicals $R_{(i)}$. Radiolysis reaction can be viewed below.

$$H_2O \rightarrow e^-_{aq}\,(2.7),\ H^+_{aq},\ OH^-_{aq},\ H^{\bullet}\,(0.6),\ OH^{\bullet}(2.7),\ O,\ H_2,\ H_2O_2 \quad (1)$$

$\mathbf{e^-_{aq}}$ is the unique reductant ($E_0$= - 2.9 V)

$$e^-_{aq} + Fe^{2+} \rightarrow Fe^+ \quad (2)$$

$$e^-_{aq} + Fe^+ \rightarrow Fe^\circ \quad (3)$$

$$Fe^\circ + Fe^+ \rightarrow Fe_2^+ \quad (4)$$

During NPs synthesis risks of possible oxidation of $Fe^+$ and other transients by radicals $OH^•$ must be considered:

$$OH + Fe^+ \rightarrow FeOH^+ \quad (5)$$

Unwanted oxidation can be avoided by $OH^•$ tra nsformation into radical products with reducing properties by introduction of the acceptors, such as alcohol, or carbonates ions (reaction 6, 7):

$$OH + (CH_3)_2CHOH \rightarrow (CH_3)_2C^•OH + H_2O \quad (k=1.6x10^9 M^{-1}s^{-1}) \quad (6)$$

$$OH + HCOO^- \rightarrow CO_2^{•-} + H_2O \quad (k=3.1x10^9 M^{-1}s^{-1}) \quad (7)$$

Thus formed reducing agents contribute to reduction reactions responsible for consequent nanoparticles yield.

Transient metal species are unstable and consequently form various aggregates $Fe_n^{m+}$ (n>m).

Conditions of the RadChem process, primarily, the radiation dose (usually chosen in 10÷50 kGy range), lab temperature, component ratios and concentrations can influence resulting NPs structure, size, shape and functionality.

## 3. Experimental

### *3.1. Chemicals*

All chemical were in analytical grade, were purchased from Aldrich (USA) and used as received, unless otherwise stated. AOT (bis(2-ethylhexyl) sulfosuccinate sodium salt) purity specification was 98%. Solvents were obtained from Component Reactive (Russia) or Aldrich.

### *3.2. Active Carbons (ACs).*

ACs from carbonized straw have been produced by pyrolytic treatment [26-27] by Dr. Mukhin, JSC ENPO "Neorganika" (Elektrostal, Russia). Table 1 shows typical specifications set for received ACs [23-25]: AC origin, surface .Received samples have been manually powered and used as is, i.e. powdered active carbons (PAC). Residues (straw) of the following cultivated plants were selected for processing into ACs: Jerusalem artichoke (*Helianthus tuberosus*), rape (*Brassica napus L. ssp. oleifera Metzg*), camelina (*Camelina sativa (L.) Crantz*), wheat (*Triticum aestivum*).

**Table 1.** Typical specifications of experimenral ACs samples

| Sample Assigned No. | AC origin | $V_\Sigma$, $cm^3/g$ | Humidity, % | Δ, $g/dm^3$ | Adsorption of iodine, % |
|---|---|---|---|---|---|
| AC1 | Jerusalem artich. | 3,00 | 1,0 | 95,3 | 34 |
| AC2 | Camelina | 2,43 | 5,7 | 140 | 43 |
| AC3 | Wheat | 3,61 | 1,9 | 66,5 | 64 |
| AC4 | Rape | 4,14 | 2,4 | 135 | 39 |

### *3.3. Spectroscopic Measurements*

UV-VIS spectroscopy was performed with a Hitachi U-3310 spectrophotometer (Japan) using a 1-mm quartz cuvette, and 0.15M AOT/ isooctane solution as reference.

### *3.4. Electrochemical characterization*

ACs electrochemical properties were studied by alternating current polarography cyclic voltammetry (CV) methods using an Intelligent Potentiostatic Control – ProMF (Russia) with IPC2000 SW in a three-electrode set-up with disposable planar working electrodes, Pt auxiliary and silver chloride reference electrodes in aqueous 0.5 M sulfuric acid electrolyte. All of the measurements were carried out at room temperature (25°C).

### *3.5. EPR measurements*

EPR spectra were recorded using EPR spectrometer PS 100.X (Russia). AC samples were placed into glass vials made of "Luch" glass, allowing irradiation of samples and further registration of paramagnetic AC particles signals without impact of irradiated glass EPR signal. Typical parameters were: middle of range – 3310 (G), scan range – 400, 2000, 4000 (G), modulation amplitude – 300 (mG) – 500, attenuation – 0 (dB), Gain: value – 3 to 9, order – 3, Phase (0/90) – 90 (deg). Measurements were carried without carbons deaeration, at room temperature.

*a. FeNPs synthesis*

Aqueous 0.3M Mohr's salt ($(NH_4)_2Fe(SO_4)_2*6H_2O$) solution was injected into 0.15M AOT/isooctane RMS in volume as determined by chosen solubilization coefficient value ω=3. RMS was, saturated by He, sealed and EB-irradiated using UELV-10-10-C-70 accelerator (10 MeV). Five samples received different absorbed radiation doses in 10-25 kGy range. UV-VIS spectroscopy records indicate, that optimal absorbed doze was achieved at 20kGy, and its further doze increase did not contribute to RadChem FeNPs formation in RMS with listed set of initial parameters, component quality and concentrations (see figure 3). FeNPs obtained in RMS with 20kGy adsorded dose were used for ACs modification series. Results of FeNPs particle size and distribution characterization performed by DSL (Zetasizer, Malvern, UK) method and corresponding AFM (BerMad 2000,Nanotec Electronica, Spain) data [28] indicated m.p.d. of ~3.5 nm of FeNPs in RSM with ω=1-3, and ~4.8 nm for FeNPs, obtained in RMS with ω=4-5.

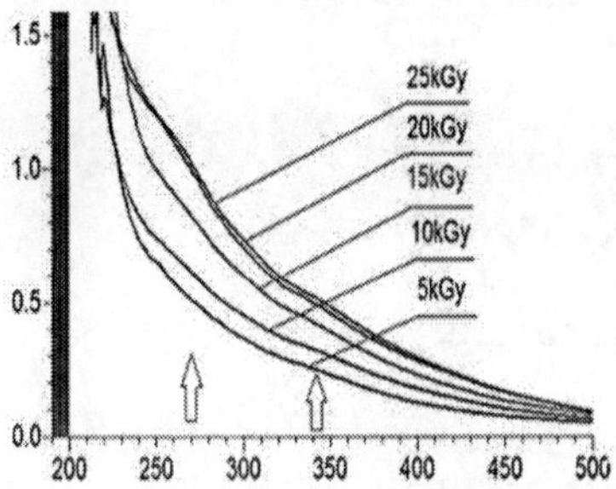


**Figure 3.** UV-VIS spectra of FeNP RMS (ω=3) samples with different adsorbed radiation dose. Dose-dependent optical absorbance increase within 5-25kGy range indicate EB-radiation feasibility level at 20 kGy.

*b. ACs activation*

Powdered AC1, AC2, AC3, AC4 were EB-irradiated using UELV-10-10-C-70 accelerator (10 MeV); absorbed radiation dose was 20 kGy for all samples: AC1-20kGy, AC2-20kGy, AC3-20kGy, AC4-20kGy. Second-step activation was performed via ACs surface modification with FeNPs. Original AC1-AC4 samples and radiated AC1-20kGy – AC4-20kGy samples were placed into in FeNPs RMS (ω=3, RadChem, D=20kGy), volume ratio of RMS and AC powder in all samples was ~ 1: 1. UV-VIS spectral control of NPs absorption process in RMS continued for 14 days.

## 4. Results and discussion

EPR spectroscopy is the standard tool for the detection of paramagnetic species and has been traditionally widely used to elucidate the nature of the carbons [29-32], it has proved existence of different types of ACs paramagnetic reaction centers, constituted by trapped electrons, and other ones - free radicals in the conjugated aromatic systems. EPR signals of these centers overlap, but have different values of the g-factor. AC radiation-induced activation appears to lead to structural changes, detected through the change of the narrow and broad signals correlation and g-factors shifts.

Obtained EPR spectra of original (xybtreated) AC1-AC4 samples have indicated the difference of the properties of detected centers dependent on the AC origin (plant). As it can be seen on figures 4 and 6, the properties of AC1 sample are characterized by three EPR signals varying in type and width: narrow signals noticed within the range of 5 ÷ 10 Oe, and two broad signals, first one ΔH is ~ up to 250 Oe, and second one has got ΔH ~ up to 600 Oe. Intensity and shape of AC4 EPR- recorded signals are notably different: the signal can be clearly seen be composed of two features: a narrow signal line, ΔH ~ 45 Oe, and second broad signal of low intensity.

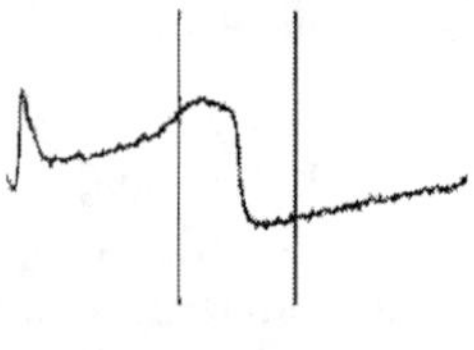

**Figure 4.** EPR spectrum of untreated AC1, derived by pyrolysis from Jerusalem artichoke straw

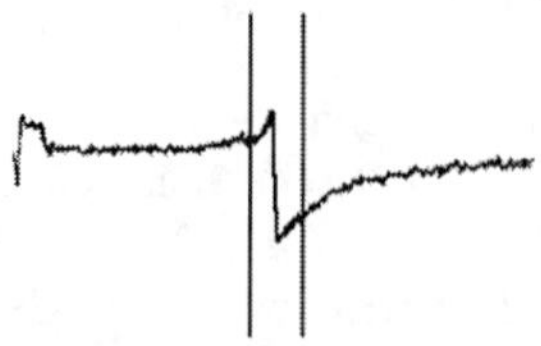

**Figure 5.** EPR spectrum of untreated AC4, derived by pyrolysis from Rape straw

After EB-radiation treatment only subtle EPR signals changes were detected for in all four radiated samples AC1-20kGy (see figure 6) – AC4-20kGy (see figure 8) . No EPR-sensitive effects of ACs irradiation, or center decomposition got revealed during reported experiment. Significant changes of paramagnetic properties of all samples got recorded after completion of the FeNPs adsorption by irradiated ACs.

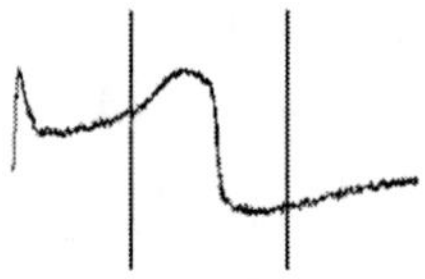

**Figure 6.** EPR spectrum of AC1-20kGy (radiated AC1, derived from Jerusalem artichoke straw).

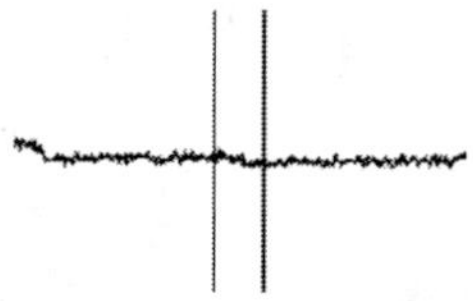

**Figure 7.** EPR spectrum of FeNPs-impregnated AC1-20kGy (derived from Jerusalem artichoke straw).

According to preliminary analysis, EPR spectra indicate that all EPR signals — narrow and broad ones – disappear from EPR spectrum of NP-modified and irradiated AC1 (see figure 7), while EPR spectrum of NP-modified and irradiated AC4 (see figure 9) clearly demonstrate intensity increase of the broad signal, going along with the weakening of the narrow one.

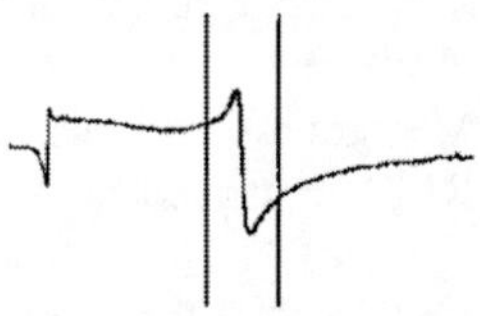

**Figure 8.** EPR spectrum of AC4-20kGy (radiated AC4, derived from Rape straw).

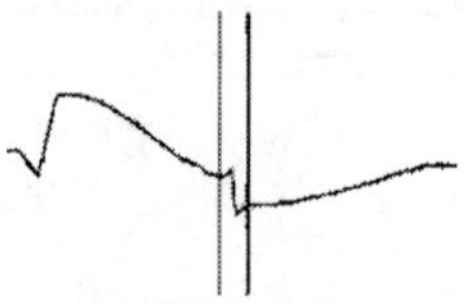

**Figure 9.** EPR spectrum of FeNPs-impregnated AC4-20kGy (derived from Rape straw).

Experimental EPR data suggest, that ACs paramagnetic centers must be involved in complex FeNP adsorption processes during ACs modification in RMS media.

Characterization achieved by CV methods consistently found that the nanomaterials based on preliminarily radiation-activated AC exhibited higher electro-catalytic activity (see figures 10-12).The shown voltammogram demonstrate that in case of FeNPs-impregnated AC1-20kGy (AC1, derived from Jerusalem artichoke straw, and FeNP-modified after being irradiated), the currents in the cathode area -750 mV at potential increase from -18 to -30 μA, while in the anode at a potential +750 mV they double in value, when compared to CV of NP-modified sample, not treated with radiation.

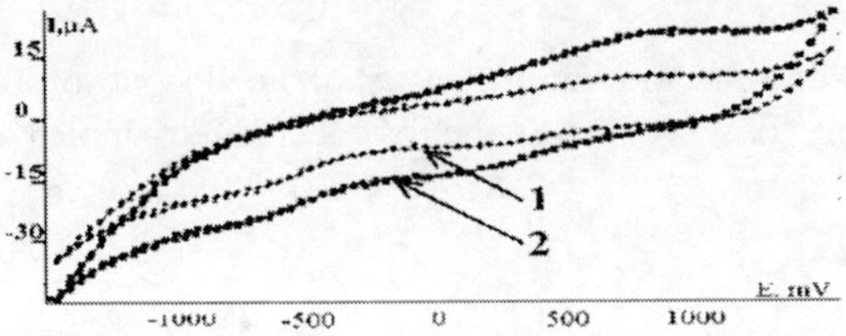


**Figure 10.** CVs of FeNPs-impregnated AC1, derived from Jerusalem artichok straw, without radiation pre-tratment (1), and FeNP-modified AC1 after radiation pre-treatment (2).

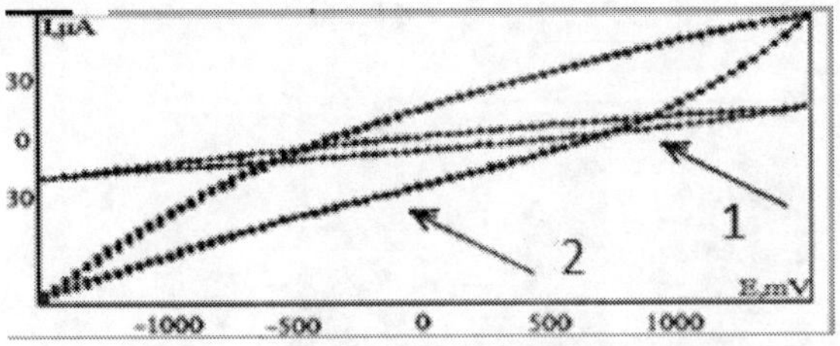


**Figure 11.** CVs of AC1-20kGy (irradiated AC1, derived from Jerusalem artichok straw), before FeNP-modification (1) , and after FeNP-modification (2).

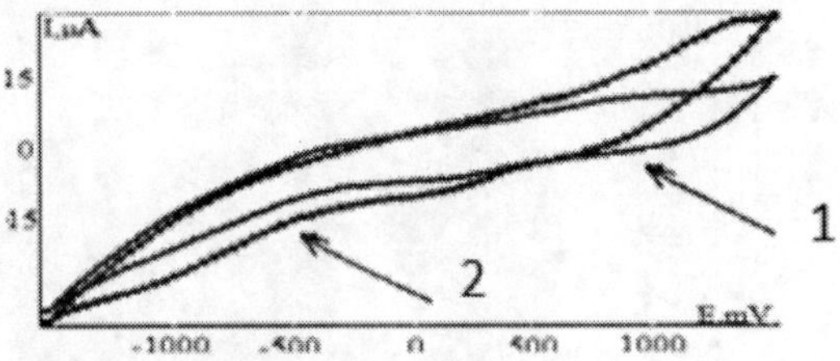


**Figure 12.** CVs of AC4-20kGy (irradiated AC4 derived from Rape straw), before FeNP-modification (1), after FeNP-modification (2).

For FeNP-modification of ACs same FeNPs RMS (ω=3, RadChem, D=20kGy) was used. UV-VIS optical absorption spectra of this FeNPs RMS demonstrate characteristic[] peak with $\lambda_{max}$~215 nm, and weakly expressed bands at 264-270 nm, and at 340 nm.

UV-VIS spectra (see figures 13, 14) recorded during modification process of AC1-20kGy illustrate multiple-stage complicated FeNPs adsorption process characteristic for all studied AC samples.

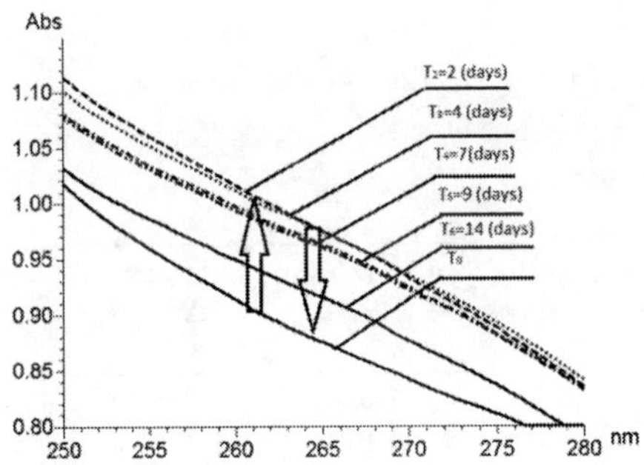


**Figure 13.** UV-VIS spectra of RMS recorded during 14-day FeNP-modification of AC4 (derived from Rape straw, untreated)

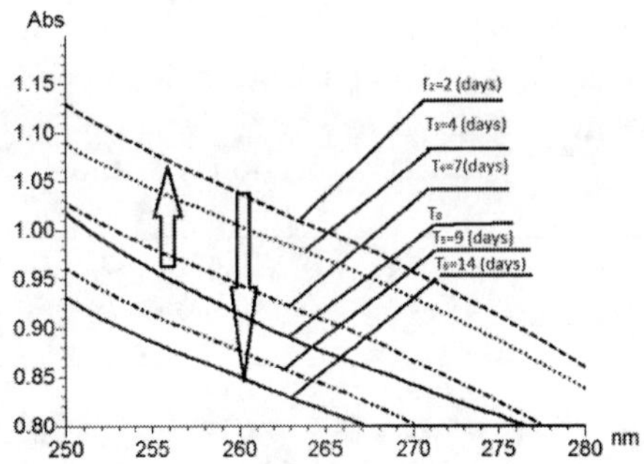


**Figure 14.** UV-VIS spectra of RMS recorded during 14-day FeNP-modification of AC4-20kGy (irradiated AC4 derived from Rape straw),

Spectrophotometric monitoring of the optical density (D) of FeNP RM modifying solutions helped providing persuasive records of initial D increase at the first stage of AC NP-modification, as well as the difference in kinetics of the absorption processes of NPs, which are accounted to ACs pre-treatment and carbon's origin.

As it has been supposed [33], this phenomena can be explained by AC-induction of further NPs formation in the RMS, being the result of the reaction between remaining metal ions, - still present in the RMC along with NPs, - and active centers found on the surface of the adsorbents. In due time, each RMSs with FeNPs, used for ACs modification, reached "FeNPs -saturation" state, followed by custom D decline. It agrees with NPs concentration decrease in liquid media due to NP adsorption within AC pores and on its surface.

A histogram, providing consolidated input on the time rate and intensity of FeNPs absorption, correlating to RMs optical density change at λ=264nm observed with time for all studied samples, can be seen on figure 15.

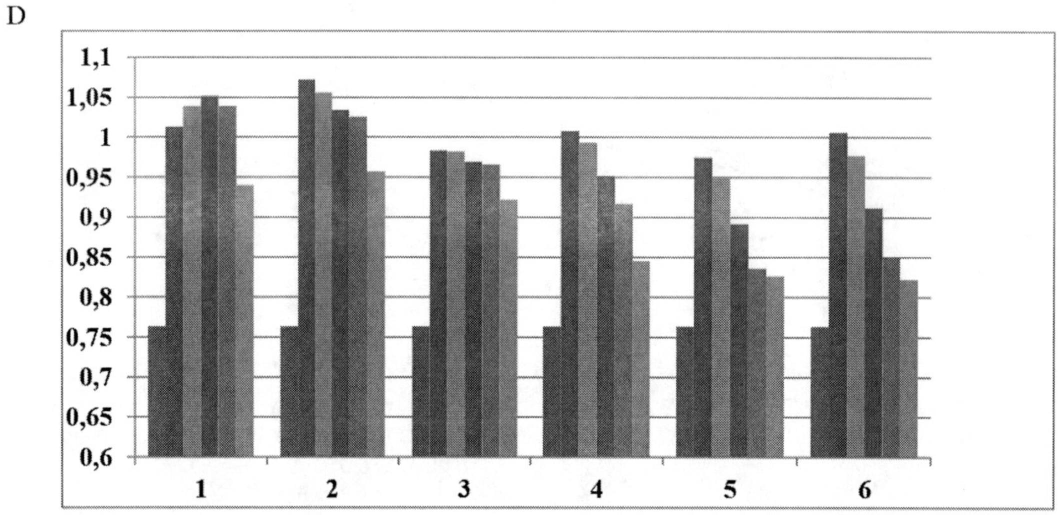


**Figure 15.** Optical density change recorded at λ=264nm for modifying FeNP RMSs with following samples: AC1– 1, AC2 – 2, AC4 – 3, AC1-20kGy – 4, AC2-20kGy – 5, AC4-20kGy – 6. Spectra Registration Time, days after adsorption start, legend:

■ Initial ■ 2 days ■ 4 days ■ 7 days ■ 9 days ■ 14 days

UV-VIS optical absorption spectra provide evidence that EB-radiation treatment of ACs make notable impact on absorbance rate and FeNP-capacity of the tested ACs ( see Table 2).

**Table 2.** Typical specifications of studied ACs samples

| Sample | $D_{\lambda264}$, a.u. ΔT=0 | $D_{\lambda264}$, a.u. ΔT=2, days | $D_{\lambda264}$, a.u. ΔT=4, days | $D_{\lambda264}$, a.u. ΔT=7, days ays | $D_{\lambda264}$, a.u. ΔT=9, days | $D_{\lambda264}$, a.u. ΔT=14, days | ΔD, % |
|---|---|---|---|---|---|---|---|
| AC1 | 0.763 | 1.013 | 1.039 | 1.052 | 1.039 | 0.940 | 11 |
| AC2 | 0.763 | 1.072 | 1.056 | 1.034 | 1.026 | 0.957 | 12 |
| AC4 | 0.763 | 0.983 | 0.982 | 0.969 | 0.966 | 0.922 | 7 |
| AC1-20kGy 4 | 0.763 | 1.008 | 0.993 | 0.952 | 0.917 | 0.845 | 19 |
| AC2-20kGy 4 | 0.763 | 0.975 | 0.951 | 0.892 | 0.836 | 0.826 | 18 |
| AC4-20kGy 4 | 0.763 | 1.006 | 0.977 | 0.912 | 0.851 | 0.822 | 22 |

FeNP-modification of AC4-20kGy (irradiated AC4 derived from Rape straw) showed this AC sample to be most active adsorbent, presumably due to inherited plant properties and EB-radiation activation, and in comparison with the untreated original AC (AC1).

FeNP-impregnated AC4-20kGy was selected for incorporation into BANchip, and biological testing of the applicability of FeNP-modified AC as the carrier for BA substance within the chips. The

IOP Conf. Series: Journal of Physics: Conf. Series **941** (2017) 012098 doi:10.1088/1742-6596/941/1/012098

crop germination tests were run at FSBSI ARRI of Rapeseed, FASO Russia, Lipetsk, Russia, within local bionanotectnological R&D program "Nanochip for the Seeds".

Bittercress (*Barbarea vulgaris*) seeds were treated with BANchips, containing varying amounts of FeNPs-impregnated AC4-20kGy carbon carrier, and tested vs untreared seeds in line with GOST 12038 – 84 test procedures.

Recorded test data [23-25] included seedling dimensions, fresh and dry weights, moisture content, relative humidity, germination total and rate values, etc. The laboratory fourfold 10-day germination tests were carried out on the filter tissues in Petri dishes at 25 ° C .

Determined germination energy and seed germination exhibited process activation due to BANchip-treatment of the seeds:

- the number of "germinated" seeds on the first day of germination, depending on the variance in their chip's composition, was 19.4-34.9% (compared to 10.1% in the control.), raising rapidly to 80.9 - 93.3% next (second) day;
- germination vigour (germinated seeds number/ total seeds ratio 3 days after seeding (DAS3), and laboratory sprouting of a seedling from a seed DAS5 have reached 100% in a number tests;
- seedlings length, their fresh and dried mass weight gain were also dependent on actual BFNchip content.

Pre-sowing treatment of plant seeds with "FeNP-impregnated AC4-20kGy" - experimental ACsapplied via BANchip technology - was recognized to be efficient practice.

## 5. Conclusion

The aborption capacity of a studied range of carbonized materials from natural straw increase in test cases, when EB pre-treatment of the sorbents was done.. According to UV-VIS spectra records, ACs EB-ionizing radiation with absorbed dose of 10-20 kGy can double the rate of FeNPs absorption from micellar solution at initial "slow" modification stage, which is complicated and contradicted in modifying micellar media, as it has been observed, by an alternative NP-formation process.

CV express registration of electrocatalytic and magnetic properties of nanomaterials, obtained as a result of absorption of FeNPs on the surface of EB radiation-treated ACs, show changes of their catalytic and magnetic activity, and suggest continuing EPR studies.

Reported experiments provide evidence, that radiation-activated and FeNPs-modified ACs from agricultural byproducts and wastes have good potential of becoming the materials of choice for further development of NPs-containing nanochip technology.

They confirm, that not only Ag and other noble metal NPs, but FeNPs, and other metal microelements should be considered for phytosanitary and safe plant seeds BAN treatment done for crop yield and productivity increase, especially under adverse conditions, which include plant diseases, insect attacks, soil and climate extremes.

Employed EB-radiation scientific tools and techniques – radiation-induced adsorbent activation, radiation-chemical metal NPs synthesis in liquid phase, and further adsorbent modifications, – are becoming an integral part of modern agrobionanotechnologies , and should eventually considerably contribute to their development, suggesting ways of effectively using materials found among renewable resources and agricultural residues. Obtained results clarify direction for further interdisciplinary studies and the path for optimization of EB accelerator-aided methods, required to be in place to meet varying needs imposed by a wide range of tasks in agriculture and food industry worldwide.

**Acknowledgement.**
We acknowledge assistance and would like to express appreciation to Dr V.M. Mukhin , RF government prize winner in the field of science, Honoured Inventor of Russia, International Academy of Ecology member, R&D head of JSC ENPO "Neorganika" (Moscow region, Elektrostal).

## References


[1] Fedorenko V F, Buklagin D C, Golubev I G and Nemenutshaya L A 2015 *Rossiiskie Nanotekhnologii* Overview of the Russian nanopreparation for processing of agricultural crops **10** № 3-4 126-131

[2] Kovalenko L V *et al.* 1998 *Proceedings of the Conference "Physicochemistry of ultrafine systems"* Ultrafine powders in the solution of the food independence of Russia Obninsk, 211

[3] Revina A A *et al.* 1998 *Patent № 2104927 (1998)* The method for producing a modified activated carbon

[4] Revina A A 2009 *Protection of Metals and Physical Chemistry of Surfaces* Adsorbtion and oxidation processes in modern nanotechnology **45** № 1 58–63

[5] Ruban I N, Voropaeva N L, Figovsky O L, Sharipov M D and Dadajanov T K 2012 *U.S. Patent № 8209902(2012)* Biologically active multifunctional nanochips and method of application for production of high-quality seed

[6] Pikaev A K, Glazunov P Y and Pavlov Yu S 1993 Radiation Center of Institute-of-Physical Chemistry in Moscow *Radiat. Phys. Chem.* **42** 887-890

[7] Revina A A 2003 *Patent RF № 2212268 (2003)* System of modification of objects with nanoparticles Bull. № 26

[8] Revina A A, Magomedbekov E P 2010 *V All-Russian conference "Physicochemical processes in condensed media and intermediate boundaries FAGRAN-2010"* Role of radiation chemistry in creation of functional nanomaterials based on metallic and bimetallic nanoparticles Voronezh, Russia 145

[9] Pavlov Yu S, Dobrohotov B B, Pavlov B A, Nepomnyaschy O N and Danilichev V A 2014 Magnetic buncher accelerator UELV-10-10-T-1 for studing fluorescence and radiation physical researches *Proceedings of XXIV Russian Particle Accelerators Conference (RuPAC'2014)* Obninsk, Russia 259-261

[10] Pavlov Yu S, *et al.* 2015 *VI Russian conference (with participation of CIS specialists) "Actual problems of high energy chemistry"* Industrial electronic-beam technologies realized by accelerator UELV-10-20-S-70-2 Moscow 268-278

[11] Pavlov Yu S, Lagov P B 2015 *15th European Conference on Radiation and Its Effects on Components and Systems (RADECS)* Magnetic Buncher Accelerator for Radiation Hardness Research and Pulse Detector Characterization 336-338

[12] Revina A A 2012 *NRNU MEPhI Scientific Session* Nanocomposites based on carbon nanotubes and metal nanoparticles **1** 172

[13] Revina A A, Lebedeva M V, Shvetsov A A and Rakov E G 2013 *Nanotechnologies: Science and Production* Electrocatalytic properties of carbon materials with iron nanoparticles based on chemical and radiation chemical reduction reactions № **1** 62–69

[14] Revina A A, *et al.* 2015 *Abstracts. I Int. scientific-practical Conf. "Graphene and related structures: synthesis, production and application"* Modified (nano)particles of silver and iron adsorbents based on activated carbons Tambov, Russia 165-167

[15] Voropaeva N L, *et al. 2015 Ecological Bulletin of the North Caucasus* Modified natural raw materials and products recycling of waste consisting of environmentally safe (nano)chips for APK **11** № 1 31-34

[16] Voropaeva N L, Mukhin V M, Revina A A, Busev S A and Karpachev V V 2015 *GeoScience Engineering* Recultivation of plant wastes (straw) and obtaining (nano) chips with bactericidal properties based on them **LXI** № 3 1-7

[17] Chekmar D V, Tanasyuk D A, Voropaeva N L and Revina A A 2015 *Materials of II All-Union. Conf. with Int. participation and Symposium "Topical problems of theory of adsorption, porosity and adsorption selectivity"* Effect of modification of carbon materials on the adsorption of NP silver Moscow-Klyazma, 76

[18] Revina A A, Voropaeva N L,Mukhin V M and Chekmar D V 2016 *Inzynieria Mineralna* The effect of iron nanoparticles on physico-chemical properties and functional activity of active

IOP Conf. Series: Journal of Physics: Conf. Series **941** (2017) 012098 doi:10.1088/1742-6596/941/1/012098

carbon from plant materials **17**(2) 247-253

[19] Pileni M P 1993 *J. Phys. Chem.* Reverse micelles as microreactors **97 (27)** 6961–6973

[20] Ershov B G 1997 *Uspekhi Khimii* Metal Ions in Unusual and unstable oxidation states in aqueous solutions: formation and properties № 2 103

[21] Revina A A 2008 *Patent RF № 2322327 (2008)* The preparation of nanostructured particles of metals and method for its production Bull. № **11**

[22] Pavlov Yu S, Souvorova O V, Kuznetsov M A and Revina A A 2016 *Nanotechnologies: Science and Production* Radiation-chemical synthesis of metal nanoparticles in reverse micellar solutions № 1 3-19

[23] Voropaeva N L, Revina A A, Mukhin V M and Chekmar D V 2016 *On the book "20th Conference on Environment and Mineral Processing"* The effect of iron nanoparticles on physico-chemical properties and functional activity of active carbon from plant materials. Ostrava, Czech Rep., 253-258

[24] Voropaeva N L, Revina A A and Karpachev V V 2016 *"Innovative Science: Past, Present and Future" a collection of articles of the International Scientific and Practical Conference* New functional (nano)materials from primary agricultural waste Ufa, Russia 2 277

[25] Shcherbatova E V, *et.al.* 2016 *"Nanomaterials and living systems" a collection of articles of the IV International Scientific and Practical Conference* Iron nanoparticles for APK combined with new functional nanocomposites based on renewable raw waste Moscow 46

[26] Mukhin V M, *et al.* 2014 *RF patent № 2527221 (2014)* Method of obtaining active carbon from plant waste

[27] Mukhin V M, *et al.* 2015 *Patent RF № 2562984 (2015)* A method of producing activated carbon from plant material – straw cruciferous oilseeds

[28] Revina A A, Bolshakova A N, Daineko S V, Yashtulov N A, Tomova Z M, Nurtdinova K F and Souvorova O V 2011 *"Science Intensive Technologies"* The comparison of optical properties and shapes for silver and iron nanosized particles synthesized by chemical reduction of metal ions in reverse micelles **12** № 6 68-73

[29] Losev B I, Bulyna E A 1959 *Reports of USSR Academy of Sciences* Paramagnetic resonance in fossil carbons **125** № 4 814-816

[30] Retcofaky H L, Stark I M and Fiedel R A 1967. Chemistry and Industry EPR g- values of coals **31** № 5 1327-1328

[31] Adashkevich S V, Lapchuk N M, Lapchuk T M, Oleschkevich A N, Steljmakh V F, Fan Suan Toan and Schilardi G O 2012 *Proceedings of the 5th Int. Scientific Conference "Materials and structures of modern electronics"* About correlation of technological and radiospectroscopic parameters of active carbons Belarus, Minsk 10-12

[32] Rusyanova N D 2003 *Nauka* Chemistry of carbons Moscow 316

[33] Revina A A, Belyakova L D, Parkaeva S A, Souvorova O V, Sergeev M O and Zolotarevski V I 2010 *Physical chemistry of surfaces and protection of materials* The study of the adsorption of palladium nanoparticles on silica by methods of spectrophotometry and atomic force microscopy **46** № 6 448-454

[34] Revina A A, Dokuchaev A G, Khaylova E B, Tedoradze M G Optical and electrical characteristics of polymer films modified with nanoaggregates *High Energy Chemistry* 2000 **43** № 1 11

# Quarter-Wave buncher for NICA project

**M Trushin[1,2], R Fatkullin[1,2], A Sitnikov[1,2], D Seleznev[2], V A Koshelev[2], A S Plastun[2], S V Barabin[2], A V Kozlov[2], V G Kuzmichev[2], G N Kropachev[2] and T Kulevoy[1,2]**

[1] National Research Nuclear University MEPhI, Moscow, Russia
[2] NRC "Kurchatov Institute" FSBI "SSC RF ITEP", Moscow, Russia

E-mail: rdfatkullin@gmail.com

**Abstract**. This paper represents the results of modeling the electrodynamic characteristics (EDC) for a quarter-wave coaxial beam buncher, simulation of thermal loads of the buncher, modeling of the mechanical changes in the geometric parameters caused by the thermal load of the buncher and modeling of the new EDC depended on this changes.

## 1. Upgrading of the injector of light ions in the NICA project

The replacement of the electrostatic pre-injector of the linear accelerator LU-20 (NICA [1]) with the RFQ accelerator was carried out in 2016 as part of the polarized light ion beams project [2].

At the moment, a new pre-injector RFQ accelerator is put into operation [3]. In order to provide the phase beam distribution prior to injection in LU-20, it is proposed to use a buncher located in the transport channel between the RFQ accelerator and LU-20 (inside the LU-20's vacuum tank). For these purposes the two-gap quarter-wave coaxial buncher (figure 1) was developed by the ITEP's engineers. The characteristics of the developed buncher are shown in Table 1.

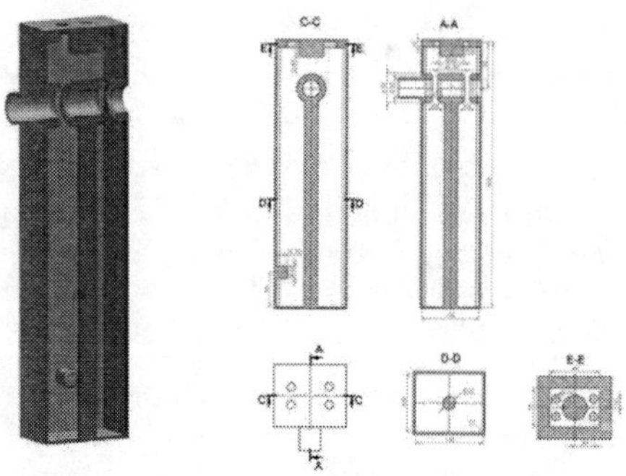

**Figure 1.** Quarter-wave coaxial buncher.

The buncher is a rectangular copper coaxial quarter-wave resonator with one drift tube, two drift half-tubes and two plungers: for manual tuning and for so-called automatic frequency control (AFC) (figure 2). The LU-20 accelerator operates in an auto-generator mode, so it is necessary to adjust the resonant frequency of the buncher. Resonant frequencies of the buncher and LU-20 accelerator must coincide for effective operation.When the resonant frequency of LU-20 is changed (for example, due to heating of the accelerator's structure), LLRF (low level RF) device receives the signal from the measuring loop and changes the frequency of the buncher by rotation of the AFC plungers.

**Table 1**. The characteristics of the buncher.

| Name | Value |
|---|---|
| Resonant frequency $f_0$, MHz | 145.2 |
| Electric field (for $C^{4+}$ ions), kV/ cm | 75 |
| Overall dimensions of the buncher, mm | 115x125x502 |

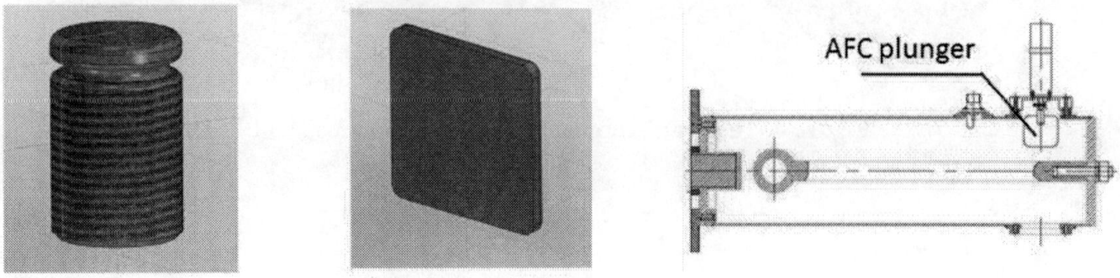


**Figure 2.** Frequency change plungers.

## 2. Simulation

Simulation of the EDC of the quarter-wave buncher was performed in CST (Student Edition). The model used in the simulation is shown in figure 3. Overall dimensions of the model is 115x125x502 mm. The material or resonator walls used for calculation is copper.

**Figure 3.** 3D model of the buncher without plungers.

Simulation of EDC includes:

- simulation of the buncher's fields distribution without plungers, at 1 J stored energy in the resonator (figure 4);

- investigation of the resonant frequency dependence based on the location of both plungers (figure 5);

Simulating of thermal losses and mechanical changes in geometric parameters of a resonator caused by a thermal load consists of:

- investigation of the temperature distribution in the buncher's structure as a function of the time operated (figure 6);
- simulation of deformations in the structure in a thermal stationary mode;
- simulation of the deviation of the resonant frequency of the deformed buncher from the frequency of the undeformed buncher.

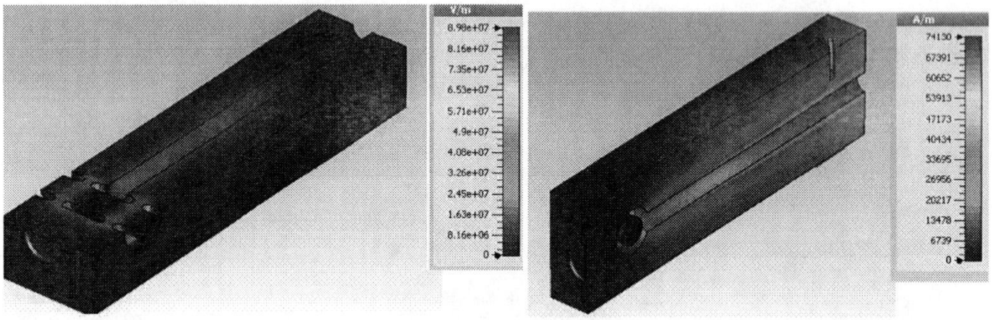


**Figure 4.** The distribution of electric (left) and magnetic (right) fields.

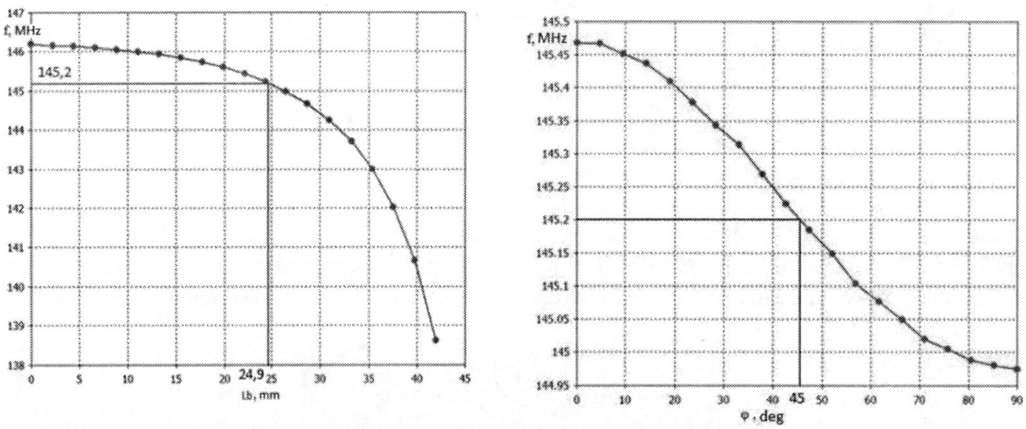


**Figure 5.** The dependence of buncher's resonant frequency on the manual (left) and AFC (right) plungers.

Thermal calculation is based on a numerical solution of the heat equation with given boundary and time conditions. The simulation was carried out with the following parameters (see Table 2).

**Table 2**. Parameters for thermal modelling.

| Name | Value |
|---|---|
| Initial temperature of the environment $t_0$, K | 293 |
| Thermal conductivity of the environment $\lambda_{en}$, $\frac{W}{m\cdot K}$ (Vacuum) | 0 |
| Heat capacity of the environment $C_{en,}$ $\frac{J}{K}$ (Vacuum) | 0 |
| Pulse duration $p$, µs | 100 |

| | |
|---|---|
| Pulse period T, s | 4 |
| Duty ratio S | 40000 |
| Peak power value W, kW | 31 |

Modeling of the mechanical changes of the resonator geometric parameters was carried out based on the calculated temperature distribution in the structure. Since we are interested in the final deviation of resonant frequency, the deformation modeling was carried out only for the stationary state of the system, when the temperature of its parts (and hence the deformation) is maximal. It was found that the maximum value of deformation caused by temperature increase is $\Delta X = 0.0194\ mm$. This value is achieved in the region located on the drift tube.

Also, the simulation of the deformed structure resonant frequency deviation was carried. The difference between the resonance frequencies is $\Delta f_{def} = 7.78 \times 10^{-4}\ MHz$. This deviation will not have a noticeable effect on the work of the buncher. Additional frequency compensation is not required.

## 3. Experimental test bench.

The buncher was manufactured in October 2016. Currently buncher passes the final stage of tuning and testing in the experimental test bench in ITEP. The buncher's vacuum chamber is tested at pressures close to the operating pressure of the linear accelerator LU-20. Figure 6 shows a schematic of an experimental vacuum installation. CT is a heated vacuum chamber; VA - vacuum shutter with electric drive; NR - high-vacuum pump TMN-1000; N - spiral pump PDV500; PT, PA – ionization vacuum meter VIT-2 with PMT-2 and PMI-2 lamps; G leak detector MSE-2000R; PM - magnetic vacuum meter PKR-251; VF – spill. Heating is carried out by heating tape ENTL-1, which is wrapped around a vacuum tank.

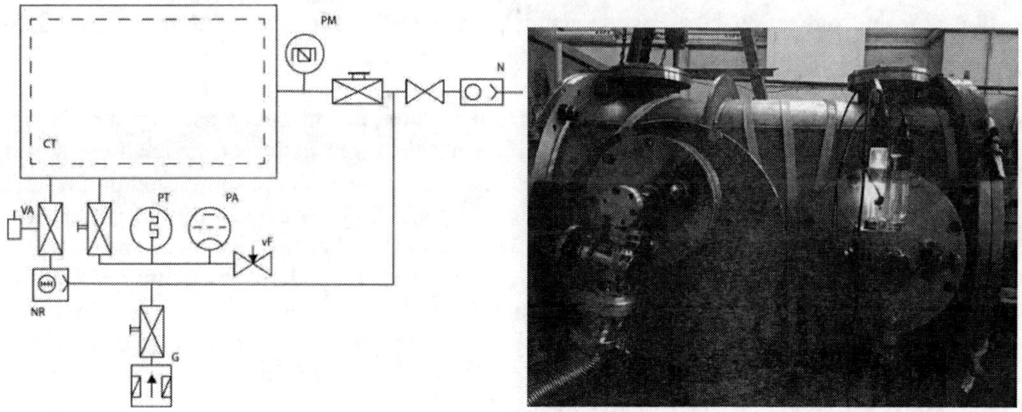


**Figure 6.** Vacuum test bench.

After passing of the necessary tests, the buncher will be transported to Dubna and installed in the "Injector Hall" of the "Nuclotron" Accelerator Complex.

**References**

[1] Trubnikov D, Agapov N, Brovko O et al 2013 NICA project at JINR *Proc. of IPAC2013* (Shanghai) p 1343

[2] Aleksandrov V, Butenko A, Golovenskiy B Commissioning of new proton and light ion injector for nuclotron-NICA et al 2016 *Proc. of IPAC2016* (Busan) p 941

[3] Butenko A, Donets E, Donets E et al 2013 Development of the NICA injection facility *Proc. of IPAC2013* (Shanghai) p 3915

# Progress towards 3-cell superconducting traveling wave cavity cryogenic test

**R Kostin[1,3†], P Avrakhov[1], A Kanareykin[1,3], V Yakovlev[2] and N Solyak[2]**

[1]Euclid Techlabs, Gaithersburg, MD, USA
[2]Fermilab, Batavia, IL, USA
[3]St. Petersburg Electrotechnical University LETI, St.Petersburg, Russia

e-mail: r.kostin@euclidtechlabs.com

**Abstract.** This paper describes a superconducting L-band *travelling wave* cavity for electron linacs as an alternative to the 9-cell superconducting standing wave Tesla type cavity. A superconducting travelling wave cavity may provide 20-40% higher accelerating gradient by comparison with conventional cavities. This feature arises from an opportunity to use a smaller phase advance per cell which increases the transit time factor and affords the opportunity to use longer cavities because of its significantly smaller sensitivity to manufacturing errors. Two prototype superconducting travelling wave cavities were designed and manufactured for a high gradient travelling wave demonstration at cryogenic temperature. This paper presents the main milestones achieved towards this test.

## 1. Introduction

A superconducting traveling wave (SC TW) cavity design was proposed [1] as a higher performance substitute for the conventional Tesla-style standing wave (SW) cavity [2]. The reason is that SC TW cavity can eventually provide a 46% higher accelerating gradient, which is critical for multi-kilometer high energy accelerators like the ILC [3]. The SC TW cavity utilizes a power feedback waveguide to recirculate the RF power at the end of the acceleration section, i.e. the cavity uses a principle similar to that of the traveling wave ring resonator (TWR). The acceleration section of the cavity was optimized [4] such that it could provide a 24% higher accelerating gradient with the same field on the surface as a conventional SW cavity. The smaller phase advance per cell (105 degrees) provides a higher transit time factor. Also, the TW cavity is significantly less sensitive to manufacturing errors[4], i.e. it could be made much longer then SW cavity, which is length-restricted to 1 meter because of field flatness degradation. The SC TW cavity is restricted only by cryomodule length and technological issues: manufacturing and cleaning of long cavities. Thus, for the structures studied here, the real estate gradient may be increased by 22% by eliminating beam pipes that contain no RF field in a cryomodule as long as only one cavity may be employed [4].

Development of this new type of cavity was started from a single cell prototype as usual. The cavity was built, cleaned and tested at the Fermilab Vertical Test Stand (VTS). This cavity was designed to check manufacturing and cleaning challenges because of its more complicated design using a superconducting feedback waveguide. The single cell prototype was tested in the SW regime. An accelerating gradient of 26 MV/m was obtained, comparable to a Tesla single cell cavity with the same surface treatment (BCP only) [4]. Thus, the presence of the feedback waveguide did not limit cavity performance and the next step could be taken towards development of a full sized SC TW cavity.

The next step is demonstration of a traveling wave in superconducting cavity. Full sized SC TW cavity testing is practically impossible because it requires a specially designed cryomodule, high power couplers and other hardware compatible with the cavity. A small prototype was proposed – a 3-cell SC TW cavity which could be tested in Fermilab VTS with the existing 500 W power supply. The three-cell prototype makes the test possible, because it is small and fits the VTS and can be excited by simple feedthroughs with small power supply. However, this makes the cavity sensitive to external loads because of its narrow bandwidth (~20 Hz). The cavity was optimized to withstand external pressure variations – stiffening ribs and rings were added to the mechanical design. Nevertheless, the cavity was still sensitive to the effects of the Lorentz force and a special tuner [5] was designed and built to compensate it. This paper presents some of the final stages towards cryogenic test of the three-cell prototype: preliminary RF measurements and final manufacturing phase.

## 2. Room temperature experimental results

### *2.1. RF Quality Control (RFQC)*

Two niobium cavities were manufactured. Routine cavity measurements which are usually referred to RF quality control (RFQC) were done, such as quality factor Q, resonant frequency f, and field distribution on the axis $E_z(z)$. Following that, one of the cavities was used for tuning studies at room temperature. The other cavity will be tested in liquid helium at Fermilab. This cavity requires additional waveguide reinforcements because of the small bandwidth during the test. The full sized traveling wave cavity was designed to use two TTF-III couplers with loaded $Q=10^6$ and does not need such reinforcement. Bead-pull measurements were made for both three-cell prototypes and are presented in Figure 1. Also shown on this figure is the field distribution along the cavity calculated by CST MWS [6] for a cavity model at room temperature before any chemical treatment (consisting of 200 μm surface etching by BCP) is applied. The integral Nb RRR300 expansion coefficient from 2K to 293K is $\alpha = 1.00143$. As one can see from Figure 2, cavity #2 has a field distribution similar to the MWS results.

Cavity spectra and Q-factors were also measured and compared to simulations (Table 1). Along with the cell modes (mode 3, 4, 5 in Table 1, bolded) the cavities also have waveguide modes (mode 1, 2, 6 in Table 1). Cell modes are the modes with EM fields concentrated in acceleration part of the resonator, WG modes are concentrated in the waveguide. Surface conductivity for simulations was chosen equal to $\sigma = 7\cdot10^6[S/m]$. The dielectric permittivity equals $\varepsilon = 1.0005[F/m]$ and corresponds to room air. The measured Q-factors differ from the simulation by less than 8%, and the resonant frequency difference of cell modes is within 5 and 7 MHz for cavities 1 and 2 respectively. Additional cavity tuning is required to get the frequencies and field distributions closer to the simulated ones. Nevertheless, the cavities were used for room temperature tuning studies and the TW mode was successfully adjusted.

**Table 1.** Cavity frequencies and Q-factors.

| | Cavity 1 | | | Cavity 2 | | | HFSS simulation | |
|---|---|---|---|---|---|---|---|---|
| Mode No. | f, MHz | Δf, MHz | $Q_0$ | f, MHz | Δf, MHz | $Q_0$ | f, MHz | $Q_0$ |
| 1 | 1044.499 | 7.40 | 2634 | 1047.47 | 10.37 | 2595 | 1037.10 | 2757 |
| 2 | 1173.649 | 8.33 | 2994 | 1173.136 | 7.82 | 2932 | 1165.32 | 3100 |
| **3** | **1277.913** | **3.63** | **5983** | **1279.697** | **5.42** | **6060** | **1274.28** | **6098** |
| **4** | **1303.469** | **2.31** | **6368** | **1304.226** | **3.07** | **6410** | **1301.16** | **6498** |
| **5** | **1305.772** | **4.53** | **6123** | **1308.04** | **6.80** | **6320** | **1301.24** | **6481** |
| 6 | 1372.615 | 8.39 | 3682 | 1378.148 | 13.93 | 3600 | 1364.22 | 3900 |

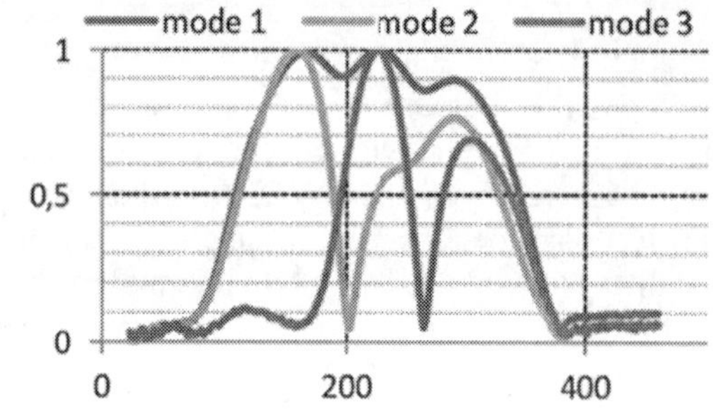


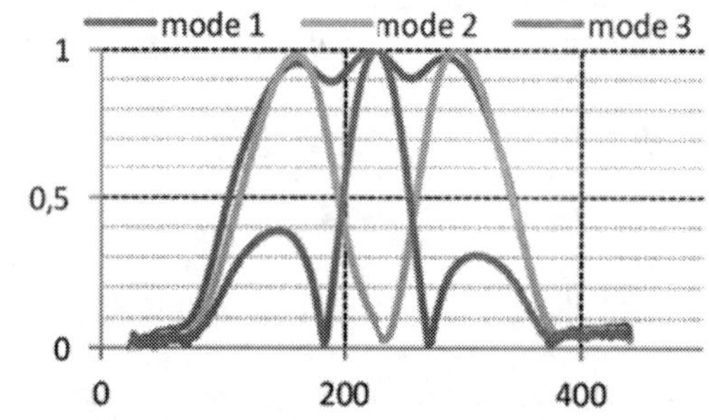


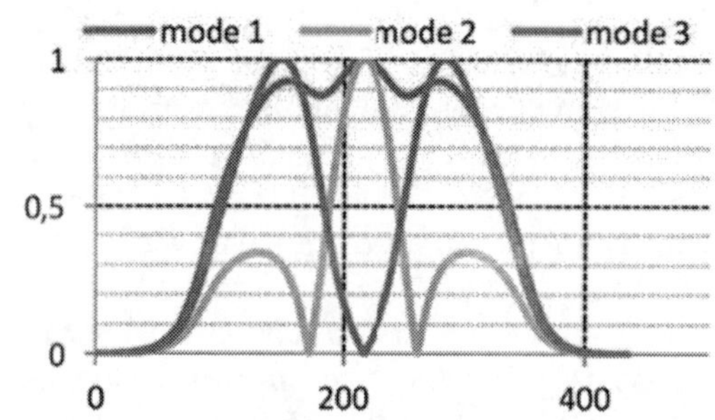


**Figure 1.** (Top to bottom) Bead-pull measurements of 3-Cell traveling wave cavities #1, #2 and simulated field distributions from CST for 3 cell modes.

### *2.2. Traveling wave adjustment*

Cavity #1 was used for bench tuning studies at room temperature. The test layout is shown in Figure 2(b) and corresponds to the schematic in Figure 2(a).

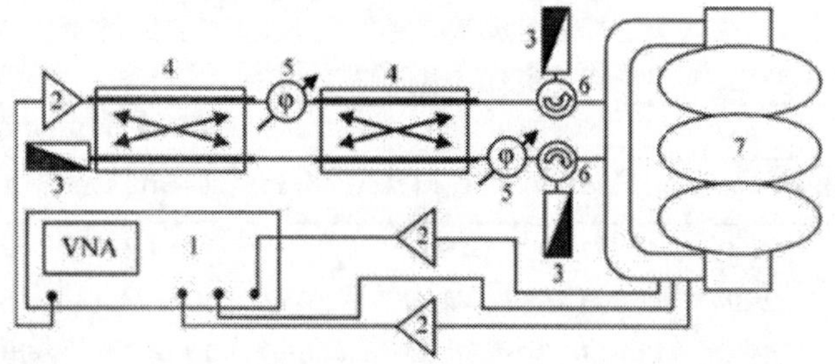


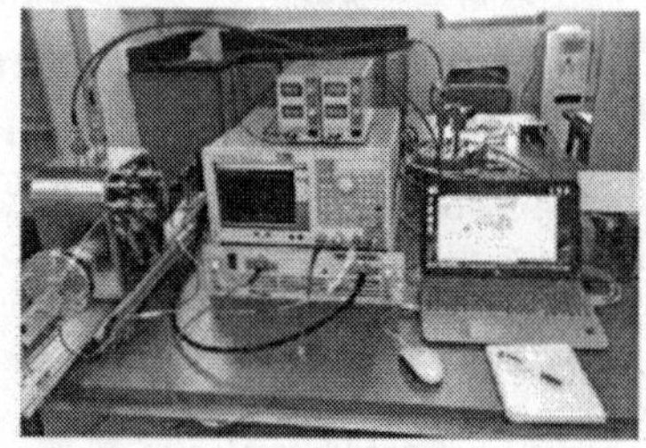

a b

**Figure 2.** 3-Cell superconducting traveling wave cavity RF feed and measurement scheme (a): 1 – Vector Network Analyzer (VNA); 2 – power amplifier; 3 – matched load; 4 – 3dB hybrid; 5 – phase shifter; 6 – circulator; 7 – resonator; and the photograph of the corresponding test layout (b).

This scheme allows amplitude and phase change in the input couplers. This is sufficient for traveling wave adjustment at room temperature but not at cryogenic temperatures because of the small bandwidth. Two hybrids and phase shifters are responsible for power and phase redistribution. A 10 W RF amplifier was used for cavity excitation. Phase shifters are the trombone type and have stepper motors for phase variation. These stepper motors were driven by a "Jova" controller [7] connected to a PC through a USB port. This allowed a quick control of cavity excitation which tremendously speeded up the traveling wave adjustment. The 3-Cell superconducting traveling wave cavity has three measurement couplers located at intervals of $\lambda/8$. The middle one is required for calibration, which consists of two steps. The first requires obtaining a minimum signal at the calibration coupler by redistributing power signals in the standing wave regime, i.e. a node is obtained at the location of the coupler. Other measurement couplers should have an equal amplitude signal, as long as they are located equidistantly from the calibration coupler. The second step of calibration is to make signals from these side couplers equal. This was done initially by adjustment of the measurement coupler loop orientation. It was later found that multiplication by a correction coefficient in the VNA (Vector Network Analyser) is more convenient. Figure 3(a) demonstrates signals from the measurement couplers after calibration.

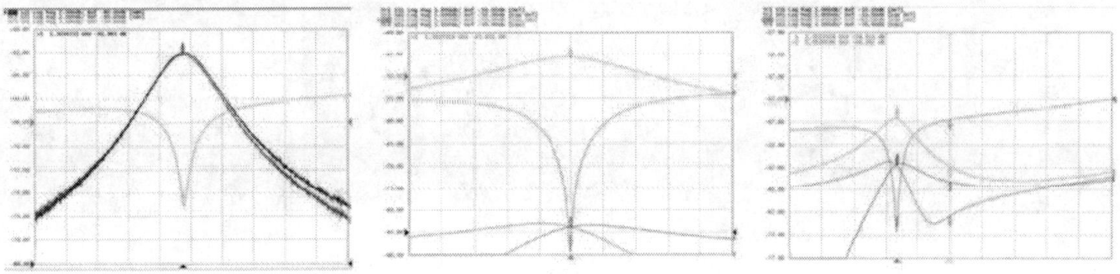

a b c

**Figure 3.** Calibrated signals from 3 measurement loops in the SW regime (a); adjusted TW in the cavity with deformed (c) and not deformed waveguide (b). Blue and red curves – signals from side measurement loops, green and pink curves (b, c) – mathematically extracted signals of waves circulating in opposite directions. Green curve (a) – middle measurement loop signal.

After calibration, the measurement coupler signals require further post-processing, i.e. extraction of information about the traveling "clock-wise" wave "b" and "anti-clockwise" wave "a". If the signal from coupler 1 is $[a+b]$, then the signal from coupler 2 is $[a\cdot e^{-i\cdot\lambda/4}+b\cdot e^{+i\cdot\lambda/4}]$. Their sum and difference will give waves "a" and "b" separately. This is possible to realize using a 90 degree hybrid or by standard functions of the VNA as in our case.

The traveling wave in the cavity #1 was adjusted by amplitude and phase redistribution at a frequency close to the 2-nd cell mode (see Table 1. Mode 4) 1303.35 MHz. The TW regime was observed at frequency range [1303.35+/-0.1] MHz which corresponds to the cavity bandwidth with full degradation at the end of this range. This result was predicted by the mathematical model [8]. Besides that, the 3-cell cavity was designed to operate at a frequency close to the 2$^{nd}$ cell mode at 2K (1300.82 MHz) because of a small reflection from the acceleration part of the waveguide. The TW at room temperature was also obtained under similar conditions. Processed signals, which corresponded to forward and backward wave amplitudes, are shown in Figure 3 (b). One wave is highly damped while the other one has a maximum at the desired frequency. The TW was ruined by compressing the waveguide (WG) with a C-clamp but was easily restored by power and phase redistribution in the input power couplers at a different frequency (see Fig. 3 (c)). WG deformations cause reflections and change of electrical length [9]. That is why the operating TW frequency was shifted to 1303.02 MHz.

## 3. Final manufacturing

The cavities require additional stiffening ribs for cryogenic tests because of the narrow bandwidth. The welding process can cause deformations of the cavities thus the tuning procedure should not be done prior to welding. Cavity #1 was chosen for cryo-test and was sent for finishing manufacturing because it required less tuning because the deviations of resonant frequencies were smaller. The cavity with a closed electromagnetic (EM) volume was measured and is shown in Figure 4(a).

**Figure 4.** The photographs at different time during the welding of stiffening ribs.

Welding stiffening ribs on the waveguide requires a flat surface for good contact. Even if the WG had a flat surface originally it would be deformed during welding. The welding places were flattened by wire EDM (electrical discharge machining) to solve this issue (see Fig. 4 (b)). After that the ribs were welded by EBW (electron beam welding) (see Fig. 4 (c-f)). Fig. 4 (f) shows the finished cavity with all the ribs attached. Flat tuner plates were also welded to three stiffening ribs and required additional CNC machining to equalize their sizes. Two Ti bars were designed and manufactured to fix the length between the beam pipes and are connected by screws to them (see Fig. 5(a)). These bars will be used to fix the cavity to special crate for testing at Fermilab. The whole assembly can be found in Figure 5(b). Holding plates are manufactured and fully compatible with the VTS.

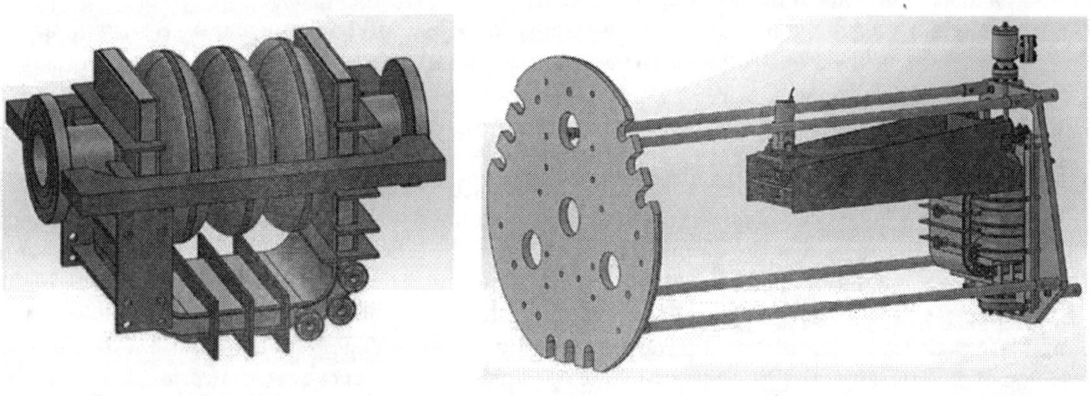

a b

**Figure 5.** 3D model of 3-cell SC TW resonator (a) and the full assembly for cryogenic test (b).

Once the cavities are fully finished, the RFQC routine will be performed. Frequencies and field distributions will be tuned as usual for any cavity. Additional tuning will take place at the operating TW frequency. The WG deformations should return the operating frequency to the expected value. After all of these procedures, the cavity will undergo the standard routine for SC cavities: chemical etching by BCP, DI water rinsing, drying process, evacuation and 2K test.

## 4. Conclusions

Two 3-cell Nb cavities with feedback waveguide were manufactured without additional stiffening required for the high power test in liquid helium. The cavities were bench tested. Quality factors, frequencies and field distribution were presented. One of the cavities was used for TW excitation studies at room temperature. As was predicted by the model, TW excitation at room temperature was obtained by power and phase redistribution in input power couplers. After room temperature tests the cavity was sent for welding of additional stiffening ribs, which has been completed. High power tests of the traveling wave prototype superconducting cavity are expected to begin in Spring 2017.

## Acknowledgments

This work was supported by USA Department of Energy contract No. DE-SC0006300 and by Ministry of Education and Science of the Russian Federation (project 3.6522.2017). Also many thanks to Fermilab Technical Division for providing us required equipment and space for testing.

## References

[1] *Avrakhov P, et al. 2005, in Proc. PAC'05, p. 4296*

[2] http://tesla.desy.de/new pages/TDR CD/start.html

[3] *http://www.linearcollider.org/ILC*

[4] *Kostin R., et al. "A high gradient test of a single-cell superconducting radio frequency cavity with a feedback waveguide." Superconductor Science and Technology 28.9 (2015): 095007.* http://dx.doi.org/10.1088/0953-2048/28/9/095007

[5] *Kostin, R., et al. "A tuner for a superconducting traveling wave cavity prototype." Journal of Instrumentation 10.10 (2015): P10038.* http://dx.doi.org/10.1088/1748-0221/10/10/P10038

[6] CST – Computer Simulation Technology, https://www.cst.com/

[7] http://www.jovasolutions.com/tims-0201-stepper-motor-controller

[8] *Kostin R. et al., 2016, in Proc. NAPAC'16,* TUPOB02. http://vrws.de/napac2016/papers/tupob02.pdf

[9] *Kostin R. et al., 2016, in Proc. LINAC'16,* MOP106021

# Influence of microwave irradiation of cement mixtures on the strength of cement stone and concrete

**M S Dmitriev[1], M V Dyakonov[1], R A Krasnokutskiy[1], A D Kolyaskin[1] and S A Dmitriev[2]**

[1] National Research Nuclear University MEPhI, 31 Kashirskoe shosse, Moscow, 115409, Russia
[2] Innovation and technology center for the «PRORYV» project, 2/8 Malaya Krasnoselskaya, Moscow, 107140, Russia

msdmitriyev@mephi.ru

**Abstract**. The influence of microwave irradiation of mixtures on the strength of cement stone and concrete was studied. Created at NRNU MEPHI experimental installation for the investigation of microwave effects on imperfect dielectrics and semiconductor materials was the source of radiation. It is shown that on the twenty-eighth day after mixing, the strength of the cement stone increases by 1.2 times, and that of concrete by 2.2 times.

## 1. Catalytic action of microwave radiation

The article is devoted to research in the field of catalytic action of microwave radiation during hydration of inorganic materials. The possibility of improving the cement stone properties by means of microwave processing of cement mixtures has been investigated.

Processes based on microwave heating have many applications in the industry. The advantages of this heating are the features of electromagnetic radiation absorption. This is a volumetric uniform heating, the possibility of concentrating high power in small volumes and the possibility of its concentration in a certain place [1]. Microwave exposure saves energy, since power is released directly in the material being processed. NRNU MEPhI has extensive experience in the development of microwave installations: an installation for a two-stage processing of radioactive waste was developed [2], nanomaterials were obtained by microwave heating [3] etc.

Also, microwave radiation can play the role of a catalyst for chemical reactions. For example, it was shown that the short-term microwave exposure to polymers which does not cause temperature increasing leads to its properties modification that can be used for practical purposes [4]. Cement is a complex mixture of minerals. Chemical reactions that occur when cement is mixed with water, lead to the formation of a structure similar to polymeric materials.

Adsorption of water by mineral particles occurs during concrete curing due to electrostatic interaction between ions and water dipoles and van der Waals forces [5]. The smaller size of the adsorption layer leads to a stronger cement particle bonds and, ultimately, the strength of the cement stone. The microwave field interferes with adsorption of additional water layers, reducing the limiting cement water-retaining capacity. Also alternating electromagnetic field facilitates mixing and therefore more compact arrangement of the cement gel particles.

## 2. Experimental installation

The experimental installation for the investigation of microwave effects on imperfect dielectrics and semiconductor materials was created at NRNU MEPhI. The installation is shown in figure 1. Its main elements are a magnetron-type microwave generator with a continuous output power, a ferrite isolator, absorbing loads and a working chamber.

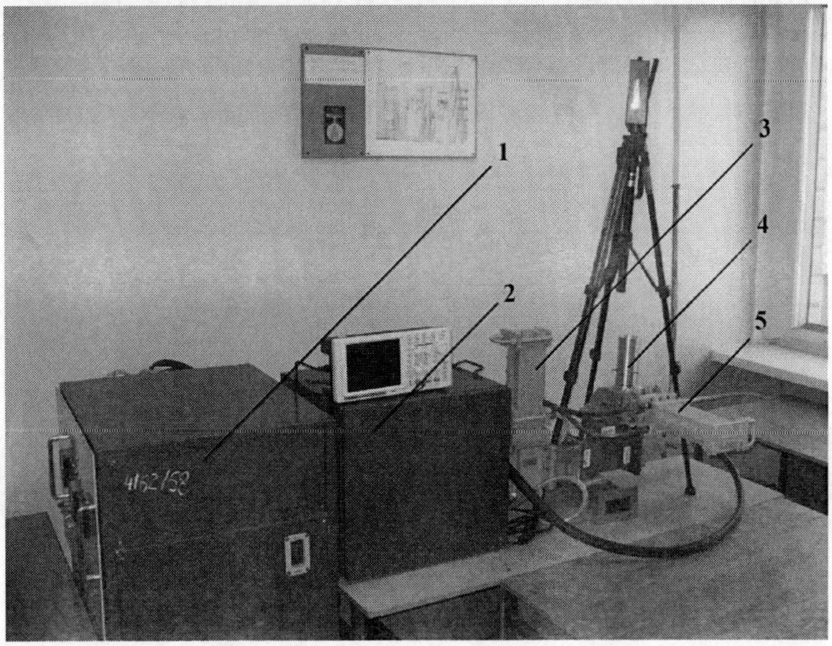


**Figure 1.** Appearance of the installation.

The installation uses a continuous generator with a power of 5 kW, based on the magnetron M-168 with an operating frequency (2450 ± 50) MHz. The generator consists of two units: a rectifier (1) forming a constant anode voltage of the magnetron and a generating unit (2). The positive quality of the magnetron is a low power level at which steady generation of microwaves begins (200 ... 300 W). This is very useful in carrying out the experiments.

Magnetron M-168 has a valid value of the standing wave ratio (VSWR) equal to 1.5. Such a VSWR value is difficult to provide in the experimental installation due to the uncertainty of the input resistance of the loaded working chamber. Therefore a ferrite isolator (3) is installed at the output of the generator to protect the magnetron from the action of the backward wave. So the magnetron is always matched to the waveguide.

A rectangular waveguide section was used as the working chamber. A circular below cutoff waveguide was installed on it (4). The microwave power not released in the test sample was dissipated in the absorbing load (5). The alundum container was placed on the bottom of the rectangular waveguide through the below cutoff waveguide.

## 3. Experiments

The container was loaded with a mixture of portland cement grade 45.5R, sand and water. In the first case the sample consisted of a water/cement mixture with a water-cement ratio of 0.4. In the second case the sample consisted of a mixture of sand and cement in a ratio of 3 to 1 with a water-cement ratio of 0.7. The container was filled with 50 $cm^3$ of mixture. After microwave processing the mixture was poured into special molds.

A series of experiments on the processing of mixture samples was carried out. The experiments were carried out at a microwave generator power of 1.2 kW. The time of microwave exposure to the samples ranged from 5 to 15 seconds. The mixture was poured into molds for subsequent testing on the seventh, fourteenth and twenty-eighth days from the moment of microwave processing. Investigation of the mechanical strength of concrete and cement stone was carried out using a hydraulic press. The cement stone obtained from the mixture processed with microwave irradiation showed an increase in strength compared to the unprocessed one by 1.2 times on the twenty-eighth day after mixing. The processed samples of the cement-sand mortar on the twenty-eighth day had strength of 2.2 times than the unprocessed samples. The best results were obtained with a microwave processing time of 10 seconds and a final temperature of the mixture of 50-55 °C. The results of the tests are shown in table 1. Rate of cement stone hardening is shown in figure 2.

**Table 1.** The results of the tests.

| Age of hardening (days) | Compressive strength (MPa) | | | |
|---|---|---|---|---|
| | Cement stone | | Concrete | |
| | Unprocessed samples | Processed samples | Unprocessed samples | Processed samples |
| 7 | 24.0 | 32.0 | 10.3 | 15.8 |
| 14 | 38.5 | 51.0 | 10.8 | 27.5 |
| 28 | 51.3 | 61.2 | 16.0 | 35.5 |

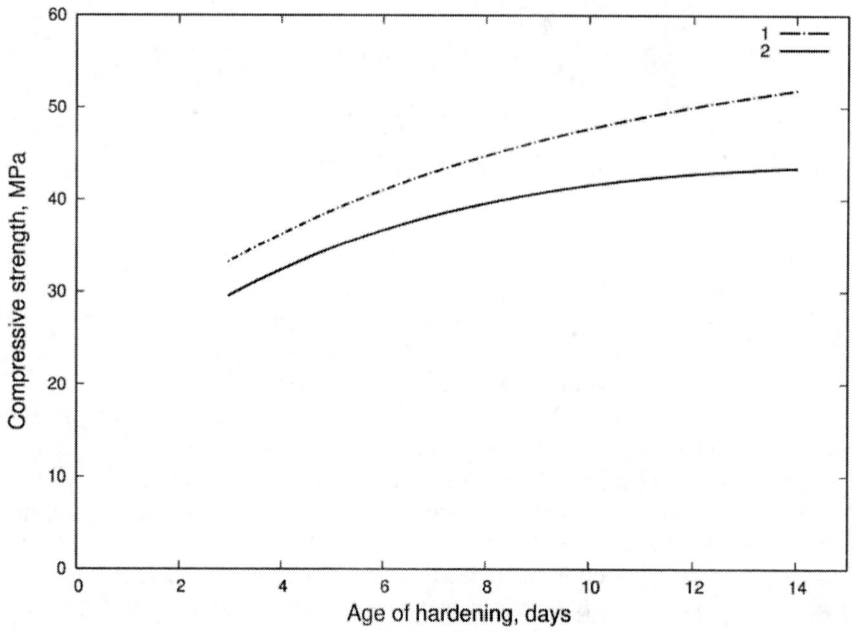


**Figure 2.** Rate of cement stone hardening: processed samples (2), unprocessed samples (1).

The obtained results give reason to believe that microwave processing of cement mixtures is a promising method to increase the mechanical strength of cement stone and concrete. This project is supported in part by the MEPhI 5/100 Program of the Russian Academic Excellence Project.

## References

[1] Didenko A N 2003 *Microwave Power Engineering: Theory and Practice* (Moscow: Nauka)

[2] Didenko A N, Dmitriev M S, Dmitriev S A, Kobelev A P, Kolyaskin A D and Lebedev V V 2010 *Nuclear Physics and Engineering* **1** 47-55

IOP Conf. Series: Journal of Physics: Conf. Series **941** (2017) 012101 doi:10.1088/1742-6596/941/1/012101

[3] Didenko A N, Dmitriev M S, Kolyaskin A D and Krasnokutskiy R A 2015 *Journal of Communications Technology and Electronics* **60** 507-11
[4] Kalganova S G and Morozova M Y 2004 *Electrical Technology Russia* **2** 91-6
[5] Akhverdov I N 1981 *Basics of Concrete Physics* (Moscow: Stroyizdat)

IOP Conf. Series: Journal of Physics: Conf. Series **941** (2017) 012102 doi:10.1088/1742-6596/941/1/012102

# Beam dynamics studies at DAΦNE: from ideas to experimental results

**M Zobov on behalf of the DAΦNE Team**

LNF-INFN, Via Enrico Fermi 40, 00044 Frascati (Rome), Italy

E-mail: mikhail.zobov@lnf.infn.it

**Abstract**. DAΦNE is the electron-positron collider operating at the energy of Φ-resonance, 1 GeV in the center of mass. The presently achieved luminosity is by about two orders of magnitude higher than that obtained at other colliders ever operated at this energy. Careful beam dynamic studies such as the vacuum chamber design with low beam coupling impedance, suppression of different kinds of beam instabilities, investigation of beam-beam interaction, optimization of the beam nonlinear motion have been the key ingredients that have helped to reach this impressive result. Many novel ideas in accelerator physics have been proposed and/or tested experimentally at DAΦNE for the first time. In this paper we discuss the advanced accelerator physics studies performed at DAΦNE.

## 1. Introduction

DAΦNE is one of the seven charged particle colliders operating in the world. It is an electron-positron collider working at the energy of the Φ resonance (1.02 GeV c.m.) aimed at providing a high rate of K mesons production [1, 2]. The DAΦNE accelerator complex (see Fig 1) consists of two main rings and an injection system composed of a full energy linear accelerator, a damping/accumulator ring and transfer lines. The presently achieved peak luminosity of $4.5 \times 10^{32}$ $cm^{-2}s^{-1}$ is by about two orders of magnitude higher than that obtained at other colliders ever operated at the same energy.

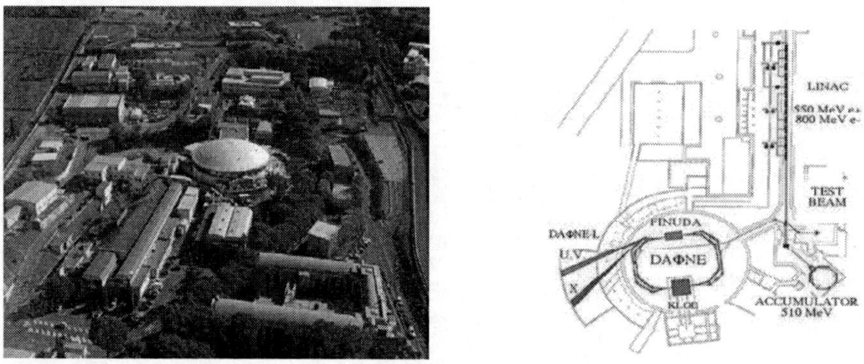


Figure 1. DAΦNE accelerator complex.

Careful beam dynamic studies such as vacuum chamber design with low beam coupling impedance, suppression of different kinds of beam instabilities, investigation of beam-beam interaction, optimization of the beam nonlinear motion have been the key ingredients that have helped to reach this impressive result. Many novel ideas in accelerator physics have been proposed and/or tested experimentally at DAΦNE for the first time. These are suppression of parasitic beam interactions with current-carrying wires [3], beam collisions with negative momentum compaction factor [4], electron cloud mitigation with dedicated electrodes [5], wigglers with "wiggling" poles [6], strong RF focusing proposal [7] and others. Besides, the innovative concept of nonlinear focusing of colliding bunches, called Crab Waist (CW) collision scheme, has been proposed at LNF INFN [8, 9] and successfully tested at DAΦNE in operational conditions providing luminosity for two different experimental detectors, SIDDHARTA [10] and KLOE-2 [11]. Considering high efficiency of the scheme for increasing collision luminosity and its relative simplicity for implementation several new collider projects have been proposed and are under development at present. These are the SuperKEKB B-factory already under commissioning in Japan [12], the SuperC-Tau factory proposed in Novosibirsk and entered in the short list of Russian mega-science projects [13], the new 100-km electron-positron Future Circular Collider (FCC-ee) under design study at CERN [14], the Higgs Factory CEPC in China [15] and some others.

In this paper we give a brief overview of some most relevant advanced accelerator physics studies performed at DAΦNE

## 2. Low impedance vacuum chamber

Since the very beginning of the DAΦNE project beam instabilities were recognized as one of the most harmful potential dangers for collider performance. For this reason the vacuum chamber was designed taking much care of its beam coupling impedance [16], both broad-band [17] and narrow-band.

New designs and novel ideas were adopted for almost all principal vacuum chamber components: RF cavities [18, 19], shielded bellows [20], longitudinal feedback kickers [21], BPMs [22], DC current monitors, injection kickers, transverse feedback kickers and others [23]. For example, the longitudinal feedback kickers based on the DAΦNE kicker design are routinely used in most high intensity lepton colliders and synchrotron radiation sources.

During the years of collider operation the vacuum chamber has been modified several times with the goal of a further impedance reduction. For example, during the collider shutdown for the SIDDHARTA experiment installation the vacuum chamber was substantially revamped [24]: two new low impedance interaction regions were designed and installed, the new fast injection kickers were implemented, the old bellows were substituted by the new ones, having Ω-shape RF sliding contacts. In addition, ion clearing electrodes that were no longer in use, have been removed from the electron ring [25]. At present the measured longitudinal broadband impedance is 0.45 Ω for the electron ring and 0.36 Ω for the positron one [24] to be compared with 13.6Ω of the EPA [26], the former CERN electron-positron accumulator ring with similar circumference and beam energy.

## 3. Collisions with negative momentum compaction factor

Since there are several potential advantages for beam dynamics and luminosity performance of a collider with a negative momentum compaction factor $\alpha_c$ [27] it was suggested to perform a dedicated machine experiment at DAΦNE with $\alpha_c < 0$. The principal results achieved during the experiment were following [4]:

- The DAΦNE lattice is shown to be very flexible and the machine optics model is proven to be reliable in providing stable collider operation with the momentum compaction factors varying in the range from -0.036 to +0.034.
- With the negative $\alpha_c$ bunches in both rings are shortened by wake fields till the microwave instability threshold as predicted by numerical simulations. Figure 2 shows the positron bunch length (left plots) for the positive (green dots) and negative alfa (blue dots) and respective charge distribution at different bunch intensities for $\alpha_c < 0$ (right plot).
- It was possible to store high bunch currents with the large negative chromaticities. No hard limit has been observed in the multibuch operations. About 1 A stable beam currents were stored in both rings.

IOP Conf. Series: Journal of Physics: Conf. Series **941** (2017) 012102 doi:10.1088/1742-6596/941/1/012102

- At beam currents up to 300 mA per beam a good specific luminosity was obtained in beam-beam collisions.

However, the microwave instability threshold is found to be somewhat lower with $\alpha_c < 0$. Since the experiment was performed before the extraction of the ion clearing electrodes [25] giving dominant contribution into the beam coupling impedance, the high current collisions were prevented by the microwave instability in the electron ring at that time.

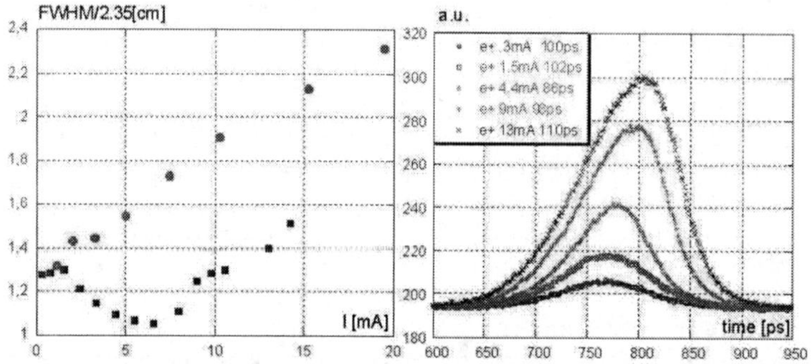


Figure 2 Bunch length (left) and charge distribution (right) for different bunch currents in the DAΦNE positron ring as measured by the streak camera (courtesy A. Stella)

## 4. Parasitic crossing compensation with wires

Long-range beam-beam interactions (parasitic crossings) were one of the sources of luminosity degradation in the DAΦNE original configuration before the collider upgrade for crab waist collisions. Due to drastic lifetime reduction the parasitic crossings put a limit on the maximum storable currents and, as a consequence, on the achievable peak and integrated luminosity. In order to mitigate this problem, numerical and experimental studies of the parasitic crossings compensation by current-carrying wires, the technique originally proposed at CERN [28], have been done [3, 29]. During the operation for the KLOE experiment two such wires were installed at both ends of the interaction region, outside the vacuum chamber. The application of the compensating wires resulted in a two fold improvement of the beam-beam induced losses for identical beam parameters, in agreement with numerical simulations [30].

## 5. e-Cloud clearing electrodes

The maximum electron beam current stored in DAΦNE was 2.5 A, i.e. the record value for the electron beam current ever stored in the modern colliders and synchrotron light sources. However, due to the electron cloud effect [31] we have not been able to exceed 1.4 A in the positron ring. In addition to this current limitation we have been suffering from other harmful e-cloud effects affecting the collider luminosity performance such as anomalous pressure rise, vertical beam size increase, tune spread along the bunch train etc. [32].

In order to cope with the strong e-cloud instabilities powerful bunch-by-bunch feedback systems have been used [33]. In addition, solenoids were installed in order to prevent the e-cloud build-up in the straight sections. However, these measures have not solved completely the problems created by the e-cloud. In particular, the strong horizontal instability still remains the worst trouble for us. Numerical simulations and experimental observations have shown that this kind of instability is triggered by the e-cloud pattern created in the wiggler and dipole magnets [34]. So it has been decided to insert special metallic electrodes [35] to suppress the e-cloud in the dipoles and the wigglers [36]

According to the dedicated numerical simulations application of the DC voltage of 200-500 V at the electrodes should decrease the electron cloud density by about 2 orders of magnitude. Indeed, the electrodes are found to be very useful to reduce the strength of the positron beam horizontal instability; to decrease the betatron tune shift and tune spread inside bunch trains; to suppress the vertical beam size blow-up of the positron beam [5]. As an example, Fig. 3 shows the betatron tune shift and spread reduction with the clearing electrodes switched on.

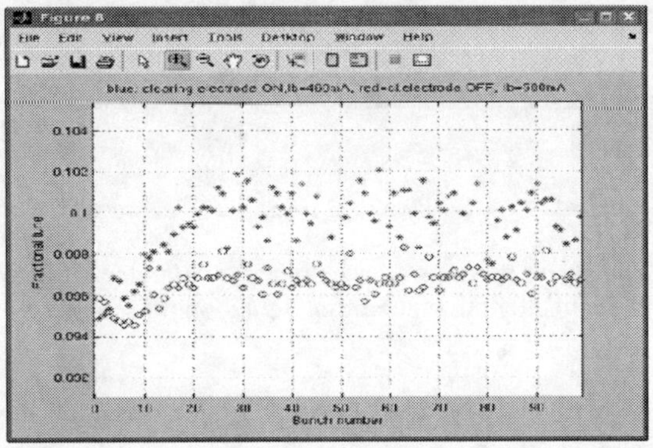


Figure 3 The horizontal betatron tune along the bunch train with the clearing electrodes off (red stars) and on (blue circles) as measured by the transverse feedback system (courtesy A.Drago).

As a result, with the electrodes switched on it is possible to store higher positron beam current, to achieve higher luminosity and to have more stable overall collider performance.

## 6. Strong RF focusing

The longitudinal strong RF focusing is an alternative way to obtain short bunches at the interaction point (IP) [37]. It consists in realizing a magnetic lattice with a large momentum compaction factor together with a strong RF gradient. In this regime the bunch length is no longer constant, but it is modulated along the ring and can be minimized at the IP. In turn, if the main impedance generating elements of the ring are located where the bunch remains long, it is possible to minimize the strength of wake fields. This helps to avoid microwave instabilities and excessive bunch lengthening due to the potential well distortion [7]. The strong RF focusing concept was proposed as one of the possible options for the DAΦNE upgrade [38].

## 7. Crab waist collision scheme

Contrary to the conventional strategies aimed at the luminosity increase, the crab waist collision scheme requires small emittance, large Piwinski angle and larger crossing angle; there is no need to decrease the bunch length and push beam currents beyond the values already achieved in the present factories. This scheme can greatly enhance the luminosity of a collider since it combines several potentially advantageous ideas: collisions with a large Piwinski angle, micro-beta insertions and suppression of beam-beam resonances using the dedicated ("crab waist") sextupoles [39]. For details of beam dynamics in crab waist collision see [9, 40, 41], for example.

The crab waist collision scheme has been successfully tested at the electron-positron collider Φ-factory DAΦNE providing luminosity increase by a factor of 3 for SIDDHARTA experiment [10], in a good agreement with numerical simulations [42]. The simulations have been also very helpful to improve the luminosity and background for the KLOE-2 experiment exploiting the crab waist scheme [43]. Figure 4 demonstrates the effect of the crab sextupoles on the beam-beam blow up and the distribution tails in DAΦNE. The advantages of the crab waist collision scheme have triggered several collider projects exploiting its potential.

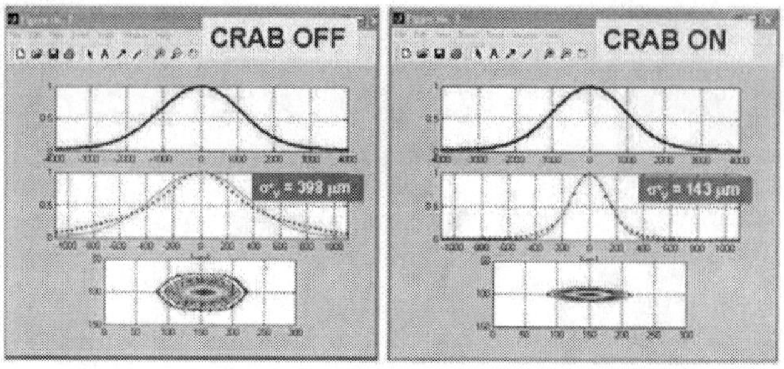


Figure 4. Transverse beam profiles with crab sextupoles off (left) and on (right) as measured by the synchrotron light monitor (courtesy A.Stella)

## 8. Conclusions

After almost 20 years of running DAΦNE is still on track providing high luminosity for physics detectors and performing world-class accelerator experiments. In the future DAΦNE can serve as a test accelerator for new collider projects.

## References

[1] Vignola G *et al* 1993 DAΦNE, the Frascati Φ-factory *Conf. Proc.* **C930517** pp.1993-1997
[2] Vignola G *et al* 1996 DAΦNE, the first Φ-factory *Conf Proc* **C960610** pp. 22-26
[3] Milardi C *et al* 2006 Wire Compensation of Parasitic Crossings in DAΦNE *Conf Proc* **C060626** pp. 2808-2810 *e-Print:* physics/0607129
[4] Zobov M *et al* 2006 DAΦNE Experience with Negative Momentum Compaction *Conf Proc* **C060626** pp. 989-991 *e-Print:* physics/0607036
[5] Alesini D *et al* 2013 DAΦNE Operation with Electron-Cloud-Clearing Electrodes *Phys. Rev. Lett.* **110** 124801
[6] Bettoni S *et al* 2014 Experimental Validation for the Compensation Method of Nonlinearities in Periodic Magnets, *Phys.Rev.ST Accel.Beams* **17** 102401
[7] Gallo A, Raimondi P and Zobov M 2003 Strong RF Focusing for Luminosity Increase: Short Bunches at the IP, *Conf Proc* **eConf C0309101** THWA003
[8] Raimondi P 2006 Status on SuperB effort *in 2nd SuperB Workshop*, Frascati, Italy
[9] Raimondi P, Shatilov D and Zobov M 2007 Beam-beam issues for colliding schemes with large Piwinski angle and crabbed waist *Preprint* physics/0702033
[10] Zobov M *et al* 2010 Test of crab-wast collisions at DAΦNE Φ-factory *Phys. Rev. Lett.* **104** 174801
[11] Milardi C *et al* 2015 DAΦNE consolidation program and operation with KLOE-2 detector *ICFA Beam Dyn. Newslett.* **67** pp. 9-27
[12] Ohnishi Y *et al* 2013 Accelerator design at SuperKEKB *Prog. Theor. Exp. Phys.* **2013** 03A011
[13] Bondar A *et al* 2013 Project of Super Charm-Tau factory at the Budker Institute of Nuclear Physics *Phys. Atom. Nucl.* **76** pp. 1072-1085.
[14] K.Oide et al 2016 Design of Beam Optics for the Future Circular Collider e+e- Collider Rings *Phys.Rev.Accel.Beams* **17** 102401
[15] Wang D *et al* 2016 CEPC partial double ring scheme and crab-waist parameters *in IAS Program on High Energy Physics* 18-21 January 2016, Hong Kong
[16] Zobov M *et al* 1995 Collective Effects and Impedance Studies for the DAΦNE Φ-Factory, *Proc. Int. Conf. on Collective Effects for B-Factories*, p. 110. *Preprint* LNF-95-041-P.
[17] Bartalucci S *et al* 1993 Broadband Model Impedance for DAΦNE Main Rings *Nucl.Instrum.Meth.* **A337** 231-241

[18] Bartalucci S *et al* 1995 Analysis of Methods for Controlling Multibunch Instabilities in DAΦNE, *Part.Accel.* **48** 213-237
[19] Alesini D *et al* 2004 Third Harmonic Cavity Design and RF Measurements for the Frascati DAΦNE Collider *Phys.Rev.ST Accel.Beams* **7** 092001
[20] Delle Monache G et al 1998 DAΦNE Shielded Bellows *Nucl.Instrum.Meth.* **A403** 185-194
[21] Boni R *et al* 1996 A Waveguide Overloaded Cavity as Longitudinal Kicker for the DAΦNE Bunch-by-Bunch Feedback System *Part.Accel.* **52** 95-113
[22] Marcellini F *et al* 1998 DAΦNE Broad-Band Button Electrodes *Nucl.Instrum.Meth.* **A402** p.27
[23] Zobov M *et al* 1998 Measures to Reduce the Impedance of Parasitic Modes in the DAΦNE Vacuum Chamber Components *Frascati Phys.Ser.* **10** 371-378
[24] Marcellini F *et al* 2008 Coupling Impedance of DAΦNE Upgraded Vacuum Chamber *Conf Proc* **C0806233** TUPP051
[25] Zobov M *et al* 2007 Impact of Ion Clearing Electrodes on Beam Dynamics in DAΦNE *JINST* **2** P08002
[26] Bartalucci S *et al* Collective Effects in the LEP Electron Positron Accumulator (EPA) *Conf. Proc.* **C870316** p. 1234
[27] Ruggiero F and Zobov M 2003 High Luminosity Issues for DAΦNE Upgrade *Conf Proc* **eConf C0309101** SAPL004
[28] Koutchouk J 2000 Principle of a Correction of Long-Range Beam-Beam Effect in LHC Using Electromagnetic Lenses *LHC-Project-Note-223*, CERN.
[29] C.Milardi *et al* 2008 DAΦNE Lifetime Optimization with Compensating Wires and Octupoles *e-Print:* arXiv:0803.1544
[30] A.Valishev et al 2015 Numerical Analysis of Parasitic Crossing Compensation with Wires in DAΦNE *6th Int. Part. Accel.Conf. (IPAC2015)* pp. 589-592 *e-Print:* arXiv:1506.07562
[31] Furman M A 2013 in *Handbook of Accelerator Physics and Engineering* pp 163-166
[32] Drago A, Vaccarezza C and Zobov M 2006 Analysis of Mechanisms Driving the Horizontal Instability in the DAΦNE Positron Ring, *DAΦNE Thechnical Note* G-**67**
[33] Drago A, Teytelman D and Zobov M 2005 Recent Observations on a Horizontal Instability in the DAΦNE Positron Ring in *Proc. of Part. Accel. Conf. (PAC2005),* pp. 1841-1843
[34] Demma T *et al* 2009 A Simulation Study of the Electron Cloud Instability at DAΦNE in *Proc. of Part. Accel. Conf. (PAC2009),* pp. 4695-4697
[35] Alesini *et al* 2010 Design and Test of the Clearing Electrodes for e-Cloud Mitigation in the e+ DAΦNE Ring *Conf. Proc.* **C100523** TUPEB002
[36] Zobov M *et al* 2012 Operating Experience with Electron Cloud Clearing Electrodes at DAΦNE *Conf. Proc.* **C1206051** pp. 259-265
[37] Gallo A, Raimondi P and Zobov M 2003 Strong RF Focusing for Luminosity Increase *e-Print:* arXiv: physics/0309066.
[38] Gallo A et al 2004 Proposal of a Strong RF Focusing Experiment at DAΦNE in *Proc. of EPAC2004*, pp. 683-685
[39] Raimondi P, Shatilov D and Zobov M 2008 Suppression of beam-beam resonances in crab waist collisions *Conf. Proc.* **C0806233** WEPP045
[40] Zobov M 2011 Next Generation of Electron-Positron Factories *Phys.Part.Nucl.* **42** pp.782-799
[41] Shatilov D, Levichev E, Simonov E and Zobov M 2011 Application of Frequency Map Analysis to Beam-Beam Effects Study in Crab Waist Collision Scheme *Phys. Rev. ST Accel. Beams* **14** 014001
[42] Zobov M 2010 Beam-beam Interaction in Novel, Very High Luminosity Parameter Regimes *Conf. Proc.* **C100523** pp. 3639-3643
[43] Zobov M *et al* 2016 Simulations of Crab Waist Collisions in DAΦNE with KLOE-2 Interaction Region *IEEE Trans.Nucl.Sci.* **63** pp.818-822

# Influence of tests results on serial production of EXFEL superconducting cavities

**D V Kostin and A A Sulimov**

Deutsches Elektronen-Synchrotron, DESY, Notkestrasse 85, 22607 Hamburg, Germany

E-mail: Alexey.Sulimov@DESY.DE

**Abstract**. The European XFEL (EXFEL) 1.3 GHz cavities were produced without any final cold RF performance guarantee for characteristics like accelerating gradient, quality factor or field emission. In the case of low-performance, DESY provided all re-treatment procedures in order to improve these characteristics getting closer to specification levels. But EXFEL cavities have very tight tolerances for the parameters like cavity length, cells eccentricity, fundamental mode frequency and field distribution on the operational mode. Usage of modern infrastructure, analytical methods and powerful tools of the database analysis allows us not only keeping these characteristics in specified ranges, but also investigate the cavity shape uncertainties and improve the damping of the higher order modes.

## 1. Introduction

European x-ray free electron laser [1] is an international project with 11 participating countries (Denmark, France, Germany, Hungary, Italy, Poland, Russia, Slovakia, Spain, Sweden and Switzerland). It is located in Germany between injector hall in DESY (Hamburg) [2] and experimental hall in Schenefeld (Schleswig-Holstein). The X-ray research laser facility is scheduled to start user operation with 1 beamline and 2 experiment stations in 2017 [3]. Electrons will be accelerated to planned energy of 17.5 GeV in superconducting radiofrequency (RF) resonators of 2 km long linear accelerator [4]. TESLA-shape 1.3 GHz Niobium cavities have 9-cells, 2 higher order mode (HOM) couplers and are operated at 2 K. But commissioning of more than 800 of those cavities requires detail tests of each accelerating element with strict requirements to its individual characteristics.

## 2. Cavity characteristics

The most important individual characteristics of main accelerator component can be separated into two classes: cryogenic parameters, which are controlled at operational temperature, and parameters at room temperature, which are measured by normal conditions.

### *2.1. Cryogenic parameters*

This set of characteristics is usually satisfying the planned values if cavities were produced according to the original recipe, which is completely described in the specification for cavity manufacturers.

*2.1.1. Unloaded quality factor ($Q_0$).* It is characterized by the power loss in cavity walls under superconducting condition. Planned value is over 1.0E+10.

*2.1.2. Field emission (x-ray).* It is characterizing the cavity dark current and identified by the appearance of radiation outside the cavity in the cryostat. Planned value is below 1.0E-2 mGy/min.

*2.1.3. Accelerating Gradient (Eacc).* It is an integral of accelerating field affected on an electron at 1 m of the cavity. There are two kinds of gradient: maximal and usable. The usable gradient is limited by criteria for previous two parameters (2.1.1. and 2.1.2) and its planned value is over 23.6 MV/m.

*2.1.4. Fundamental mode frequency (Fpi @ 2K).* It is a frequency of mechanically relaxed cavity before tuning to the exact value of 1.3 GHz. Planned value for cavity under this condition is (1299.700 ± 0.100) MHz.

*2.1.5. HOM spectra (HOM SP).* This is a set of frequencies and loaded quality factors for each mode and polarization in three passbands TE111, TM110 and TM011. Some of them require a special attention [5, 6].

*2.2. Parameters at room temperature*
The EXFEL cavities have very tight tolerances for the parameters, which are describing the cavity condition before a cryogenic test.

*2.2.1. Field flatness (FF).* It is a ratio between the minimal and maximal amplitude of longitudinal E-field on cavity axis in the middle of cells for the fundamental mode. Planned value after cavity integration in helium vessel and the pressure test is over 90 %.

*2.2.2. Fundamental mode spectrum (FM SP).* It is nine frequencies of the TM010 mode. Stability of this parameter indicates the level of non-uniform deformations and risk of FF changes. Planned value of the highest frequency of FM SP (Fpi) for cavity under vacuum is (1297.750 ± 0.100) MHz.

*2.2.3. Cavity eccentricity (ECC).* It is a maximal value of cells and flanges centers displacement relative to axis crossing the middles of reference rings [7]. Planned value before cavity integration in a helium vessel is below 0.4 mm.

*2.2.4. Length (L).* It is a longitudinal dimension, which is critical for cavity integration in accelerating module. This parameter is measured (or calculated) either between the end flanges or between reference rings [8]. Planned value of cavity length between reference rings is (1059 ± 3) mm.

*2.2.5. Shape deviation of the inner surface.* This parameter allows us to keep the TESLA shape geometry close to drawings and simulated cavity shape. Planned value for cavities components in original specification is ±0.20 mm. It was an agreement between DESY and cavity manufacturers that deviation up to ±0.30 mm can be accepted only for 10 % of cells surface.

## 3. Cavity tests and influence of their results

Cavity tests – it is a set of measurements under cryogenic and room temperatures. Cryogenic tests are expensive and time-consuming. Their amount has to be reduced for serial production of more than 800 cavities. This problem was solved by the founding of the exact procedure of cavity manufacturing, including all details of chemical surface treatment, and its implementation in an industrial process. The quality assurance procedure at room temperature provides us with a possibility of supervising the complete production sequence.

The Engineering Data Management System [9] and EXFEL cavity database [10, 11] automatize the documentation flow, collecting the data and results control during the cavities manufacturing and both types of tests.

### *3.1. Cryogenic tests*

Cavity performance measurements at 2K are done for the separate cavities in a vertical cryostat for 4 cavities in CW mode and in accelerating module in the module test stand after the tuning to the operating frequency [12-18]. In the module test, each of 8 cavities is tested first separately by tuning it to the operating frequency of 1.3 GHz up to a breakdown (quench) or RF power limit. Operating gradient limit is given by described limitations from a breakdown and gamma radiation (field emission – dark current). A general RF power limit is set by 31MV/m for a cavity. After a single cavities tests, all 8 cavities are tested together in the module to get a cryogenic performance as well as combined field emission properties. Obtained operating gradients are realized as a fixed tailored RF waveguides power distribution to maximize a module acceleration with given limitations.

Unloaded quality factor, field emission and accelerating gradient are generally satisfying the EXFEL specification requirements. The test results of these parameters have no impact on serial production. In the case of necessity, these characteristics can be improved by re-treatment with an individual plan for each cavity. Average value of the accelerating gradient for all EXFEL cavities is 27 MV/m and estimated final energy of the accelerated beam is over the planned value [19, 20].

Fundamental mode frequency at 2K is well correlated with room temperature measurements. The planned difference in EXFEL specification is 1.95 MHz, but statistical analysis of first twelve modules detected a systematic error and real frequency changes during cooling down is $(2.00 \pm 0.05)$ MHz.

During cavity vertical tests it was observed that the damping of the second monopole mode (TM011) showed the largest variation and sometimes up to 2-3 times lower than the originally allowed limit. It was concluded that this TM011-damping degradation was caused by cavity geometry deviation within the specified mechanical tolerances [6].

Solving the problem with HOM SP and adjusting of Fpi @ 2K had the most influence on cavity serial production. These aspects were integrated into quality assurance procedure at room temperature.

### *3.2. Quality assurance at room temperature*

Control of cavity parameters under the normal conditions [7] begins by the cavity manufacturers RI Research Instruments GmbH (RI) and Ettore Zanon S.p.A. (EZ) under the supervision of DESY (Hamburg) and INFN (Milano) [21].

The Cavity Tuning Machine (CTM) and Half-Cell Measurement Machine (HAZEMEMA) [22] were used for room temperature measurements and adjustment of corresponding cavity characteristics. The routine tuning on the CTM allowed us to keep cavity eccentricity and field flatness inside the tolerances. The final statistic shows that more than 70 % cavities have FF over 95 % and ECC for 75 % cavities is below 0.3 mm.

More efforts were required to adjust the CTM for correcting of the pi-mode frequency of FM SP. This characteristic depends on production step, cavity filling (air, argon or vacuum), and the presence of field measurement system (FMS). Fluctuation of all parameters of the chemical treatment procedure, which are individual for RI and EZ, change Fpi value. It was determined at the beginning of production that Fpi for full equipped cavity under vacuum is below planned value. After collecting and analysis of some statistics, the reasons of Fpi deviations were identified and the corresponding correction for the CTM parameters was done. The same algorithm was used, when EZ installed their own infrastructure for main electropolishing process and the ratio of removed material from equator and iris areas [23] caused an increase of Fpi value.

Measurements of cavity components on the HAZEMEMA [24] allow predicting of final cavity length before equator welding [25]. The accuracy of this prediction is 0.4 mm [8]. This is several times lower than tolerance for this characteristic. It provides us with a possibility to variate the cavity length during production being always inside the specified tolerances. In the middle of serial production, some limitation of usage the cavities on the first position in an accelerating module were determined. This problem was solved by reduction an average value of L till 1058 mm.

Quality assurance of cavity production is finalizing by comparison of incoming inspection results [26] with full documentation from cavity manufacturer. Non-uniform cavity deformation and risk of

FF changes are estimated by FM SP deviations. In case this value overcomes a specific limit the cavity had to be sent back to the manufacturer to repeat some reference measurements and repair the cavity if it was necessary.

The impact of cells shape uncertainties on the degradation of HOM damping efficiency was analyzed [27, 28]. Additional RF measurements and calculations of cells' eigenfrequencies were integrated into serial production procedure. They were used to identify equator welding instabilities and influence of electropolishing processes on cavity FM SP [23, 25]. Finally, the cells shape uncertainties were compensated by asymmetrical field (length) correction of cavity end cells.

## 4. Conclusion

Finally, we can select the cavity characteristics which needed the most concentration during EXFEL serial cavity production.

Target values for fundamental mode frequency and cavity length have to be changed several times. So the statistics of test results for already produced cavities was used for adjusting of manufacturing parameters on the CTM and HAZEMEMA.

The most challenging problem was increasing of HOM damping efficiency. It requires a special attention on data analysis of inner surface shape deviations, preparation of additional measurements and calculations before implementation of necessary corrections in the process of serial production.

Collaborations between participants of the EXFEL project, elaborately prepared specification for cavity manufacturing and taking into consideration of the tests results guaranteed the success of EXFEL cavity mass production.

## References

[1] H. Weise, "The TESLA X-FEL Project", in *Proceedings of EPAC 2004*, Lucerne, Switzerland.
[2] www.xfel.desy.de
[3] www.xfel.eu
[4] "XFEL: The European X-Ray Free-Electron Laser - Technical Design Report", DESY, Hamburg, Germany, DESY_06-097, 2006, doi:10.3204/DESY_06-097
[5] J. Sekutowicz, "HOM Damping and Power Extraction from Superconducting Cavities", in *Proceedings of LINAC 2006*, August 21-25, 2006, Knoxville, Tennessee USA, pp.507-510.
[6] A. Sulimov et al., "Efficiency of High Order Modes Extraction in the European XFEL Linac", in *Proceedings of LINAC 2014*, Geneva, Switzerland, 2014, pp. 883-885.
[7] A. Sulimov, "RF Measurements for Quality Assurance During Cavity Mass Production", in *Proceedings of SRF2015*, Whistler, BC, Canada, 2015, pp. 955-960.
[8] A. Sulimov et al., "The Statistics of Industrial XFEL Cavities Fabrication at Research Instruments", in *Proceedings of SRF 2013*, Paris, France, 2013, pp. 234-236.
[9] J. Iversen et al., "Using an Engineering Data Management System for Series Cavity Production for The European XFEL", in *Proceedings of SRF 2013*, Paris, France, 2013, pp. 183-185.
[10] P.D. Gall et al., "XFEL Database Structure & Loading System", in *Proceedings of SRF2015*, Whistler, Canada, 2015, pp. 1166-1167.
[11] P.D. Gall et al., "XFEL Database User Interface", in *Proceedings of SRF2015*, Whistler, Canada, 2015, pp. 1168-1170.
[12] D. Kostin et al., "Update on Module Measurements for the XFEL Prototype Modules", in *Proceedings of SRF2011*, Chicago, IL USA, 2011. pp. 65-68.
[13] D. Reschke, for the European XFEL Accelerator Consortium, "Performance Of Superconducting Cavities for the European XFEL", in *Proceedings of IPAC2016*, Busan, Korea, 2016, pp. 3186-3191.
[14] J. Swierblewski, "Large Scale Testing of SRF Cavities and Modules", in *Proceedings of LINAC 2014*, Geneva, Switzerland, 2014, pp. 426-430.
[15] M. Wiencek et al., "Improvements of the RF test procedure for European XFEL cryomodules", in *Proceedings of SRF2015*, Whistler, Canada, 2015, pp. 914-918.

IOP Conf. Series: Journal of Physics: Conf. Series **941** (2017) 012103 doi:10.1088/1742-6596/941/1/012103

[16] D. Reschke, "Infrastructure, Methods and Test Results for the Testing of 800 Series Cavities for the European XFEL", in *Proceedings of SRF2013*, Paris, France, 2013, pp. 812-815.
[17] M. Wiencek et al., "Update and status of the test results of the XFEL series accelerator modules", in *Proceedings of SRF2015*, Whistler, Canada, 2015, pp. 319-323.
[18] O. Napoly, "Module performance in XFEL cryomodule mass production", in *Proceedings of SRF2015*, Whistler, Canada, 2015, FRAA02.
[19] H. Weise, "Status of the European XFEL", in *Proceedings of LINAC2016*, East Lansing, MI, USA, 2016, MO1A02
[20] N. Walker et al., "Performance Analysis of the European XFEL SRF Cavities, from Vertical Test to Operation in Modules", in *Proceedings of LINAC2016*, East Lansing, MI, USA, 2016, WE1A04
[21] W. Singer et al., "Production of superconducting 1.3-GHz cavities for the European X-ray Free Electron Laser", Phys. Rev. Accel. Beams 19, 092001 – Published 9 September 2016
[22] J.-H. Thie et al., "Operation Experience with Half Cell Measurement Machine & Cavity Tuning Machine in 3 Years of the European XFEL Cavity Series Production", in *Proceedings of SRF2015*, Whistler, Canada, 2015, pp. 1149-1153.
[23] A. Sulimov et al., "RF Analysis of Electropolishing for EXFEL Cavities Production at Ettore Zanon Spa", presented at LINAC 16, East Lansing, MI, USA, 2016, paper TUPL035.
[24] A. Sulimov et al., "Description and First Experience with the RF Measurement Procedure for the European XFEL SC Cavity Production", in *Proceedings of IPAC'11*, San Sebastian, Spain, 2011, pp. 277-279.
[25] A. Sulimov, "RF Analysis of Equator Welding Stability for the European XFEL Cavities", in *Proceedings of SRF2015*, Whistler, BC, Canada, 2015, p. 1272-1273.
[26] K. Krzysik et al., "Tests of the 1.3 GHz Superconducting Cavities for the European X-Ray Free Electron Laser", in *Proceedings of SRF2013*, Paris, France, 2013, pp. 191-193.
[27] A. Sulimov et al., "HOM Suppression Improvement for Mass Production of EXFEL Cavities at RI", in *Proceedings of LINAC2016*, East Lansing, MI, USA, 2015, THPLR018
[28] A. Sulimov et al., "Practical Aspects of HOM Suppression Improvement for TM011", in *Proceedings of SRF2015*, Whistler, BC, Canada, 2015, pp. 1277-1278.

# The instrument for investigating magnetic fields of isochronous cyclotrons

**N V Avreline**

TRIUMF, Vancouver, CANADA

E-mail: navreline@triumf.ca

**Abstract**. A new instrument was designed and implemented in order to increase the measurement accuracy of magnetic field maps for isochronous Cyclotrons manufactured by Advanced Cyclotron Systems Inc. This instrument uses the Hall Probe (HP) from New Zealand manufacturer Group3. The specific probe used is MPT-141 HP and can measure magnetic field in the range from 2G to 21kG. Use of a fast ADC NI9239 module and error reduction algorithms, based on a polynomial regression method, allowed to reduce the noise to 0.2G. The design of this instrument allows to measure high gradient magnetic fields, as the resolution of the HP arm angle is within 0.0005° and the radial position resolution is within 25μm. A set of National Instrument interfaces connected to a desktop computer through a network are used as base control and data acquisition systems.

## 1. Instrument Overview

The instrument consists of the Hall Probe positioning mechanism, the electronics module based on compact National Instruments NI cDAQ-9188 chassis and a desktop computer equipped with custom-made LabVIEW 2011 software [1]. This software performs control, data acquisition and post processing tasks.

**Table 1.** Technical characteristics of the instrument

| | |
|---|---|
| Magnetic field accuracy | $5 * 10^{-5}$ T (in hills) |
| Azimuthal, radial resolution | 0.0005°, 25μm |
| Magnetic field range | 0.4 – 2.2 T |
| Speed of radial motion | 75 – 500 mm/s |
| Measurement time | 70 min |
| Number of samples per one radial scan | 52000 |

### *1.1. HP Positioning Mechanism (HPM)*

HPM is designed to move the HP in the radial direction and to change the scan angle. HPM allows to move the HP to any coordinate of the middle plane, the plane mid-way between the cyclotron magnet's poles. To provide this two-dimensional motion, two stepper motors with coaxial shafts are used. For azimuthal rotation, the Harmonic Drive LLC (FHA-25C-100-US250-A) is used [2]. The advantage of this motor is that it uses a harmonic gear, has a hollow shaft, and has turning accuracy of $2.8 * 10^{-4}$ °. Its transmission consists of specially designed disc brakes that allow to overcome inertia and allow to stop the shaft at the required angle with high precision. An inductosyn encoder Farrand Controls, Inc. [3] with a resolution of 0.0001° is used for angle measurements.

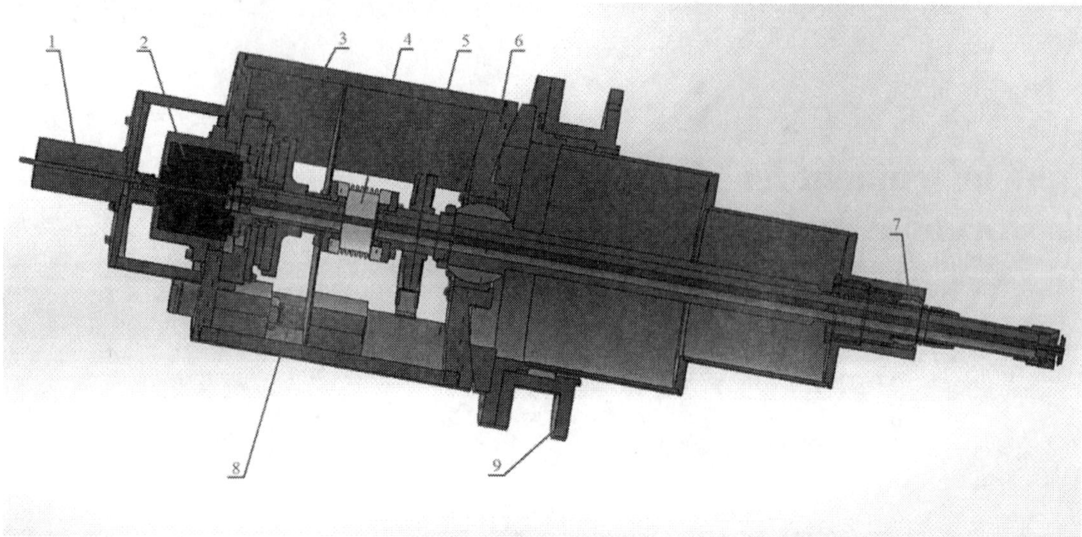


**Figure 1.** The design of the instrument's transmission

The transmission includes: 1 - Stepper motor for the radial direction, 2 - Stepper motor with the harmonic gear for azimuthal rotation, 3 - The inductosyn, 4 - The bellows made of stainless steel, 5 - The clutch, 6 - The big spherical bearing sitting on an aluminum sphere, 7 - The small ball bearing, 8 - The pneumatic disk brakes, 9 - Flange for the attachment of the transmission to a cyclotron magnet.

The instrument's transmission is inserted in the central hole of the cyclotron magnet and rotates the HP cart's plastic panel in the plane between the magnet's poles. The plastic panel was manufactured as a lamination of two carbon fiber layers (Illstreet Composites) with one layer of polyurethane foam (LAST-A-FOAM FR-7106, manufactured by Fiberlay, Inc.) placed in-between. This allows for the panel to be lightweight and rigid. To ensure precise mechanical tuning, lamination was done in a vacuum press and on a granite table. The panel also contains rails that provide radial motion of the HP cart along its length. The rails are made of Delrin (polyoxymethylene) and were machined in-place after the lamination process.

**Figure 2.** The HP cart panel as installed inside the magnet of the TR-24 Cyclotron

An experimental study showed that the mechanical vibration, recalculated in magnetic field, was 4G if the HP cart was moving on wheels and 0.5G if it was sliding on rails. Thus, the later design was chosen. The friction sliders were made of the self-lubricated material SP-21 manufactured by DuPont VESPEL.

The Parker-Compumotor ES23B stepper motor pulls the HP cart via a timing belt and a timing pulley located at the end of motor's shaft. The radial coordinate readings are made using optical transmissive sensors (HOA1872-012) and a transparent plastic film. The film contains 2mm black transverse strips, placed with a 4mm period. These strips interrupt the sensor's light beam and generate voltage pulses as a result (Fig. 3). When the voltage reading is zero, the HP cart is located precisely at the black strip's edge. Linear interpolation is used to determine those exact points and adjust magnetic field readings accordingly.

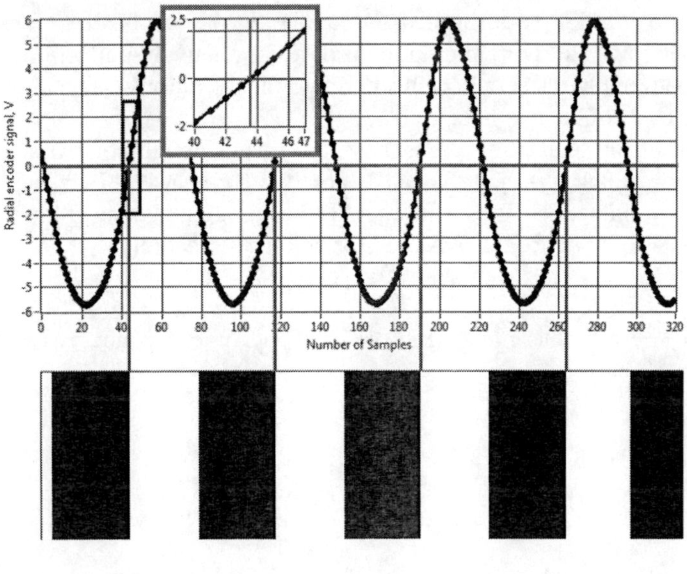


**Figure 3.** Determination of HP's radial coordinate

IOP Conf. Series: Journal of Physics: Conf. Series **941** (2017) 012104 doi:10.1088/1742-6596/941/1/012104

The total error of coordinates depends on the tolerance of etching strips, dirtiness of film, scratches of the film and on deflections in the position of the film from the center of optical sensor's gap. The experimental study of this error showed that the error is less than 25μm.

This instrument also contains an optical sensor to determine a reference point for the panel's angles (i.e. a home position). Reproducibility of the home position was experimentally checked by comparing home position angles before and after 360° mapping cycles. The initial and final angles were within 0.002°.

### *1.2. Acquisition and control electronics*

A NI cDAQ-9188 CompactDAQ Ethernet chassis with five modules is the main interface between the CPU, the sensors and the actuators. Most of the software is written in LabVIEW 2011. This allows to control all actuators and sensors and to carry out live data processing.

## 2. Alignment and Calibration

Prior to any data collection, the main shaft of the instrument as well as the HP height are aligned so that the HP travels within 100μm of the mid-plane at the all angles and radii. A spacer is used to position the HP panel in the middle plane. After installation, the HP panel's position is verified by a Mitutoyo dial gauge (25μm resolution). Three tuning screws are used to adjust the angle of the rotational axis. Those screws move the transmission assembly with respect to the aluminum sphere which is attached to the magnet. The results of this alignment are checked by attaching an aluminum bar to the main hollow shaft and placing the same dial gauge at the outside end of the bar, with the tip pointing down to the magnet's surface. Then the main shaft is manually rotated and dial gauge readings are checked to ensure that the angle of the rotational axis is within ±0.0025°.

The inductosyn is used to measure the panel's angle. The inductosyn is a transformer with the primary winding installed on the rotor and two secondary windings installed on the stator. The required amplitude of the secondary winding's signal is adjusted by changing the gap between the stator and the rotor. According to the datasheet for this inductosyn, this gap should be in the range from 0.1mm to 0.3mm. The error is caused by the shifts in the stator and rotor axis. To minimize the error, the rotor's position is tuned relative to the hollow shaft's axis. The shift between the stator and the rotor is measured in two steps. The dial gauge is attached to the stator in the first step and to the rotor in the second step. The hollow shaft is rotated during those measurements. The inductosyn signal is converted to the quadrature signal and is then processed by the NI 9411 module to obtain the digital value of HP panel's angle.

The calibration of the HP is performed by comparing the HP readings with the readings of the Drusch Gaussmeter NMR 20. For calibration purposes, a relatively uniform magnetic field is formed in the small gap between two steel plugs inserted into the cyclotron's magnet. The calibration is done at seven different levels of the magnetic field.

The calibration of the temperature sensor is performed through comparison of HP's voltage and temperature sensor's voltage. During the calibration, the HP is heated by a fan for ten minutes and is then cooled down by ambient air for 20 minutes. The calibration is done at three different levels of the magnetic field. The following formulas is used for voltage correction (1):

$$\Delta V_{corr}(B, V_{temp}) = \frac{dV_{hp}}{dV_{temp}}(B)\left(V_{temp\ initial} - V_{temp}\right) \tag{1}$$

Coefficients $\frac{dV_{hp}}{dV_{temp}}$ are determined (Fig. 4) from the time dependences of HP's voltage and of temperature sensor's voltage.

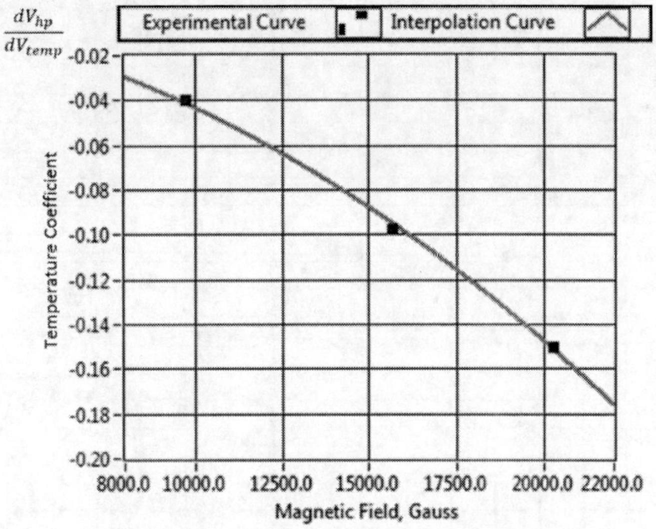


**Figure 4.** Coefficient of temperature calibration

## 3. Data Collection and Analysis

### *3.1. Data acquisition*

Usually, the 360° mapping process starts in the middle of the first hill and consists of 181 radial scans of 2° each (last scan overlaps the first scan to verify consistency). If necessary, the starting position and the step size are adjustable. To determine the initial angle of the HP panel, an optical sensor is used. All trajectories of the HP motion pass through the centre of the magnet to obtain field samples from the opposite part of the magnet. These samples are used to correct for the magnetic field errors due to mapper alignment, air temperature drift and HP noise. These samples are also used to determine the radial direction's midpoint and magnet's centre position (where the magnetic field reaches its minimum). To reduce the data acquisition cycle, all data from HP, two optical coordinate encoders and the temperature sensor are recorded in the memory of ADC NI 9239. Then, when the HP cart moves backward, the memory contents are transferred to the computer.

### *3.2. Magnetic field measurements data treatment*

The post-processing of data reduces the data's noise level and is done as the HP cart moves backwards. This correction is performed using sliding 100-point data intervals. For each data interval, a 3rd order polynomial curve is fit using the method of least squares. For the interval's mid-point, the interpolated value becomes the new true value of the magnetic field at that point (Fig. 5).

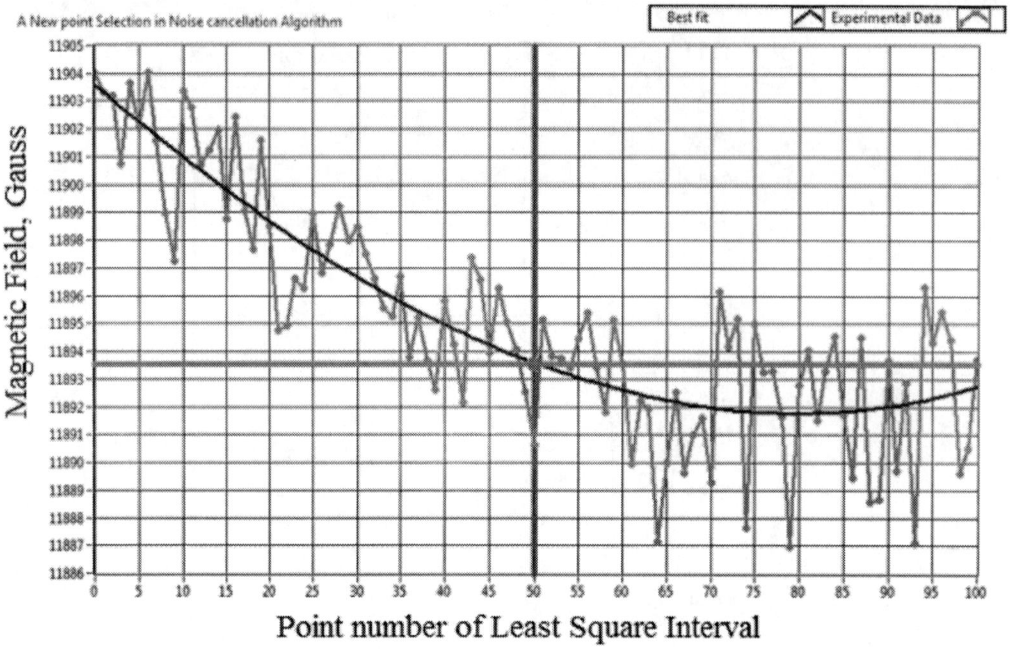


Figure 5. HP Noise Reduction

An investigation of magnetic field measurement errors determined that errors of at most 0.2G are found in the areas of large magnetic field gradients (210G/mm). It was also experimentally determined that the vibration of HP cart increases the error by 20μG in the same areas of large magnetic field gradients. This increase was negligible in comparison to the 0.2G error measured otherwise.

### *3.3. Magnetic field measurement equipment*

Group3's Hall Effect Digital Teslameter DTM-133 with HP cable MPT-141 [4] were modified for use in this instrument. Specifically, the amplifier was modified to have higher gain and components were added to read temperature. This field measurement equipment has some advantages in comparison to similar equipment offered by other companies. Specifically, to reduce noise and instrument drift, this equipment uses amplitude-modulated signal from the HP instead of the DC signal used in other instruments. This modulation has frequency of 20 kHz. Another advantage is that Group3's equipment uses double twisted pair cable for the HP signal and single twisted pair cable for the BIAS and the temperature sensor. The Group3 cables were compared to the cables from SENIS through a study of induced voltages, when the cables were loaded by a 1.1Ω resistor instead of the HP. The results show that the maximal error for the SENIS's HP cable is 45G and for Group3's HP cable is 0.2G.

### *3.4. Dynamic errors of measurement of magnetic fields*

To suppress parasitic noise and the other parasitic signals, DTM-133 has two low-pass filters with cutoff frequencies of 16 kHz and 1746 Hz. To improve the signal-noise ratio, an additional low-pass filter with lower cutoff frequency could be installed, but in this case additional errors appeared to be caused by this filter. Specifically, as the HP moves in the static magnetic field that is variable in the radial direction, the signal at the amplifier's input becomes the signal changing over time. This dynamic error depends on the filter's cutoff frequency as illustrated in Fig.6.

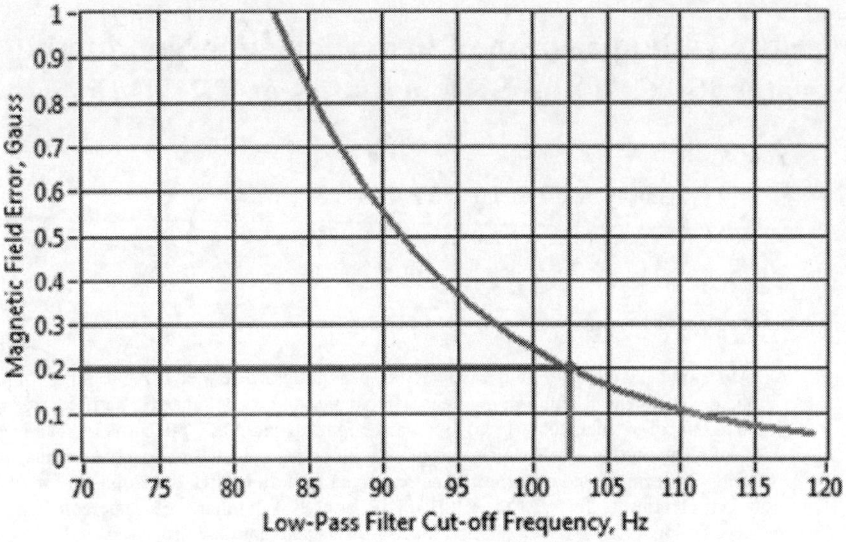


**Figure 6.** Analysis of the error caused by the use of a low-pass filter

## 4. Conclusion

The new instrument was verified through comparison of the magnetic field maps it produced to the ones produced by an instrument which was successfully used by Advanced Cyclotron Systems Inc. for over 20 years. The verification was performed at HP speeds of 75 mm/s, 150 mm/s and 250 mm/s. At the present time, more than 12 cyclotrons were built and commissioned using this instrument. The instrument demonstrated reliability, accuracy, and high performance throughout.

## References

[1] Avreline N V, Gyles W, Watt R L, Methods of increasing Accuracy in Precision Magnetic Field Measurements of Cyclotron Magnets, 20th International Conference on Cyclotrons and their Applications, Vancouver, Canada, 2013, p.283

[2] Harmonic Drive, http://www.harmonicdrive.net

[3] Ruhle Companies Inc., http//www.ruhle.com

[4] Group 3 Tech., http://www.group3technology.com

# Experience with operation of heavy ion superconducting accelerator ISAC-II and SRF activities at TRIUMF

**V Zvyagintsev, R E Laxdal and Z Yao**

4004 Wesbrook Mall, Vancouver, BC, Canada, V6T 2A3

zvyagint@triumf.ca

**Abstract**. The history of superconducting heavy ion accelerator ISAC-II started in 2000 with the development of niobium quarter wave cavities and cryomodules. The first stage of ISAC-II is in operation since 2006. In 2010 it was completed and made TRIUMF ISAC a leading ISOL facility supporting a full physics program with both stable and radioactive beams. TRIUMF is using its accumulated experience and resources for farther SRF development for ELINAC and external projects for VECC, RISP, FRIB and SLAC. Status of Superconducting ISAC-II accelerator and SRF development aspects, results and plans are discussed.

## 1. ISAC-II

SRF at TRIUMF began in 2000 with cavity and infrastructure development in support of the ISAC-II heavy ion linac as an extension of ISAC facility for ISOL based radioactive ion beam production and acceleration. In 2006 Phase-I of ISAC-II with an acceleration voltage of 20 MV was commissioned for operation [1]. In 2010 the design goal of ISAC-II for 40 MV of acceleration voltage was achieved with completion of Phase-II [2]. ISAC became a leading ISOL facility supporting a full physics program with both stable and radioactive beams being delivered including stable beams of 16O5+, 15N4+, 20Ne5+ and radioactive beams (and their stable pilot beams) of 26Na, 26Al6+, (26Mg6+), 6He1+, (12C2+), 24Na5+, (24Mg5+), 11Li2+, (22Ne4+) with 74Br14+ and other high mass beams from the charge state booster.

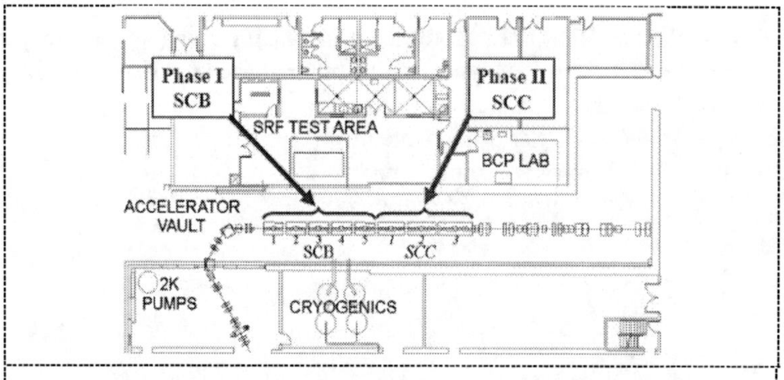


**Figure 1.** Layout of ISAC-II linac and SRF infrastructure.

The Phase-I segment (SCB section of Fig. 1) consists of twenty 106 MHz quarter wave cavities housed in five cryomodules with four cavities per cryomodule. The Phase-II (SCC section) consists of twenty 141 MHz QWR cavities at $\beta$=0.11 in three cryomodules with six cavities in each of the first two modules and eight cavities in the third (SCC section in Fig. 1). Both Phase-I and Phase-II cryomodules have one 9T superconducting solenoid symmetrically placed in the cryomodule.

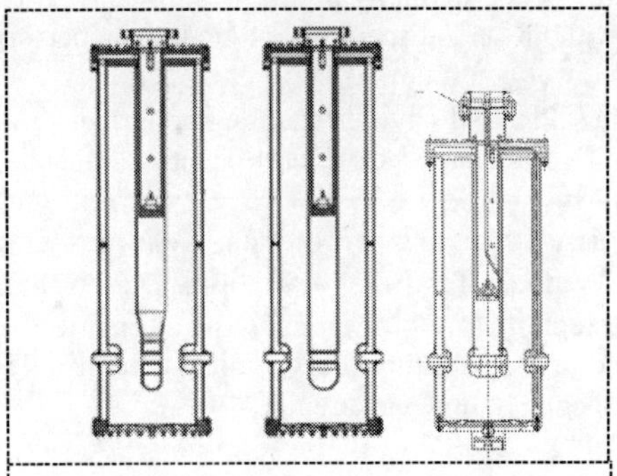

**Figure 2.** ISAC-II cavities for β=0.057, 0.071 and 0.11.

### *1.1. Cavities*

The first eight Phase-I cavities have a geometric β of 0.057 and the remainder a geometric β of 0.071 (Fig. 2). The cavity design was conducted in collaboration with INFN-LNL (Italy) with adoption of ALPI INFN-LNL coaxial bulk Nb cavities concept. The cryomodule design developed at TRIUMF is based on the rf vacuum volume open to the cryomodule isolation vacuum. The cavities are specified to operate at 106MHz and to provide an effective acceleration of 1.1MV for a cavity power of 7W at 4.2K with corresponding peak surface fields of 30MV/m and 60mT [3]. 20 Phase-I cavities were fabricated at Zanon (Italy) and assembled in 5 cryomodules designed and fabricated at TRIUMF.

**Table 1.** Parameters of ISAC-II cavities

| β | $f_o$ (MHz) | $R_sQ$ (Ohm) | $U/E_a^2$ (J/(MV/m)$^2$) | $E_a/E_p$ | $B_p/E_a$ (mT/(MV/m)) |
|---|---|---|---|---|---|
| 0.057 | 106.08 | 20.1 | 0.100 | 5.2 | 10.3 |
| 0.071 | 106.08 | 19.1 | 0.094 | 4.7 | 10.1 |
| 0.110 | 141.44 | 26.0 | 0.067 | 4.9 | 10.0 |

The Phase-II 141 MHz superconducting cavity with β of 0.11 is shown in Fig. 2. It was developed at TRIUMF and has a similar structure to the ISAC-II Phase-I linac cavity. The chief difference here besides the frequency is the inner conductor beam port region is outfitted with a donut style drift tube to improve the transit time factor. The Phase-II cavity has the same specification as the Phase-I cavities [4]. Twenty Phase-II cavities were produced by PAVAC Industries in Canada. Three cryomodules with high beta cavities were successfully commissioned in April 2010 [2]. An SRF infrastructure for SC development including SRF test area, clean room and chemical laboratory (Fig. 1) was created at TRIUMF. The parameters of Phase-I and II cavities are presented in the Table 1.

Tuning of the cavities is provided with deformation of Nb plates bolted to the bottom flange. A mechanical damper installed inside of the inner conductor provides >10 dB attenuation of microphonics noise. The cavities operate in strong overcoupled regime (coupling ~50-100) to provide enough bandwidth to maintain stable operation from microphonics. The LN2 cooled coupling loop produces <0.25 W power dissipation to the helium system at 200W forward power.

### *1.2. Operation experience.*

The performance of the cavities is monitored periodically during start-up after shutdown. The linac is warmed up once per year for three months as part of the site maintenance shutdown. The linac cavities operate with an average gradient $E_a$ corresponding to a peak surface field $E_p$ of 32 MV/m for Phase-I and 28.5 MV/m for Phase-II without any discernible reduction in performance. The reasons why Phase-II cavities performance is lower could be:

- Q-disease. Tests show that Phase-II cavities start degrading performance after 1h in the range of temperatures 200-100K, for Phase-I cavities it occurs after 10h. It could be due to higher hydrogen content in the Nb.
- During production of Phase-II cavities two of them were rejected due to vacuum leaks that opened in the donut weld after a final BCP of 100μm. Due to the tight schedule we limited BCP (in the beam tube region) to 60μm on subsequent cavities. All leaking cavities were successfully repaired and tested. Another four cavities were installed in cryomodules without single cavity cryostat tests due to time constraints.

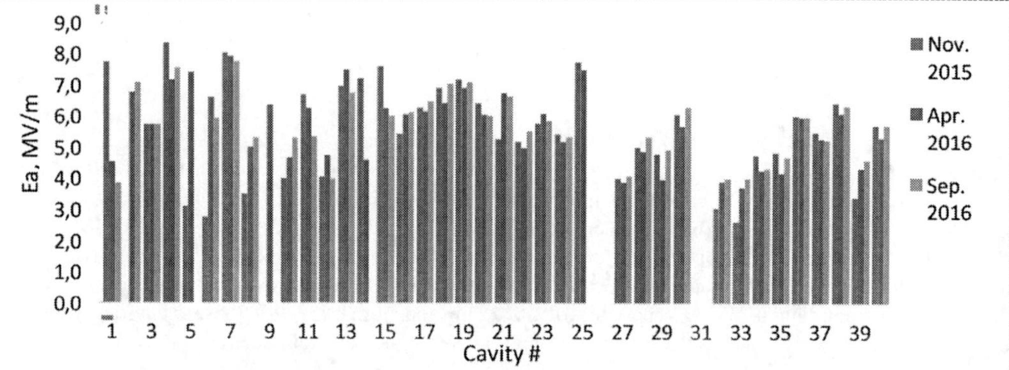


**Figure 3.** ISAC-II cavities performance as acceleration gradient $E_a$ measured at power dissipation of 7W on cavities walls during recent in situ cavity tests.

We experience some cavity failures for different reasons but it doesn't stop operation. Figure 3 shows the change of cavity performance and failures as a function of the time of operation; it could be restored after conditioning or MRO, dropped and even failed. Since every cavity has an independent RF system, we can compensate the performance of the unavailable cavities by increasing the gradient in other cavities (at power dissipation >7W).

We are conducting continuous development to upgrade the cavities systems and mitigate failures. These developments include:

- Replacement of coupler loop mechanical joints.
- Due to several failures of internal cables due to RF glow discharge in vacuum since 2015 we started replacing 3/8" for ½" ANDREW HELIAX cables with positive results.
- Replacing Phase-I couplers that use a rack and pinion mechanical arrangement and Teflon guide bearing with Phase-II couplers with a more robust design that includes non-magnetic cross-roller bearings and symmetric loading. This has improved the mechanical motion which is important because the ½" cable is more rigid and provides more side load for the coupler mechanism.
- During maintenance we are doing high pressure rinsing and sometimes light etch to recover the cavities performance.

During operation cryogenics failures can cause cavity degradation after recovery including the effects of:

IOP Conf. Series: Journal of Physics: Conf. Series **941** (2017) 012105 doi:10.1088/1742-6596/941/1/012105

- Trapped magnetic flux from short interruption of LHe supply. Full recovery (~two hours activity) involves degaussing the solenoid and environs, then warming cavities and solenoid to 30K to quench all solenoid trapped flux, then recooling the cold mass [5].
- Q-disease due to long interruption of LHe supply. Full recovery requires cavity warmup to room temperature.

Low level multipacting in some cavities is responsible for the delay of start-up and tuning. It is three orders of magnitude less than the operational field level and doesn't affect performance. Pulse RF conditioning in self-excited loop is required to start these cavities and sometimes it takes a significant time delaying beam delivery. We implemented a driven option for multipacting conditioning from a signal generator and apparently we see that it is more efficient. Multipacting disappears during cavity operation and reappears after warmup.

The Phase-I system uses tube amplifiers and they have been a source of downtime due to tube aging issues causing phase drift and non-linear output affecting LLRF operation. Replacement solid state amplifiers have been developed and successfully tested in operation. All tube amplifiers will be replaced with new solid state amplifiers in the summer of 2018

For the future we intend to make a development for degassing of Phase-II cavities to eliminate Q-disease issue and increase the cavities operational gradient.

## 2. Other SRF activities at TRIUMF

The accumulated experience and resources from ISAC-II linac development were used for farther SRF development at TRIUMF including ELINAC and external VECC, RISP, FRIB and SLAC projects.

### *2.1. ELINAC*

The ARIEL project [6] will allow an increase in the radioactive ion beam (RIB) hours with the addition of a new electron linac (ELINAC) driver of 50 MeV (0.5 MW), a new proton line from the 500MeV cyclotron and new production target stations. Accelerated electrons can be used to generate RIBs via the photo-fission process. The electrons are stopped in a converter to generate bremsstrahlung photons for fission in actinide target material. An electron beam intensity and energy of 10 mA and 50 MeV is required for a fission rate of $10^{13}$ fissions/sec.

The electron linac is housed in a pre-existing shielded experimental hall adjacent to the TRIUMF 500 MeV cyclotron that has been re-purposed as an accelerator vault. The e-Linac is being installed in a phased way with stages shown schematically in Fig. 4.

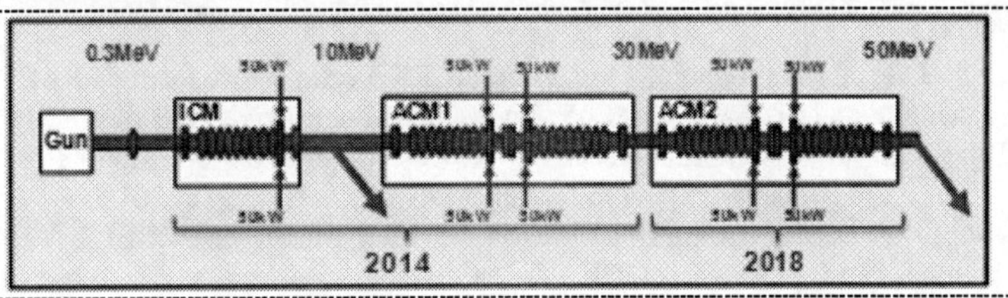


**Figure 4.** The stages of the ELINAC project.

A first phase consisting of a 300 kV 16 mA electron gun, an injector cryomodule, ICM, containing one 1.3 GHz nine-cell cavity and an accelerating cryomodule, ACM1, that now contains one 1.3 GHz nine-cell cavity (and eventually two cavities) plus associated beamlines is now installed and is being commissioned with one cavity in ACM1 at energy of 23 MeV[7]. This first phase is designed to accelerate CW up to 10 mA of electrons at 30 MeV. The second cavity for ACM1 finally was fabricated and installed for commissioning in 2017. The initial beam dumps and production targets will only be compatible with 100 kW operation. A second phase, dependent on funding, will see the addition of a second accelerating module, ACM2, and a ramp up in beam intensity to the full capability of 50 MeV 0.5 MW.

### *2.2. External projects*
In frame of collaboration agreement with VECC (India) TRIUMF developed and successfully commissioned a copy of the ELINAC injector cryomodule ICM2 and 30 kW CW IOT Transmitter which was used for ELINAC power coupler conditioning. The IOT Transmitter was used also for 4kW 1.3GHz couplers conditioning for the SLAC LCLS-II project.

TRIUMF SRF infrastructure is extensively involved in testing of SC QWR and HWR cavities for RISP. We are conducting development of a novel Spoke cavity for RISP. We developed and fabricated variable test couplers for SRF tests of FRIB SC QWR and HWR cavities.

### *2.3. SRF development*
TRIUMF is developing design and 'in house' fabrication of SC Deflecting cavity for ELINAC ERL separator. We are conducting a series of developments for SRF technology: μSR material samples study, vertical electro polishing, induction oven for cavities degassing and doping, T-map for SRF cavities tests.

## 3. Summary
In 2017 we are going to complete Phase-I of ELINAC with installation and commissioning of second cavity in ACM cryomodule and completion of the RF System.

We are going to proceed with development for ISAC-II cavities performance and reliability.
External projects and collaborations help to raise expertise and extend competency of TRIUMF SRF team.

## References
[1] Laxdal R E, "Commissioning and Early Experiments With ISAC-II," PAC'07, Albuquerque, New Mexico, USA, June 2007, THXAB01, p. 2593-2597 (2007)

[2] Laxdal R E et al., "Operating Experience of the 20MV Upgrade Linac," LINAC'10, Tsukuba, Japan, September 2010, MO202, p. 21-25 (2010)

[3] Facco A et al., "The Superconducting Medium Beta Prototype for Radioactive Beam Acceleration at TRIUMF," PAC'01, Chicago, USA, June 2001, MPPH134, p. 1092-1094 (2001)

[4] Zvyagintsev V et al., "Production and Testing Results of Superconducting Cavities for ISAC-II High Beta Section," PAC'09, Vancouver, Canada, June 2009, TU4PBC04, p. 786-788 (2009)

[5] Laxdal R E, et al, "Cryogenic, Magnetic and RF Performance of the ISAC-II Medium Beta Cryomodule at TRIUMF", PAC'05, Knoxville, May 2005, TPPT052, p. 3191-3193 (2005)

[6] Merminga L et al., "ARIEL: TRIUMF's Advanced Rare Isotope Laboratory", WEOBA001, IPAC'11, San Sebastian, Spain, September 2011, p. 1917-1918 (2011)

[7] Zvyagintsev V et al., "Commissioning of the SRF Linac for ARIEL", TUAA02, Whistler, September 2015, p. 457-461 (2015)

IOP Conf. Series: Journal of Physics: Conf. Series **941** (2017) 012106 doi:10.1088/1742-6596/941/1/012106

# Cloud-based design of high average power traveling wave linacs

**S V Kutsaev[1], Y Eidelman[2], D L Bruhwiler[2], P Moeller[2], R Nagler[2] and J Barbe Welzel[2]**

[1]RadiaBeam Technologies LLC, 1717 Stewart St, Santa Monica, CA 90403, USA,
[2]RadiaSoft LLC, 1348 Redwood Ave, Boulder, CO 80304, USA

kutsaev@radiabeam.com

**Abstract.** The design of industrial high average power traveling wave linacs must accurately consider some specific effects. For example, acceleration of high current beam reduces power flow in the accelerating waveguide. Space charge may influence the stability of longitudinal or transverse beam dynamics. Accurate treatment of beam loading is central to the design of high-power TW accelerators, and it is especially difficult to model in the meter-scale region where the electrons are nonrelativistic. Currently, there are two types of available codes: tracking codes (e.g. PARMELA or ASTRA) that cannot solve self-consistent problems, and particle-in-cell codes (e.g. Magic 3D or CST Particle Studio) that can model the physics correctly but are very time-consuming and resource-demanding. Hellweg is a special tool for quick and accurate electron dynamics simulation in traveling wave accelerating structures. The underlying theory of this software is based on the differential equations of motion. The effects considered in this code include beam loading, space charge forces, and external magnetic fields. We present the current capabilities of the code, provide benchmarking results, and discuss future plans. We also describe the browser-based GUI for executing Hellweg in the cloud.

## 1. Introduction

High power travelling wave (TW) linear accelerators are the most promising solution for high-current applications that include: treatment of potable and waste water, removal of pollutants from stack gases, increased efficiency of material processing; and replacement of radioactive sources in sterilization application. The pulsed current in such linacs can be as high as tens of amperes. Such high currents can cause a number of specific effects, which must be accurately considered in the design stage. One of the most important of them is a beam loading effect, when a high current beam reduces the accelerating electric field amplitude and, hence, the output beam energy. Another important effect is the space charge forces acting in the bunch, which can be comparable to the single-particle electromagnetic forces at low energy. The space charge algorithm should also include the influence of neighbouring bunches, since in industrial linacs the beam typically consists of a train of bunches filling every RF period.

There is no commercially available software that provides fast, accurate calculations of beam loading effects in high power TW linear accelerators. For example, the commercial GPT code can only model beam loading for relativistic beams. The popular PARMELA code does not model beam loading at all, so the correct amplitude and phase in each accelerating cell must be calculated externally. Commercial codes like Magic 3D and CST Particle Studio can model the physics correctly;

however, they are expensive and the required run time is long, so multiple parameter optimization for system design is impractical.

Meanwhile, a rather simple and convenient method for intensive beam dynamics in TW accelerating structures analysis has been proposed [1] and realized in the Hellweg code [2]. This code is based on differential equations that can accurately simulate beam dynamics with beam loading, space charge, and external magnetic fields. We have recently enhanced the Hellweg code with improved algorithms and a cloud-based Graphical User Interface (GUI) to make this code available to the accelerator community.

## 2. Hellweg code overview

Hellweg was originally developed at the RF Technology Laboratory of Moscow Engineering-Physics Institute. Hellweg is a C++ application that runs on Microsoft Windows with a GUI. The kernel can be executed from the command line on GNU/Linux. The input data for this program is a plain text file that specifies the beam (initial phase, energy and phase distributions, input current, Twiss parameters), the traveling wave structure (input power, frequency and cell parameters), and computational parameters (number of macroparticles, mesh, etc.).

The numerical model is based on self-consistent equations describing electron motion in waveguide structures with variable dimensions. This system includes 2D dynamics in an axisymmetric structure; an equation for a self-consistent RF-field amplitude created by the beam; and an equation for a particle's phase in a self-consistent field. The code used the simplified ellipsoid-based single-bunch space charge model [3] and simple approximations of external magnetic fields.

Modern accelerator design, however, can require non-axially-symmetric field profiles such as multipole magnetic fields or accelerating structures with a cyclic variation of azimuthal asymmetries, which requires fully 3D dynamics. This then required the space charge model to be improved to the general 3D ellipsoid form for arbitrary aspect ratio, consideration of the space charge field for particles outside the beam, and adding fields from the neighboring bunches (which are not independently simulated but must be included approximately). Some applications (e.g. novel cargo inspection systems based on energy-ramping accelerators) require the consideration of realistic external magnetic fields, usually imported from pre-calculated models, and the ability to properly model the initial beam distribution including spherical cathodes and asymmetric 4D Twiss parameters. Some users require the ability to use Hellweg for only part of the accelerator system, for which robust import/export of multiple file formats is required.

A browser-based GUI for cloud computing is necessary to attract a significant user base in both the commercial/engineering and research communities. Scientific cloud computing is the only practical way to sustainably support software and users in the absence of significant funding. As described below, this approach also enables instantaneous collaboration and computational reproducibility.

We have recently made the following improvements to the code:

- converted the code to 3D model
- revised the beam initialization
- added new results for visualization
- enhanced the parametrization of solenoid fields and added fringe field effects
- improved the space charge model to 3D ellipsoid, including fields outside of the bunch
- added multi-bunch space charge model
- added import/export of beam to/from CST Particle Studio and ASTRA
- added import of 1D, 2D and 3D magnetic field maps
- added quadrupole elements with predefined 2D field maps
- developed a browser-based GUI, which exports Python script for command-line use

In the following section, we briefly discuss some of these enhancements.

## 3. New three-dimensional model

Hellweg was originally developed to solve beam dynamics problems with 2D cylindrical symmetry. This approach is sufficient for standard DLS-type accelerating structures with only the fundamental accelerating cavity mode; however, it is not adequate for the wide range of 3D systems, nor for problems with broken 2D symmetry: solenoids, quadrupole optical elements, higher order modes (HOM), asymmetrical structures etc.

We have derived from first principles the 3D differential equations for all velocity components ($\beta_z$, $\beta_r$ and $\beta_\theta$), where $\beta=v/c$. The equations include all six RF field components ($E_z$, $E_r$, $E_\theta$, $H_z$, $H_r$, $H_\theta$) and three external magnetic field components ($H_z^{ext}$, $H_r^{ext}$, $H_\theta^{ext}$):

$$\frac{d\beta_\zeta}{d\zeta} = \frac{1}{\gamma\beta_\zeta}\left[(1-\beta_\zeta^2)A_\zeta + \beta_\eta(H_\theta - \beta_\zeta A_\eta) - \beta_\theta(H_\eta + \beta_\zeta A_\theta) + \beta_\eta H_\theta^{ext} - \beta_\theta H_\eta^{ext}\right] \tag{1}$$

$$\frac{d\beta_\eta}{d\zeta} = \frac{1}{\gamma\beta_\zeta}\left[(1-\beta_\eta^2)A_\eta + \beta_\theta(H_\zeta - \beta_\eta A_\theta) - \beta_\zeta(H_\theta + \beta_\eta A_\zeta) + \beta_\theta H_\zeta^{ext} - \beta_\zeta H_\theta^{ext}\right] + \frac{\beta_\theta^2}{\eta\beta_\zeta} \tag{2}$$

$$\frac{d\beta_\theta}{d\zeta} = \frac{1}{\gamma\beta_\zeta}\left[(1-\beta_\theta^2)A_\theta + \beta_\zeta(H_\eta - \beta_\theta A_\zeta) - \beta_\eta(H_\zeta + \beta_\theta A_\eta) + \beta_\zeta H_\eta^{ext} - \beta_\eta H_\zeta^{ext}\right] - \frac{\beta_\theta\beta_\eta}{\eta\beta_\zeta} \tag{3}$$

Here all values are normalized by the wavelength λ or the particle rest energy $W_0$ as follows:

$$\zeta = \frac{z}{\lambda},\ \eta = \frac{r}{\lambda},\ \alpha = \frac{r}{\lambda}\dot{\theta} = \eta\dot{\theta},\ A = \frac{E\lambda}{W_o},\ H = c\frac{B\lambda}{W_o} \tag{4}$$

These equations were derived from the following fundamental equation of motion:

$$\frac{d\vec{v}}{dt} = \frac{e}{m_0}\sqrt{1-\frac{v^2}{c^2}}\left(\vec{E} + \frac{\vec{v}\times(\vec{B}+\overrightarrow{B_{ext}})}{c} - \frac{\vec{v}(\vec{v}\cdot\vec{E})}{c^2}\right) \tag{5}$$

To take into account the space charge of the particle we have solved the Poisson equation for the ellipsoid beam and obtained the following adjustment component values for the electric field:

$$E_r = \frac{\rho}{2\varepsilon_0\gamma^2}\left(M_x(\lambda)\cdot x\cos\theta + M_y(\lambda)\cdot y\sin\theta\right) \tag{6}$$

$$E_\theta = \frac{\rho}{2\varepsilon_0\gamma^2}\left(M_y(\lambda)\cdot y\cos\theta - M_x(\lambda)\cdot x\sin\theta\right) \tag{7}$$

$$E_z = \frac{\rho}{2\varepsilon_0}M_z(\lambda)z \tag{8}$$

Here $M_{x,y,z}(\lambda)$ is a form factor of the beam defined as follows:

$$M_u(\lambda) = R_xR_yR_zu\int_\lambda^\infty \frac{ds}{(R_u^2+s)\sqrt{(R_x^2+s)(R_y^2+s)(R_z^2+s)}} \tag{9}$$

In Eq. (9), $R_x$, $R_y$, and $R_z$ are the bunch dimensions, $u$ represents a coordinate $x$, $y$ or $z$, and $\lambda$ is a root of a characteristic equation:

$$\frac{x^2}{R_x^2+\lambda} + \frac{y^2}{R_y^2+\lambda} + \frac{z^2}{R_z^2+\lambda} = 1 \tag{10}$$

A multi-bunch space charge effect can be included by solving the characteristic equation for two neighbouring bunches traveling ahead and behind of the reference bunch, at a distance L:

$$E_z = \frac{\rho}{2\varepsilon_0}\left[M_z(\lambda)z + M_z(\lambda_1)(z-L) + M_z(\lambda_2)(z+L)\right] \tag{11}$$

The detailed derivation and analysis of the formulas (1)-(11) will be presented elsewhere.

Support for 3D dynamics required major revision of the particle initialization algorithm. Previously, the beam's transverse coordinates could only be initialized via three Twiss parameters: $\alpha$, $\beta$ and ε. Moreover, the distribution was Gaussian in the $x$-coordinate, not in the radial variable, which can be technically incorrect for realistic beams. We have recently added the following options of beam initialization, which cover many cases of realistic 3D distributions:

- 4D Twiss parameters: $\alpha_x$, $\beta_x$ [cm/rad], $\varepsilon_x$ [cm*rad] $\alpha_y$, $\beta_y$ [cm/rad], $\varepsilon_y$ [cm*rad]
- Gaussian longitudinal distribution; W0 [MeV] ΔW [MeV] $\sigma_W$ [MeV] $\varphi_0$ [deg] Δφ [deg] $\sigma_\varphi$ [deg]

- Emission from a spherical cathode: $R_{cath}$ [cm], $R_{sph}$ [cm], kT [eV]
- Kapchinsky-Vladimisky distribution: $a_x$ [cm], $b_y$ [cm], φ [deg], h

Generic beam initialization does not always provide the required particle distribution, especially when the electrons come from an external simulation of a particle source, such as a gun or another accelerator. Also, users often need to track a single particle with defined parameters, or a beam with some exotic distribution. For these cases, we added the capability to import a sequence of particles from a text file.

## 4. Example of industrial linac design

We benchmarked the code on simple problems that allow for analytical calculation. Such tests included:

- Beam in a drift pipe with and without magnetic field
- Beam in RF accelerating structure
- Bunching
- Acceleration
- Beam loading
- Emittance conservation

Also, we have designed an industrial linac with Hellweg and compared the simulations with the PIC module of CST Particle Studio [4]. The linac is designed for a maximum power of 9 MeV, with a 2π/3 constant impedance travelling wave structure. The accelerator is powered by a klystron with 5 MW peak power at operating frequency of 2856 MHz. The accelerating structure consists of 25 cells, including lower-beta bunching cells. Beam focusing is performed with three 0.1T solenoids. The linac design is presented in Figure 1.

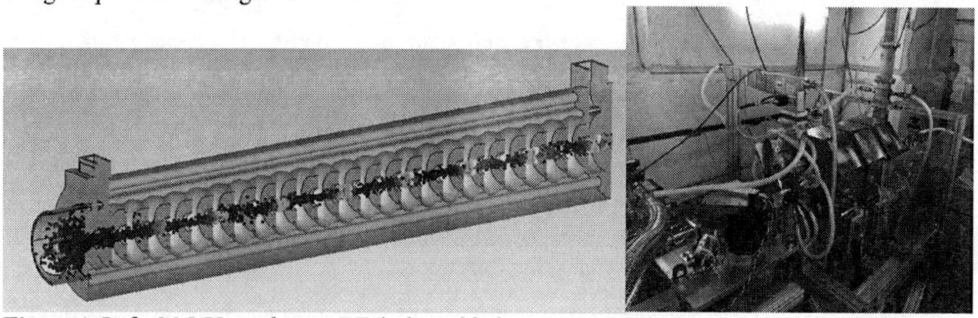

**Figure 1.** Left: 9 MeV accelerator RF design with the propagating beam (color represents the energy up to 9 MeV). Right: The actual picture of accelerator.

The comparison of the output beam parameters from Hellweg and CST are presented in Table 1 and Figure 2 and demonstrate a very good agreement.

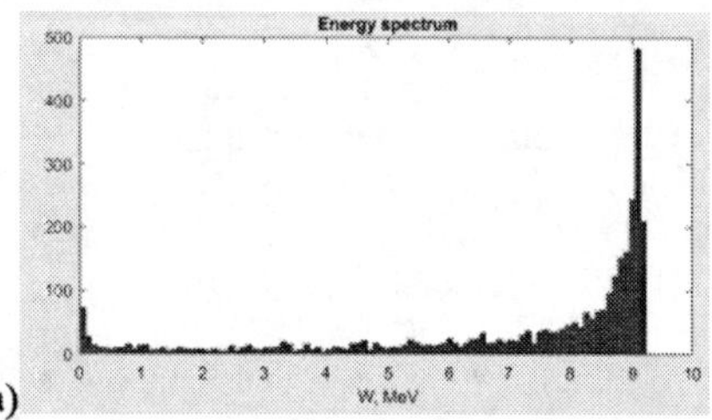


a)

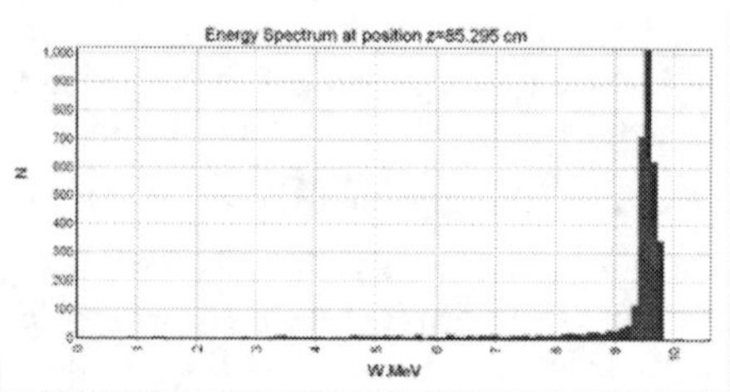


b)

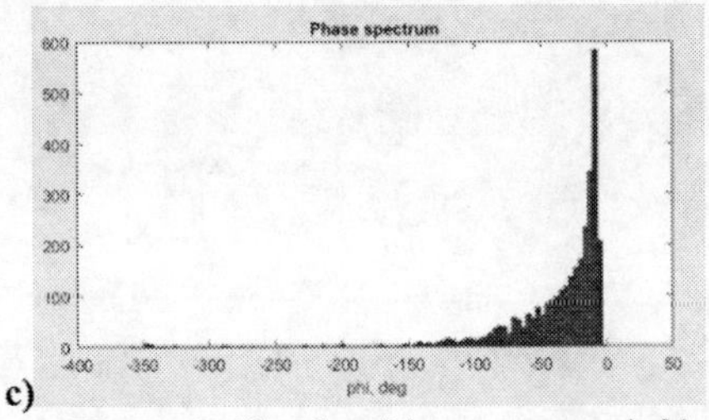


c)

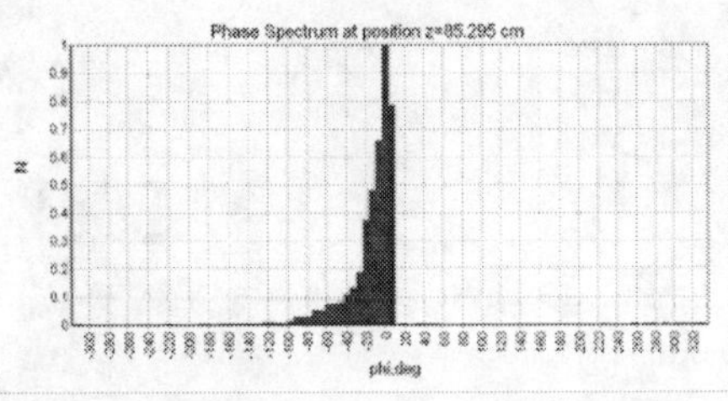


d)

**Figure 2.** Simulated beam energy (a,b) and phase (c,d) spectra in CST Particle Studio (a,c) and Hellweg (b,d).

**Table 1.** Comparison of Hellweg and CST simulations.

| Code | Hellweg | CST |
|---|---|---|
| Input current | 250 mA | 250 mA |
| Output current | 124 mA | 115 mA |
| Beam energy | 9.6 MeV | 9.2 MeV |
| Input power | 4.6 MW | 4.6 MW |
| Output power | 1.73 MW | 1.72 MW |
| Simulation time | ~30 sec | ~8 hrs |

## 5. Browser-based GUI for cloud computing

We have ported the Hellweg physics kernel from Windows to the gcc compiler on GNU/Linux, including an initial Python framework for unit and regression testing, while retaining the flexibility to compile with the original GUI for Windows. [5] The source structure is designed to enable delivery of the Hellweg kernel via PyPI [6] in the future, from the command line via 'pip install rslinac'. Using the open source framework Sirepo [7], we have developed an initial browser-based GUI for Hellweg and made it publicly available for beta testing. [8] Sirepo allows easy integration of legacy, open source simulation codes. [9] The platform is designed to encourage reproducible research where sharing of simulations is crucial. [10] Figure 3 shows a screen shot of dynamics in an RF cavity.

Sirepo GUIs invoke JavaScript in the browser, while Python is invoked on a remote Linux server to execute the requested scientific kernel (e.g. Hellweg) within a Docker [11] container. Sirepo apps work on a tablet as well as a desktop computer, and they enable instantaneous collaboration – one can send a link to anyone, allowing them to immediately copy the full simulation and start working with it. One can also export a working Python script with all the associated input files, so the GUI does not prevent expert users from working on the command line.

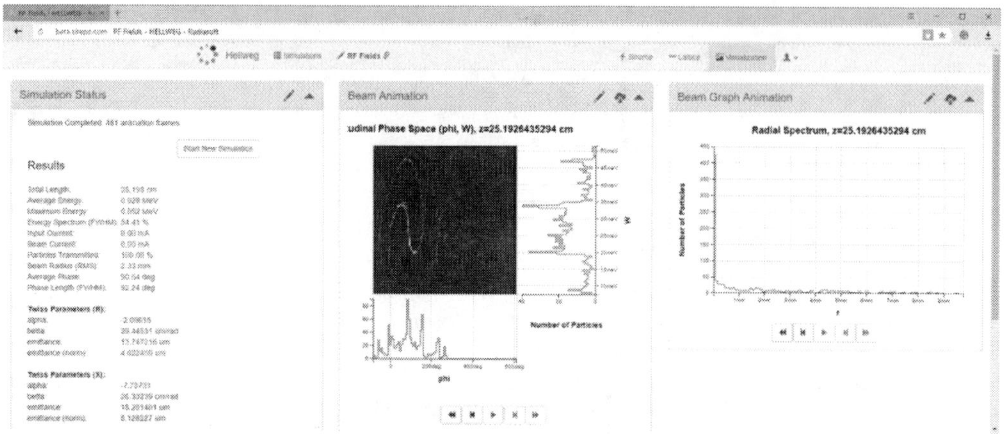


**Figure 3.** Screenshot of browser-based Sirepo GUI, running an example Hellweg simulation.

## 6. Conclusion

The updated Hellweg presents the scientific and commercial opportunity to enable cloud-based design of high average power traveling wave linacs. The improved physics of Hellweg and the implementation it within Sirepo as an open source framework provides a fast, accurate and convenient tool for cloud computing of electron beam dynamics problems to accelerator physicists community.

## References

[1] Masunov E.S., "Bunching and Acceleration of the Intensive Beam in the Waveguide Accelerating Structure with a Presence of an External Magnetic Field", Journal of Technical Physics 49, pp. 1462-1463 (1979).

[2] Kutsaev S.V., "Electron dynamics simulations with Hellweg 2D code", Nuclear Instruments and Methods in Physics Research A618, pp. 298–305 (2010).

[3] Lapostolle P. M., "Effets de la charge d'espace dans un accélérateur linéaire `a protons," CERN, Geneva, Tech. Rep. CERN-AR-Int-SG-65-15, Jul 1965, in French

[4] Kutsaev S.V. et al., "Design, Simulations and Experimental Demonstration of an Intra-Pulse Ramped-Energy Travelling Wave Linac for Cargo Inspection", Proc. of NAPAC'16 Conference, 2016

[5] The Hellweg source repository on GitHub, https://github.com/radiasoft/rslinac

[6] PyPI – the Python Package Index, https://pypi.python.org/pypi

[7] The Sirepo framework for cloud computing, http://sirepo.com

[8] Open beta test of Hellweg in the cloud, https://beta.sirepo.com/hellweg

[9] Nagler R., Moeller P. and Bruhwiler, D.L. "Sirepo: An Open Source Platform for Portable Reproducible Simulations," https://doi.org/10.6084/m9.figshare.4688566.v1

[10] Nagler R., Bruhwiler D.L., Moeller P. and Webb S.D., "Sustainability and Reproducibility via Containerized Computing," https://arxiv.org/abs/1509.08789

[11] What is Docker," https://www.docker.com/what-docker

**Acknowledgement**

This material is based upon work supported by the U.S. Department of Energy, Office of Science, Office of High Energy Physics, under Award Number DE-SC0015897.

# Method of the production and trapping of thorium ions for nuclear transition investigation

**P V Borisyuk [1], S P Derevyashkin[1], K Yu Khabarova [2,3], N N Kolachevsky[1,2,3], Yu Yu Lebedinsky[1], S S Poteshin[1], A A Sysoev[1], E V Tkalya[1,4,5], D O Tregubov [1,2,3], V I Troyan [1], O S Vasiliev[1] and V P Yakovlev[1]**

[1]National Research Nuclear University MEPhI (Moscow Engineering Physics Institute), 115409, Kashirskoe sh. 31, Moscow, Russia
[2]P.N. Lebedev Physical Institute of the Russian Academy of Sciences, 119991, 53 Leninskiy pr., Moscow, Russia
[3]Moscow Institute of Physics and Technology (State University), 141700, 9 Institutskiy per., Dolgoprudny, Moscow region, Russia
[4]Lomonosov Moscow State University Skobeltsyn Institute of Nuclear Physics (MSU SINP), 119991, 1b2 Leninskie gory, Moscow, Russia
[5]Nuclear Safety Institute of the Russian Academy of Sciences (IBRAE RAN), 115191, 52 Bolshaya Tulskaya st., Moscow, Russia

E-mail: sp.derevyashkin@gmail.com

**Abstract**. This paper describes the techniques of production and trapping of thorium ions required to perform spectroscopic studies of thorium nuclear transition.

## 1. Introduction

Laser spectroscopy plays a key role in the solving fundamental physics problem such as the testing of the general theory of relativity [1], the determination of fundamental constants [2], the development of new quantum technologies [3], and the creation of precise frequency standards [4].

The currently achieved fractional frequency uncertainty does not allow to perform some fundamental physics investigation, such as testing the theory of fundamental constants variation. The usage of nuclear transitions instead of atomic ones would allow us to increase the precision of measurements by several orders of magnitude [5]. Nucleus is characterized by extremely low polarizability and is shielded from external perturbation by an electron cloud. The narrowness of the spectral line and the low transition intensity make nuclear transitions a good candidate for creation new generation of frequency standards. Unfortunately, the typical energies of nuclear transitions are within the keV to MeV range, no common sources of coherent radiation are available at this spectral range. The unique exception is the isomeric transition in the thorium-229 isotope nucleus, its energy, according to the latest data, is $7.8 \pm 0.5$ eV [6].

To study the isomeric state of thorium-229, $^{229}Th^{+}$ and $^{229}Th^{3+}$ ions are used, the former is relatively easy to obtain, it has a complex system of electronic levels, so it can be used in the realization of the electronic bridge mechanism for excitation of the isomeric state [7], and the latter is most convenient for investigate isomeric transition energy, due to the high ionization potential (27 eV) and the simple electronic levels structure [8].

In addition to the enormous technical difficulties associated with the measurement of an isomeric nuclear transition energy, production of the thorium ions and trapping it for subsequent laser cooling and spectroscopic studies, are also a serious problem. The isotope of thorium-229 can only be produced artificially and only in very small quantities. In addition, thorium-229 isotope is very radioactive. In the experiments to determine the energy of the isomeric transition in the thorium-229 nucleus researchers use, either individual ions are placed in an ion trap [9], or thorium-229 ions are embedded into a crystal transparent in the vacuum ultraviolet range [10]. In this paper, the methods of production and trapping thorium ions in a quadrupole linear ion trap are reported.

## 2. Production of ions

Thorium ions are generated by laser ablation. During short laser pulse plasma plume containing large number of particles (~$10^{14}$) is generated. Large number of ions, inside trap at the same time, allows you to trap sufficient number of ions. In addition, the laser ablation provides local evaporation of the target [11], which is good for working with a limited target material. This is important because of $^{229}$Th radioactivity. Allowed sample amount of $^{229}$Th in laboratory is nanograms of matter.

Due to the high radioactivity of the isotope $^{229}$Th initial stages of work were carried out using a model sample of monoisotopic and chemically inert gold 197Au and stable thorium isotope $^{232}$Th.

Laser ablation system is based on Q-switched Nd:YAG laser. Laser pulse parameters are following: duration of the laser pulse $\tau$ = 25 ns, pulse energy $E$ = 50 mJ, the laser spot radius on the sample $r$ = 100 μm. Thus the obtained power density is about $I$ ~ 1 GW/cm$^2$. Mass spectra of gold ions, generated via laser ablation, obtained by quadrupole mass filter is shown in Fig. 1.

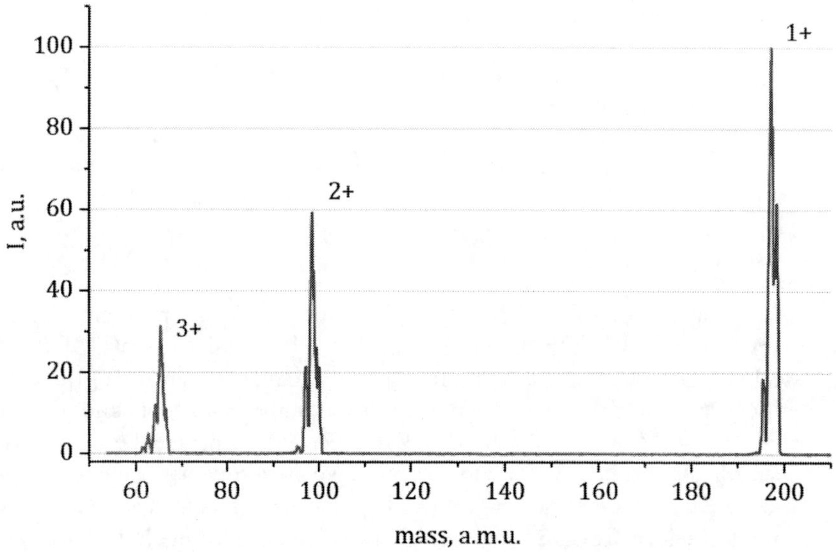


**Figure 1.** Mass spectra of gold ions generated by laser ablation.

## 3. The trapping of ions

Multisectional quadrupole linear ion trap [12] was used for trapping ions. The trap (Fig. 2) consists of five separated quadrupole segments, which makes it possible to vary the potential of each quadrupole individually and thus forming a potential profile for capturing and keeping ions in a wide energy range of 1 – 500 eV. Each quadrupole section consists of four stainless steel cylindrical rods with a diameter of 8 mm mounted at a diagonal distance of 7.1 mm between the surfaces of the rods.

Trap's rods are supplied with voltage oscillating at a frequency 1.22 MHz. In addition, a DC bias voltage can be applied to one of the sections to perform mass filtering. Power supplies allow mass filtering a range of 2 – 250 Da, with a resolution ~1 Da.

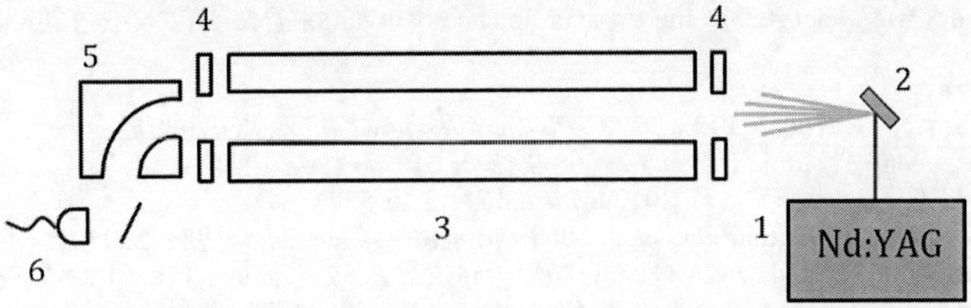


**Figure 2.** Scheme of experimental setup. 1 – ablation Nd:YAG laser, 2 – sample, 3 – quadrupole linear Paul ion trap, 4 – end cap electrodes, 5 – energy analyser, 6 – secondary electron multiplier.

Trapping process was implemented by supplying DC voltage to end cap electrodes, synchronized (synchronization time 1 μs) with the laser pulse. End cap electrodes have 2 mm diameter apertures coaxial with the ion trap axis. Value of "trapping potentials" applied to the electrodes ~100 V.

Trapping time of up to 100 seconds was demonstrated by registration of secondary electron multiplier signal, corresponded to ions released from the trap. In Fig. 3, dependence of the ions number vs the time of their registration by secondary electron multiplier for different trapping times are presented.

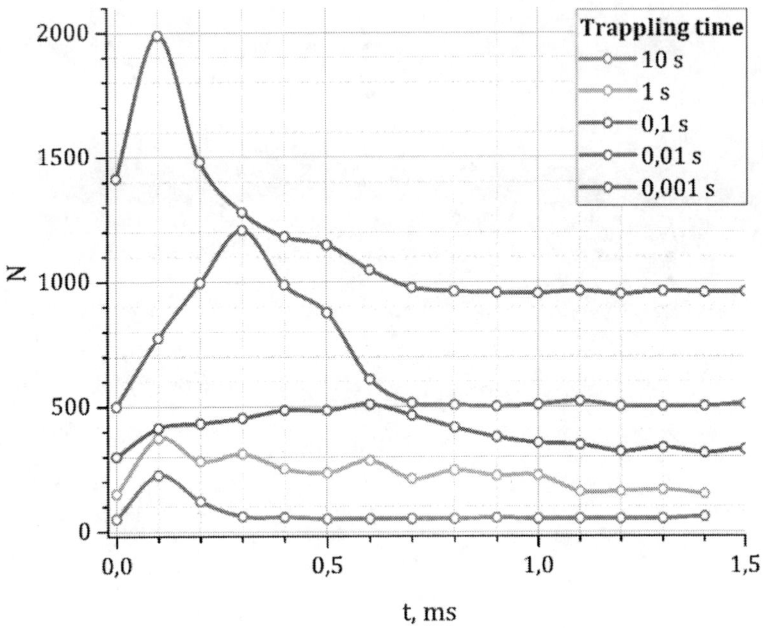


**Figure 3.** Dependence of the ions number vs the time of their registration by secondary electron multiplier, for different trapping times.

Maximum number of ions are registered about 0.3 ms. The appearance of slower ions at long trapping times ~ 1 s can be explained by ions interaction with gas molecules in a vacuum chamber evacuated at a pressure of ~ $10^{-7}$ Torr.

**Acknowledgments**
The work was supported by the Russian Science Foundation (project No. 16-12-00001).

**References**
[1] Ashby N, Heavner T P et al. 2007 *Physical Review Letters* **98** 070802.
[2] Rosenband T, Hume D B et al. 2008 *Science* **319** pp 1808-12.
[3] Ladd T D, Jelezko F et al. 2010 *Nature* **464** pp 45-53.
[4] Rosenband T, Schmidt P O et al. 2007 *Physical Review Letters* **98** p 220801.
[5] Campbell C J, Radnaev A G et al. 2012 *Physical Review Letters* **108** p 120802.
[6] Beck B R, Becker J A et al. 2007 *Physical Review Letters* **98** p 142501.
[7] Porsev S G and Flambaum V V 2010 *Physical Review A* **81** p 042516.
[8] Peik E and Tamm C 2003 *Europhys. Lett.* **61** p 181.
[9] Campbell C J, Steele A V et al. 2009 *Physical Review Letters* **102** p 233004.
[10] Rellergert W G, DeMille D et al. 2010 *Physical Review Letters* **104** p 200802.
[11] Troyan V I, Borisyuk P V et al. 2013 *Physical Review Letters* **10** p 105301.
[12] Troyan V I, Borisyuk P V et al. 2015 *European Journal of Mass Spectrometry* **21** pp 1-12.

IOP Conf. Series: Journal of Physics: Conf. Series **941** (2017) 012108 doi:10.1088/1742-6596/941/1/012108

# Methodical errors caused by superposition of electrical and optical signals in a heterodyne laser interferometer

**T V Kazieva[1], A P Kuznetsov[1], M V Ponarina[1], K L Gubskiy[2], S V Norkin[1], K S Lukyanov[1], A A Rusakov[3], K O Baranova[3] and I I Maslenikov[3]**

[1]National Research Nuclear University MEPhI (Moscow Engineering Physics Institute), Kashirskoe sh. 31, 115409, Moscow, Russia
[2]LaserEye LLC, 115409, 31 Kashirskoye sh., Moscow, Russia
[3]Technological Institute for Superhard and Novel Carbon Materials, 142190, 7a Centralnaya st., Troitsk, Moscow, Russia

E-mail: glizerogen@gmail.com

**Abstract.** Distortions of the displacements measured by the heterodyne interferometer due to penetration of the electric excitation signal of the acousto-optic modulator into the path of registration and processing of optical signals are considered. The level of this type of noise is estimated and the ways of its elimination from data obtained using a three-coordinate heterodyne interferometer are proposed.

## 1. Introduction

Nowadays, the metrological certification of measuring equipment operating in the nanometer and micron range of displacements is performed in several ways. One of them is the development of special equipment combining the capabilities of atomic force microscopy and laser interferometry. The combination of a three-axis heterodyne laser interferometer and a nanopositioner makes it possible to create metrological atomic force microscopes (AFM) that measure the vertical and horizontal dimensions of microscopic objects under study with nanometer uncertainty and ensure the traceability of such measurements to the standard meter [1–2].

## 2. Interferometer

Laser interferometers for linear displacement measurement can be divided into two main types by the method of information signal generation: homodyne and heterodyne ones. The peculiarity of the heterodyne interferometer is the transfer of the measured optical phase shift $\delta$ to the radio frequency range. The transfer of an informative signal to an operating frequency significantly exceeding the flicker-noise band of a photodetector and a laser, and the subsequent narrow-band filtering allow the signal-to-noise ratio to be increased by several orders of magnitude.

An example of optical scheme of a three-coordinate heterodyne laser interferometer with a scanning field up to 100×100 μm in the sample plane and up to 10 μm vertically is shown in Fig. 1. The heterodyne frequency shift in this interferometer is realized with the acousto-optic modulator (AOM). Mirrors M1–M11 form three heterodyne interferometers that measure the movement of a nanopositioner, on which triple prisms (TP) are mounted in three orthogonal planes serving as reflectors [3]. The polarization is matched by means of a half-wave phase plate $\lambda/2$. After optical mixing on mirrors M9, M10, and M11, radiation with the help of collimators K is introduced into optical fibers, through which it is transported to an analog-digital module for photoregistration and processing of interferometer signals.

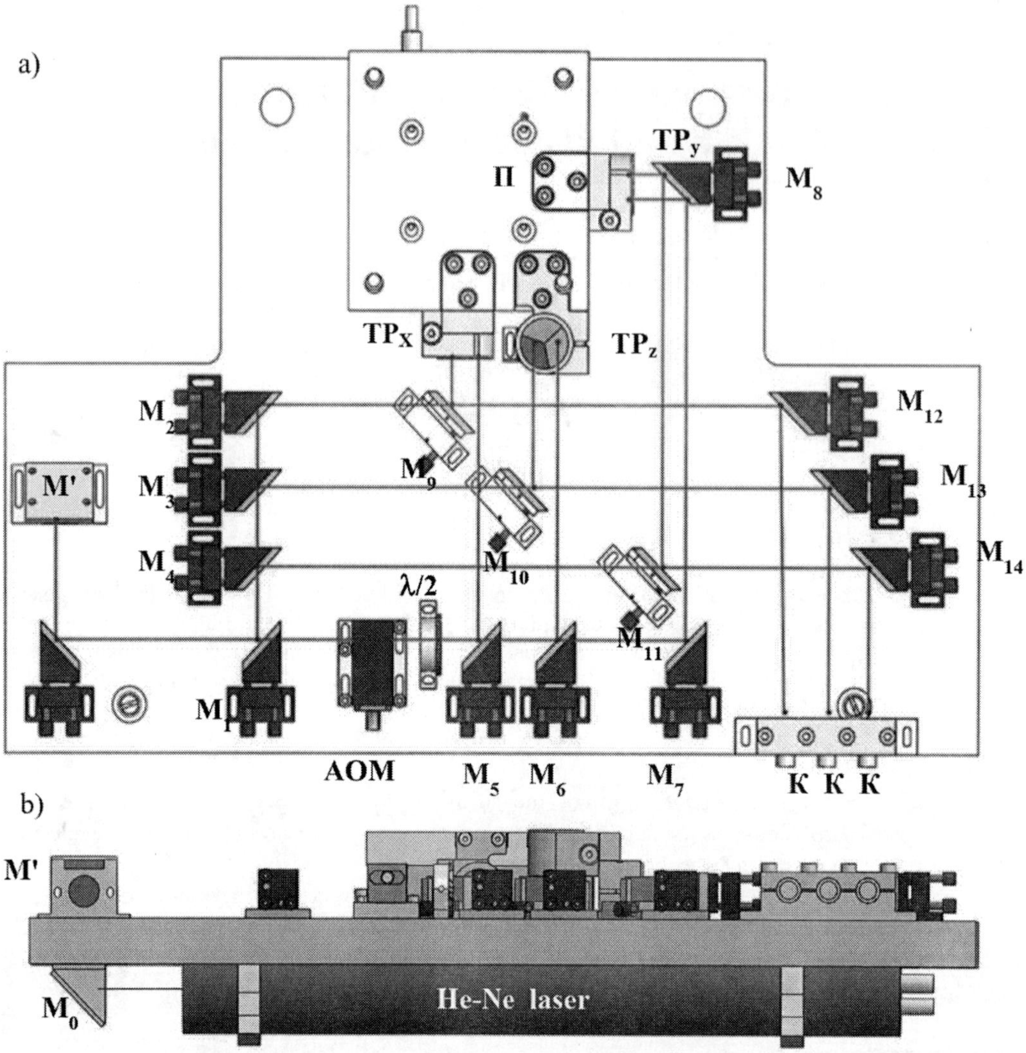


**Figure 1.** Optical scheme of a three-coordinate heterodyne interferometer: a) top view, b) side view (P—piezoelectric table, M0—mirror for laser input to the circuit from the laser, AOM—acousto-optic modulator, M1–M14—optical mirrors, λ/2—half-wave plate, K—collimator, X, Y, Z—triple prisms).

The principle of the phase meter is based on the quadrature-differential scheme of direct analog-digital conversion of the phase-modulated signal with the subsequent digital processing of informative counts in the programmable logic integrated circuit (FPGA) [4].

The clock frequency *f* for the synchronization of the operation of the entire circuit is determined by a quartz generator (G). After passing the divider (*f*/4), which reduces the frequency of the signal fourfold, subsequent filtering and amplifying, the control signal from the generator goes to the input of the AOM driver (Fig. 2).

Since the generator and the interference signal recording system are located in the same electronic unit, there is a certain level of penetration of the electric signal that supplies the AOM into the path of receiving and processing signals from the photodetectors, which leads to an error in the phase measurement. The sign and magnitude of this error depend on the current value of the phase difference between the optical and electrical signals. The nature of mutual influence of these signals becomes clear when analyzing the electrical circuit of a three-coordinate heterodyne interferometer. In Fig. 2 electrical noise pickup is shown as a dashed wavy arrow.

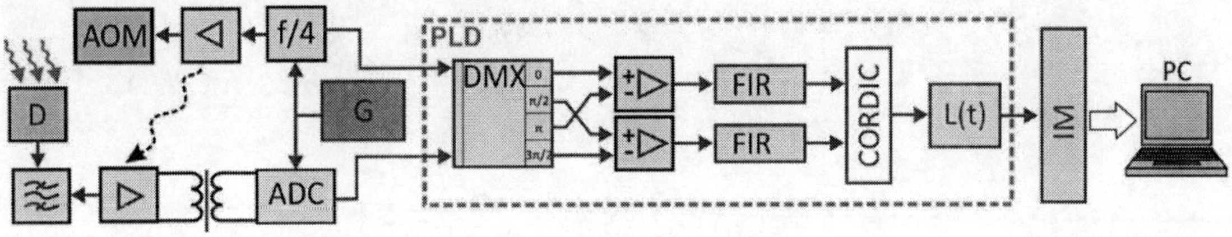


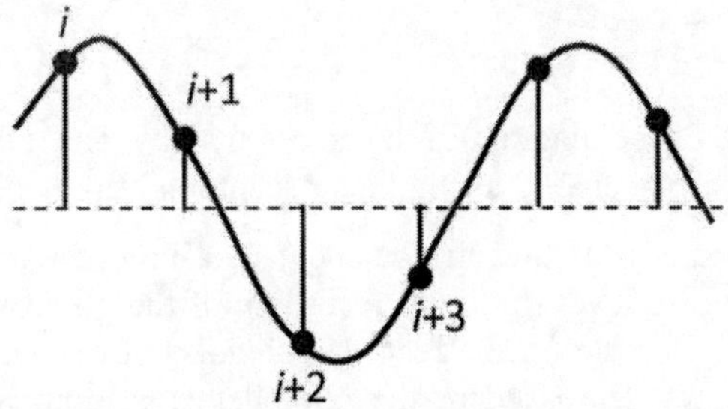


**Figure 2.** Functional scheme of recording and analog-digital processing of signals of the interferometer (a) and time diagram of sample reading of ADC (b). D – photodetector, AOM – acousto-optic modulator, G – control oscillator, ADC – analog-to-digital converter, *f*/4 – frequency divider, DMX – demultiplexer, FIR – low pass filter with finite impulse response, PLD – programmable logic integrated circuit, CORDIC – module that implements the CORDIC algorithm for phase calculation, *L*(*t*) – phase shift length modulation module, IM – interface module, PC – personal computer.

## 3. Error of registered signals

In the ideal case, the signals detected by the photodetectors are:

$$A(t) = A_0(t)\cos(2\pi f_o t+\delta(t)), \quad (1)$$

where $A_0(t)$ is the slowly varying photocurrent amplitude, $f_0$ is the working frequency of the received signal, $\delta(t)$ is the phase difference of the interfering beams. In the presence of electrical noise pickup from the AOM supply channel to the amplifiers recording the photocurrent, the signal arriving at the digital quadrature detection system is distorted and becomes:

$$A'(t) = A_0(t)\cos(2\pi f_0 t+\delta(t))+a_2\sin(2\pi f_0 t)+a_1\cos(2\pi f_0 t), \quad (2)$$

where $a_2\sin(2\pi f_0 t)$ and $a_1\cos(2\pi f_0 t)$ describe the parasitic electrical signal that penetrates the photorecording path. If $A_0 >> a_0$, then after quadrature detection, i. e. multiplying by $\sin(2\pi f_0 t)$ and $\cos(2\pi f_0 t)$ (hardware, synchronous detection takes place in digital form, by multiplying the current sample of the recorded signal by the values $\sin(2\pi f_0 t)$ and $\cos(2\pi f_0 t)$ at the sampling points), we receive not the expected signals $A(t)\sin\delta(t)$ and $A(t)\cos\delta(t)$, but the distorted $A(t)\sin\delta(t)+a_2$ and $A(t)\cos\delta(t)+a_1$ instead. By choosing the initial phase, one can ensure $a_2 = 0$ and simplify further analysis. Such pattern corresponds to the presence of parasitic phase modulation of the detected signal, and when the interferometer signal is recorded during linear scanning of the measured displacement of the AFM positioning system, it appears as specific sinusoidal deviations of the recorded displacement from the linear graph (see Fig. 3). Shifting one of the quadrature components results in distortion of the signal recorded by the CORDIC algorithm (COordinate Rotation DIgital Computer). As a result, the phase angle $\delta_i(t)$ will be described by the formula: $\delta_i(t) = \delta(t)+(a_1/A_0)\cos\delta(t)$.

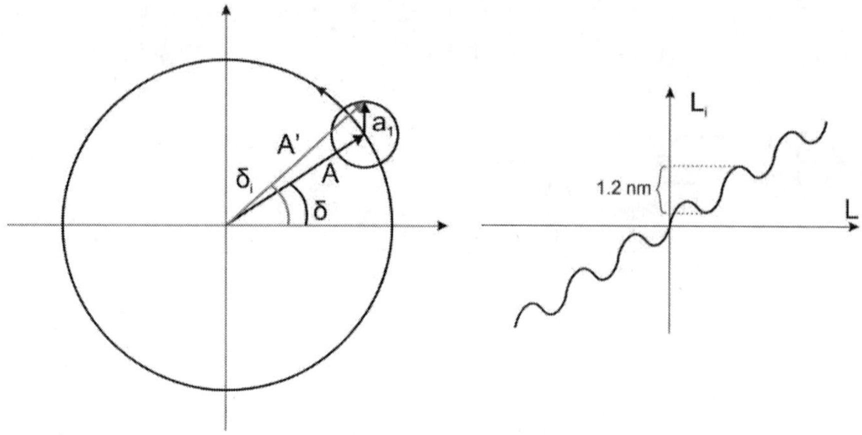


**Figure 3.** Phase diagram of the detected signal and a typical view of the visible displacement during linear scanning of the positioning system.

Since the achieved accuracy of the displacement measurement for the NanoScan-3Di device is better than 1 nm and the noise level in digital recording of the phase shift signal is less than 0.1 nm, these distortions are clearly visible on the recorded signals. Moreover, they clearly appear when the level of the optical signal is decreased at the photodetector input and lead to an increase in the measurement error.

Because the behavior of this kind of distortion is periodic in relation to the phase, it allows us to offer a fairly simple mechanism for suppressing this type of noise by Fourier filtering of the measured displacements. Such an approach is applicable when the range of recorded movements significantly exceeds several wavelengths of the laser radiation used. An example of the sinusoidal distortions of the recorded displacement is shown in Fig. 4. These data were obtained with a deliberate attenuation of the signal at the photodetector input by a factor of 2, for greater intensity of the effect. With qualitative adjustment of the entire interferometer and the effective introduction of laser radiation into fiber-optic communication lines (transport fiber), the sinusoidal form of the interference is distorted by noise originating from temperature fluctuations in the refractive index of air in the interferometer's operating and reference arms.

As can be seen from the experimental data, this type of error can reach several nanometers and strongly resembles the so-called non-quadrature error caused by ellipticity of the polarization of the interfering light and the incorrect tuning of the quadrature detector (the difference in the phase shift between the reference signals and/or the difference in signal amplitudes, etc.). The ellipticity of the polarization of the interfering beams, in the case of heterodyne interferometry, does not contribute to the detected signals. In the registration scheme considered, the error associated with the tuning of the quadrature detector is negligible, since digital quadrature heterodyning is used to generate quadrature signals by processing four "instantaneous" samples over the period of the detected signal (see Fig. 2b). Formation of the phases of the digitizing signals and the signal going to the AOM occurs by the method of digital division of the initial highly stable frequency (144 MHz) signal (Fig. 2). The level of phase instability (jitter) inherent in the utilized microcircuits allows us to assert that the error caused by non-quadraturity of the digitization does not exceed 0.3 mrad, and the resulting geometric error is be less than 0.1 nm.

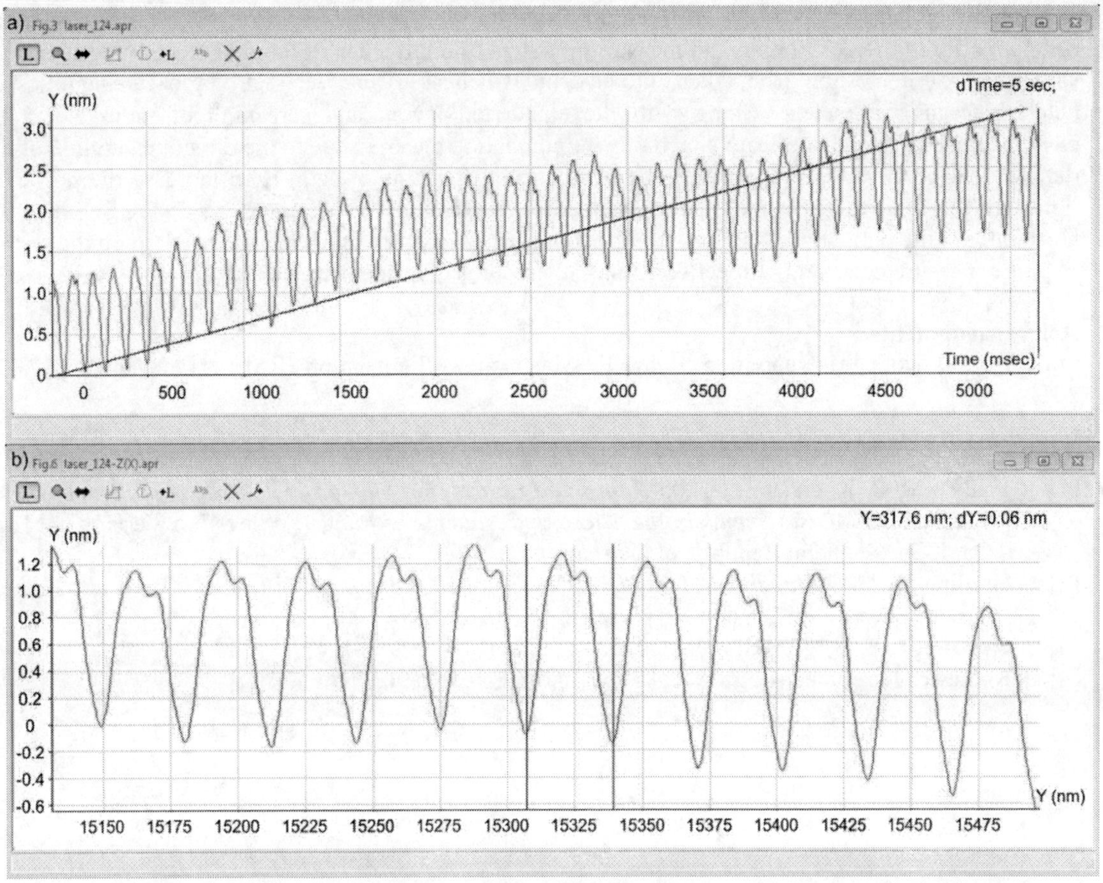


**Figure 4.** Examples of periodic interference caused by penetration of the electrical signal of the AOM supply into the photocurrent registration path, a) the red line is the recorded displacement vs. time, obtained by subtracting the approximated line from the received signal (black line, scale in arbitrary units); b) the period of the resulting sine is 317.6 nm.

When studying the size and shape of real structures in an AFM, height differences often do not exceed the wavelength of the laser radiation used. In a situation where the shape of the surface is not plane, the effect of the error caused by the penetration of the signal of the AOM electrical supply into the optical signal processing channel of the interferometer results in a systematic error depending on the current phase shift of the interference signal with respect to the AOM signal. It is impossible to eliminate this error without knowing the current value of the phase shift between optical and electrical signals.

Thus, one of the requirements for radio electronic equipment used in heterodyne laser interferometers with direct digital quadrature heterodyning is to minimize the level of electrical interference induced by the AOM circuit in the circuits that record the interference signals. In the Michelson interferometer used in the NanoScan 3Di, the phase shift of $2\pi$ occurs when the scanner is displaced by 316 nm and, correspondingly, at a level of penetration of the electric signal into the optical path of the order of 1%, the relative error in displacement measurements in the 50 nm range (~ 1 rad of the phase shift) would not exceed 1%, while the absolute error in the entire range of the scanner can reach 1.2 nm. Hence, in order to realize the possibility of measurements with an accuracy better than 0.5 nm, the radio electronic system should guarantee the level of the electronic noise under discussion below 0.1%, i. e. –60 dB, which is quite feasible in the frequency range of tens and hundreds of MHz commonly used in radio electronic systems of heterodyne interferometers.

## 4. Conclusion

The discussed phase-dependent error along with the well-studied Abbe errors, the mutual non-orthogonality of the positioning and measuring system, the influence of thermal electronics noise and shot noise of light should be taken into account when assessing the metrological capabilities of heterodyne interferometers. Along with the elaborated layout and screening of radio electronic elements, one of the most effective methods for combating the error due to the direct penetration of the radio signal of the AOM supply into the photocurrent registration path, is to increase the power of light interfering in each channel and to struggle for the contrast of the interference image, since the more intense the light signal, the less the effect of the carrier frequency electrical pickup on the results of phase measurements and, respectively, on the measured displacement.

## Acknowledgments

Funding: this work was supported by the Russian Science Foundation (RSF) (grant number 17-72-20164).

## References

[1] Wilkening G, Koenders L, 2005 *Nanoscale Calibration Standards and Methods: Dimensional and Related Measurements in the Micro and Nanometer Range* (Wiley-Vch Verlag GmbH & Co. KGaA, Federal Republic of Germany).

[2] Harding K, 2013 *Handbook of Optical Dimensional Metrology* (CRC Press Taylor & Francis Group, Boca Raton London New York).

[3] Gogolinskii K V, Gubskii K L et al. 2012 *Meas. Tech.*. **55** pp 1–6.

[4] Kazieva T V, Kuznetsov A P et.al *Proc. of SPIE* **9636** p 96360.

IOP Conf. Series: Journal of Physics: Conf. Series **941** (2017) 012109 doi:10.1088/1742-6596/941/1/012109

# Formation and study of properties of Ta and Mo nanocluster films

**T I Kozlova[1], O S Vasilyev[1], P V Borisyuk[1] and Yu Yu Lebedinskii[1,2]**

[1]National Research Nuclear University MEPhI (Moscow Engineering Physics Institute), Kashirskoe sh. 31, 115409, Moscow, Russia
[2]Moscow Institute of Physics and Technology (State University), Institutskiy per. 9, 141701, Dolgoprudny, Moscow Region, Russia

E-mail: kozlovati@ro.ru

**Abstract.** In the work the results of studies of thin-film samples of nanoclusters of tantalum and molybdenum metals on the surface of silicon dioxide $SiO_2$/Si (001) at room temperature are presented. The chemical composition and electronic structure of the obtained nanocluster films of Ta and Mo were controlled *in situ* by X-ray photoelectron spectroscopy (XPS). Susceptibility to oxidation during the exposure to the atmosphere of then nanocluster films, as well as their thermal stability when heated in a vacuum to 600 °C were studied *ex situ* by the XPS method. The size and shape of the nanoclusters composing the film were estimated *ex situ* by analyzing images obtained with a scanning electron microscope. The band structure before and after oxidation was studied by measuring the bandgap of the formed Ta and Mo films by the method of electron energy loss characteristic spectroscopy (REELS). Conclusions about thermoelectric properties of the formed nanocluster films of Ta and Mo were made.

## 1. Introduction

Study of the properties of nanostructured materials is of current interest due to the development of nanoelectronics, in particular, micro- and nanoelectromechanical systems (MEMS and NEMS) [1]. However, one of the main constraints to wide-scale introduction of the systems is the problem of producing miniature power supplies for MEMS and NEMS [2]. One of the perspective miniature power supply units for MEMS and NEMS is radioisotope-based ones [3]. Their advantages are long, stable operation, the lack of maintenance, and the possibility of forming radioisotope sources of micron sizes. The most efficient way to convert the energy released during nuclear decay into electric one is based on thermoelectric conversion [4] and is known as Seebeck effect. However, modern semiconductor thermoelectric materials don't have sufficient efficiency for use in such systems. The efficiency of thermoelectric materials is described by the dimensionless coefficient ZT, which is directly proportional to the electrical conductivity $\sigma$, the square to the value of the thermoelectric power S, and inversely proportional to the thermal conductivity $\kappa$ [5]. Currently the maximum value of the efficiency ZT ~ 1.3 is obtained for thermoelectric material $AgSbTe_2$ [6] for the operating temperature range of radioisotope power sources 400-600K (according to [7]). The main attempts to increase the thermoelectric efficiency of the material are based on the use of semiconductor nanostructures [8]. In order to increase the efficiency of thermoelectric materials, particularly, by changing the value of the thermoelectric power, it is necessary to reduce the size of the particles making up the thermoelectric material. However, if the size of individual semiconductor nanostructures is less than 50 nm, then materials based on such structures lose electrical conductivity up to the manifestation of dielectric properties [9], and as result they become less effective.

One of approaches to solve this problem is to use porous metal nanostructures, in particular, ensembles of nanoclusters in the form of films for this purpose [7]. Metal nanoclusters with decrease of their size are less susceptible to changes of electrical conductivity, manifesting semiconductor properties at characteristic size less than 3 nm. It is expected that the electrical conductivity of such films will be close to metallic one due to carriers tunneling between nanoclusters and percolation effects. At the same time, with the decrease of size of metallic nanoclusters, a strong increase of thermoelectric power value of nanocluster films is expected [10], since the latter depends on the density of the states at the Fermi level. Moreover, the porous structure is expected to result in a decrease of the phonon thermal conductivity due to scattering at the boundaries of nanoparticles [8]. Thus, such materials as porous films based on metal nanoclusters may be potential candidates for the role of a highly effective thermoelectric material.

In this work thin nanocluster films of tantalum and molybdenum were formed on surface of dielectric substrates with different sizes of nanoclusters. Analysis of the chemical composition of these films was carried out. Their susceptibility to oxidation after exposure to the atmosphere, as well as their thermal stability under heating, were studied. Evaluation of the size and shape of large-sized nanoclusters constituting the film were performed. The band structure of the films before and after oxidation was studied. Conclusions on the thermoelectric properties of the formed films were made.

## 2. Materials and methods

Silicon dioxide substrate $SiO_2/Si$ (001) with a size of 10 mm x 10 mm was chosen for the samples, since a heat and electrical insulating material is required for the effective operation of the thermoelectric material. Tantalum and molybdenum were chosen, since these materials have different signs of thermoelectric power value, they are high-melting and well evaporated with the magnetron sputtering method.

Thin films were formed using a magnetron deposition source NanoGen 50 (Mantis Deposition, UK) in the preparation chamber of ultrahigh vacuum system Multiprobe MXPS RM VT AFM-25 (Omicron NanoTechnology GmbH, Germany). In the preparation chamber the pressure was $p \approx 10^{-3}$ Torr during deposition process. The working parameters of the cluster source were: buffer gases fluxes: Ar from 15 to 30 sccm, He – 0-50 sccm; magnetron power – 20-90 W; aggregation zone – 2-10 cm. In the process of nucleation in this source, nanoclusters acquire a single charge, which allowed the size filtration with a dispersion of 2% in the quadrupole mass spectrometer MesoQ (Mantis Deposition, UK). Positive voltage of 800 V was applied to the substrate for better adhesion and formation of porous films consisting of individual nanoclusters. When applying voltage above this value, a continuous homogeneous film was formed due to the "hard" deposition of nanoclusters. Otherwise, when the voltage on the substrate is less than 800 V, the film does not form due to the poor adhesion of the nanoclusters of these metals to the surface of the silicon dioxide. The chemical composition of the surface was controlled *in situ* (at a level of 0.1%) by XPS (the source of radiation MgKα, 1253.6 eV). The thickness and morphology of Ta and Mo films consisting of large clusters was studied *ex situ* by scanning electron microscopy using a microscopes JSM 7001F (JEOL, Japan) and HELIOS 660 (FEI, USA). Samples were then again analyzed in the XPS spectrometer Theta Probe (Thermo Fisher Scientific Inc.; the source of radiation $MgK_\alpha$) after exposure to the atmosphere in order to study their susceptibility to oxidation and thermal stability when heating in vacuum ($5 \cdot 10^{-9}$) up to 600 °C. The band structure of nanocluster films before and after exposure to the atmosphere was studied using REELS method realized in the UHV analysis chamber Multiprobe MXPS RM VT AFM-25 (initial energy of electrons was 500 eV).

## 3. Results and Discussion

As a result of the work, two series of samples with monodisperse tantalum films consisting of nanoclusters of various sizes in the range from 1.3 to 5.8 nm, and molybdenum films with sizes of nanoclusters from 1.4 to 7.5 nm were formed.

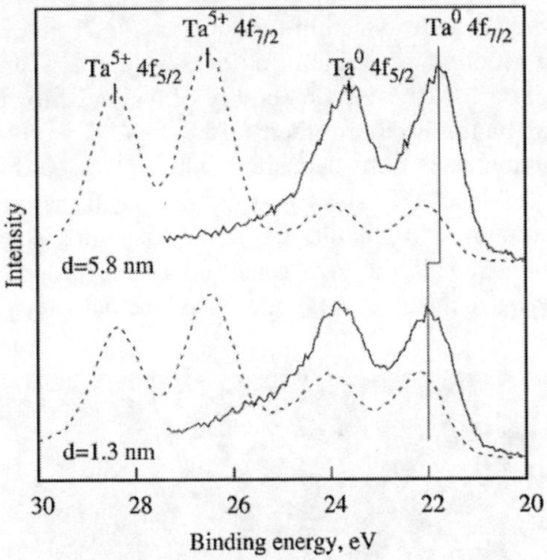


**Figure 1**.-spectra of the core levels of Ta films consisted of nanoclusters of 1.3 nm (bottom spectra) and 5.8 nm (upper spectra) in diameter obtained *in situ* (solid line) and *ex situ* (dashed line) with the XPS method.

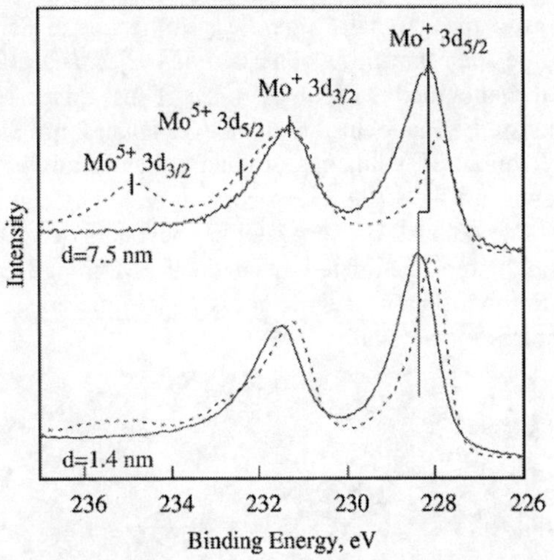


**Figure 2**. Typical X-ray spectra of the core levels of Mo films consisted of nanoclusters of 1.4 nm (bottom spectra) and 7.5 nm (upper spectra) in diameter obtained *in situ* (solid line) and *ex situ* (dashed line) with the XPS method.

Typical X-ray spectra of the core levels of monodisperse films consisted of Ta nanoclusters of sizes 1.3 nm and 5.8 nm in diameter, as well as Mo nanoclusters sizes of 1.4 nm and 7.5 nm in diameter before and after exposure to the atmosphere are shown in figures 1 and 2. The analysis of the *in situ* X-ray spectra showed that all samples immediately after the deposition process are of a metallic state (Ta or Mo) with low oxygen and carbon content (<10%). According to the absence of peaks in the *in situ* spectra (fig. 1 and 2) corresponding to the oxide states, it can be concluded that oxygen is adsorbed into the near-surface layers of the films without forming a chemical bond with the atoms of the nanoclusters.

It is shown that with decrease of characteristic sizes of the nanoclusters composing the films, the binding energy of the core levels (Ta*4f* and Mo*3d*) shifts toward larger values (see fig. 1 and 2), which may be related to the size effect [11]. Indeed, let us consider an electron located on the core level of an atom in a nanocluster. The energy of the initial state of the electron (before the photoionization process) in absolute value coincides with the orbital energy of an electron on a given atomic shell. Photoionization leads to the formation of an uncompensated positive charge (hole) in the atom acting as a potential of disturbance on the electrons remaining in the solid. As a result of the relaxation of the electronic subsystem (screening of the core hole) the relaxation energy is released. Since the time of intraatomic relaxation is comparable with the time of the photoelectron leaving the atom, then the released energy is acquired by the photoelectron, leading to an increase of its kinetic energy and, consequently, to a decrease of a binding energy measured by the XPS. The weakening of the shielding of the core hole by the remaining electrons occurs due to the narrowing of the valence band with respect to the bulk metal in the nanocluster with a decrease of its size. This leads to the fact that the absolute screening decreases, and its difference with the bulk screening increases, thus the energy of the core level recorded with XPS method shifts toward higher values.

An analysis of the *ex situ* spectra of tantalum and molybdenum showed that exposure the samples to the atmosphere leads to strong oxidation of the films (see Figures 1 and 2). In the X-ray spectra for nanocluster tantalum and molybdenum films there are peaks corresponding to two phases: oxide $Ta_2O_5$ (or $Mo_2O_5$) and metallic Ta (or Mo), which is confirmed by stoichiometric analysis. At the same time, analogously to the samples not exposed to oxidation, the positions of Ta4f (and Mo3d) peaks corresponding to the metallic phase also exhibit a dimensional shift of the binding energy for all samples of nanocluster films (see Fig. 1 and 2). Also, the analysis of the X-ray spectra of oxidized samples of nanocluster films showed the presence of typical oxide and metal phases of the valence band spectra of all samples of Ta and Mo. This may indicate the preservation of the conductive properties of nanocluster films of these metals, regardless of the size of nanoclusters. In addition, from the spectra (Fig. 1 and 2), it can be seen that films consisted of larger nanoclusters are more susceptible to oxidation than films made from nanoclusters of smaller size. Thus, this may indicate a lower catalytic activity of Ta and Mo films consisted of nanoclusters of size less than 2 nm and it may be important for their further use.

Typical SEM images of tantalum and molybdenum nanocluster films deposited onto $SiO_2$/Si (001) are shown in Fig. 3 (the view from above). The figures 3a and 3b clearly show that the formed films on the surface consist of separate nanoclusters and are continuous. It should be noted that the size of nanoclusters, estimated from the SEM images obtained, exceeds the size specified in the process of filtration in a quadrupole mass spectrometer. This may be a result of coagulation of nanoclusters and requires further study.

**Figure 3**. Typical SEM-images of films consisted of nanoclusters of Mo (left side) and Ta (right side) (top view).

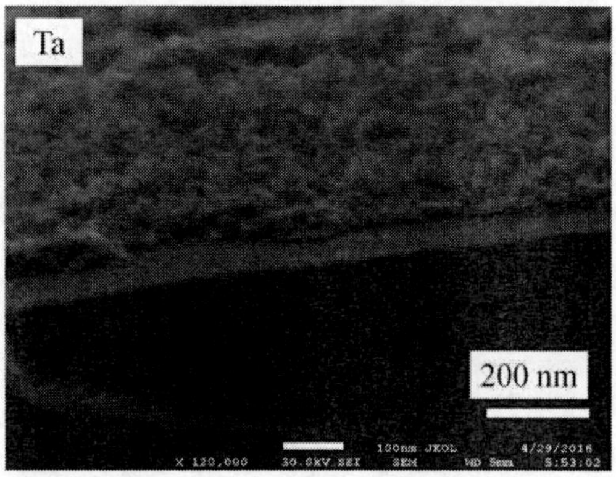


**Figure 4**. SEM-image of substrate $SiO_2$/Si (001) with nanocluster films of Ta (side view).

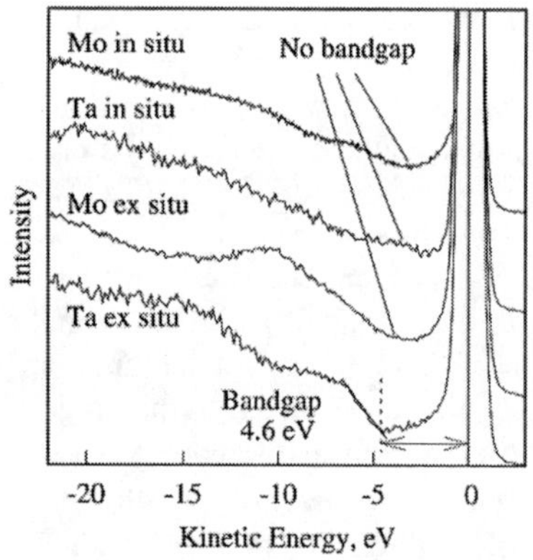


**Figure 5**. Spectra of characteristic electron energy loss obtained *in situ* and *ex situ* for Ta and Mo nanocluster films.

The cleavage of the substrate $SiO_2$/Si (001) with nanocluster films was made to estimate the thickness and internal structure of the films deposited. Typical SEM image of the cleavage, which is made for a sample with Ta nanocluster film on the $SiO_2$/Si (001) surface, is shown in Fig. 4. From the analysis of the image (see Fig. 4) we can conclude that the thickness of the nanocluster film is of the order of 10 nm. In addition, it is seen that the shape and size of the clusters contacting the substrate and lying on the surface coincide.

The analysis of the band structure was carried out with the REELS method by measuring the width of bandgap of the formed Ta and Mo nanocluster films. Fig. 5 shows the spectra of the characteristic losses of the electron energy obtained *ex situ* and *in situ*. It was found that the samples of Ta nanocluster films after exposure to the atmosphere have a bandgap about 4.6 eV. An XPS analysis with angular resolution was additionally carried out to determine the depth of oxidation of the resulting films. It was concluded that the formed nanocluster films of Ta after exposure to the atmosphere show only surface oxidation, remaining metallic in the depth (more than 2-3 nm from the surface). It is possible that Mo nanocluster film samples are conductive, because it is difficult to conclude that there is bandgap from the loss spectra for Mo. Since while using these films as thermoelectric materials, their heating will occur, the thermal stability of the obtained samples was studied by heating in a vacuum to 600 ° C, followed by an analysis of changes in chemical composition by the XPS method. Differences in the spectra obtained before and after heating were not found. Thus, due to the absence of shifts in the binding energy of the core levels, it can be concluded that the size of the nanoparticles don't change after heating, and the resulting nanocluster films do not lose their porous structure due to heating. This fact is important for solving the problem of creating a highly effective thermoelectric material.

Thus the nanocluster films obtained have a clear porous structure, consist of individual nanoclusters, are conductive and resistant to heating up to 600 ° C. The following work will be devoted to the study of thermoelectric properties of the nanocluster films obtained.

## 4. Conclusion

Thus, deposition parameters of high-melting metals Ta and Mo nanocluster films deposition with a clear porous structure on the surface of the $SiO_2$/Si (001) dielectric substrate by magnetron sputtering were determined. The analysis of the chemical state of the nanocluster films obtained was made. Their susceptibility to oxidation after exposure to the atmosphere, as well as their thermal stability under heating were studied. Evaluation of the size and shape of large-sized nanoclusters constituting the film were performed. The band structure of the films before and after oxidation was studied.

**Acknowledgements**

This work was financially supported by Grant of the Russian Science Foundation (project No. 16-19-00168).

## References

[1] Kim K, Guo J et al. 2015 *ACS Nano* **9** pp 548-54.
[2] Holmberg S, Perebikovsky A et al. 2014 *Micromachines* **5** pp 171-203.
[3] Lal A and Blanchard J 2004 *IEEE Spectrum* **41** pp 36-41.
[4] Prelas M A, Weaver C L et al. 2014 *Progress in Nuclear Energy* **75** pp 117-48.
[5] Ziman J M 1961 *Advances in Physics* **10** pp 1-56
[6] Gayner C and Kar K 2016 *Progress in Materials Science* **83** pp 330-82.
[7] Borisyuk P V, Vasilyev O S et al. 2016 *Chemical Physics* **478** pp 2-7.
[8] Hilaal A and Ramakrishna S 2013 *Nano Energy* **2** pp 190-212.
[9] Lee H, Vashaee D et al. 2010 *Applied Physics* **107** pp 1-7.
[10] Borisyuk P V, Vasilyev O S et al. 2015 *Applied Surface Science* **336** pp 359-63.
[11] Borman V D, Borisyuk P V et al. 2006 *Journal of Experimental and Theoretical Physics* **129** pp 343-55 (in Russian).

IOP Conf. Series: Journal of Physics: Conf. Series **941** (2017) 012110 doi:10.1088/1742-6596/941/1/012110

# Investigation of the electronic properties of the surface and bulk forms of gold and palladium

**U N Kurelchuk, P V Borisyuk, O S Vasilyev, Yu Yu Lebedinsky**

[1]National Research Nuclear University MEPhI (Moscow Engineering Physics Institute), 115409, 31 Kashirskoe sh., Moscow, Russia
[2]Moscow Institute of Physics and Technology (State University), 141701, 9 Institutskiy per., Dolgoprudny, Moscow Region, Russia

E-mail: unkurelchuk@mephi.ru

**Abstract.** The density of electronic states for bulk metals Au and Pd, their surfaces in the form of polycrystalline surface layers of nanometer thickness is investigated. The calculations were performed using density functional theory with pseudopotential in full relativistic approximation. Approximations have been found that provide calculations the density of electronic states of noble metal surfaces that describe the experimentally observed features of XPS spectra of the valence band of these metals.

## 1. Introduction

Nanoporous films made of noble metallic nanoclusters are promising materials for high-efficiency thermoelectric elements. Modern computational modeling and numeric methods allow solving such problems as: prediction of stable nanoclusters from, predetermined materials with certain dimensions, theoretical study of their electronic, thermal, magnetic and other properties; computational experiments for planning and optimizing physical experiments and interpreting their results.

In this article we consider the features of the experimental and theoretical investigation of the density of electron states (DOS), and we develop an approach to the analysis of DOS obtained theoretically and experimentally for noble transition metals, for example, the surfaces of polycrystalline samples of d-metals Au, Pd. In the future, it is planned to develop an approach to implement for the analysis of metallic nanoclusters and porous nanocluster films.

One of the most common and informative experimental methods for studying electronic properties is X-ray photoelectron spectroscopy (XPS) [1]. Its essence is in obtaining the distribution of the number of emitted photoelectrons depending on their binding energy, and this distribution describes the density of occupied electronic states in the material. It should be noted that photoelectrons can be collected from the depth no more than 2-3 nm or 10-20 monoatomic layers, so the method is sensitive only to the upper surface layer [1]. The analysis of publications concerning DOS, started from the earliest studies of the bulk state, and up to the modern calculations of nanoclusters, shows that the comparison of theoretical DOS with XPS spectra has for the most part a qualitative similarity. Generally, the comparison of volumetric DOS and XPS spectrum is not a comparison of two identical systems, since XPS is dealing with a surface, whose state is changed in comparison with the bulk structure of the material, even the ideally prepared surface – avoiding adsorption, oxidation, etc. It is necessary to compare with the experimental spectrum the theoretically calculated surface spectrum of the investigated material of that thickness and with those surface features specific for the sample and experiment.

A feature of nanoscale material modeling techniques is presence of both quantum (ab initio) and many-body approaches. For structures with a big (>$10^3$) number of atoms very powerful approach is density functional theory (DFT) with pseudopotential. In this work we use plane-wave basis set and pseudopotential DFT technique, with geometric optimization, implemented in PW DFT code Quantum Espresso [2]. Calculations was performed using resources of NRNU MEPhI high-performance computing center.

## 2. DFT DOS calculation for *d* nobble metals, particularly Au and Pd.

The properties of noble metals are due to their d-band filled and localized near the Fermi level. The width of the *d*-band, it position relative to the Fermi level, spin-orbit splitting is DOS characteristics which cause physical properties, and it all can be measured experimentally by the XPS spectrum [4].

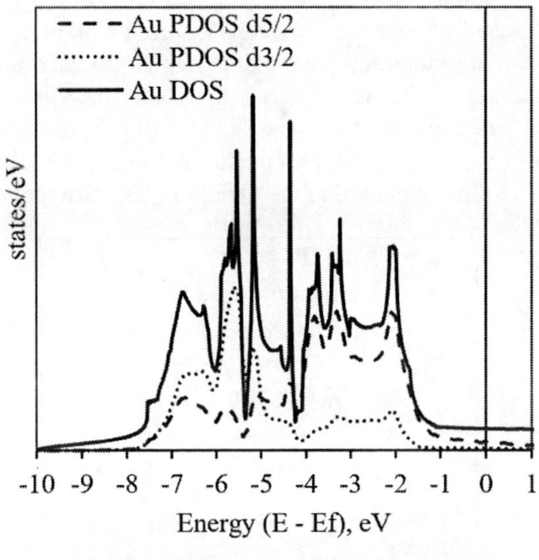


**Fig. 1a.** Total and *d*-DOS of bulk Au

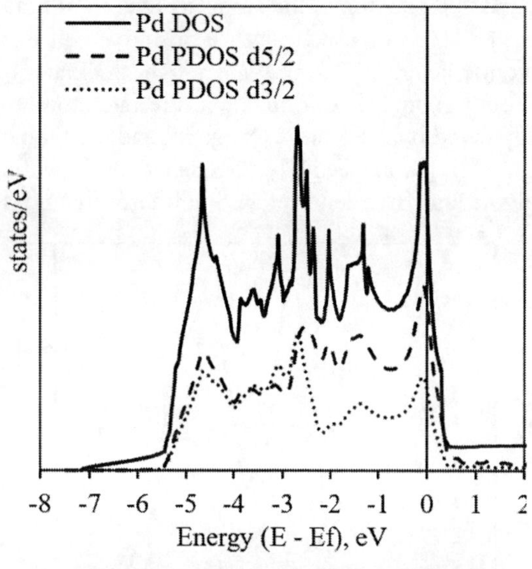


**Fig. 1b.** Total and *d*-DOS of bulk Pd

The DFT study of the electronic structure of bulk gold and palladium was provided in DFT GGA (PBE) approximation [3] of the Ex, for the noncollinear spin orientation with the spin-orbit interaction and nonmagnetic state. The interaction of valence electrons with the core is described with the full-relativistic ultrasoft Vanderbild pseudopotentials [5]. The obtained DOS show a more accurate correspondence to the experimental data, spin-orbit splitting, rather than the results of DOS from databases for bulk metals [6,7]. Figures 1a, 1b show the calculated total and d-projected DOS. *PBE).*, This approximation of the DFT correctly describes the valence band, its distance to the Fermi level and the position of the peaks, the spin-orbit splitting of $d_{3/2}$, $d_{5/2}$ on atomistic scale, and was therefore also used for modeling nanoscaled surfaces.

## 3. Metal Au, Pd surfaces modeling

The structure with which photoelectrons are analyzed in XPS of these noble metals is a polycrystalline surface with a thickness of about 1 nm [1]. An arbitrarily cleaved surface consists of sections of surfaces of different crystalline faces. Like the growth of fcc clusters and spontaneous surface formation, it is most likely that the most closely packed faces with the lowest surface energy are [111], 100 ([200]). Analysis of diffraction patterns of polycrystalline gold shows that the greatest intensity is given by the faces [111]: [200]: [220] in the ratio ~ 2:1:0.8. Similar pictures are observed for thin films, surfaces, fcc-nanoclusters Au, Pt, Pd. [8].

In the XPS, the shape of the spectral line is related to the density of electronic states $n\ (E)$ as:

$$I(E) = I_0 \int I_{DS}(E - E')\, n(E')G(E', \sigma)dE' \quad (1)$$

$G\ (E'; \sigma)$ is the total instrumental broadening, which is described by the Gaussian function, and the Dons-Schünich $I_{DS}$-function [9]. The fluxes of photoelectrons from each part of the surface (with information on the occupied electronic states in it) are summed in the detector, therefore, because of the transformation (1) is linear, the spectrum can be represented as a superposition of the spectra $I\ (n_i\ (E))$ from the sites of the most probable *hkl* configuration - [111], [200], [220] weighed with Ci corresponding to the peaks of X-ray diffraction.

$$I(E) = \sum_{i=[hkl]_i} C_i \cdot I(n_i(E)) \quad (2)$$

An infinite film with a thickness of 1 nm, cut in the [*hkl*] direction in a fcc crystal, is considered. To solve the periodic problem for such a structure, such a Bravais pseudo-lattice is constructed so that the periodicity in all directions is preserved, but the films do not interact. At the same time, the upper surface of the film remains free, and the lower surface is part of the bulk. In the absence of part of the bonds, the atoms on the surface are rearranged so as to minimize the energy of the system, on the other hand, the position of the atoms inside the film must be equal to the bulk one.

The Bravais pseudo-lattice primitive cell is composed of 2×2 minimal cells in the *XY* plane and 5 atomic layers in *Z* to provide a layer thickness of 1 nm and the motion of atoms relative to each other during geometric optimization, the interatomic distances in the first 2 layers are fixed, the films are separated by a 1.5 nm. Hexagonal and tetragonal Bravais lattices were constructed for [111] and [200] [220] films, respectively. Geometric optimization of these structures was performed using the BFGS algorithm (Broyden-Fletcher-Goldfarb-Shanno) [10-13] implemented in the Quantum Espresso code.

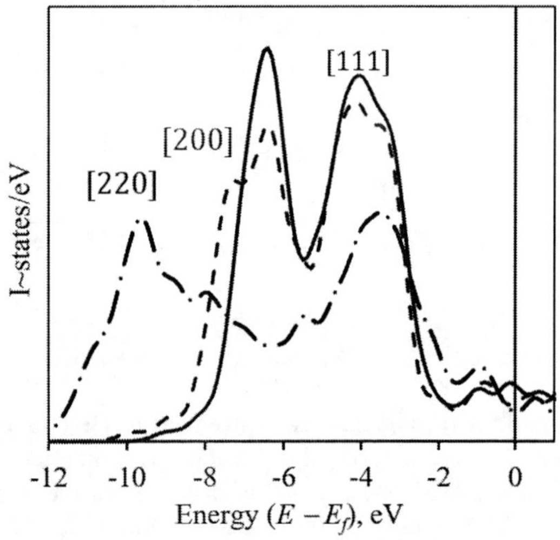


**Fig. 2a**. Smeared DOS of Au model surfaces.

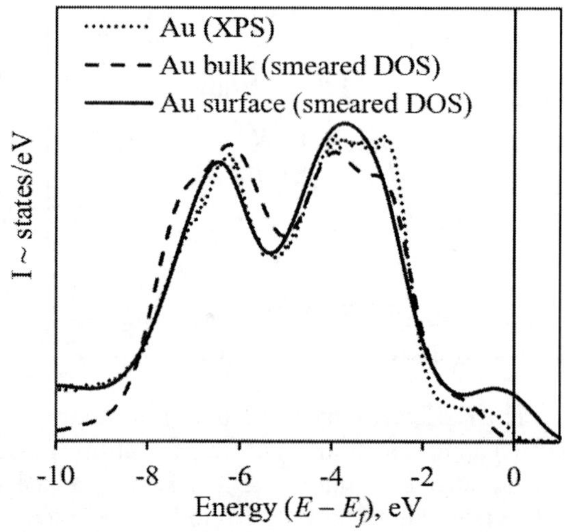


**Fig. 2b**. . Au valence band lines: experimental XPS, calculated from model surface DOS, and from bulk DOS..

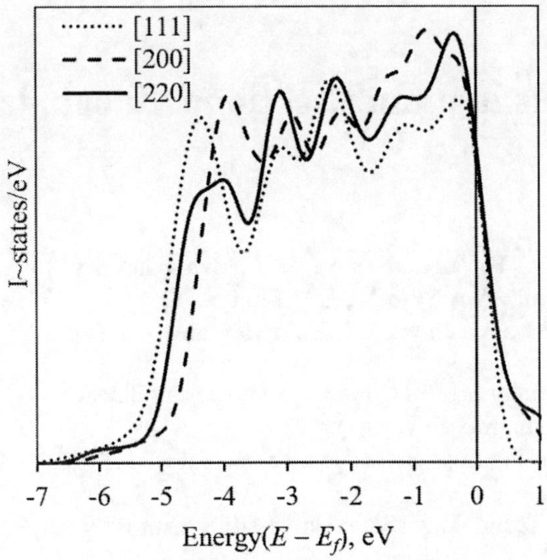


**Fig. 3a**. Smeared DOS of Pd model surfaces.

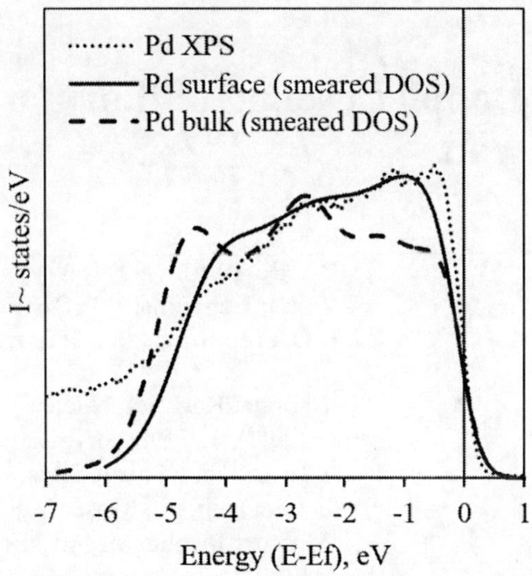


**Fig. 3b** . Pd valence band lines: experimental XPS, calculated from model surface DOS, and from bulk DOS.

In Fig. 2a showed the smeared DOS of model areas of the surface $I_{[hkl]i}$ and in Fig. 2b their superposition according to rule (2) with instrumental broadening, describing the XPS of the valence band of gold. The same for palladium is shown in Fig. 3a, 3b. It can be seen that the model of a surface more accurately describes the filled electronic states of the valence band (XPS) than the model of a bulk sample.

Thus, this approach reveals the suitability for modeling the surface DOS in the aggregate with an experimental study, and will be used for the analysis of nanoclusters and porous films formed from them.

**Acknowledgements**
This work was financial supported by the Russian Federation President Grant to support young scientists (№ 14.Y30.17.2948-MK)

**References**

[1] Bormann V D 2014 *Physical foundations of methods for studying nanostructures and solids* (NRNU MEPhI).
[2] Paolo G; Baroni S et al. 2009 *Journal of Physics: Condensed Matter*. **21** p 395502.
[3] John P P, Kieron B et al. *Phys. Rev. Lett.* **77** p 3865.
[4] Irkhin V Yu and Irkhin Yu P 2007 *Cambridge international science publishing*
[5] http://www.quantum-espresso.org/wp-content/uploads/upf_files/
[6] NIMS Material database http://mits.nims.go.jp/index_en.html
[7] NIST Standard reference database http://webbook.nist.gov/chemistry/
[8] Asadi Asadabad M, Jafari Eskandari M 2015 *Synthesis and Reactivity in Inorganic, Metal-Organic, and Nano-Metal Chemistry* **45** pp 323-326
[9] Doniach S and Sunjic M 1970 *Journal Phys. C* **3** p 284
[10] Broyden C C 1970 *Journal Inst. Math. Appl.* **6** p 64
[11] Fletcher R 1970 *Journal Comput.* **13** p 317.
[12] Goldfarb D 1970 *Math.Comput.* **24** p 23.
[13] Shanno D F 1970 *Math.Comput.* **24** p 647.

IOP Conf. Series: Journal of Physics: Conf. Series **941** (2017) 012111 doi:10.1088/1742-6596/941/1/012111

# Doppler cooling of thorium ions in a multisectional linear Paul trap

**P V Borisyuk[1], S P Derevyashkin[1], K YuKhabarova[2,3], N N Kolachevsky[1,2,3], Yu Yu Lebedinsky[1,3], S S Poteshin[1], A A Sysoev[1], E V Tkalya[1,4,5], D O Tregubov[1,2,3], V I Troyan[1], O S Vasiliev[1], V P Yakovlev[1]and V I Yudin[6]**

[1]National Research Nuclear University MEPhI (Moscow Engineering Physics Institute), 115409, 31 Kashirskoe sh., Moscow, Russia
[2]P.N. Lebedev Physical Institute of the Russian Academy of Sciences, 119991, 53 Leninskiy pr., Moscow, Russia
[3]Moscow Institute of Physics and Technology (State University), Institutskiy per. 9, 141700, Dolgoprudny, Moscow Region, Russia
[4]Lomonosov Moscow State University Skobeltsyn Institute of Nuclear Physics (MSU SINP), 119991, 1b2 Leninskie gory, Moscow, Russia
[5]Nuclear Safety Institute of the Russian Academy of Sciences (IBRAE RAN), 115191, 52 Bolshaya Tulskaya st., Moscow, Russia
[6]Novosibirsk state university, 630090, 2 Pirogova st., Novosibirsk, Russia

E-mail: treg.dim@gmail.com

**Abstract**. Doppler cooling of thorium ions is described. The technique is to be applied for cooling of ions in a multisectional linear Paul trap for precise spectroscopy of ions and eventually detecting an isomeric nuclear transition.

## 1. Introduction

Today the second is defined according to the transition between the two hyperfine levels of the ground state of the caesium-133, and the best atomic clocks based on $^{133}$Cs reach the uncertainty $2\times10^{-16}$ [1]. However, recent achievements in laser spectroscopy paved the way towards more precise and stable frequency standards [2], which lead to a significant breakthrough in both applied and fundamental physics, e.g. problems of the theory of relativity or constant variance.

One of the ways to increase the uncertainty of a standard is to use higher frequency of oscillations, i.e. move from microwave to optical frequency. It has been done for the neutral atomic ensembles and single ions, with relative uncertainty being as low as $1.5\times10^{-18}$ [3]. Further progress, however, is complicated by systematic frequency shifts due to the sensitivity of electrons to external fields. New approach is therefore necessary.

One of the possible solutions is to use a unique nuclear isomeric transition in thorium-229 with an excitation energy similar to optical frequencies of electron transitions. Since nucleus has low polarizability and is shielded by electron shell, nuclear transitions are far less sensitive to external perturbations, significantly increasing the precision of measurements and making it possible to reach the next level of relative uncertainty $10^{-19}$ or even lower [4]. The energy of the transition corresponds to the wavelength ≈ 170 nm, which is in the range of such an important tool for frequency standard development as frequency combs.

Thorium-229 is a product of uranium-233 alpha-decay with a half-life 7880 years. Early research [5] of $^{229}$Th nuclear energy levels using gamma-spectroscopy showed that two rotational bands (main and isomer) of the nucleus are close to each other. The difference was estimated to be 1 ± 4 eV at first [6], and then 3.5 ± 1.0 eV later [7]. Further research assisted with understanding the probability of the transition and different decay channels [8]. For example, recent experiment on measuring the lifetime of the upper level of the transition through the electron conversion decay channel (7±1 μs) [9] is in good agreement with theory (10 μs) [10].

Although laser cooling of thorium ions is an essential step towards spectroscopy of the nuclear isomeric level, the excitation of this level is a challenge. The direct excitation has not yet been demonstrated due to a large uncertainty of the energy and long lifetime of the excited state. Therefore, it may be more feasible to use a method of reverse electron bridge [11]. This mechanism, if achievable, will make use of well-known electronic transitions and will have a higher cross-section compared to the direct excitation.

Despite many efforts, precise spectroscopy of the isomeric level has not been done yet. Nevertheless, we are now at the important stage of research towards thorium frequency standard. Our group has recently joined this research as well by developing a multisectional linear Paul trap for thorium ions [12]. Now we have plans to make use of Doppler cooling for spectroscopy of ions and researching ways of excitation of the isomeric nuclear level. In this work our plans of Doppler cooling of $^{229}Th^{3+}$ ions together with $^{232}Th^{3+}$ are presented.

## 2. Level structure of ions

In order to successfully laser cool ions, precise knowledge of their energy structure is necessary. Spectroscopy of energy levels suitable for laser cooling was performed by the group from the USA [13]. Energy level diagrams for both ions are presented on Fig 1.

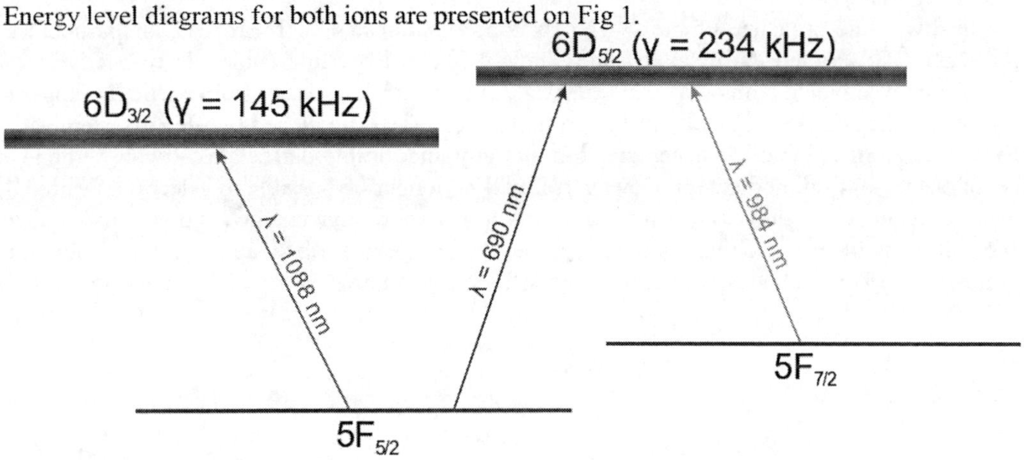


**Figure 1.** Energy levels of both $^{232}Th^{3+}$ and $^{229}Th^{3+}$ (hyperfine sublevels omitted) used for cooling

Since nuclear spin of the isotope $^{232}Th^{3+}$ is zero, it has no hyperfine structure. Transition $5F_{5/2}\rightarrow 6D_{3/2}$ has a wavelength 1088 nm, and a natural linewidth of the upper level $\gamma_{3/2}$ = 145 kHz, according to calculations [14]. Level $6D_{3/2}$ has a linewidth $\gamma_{3/2}$ = 234 kHz, and decays to the two fine sublevels of the ground state $5F_{5/2}$ and $5F_{7/2}$ with a branching ratio 1:8. Corresponding wavelengths of the transitions are 690 nm and 984 nm.

Unlike $^{232}Th^{3+}$ isotope $^{229}Th^{3+}$ has nuclear spin $I$ = 5/2, which leads to a rather complex hyperfine structure of energy levels. It can be described as following [15]:

$$\delta E_i = \Delta_i + \frac{K_i}{2}A_i + \frac{3K_i(K_i+1)/2 - 2I(I+1)J_i(J_i+1)}{4I(2I-1)J_i(2J_i-1)}B_i\,,$$

Where $K_i = F_i (F_i + 1) - I (I + 1) - J_i (J_i + 1)$, $I$– nuclear spin, $J_i$ – total electron momentum, $F$– total atomic momentum, $\delta E_i$ – difference between level of $^{229}Th^{3+}$ and the same level of $^{232}Th^{3+}$, $i$ – index of the corresponding level $e$ (excited) or $g$ (ground), $A$, $B$, $\Delta$ – coefficients that are not dependent on $F$. All coefficients $A$, $B$ and $\Delta_e$ - $\Delta_g$ of all transitions were measured in the work [13].

## 3. Doppler cooling

The choice of isotope $^{232}Th^{3+}$ was primary dictated by its simple level structure, which facilitates Doppler cooling of these ions compared to $^{229}Th^{3+}$, and can later assist with cooling of $^{229}Th^{3+}$ through the mechanism of sympathetic cooling [12]. Closed transition 1088 nm and three-level lambda system 690 nm – 984 nm can both be used for Doppler cooling. Doppler limit of temperature in this case can be estimated as $T_D = \hbar\gamma/(2k_B) \approx 1$ μK, which is low enough to form ion Coulomb crystals [16].

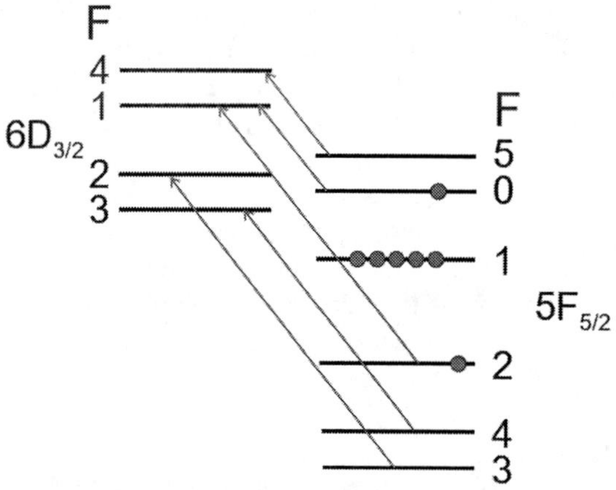


**Figure 2.** The first stage of Doppler cooling of $^{229}Th^{3+}$ ions using 1088 nm transition.

The hyperfine structure of the $^{229}Th^{3+}$ isotope complicates such process. In particular, for the transition 1088 ground and excited states have 4 and 6 hyperfine sublevels respectively, with the distance between them being 100 ÷ 1000 MHz. Since an ion can decay from particular upper sublevel $F$ to any of the sublevels $F$, $F \pm 1$ of the ground state, there is no closed two-level system. Addressing many transitions is therefore necessary to avoid any undesirable dark states. It can be done with the use of electro-optical modulators. However, it will no longer be possible to select the desired detuning of laser frequency in this case, which forces us to use more convenient cooling scheme as a second stage. It is worth mentioning that during the first stage described above, the driving transitions optically pump ions to the certain ground hyperfine sublevel $5F_{5/2}$ (F = 1) (Fig. 2).

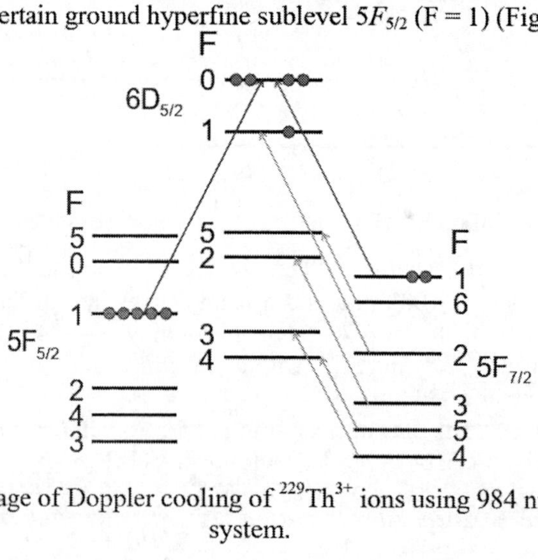


**Figure 3.** The second stage of Doppler cooling of $^{229}Th^{3+}$ ions using 984 nm ↔ 690 nm lambda system.

This particular state is chosen so that the three-level lambda scheme $5F_{5/2}(F = 1) \leftrightarrow 5D_{5/2}(F = 0) \leftrightarrow 5F_{7/2}(F = 1)$ can be used for the second stage cooling (Fig. 3). It is a closed system because the transition $F = 0 \leftrightarrow F = 0$ is forbidden (left branch on Fig. 3). Since only two frequency detunings are to be controlled in this case, efficient Doppler cooling is feasible. Still, non-resonance excitation to another sublevels remains possible, which makes driving repumper transitions necessary.

The two described stages of Doppler cooling of ions can be used simultaneously. At first, when the temperature of ions is high, cooling on the 1088 nm works inefficiently due to high possibility of non-resonance excitation. Still, on average the temperature is decreasing and ions are getting pumped to the $5F_{5/2}$ ($F = 1$) sublevel. When the temperature becomes <10 K, the Doppler broadening of all the sublevels becomes <200 MHz, thus allowing to address certain hyperfine sublevels separately with low probability of non-resonance excitation. The second stage cooling becomes relevant at this point, while the 1088 nm transitions remains working as a repumper.

## 4. Conclusion

We have created a multisectional linear Paul trap for thorium ions and are planning to apply laser cooling in order to achieve more precise spectroscopy of ions, the isomeric nuclear transition in particular. The technique of laser cooling is in accordance with that of a group in the USA, and can be used to cool ions to crystallization.

**Acknowledgements**
The work was supported by the Russian Science Foundation (project number 16-12-00001).

**References**

[1] Levi F, Calonico D et al. 2014 *Metrologia* **51** p 270.
[2] Riehle F 2006 *Frequency standards: basics and applications* John Wiley & Sons.
[3] Schioppo M, Brown R C et al. 2017 *Nature Photonics* **11** pp 48-52.
[4] Campbell C J, Radnaev A G et al. 2012 *Physical Review Letters* **108** p 120802.
[5] Kroger L and Reich C 1976 *Nuclear Physics A* **259** pp 29-60.
[6] Reich C and Helmer R 1990 *Physical Review Letters* **64** p 271.
[7] Helmer R and Reich C 1994 *Physical Review C* **49** p 1845.
[8] Tkalya E V, Zherikhin A N et al. 2000 *Physical Review C* **61** p 064308.
[9] Seiferle B, von der Wense L et al. 2017 *Physical Review Letters* **118** p 042501.
[10] Tkalya E, Schneider C et al. 2015 *Physical Review C* **92** p 054324.
[11] Porsev S, Flambaum V et al. 2010 *Physical review letters* **105** p 182501.
[12] Troyan V I, Borisyuk P V et al. 2015 *European Journal of Mass Spectrometry* **21** 1-12.
[13] Campbell C, Radnaev A and Kuzmich A 2011 *Physical Review Letters* **106** p 223001.
[14] Safronova U I, Johnson W R et al. 2006 *Physical Review A* **74** p 042511.
[15] Sobelman I I 1963 *Vvedenie v teoriyu atomnykh spectrov* (M.: Fizmatgiz) par. 23 (in Russian).
[16] Wineland D J, Bergquist J C et al. 1987 *Physical review letters* **59** p 2935.

# Estimation of the operating parameters of miniature radioisotope thermoelectric power unit based on the Th-228 isotope

**V V Fetisov[1], O S Vasilyev[1], P V Borisyuk[1]and Yu YuLebedinskii[1,2]**

[1]National Research Nuclear University MEPhI (Moscow Engineering Physics Institute), 115409, 31 Kashirskoe sh., Moscow, Russia
[2]Moscow Institute of Physics and Technology (State University), 141701, 9 Institutskiy per., Dolgoprudny, Moscow Region,Russia

E-mail: nanovova@mail.ru

**Abstract.**The paper considersthe construction of a miniature radioisotope power unit based on thermoelectric conversion of thermal energy released during nuclear decay. It is proposed to use thin fluoropolymer films (membranes) as a dielectric heat-insulating material. The results of numerical simulation of a prototype of a miniature radioisotope thermoelectric battery unit based on the thorium-228 isotope in the ANSYS program are presented. The geometry of the system has been optimized. It was established that the temperature of the source can reach about 1033 K, and the efficiency of the considered battery unit can reach 16.8%, which corresponds to modern power supplies of this type.

## 1. Introduction

The development of nanoelectronics, in particular, micro- and nanoelectromechanical systems (MEMS and NEMS) [1] is currently limited by the lack of effective commensurate power supplies [2]. Power sources based on radioisotope source are one of the perspective miniature power supply units for MEMS and NEMS [3]. Their advantages include long, stable work, lack of maintenance, noiselessness. One of the shortest ways to convert the energy of nuclear decay into electrical one is the thermoelectric conversion [4]. So, it is necessary to consider the process of conversion of the thermal energy of a radioisotope into electrical one at the characteristic scales of the system in order to prevent energy losses.

In this paper, the possibility of creating miniature thermoelectric radioisotope battery is considered. The design of the power source is shown in Figure 1 (a, b).

In the figure, the element of a disk form (1) located in the center of the upper surface of the unit represents the radioisotope source of thermal energy, that initiates local heating of the junction point of a pair of thermoelectric material tracks (2), (3). The thermal insulation layer (4) reduces the energy loss on the substrate (5) that operates as the thermostat. The substrate has a through hole, which is necessary for a high thermal insulation of the system. When using this type of power source design, a number of requirements on the geometry and materials arise. All they are described in this paper.

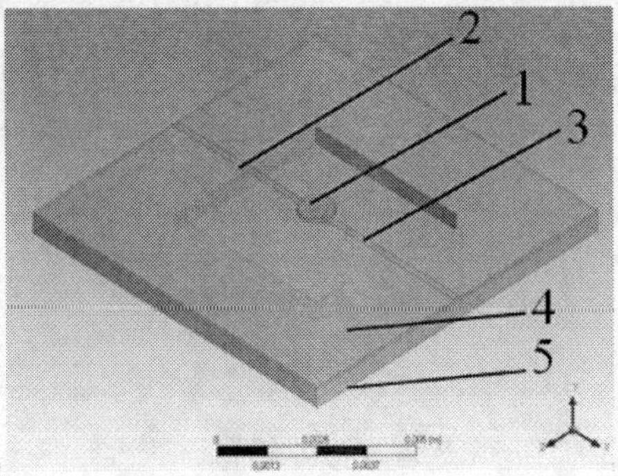


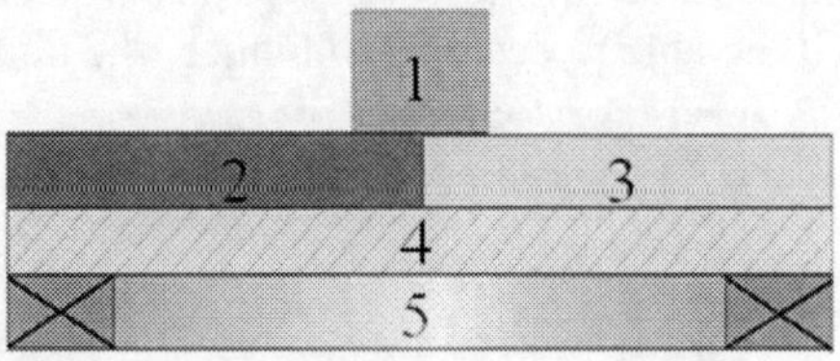


**Figure 1.** a) Isometric view of the construction of a radioisotope thermoelectric power source. b) Schematic representation of the design of a radioisotope thermoelectric power source (side view of a section) (not in scale)

## 2. Materials and experimental procedure

Let us consider in detail all the elements of the nuclear power source design shown in figure 1a radioisotope that produces $\alpha$- or $\beta$-particles with a high specific power is needed for use as a fuel element in a miniaturized power source. The lifetime of the radioisotope is an important criterion here. Typically, the specific power is inversely proportional to the lifetime, i.e., the higher the value of the half-life is, the lower is the specific power of the material. A pure α-source of polonium-210 has high specific power ($N$ = 210 W/g) and short half-life ($T_{1/2}$ = 134 days), so it is convenient to use for energy-intensive but not time-consuming tasks. In turn, such element as nickel-63 ($\beta$-decay, $N$ = 0.5 W/g, $T_{1/2}$ = 100.1 years) is suitable for long life system supplying, that doesn't not require high electrical power. In laboratory conditions it is convenient to use a radioisotope that has parameters of specific power and lifetime, such as thorium-228 has ($\alpha$-decay, $N$ = 26 W/g, $T_{1/2}$ = 1.9 years). It should be noted that metallic thorium is hardly accessible, therefore it is advisable to use thorium dioxide, which is easy to synthesize and, in addition, is much less susceptible to self-destruction and self-dispersion. To ensure maximum heating of nuclear material and minimize heat losses, the radioisotope source and hot ends of thermoelectric materials are located on a thin membrane of the heat insulator. Calculations show that silicon dioxide $SiO_2$ thermal insulation membrane with a thermal conductivity of $k$ = 1.5 W/(m·K) the effective thickness is about 1-5 μm. At the same time, with the use of PVdF (thermal conductivity $k$ = 0.2 W/(m·K)) as thermal insulation material the efficiency of the radioisotope / membrane system is significantly increased. We have developed a technique for depositing thin PVdF films and membranes on $SiO_2$/Si(001) substrates of 1x1 cm in size with a central through hole of 6 mm in diameter. Deposition process was the following: the pre-cleaned substrates were placed into a Petri dish and poured out of 5% acetone PVdF solution and then the Petri dish was closed with a lid and held for 20 hours until completely dried. This method provides formation of plain and smooth heat-insulating membranes with an average thickness of ~ 20 μm. The SEM image of such film is shown in Fig. 2a.

Figure 2b shows a SEM image of a similar PVdF membrane with a thermoelectric material deposited onto its surface in the form of a track with dimensions of 8 × 0.5 × 0.001 (length × width × height, mm).

The geometry of thermoelectric tracks was selected from the following considerations. The peak electrical power of the power unit is determined by the short-circuit current $I_{sc}$ multiplied by the open-circuit power supply voltage $U$. The value of $U$ in this case is determined by the contact potential difference (Seebeck effect) and, consequently, depends on the temperature gradient along the thermoelectric tracks. It is important to note that the temperature gradient, in turn, depends on the cross-section of the thermoelectric track $s$. So the smaller the cross-section of the thermoelectric track is, the higher the temperature can be obtained. Heat transfer in metals is carried to the greatest extent by the flux of free electrons, therefore the maximum current of the power source will be proportional to the heat fluxes (W/cm$^2$) passing along the thermoelectric tracks multiplied by the cross-section. Thus we can introduce the parameter $\alpha$, which is obtained by multiplying the heat fluxes by the temperature difference between the hot and cold ends of thermoelectric tracks. In this case, the parameter $\alpha$ is a function that depends on the cross-section of thermoelectric tracks, and determines the efficiency of the power source. As the cross section decreases, the temperature of the hot end will increase, but the heat flux will decrease, so if we find the maximum of the function of parameter $\alpha$, we will be able to determine uniquely the optimal geometry of the system.

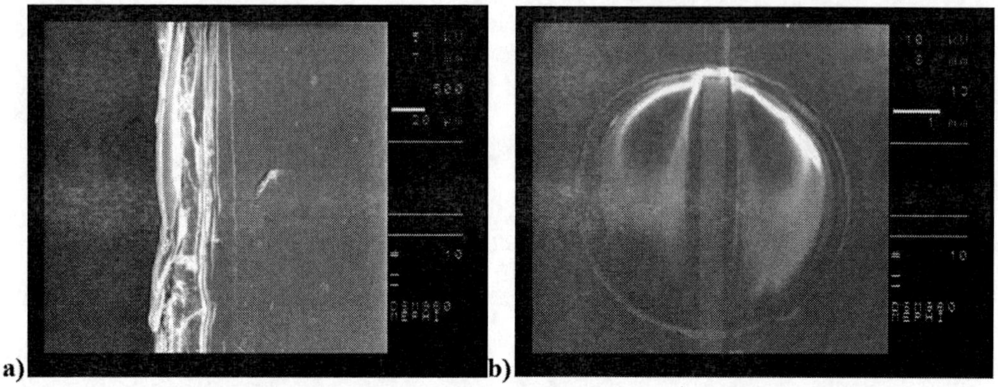


**Figure 2.**a) SEM image of a PVdF film on silicon substrate (side view).
b) SEM image of gold thermoelectric track on the PVdF membrane.

## 3. Results

A number of model experiments were carried out in the software package ANSYS, in which the temperature and heat flux distributions along thermoelectric tracks were found for different cross-sections. For each calculation, the geometric parameters and materials of the radioisotope element and heat insulator remained unchanged, but the width and height of thermoelectric tracks varied (Table 1).

**Table 1.** Geometric and thermophysical parameters of the elements of the construction of a radioisotope thermoelectric power source

| | |
|---|---|
| Substrate characteristics | |
| Length × width × height, cm × cm × cm | 1×1×0.05 |
| Hole diameter, cm | 0.6 |
| Thermal conductivity (Si), W / (m · K) | 148 |
| Characteristics of the thermal insulation layer | |
| Length × width × height, cm × cm × cm | 1×1×0.002 |
| Thermal conductivity (PVdF), W / (m · K) | 0.2 |
| Characteristics of thermoelectric tracks | |
| Width, μm | 100÷ 500 |
| Height, μm | 10 ÷ 100 (10) |
| Thermal conductivity (Au), W / (m · K) | 204 |
| Thermal conductivity (Pt), W / (m · K) | 72 |
| Characteristics of the radioisotope source | |
| Radius, μm | 500 |
| Height, μm | 10 |
| Материал | $ThO_2$(Thoriumdioxide-228) |
| Specific power, W / cm$^3$ | 221.7 |
| Power, mV | 17.8 |
| Thermal conductivity, W / (m · K) | 9.7 |

IOP Conf. Series: Journal of Physics: Conf. Series **941** (2017) 012112 doi:10.1088/1742-6596/941/1/012112

The analysis of the obtained data resulted in the dependence of the parameter α on the cross-section of thermoelectric tracks. For convenience, the heat fluxes were normalized to the total power released from the nuclear element. The plot of the parameter α is shown in the figure 3.

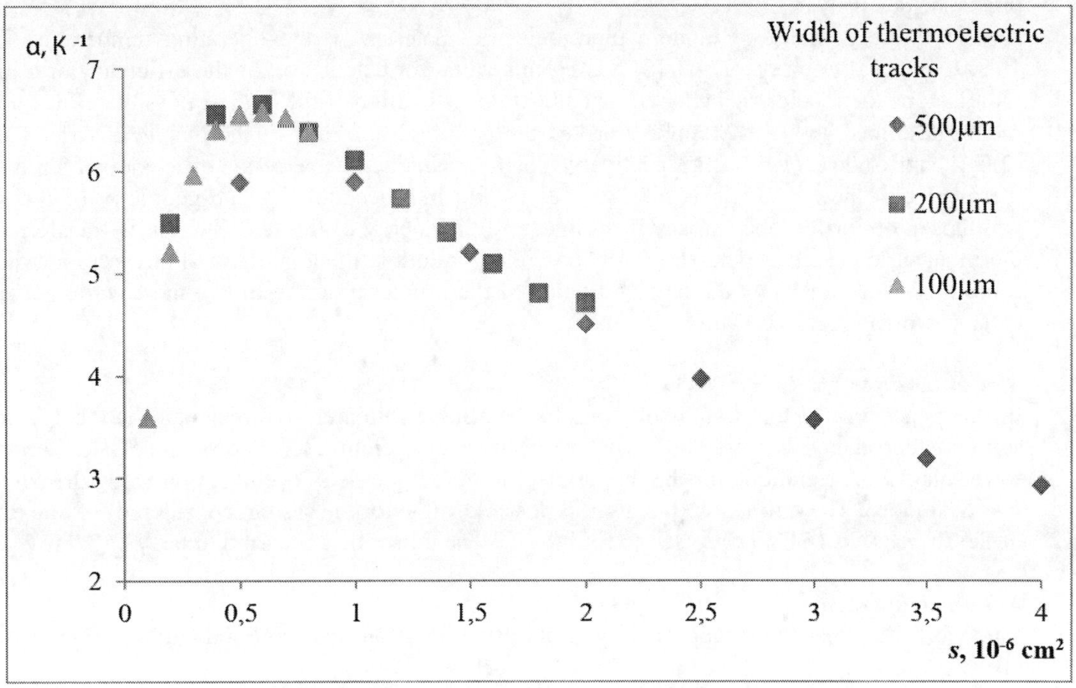


**Figure 3.** Dependence of the parameter α ($K^{-1}$) on the cross-section of thermoelectric tracks ($10^{-6}$ $cm^2$)

The graph has a bell shape. One can clearly see the maximum of the function α, which is achieved at a cross section of $0.6 \times 10^{-6}$ $cm^2$. At this point, the temperature of the hot end is approximately 1033 K. That cross-section of thermoelectric trackis the optimal one. These calculations were carried out for metallic gold and platinum materials of the thermoelectric tracks. Such materials were chosen because gold and platinum are two noble metals with thermoelectric power value of a different sign. However, it must be said that these materials are suitable only for testing the technique. Nevertheless, even for a such test materials the efficiency of such radioisotope thermoelectric power unit can be compared with modern power units of this type. Indeed, the efficiency of thermoelectric materials is usually described by the *ZT* factor, which is directly proportional to the square of the Seebeck coefficient $S^2$, the electrical conductivity of the material σ, and inversely proportional to the thermal conductivity κ

$$ZT = T\frac{S^2\sigma}{\kappa} \quad (1)$$

In order to make this parameter higher, our scientific group suggests using porous films made of metal nanoclusters as thermoelectric materials [5]. It is expected that the porous structure of the material will lead to a decrease of the phonon thermal conductivity due to the scattering of phonons at the interfaces of the nanoclusters. In case of metal porous film, the electrical conductivity willbe comparable with the metallic one due to electron tunneling between nanoclusters and percolation effects. A significant increase of the Seebeck coefficient is also expected [6].

The principle of operation of thermoelectric materials is similar to the principle of operation of a thermal engine that uses electrons and positively charged holes as energy carriers. The efficiency of the thermoelectric conversion is determined by the Carnot cycle and the material factor *ZT* [7]. The value of the efficiency of thermoelectric converters η can be found by the formula

$$\eta = \frac{\Delta T}{T_h}\left(\frac{\sqrt{1+Z\bar{T}}-1}{\sqrt{1+Z\bar{T}}+\frac{T_c}{T_h}}\right), \tag{2}$$

where $\bar{T} = (T_h - T_c)/2$ and $\Delta T = T_h - T_c$, $T_h$ and $T_c$ – temperature of hot and cold ends of thermoelectric tracks, respectively.

The figure of merit of modern thermoelectric materials at the operating temperature of about 1064 K is equal to unity $ZT = 1$ [8]. Using this value for calculation of the efficiency of a miniature radioisotope thermoelectric battery, one can get the estimation of the highest possible efficiency. In the model described above, the temperature of the cold and hot ends of thermoelectric tracks is 22 °C (295 K) and 760 °C (1033 K) respectively. Then the obtained value of the efficiency of a radioisotope thermoelectric power source is $\eta$ = 16.8%. Taking into account that the geometry of the structure waschosen optimally, one can say that almost all the energy of the fuel element is transferred by the thermoelectric materials. Therefore, it is possible to multiply the total thermal power of a radioisotope source ($N$ = 17.8 mW) by the obtained value of the efficiency and estimate the electric power of the battery, which in this case will be 2.9 mW.

## 4. Conclusion

In this paper, one of the design solutions for creating a miniature radioisotope thermoelectric power source is considered. Materials of structural elements (thorium-228, PVdF, nanocluster porous films) were selected. A technique for the thermoelectric tracks geometry optimization using the parameter α was developed. The estimation (the highest possible value for the system considered) of the efficiency of the source ($\eta$ = 16.8%) was used to find the electric power that occurred to be $N_e$ = 2.9 mW.

**Acknowledgements**

This work was financially supported by Grant of the Russian Science Foundation (project No. 16-19-00168).

## References

[1] Kim K, Guo J et al. 2015 *ACS Nano* **9** pp 548-54.
[2] Holmberg S, Perebikovsky A et al .2014 *Micromachines* **5** pp 171-203.
[3] Lal A and Blanchard J 2004 *IEEE Spectrum* **41** pp 36-41.
[4] Prelas M A, Weaver C L et al. 2014 *Progress in Nuclear Energy* **75** pp 117-48.
[5] Borisyuk P V, Vasilyev O S et al. 2016 *Chemical Physics* **478** pp 2-7.
[6] Ziman J M 1961 *Advances in Physics* **10** pp 1-56.
[7] Macia-Barber E 2015 *Thermoelectric Materials Advances and Applications (*Boca Raton: Taylor & Francis Group).
[8] Gayner C and Kar K 2016 *Progress in Materials Science* **83** pp 330-82.

IOP Conf. Series: Journal of Physics: Conf. Series **941** (2017) 012113 doi:10.1088/1742-6596/941/1/012113

# Microwave frequency standard based on $^{25}Mg^{+}$ ions

**I Zalivako[1, 2, 3], I Semerikov[1, 2], A Borisenko[1, 2, 3], K Khabarova[1, 2, 4], V Sorokin[1] and N Kolachevsky[1, 2, 4, 5]**

[1]P. N. Lebedev Physical Institute of the Russian Academy of Sciences, 119991, 53 Leninsky prospekt, Moscow, Russia
[2]Russian Quantum Center, Business-Center Ural, 143025, 100A Novaya st., Skolkovo, Moscow, Russia
[3] Moscow Institute of Physics and Technology (State University), 141700, 9 Institutskiy per., Dolgoprudny, Moscow region, Russia
[4]All-Russian Scientific Research Institute of Physical-Technical and Radiotechnical Measurements, 141570, Mendeleevo, Moscow region, Russia
[5]National Research Nuclear University MEPhI (Moscow Engineering Physics Institute), 115409, 31 Kashirskoe sh., Moscow, Russia

E-mail: zalivako.ilya@yandex.ru

**Abstract.** Development of compact transportable frequency standards is an important goal of the modern applied physics. Despite of the great progress in the field of optical frequency standards with their ultimate accuracy and stability, they remain bulky, complex and sensitive systems. At the other hand, the best microwave frequency standards like cesium fountains are also bulky and sensitive to adjustments. Combination of techniques developed for optical and microwave standards can result in compact and robust transportable frequency standard. Here we suggest the microwave transition in $^{25}Mg^{+}$ laser cooled ion cloud for realization of highly accurate microwave frequency reference. We describe its design and experimental sequence, estimate its expected accuracy and stability.

## 1. Introduction

The progress in the field of frequency standards impacted wide range of technologies and opened new possibilities for modern fundamental research. Highly stable frequency standards are used for global navigation satellite systems (GNSS) [1] and fast data transfer using telecommunication links [2]. They are applied in radioastronomy and tests of fundamental theories like Einstein equivalence principle [3], searching for exoplanets and gravitational waves detection.

Most of today's ultimately accurate and stable clocks utilize optical transitions in neutral atoms confined in optical lattices or transitions in single ions confined in radio-frequency traps. Such clocks show short-term relative instability on the level of $2\times10^{-18}$ and fractional inaccuracy reaching $2.1\times10^{-18}$ [4, 5]. However these systems are still very complex, sensitive to misalignments and vibrations, heavy and bulky. Also optical frequency standards require femtosecond frequency combs for frequency conversion from optical to radio-frequency domain. All these factors restrict their applications in many applied fields. There are several groups working on development of compact and transportable optical clocks [6, 7] but the volume and mass still appear to be inappropriate for e.g. practical usage on the GNSS satellites.

Another approach relies solely on highly stable Fabry-Perot cavity as a frequency reference. The laser stabilized to e.g. a cavity made from Ultra-Low Expansion (ULE) glass or crystalline silicone provides prominent short-term frequency instability of order $10^{-16}$ [8, 9]. However, for the long integration time (more than $10^{3}$ s) the frequency stability is insufficient for most of today's applications due to long-term drifts coming from temperature fluctuations and recrystallization of cavity's material.

At the moment the best tradeoff between frequency stability, mass and robustness is achieved for radio-frequency clocks based on clouds of neutral atoms or ions. Today all GNSSs use such standards as on-board timekeepers. The most widespread spaceborn atomic clocks (installed on GPS, GLONASS, GALILEO, BEIDOU, IRNSS and QZSS satellites) are based on Rb gas cells and Cs atomic beams. These Rb clocks have mass of around 3 kg and their relative instability reaches $10^{-14}$ level after 8 hours of averaging [10]. Another promising on-board frequency standard which is currently being installed on GALILEO satellites is a passive hydrogen maser. While having higher mass, close to 12 kg, H-maser exhibit higher both short- and long-term stability [10]. Corresponding Allan deviation reaches $2\times10^{-15}$ in relative units. Along with better stability, passive hydrogen masers show lower frequency sensitivity to temperature fluctuations. However they are subjected to rare but significant frequency jumps caused by degradation of a hydrogen cell.

Ion radio-frequency clocks are also of great interest for space applications. In contrast to clocks discussed in previous paragraph, which are based on neutral atoms in gas cells or atomic beams, ions are stored in radio-frequency Paul traps. This ensures long interaction time between ions and interrogation field, absence of collisions between ions and cell walls which provides better frequency stability and accuracy. At the same time Paul traps are very robust and have trap depth on the order of several eV. At the moment several groups worldwide are developing atomic clocks where microwave transitions between hyperfine components of ground state in ions are used as a frequency reference. $^{199}Hg^{+}$ [11], $^{137}Ba^{+}$ [12], $^{9}Be^{+}$ [13], $^{171}Yb^{+}$ [14] and $^{113}Cd^{+}$ [15] ions are under investigation. The Deep Space Atomic Clock developed in JPL [16] and based on $^{199}Hg^{+}$ ions trapped in combined quadrupole and 16-pole Paul trap worth special attention. It was designed for spacecraft navigation in deep space and exhibit stability by the order of magnitude better than clocks currently used in GNSSs, however they are still more heavy (17.5 kg). Relative instability of this clock after 24 hours of averaging is on the level of $2\text{x}10^{-15}$. Space tests of this clock are scheduled to be carried out this year.

Here we consider design and estimate characteristics of a new transportable radio-frequency atomic clock based on a laser cooled cloud of $^{25}Mg^{+}$ ions trapped in a linear quadrupole Paul trap. Magnetic dipole transition at 1.789 GHz between hyperfine components of ground state serves as the clock transition. The clock will need only one diode laser system for cooling, ions state preparation and detection which makes this clock quite compact and robust.

## 2. The level scheme and the experimental sequence

Since $^{25}Mg^{+}$ has nuclear spin of $I = 5/2$, each energy level possesses the hyperfine structure. The hyperfine splitting of the $^{25}Mg^{+}$ ion ground state equals to 1.7 GHz and can be chosen as a clock transition for the frequency standard (Fig. 1). Due to relatively simple energy level structure one could use single laser system for both cooling and state preparation. Experimental cycle consists of 4 main stages: (i) laser cooling, (ii) state preparation, (iii) interrogation with RF field, (iv) state detection.

(i) The laser cooling on a broadband 280 nm $^{2}S_{1/2}$ ($F = 3$, $m_F = 3$) $\rightarrow$ $^{2}P_{3/2}$ ($F = 4$, $m_F = 4$) transition with natural linewidth $\Gamma = 2\pi\times41.4$ MHz and saturation intensity $I_{sat} = 250$ mW/cm$^{2}$ can be performed. Since the hyperfine splitting of the $^{2}P_{3/2}$ state has the same order of magnitude as line width of cooling transition, the transition $^{2}S_{1/2}$ ($F = 3$, $m_F = 3$) $\rightarrow$ $^{2}P_{3/2}$ ($F = 3$, $m_F = 3$) will be also excited, such that after some time all population will be transferred to the $^{2}S_{1/2}$ ($F = 2$, $m_F = 2$) metastable state. To prevent ions loss from the cooling cycle a repumping beam exciting $^{2}S_{1/2}$ ($F = 2$, $m_F = 2$) $\rightarrow$ $^{2}P_{3/2}$ ($F = 3$, $m_F = 3$) transition is required (Fig. 2).

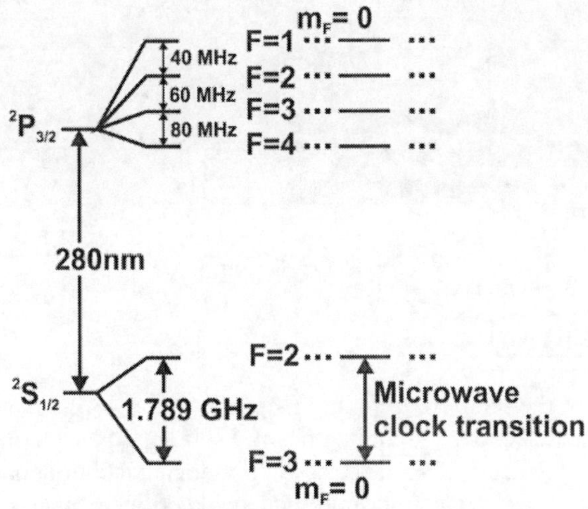


**Figure 1.** $^{25}Mg^{+}$ ion levels involved.

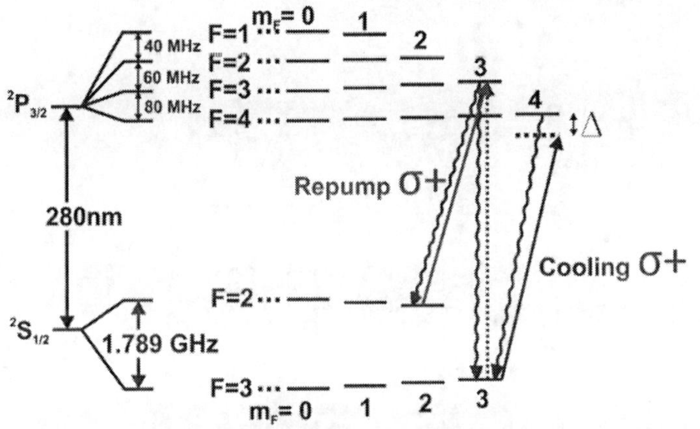


**Figure 2.** Cooling scheme. Δ – detuning of the laser frequency from the transition.

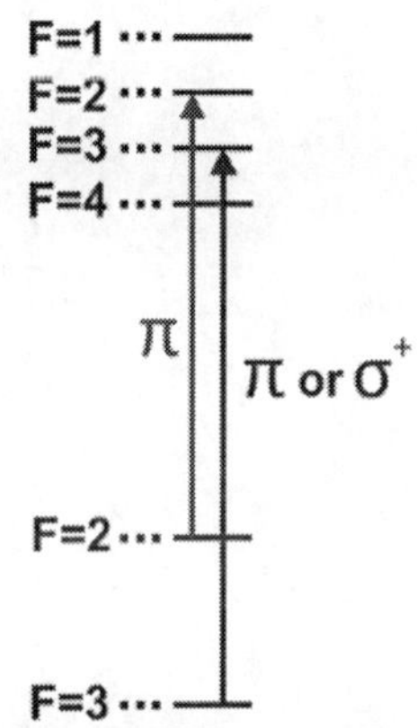


**Figure 3.** Pumping to the $^2S_{1/2}$ ($F = 2$, $m_F = 0$) state.

(ii) For the $^2S_{1/2}$ ($F$ = 2, $m_F$ = 0) → $^2S_{1/2}$ ($F$ = 3, $m_F$ = 0) transition with the frequency of $\nu_0$ = 1.789 GHz the linear Zeeman shift is suppressed, and measured transition frequency can be expressed as $\nu_{clock} = \nu_0 + \eta B^2$, where $\eta$ = 2.199 kHz/G² and $B$ – magnetic induction. Thus pumping to the $^2S_{1/2}$ ($F$ = 2, $m_F$ = 0) state before clock transition excitation is necessary. For this purpose a $\pi$-polarized beam exciting $^2S_{1/2}$ ($F$ = 2) → $^2P_{3/2}$ ($F$ = 2) transition and a beam for $^2S_{1/2}$ ($F$ = 3) → $^2P_{3/2}$ ($F$ = 3) transition excitation with polarization changing periodically between $\pi$ and $\sigma^+$ should be used. Since transitions between $\Delta F = 0$ and $\Delta m_F = 0$ are forbidden, the $^2S_{1/2}$ ($F = 2$, $m_F = 0$) becomes a dark state (Fig. 3).

(iii) To probe the clock transition the Ramsey spectroscopy with two $\pi/2$ microwave pulses with $T_{rams} \approx 1$ s pause is suggested. Duration of the pulses should be defined experimentally.

(iv) To detect the part of excited ions the fluorescence rate of the ion cloud can be measured.

## 3. The setup

Basic scheme of setup is presented at the Fig. 4 and consists of 4 main parts: (i) the laser system, (ii) the ion trapping system, (iii) the microwave interrogation system, (iv) the detection system.

(i) Scheme of the laser setup is shown at the Fig. 5. Commercial Toptica TA pro laser provides 0.8 W power at 1120 nm. It is frequency stabilized with several MHz accuracy to a wavemeter (WS - U, Angstrom). The fourth harmonic of the laser irradiation is obtained by two consequent frequency

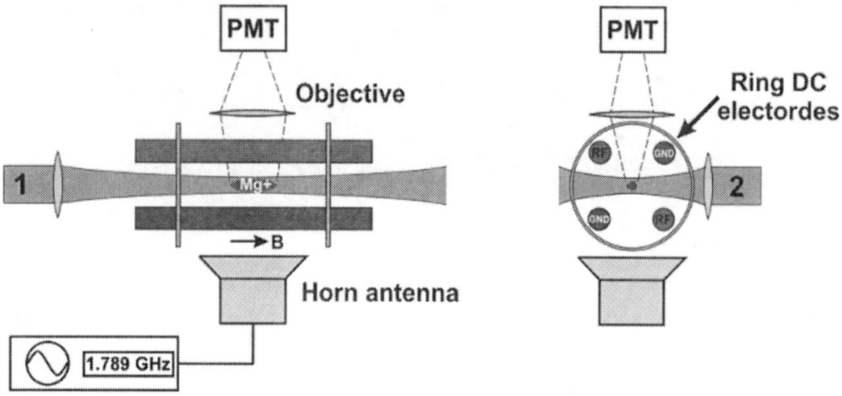


**Figure 4.** Sketch of the setup for trapping $^{25}Mg^+$ ions, laser cooling, clock transition excitation and fluorescence detection. 1 – cooling beam, 2 – pumping beam, PMT – photomultiplier, B – magnetic field. RF and GND electrodes provide radial confinement of ions and ring DC electrodes provide axial confinement.

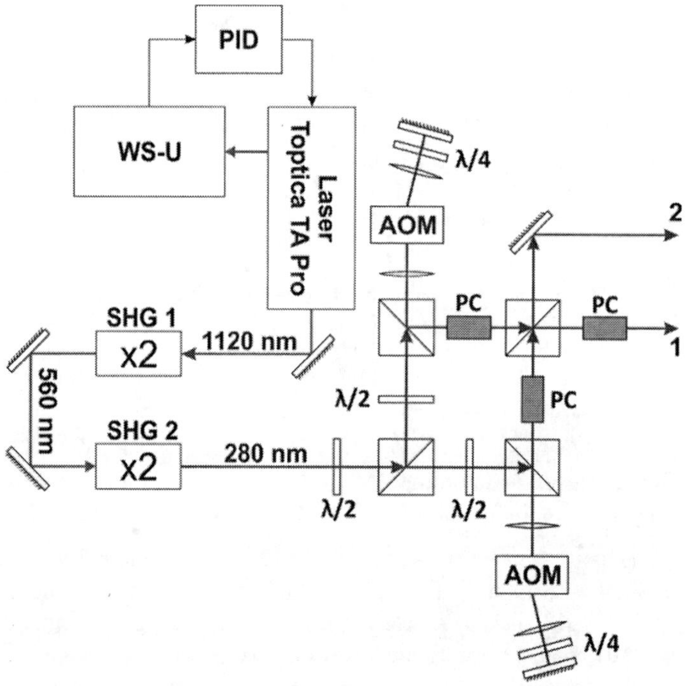


**Figure 5.** Laser system scheme. WS-U – wavelength meter, SHG 1 and 2 – home-built second harmonic generators based on LBO and BBO crystals respectively, AOM – acoustooptical modulators, PC – Pockels cells, 1 and 2 – cooling and pumping beams going into the trap.

doublings in two home-built second harmonic generators based on LBO and BBO crystals respectively. The crystals are housed in bow-tie cavities. The output power of about 5 mW at 280 nm is provided. Two double-pass acousto-optical modulators (AOM) allow for fast frequency tuning of laser irradiation. For polarization switching we suggest Pockels cells.

(ii) The linear Paul trap with four cylindrical rods is suitable for storage up to $10^6$ ions. The trap is described in details in [17]. Secular frequencies for $^{25}Mg^+$ ions in the trap are $\omega_{rad} = 2\pi \times 500$ kHz for radial movement and $\omega_{ax} = 2\pi \times 100$ kHz for axial. Trap depth corresponds to $D = 3$ eV. Ions are generated by electron impact ionization of magnesium atomic beam in the center of the trap. Ion trap, electron gun and atomic oven are situated inside the vacuum chamber under pressure less than $10^{-10}$

mbar. To separate $^{25}Mg^+$ isotope of magnesium from other particles the mass-selective properties of the trap can be used.

(iii) The microwave interrogation is provided by SRS DS345 generator stabilized to a passive hydrogen maser, which radiation will be delivered to the ion cloud via home-made horn antenna. To remove degeneracy of magnetic sublevels magnetic field with induction on the order of $B \approx 10$ mG along the trap axis should be applied.

(iv) The fluorescence signal at 280 nm is collected by high aperture lens assembly made from UV grade fused silica with antireflection coating and detected by a photomultiplier tube (PMT).

## 4. Expected performance

Since the ion cloud size is by two orders of magnitude less than clock transition wavelength, Lamb-Dicke regime is realized and the first order Doppler effect is suppressed. The main systematics are expected to be the second order Doppler and the quadratic Zeeman effect. Other systematics as the Stark effect, caused by mutual interaction between ions are negligible. The second order Doppler effect can be expressed as [18]:

$$\frac{\delta\nu_{dop}}{\nu} = -\frac{3k_BT}{2mc^2}\left(1+\frac{2}{3}N_d^k\right),$$

where $k_B$ – the Boltsman constant, $T$ – the temperature of the ion cloud, $m$ – the ion mass, $c$ – the speed of light in vacuum and $N_d^k$ characterizes the second-order Doppler shift caused by micromotion. For this type of traps and in presence of laser cooling $N_d^k \approx 3$. For estimated 1 K temperature uncertainty, relative uncertainty of the clock transition frequency becomes $\delta\nu_{dop}/\nu = 2\times10^{-14}$.

Uncertainty due to the quadratic Zeeman effect can be expressed as

$$\frac{\delta\nu_{Zeeman}}{\nu} = \frac{2\eta B\delta B}{\nu_{HFS}}.$$

If $\delta B \approx 1$ μG, relative frequency uncertainty of the clock transition frequency is estimated to be $\Delta\nu_{Zeeman}/\nu = 3\times10^{-14}$.

The relative instability can be estimated as [19]

$$\sigma_y(\tau) = \frac{1}{2\pi\nu_0\Delta T_{rams}\mathrm{SNR}}\sqrt{\frac{T_c}{\tau}},\ \tau \geq T_c \geq 2\,\Delta T_{rams},$$

where SNR – signal to noise ratio, – measurement cycle time, $\tau$ – averaging time. If the only noise presented is the quantum projection noise, then SNR = $\sqrt{N_{ions}}$, where $N_{ions}$ – number of ions in the trap. With $\nu_0$=1.789 GHz, $\Delta T_{rams}$ = 1 s, $T_c$ = 2 $\Delta T_{rams}$, $N_{ions}$ = $10^6$ the estimated relative instability is $1.3\times10^{-13}\,\tau^{-1/2}$.

## 5. Conclusion

Design and expected performance of the microwave frequency standard based on laser-cooled $^{25}Mg^+$ ions are presented. The estimated fractional inaccuracy of the standard is estimated to be on the level of $3\times10^{-14}$ and mostly limited by the quadratic Zeeman effect. The fractional instability at the short averaging time is estimated to be of $1.3\times10^{-13}\,\tau^{-1/2}$. Experiments aimed for study the microwave transition $^{25}Mg^+$ ions are important for new generation of frequency references as well as for elements of quantum logic, where ions can play as an interface between microwave and optical Q-bits.

**Acknowledgments**
The work is supported by the Russian Science Foundation (Grant 16-12-00096).

**References**

[1] Dow J M, Neilan R E et al. 2009 *Journal of Geodesy* **83** p 191.
[2] Gilbert S L, Swann W C et al. 2001 *Proc. SPIE* **4269** p 184.
[3] Rosenband T, Hume D B et al. 2008 *Science* **319** p 1808.
[4] Nicholson T L, Campbell S L et al. 2015 *Nature Communications* **6** p 6896.

IOP Conf. Series: Journal of Physics: Conf. Series **941** (2017) 012113 doi:10.1088/1742-6596/941/1/012113


[5] Chou C W, Hume D et al. 2010 *Phys. Rev. Lett.* **104** p 070802.
[6] Cao J, Zhang P et al. 2016 *arXiv preprint* arXiv:1607.03731.
[7] Falke S, Lemke N et al. 2014 *New J. Phys.* **16** p 073023.
[8] Häffner S, Falke S et al. 2015 *Optics Letters* **40** p 2112.
[9] Matei D G, Legero T et al. 2016 *J. Phys.: Conf. Series* **723** p 012031.
[10] Rochat P, Droz F et al. 2012 *Proc.of European Nav. Conf.* 25.
[11] Burt E A, Diener W A et al. 2008 *IEEE transactions on ultrasonics, ferroelectrics and frequencycontrol* **55** p 2586.
[12] Knab H, Niebing K et al. 1985 *IEEE Transactions on Instrumentation and Measurement* **34**(2) 242.
[13] Bollinger J J, Heinzen D J et al. 1991 *IEEE TRANSACTIONS ON INSTRUMENTATION AND MEASUREMENT* **40** p 126.
[14] Schwindt P D D, Jau Y et al. 2016 *Rev. Sci. Instrum.* **87** p 053112.
[15] Miao K, Zhang J W et al. 2015 *Optics Letters* **40** p 4249.
[16] Tjoelker R L, Prestage J D et al. 2016 *IEEE transactions on ultrasonics, ferroelectrics, and frequency control* **63** p 1034.
[17] Semerikov I A, Zalivako I V et al. 2016 *Quantum Electron* **46** p 935.
[18] Prestage J D, Tjoelker R L et al. 1999 *Joint Meeting EFTF - IEEE IFCS* **124** p 121.
[19] Fisk P T H 1997 *Rep. Prog. Phys.* **60** p 761.

IOP Conf. Series: Journal of Physics: Conf. Series **941** (2017) 012114 doi:10.1088/1742-6596/941/1/012114

# Measurement of the upper clock level lifetime in $^{169}$Tm

**A A Golovizin[1,2,3], E S Kalganova[1,2,3], D O Tregubov[1,2,4], G A Vishnyakova[1,5], D D Sukachev[1,3,6] and V N Sorokin[1,3]**

[1]P.N. Lebedev Physical Institute of the Russian Academy of Sciences, 119991, 53 Leninsky pr., Moscow, Russia
[2]Moscow Institute of Physics and Technology (State University), 141700, 9 Institutskiy per., Dolgoprudny, Moscow region, Russia
[3]Russian Quantum Center, Business-Center Ural, 143025, 100A Novaya st., Skolkovo, Moscow, Russia
[4]National Research Nuclear University MEPhI (Moscow Engineering Physics Institute), 115409, 31 Kashirskoe sh., Moscow, Russia
[5]Max-Planck-Institut fur Quantcnoptik, 85748, 1 Hans-Kopfermann-Strasse, Garching, Germany
[6]Department of Physics, Harvard University, 02138, 17 Oxford st., Cambridge, Massachusetts, USA

E-mail: artem.golovizin@gmail.com

**Abstract.** We determine a lifetime of the $4f^{13}6s^2$ ($J = 5/2$) clock level in $^{169}$Tm atoms trapped in an optical lattice at 532 nm directly exciting a 1.14 μm magneto-dipole transition and measuring a decay rate from this level to the ground ($J = 7/2$) state. The measured lifetime of 112(4) ms corresponds to a transition spectral line width of 1.4 Hz which is in a good agreement with a theoretical prediction of 1.14 Hz.

## 1. Introduction

Precise timekeeping is an important technology for industry and science. The realization of the optical lattice clocks brings the fractional instability to the $10^{-18}$ range [1, 2], and enables new fundamental research, like testing a drift of the fundamental constants [3, 4] and probing the general relativity [5]. The main limiting factor in optical clocks is the clock transition shift caused by the black body radiation (BBR) [6, 7, 8]. Magneto-dipole transitions between fine-structure states of the lanthanides correspond to flipping of spin of the unpaired electrons in the submerged $4f$ electron shell. This shell is shielded from an external electric field by the outer $5s^2$ and $6s^2$ shells [9]. This shielding results in a small BBR shift of these magneto-dipole transitions and renders them as promising candidates for new optical clocks [10, 11].

Thulium is a lanthanide with one vacancy in the inner $4f$ electron shell. Similar to other lanthanides [12, 13, 13, 14], laser cooling of Tm is achieved in two stages. The first-stage cooling is done at a strong 410.6 nm transition which routinely allows reaching a sub-Doppler temperature of 80 μK in a cloud of $2\text{x}10^6$ atoms [15]. The second cooling stage at a weak 530.7 nm transition results in a Doppler-limited temperature of 9 μK [16]. This temperature is low enough to load atoms in a shallow optical trap or a lattice using 532 nm laser radiation [17]. Relevant Tm levels are shown in Fig. 1a.

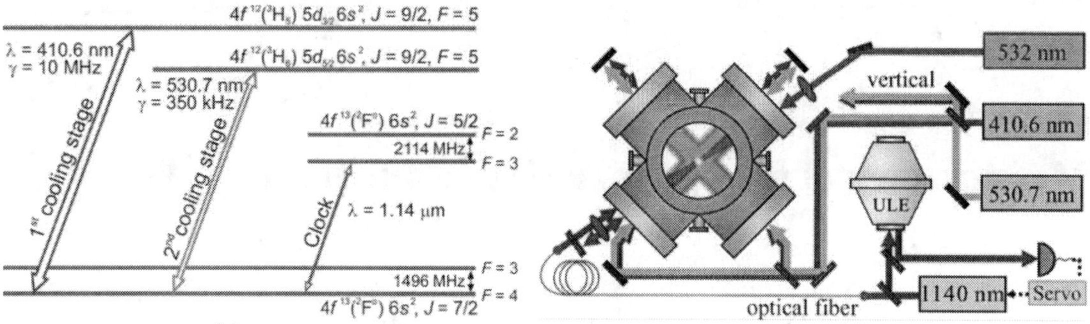


(a) (b)

**Figure 1.** (a) Relevant energy levels of $^{169}$Tm. The strong transition at 410.6 nm is used for the first-stage laser cooling, and detecting the ground state populations and the weak transition at 530.7 nm is used for the second-stage cooling. The proposed clock transition $4f^{13}6s^2$ ($J = 7/2$, $F = 4$) → $4f^{13}6s^2$ ($J = 5/2$, $F = 3$) is at the wavelength of 1.14 μm.

(b) A Schematic diagram of the experimental setup: blue and light green lines are the cooling beams with 410.6 nm and 530.7 nm wavelength, respectively, dark green is optical lattice beam. 1.14 μm laser is stabilized to a high-finesse ULE cavity and is guided to the MOT region via optical fiber.

In this paper, we present measurement of the upper $|J = 5/2\rangle$ clock level lifetime of Tm atoms trapped in the optical lattice at 532 nm. We also show that after loading atoms to the optical lattice only a small fraction of atoms are in the $|m_F = 0\rangle$ and additional pumping is required.

## 2. Lifetime of the $|J = 5/2\rangle$ clock level

The lifetime of the $|J = 5/2\rangle$ clock level is measured by excitation of the magnetic dipole transition at 1.14 μm in a 1D optical lattice. An experimental setup is shown in Fig. 1b and described elsewhere [16, 17]. About 106 thulium atoms are laser cooled down to 20 μK in a narrow-line magneto-optical trap (MOT) operating at 530.7 nm and then are recaptured by the 1D optical lattice. The lattice is formed by a retro-reflected focused 532nm cw laser beam (waistradius is 50 μm, laser power is 3W) and superimposed with the atomic cloud. The trap depth is calculated to be 400 μK for the ground $|J = 7/2, F = 4\rangle$ state which together with good spatial overlapping provides a recapture efficiency of 40%.

After recapture, we switch the MOT off and wait for 20 ms to let uncaptured atoms escape. Then, we excite atoms to the $|J = 5/2, F = 3\rangle$ level applying a resonant 1.14 μm laser pulse for 30 ms [18]. The laser is actively stabilized to a high-finesse ultralow expansion (ULE) cavity [19] which narrows the laser spectral line width down to ~10 Hz. After the interrogation pulse, a resonant 410.6 nm laser pulse of 1 ms is applied to remove atoms from the $|J = 7/2\rangle$ ground state (Fig. 1a). Atoms exited to the $|J = 5/2, F = 3\rangle$ decay back to the $|J = 7/2, F = 4\rangle$ ground state, which population is monitored by means of a fluorescence induced by a delayed 410.6 nm probe pulse.

The increase of the population of the $|J = 7/2, F = 4\rangle$ ground state is described by the exponential function

$$N(t) = N_0\,(1 - \exp(-t/\tau)), \quad (1)$$

where $\tau$ is the lifetime of the the excited $|J = 5/2, F = 3\rangle$ state and $N_0$ is a total number of atoms returned to the ground state. By fitting the experimental data presented in Fig. 2a, we measure $\tau = 112(4)$ ms. It is the lower bound for the $|J = 5/2\rangle$ level natural lifetime since the measured lifetime can be reduced by additional weak losses from the $|J = 5/2\rangle$ evel in the optical lattice. Thus, the natural line width of the clock transition $|J = 7/2\rangle \rightarrow |J = 5/2\rangle$ is not broader than 1.4 Hz and is consistent with a previous measurement in $^4$He matrix [20] and a theoretical prediction of 1.14 Hz [21].

IOP Conf. Series: Journal of Physics: Conf. Series **941** (2017) 012114 doi:10.1088/1742-6596/941/1/012114

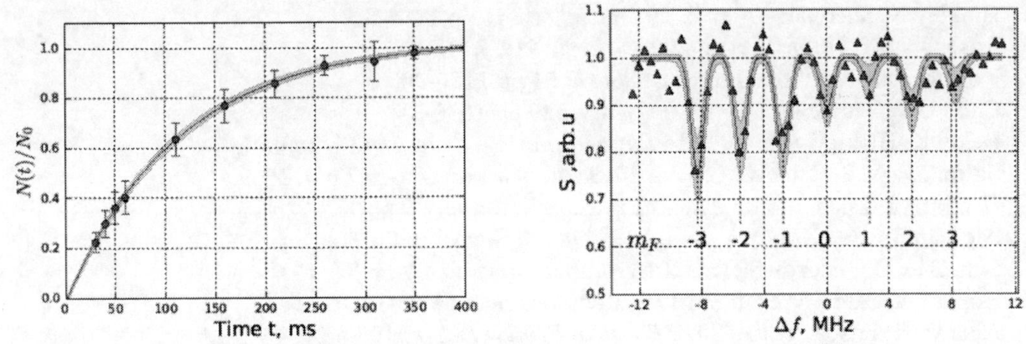

(a) (b)

**Figure 2.** (a) Measurement of the lifetime of the $|J = 5/2\rangle$ evel. Dots are the normalized number of atoms decayed to the ground $|J = 7/2, F = 4\rangle$ level averaged over 9 series of measurements. A solid curve is a fit by (1) with a statistical $1\sigma$ standard deviation shown as gray area.

(b) Normalized Zeeman spectrum of $|J = 7/2, F = 4\rangle \rightarrow |J = 5/2, F = 3\rangle$ transition in $\pi$-polarized 1.14 μm radiation. $\Delta f$ is probe beam frequency detuning from $|m_F = 0\rangle \rightarrow |m_F = 0\rangle$ transition, $S$ is normalized number of atoms remained in the $|J = 7/2, F = 4\rangle$ ground state after excitation by 30 ms probe pulse at 1.14 μm. Experimental data (triangles) are fitted with a sum of equidistantly separated Gaussian functions with equal widths (solid line). Gray area shows statistical $1\sigma$ standard deviation of the fit.

## 3. Ground state magnetic sublevels population distribution

To minimize sensetivity of the $|J = 7/2, F = 4\rangle \rightarrow |J = 5/2, F = 3\rangle$ clock transition frequency to interatomic dipole-dipole interaction, vector Stark shift, and linear Zeeman shift we plan to interrogare $|m_F = 0\rangle \rightarrow |m_F = 0\rangle$ of its Zeeman manifold transition in future Tm optical frequency standard [10]. To improve signal-to-noise ratio and to decrease systematic errors, atomic ensemble should be prepared in the $|m_F = 0\rangle$ state. In Fig. 2b we show spectrum of the 1.14 μm transition under applied an external magnetic field $B_0 = 6.6$ Gs in a $\pi$-polarized probe light where 7 transitions with $m_F$ from -3 to +3 are clearly distinguishable. There are no $\pi$-transitions from $m_F = -4$ or $m_F = 4$ to upper $|J = 5/2, F = 3\rangle$ level. A spectral line width of the individual transition is approximately 0.55 MHz and is close to expected inhomogeneous power broadening of ~0.4 MHz caused by different dynamic polarizabilities of the $|J = 7/2, F = 4\rangle$ and $|J = 5/2, F = 3\rangle$ levels at 532 nm. Taking into account populations of $m_F = -4$ and $m_F = 4$ sublevels we estimate that only about 10% of all atoms in the optical lattice are in the $|m_F = 0\rangle$ state. Required optical pumping to the $|F = 4, m_F = 0\rangle$ sublevel can be performed using a $\pi$-polarized light resonant with a $|J = 7/2, F = 4\rangle \rightarrow |J = 9/2, F = 4\rangle$ transition at 410.6 nm since the $|F = 4, m_F = 0\rangle$ state is a dark state in this configuration [22, 23].

In conclusion, we measured the lifetime of the upper clock level $4f^{13}6s^2$ ($J = 7/2$) of Tm in the 1D optical lattice at 532 nm to be 112(4) ms which sets the upper limit to a natural spectral linewidth of 1.4 Hz of the magneto-dipole transition $|J = 7/2, F = 4\rangle \rightarrow |J = 5/2, F = 3\rangle$. Optical pumping to the $|F = 4, m_F = 0\rangle$ state is required since only a small fraction of atoms in optical lattice is initially in the $|m_F = 0\rangle$ state. Finally, narrow line width and small sensitivity to BBR shift [10] of the $|J = 7/2, F = 4\rangle \rightarrow |J = 5/2, F = 3\rangle$ transition as well as simple experimental scheme make the Tm atom a promising candidate for optical frequency standards.

**Acknowledgements**
Authors acknowledge K.Yu. Khabarova and N.N. Kolachevsky for supporting the project and fruitful discussions.

IOP Conf. Series: Journal of Physics: Conf. Series **941** (2017) 012114 doi:10.1088/1742-6596/941/1/012114

**References**


[1] Bloom B J, Nicholson T L et al. 2014 *Nature* **506** pp 71-5.
[2] Hinkley N, Sherman J A et al. 2013 *Science* **341** pp 1215-8.
[3] Rosenband T, Hume D et al. 2008 *Science* **319** pp 1808-12.
[4] Kolachevsky N N 2008 *Physics-Uspekhi* **51** pp 1180-90.
[5] Chou C W, Hume D et al. 2010 *Science* **329** pp 1630-3.
[6] Beloy K, Hinkley N et al., 2014 *Physical Review Letters* **113** p 260801.
[7] Safronova M S, Porsev S G et al. 2013 *Physical Review A* **87** p 012509.
[8] Ushijima I, Takamoto M et al. 2015 *Nature Photonics* **9** pp 185-9.
[9] Aleksandrov E, Kotylev V et al. 1983 *Opt. Spektrosc.* **54** p 3.
[10] Sukachev D, Fedorov S et al. 2016 *Physical Review A* **94** p 022512.
[11] Kozlov A, Dzuba V et al. 2013 *Physical Review A* **88** p 032509.
[12] Aikawa K, Frisch A et al. 2012 *Physical Review Letters* **108** p 210401.
[13] Lu M, Burdick N Q et al. 2011 *Physical Review Letters* **107** p 190401.
[14] Miao J, Hostetter J et al. 2014 *Physical Review A* **89** p 041401.
[15] Sukachev D, Sokolov A et al. 2010 *Physical Review A* **82**.
[16] Vishnyakova G A, Golovizin A A et al. 2016 *Physics-Uspekhi* **59** p168.
[17] Sukachev D D, Kalganova E S et al. 2014 *Quantum Electronics* **44** pp 515-20.
[18] Golovizin A A, Kalganova E S et al. 2015 *Quantum Electronics* **45** pp 482-5.
[19] Alnis J, Matveev A et al. 2008 *Physical Review A* **77** p 053809.
[20] Ishikawa K, Hatakeyama A et al., 1997 *Physical Review B* **56** pp 780-7.
[21] Kolachevsky N N, Akimov A V et al. 2007 *Applied Physics B* **89** pp 589-94.
[22] Magunov A I and Palchikov V G 2014 *Journal of Exper. and Theor.Physics* **118** pp 687-700.
[23] Martin J, Shore B W et al. 1996 *Physical Review A* **54** pp 1556-69.